STUDENT SOLUTIONS MANUAL

Mark McCombs

PRECALCULUS

SIXTH EDITION

Michael Sullivan

Prentice
Hall

Upper Saddle River, NJ 07458

Editor in Chief: Sally Yagan
Associate Editor: Dawn Murrin
Assistant Managing Editor, Math Media Production: John Matthews
Production Editor: Donna Crilly
Supplement Cover Manager: Paul Gourhan
Supplement Cover Designer: PM Workshop Inc.
Manufacturing Manager: Trudy Pisciotti

Printed in the United States of America

10 9 8 7 6 5 4 3 2

ISBN   0-13-041221-X

Prentice-Hall International (UK) Limited, London
Prentice-Hall of Australia Pty. Limited, Sydney
Prentice-Hall Canada, Inc., Toronto
Prentice-Hall Hispanoamericana, S.A., Mexico City
Prentice-Hall of India Private Limited, New Delhi
Pearson Education Asia Pte. Ltd., Singapore
Prentice-Hall of Japan, Inc., Tokyo
Editora Prentice-Hall do Brazil, Ltda., Rio de Janeiro

Contents

Chapter 5   Trigonometric Functions

Chapter 6   Analytic Trigonometry

Chapter 7   Applications of Trigonometric Functions

Chapter 8   Polar Coordinates; Vectors

Chapter 9   Analytic Geometry

Appendix B Graphing Utilities

Preface

The Student's Solutions Manual to accompany Precalculus, 6th Edition by Michael Sullivan contains detailed solutions to all of the problems in the textbook. TI-83 graphing calculator screens have been included to demonstrate the use of the graphics calculator in solving and in checking solutions to the problems where requested. Every attempt has been made to make this manual as error free as possible. If you have suggestions, error corrections, or comments please feel free to write to me about them.

A number of people need to be recognized for their contributions in the preparation of this manual. Thanks go to Sally Yagan, Dawn Murrin and Audra Walsh at Prentice Hall. Thanks also to Rachelle DeCoste, Kyle Kneisl, Laura Stevens and Scott Young for thoroughly checking the solutions for errors.

I especially wish to thank my mother, Sarah, and my brothers, Kirk and Doug, for their unwavering support and encouragement.

I am also greatly indebted to Neil Young and Blind Willie McTell for helping me endure the many hours of editing the manuscript.

Finally, I want to thank Lily for her invaluable help in formatting the manual.

Mark A. McCombs
Department of Mathematics
Campus Box 3250
University of North Carolina at Chapel Hill
Chapel Hill, NC 27599
mccombs@math.unc.edu

Graphs

1.1 Rectangular Coordinates

1. (a) Quadrant II
   (b) Positive x-axis
   (c) Quadrant III
   (d) Quadrant I
   (e) Negative y-axis
   (f) Quadrant IV

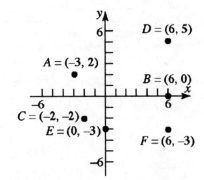

3. The points will be on a vertical line that is two units to the right of the y-axis.

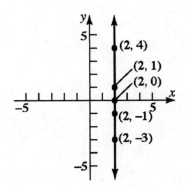

5. $d(P_1, P_2) = \sqrt{(2-0)^2 + (1-0)^2} = \sqrt{4+1} = \sqrt{5}$

7. $d(P_1, P_2) = \sqrt{(-2-1)^2 + (2-1)^2} = \sqrt{9+1} = \sqrt{10}$

9. $d(P_1, P_2) = \sqrt{(5-3)^2 + (4-(-4))^2} = \sqrt{2^2 + 8^2} = \sqrt{4+64} = \sqrt{68} = 2\sqrt{17}$

11. $d(P_1, P_2) = \sqrt{(6-(-3))^2 + (0-2)^2} = \sqrt{9^2 + (-2)^2} = \sqrt{81+4} = \sqrt{85}$

13. $d(P_1, P_2) = \sqrt{(6-4)^2 + (4-(-3))^2} = \sqrt{2^2 + 7^2} = \sqrt{4+49} = \sqrt{53}$

15. $d(P_1, P_2) = \sqrt{(2.3-(-0.2))^2 + (1.1-0.3)^2} = \sqrt{(2.5)^2 + (0.8)^2}$
    $= \sqrt{6.25+0.64} = \sqrt{6.89} \approx 2.625$

17. $d(P_1, P_2) = \sqrt{(0-a)^2 + (0-b)^2} = \sqrt{a^2 + b^2}$

19. $A = (-2,5), \quad B = (1,3), \quad C = (-1,0)$

$d(A,B) = \sqrt{(1-(-2))^2 + (3-5)^2} = \sqrt{3^2 + (-2)^2}$

$\quad = \sqrt{9+4} = \sqrt{13}$

$d(B,C) = \sqrt{(-1-1)^2 + (0-3)^2} = \sqrt{(-2)^2 + (-3)^2}$

$\quad = \sqrt{4+9} = \sqrt{13}$

$d(A,C) = \sqrt{(-1-(-2))^2 + (0-5)^2} = \sqrt{1^2 + (-5)^2}$

$\quad = \sqrt{1+25} = \sqrt{26}$

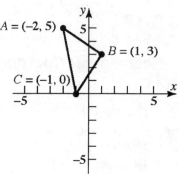

Verifying that $\triangle$ ABC is a right triangle by the Pythagorean Theorem:

$$[d(A,B)]^2 + [d(B,C)]^2 = [d(A,C)]^2$$

$$\left(\sqrt{13}\right)^2 + \left(\sqrt{13}\right)^2 = \left(\sqrt{26}\right)^2$$

$$13 + 13 = 26$$

$$26 = 26$$

The area of a triangle is $A = \dfrac{1}{2} \cdot bh$. In this problem,

$$A = \frac{1}{2} \cdot [d(A,B)] \cdot [d(B,C)] = \frac{1}{2} \cdot \sqrt{13} \cdot \sqrt{13} = \frac{1}{2} \cdot 13 = \frac{13}{2} \text{ square units}$$

21. $A = (-5,3), \quad B = (6,0), \quad C = (5,5)$

$d(A,B) = \sqrt{(6-(-5))^2 + (0-3)^2} = \sqrt{11^2 + (-3)^2}$

$\quad = \sqrt{121+9} = \sqrt{130}$

$d(B,C) = \sqrt{(5-6)^2 + (5-0)^2} = \sqrt{(-1)^2 + 5^2}$

$\quad = \sqrt{1+25} = \sqrt{26}$

$d(A,C) = \sqrt{(5-(-5))^2 + (5-3)^2} = \sqrt{10^2 + 2^2}$

$\quad = \sqrt{100+4} = \sqrt{104}$

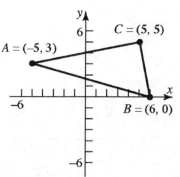

Verifying that $\triangle$ ABC is a right triangle by the Pythagorean Theorem:

$$[d(A,C)]^2 + [d(B,C)]^2 = [d(A,B)]^2$$

$$\left(\sqrt{104}\right)^2 + \left(\sqrt{26}\right)^2 = \left(\sqrt{130}\right)^2$$

$$104 + 26 = 130$$

$$130 = 130$$

The area of a triangle is $A = \dfrac{1}{2} \cdot bh$. In this problem,

$$A = \frac{1}{2} \cdot [d(A,C)] \cdot [d(B,C)] = \frac{1}{2} \cdot \sqrt{104} \cdot \sqrt{26} = \frac{1}{2} \cdot \sqrt{2704} = \frac{1}{2} \cdot 52 = 26 \text{ square units}$$

23. $A = (4, -3), \ B = (0, -3), \ C = (4, 2)$

$$d(A,B) = \sqrt{(0-4)^2 + (-3-(-3))^2} = \sqrt{(-4)^2 + 0^2}$$
$$= \sqrt{16+0} = \sqrt{16} = 4$$
$$d(B,C) = \sqrt{(4-0)^2 + (2-(-3))^2} = \sqrt{4^2 + 5^2}$$
$$= \sqrt{16+25} = \sqrt{41}$$
$$d(A,C) = \sqrt{(4-4)^2 + (2-(-3))^2} = \sqrt{0^2 + 5^2}$$
$$= \sqrt{0+25} = \sqrt{25} = 5$$

Verifying that $\triangle$ ABC is a right triangle by the Pythagorean Theorem:

$$[d(A,B)]^2 + [d(A,C)]^2 = [d(B,C)]^2$$

$$4^2 + 5^2 = \left(\sqrt{41}\right)^2 \rightarrow 16 + 25 = 41 \rightarrow 41 = 41$$

The area of a triangle is $A = \dfrac{1}{2} \cdot bh$. In this problem,

$$A = \frac{1}{2} \cdot [d(A,B)] \cdot [d(A,C)] = \frac{1}{2} \cdot 4 \cdot 5 = 10 \text{ square units}$$

25. All points having an x-coordinate of 2 are of the form (2, y). Those which are 5 units from (–2, –1) are:

$$\sqrt{(2-(-2))^2 + (y-(-1))^2} = 5 \rightarrow \sqrt{4^2 + (y+1)^2} = 5$$

$$\text{Squaring both sides}: \ 4^2 + (y+1)^2 = 25$$

$$16 + y^2 + 2y + 1 = 25$$

$$y^2 + 2y - 8 = 0$$

$$(y+4)(y-2) = 0$$

$$y = -4 \ \text{ or } \ y = 2$$

Therefore, the points are (2, –4) or (2, 2).

27. All points on the x-axis are of the form (x, 0). Those which are 5 units from (4, –3) are:

$$\sqrt{(x-4)^2 + (0-(-3))^2} = 5 \rightarrow \sqrt{(x-4)^2 + 3^2} = 5$$

$$\text{Squaring both sides}: \ (x-4)^2 + 9 = 25$$

$$x^2 - 8x + 16 + 9 = 25 \rightarrow x^2 - 8x = 0$$

$$x(x-8) = 0 \rightarrow x = 0 \ \text{ or } \ x = 8$$

Therefore, the points are (0, 0) or (8, 0).

29. The coordinates of the midpoint are:

$$(x,y) = \left(\frac{x_1 + x_2}{2}, \frac{y_1 + y_2}{2}\right) = \left(\frac{5+3}{2}, \frac{-4+2}{2}\right) = \left(\frac{8}{2}, \frac{-2}{2}\right) = (4, -1)$$

3

Chapter 1   Graphs

31. The coordinates of the midpoint are:

$(x,y) = \left(\dfrac{x_1+x_2}{2}, \dfrac{y_1+y_2}{2}\right) = \left(\dfrac{-3+6}{2}, \dfrac{2+0}{2}\right) = \left(\dfrac{3}{2}, \dfrac{2}{2}\right) = \left(\dfrac{3}{2}, 1\right)$

33. The coordinates of the midpoint are:

$(x,y) = \left(\dfrac{x_1+x_2}{2}, \dfrac{y_1+y_2}{2}\right) = \left(\dfrac{4+6}{2}, \dfrac{-3+1}{2}\right) = \left(\dfrac{10}{2}, \dfrac{-2}{2}\right) = (5,-1)$

35. The coordinates of the midpoint are:

$(x,y) = \left(\dfrac{x_1+x_2}{2}, \dfrac{y_1+y_2}{2}\right) = \left(\dfrac{-0.2+2.3}{2}, \dfrac{0.3+1.1}{2}\right) = \left(\dfrac{2.1}{2}, \dfrac{1.4}{2}\right) = (1.05, 0.7)$

37. The coordinates of the midpoint are:

$(x,y) = \left(\dfrac{x_1+x_2}{2}, \dfrac{y_1+y_2}{2}\right) = \left(\dfrac{a+0}{2}, \dfrac{b+0}{2}\right) = \left(\dfrac{a}{2}, \dfrac{b}{2}\right)$

39. The midpoint of AB is: $D = \left(\dfrac{0+0}{2}, \dfrac{0+6}{2}\right) = (0,3)$

The midpoint of AC is: $E = \left(\dfrac{0+4}{2}, \dfrac{0+4}{2}\right) = (2,2)$

The midpoint of BC is: $F = \left(\dfrac{0+4}{2}, \dfrac{6+4}{2}\right) = (2,5)$

$d(C,D) = \sqrt{(0-4)^2 + (3-4)^2} = \sqrt{(-4)^2 + (-1)^2} = \sqrt{16+1} = \sqrt{17}$

$d(B,E) = \sqrt{(2-0)^2 + (2-6)^2} = \sqrt{2^2 + (-4)^2} = \sqrt{4+16} = \sqrt{20} = 2\sqrt{5}$

$d(A,F) = \sqrt{(2-0)^2 + (5-0)^2} = \sqrt{2^2 + 5^2} = \sqrt{4+25} = \sqrt{29}$

41. $d(P_1,P_2) = \sqrt{(-4-2)^2 + (1-1)^2} = \sqrt{(-6)^2 + 0^2} = \sqrt{36} = 6$

$d(P_2,P_3) = \sqrt{(-4-(-4))^2 + (-3-1)^2} = \sqrt{0^2 + (-4)^2} = \sqrt{16} = 4$

$d(P_1,P_3) = \sqrt{(-4-2)^2 + (-3-1)^2} = \sqrt{(-6)^2 + (-4)^2} = \sqrt{36+16} = \sqrt{52} = 2\sqrt{13}$

Since $[d(P_1,P_2)]^2 + [d(P_2,P_3)]^2 = [d(P_1,P_3)]^2$, the triangle is a right triangle.

43. $d(P_1,P_2) = \sqrt{(0-(-2))^2 + (7-(-1))^2} = \sqrt{2^2 + 8^2} = \sqrt{4+64} = \sqrt{68} = 2\sqrt{17}$

$d(P_2,P_3) = \sqrt{(3-0)^2 + (2-7)^2} = \sqrt{3^2 + (-5)^2} = \sqrt{9+25} = \sqrt{34}$

$d(P_1,P_3) = \sqrt{(3-(-2))^2 + (2-(-1))^2} = \sqrt{5^2 + 3^2} = \sqrt{25+9} = \sqrt{34}$

Since $d(P_2,P_3) = d(P_1,P_3)$, the triangle is isosceles.

Since $[d(P_1,P_3)]^2 + [d(P_2,P_3)]^2 = [d(P_1,P_2)]^2$, the triangle is also a right triangle.

Therefore, the triangle is an isosceles right triangle.

45. $P_1 = (1,3), \ P_2 = (5,15)$

$d(P_1,P_2) = \sqrt{(5-1)^2 + (15-3)^2} = \sqrt{4^2 + 12^2} = \sqrt{16+144} = \sqrt{160} = 4\sqrt{10}$

47. $P_1 = (-4,6), \ P_2 = (4,-8)$

$$d(P_1, P_2) = \sqrt{(4-(-4))^2 + (-8-6)^2} = \sqrt{8^2 + (-14)^2} = \sqrt{64+196} = \sqrt{260} = 2\sqrt{65}$$

49.   Plot the vertices of the square at
      $(0, 0), (0, s), (s, s),$ and $(s, 0)$.

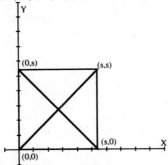

Find the midpoints of the diagonals.

$$M_1 = \left(\frac{0+s}{2}, \frac{0+s}{2}\right) = \left(\frac{s}{2}, \frac{s}{2}\right)$$

$$M_2 = \left(\frac{0+s}{2}, \frac{s+0}{2}\right) = \left(\frac{s}{2}, \frac{s}{2}\right)$$

Since the coordinates of the midpoints
are the same, the diagonals of a square
intersect at their midpoints.

51.   Using the Pythagorean Theorem:

$$90^2 + 90^2 = d^2$$
$$8100 + 8100 = d^2$$
$$16200 = d^2$$
$$d = \sqrt{16200}$$
$$d = 90\sqrt{2} \approx 127.28 \text{ feet}$$

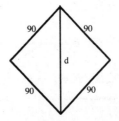

53.   (a)   First: (90, 0), Second: (90, 90)
            Third: (0, 90)

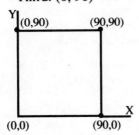

      (b)   Using the distance formula:
            $$d = \sqrt{(310-90)^2 + (15-90)^2}$$
            $$= \sqrt{220^2 + (-75)^2}$$
            $$= \sqrt{54025} \approx 232.4 \text{ feet}$$

      (c)   Using the distance formula:
            $$d = \sqrt{(300-0)^2 + (300-90)^2}$$
            $$= \sqrt{300^2 + 210^2}$$
            $$= \sqrt{134100} \approx 366.2 \text{ feet}$$

55.   The Intrepid heading east moves a distance $30t$ after t
      hours. The truck heading south moves a distance $40t$
      after t hours. Their distance apart after t hours is:

$$d = \sqrt{(30t)^2 + (40t)^2}$$
$$= \sqrt{900t^2 + 1600t^2}$$
$$= \sqrt{2500t^2}$$
$$= 50t$$

Graphs

1.2   Graphs of Equations

1.

3.

5.

7.

9.

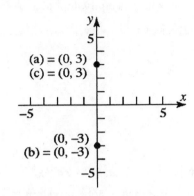

(a) = (0, 3)
(c) = (0, 3)

(0, –3)
(b) = (0, –3)

11. (a) $(-1, 0), (1, 0)$ (b) symmetric to the x-axis, y-axis and origin

13. (a) $\left(-\frac{\pi}{2}, 0\right), \left(\frac{\pi}{2}, 0\right), (0, 1)$ (b) symmetric to the y-axis

15. (a) $(0, 0)$ (b) symmetric to the x-axis

17. (a) $(1, 0)$ (b) not symmetric to x-axis, y-axis, or origin

19. (a) $(-1.5, 0), (1.5, 0), (0, -2)$ (b) symmetric to the y-axis

21. (a) none   (b) symmetric to the origin

23. $y = x^4 - \sqrt{x}$

$0 = 0^4 - \sqrt{0}$ $1 = 1^4 - \sqrt{1}$ $0 = (-1)^4 - \sqrt{-1}$

$0 = 0$ $1 \neq 0$ $0 \neq 1 - \sqrt{-1}$

$(0, 0)$ is on the graph of the equation.

25. $y^2 = x^2 + 9$

$3^2 = 0^2 + 9$ $0^2 = 3^2 + 9$ $0^2 = (-3)^2 + 9$

$9 = 9$ $0 \neq 18$ $0 \neq 18$

$(0, 3)$ is on the graph of the equation.

27. $x^2 + y^2 = 4$

$0^2 + 2^2 = 4$ $(-2)^2 + 2^2 = 4$ $\left(\sqrt{2}\right)^2 + \left(\sqrt{2}\right)^2 = 4$

$4 = 4$ $8 \neq 4$ $4 = 4$

$(0, 2)$ and $\left(\sqrt{2}, \sqrt{2}\right)$ are on the graph of the equation.

29. $x^2 = y$

y - intercept: Let $x = 0$, then $y = 0$ $(0, 0)$
x - intercept: Let $y = 0$, then $x = 0$ $(0, 0)$

Test for symmetry:

x - axis: Replace y by $-y$ so $x^2 = -y$, which is not equivalent to $x^2 = y$.

7

y - axis: Replace x by $-x$ so $(-x)^2 = y$ or $x^2 = y$, which is equivalent to $x^2 = y$.

Origin :   Replace x by $-x$ and y by $-y$ so $(-x)^2 = -y$ or $x^2 = -y$,

which is not equivalent to $x^2 = y$.

Therefore, the graph is symmetric with respect to the y - axis.

31. $y = 3x$
   y - intercept: Let $x = 0$, then $y = 0$ $(0,0)$
   x - intercept: Let $y = 0$, then $x = 0$ $(0,0)$
   Test for symmetry:
    x - axis: Replace y by $-y$ so $-y = 3x$, which is not equivalent to $y = 3x$.
    y - axis :   Replace x by $-x$ so $y = 3(-x)$ or $y = -3x$, which is not equivalent to $y = 3x$.
    Origin :   Replace x by $-x$ and y by $-y$ so $-y = 3(-x)$ or $y = 3x$,

       which is equivalent to $y = 3x$.
   Therefore, the graph is symmetric with respect to the origin.

33. $x^2 + y - 9 = 0$
   y - intercept : Let $x = 0$, then $y = 9$ $(0,9)$
   x - intercept : Let $y = 0$, then $x = \pm 3$ $(-3,0),(3,0)$
   Test for symmetry:
    x - axis: Replace y by $-y$ so $x^2 + (-y) - 9 = 0$ or $x^2 - y - 9 = 0$,

       which is not equivalent to $x^2 + y - 9 = 0$.
    y - axis: Replace x by $-x$ so $(-x)^2 + y - 9 = 0$ or $x^2 + y - 9 = 0$,

       which is equivalent to $x^2 + y - 9 = 0$.
    Origin: Replace x by $-x$ and y by $-y$ so $(-x)^2 + (-y) - 9 = 0$ or $x^2 - y - 9 = 0$,

       which is not equivalent to $x^2 + y - 9 = 0$.
   Therefore, the graph is symmetric with respect to the y-axis.

35. $9x^2 + 4y^2 = 36$
   y - intercept : Let $x = 0$, then $y = \pm 3$ $(0,-3),(0,3)$
   x - intercept : Let $y = 0$, then $x = \pm 2$ $(-2,0),(2,0)$
   Test for symmetry:
x - axis: Replace y by $-y$ so $9x^2 + 4(-y)^2 = 36$ or $9x^2 + 4y^2 = 36$,

       which is equivalent to $9x^2 + 4y^2 = 36$.

y - axis :   Replace x by $-x$ so $9(-x)^2 + 4y^2 = 36$ or $9x^2 + 4y^2 = 36$,

       which is equivalent to $9x^2 + 4y^2 = 36$.

Origin: Replace x by $-x$ and y by $-y$ so $9(-x)^2 + 4(-y)^2 = 36$ or $9x^2 + 4y^2 = 36$,

       which is equivalent to $9x^2 + 4y^2 = 36$.
   Therefore, the graph is symmetric with respect to the x-axis, the y-axis and the origin.

37. $y = x^3 - 27$

y - intercept: Let $x = 0$, then $y = 0^3 - 27$
$$y = -27 \qquad (0, -27)$$

x - intercept: Let $y = 0$, then $0 = x^3 - 27$
$$x^3 = 27$$
$$x = 3 \qquad (3, 0)$$

Test for symmetry:

x - axis: Replace y by $-y$ so $-y = x^3 - 27$, which is not equivalent to $y = x^3 - 27$.

y - axis: Replace x by $-x$ so $y = (-x)^3 - 27$ or $y = -x^3 - 27$,

which is not equivalent to $y = x^3 - 27$.

Origin: Replace x by $-x$ and y by $-y$ so $-y = (-x)^3 - 27$ or

$y = x^3 + 27$, which is not equivalent to $y = x^3 - 27$.

Therefore, the graph is not symmetric to the x-axis, the y-axis, or the origin.

39. $y = x^2 - 3x - 4$

y - intercept: Let $x = 0$, then $y = 0^2 - 3(0) - 4$
$$y = -4 \qquad (0, -4)$$

x - intercept: Let $y = 0$, then $0 = x^2 - 3x - 4$
$$(x - 4)(x + 1) = 0$$
$$x = 4 \quad x = -1 \quad (4, 0), (-1, 0)$$

Test for symmetry:

x - axis: Replace y by $-y$ so $-y = x^2 - 3x - 4$, which is not

equivalent to $y = x^2 - 3x - 4$.

y - axis: Replace x by $-x$ so $y = (-x)^2 - 3(-x) - 4$ or $y = x^2 + 3x - 4$,

which is not equivalent to $y = x^2 - 3x - 4$.

Origin: Replace x by $-x$ and y by $-y$ so $-y = (-x)^2 - 3(-x) - 4$ or

$y = -x^2 - 3x + 4$, which is not equivalent to $y = x^2 - 3x - 4$.

Therefore, the graph is not symmetric to the x-axis, the y-axis, or the origin.

41. $y = \dfrac{3x}{x^2 + 9}$

y - intercept : Let $x = 0$, then $y = \dfrac{0}{0 + 9}$
$$y = 0 \qquad (0, 0)$$

x - intercept : Let $y = 0$, then $0 = \dfrac{3x}{x^2 + 9}$
$$3x = 0 \rightarrow x = 0 \quad (0, 0)$$

Test for symmetry:

x - axis: Replace y by $-y$ so $-y = \dfrac{3x}{x^2 + 9}$, which is not equivalent to $y = \dfrac{3x}{x^2 + 9}$.

y - axis: Replace x by $-x$ so $y = \dfrac{3(-x)}{(-x)^2 + 9}$ or $y = \dfrac{-3x}{x^2 + 9}$,

which is not equivalent to $y = \dfrac{3x}{x^2 + 9}$.

Origin :   Replace x by $-x$ and y by $-y$ so $-y = \dfrac{-3x}{(-x)^2 + 9}$ or

$y = \dfrac{3x}{x^2 + 9}$, which is equivalent to $y = \dfrac{3x}{x^2 + 9}$.

Therefore, the graph is symmetric with respect to the origin.

43. $y = \dfrac{-x^3}{x^2 - 9}$

y - intercept :   Let $x = 0$, then $y = \dfrac{0}{-9} = 0$

$$(0,0)$$

x - intercept :   Let $y = 0$, then $0 = \dfrac{-x^3}{x^2 - 9} \rightarrow -x^3 = 0$

$$x = 0 \qquad (0,0)$$

Test for symmetry:

x - axis: Replace y by $-y$ so $-y = \dfrac{-x^3}{x^2 - 9}$, which is not equivalent to $y = \dfrac{-x^3}{x^2 - 9}$.

y - axis: Replace x by $-x$ so $y = \dfrac{-(-x)^3}{(-x)^2 - 9}$ or $y = \dfrac{x^3}{x^2 - 9}$,

which is not equivalent to $y = \dfrac{-x^3}{x^2 - 9}$.

Origin :   Replace x by $-x$ and y by $-y$ so $-y = \dfrac{-(-x)^3}{(-x)^2 - 9}$ or

$-y = \dfrac{x^3}{x^2 - 9}$, which is equivalent to $y = \dfrac{-x^3}{x^2 - 9}$.

Therefore, the graph is symmetric with respect to the origin.

45. $y = x^3$

47. $y = \sqrt{x}$

49. $y = 3x + 5$
$2 = 3a + 5$
$3a = -3$
$a = -1$

51. $2x + 3y = 6$
$2a + 3b = 6 \rightarrow b = 2 - \dfrac{2}{3}a$

53. Center = (2, 1)
Radius = distance from (0,1) to (2,1)
$= \sqrt{(2-0)^2 + (1-1)^2}$
$= \sqrt{4} = 2$
$(x-2)^2 + (y-1)^2 = 4$

55. Center = midpoint of (1,2) and (4,2)
$= \left(\dfrac{1+4}{2}, \dfrac{2+2}{2}\right) = \left(\dfrac{5}{2}, 2\right)$
Radius = distance from $\left(\dfrac{5}{2},2\right)$ to (4,2)
$= \sqrt{\left(4-\dfrac{5}{2}\right)^2 + (2-2)^2}$
$= \sqrt{\dfrac{9}{4}} = \dfrac{3}{2}$
$\left(x-\dfrac{5}{2}\right)^2 + (y-2)^2 = \dfrac{9}{4}$

57. $(x-h)^2 + (y-k)^2 = r^2$
$(x-0)^2 + (y-0)^2 = 2^2$
$x^2 + y^2 = 4$
General form:
$x^2 + y^2 - 4 = 0$

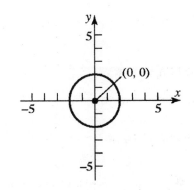

59. $(x-h)^2 + (y-k)^2 = r^2$
$(x-1)^2 + (y-(-1))^2 = 1^2$
$(x-1)^2 + (y+1)^2 = 1$
General form:
$x^2 - 2x + 1 + y^2 + 2y + 1 = 1$
$x^2 + y^2 - 2x + 2y + 1 = 0$

61. $(x-h)^2 + (y-k)^2 = r^2$
$(x-0)^2 + (y-2)^2 = 2^2$
$x^2 + (y-2)^2 = 4$
General form:
$x^2 + y^2 - 4y + 4 = 4$
$x^2 + y^2 - 4y = 0$

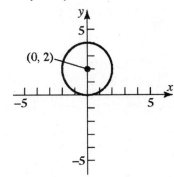

63. $(x-h)^2 + (y-k)^2 = r^2$
     $(x-4)^2 + (y-(-3))^2 = 5^2$
       $(x-4)^2 + (y+3)^2 = 25$
   General form:
   $x^2 - 8x + 16 + y^2 + 6y + 9 = 25$
        $x^2 + y^2 - 8x + 6y = 0$

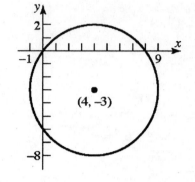

65. $x^2 + y^2 = 4$
   $x^2 + y^2 = 2^2$
   Center: (0,0)
   Radius = 2

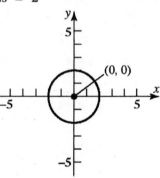

67. $2(x-3)^2 + 2y^2 = 8$
     $(x-3)^2 + y^2 = 4$
   Center: (3,0)
   Radius = 2

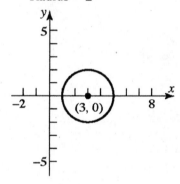

69. $x^2 + y^2 + 4x - 4y - 1 = 0$
        $x^2 + 4x + y^2 - 4y = 1$
   $(x^2 + 4x + 4) + (y^2 - 4y + 4) = 1 + 4 + 4$
        $(x+2)^2 + (y-2)^2 = 3^2$
   Center: (−2,2)
   Radius = 3

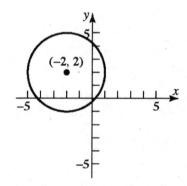

71. $x^2 + y^2 - x + 2y + 1 = 0$

$$x^2 - x + y^2 + 2y = -1$$

$$\left(x^2 - x + \frac{1}{4}\right) + (y^2 + 2y + 1) = -1 + \frac{1}{4} + 1$$

$$\left(x - \frac{1}{2}\right)^2 + (y + 1)^2 = \left(\frac{1}{2}\right)^2$$

Center: $\left(\frac{1}{2}, -1\right)$

Radius = $\frac{1}{2}$

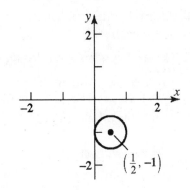

73. $2x^2 + 2y^2 - 12x + 8y - 24 = 0$

$$x^2 + y^2 - 6x + 4y = 12$$

$$x^2 - 6x + y^2 + 4y = 12$$

$$(x^2 - 6x + 9) + (y^2 + 4y + 4) = 12 + 9 + 4$$

$$(x - 3)^2 + (y + 2)^2 = 5^2$$

Center: $(3, -2)$

Radius = 5

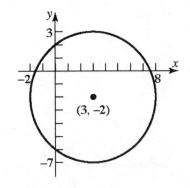

75. Center at (0,0); containing point (- 3, 2).

$$r = \sqrt{(-3 - 0)^2 + (2 - 0)^2} = \sqrt{9 + 4} = \sqrt{13}$$

Equation: $(x - 0)^2 + (y - 0)^2 = \left(\sqrt{13}\right)^2$

$$x^2 + y^2 = 13 \rightarrow x^2 + y^2 - 13 = 0$$

77. Center at (2,3); tangent to the x-axis.

$r = 3$

Equation: $(x - 2)^2 + (y - 3)^2 = 3^2$

$$x^2 - 4x + 4 + y^2 - 6y + 9 = 9 \rightarrow x^2 + y^2 - 4x - 6y + 4 = 0$$

79. Endpoints of a diameter are (1,4) and (−3,2).
The center is at the midpoint of that diameter:

Center: $\left(\frac{1 + (-3)}{2}, \frac{4 + 2}{2}\right) = (-1, 3)$

Radius: $r = \sqrt{(1 - (-1))^2 + (4 - 3)^2} = \sqrt{4 + 1} = \sqrt{5}$

Equation: $(x - (-1))^2 + (y - 3)^2 = \left(\sqrt{5}\right)^2$

$$x^2 + 2x + 1 + y^2 - 6y + 9 = 5 \rightarrow x^2 + y^2 + 2x - 6y + 5 = 0$$

81. (c)   83.   (b)   85. (b), (c), (e) and (g)

13

87. $x^2 + y^2 + 2x + 4y - 4091 = 0$
     $x^2 + 2x + y^2 + 4y - 4091 = 0$

$x^2 + 2x + 1 + y^2 + 4y + 4 = 4091 + 5 \rightarrow (x+2)^2 + (y+2)^2 = 4096$

The circle representing Earth has center $(-2, -2)$ and radius $= \sqrt{4096} = 64$
So the radius of the satellite's orbit is $64 + 0.6 = 64.06$ units.

The equation of the orbit is $(x+2)^2 + (y+2)^2 = (64.06)^2$.

89. – 91.   Answers will vary

Graphs

1.3 Lines

1. (a) Slope $= \dfrac{1-0}{2-0} = \dfrac{1}{2}$

(b) If x increases by 2 units, y will increase by 1 unit.

3. (a) Slope $= \dfrac{1-2}{1-(-2)} = -\dfrac{1}{3}$

(b) If x increases by 3 units, y will decrease by 1 unit.

5. (x_1, y_1) (x_2, y_2)
    $(2,3)$ $(4,0)$

Slope $= \dfrac{y_2 - y_1}{x_2 - x_1} = \dfrac{0-3}{4-2} = -\dfrac{3}{2}$

7. (x_1, y_1) (x_2, y_2)
    $(-2,3)$ $(2,1)$

Slope $= \dfrac{y_2 - y_1}{x_2 - x_1} = \dfrac{1-3}{2-(-2)} = \dfrac{-2}{4} = -\dfrac{1}{2}$

9. (x_1, y_1) (x_2, y_2)
    $(-3,-1)$ $(2,-1)$

Slope $= \dfrac{y_2 - y_1}{x_2 - x_1} = \dfrac{-1-(-1)}{2-(-3)} = \dfrac{0}{5} = 0$

11. (x_1, y_1) (x_2, y_2)
    $(-1,2)$ $(-1,-2)$

Slope $= \dfrac{y_2 - y_1}{x_2 - x_1} = \dfrac{-2-2}{-1-(-1)} = \dfrac{-4}{0}$

Slope is undefined.

13.

15.

17.

19.

21. (0,0) and (2,1) are points on the line.

Slope $= \dfrac{1-0}{2-0} = \dfrac{1}{2}$

y- intercept is 0; using $y = mx + b$:

$y = \dfrac{1}{2}x + 0 \rightarrow 2y = x \rightarrow 0 = x - 2y$

$x - 2y = 0$ or $y = \dfrac{1}{2}x$

23. (–2,2) and (1,1) are points on the line.

Slope $= \dfrac{1-2}{1-(-2)} = -\dfrac{1}{3}$

Using $y - y_1 = m(x - x_1)$

$y - 1 = -\dfrac{1}{3}(x-1)$

$y - 1 = -\dfrac{1}{3}x + \dfrac{1}{3} \rightarrow y = -\dfrac{1}{3}x + \dfrac{4}{3}$

$x + 3y = 4$ or $y = -\dfrac{1}{3}x + \dfrac{4}{3}$

25. $y - y_1 = m(x - x_1),\ \ m = 2$

$y - 3 = 2(x - 3)$

$y - 3 = 2x - 6 \rightarrow y = 2x - 3$

$2x - y = 3$ or $y = 2x - 3$

27. $y - y_1 = m(x - x_1),\ \ m = -\dfrac{1}{2}$

$y - 2 = -\dfrac{1}{2}(x - 1)$

$y - 2 = -\dfrac{1}{2}x + \dfrac{1}{2} \rightarrow y = -\dfrac{1}{2}x + \dfrac{5}{2}$

$x + 2y = 5$ or $y = -\dfrac{1}{2}x + \dfrac{5}{2}$

29. Slope = 3; containing (–2,3)

$y - y_1 = m(x - x_1)$

$y - 3 = 3(x - (-2))$

$y - 3 = 3x + 6 \rightarrow y = 3x + 9$

$3x - y = -9$ or $y = 3x + 9$

31. Slope $= -\dfrac{2}{3}$; containing $(1,-1)$

$y - y_1 = m(x - x_1)$

$y - (-1) = -\dfrac{2}{3}(x - 1)$

$y + 1 = -\dfrac{2}{3}x + \dfrac{2}{3} \rightarrow y = -\dfrac{2}{3}x - \dfrac{1}{3}$

$2x + 3y = -1$ or $y = -\dfrac{2}{3}x - \dfrac{1}{3}$

33. Containing $(1,3)$ and $(-1,2)$

$m = \dfrac{2-3}{-1-1} = \dfrac{-1}{-2} = \dfrac{1}{2}$

$y - y_1 = m(x - x_1)$

$y - 3 = \dfrac{1}{2}(x - 1)$

$y - 3 = \dfrac{1}{2}x - \dfrac{1}{2} \rightarrow y = \dfrac{1}{2}x + \dfrac{5}{2}$

$x - 2y = -5$ or $y = \dfrac{1}{2}x + \dfrac{5}{2}$

35. Slope $= -3$; y-intercept $= 3$

$y = mx + b$

$y = -3x + 3$

$3x + y = 3$ or $y = -3x + 3$

37. x-intercept $= 2$; y-intercept $= -1$

Points are $(2,0)$ and $(0,-1)$

$m = \dfrac{-1-0}{0-2} = \dfrac{-1}{-2} = \dfrac{1}{2}$

$y = mx + b$

$y = \dfrac{1}{2}x - 1$

$x - 2y = 2$ or $y = \dfrac{1}{2}x - 1$

39. Slope undefined; passing through $(2,4)$

This is a vertical line.

$x = 2$

No slope intercept form.

41. Parallel to $y = 2x$; Slope $= 2$

Containing $(-1,2)$

$y - y_1 = m(x - x_1)$

$y - 2 = 2(x - (-1))$

$y - 2 = 2x + 2 \rightarrow y = 2x + 4$

$2x - y = -4$ or $y = 2x + 4$

43. Parallel to $2x - y = -2$; Slope $= 2$

Containing the point $(0,0)$

$y - y_1 = m(x - x_1)$

$y - 0 = 2(x - 0)$

$y = 2x$

$2x - y = 0$ or $y = 2x$

45. Parallel to $x = 5$;

Containing $(4,2)$

This is a vertical line.

$x = 4$

No slope intercept form.

47. Perpendicular to $y = \dfrac{1}{2}x + 4$;

Slope of perpendicular $= -2$

Containing $(1,-2)$

$y - y_1 = m(x - x_1)$

$y - (-2) = -2(x - 1)$

$y + 2 = -2x + 2 \rightarrow y = -2x$

$2x + y = 0$ or $y = -2x$

49. Perpendicular to $2x + y = 2$;

Containing $(-3,0)$

Slope of perpendicular $= \dfrac{1}{2}$

$y - y_1 = m(x - x_1)$

$y - 0 = \dfrac{1}{2}(x - (-3)) \rightarrow y = \dfrac{1}{2}x + \dfrac{3}{2}$

$x - 2y = -3$ or $y = \dfrac{1}{2}x + \dfrac{3}{2}$

51. Perpendicular to $x = 8$;
     Slope of perpendicular $= 0$
     Containing $(3,4)$
     $y - y_1 = m(x - x_1)$
     $y - 4 = 0(x - 3) \rightarrow y - 4 = 0 \rightarrow y = 4$
     $y = 4$ or $y = 0x + 4$

53. $y = 2x + 3$
     Slope $= 2$
     y-intercept $= 3$

55. $\frac{1}{2}y = x - 1$
     $y = 2x - 2$
     Slope $= 2$
     y-intercept $= -2$

57. $y = \frac{1}{2}x + 2$
     Slope $= \frac{1}{2}$
     y-intercept $= 2$

59. $x + 2y = 4$
     $2y = -x + 4 \rightarrow y = -\frac{1}{2}x + 2$
     Slope $= -\frac{1}{2}$
     y-intercept $= 2$

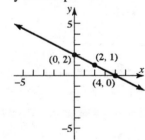

61. $2x - 3y = 6$
     $-3y = -2x + 6 \rightarrow y = \frac{2}{3}x - 2$
     Slope $= \frac{2}{3}$
     y-intercept $= -2$

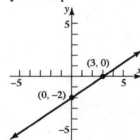

63. $x + y = 1$
$y = -x + 1$
Slope $= -1$
y-intercept $= 1$

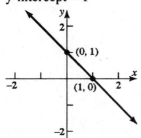

65. $x = -4$
Slope is undefined
y-intercept - none

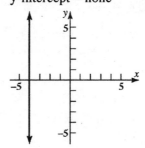

67. $y = 5$
Slope $= 0$
y-intercept $= 5$

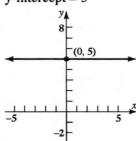

69. $y - x = 0$
$y = x$
Slope $= 1$
y-intercept $= 0$

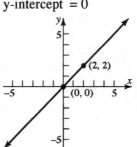

71. $2y - 3x = 0$
$2y = 3x \rightarrow y = \dfrac{3}{2}x$
Slope $= \dfrac{3}{2}$
y-intercept $= 0$

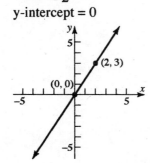

73. The equation of the x-axis is $y = 0$.
(The slope is 0 and the y-intercept is 0.)

75. Consider the points
$A(-2,5),\; B(1,3)$ and $C(-1,0)$
slope of $\overline{AB} = \dfrac{3-5}{1-(-2)} = -\dfrac{2}{3}$

slope of $\overline{AC} = \dfrac{0-5}{-1-(-2)} = -\dfrac{5}{3}$

slope of $\overline{BC} = \dfrac{0-3}{-1-1} = \dfrac{3}{2}$

Therefore, $\triangle ABC$ has a right angle at vertex B since

slope $\overline{AB} = -\dfrac{2}{3}$ and slope $\overline{BC} = \dfrac{3}{2}$

$\Rightarrow \overline{AB} \perp \overline{BC}$

19

77. Consider the points $A(-1,0)$, $B(2,3)$, $C(1,-2)$ and $D(4,1)$

slope of $\overline{AB} = \dfrac{3-0}{2-(-1)} = 1$; slope of $\overline{CD} = \dfrac{1-(-2)}{4-1} = 1$

slope of $\overline{AC} = \dfrac{-2-0}{1-(-1)} = -1$; slope of $\overline{BD} = \dfrac{1-3}{4-2} = -1$

Therefore, the quadrilateral $ACDB$ is a parallelogram since
slope $\overline{AB} = 1$ and slope $\overline{CD} = 1 \rightarrow \overline{AB}$ is parallel to $\overline{CD}$

slope $\overline{AC} = -1$ and slope $\overline{BD} = -1 \rightarrow \overline{AC}$ is parallel to $\overline{BD}$

Furthermore,
slope $\overline{AB} = 1$ and slope $\overline{BD} = -1 \rightarrow \overline{AB} \perp \overline{BD}$

slope $\overline{AC} = -1$ and slope $\overline{CD} = 1 \rightarrow \overline{AC} \perp \overline{CD}$

So the quadrilateral $ACDB$ is a rectangle.

79. $(^\circ C, ^\circ F) = (0, 32)$; $(^\circ C, ^\circ F) = (100, 212)$

slope $= \dfrac{212 - 32}{100 - 0} = \dfrac{180}{100} = \dfrac{9}{5}$

$^\circ F - 32 = \dfrac{9}{5}(^\circ C - 0)$

$^\circ F - 32 = \dfrac{9}{5}(^\circ C)$

$^\circ C = \dfrac{5}{9}(^\circ F - 32)$

If $^\circ F = 70$, then

$^\circ C = \dfrac{5}{9}(70 - 32) = \dfrac{5}{9}(38)$

$^\circ C \approx 21^\circ$

81. (a) Since there is only a profit of
$0.50 per copy and the expense
of $100 must be deducted, the
profit is:

$P = 0.50x - 100$

(b) $P = 0.50(1000) - 100$

$= 500 - 100 = \$400$

(c) $P = 0.50(5000) - 100$

$= 2500 - 100 = \$2400$

83. $C = 0.06543x + 5.65$

For 300 kWh,

$C = 0.06543(300) + 5.65 = \$25.28$

For 750 kWh,

$C = 0.06543(750) + 5.65 = \$54.72$

85. (b)   87. (d)

89. Slope = 1; y-intercept = 2
$y = x + 2$ or $x - y = -2$

91. Slope $= -\dfrac{1}{3}$; y-intercept $= 1$

$y = -\dfrac{1}{3}x + 1$ or $x + 3y = 3$

93. (b), (c), (e) and (g)

95. (c)

97. (a) $x^2 + (mx + b)^2 = r^2$

$x^2 + m^2x^2 + 2bmx + b^2 = r^2$

$(1 + m^2)x^2 + 2bmx + b^2 - r^2 = 0$

There is one solution if and only if the discriminant is zero.

$$(2bm)^2 - 4(1 + m^2)(b^2 - r^2) = 0$$

$$4b^2m^2 - 4b^2 + 4r^2 - 4b^2m^2 + 4m^2r^2 = 0$$

$$-4b^2 + 4r^2 + 4m^2r^2 = 0$$

$$-b^2 + r^2 + m^2r^2 = 0$$

$$r^2(1 + m^2) = b^2$$

(b) Using the quadratic formula, knowing that the discriminant is zero:

$$x = \frac{-2bm}{2(1 + m^2)} = \frac{-bm}{\dfrac{b^2}{r^2}} = \frac{-bmr^2}{b^2} = \frac{-mr^2}{b}$$

$$y = m\left(\frac{-mr^2}{b}\right) + b = \frac{-m^2r^2}{b} + b = \frac{-m^2r^2 + b^2}{b} = \frac{r^2}{b}$$

(c) The slope of the tangent line is m.

The slope of the line joining the point of tangency and the center is:

$$\frac{\left(\dfrac{r^2}{b} - 0\right)}{\left(\dfrac{-mr^2}{b} - 0\right)} = \frac{r^2}{b} \cdot \frac{b}{-mr^2} = -\frac{1}{m}$$

99. $x^2 + y^2 - 4x + 6y + 4 = 0$

$(x^2 - 4x + 4) + (y^2 + 6y + 9) = -4 + 4 + 9$

$(x - 2)^2 + (y + 3)^2 = 9$

Center: $(2, -3)$

Slope from center to $\left(3, 2\sqrt{2} - 3\right)$ is $\dfrac{2\sqrt{2} - 3 - (-3)}{3 - 2} = \dfrac{2\sqrt{2}}{1} = 2\sqrt{2}$

Slope of the tangent line is: $\dfrac{-1}{2\sqrt{2}} = \dfrac{-\sqrt{2}}{4}$

Equation of the tangent line:

$$y - \left(2\sqrt{2} - 3\right) = \frac{-\sqrt{2}}{4}(x - 3) \rightarrow y - 2\sqrt{2} + 3 = \frac{-\sqrt{2}}{4}x + \frac{3\sqrt{2}}{4}$$

$$4y - 8\sqrt{2} + 12 = -\sqrt{2}x + 3\sqrt{2} \rightarrow \sqrt{2}x + 4y = 11\sqrt{2} - 12$$

101. Find the centers of the two circles:

$$x^2 + y^2 - 4x + 6y + 4 = 0 \qquad\qquad x^2 + y^2 + 6x + 4y + 9 = 0$$

$$(x^2 - 4x + 4) + (y^2 + 6y + 9) = -4 + 4 + 9 \qquad (x^2 + 6x + 9) + (y^2 + 4y + 4) = -9 + 9 + 4$$

$$(x - 2)^2 + (y + 3)^2 = 9 \qquad\qquad (x + 3)^2 + (y + 2)^2 = 4$$

Center: (2, -3)   Center: (-3, -2)

    Find the slope of the line containing the centers:

$$m = \frac{-2 - (-3)}{-3 - 2} = -\frac{1}{5}$$

    Find the equation of the line containing the centers:

$$y + 3 = -\frac{1}{5}(x - 2) \rightarrow 5y + 15 = -x + 2 \rightarrow x + 5y = -13$$

103. $2x - y = C$

    Graph the lines:

        $2x - y = -2$

        $2x - y = 0$

        $2x - y = 4$

    All the lines have the same slope, 2.

    The lines are parallel.

105. $y = 2$

107. A vertical line cannot be written in slope-intercept form since the slope is undefined.

109. Two lines that have equal slopes and equal y-intercepts must be represented by equivalent equations.

111. Two lines that have the same slope but different x-intercepts cannot have the same y-intercept.

        Assume Line 1 has equation $y = mx + b_1$ and Line 2 has equation $y = mx + b_2$,

        Line 1 has x-intercept $-\dfrac{b_1}{m}$ and y-intercept b_1.

        Line 2 has x-intercept $-\dfrac{b_2}{m}$ and y-intercept b_2.

        Assume also that Line 1 and Line 2 have unequal x-intercepts.

        If the lines have the same y-intercept, then $b_1 = b_2$.

$$b_1 = b_2 \Rightarrow \frac{b_1}{m} = \frac{b_2}{m} \Rightarrow -\frac{b_1}{m} = -\frac{b_2}{m}$$

    But $-\dfrac{b_1}{m} = -\dfrac{b_2}{m} \Rightarrow$ Line 1 and Line 2 have the same x-intercept, which contradicts the

    original assumption that the lines have unequal x-intercepts.

    Therefore, Line 1 and Line 2 cannot have the same y-intercept.

113 – 115. Answers will vary.

Graphs

1.4   Scatter Diagrams; Linear Curve Fitting

1.   Linear, $m > 0$ 3.   Linear, $m < 0$ 5.   Nonlinear

7.   (a)

   (b)   Answers will vary. We select (4,6) and (8,14). The slope of the line containing these points is:
$$m = \frac{14-6}{8-4} = \frac{8}{4} = 2$$
The equation of the line is:
$$y - y_1 = m(x - x_1)$$
$$y - 6 = 2(x - 4)$$
$$y - 6 = 2x - 8$$
$$y = 2x - 2$$

   (c)

   (d)   Using the LINear REGresssion program, the line of best fit is:
$$y = 2.0357x - 2.3571$$

   (e)

9.   (a)

   (b)   Answers will vary. We select (–2,–4) and (1,4). The slope of the line containing these points is:
$$m = \frac{4-(-4)}{1-(-2)} = \frac{8}{3}$$
The equation of the line is:
$$y - y_1 = m(x - x_1)$$
$$y - (-4) = \frac{8}{3}(x - (-2))$$
$$y + 4 = \frac{8}{3}x + \frac{16}{3}$$
$$y = \frac{8}{3}x + \frac{4}{3}$$

(c)

(d) Using the LINear REGresssion program, the line of best fit is:
$$y = 2.2x + 1.2$$

(e)

11. (a)

(b) Answers will vary. We select (30,95) and (60,70). The slope of the line containing these points is:
$$m = \frac{70 - 95}{60 - 30} = \frac{-25}{30} = -\frac{5}{6}$$
The equation of the line is:
$$y - y_1 = m(x - x_1)$$
$$y - 95 = -\frac{5}{6}(x - 30)$$
$$y - 95 = -\frac{5}{6}x + 25$$
$$y = -\frac{5}{6}x + 120$$

(c)

(d) Using the LINear REGresssion program, the line of best fit is:
$$y = -0.72x + 116.6$$

(e)

13. (a)

(b) Answers will vary. We select
(−20,100) and (−15,118). The slope
of the line containing these points is:
$$m = \frac{118 - 100}{-15 - (-20)} = \frac{18}{5}$$
The equation of the line is:
$$y - y_1 = m(x - x_1)$$
$$y - 100 = \frac{18}{5}(x - (-20))$$
$$y - 100 = \frac{18}{5}x + 72$$
$$y = \frac{18}{5}x + 172$$

(c)

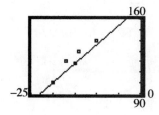

(d) Using the LINear REGresssion
program, the line of best fit is:
$$y = 3.8613x + 180.2920$$

(e)

15. (a)

(b) using points $(20,16)$ and $(50,39)$,
$$slope = \frac{39 - 16}{50 - 20} = \frac{23}{30}$$

the point slope formula yields
$$C - 16 = \frac{23}{30}(I - 20)$$
$$C = \left(\frac{23}{30}\right)I - \frac{460}{30} + 16$$
$$C = 0.76I + 0.67$$

(c) As disposable income increases by
$1, consumption increases by
$0.76.

(d) $C = 0.77(42) + 0.67 = \$33.01$
A family with disposable income of
$42,000 consumes about $33,010.

(e) $C = 0.7549I + 0.6266$

17. (a)   (Data used in graphs is in
                 thousands.)

     (b) $L = 2.9814I - 0.0761$
     (c)

     (d)   As annual income increases by
             $1, the loan amount increases
             by $2.9814.

     (e) $L = 2.9814(42) - 0.0761 = 125.143$
             A person with an annual income of
             $42,000 would qualify for a loan of
             about $125,143.

19. (a)

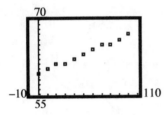

     (b) $T = 0.0782h + 59.0909$
     (c)

     (d)   As relative humidity increases
             by 1%, the apparent temperature
             increases by 0.0782°.

     (e) $T = 0.0782(75) + 59.0909 = 64.96$
             A relative humidity of 75% would
             give an apparent temperature of 65°.

21. (a)

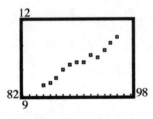

     (b) $M = 0.1633x - 4.4691$
     (c)

     (d)   As the year increases by 1, the
             average miles per car (in thousands)
             increases by 0.1633.

     (e) $M = 0.1633(97) - 4.4691 = 11.371$
             In 1997, the average number of miles
             driven per car is 11,371.

Graphs

1.R   Chapter Review

1. Intercepts: (0,0)
   Test for symmetry:
   x - axis: Replace y by $-y$ so $2x = 3(-y)^2$ or $2x = 3y^2$, which is
                   equivalent to $2x = 3y^2$.

   y - axis: Replace x by $-x$ so $2(-x) = 3y^2$ or $-2x = 3y^2$,
                   which is not equivalent to $2x = 3y^2$.

   Origin :   Replace x by $-x$ and y by $-y$ so $2(-x) = 3(-y)^2$ or
                   $-2x = 3y^2$, which is not equivalent to $2x = 3y^2$.

   Therefore, the graph is symmetric with respect to the x - axis.

3. Intercepts: (0, 2), (0, -2), (4, 0), (-4, 0)
   Test for symmetry:
   x - axis: Replace y by $-y$ so $x^2 + 4(-y)^2 = 16$ or $x^2 + 4y^2 = 16$,
                   which is equivalent to $x^2 + 4y^2 = 16$.

   y - axis: Replace x by $-x$ so $(-x)^2 + 4y^2 = 16$ or $x^2 + 4y^2 = 16$,
                   which is equivalent to $x^2 + 4y^2 = 16$.

   Origin :   Replace x by $-x$ and y by $-y$ so $(-x)^2 + 4(-y)^2 = 16$ or $x^2 + 4y^2 = 16$,
                   which is equivalent to $x^2 + 4y^2 = 16$.

   Therefore, the graph is symmetric with respect to the x-axis, the y-axis and the origin.

5. Intercepts: (0, 1)
   Test for symmetry:
   x - axis: Replace y by $-y$ so $-y = x^4 + 2x^2 + 1$,
                   which is not equivalent to $y = x^4 + 2x^2 + 1$.

   y - axis: Replace x by $-x$ so $y = (-x)^4 + 2(-x)^2 + 1$ or $y = x^4 + 2x^2 + 1$,
                   which is equivalent to $y = x^4 + 2x^2 + 1$.

   Origin :   Replace x by $-x$ and y by $-y$ so $-y = (-x)^4 + 2(-x)^2 + 1$ or $-y = x^4 + 2x^2 + 1$,
                   which is not equivalent to $y = x^4 + 2x^2 + 1$.

   Therefore, the graph is symmetric with respect to the y-axis.

7.   Intercepts: (0,0), (0,–2), (–1,0)
      Test for symmetry:
      x - axis: Replace y by $-y$ so $x^2 + x + (-y)^2 + 2(-y) = 0$ or $x^2 + x + y^2 - 2y = 0$,
                 which is not equivalent to $x^2 + x + y^2 + 2y = 0$.
      y - axis: Replace x by $-x$ so $(-x)^2 + (-x) + y^2 + 2y = 0$ or $x^2 - x + y^2 + 2y = 0$,
                 which is not equivalent to $x^2 + x + y^2 + 2y = 0$.
      Origin :   Replace x by $-x$ and y by $-y$ so $(-x)^2 + (-x) + (-y)^2 + 2(-y) = 0$ or
                 $x^2 - x + y^2 - 2y = 0$, which is not equivalent to $x^2 + x + y^2 + 2y = 0$.
      Therefore, the graph is not symmetric to the x-axis, the y-axis, or the origin.

9.   Slope = –2; containing (3,–1)
      $y - y_1 = m(x - x_1)$

      $y - (-1) = -2(x - 3)$

      $y + 1 = -2x + 6 \rightarrow y = -2x + 5$

      $2x + y = 5$ or $y = -2x + 5$

11.   Slope undefined; containing (–3,4)
      This is a vertical line.
      $x = -3$
         No slope intercept form.

13.   y-intercept = –2; containing (5,–3)
      Points are (5,–3) and (0,–2)
      $m = \dfrac{-2 - (-3)}{0 - 5} = \dfrac{1}{-5} = -\dfrac{1}{5}$
      $y = mx + b$

      $y = -\dfrac{1}{5}x - 2$

      $x + 5y = -10$ or $y = -\dfrac{1}{5}x - 2$

15.   Parallel to $2x - 3y = -4$;
      Slope $= \dfrac{2}{3}$; containing (–5,3)
         $y - y_1 = m(x - x_1)$

         $y - 3 = \dfrac{2}{3}(x - (-5)) \rightarrow y - 3 = \dfrac{2}{3}x + \dfrac{10}{3}$

         $y = \dfrac{2}{3}x + \dfrac{19}{3}$

      $2x - 3y = -19$ or $y = \dfrac{2}{3}x + \dfrac{19}{3}$

17.   Perpendicular to $x + y = 2$;
      Containing (4,–3)
      Slope of perpendicular = 1
         $y - y_1 = m(x - x_1)$
      $y - (-3) = 1(x - 4)$
         $y + 3 = x - 4$
            $y = x - 7$
         $x - y = 7$ or $y = x - 7$

19. $4x - 5y = -20$
      x-intercept = –5; y-intercept = 4

21. $\frac{1}{2}x - \frac{1}{3}y = -\frac{1}{6}$

x - intercept $= -\frac{1}{3}$;   y - intercept $= \frac{1}{2}$

23. $\sqrt{2}x + \sqrt{3}y = \sqrt{6}$

x-intercept $= \sqrt{3}$;   y-intercept $= \sqrt{2}$

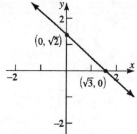

25. $x^2 + (y-1)^2 = 4$
Center: (0,1)
Radius $= 2$

27. $x^2 + y^2 - 2x + 4y - 4 = 0$
$$x^2 - 2x + y^2 + 4y = 4$$
$$(x^2 - 2x + 1) + (y^2 + 4y + 4) = 4 + 1 + 4$$
$$(x-1)^2 + (y+2)^2 = 3^2$$
Center: (1,–2)   Radius $= 3$

29. $3x^2 + 3y^2 - 6x + 12y = 0$
$$x^2 + y^2 - 2x + 4y = 0$$
$$x^2 - 2x + y^2 + 4y = 0$$
$$(x^2 - 2x + 1) + (y^2 + 4y + 4) = 1 + 4$$
$$(x-1)^2 + (y+2)^2 = \left(\sqrt{5}\right)^2$$
Center: (1,–2)   Radius $= \sqrt{5}$

31. Given the points (7,4) and (–3,2).

Slope: $m = \dfrac{2-4}{-3-7} = \dfrac{-2}{-10} = \dfrac{1}{5}$

Distance: $d = \sqrt{(-3-7)^2 + (2-4)^2} = \sqrt{100+4} = \sqrt{104} = 2\sqrt{26}$

Midpoint: $\left(\dfrac{7+(-3)}{2}, \dfrac{4+2}{2}\right) = (2,3)$

33. Given the points $A = (-2, 0)$, $B = (-4, 4)$, $C = (8, 5)$.
  (a) Find the distance between each pair of points.

$$d_{A,B} = \sqrt{(-4-(-2))^2 + (4-0)^2} = \sqrt{4+16} = \sqrt{20} = 2\sqrt{5}$$

$$d_{B,C} = \sqrt{(8-(-4))^2 + (5-4)^2} = \sqrt{144+1} = \sqrt{145}$$

$$d_{A,C} = \sqrt{(8-(-2))^2 + (5-0)^2} = \sqrt{100+25} = \sqrt{125} = 5\sqrt{5}$$

$$\left(\sqrt{20}\right)^2 + \left(\sqrt{125}\right)^2 = \left(\sqrt{145}\right)^2 \quad \rightarrow \quad 20+125 = 145 \quad \rightarrow \quad 145 = 145$$

The Pythagorean theorem is satisfied, so this is a right triangle.

  (b) Find the slopes:

$$m_{AB} = \frac{4-0}{-4-(-2)} = \frac{4}{-2} = -2; \; m_{BC} = \frac{5-4}{8-(-4)} = \frac{1}{12}; \; m_{AC} = \frac{5-0}{8-(-2)} = \frac{5}{10} = \frac{1}{2}$$

$$m_{AB} \cdot m_{AC} = -2 \cdot \frac{1}{2} = -1$$

Since the product of the slopes is –1, the sides of the triangle are perpendicular and the triangle is a right triangle.

35. slope of $\overline{AB} = \dfrac{1-5}{6-2} = -1$; slope of $\overline{AC} = \dfrac{-1-5}{8-2} = -1$; slope of $\overline{BC} = \dfrac{-1-1}{8-6} = -1$

therefore the points are collinear.

37. (a)

  (b) $\text{slope} = \dfrac{5.87 - 6.69}{1990 - 1987} = -0.27\overline{3}$

  (c) for each 1 year increase, the concentration decreases by $0.27\overline{3}$ ppm.

  (d) $\text{slope} = \dfrac{4.88 - 5.87}{1993 - 1990} = -0.33$

  (e) for each 1 year increase, the concentration decreases by 0.33 ppm.

  (f) $y = 618.477 - 0.308x$

  (g) for each 1 year increase, the concentration decreases by 0.308 ppm.

  (h) Answers will vary.

  (i) As time passes, the average level of carbon monoxide is decreasing more rapidly.

39. Answers will vary.

41. Set the axes so that the field's maximum dimension is along the x-axis.

Let $2h = width$, $2k = height$, therefore the point farthest from the origin has coordinates $P(h,k)$. So the distance from the origin to point P is $r = \sqrt{h^2 + k^2}$ = the radius of the circle

Using 1 sprinkler arm:

If we place the sprinkler at the origin, we get a circle with equation $x^2 + y^2 = r^2$, where $r = \sqrt{h^2 + k^2}$. So how much water is wasted ? The area of the field $= A_F = 4hk$.

The area of the circular water pattern $A_C = \pi \cdot r^2 = \pi \cdot \left(\sqrt{h^2 + k^2} \right)^2 = \pi \cdot \left(h^2 + k^2 \right)$.

Therefore the amount of water wasted $= A_C - A_F = \pi \cdot \left(h^2 + k^2 \right) - 4hk$.

Using 2 sprinkler arms:

Place sprinkler 1 at $\frac{1}{4} \cdot (width)$ and sprinkler 2 at $\frac{3}{4} \cdot (width)$. Let the origin correspond to the center of the circle formed by sprinkler 1, the arm must be set so that the water will hit the point $Q\left(\frac{1}{2}h, k \right)$.

The distance from the origin to point Q is given by $r = \sqrt{\frac{1}{4}h^2 + k^2}$ = the radius of circle 1.

We need the same settings for sprinkler 2.

So the total area of the 2 circles $= 2\pi \cdot r^2 = 2\pi\left(\sqrt{\frac{1}{4}h^2 + k^2} \right)^2 = 2\pi\left(\frac{1}{4}h^2 + k^2 \right)$

Therefore the amount of water wasted $= A_C - A_F = 2\pi\left(\frac{1}{4}h^2 + k^2 \right) - 4hk$.

In order to decide when to switch from Case 1 to Case 2, we want to determine when the waste in Case 2 is less than the waste in Case 1.

That is, we want to solve: $2\pi\left(\frac{1}{4}h^2 + k^2 \right) - 4hk < \pi \cdot \left(h^2 + k^2 \right) - 4hk$

$2\pi\left(\frac{1}{4}h^2 + k^2 \right) - 4hk < \pi \cdot \left(h^2 + k^2 \right) - 4hk$

$\frac{\pi}{2}h^2 + 2\pi k^2 - 4hk < \pi h^2 + \pi k^2 - 4hk$

$\frac{\pi}{2}h^2 + 2\pi k^2 < \pi h^2 + \pi k^2$

$\frac{1}{2}h^2 + 2k^2 < h^2 + k^2 \rightarrow k^2 < \frac{1}{2}h^2 \rightarrow k < \sqrt{\frac{1}{2}} \cdot h$

So Case 2 is the better choice when the rectangle's dimensions obey the inequality $\frac{k}{h} < \frac{\sqrt{2}}{2}$.

Functions and Their Graphs

2.1   Functions

1. Function
    Domain: {Dad, Colleen, Kaleigh,
    Marissa}
    Range: {Jan. 8, Mar. 15, Sept. 17}

3. Not a function

5. Function
    Domain: {2, –3, 4, 1}
    Range: {6, 9, 10}

7. Function
    Domain: {1, 2, 3, 4}
    Range: {3}

9. Not a function

11. Function
    Domain: {–2, –1, 0, 1}
    Range: {4, 1, 0}

13. $f(x) = 3x^2 + 2x - 4$
    (a) $f(0) = 3(0)^2 + 2(0) - 4 = -4$
    (b) $f(1) = 3(1)^2 + 2(1) - 4 = 3 + 2 - 4 = 1$
    (c) $f(-1) = 3(-1)^2 + 2(-1) - 4 = 3 - 2 - 4 = -3$
    (d) $f(-x) = 3(-x)^2 + 2(-x) - 4 = 3x^2 - 2x - 4$
    (e) $-f(x) = -(3x^2 + 2x - 4) = -3x^2 - 2x + 4$
    (f) $f(x + 1) = 3(x + 1)^2 + 2(x + 1) - 4 = 3(x^2 + 2x + 1) + 2x + 2 - 4$
         $= 3x^2 + 6x + 3 + 2x + 2 - 4 = 3x^2 + 8x + 1$
    (g) $f(2x) = 3(2x)^2 + 2(2x) - 4 = 12x^2 + 4x - 4$
    (h) $f(x + h) = 3(x + h)^2 + 2(x + h) - 4 = 3(x^2 + 2xh + h^2) + 2x + 2h - 4$
         $= 3x^2 + 6xh + 3h^2 + 2x + 2h - 4$

15. $f(x) = \dfrac{x}{x^2 + 1}$
    (a) $f(0) = \dfrac{0}{0^2 + 1} = \dfrac{0}{1} = 0$
    (b) $f(1) = \dfrac{1}{1^2 + 1} = \dfrac{1}{2}$
    (c) $f(-1) = \dfrac{-1}{(-1)^2 + 1} = \dfrac{-1}{1 + 1} = -\dfrac{1}{2}$
    (d) $f(-x) = \dfrac{-x}{(-x)^2 + 1} = \dfrac{-x}{x^2 + 1}$

(e) $-f(x) = -\dfrac{x}{x^2+1} = \dfrac{-x}{x^2+1}$

(f) $f(x+1) = \dfrac{x+1}{(x+1)^2+1} = \dfrac{x+1}{x^2+2x+1+1} = \dfrac{x+1}{x^2+2x+2}$

(g) $f(2x) = \dfrac{2x}{(2x)^2+1} = \dfrac{2x}{4x^2+1}$

(h) $f(x+h) = \dfrac{x+h}{(x+h)^2+1} = \dfrac{x+h}{x^2+2xh+h^2+1}$

17. $f(x) = |x|+4$

(a) $f(0) = |0|+4 = 0+4 = 4$

(b) $f(1) = |1|+4 = 1+4 = 5$

(c) $f(-1) = |-1|+4 = 1+4 = 5$

(d) $f(-x) = |-x|+4 = |x|+4$

(e) $-f(x) = -(|x|+4) = -|x|-4$

(f) $f(x+1) = |x+1|+4$

(g) $f(2x) = |2x|+4 = 2|x|+4$

(h) $f(x+h) = |x+h|+4$

19. $f(x) = \dfrac{2x+1}{3x-5}$

(a) $f(0) = \dfrac{2(0)+1}{3(0)-5} = \dfrac{0+1}{0-5} = -\dfrac{1}{5}$

(b) $f(1) = \dfrac{2(1)+1}{3(1)-5} = \dfrac{2+1}{3-5} = \dfrac{3}{-2} = -\dfrac{3}{2}$

(c) $f(-1) = \dfrac{2(-1)+1}{3(-1)-5} = \dfrac{-2+1}{-3-5} = \dfrac{-1}{-8} = \dfrac{1}{8}$

(d) $f(-x) = \dfrac{2(-x)+1}{3(-x)-5} = \dfrac{-2x+1}{-3x-5} = \dfrac{2x-1}{3x+5}$

(e) $-f(x) = -\dfrac{2x+1}{3x-5} = \dfrac{-2x-1}{3x-5}$

(f) $f(x+1) = \dfrac{2(x+1)+1}{3(x+1)-5} = \dfrac{2x+2+1}{3x+3-5} = \dfrac{2x+3}{3x-2}$

(g) $f(2x) = \dfrac{2(2x)+1}{3(2x)-5} = \dfrac{4x+1}{6x-5}$

(h) $f(x+h) = \dfrac{2(x+h)+1}{3(x+h)-5} = \dfrac{2x+2h+1}{3x+3h-5}$

21. Graph $y = x^2$. The graph passes the vertical line test. Thus, the equation represents a function.

23. Graph $y = \dfrac{1}{x}$. The graph passes the vertical line test. Thus, the equation represents a function.

25. $y^2 = 4 - x^2$

Solve for y: $y = \pm\sqrt{4 - x^2}$

For $x = 0, y = \pm 2$. Thus, (0,2) and (0,–2) are on the graph. This is not a function, since a distinct x corresponds to two different y's.

27. $x = y^2$

Solve for y: $y = \pm\sqrt{x}$

For $x = 1, y = \pm 1$. Thus, (1,1) and (1,–1) are on the graph. This is not a function, since a distinct x corresponds to two different y's.

29. Graph $y = 2x^2 - 3x + 4$. The graph passes the vertical line test. Thus, the equation represents a function.

31. $2x^2 + 3y^2 = 1$

Solve for y:

$$2x^2 + 3y^2 = 1 \rightarrow 3y^2 = 1 - 2x^2 \rightarrow y^2 = \frac{1 - 2x^2}{3}$$

$$y = \pm\sqrt{\frac{1 - 2x^2}{3}}$$

For $x = 0, y = \pm\sqrt{\frac{1}{3}}$. Thus, $\left(0, \sqrt{\frac{1}{3}}\right)$ and $\left(0, -\sqrt{\frac{1}{3}}\right)$ are on the graph. This is not a function, since a distinct x corresponds to two different y's.

33. $f(x) = -5x + 4$

Domain: {Real Numbers}

35. $f(x) = \frac{x}{x^2 + 1}$

Domain: {Real Numbers}

37. $g(x) = \frac{x}{x^2 - 16}$

$x^2 - 16 \neq 0$

$x^2 \neq 16 \rightarrow x \neq \pm 4$

Domain: $\{x | x \neq -4, x \neq 4\}$

39. $F(x) = \frac{x - 2}{x^3 + x}$

$x^3 + x \neq 0$

$x(x^2 + 1) \neq 0$

$x \neq 0, \quad x^2 \neq -1$

Domain: $\{x | x \neq 0\}$

41. $h(x) = \sqrt{3x - 12}$

$3x - 12 \geq 0$

$3x \geq 12$

$x \geq 4$

Domain: $\{x | x \geq 4\}$

43. $f(x) = \frac{4}{\sqrt{x - 9}}$

$x - 9 > 0$

$x > 9$

Domain: $\{x | x > 9\}$

45. $p(x) = \sqrt{\dfrac{2}{x-1}}$

$\dfrac{2}{x-1} > 0 \quad \rightarrow \quad x-1 > 0 \quad \rightarrow \quad x > 1$

Domain: $\left\{x \mid x > 1\right\}$

47. (a) $f(0) = 3$ since $(0,3)$ is on the graph.

$f(-6) = -3$ since $(-6,-3)$ is on the graph.

(b) $f(6) = 0$ since $(6, 0)$ is on the graph.

$f(11) = 1$ since $(11, 1)$ is on the graph.

(c) $f(3)$ is positive since $f(3) \approx 3.7$.

(d) $f(-4)$ is negative since $f(-4) = -1$.

(e) $f(x) = 0$ when $x = -3$, $x = 6$, and $x = 10$.

(f) $f(x) > 0$ when $-3 < x < 6$, and $10 < x \leq 11$.

(g) The domain of f is $\left\{x \mid -6 \leq x \leq 11\right\}$ or $\left[-6, 11\right]$

(h) The range of f is $\left\{y \mid -3 \leq y \leq 4\right\}$ or $\left[-3, 4\right]$

(i) The x-intercepts are $(-3, 0), (6, 0)$, and $(11, 0)$.

(j) The y-intercept is $(0, 3)$

(k) The line $y = \dfrac{1}{2}$ intersects the graph 3 times.

(l) The line $x = 5$ intersects the graph 1 time.

(m) $f(x) = 3$ when $x = 0$ and $x = 4$.

(n) $f(x) = -2$ when $x = -5$ and $x = 8$.

49. Not a function since vertical lines will intersect the graph in more than one point.

51. Function   (a) Domain: $\left\{x \mid -\pi \leq x \leq \pi\right\}$;   Range: $\left\{y \mid -1 \leq y \leq 1\right\}$

(b) $\left(-\dfrac{\pi}{2}, 0\right)$, $\left(\dfrac{\pi}{2}, 0\right)$, $(0,1)$ (c) y-axis

53. Not a function since vertical lines will intersect the graph in more than one point.

55. Function   (a) Domain: $\left\{x \mid x > 0\right\}$;   Range: $\left\{y \mid y \in \text{Real Numbers}\right\}$

(b) $(1, 0)$

(c) No symmetry to the x-axis, y-axis, or origin.

57. Function   (a) Domain: $\left\{x \mid x \in \text{Real Numbers}\right\}$;   Range: $\left\{y \mid y \leq 2\right\}$

(b) $(-3,0), (3,0), (0,2)$

(c) y-axis

59. Function   (a) Domain: $\left\{x \mid x \in \text{Real Numbers}\right\}$;   Range: $\left\{y \mid y \geq -3\right\}$

(b) $(1,0), (3,0), (0,9)$

(c) No symmetry to the x-axis, y-axis, or origin.

61.
$f(x) = 2x^2 - x - 1$

(a) $f(-1) = 2(-1)^2 - (-1) - 1 = 2$ (-1,2) is on the graph of f.

(b) $f(-2) = 2(-2)^2 - (-2) - 1 = 9$ (-2,9) is on the graph of f.

(c) Solve for x:
$$-1 = 2x^2 - x - 1$$
$$0 = 2x^2 - x$$
$$0 = x(2x - 1) \to x = 0, x = \frac{1}{2}$$

(0, -1) and $\left(\frac{1}{2}, -1\right)$ are points on the graph of f.

(d) The domain of f is: $\{x \mid x \text{ is any real number}\}$.

(e) x-intercepts:
$$f(x) = 0$$
$$2x^2 - x - 1 = 0 \to (2x + 1)(x - 1) = 0$$
$$\to x = -\frac{1}{2}, x = 1 \to \left(-\frac{1}{2}, 0\right) \text{ and } (1,0)$$

(f) y-intercept: $f(0) = 2(0)^2 - 0 - 1 = 0 \to (0,0)$

63. $f(x) = \dfrac{x + 2}{x - 6}$

(a) $f(3) = \dfrac{3 + 2}{3 - 6} = -\dfrac{5}{3} \neq 14$ (3,14) is not on the graph of f.

(b) $f(4) = \dfrac{4 + 2}{4 - 6} = \dfrac{6}{-2} = -3$ (4,-3) is the point on the graph of f.

(c) Solve for x:
$$2 = \frac{x + 2}{x - 6}$$
$$2x - 12 = x + 2 \qquad\qquad (14, 2) \text{ is a point on the graph of } f.$$
$$x = 14$$

(d) The domain of f is $\{x \mid x \neq 6\}$.

(e) x-intercepts:
$$f(x) = 0$$
$$\frac{x + 2}{x - 6} = 0 \to x + 2 = 0$$
$$\to x = -2 \to (-2,0)$$

(f) y-intercept: $f(0) = \dfrac{0 + 2}{0 - 6} = -\dfrac{1}{3} \to \left(0, -\dfrac{1}{3}\right)$

65. $f(x) = \dfrac{2x^2}{x^4 + 1}$

(a) $f(-1) = \dfrac{2(-1)^2}{(-1)^4 + 1} = \dfrac{2}{2} = 1$ (-1,1) is a point on the graph of f.

(b) $f(2) = \dfrac{2(2)^2}{(2)^4 + 1} = \dfrac{8}{17}$ $\left(2, \dfrac{8}{17}\right)$ is a point on the graph of f.

(c) Solve for x:

$$1 = \frac{2x^2}{x^4 + 1}$$

$$x^4 + 1 = 2x^2$$

$$x^4 - 2x^2 + 1 = 0 \qquad\qquad (1,1) \text{ and } (-1,1) \text{ are points on the graph of } f.$$

$$(x^2 - 1)^2 = 0$$

$$x^2 - 1 = 0 \rightarrow x = \pm 1$$

(d) The domain of f is $\{\text{Real Numbers}\}$.

(e) x-intercepts:

$$f(x) = 0$$

$$\frac{2x^2}{x^4 + 1} = 0 \rightarrow 2x^2 = 0 \rightarrow x = 0 \rightarrow (0,0)$$

(f) y-intercept: $f(0) = \dfrac{2x^2}{x^4 + 1} = \dfrac{0}{0 + 1} = 0 \rightarrow (0,0)$

67. Solving for A:

$f(x) = 2x^3 + Ax^2 + 4x - 5$ and $f(2) = 5$

$f(2) = 2(2)^3 + A(2)^2 + 4(2) - 5$

$5 = 16 + 4A + 8 - 5$

$5 = 4A + 19$

$-14 = 4A \rightarrow A = -\dfrac{7}{2}$

69. Solving for A:

$f(x) = \dfrac{3x + 8}{2x - A}$ and $f(0) = 2$

$f(0) = \dfrac{3(0) + 8}{2(0) - A}$

$2 = \dfrac{8}{-A}$

$-2A = 8$

$A = -4$

71. Solving for A:

$f(x) = \dfrac{2x - A}{x - 3}$ and $f(4) = 0$

$f(4) = \dfrac{2(4) - A}{4 - 3}$

$0 = \dfrac{8 - A}{1}$

$0 = 8 - A$

$A = 8$

f is undefined when $x = 3$.

73. (a) III   (b) IV   (c) I   (d) V   (e) II

75.

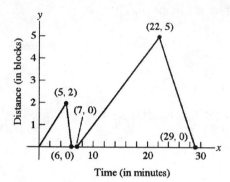

77. (a) Two times
     (b) Kevin's distance from home increased steadily at a rate of 1.5 miles per hour.
     (c) Kevin's distance from home did not change.
     (d) Kevin's distance from home decreased steadily at a rate of 10 miles per hour.
     (e) Kevin stayed at home for 0.2 hours.
     (f) Kevin's distance from home increased rapidly at the beginning and then tapered off to a very slow change in distance from home.
     (g) Kevin's distance from home did not change.
     (h) Kevin's distance from home decreased rapidly at the beginning of the interval and then tapered off as he got closer to home.
     (i) The furthest distance Kevin is from home is 3 miles.

79. (a) $H(1) = 20 - 4.9(1)^2 = 20 - 4.9 = 15.1$ meters
         $H(1.1) = 20 - 4.9(1.1)^2 = 20 - 4.9(1.21) = 20 - 5.929 = 14.071$ meters
         $H(1.2) = 20 - 4.9(1.2)^2 = 20 - 4.9(1.44) = 20 - 7.056 = 12.944$ meters
         $H(1.3) = 20 - 4.9(1.3)^2 = 20 - 4.9(1.69) = 20 - 8.281 = 11.719$ meters

     (b) $H(x) = 15$ $H(x) = 10$ $H(x) = 5$
         $15 = 20 - 4.9x^2$ $10 = 20 - 4.9x^2$ $5 = 20 - 4.9x^2$
         $-5 = -4.9x^2$ $-10 = -4.9x^2$ $-15 = -4.9x^2$
         $x^2 = 1.0204$ $x^2 = 2.0408$ $x^2 = 3.0612$
         $x = 1.01$ seconds $x = 1.43$ seconds $x = 1.75$ seconds

     (c) $H(x) = 0$
         $0 = 20 - 4.9x^2$
         $-20 = -4.9x^2 \rightarrow x^2 = 4.0816 \rightarrow x = 2.02$ seconds

81. $h(x) = -\dfrac{32x^2}{130^2} + x$

     (a) $h(100) = -\dfrac{32(100)^2}{130^2} + 100 = -\dfrac{320000}{16900} + 100 = -18.93 + 100 = 81.07$ feet

     (b) $h(300) = -\dfrac{32(300)^2}{130^2} + 300 = -\dfrac{2880000}{16900} + 300 = -170.41 + 300 = 129.59$ feet

(c) $h(500) = -\dfrac{32(500)^2}{130^2} + 500 = -\dfrac{8000000}{16900} + 500 = -473.37 + 500 = 26.63$ feet

(d)   Solve $h(x) = -\dfrac{32x^2}{130^2} + x = 0$

$x\left(-\dfrac{32x}{130^2} + 1\right) = 0$

$x = 0$ or $-\dfrac{32x}{130^2} + 1 = 0 \rightarrow 1 = \dfrac{32x}{130^2} \rightarrow x = \dfrac{130^2}{32} = 528.125$ feet

83. $C(x) = 100 + \dfrac{x}{10} + \dfrac{36000}{x}$

(a) $C(500) = 100 + \dfrac{500}{10} + \dfrac{36000}{500} = 100 + 50 + 72 = \$222$

(b) $C(450) = 100 + \dfrac{450}{10} + \dfrac{36000}{450} = 100 + 45 + 80 = \$225$

(c) $C(600) = 100 + \dfrac{600}{10} + \dfrac{36000}{600} = 100 + 60 + 60 = \$220$

(d) $C(400) = 100 + \dfrac{400}{10} + \dfrac{36000}{400} = 100 + 40 + 90 = \$230$

(e)   Graphing:

(f)   As x varies from 400 to 600 mph, the cost decreases from \$230 to \$220.

85.   Let x represent the length of the rectangle.

Then $\dfrac{x}{2}$ represents the width of the rectangle, since the length is twice the width.

The function for the area is: $A(x) = x \cdot \dfrac{x}{2} = \dfrac{x^2}{2} = \dfrac{1}{2}x^2$

87.   Let x represent the number of hours worked.
The function for the gross salary is: $G(x) = 10x$

89.   (a)   The relation is not a function   (b)
because 23 is paired with both
56 and 53.

(c) Using the points (20,60) and
(30, 44) we get
$D = -1.6p + 92$

(d) As the price of the jeans increases by
$1, the demand for the jeans decreases
by 1.6.

(e) $D(p) = -1.6p + 92$

(f) Domain: $\{p \mid p > 0\}$

(g) $D(28) = -1.6(28) + 92$
$= 47.2 \approx 47$
Demand is about 47 pairs.

(h) $D = -1.3355p + 86.1974$

91. (a) The relation is a function.
Each time value is paired with
exactly one distance value.

(b)

(c) Using the points (0,0) and
(8, 300) we get
$s = 37.5t$

(d) As the time increases by 1 hour, the
distances increases by 37.5 miles.

(e) $s(t) = 37.5t$

(f) Domain: $\{t \mid t \geq 0\}$

(g) $s(11) = 37.5(11)$
$= 412.5$ miles

(h) $s = 37.7833t - 19.1333$

93. (a) $h(x) = 2x$
$h(a + b) = 2(a + b) = 2a + 2b = h(a) + h(b); \quad h(x) = 2x$ has the property.

(b) $g(x) = x^2$
$g(a + b) = (a + b)^2 = a^2 + 2ab + b^2 \neq a^2 + b^2 = h(a) + h(b)$
$g(x) = x^2$ does not have the property.

(c) $F(x) = 5x - 2$
$F(a + b) = 5(a + b) - 2 = 5a + 5b - 2 \neq 5a - 2 + 5b - 2 = h(a) + h(b)$
$F(x) = 5x - 2$ does not have the property.

(d) $G(x) = \dfrac{1}{x}$
$G(a + b) = \dfrac{1}{a + b} \neq \dfrac{1}{a} + \dfrac{1}{b} = h(a) + h(b)$
$G(x) = \dfrac{1}{x}$ does not have the property.

95. No, $x = -1$ is not in the domain of g, but it is in the domain of f.

97. A function may have any number of x-intercepts; it can have only one y-intercept.

99. The only such function is $f(x) = 0$.

Chapter 2

Functions and Their Graphs

2.2   Properties of Functions

1. Yes

3. No   It only increases on (5, 10).

5. f is increasing on the intervals: $(-8, -2), (0, 2), (5, 10)$.

7. Yes. The local maximum at $x = 2$ is 10.

9. f has local maxima at $x = -2$ and $x = 2$. The local maxima are 6 and 10, respectively.

11. (a) Intercepts: $(-2,0), (2,0)$, and $(0,3)$.
    (b) Domain: $\{x \mid -4 \le x \le 4\}$; Range: $\{y \mid 0 \le y \le 3\}$.
    (c) Interval notation: Increasing: $(-2, 0)$ and $(2, 4)$; Decreasing: $(-4, -2)$ and $(0, 2)$.
         Inequality notation: Increasing: $-2 < x < 0$ and $2 < x < 4$
                                     Decreasing: $-4 < x < -2$ and $0 < x < 2$
    (d) Since the graph is symmetric to the y-axis, the function is <u>even</u>.

13. (a) Intercepts: $(0,1)$.
    (b) Domain: { Real Numbers }; Range: $\{y \mid y > 0\}$.
    (c) Interval notation: Increasing: $(-\infty, +\infty)$; Decreasing: never.
         Inequality notation: Increasing: $-\infty < x < +\infty$
                                     Decreasing: never
    (d) Since the graph is not symmetric to the y-axis or the origin, the function is <u>neither</u> even nor odd.

15. (a) Intercepts: $(-\pi,0), (0,0)$, and $(\pi,0)$.
    (b) Domain: $\{x \mid -\pi \le x \le \pi\}$; Range: $\{y \mid -1 \le y \le 1\}$.
    (c) Interval notation: Increasing: $\left(-\dfrac{\pi}{2}, \dfrac{\pi}{2}\right)$; Decreasing: $\left(-\pi, -\dfrac{\pi}{2}\right)$ and $\left(\dfrac{\pi}{2}, \pi\right)$.

         Inequality notation: Increasing: $-\dfrac{\pi}{2} < x < \dfrac{\pi}{2}$

                                     Decreasing: $-\pi < x < -\dfrac{\pi}{2}$ and $\dfrac{\pi}{2} < x < \pi$

    (d) Since the graph is symmetric to the origin, the function is <u>odd</u>.

17. (a) Intercepts: $\left(0,\frac{1}{2}\right),\left(\frac{1}{3},0\right)$, and $\left(\frac{5}{2},0\right)$.

(b) Domain: $\left\{x\middle|-3 \le x \le 3\right\}$; Range: $\left\{y\middle|-1 \le y \le 2\right\}$.

(c) Interval notation: Increasing: $(2, 3)$; Decreasing: $(-1, 1)$;

Constant: $(-3, -1)$ and $(1, 2)$.

Inequality notation: Increasing: $2 < x < 3$; Decreasing: $-1 < x < 1$;

Constant: $-3 < x < -1$ and $1 < x < 2$.

(d) Since the graph is not symmetric to the y-axis or the origin, the function is <u>neither</u> even nor odd.

19. (a) Intercepts: $(0, 2), (-2, 0)$, and $(2, 0)$.

(b) Domain: $\left\{x\middle|-4 \le x \le 4\right\}$; Range: $\left\{y\middle|0 \le y \le 2\right\}$.

(c) Interval notation: Increasing: $(-2, 0)$ and $(2, 4)$;

Decreasing: $(-4, -2)$ and $(0, 2)$.

Inequality notation: Increasing: $-2 < x < 0$ and $2 < x < 4$;

Decreasing: $-4 < x < -2$ and $0 < x < 2$.

(d) Since the graph is symmetric to the y-axis, the function is <u>even</u>.

21. (a) f has a local maximum of 3 at $x = 0$.

(b) f has a local minimum of 0 at both $x = -2$ and $x = 2$.

23. (a) f has a local maximum of 1 at $x = \dfrac{\pi}{2}$.

(b) f has a local minimum of -1 at $x = -\dfrac{\pi}{2}$.

25. $f(x) = 5x$

(a) $\dfrac{f(x)-f(1)}{x-1} = \dfrac{5x-5}{x-1} = \dfrac{5(x-1)}{x-1} = 5$

(b) $\dfrac{f(2)-f(1)}{2-1} = \dfrac{10-5}{2-1} = \dfrac{5}{1} = 5$

Slope = 5; Containing (1, 5):

$y - 5 = 5(x-1)$

(c) $y - 5 = 5x - 5$

$\quad\quad y = 5x$

27. $f(x) = 1 - 3x$

(a) $\dfrac{f(x)-f(1)}{x-1} = \dfrac{1-3x-(-2)}{x-1}$

$= \dfrac{-3x+3}{x-1} = \dfrac{-3(x-1)}{x-1} = -3$

(b) $\dfrac{f(2)-f(1)}{2-1} = \dfrac{1-3(2)-(-2)}{2-1} = \dfrac{-3}{1} = -3$

Slope = -3; Containing (1, -2):

$y - (-2) = -3(x-1)$

(c) $\quad y + 2 = -3x + 3$

$\quad\quad\quad y = -3x + 1$

29. $f(x) = x^2 - 2x$

(a) $\dfrac{f(x) - f(1)}{x - 1} = \dfrac{x^2 - 2x - (-1)}{x - 1}$

$= \dfrac{x^2 - 2x + 1}{x - 1} = \dfrac{(x-1)^2}{x - 1} = x - 1$

(b) $\dfrac{f(2) - f(1)}{2 - 1} = \dfrac{2^2 - 2(2) - (-1)}{2 - 1} = \dfrac{1}{1} = 1$

Slope = 1; Containing $(1, -1)$:

(c) $y - (-1) = 1(x - 1)$

$y + 1 = 1x - 1$

$y = x - 2$

31. $f(x) = x^3 - x$

(a) $\dfrac{f(x) - f(1)}{x - 1} = \dfrac{x^3 - x - 0}{x - 1} = \dfrac{x^3 - x}{x - 1}$

$= \dfrac{x(x-1)(x+1)}{x - 1} = x^2 + x$

(b) $\dfrac{f(2) - f(1)}{2 - 1} = \dfrac{2^3 - 2 - 0}{2 - 1} = \dfrac{6}{1} = 6$

(c) Slope = 6; Containing $(1, 0)$:

$y - 0 = 6(x - 1)$

$y = 6x - 6$

33. $f(x) = \dfrac{2}{x + 1}$

(a) $\dfrac{f(x) - f(1)}{x - 1} = \dfrac{\left(\dfrac{2}{x+1} - 1\right)}{x - 1} = \dfrac{\left(\dfrac{2 - x - 1}{x + 1}\right)}{x - 1}$

$= \dfrac{1 - x}{(x - 1)(x + 1)} = \dfrac{-1}{x + 1}$

(b) $\dfrac{f(2) - f(1)}{2 - 1} = \dfrac{\left(\dfrac{2}{2 + 1} - 1\right)}{2 - 1}$

(c) $= \dfrac{\left(-\dfrac{1}{3}\right)}{1} = -\dfrac{1}{3}$

Slope $= -\dfrac{1}{3}$; Containing $(1, 1)$:

$y - 1 = -\dfrac{1}{3}(x - 1)$

$y - 1 = -\dfrac{1}{3}x + \dfrac{1}{3} \rightarrow y = -\dfrac{1}{3}x + \dfrac{4}{3}$

35. $f(x) = \sqrt{x}$

(a) $\dfrac{f(x) - f(1)}{x - 1} = \dfrac{\sqrt{x} - 1}{x - 1}$

(b) $\dfrac{f(2) - f(1)}{2 - 1} = \dfrac{\sqrt{2} - 1}{1} = \sqrt{2} - 1$

Slope $= \sqrt{2} - 1$; Containing $(1, 1)$:

(c) $y - 1 = \left(\sqrt{2} - 1\right)(x - 1)$

$y - 1 = \left(\sqrt{2} - 1\right)x - \left(\sqrt{2} - 1\right)$

$y = \left(\sqrt{2} - 1\right)x - \sqrt{2} + 2$

37. $f(x) = 4x^3$

$f(-x) = 4(-x)^3 = -4x^3$

f is odd.

39. $g(x) = -3x^2 - 5$

$g(-x) = -3(-x)^2 - 5 = -3x^2 - 5$

g is even.

41. $F(x) = \sqrt[3]{x}$

$F(-x) = \sqrt[3]{-x} = -\sqrt[3]{x}$

F is odd.

43. $f(x) = x + |x|$

$f(-x) = -x + |-x| = -x + |x|$

f is neither even nor odd.

45. $g(x) = \dfrac{1}{x^2}$

$$g(-x) = \dfrac{1}{(-x)^2} = \dfrac{1}{x^2}$$

g is even.

47. $h(x) = \dfrac{-x^3}{3x^2 - 9}$

$$h(-x) = \dfrac{-(-x)^3}{3(-x)^2 - 9} = \dfrac{x^3}{3x^2 - 9}$$

h is odd.

49. $f(x) = 2x + 5$

$$m_{\text{sec}} = \dfrac{f(x+h) - f(x)}{h} = \dfrac{2(x+h) + 5 - 2x - 5}{h} = \dfrac{2h}{h} = 2$$

51. $f(x) = x^2 + 2x$

$$m_{\text{sec}} = \dfrac{f(x+h) - f(x)}{h} = \dfrac{(x+h)^2 + 2(x+h) - (x^2 + 2x)}{h}$$

$$= \dfrac{x^2 + 2xh + h^2 + 2x + 2h - x^2 - 2x}{h} = \dfrac{2xh + h^2 + 2h}{h} = 2x + h + 2$$

53. $f(x) = 2x^2 - 3x + 1$

$$m_{\text{sec}} = \dfrac{f(x+h) - f(x)}{h} = \dfrac{2(x+h)^2 - 3(x+h) + 1 - \left(2x^2 - 3x + 1\right)}{h}$$

$$= \dfrac{2(x^2 + 2xh + h^2) - 3x - 3h + 1 - 2x^2 + 3x - 1}{h}$$

$$= \dfrac{2x^2 + 4xh + 2h^2 - 3x - 3h + 1 - 2x^2 + 3x - 1}{h} = \dfrac{4xh + 2h^2 - 3h}{h} = 4x + 2h - 3$$

55. $f(x) = \dfrac{1}{x}$

$$m_{\text{sec}} = \dfrac{f(x+h) - f(x)}{h} = \dfrac{\left(\dfrac{1}{x+h} - \dfrac{1}{x}\right)}{h}$$

$$= \dfrac{\left(\dfrac{x - (x+h)}{(x+h)x}\right)}{h} = \left(\dfrac{x - x - h}{(x+h)x}\right)\left(\dfrac{1}{h}\right) = \left(\dfrac{-h}{(x+h)x}\right)\left(\dfrac{1}{h}\right) = \dfrac{-1}{(x+h)x}$$

57. (a), (b), (e)

(c)   Average rate of change $= \dfrac{28000 - 0}{25 - 0} = \dfrac{28000}{25} = 1120$

(d)   For each additional bicycle sold between 0 and 25, the total revenue increases by $1120.

(f)   Average rate of change $= \dfrac{64835 - 62360}{223 - 190} = \dfrac{2475}{33} = 75$

(g)   For each additional bicycle sold between 190 and 223, the total revenue increases by $75.

59.   (a), (b), (e)

(c)   Average rate of change $= \dfrac{153 - 50}{1 - 0} = \dfrac{103}{1} = 103$

(d)   The population is increasing at a rate of 103 per day between day 0 and day 1.

(f)   Average rate of change $= \dfrac{1280 - 839}{6 - 5} = \dfrac{441}{1} = 441$

(g)   The population is increasing at a rate of 441 per day between day 5 and day 6.

(h)   As time passes, the average rate of change of the population is increasing.

61.   One at most because if f is increasing it could only cross the x-axis at most one time. It could not "turn" and cross it again or it would start to decrease.

Functions and Their Graphs

2.3   Library of Functions; Piecewise-Defined Functions

1.   C   3.   E   5.   B   7.   F

9.

11.

13.

15.   (a) $f(-2) = (-2)^2 = 4$
      (b) $f(0) = 2$
      (c) $f(2) = 2(2) + 1 = 5$

17.   (a) $f(1.2) = \text{int}(2(1.2)) = \text{int}(2.4) = 2$
      (b) $f(1.6) = \text{int}(2(1.6)) = \text{int}(3.2) = 3$
      (c) $f(-1.8) = \text{int}(2(-1.8))$
      $\qquad = \text{int}(-3.6) = -4$

19. $f(x) = \begin{cases} 2x & \text{if } x \neq 0 \\ 1 & \text{if } x = 0 \end{cases}$

      (a)   Domain: {Real Numbers}
      (b)   x-intercept: none
            y-intercept: (0,1)

(c)

      (d)   Range: $\left\{ y \mid y \neq 0 \right\}$

21. $f(x) = \begin{cases} -2x + 3 & \text{if } x < 1 \\ 3x - 2 & \text{if } x \geq 1 \end{cases}$

    (a)   Domain: {Real Numbers}

    (b)   x-intercept: none

            y-intercept: (0,3)

    (c)

    (d)   Range: $\{y \mid y \geq 1\}$

23. $f(x) = \begin{cases} x + 3 & \text{if } -2 \leq x < 1 \\ 5 & \text{if } x = 1 \\ -x + 2 & \text{if } x > 1 \end{cases}$

    (a)   Domain: $\{x \mid x \geq -2\}$

    (b)   x-intercept: (2, 0)

            y-intercept: (0, 3)

    (c)

    (d)   Range: $\{y \mid y < 4\} \cup \{5\}$

25. $f(x) = \begin{cases} 1 + x & \text{if } x < 0 \\ x^2 & \text{if } x \geq 0 \end{cases}$

    (a)   Domain: {Real Numbers}

    (b)   x-intercept: (–1,0), (0,0)

            y-intercept: (0,0)

    (c)

    (d)   Range: {Real Numbers}

27. $f(x) = \begin{cases} |x| & \text{if } -2 \leq x < 0 \\ 1 & \text{if } x = 0 \\ x^3 & \text{if } x > 0 \end{cases}$

    (a)   Domain: $\{x \mid x \geq -2\}$

    (b)   x-intercept: none

            y-intercept: (0, 1)

    (c)

    (d)   Range: $\{y \mid y > 0\}$

29. $h(x) = 2\text{int}(x)$
   (a) Domain: {Real Numbers}
   (b) x-intercept: all ordered pairs $(x,0)$ when $0 \le x < 1$.
       y-intercept: $(0,0)$
   (c)

31. $f(x) = \begin{cases} -x & \text{if } -1 \le x \le 0 \\ \dfrac{1}{2}x & \text{if } 0 < x \le 2 \end{cases}$

33. $f(x) = \begin{cases} -x & \text{if } x \le 0 \\ -x+2 & \text{if } 0 < x \le 2 \end{cases}$

   (d) Range: {Even Integers}

35. (a) Charge for 50 therms: $C = 9.45 + 0.36375(50) + 0.3128(50) = \$43.28$
   (b) Charge for 500 therms:
       $$C = 9.45 + 0.36375(50) + 0.11445(450) + 0.3128(500) = \$235.54$$
   (c) The monthly charge function:

   $$C = \begin{cases} 9.45 + 0.36375x + 0.3128x & \text{for } 0 \le x \le 50 \\ 9.45 + 0.36375(50) + 0.11445(x-50) + 0.3128x & \text{for } x > 50 \end{cases}$$

   $$= \begin{cases} 9.45 + 0.67655x & \text{for } 0 \le x \le 50 \\ 9.45 + 18.1875 + 0.11445x - 5.7225 + 0.3128x & \text{for } x > 50 \end{cases}$$

   $$= \begin{cases} 9.45 + 0.67655x & \text{for } 0 \le x \le 50 \\ 21.915 + 0.42725x & \text{for } x > 50 \end{cases}$$

   (d)

37. (a) $W = 10°C$
   (b) $W = 33 - \dfrac{(10.45 + 10\sqrt{5} - 5)(33 - 10)}{22.04} = 3.98°C$
   (c) $W = 33 - \dfrac{(10.45 + 10\sqrt{15} - 15)(33 - 10)}{22.04} = -2.67°C$
   (d) $W = 33 - 1.5958(33 - 10) = -3.7°C$
   (e) When $0 \le v < 1.79$, the wind speed is so small that there is no effect on the temperature.

Section 2.3   Library of Functions; Piecewise-Defined Functions

(f)   For each drop of 1° in temperature, the wind chill factor drops approximately 1.6°C.
       When the wind speed exceeds 20, there is a constant drop in temperature.

39.   Each graph is that of $y = x^2$, but shifted vertically. If $y = x^2 + k$, $k > 0$, the shift is up k
       units; if $y = x^2 + k$, $k < 0$, the shift is down $|k|$ units. The graph of $y = x^2 - 4$ is the same
       as the graph of $y = x^2$, but shifted down 4 units. The graph of $y = x^2 + 5$ is the graph of
       $y = x^2$, but shifted up 5 units.

41.   Each graph is that of $y = |x|$, but either compressed or stretched. If $y = k|x|$ and $k > 1$,
       the graph is stretched; if $y = k|x|$ and $0 < k < 1$, the graph is compressed. The graph of
       $y = \frac{1}{4}|x|$ is the same as the graph of $y = |x|$, but compressed. The graph of $y = 5|x|$ is
       the same as the graph of $y = |x|$, but stretched.

43.   The graph of $y = \sqrt{-x}$ is the reflection about the y-axis of the graph of $y = \sqrt{x}$. The same
       type of reflection occurs when graphing $y = 2x + 1$ and $y = 2(-x) + 1$. The conclusion is
       that the graph of $y = f(-x)$ is the reflection about the y-axis of the graph of $y = f(x)$.

45.   For the graph of $y = x^n$, n a positive even integer, as n increases, the graph of the function
       is narrower for $|x| > 1$ and flatter for $|x| < 1$.

47. $f(x) = \begin{cases} 1 & \text{if } x \text{ is rational} \\ 0 & \text{if } x \text{ is irrational} \end{cases}$ Domain = { all real numbers}   Range = {0,1}

       y-intercept: $x = 0 \rightarrow x$ is rational $\rightarrow y = 1$, so the y-intercept is (0, 1).
       x-intercept: $y = 0 \rightarrow x$ is irrational, so the graph has infinitely many x-intercepts, namely,
       there is an x-intercept at each irrational value for x.
       $f(-x) = 1 = f(x)$ when x is rational; $f(-x) = 0 = f(x)$ when x is irrational $\therefore$ f is even.
       The graph of f consists of 2 infinite clusters of distinct points, extending
       horizontally in both directions.
       One cluster is located 1 unit above the x-axis, and the other is located along the x-axis.

Chapter 2

Functions and Their Graphs

2.4   Graphing Techniques: Transformations

1. B   3. H   5. I   7. L   9.   F

11. G   13. $y = (x - 4)^3$ 15. $y = x^3 + 4$

17. $y = -x^3$ 19. $y = 4x^3$

21. (1) $y = \sqrt{x} + 2$ 23. (1) $y = -\sqrt{x}$
    (2) $y = -\left(\sqrt{x} + 2\right)$ (2) $y = -\sqrt{x} + 2$
    (3) $y = -\left(\sqrt{-x} + 2\right)$ (3) $y = -\sqrt{x + 3} + 2$

25. C   27. C

29. $f(x) = x^2 - 1$ 31. $g(x) = x^3 + 1$
Using the graph of $y = x^2$, vertically   Using the graph of $y = x^3$, vertically
shift downward 1 unit.   shift upward 1 unit.

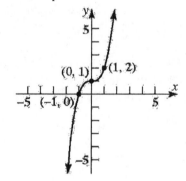

50

33. $h(x) = \sqrt{x-2}$

Using the graph of $y = \sqrt{x}$, horizontally shift to the right 2 units.

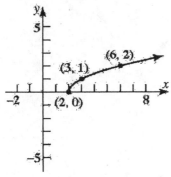

35. $f(x) = (x-1)^3 + 2$

Using the graph of $y = x^3$, horizontally shift to the right 1 unit, then vertically shift up 2 units.

37. $g(x) = 4\sqrt{x}$

Using the graph of $y = \sqrt{x}$, vertically stretch by a factor of 4.

39. $h(x) = \dfrac{1}{2x}$

Using the graph of $y = \dfrac{1}{x}$, vertically compress by a factor of $\dfrac{1}{2}$.

41. $f(x) = -|x|$
     Reflect the graph of $y = |x|$, about the x-axis.

43. $g(x) = |-x|$
     Reflect the graph of $y = |x|$, about the y-axis.

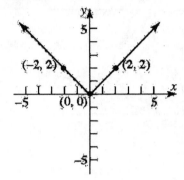

45. $h(x) = -x^3 + 2$
     Reflect the graph of $y = x^3$ on the x-axis, vertically shift upward 2 units.

47. $f(x) = 2(x + 1)^2 - 3$
     Using the graph of $y = x^2$, horizontally shift to the left 1 unit, vertically stretch by a factor of 2, and vertically shift downward 3 units.

49. $g(x) = \sqrt{x - 2} + 1$
     Using the graph of $y = \sqrt{x}$, horizontally shift to the right 2 units and vertically shift upward 1 unit.

51. $h(x) = \sqrt{-x} - 2$
     Reflect the graph of $y = \sqrt{x}$, about the y-axis and vertically shift downward 2 units.

53. $f(x) = -(x+1)^3 - 1$

Using the graph of $y = x^3$, horizontally shift to the left 1 units, reflect the graph on the x-axis, and vertically shift downward 1 unit.

55. $g(x) = 2|1-x| = 2|x-1|$

Using the graph of $y = |x|$, horizontally shift to the right 1 unit, and vertically stretch by a factor or 2.

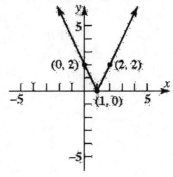

57. $h(x) = 2\text{int}(x-1)$

Using the graph of $y = \text{int}(x)$, horizontally shift to the right 1 unit, and vertically stretch by a factor of 2.

59. (a) $F(x) = f(x) + 3$
        Shift up 3 units.

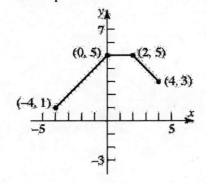

(b) $G(x) = f(x+2)$
        Shift left 2 units.

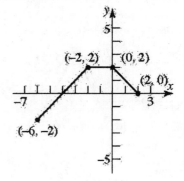

(c) $P(x) = -f(x)$
Reflect about the x-axis.

(d) $H(x) = f(x+1) - 2$
Shift left 1 unit and shift down 2 units.

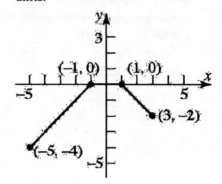

(e) $Q(x) = \dfrac{1}{2} f(x)$

Compress vertically by a factor of $\dfrac{1}{2}$.

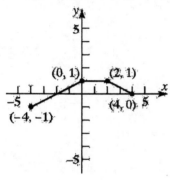

(f) $g(x) = f(-x)$
Reflect about y-axis.

(g) $h(x) = f(2x)$
Compress horizontally by a factor of $\dfrac{1}{2}$.

61. (a) $F(x) = f(x) + 3$
          Shift up 3 units.

(b) $G(x) = f(x + 2)$
      Shift left 2 units.

(c) $P(x) = -f(x)$
      Reflect about the x-axis.

(d) $H(x) = f(x + 1) - 2$
      Shift left 1 unit and shift down 2 units.

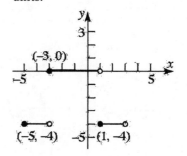

(e) $Q(x) = \dfrac{1}{2} f(x)$

Compress vertically by a factor of $\dfrac{1}{2}$.

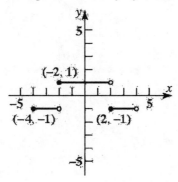

(f) $g(x) = f(-x)$
      Reflect about y-axis.

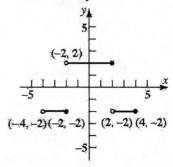

(g) $h(x) = f(2x)$
Compress horizontally by a factor
of $\dfrac{1}{2}$.

63.   (a) $F(x) = f(x) + 3$
              Shift up 3 units.

(b) $G(x) = f(x + 2)$
       Shift left 2 units.

(c) $P(x) = -f(x)$
       Reflect about the x-axis.

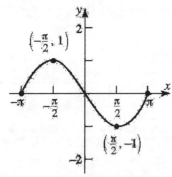

(d) $H(x) = f(x + 1) - 2$
       Shift left 1 unit and shift down 2
       units.

(e) $Q(x) = \dfrac{1}{2} f(x)$

Compress vertically by a factor of $\dfrac{1}{2}$.

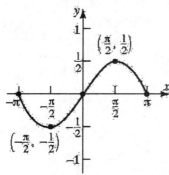

(f) $g(x) = f(-x)$
Reflect about y-axis.

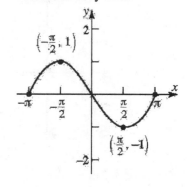

(g) $h(x) = f(2x)$
Compress horizontally by a factor of $\dfrac{1}{2}$.

65. (a) $y = |x + 1|$

(b) $y = |4 - x^2|$

(c) $y = |x^3 + x|$

(d) Any part of the graph of $y = f(x)$ that lies below the x-axis is reflected about the x-axis to obtain the graph of $y = |f(x)|$.

67.   (a) $y = |f(x)|$

(b) $y = f(|x|)$

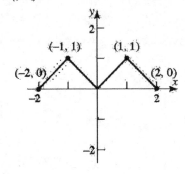

69. $f(x) = x^2 + 2x$
$f(x) = (x^2 + 2x + 1) - 1$
$f(x) = (x + 1)^2 - 1$
Using $f(x) = x^2$, shift left 1 unit and shift
   down 1 unit.

71. $f(x) = x^2 - 8x + 1$
$f(x) = (x^2 - 8x + 16) + 1 - 16$
$f(x) = (x - 4)^2 - 15$
Using $f(x) = x^2$, shift right 4 units and
   shift down 15 units.

73. $f(x) = x^2 + x + 1$
$f(x) = \left(x^2 + x + \dfrac{1}{4}\right) + 1 - \dfrac{1}{4}$
$f(x) = \left(x + \dfrac{1}{2}\right)^2 + \dfrac{3}{4}$

Using $f(x) = x^2$, shift left $\dfrac{1}{2}$ unit and

shift up $\dfrac{3}{4}$ unit.

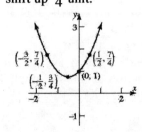

75. $y = (x - c)^2$
If $c = 0$, $y = x^2$.
If $c = 3$, $y = (x - 3)^2$; shift right 3 units.
If $c = -2$, $y = (x + 2)^2$; shift left 2 units.

77. $F = \dfrac{9}{5}C + 32$

$F = \dfrac{9}{5}(K - 273) + 32$

Shift the graph 273 units to the right.

79.   **(a)**

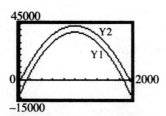

(b)   Select the 10% tax since the profits are higher.

(c)   The graph of Y1 is obtained by shifting the graph of $p(x)$ vertically down 10,000. The graph of Y2 is obtained by multiplying the y-coordinate of the graph of $p(x)$ by 0.9. Thus, Y2 is the graph of $p(x)$ vertically compressed by a factor of 0.9.

(d)   Select the 10% tax since the graph of $Y1 = 0.9p(x) \geq Y2 = -0.05x^2 + 100x - 6800$ for all x in the domain.

Chapter 2

Functions and Their Graphs

2.5 Operations on Functions; Composite Functions

1. $f(x) = 3x + 4 \qquad g(x) = 2x - 3$
   (a) $(f + g)(x) = 3x + 4 + 2x - 3 = 5x + 1$ The domain is all real numbers.
   (b) $(f - g)(x) = (3x + 4) - (2x - 3) = 3x + 4 - 2x + 3 = x + 7$
        The domain is all real numbers.
   (c) $(f \cdot g)(x) = (3x + 4)(2x - 3) = 6x^2 - 9x + 8x - 12 = 6x^2 - x - 12$
        The domain is all real numbers.
   (d) $\left(\dfrac{f}{g}\right)(x) = \dfrac{3x + 4}{2x - 3}$ The domain is all real numbers except $\dfrac{3}{2}$.

3. $f(x) = x - 1 \qquad g(x) = 2x^2$
   (a) $(f + g)(x) = x - 1 + 2x^2 = 2x^2 + x - 1$ The domain is all real numbers.
   (b) $(f - g)(x) = (x - 1) - (2x^2) = x - 1 - 2x^2 = -2x^2 + x - 1$
        The domain is all real numbers.
   (c) $(f \cdot g)(x) = (x - 1)(2x^2) = 2x^3 - 2x^2$ The domain is all real numbers.
   (d) $\left(\dfrac{f}{g}\right)(x) = \dfrac{x - 1}{2x^2}$ The domain is all real numbers except 0.

5. $f(x) = \sqrt{x} \qquad g(x) = 3x - 5$
   (a) $(f + g)(x) = \sqrt{x} + 3x - 5$ The domain is $\{x \mid x \geq 0\}$.
   (b) $(f - g)(x) = \sqrt{x} - (3x - 5) = \sqrt{x} - 3x + 5$ The domain is $\{x \mid x \geq 0\}$.
   (c) $(f \cdot g)(x) = \sqrt{x}(3x - 5) = 3x\sqrt{x} - 5\sqrt{x}$ The domain is $\{x \mid x \geq 0\}$.
   (d) $\left(\dfrac{f}{g}\right)(x) = \dfrac{\sqrt{x}}{3x - 5}$ The domain is $\left\{x \mid x \geq 0 \text{ and } x \neq \dfrac{5}{3}\right\}$.

7. $f(x) = 1 + \dfrac{1}{x} \qquad g(x) = \dfrac{1}{x}$
   (a) $(f + g)(x) = 1 + \dfrac{1}{x} + \dfrac{1}{x} = 1 + \dfrac{2}{x}$ The domain is $\{x \mid x \neq 0\}$.
   (b) $(f - g)(x) = 1 + \dfrac{1}{x} - \dfrac{1}{x} = 1$ The domain is $\{x \mid x \neq 0\}$.
   (c) $(f \cdot g)(x) = \left(1 + \dfrac{1}{x}\right)\dfrac{1}{x} = \dfrac{1}{x} + \dfrac{1}{x^2}$ The domain is $\{x \mid x \neq 0\}$.

(d) $\left(\dfrac{f}{g}\right)(x) = \dfrac{\left(1 + \dfrac{1}{x}\right)}{\left(\dfrac{1}{x}\right)} = \dfrac{\left(\dfrac{x+1}{x}\right)}{\left(\dfrac{1}{x}\right)} = \dfrac{x+1}{x} \cdot \dfrac{x}{1} = x + 1$ The domain is $\left\{x \mid x \neq 0\right\}$.

9. $f(x) = \dfrac{2x + 3}{3x - 2}$ $g(x) = \dfrac{4x}{3x - 2}$

(a) $(f + g)(x) = \dfrac{2x + 3}{3x - 2} + \dfrac{4x}{3x - 2} = \dfrac{2x + 3 + 4x}{3x - 2} = \dfrac{6x + 3}{3x - 2}$

The domain is $\left\{x \mid x \neq \dfrac{2}{3}\right\}$.

(b) $(f - g)(x) = \dfrac{2x + 3}{3x - 2} - \dfrac{4x}{3x - 2} = \dfrac{2x + 3 - 4x}{3x - 2} = \dfrac{-2x + 3}{3x - 2}$

The domain is $\left\{x \mid x \neq \dfrac{2}{3}\right\}$.

(c) $(f \cdot g)(x) = \left(\dfrac{2x + 3}{3x - 2}\right)\left(\dfrac{4x}{3x - 2}\right) = \dfrac{8x^2 + 12x}{(3x - 2)^2}$

The domain is $\left\{x \mid x \neq \dfrac{2}{3}\right\}$.

(d) $\left(\dfrac{f}{g}\right)(x) = \dfrac{\left(\dfrac{2x + 3}{3x - 2}\right)}{\left(\dfrac{4x}{3x - 2}\right)} = \dfrac{2x + 3}{3x - 2} \cdot \dfrac{3x - 2}{4x} = \dfrac{2x + 3}{4x}$

The domain is $\left\{x \mid x \neq \dfrac{2}{3} \text{ and } x \neq 0\right\}$.

11. $f(x) = 3x + 1$ $(f + g)(x) = 6 - \dfrac{1}{2}x$

$6 - \dfrac{1}{2}x = 3x + 1 + g(x) \rightarrow 5 - \dfrac{7}{2}x = g(x) \rightarrow g(x) = 5 - \dfrac{7}{2}x$

13. $f(x) = 2x$ $g(x) = 3x^2 + 1$

(a) $(f \circ g)(4) = f(g(4)) = f\left(3(4)^2 + 1\right) = f(49) = 2(49) = 98$

(b) $(g \circ f)(2) = g(f(2)) = g(2 \cdot 2) = g(4) = 3(4)^2 + 1 = 48 + 1 = 49$

(c) $(f \circ f)(1) = f(f(1)) = f(2(1)) = f(2) = 2(2) = 4$

(d) $(g \circ g)(0) = g(g(0)) = g\left(3(0)^2 + 1\right) = g(1) = 3(1)^2 + 1 = 4$

15. $f(x) = 4x^2 - 3$ $g(x) = 3 - \dfrac{1}{2}x^2$

(a) $(f \circ g)(4) = f(g(4)) = f\left(3 - \dfrac{1}{2}(4)^2\right) = f(-5) = 4(-5)^2 - 3 = 97$

(b) $(g \circ f)(2) = g(f(2)) = g(4(2)^2 - 3) = g(13) = 3 - \dfrac{1}{2}(13)^2 = 3 - \dfrac{169}{2} = -\dfrac{163}{2}$

(c) $(f \circ f)(1) = f(f(1)) = f(4(1)^2 - 3) = f(1) = 4(1)^2 - 3 = 1$

(d) $(g \circ g)(0) = g(g(0)) = g\left(3 - \frac{1}{2}(0)^2\right) = g(3) = 3 - \frac{1}{2}(3)^2 = 3 - \frac{9}{2} = -\frac{3}{2}$

17. $f(x) = \sqrt{x}$ $g(x) = 2x$
   (a) $(f \circ g)(4) = f(g(4)) = f(2(4)) = f(8) = \sqrt{8} = 2\sqrt{2}$
   (b) $(g \circ f)(2) = g(f(2)) = g\left(\sqrt{2}\right) = 2\sqrt{2}$
   (c) $(f \circ f)(1) = f(f(1)) = f\left(\sqrt{1}\right) = f(1) = \sqrt{1} = 1$
   (d) $(g \circ g)(0) = g(g(0)) = g(2(0)) = g(0) = 2(0) = 0$

19. $f(x) = |x|$ $g(x) = \frac{1}{x^2 + 1}$
   (a) $(f \circ g)(4) = f(g(4)) = f\left(\frac{1}{4^2 + 1}\right) = f\left(\frac{1}{17}\right) = \left|\frac{1}{17}\right| = \frac{1}{17}$
   (b) $(g \circ f)(2) = g(f(2)) = g(|2|) = g(2) = \frac{1}{2^2 + 1} = \frac{1}{5}$
   (c) $(f \circ f)(1) = f(f(1)) = f(|1|) = f(1) = |1| = 1$
   (d) $(g \circ g)(0) = g(g(0)) = g\left(\frac{1}{0^2 + 1}\right) = g(1) = \frac{1}{1^2 + 1} = \frac{1}{2}$

21. $f(x) = \frac{3}{x + 1}$ $g(x) = \sqrt{x}$
   (a) $(f \circ g)(4) = f(g(4)) = f\left(\sqrt{4}\right) = f(2) = \frac{3}{2 + 1} = \frac{3}{3} = 1$
   (b) $(g \circ f)(2) = g(f(2)) = g\left(\frac{3}{2 + 1}\right) = g\left(\frac{3}{3}\right) = \sqrt{1} = 1$
   (c) $(f \circ f)(1) = f(f(1)) = f\left(\frac{3}{1 + 1}\right) = f\left(\frac{3}{2}\right) = \frac{3}{\left(\frac{3}{2}\right) + 1} = \frac{3}{\left(\frac{5}{2}\right)} = \frac{6}{5}$
   (d) $(g \circ g)(0) = g(g(0)) = g\left(\sqrt{0}\right) = g(0) = \sqrt{0} = 0$

23. The domain of g is $\{x \mid x \neq 0\}$. The domain of f is $\{x \mid x \neq 1\}$.
   Thus, $g(x) \neq 1$, so we solve:
$$g(x) = 1 \rightarrow \frac{2}{x} = 1 \rightarrow x = 2$$
   Thus, $x \neq 2$; so the domain of $f \circ g$ is $\{x \mid x \neq 0, x \neq 2\}$.

25. The domain of g is $\{x \mid x \neq 0\}$. The domain of f is $\{x \mid x \neq 1\}$.
   Thus, $g(x) \neq 1$, so we solve:
$$g(x) = 1 \rightarrow -\frac{4}{x} = 1 \rightarrow x = -4$$
   Thus, $x \neq -4$; so the domain of $f \circ g$ is $\{x \mid x \neq -4, x \neq 0\}$.

27. The domain of g is {Real Numbers}. The domain of f is $\{x \mid x \geq 0\}$.

Thus, $g(x) \geq 0$, so we solve:

$$g(x) \geq 0$$
$$2x + 3 \geq 0 \qquad \text{Thus, the domain of } f \circ g \text{ is } \left\{x \mid x \geq -\frac{3}{2}\right\}.$$
$$x \geq -\frac{3}{2}$$

29. The domain of g is $\{x \mid x \geq 1\}$. The domain of f is {Real Numbers}.

Thus, the domain of $f \circ g$ is $\{x \mid x \geq 1\}$.

31. $f(x) = 2x + 3 \qquad g(x) = 3x$

The domain of f is all real numbers. The domain of g is all real numbers.

(a) $(f \circ g)(x) = f(g(x)) = f(3x) = 2(3x) + 3 = 6x + 3$ Domain: All real numbers.

(b) $(g \circ f)(x) = g(f(x)) = g(2x + 3) = 3(2x + 3) = 6x + 9$

Domain: All real numbers.

(c) $(f \circ f)(x) = f(f(x)) = f(2x + 3) = 2(2x + 3) + 3 = 4x + 6 + 3 = 4x + 9$

Domain: All real numbers.

(d) $(g \circ g)(x) = g(g(x)) = g(3x) = 3(3x) = 9x$ Domain: All real numbers.

33. $f(x) = 3x + 1 \qquad g(x) = x^2$

The domain of f is all real numbers. The domain of g is all real numbers.

(a) $(f \circ g)(x) = f(g(x)) = f(x^2) = 3x^2 + 1$ Domain: All real numbers.

(b) $(g \circ f)(x) = g(f(x)) = g(3x + 1) = (3x + 1)^2 = 9x^2 + 6x + 1$

Domain: All real numbers.

(c) $(f \circ f)(x) = f(f(x)) = f(3x + 1) = 3(3x + 1) + 1 = 9x + 3 + 1 = 9x + 4$

Domain: All real numbers.

(d) $(g \circ g)(x) = g(g(x)) = g(x^2) = (x^2)^2 = x^4$ Domain: All real numbers.

35. $f(x) = x^2 \qquad g(x) = x^2 + 4$

The domain of f is all real numbers. The domain of g is all real numbers.

(a) $(f \circ g)(x) = f(g(x)) = f(x^2 + 4) = (x^2 + 4)^2 = x^4 + 8x^2 + 16$

Domain: All real numbers.

(b) $(g \circ f)(x) = g(f(x)) = g(x^2) = (x^2)^2 + 4 = x^4 + 4$ Domain: All real numbers.

(c) $(f \circ f)(x) = f(f(x)) = f(x^2) = (x^2)^2 = x^4$ Domain: All real numbers.

(d) $(g \circ g)(x) = g(g(x)) = g(x^2 + 4) = (x^2 + 4)^2 + 4 = x^4 + 8x^2 + 16 + 4$

$$= x^4 + 8x^2 + 20 \text{ Domain: All real numbers.}$$

37. $f(x) = \dfrac{3}{x - 1} \qquad g(x) = \dfrac{2}{x}$ The domain of f is $\{x \mid x \neq 1\}$.

The domain of g is $\{x \mid x \neq 0\}$.

(a) $\quad (f \circ g)(x) = f(g(x)) = f\left(\dfrac{2}{x}\right) = \dfrac{3}{\left(\dfrac{2}{x} - 1\right)} = \dfrac{3}{\left(\dfrac{2 - x}{x}\right)} = \dfrac{3x}{2 - x}$

Domain of $f \circ g$ is $\{x \mid x \neq 0,\ x \neq 2\}$.

(b) $\quad (g \circ f)(x) = g(f(x)) = g\left(\dfrac{3}{x - 1}\right) = \dfrac{2}{\left(\dfrac{3}{x - 1}\right)} = \dfrac{2(x - 1)}{3}$

Domain of $g \circ f$ is $\{x \mid x \neq 1\}$

(c) $\quad (f \circ f)(x) = f(f(x)) = f\left(\dfrac{3}{x - 1}\right) = \dfrac{3}{\left(\dfrac{3}{x - 1} - 1\right)} = \dfrac{3}{\left(\dfrac{3 - (x - 1)}{x - 1}\right)} = \dfrac{3(x - 1)}{4 - x}$

Domain of $f \circ f$ is $\{x \mid x \neq 1,\ x \neq 4\}$.

(d) $\quad (g \circ g)(x) = g(g(x)) = g\left(\dfrac{2}{x}\right) = \dfrac{2}{\left(\dfrac{2}{x}\right)} = \dfrac{2x}{2} = x$

Domain of $g \circ g$ is $\{x \mid x \neq 0\}$.

39. $\quad f(x) = \dfrac{x}{x - 1} \qquad g(x) = -\dfrac{4}{x}$

The domain of f is $\{x \mid x \neq 1\}$. The domain of g is $\{x \mid x \neq 0\}$.

(a) $\quad (f \circ g)(x) = f(g(x)) = f\left(-\dfrac{4}{x}\right) = \dfrac{\left(-\dfrac{4}{x}\right)}{\left(-\dfrac{4}{x} - 1\right)} = \dfrac{\left(-\dfrac{4}{x}\right)}{\left(\dfrac{-4 - x}{x}\right)} = \dfrac{-4}{-4 - x} = \dfrac{4}{4 + x}$

Domain of $f \circ g$ is $\{x \mid x \neq -4,\ x \neq 0\}$.

(b) $\quad (g \circ f)(x) = g(f(x)) = g\left(\dfrac{x}{x - 1}\right) = -\dfrac{4}{\left(\dfrac{x}{x - 1}\right)} = \dfrac{-4(x - 1)}{x}$

Domain of $g \circ f$ is $\{x \mid x \neq 0,\ x \neq 1\}$.

(c) $\quad (f \circ f)(x) = f(f(x)) = f\left(\dfrac{x}{x - 1}\right) = \dfrac{\left(\dfrac{x}{x - 1}\right)}{\left(\dfrac{x}{x - 1} - 1\right)} = \dfrac{\left(\dfrac{x}{x - 1}\right)}{\left(\dfrac{x - (x - 1)}{x - 1}\right)} = \dfrac{x}{1} = x$

Domain of $f \circ f$ is $\{x \mid x \neq 1\}$.

(d) $\quad (g \circ g)(x) = g(g(x)) = g\left(\dfrac{-4}{x}\right) = -\dfrac{4}{\left(-\dfrac{4}{x}\right)} = \dfrac{-4x}{-4} = x$

Domain of $g \circ g$ is $\{x \mid x \neq 0\}$.

41. $f(x) = \sqrt{x}$ $g(x) = 2x + 3$
The domain of f is $\{x \mid x \geq 0\}$. The domain of g is {Real Numbers}.

(a) $(f \circ g)(x) = f(g(x)) = f(2x + 3) = \sqrt{2x + 3}$ Domain of $f \circ g$ is $\left\{x \mid x \geq -\dfrac{3}{2}\right\}$.

(b) $(g \circ f)(x) = g(f(x)) = g\left(\sqrt{x}\right) = 2\sqrt{x} + 3$ Domain of $g \circ f$ is $\{x \mid x \geq 0\}$.

(c) $(f \circ f)(x) = f(f(x)) = f\left(\sqrt{x}\right) = \sqrt{\sqrt{x}} = x^{1/4} = \sqrt[4]{x}$
Domain of $f \circ f$ is $\{x \mid x \geq 0\}$.

(d) $(g \circ g)(x) = g(g(x)) = g(2x + 3) = 2(2x + 3) + 3 = 4x + 6 + 3 = 4x + 9$
Domain of $g \circ g$ is {Real Numbers}.

43. $f(x) = x^2 + 1$ $g(x) = \sqrt{x - 1}$
The domain of f is {Real Numbers}. The domain of g is $\{x \mid x \geq 1\}$.

(a) $(f \circ g)(x) = f(g(x)) = f\left(\sqrt{x - 1}\right) = \left(\sqrt{x - 1}\right)^2 + 1 = x - 1 + 1 = x$
Domain of $f \circ g$ is $\{x \mid x \geq 1\}$.

(b) $(g \circ f)(x) = g(f(x)) = g\left(x^2 + 1\right) = \sqrt{x^2 + 1 - 1} = \sqrt{x^2} = |x|$
Domain of $g \circ f$ {Real Numbers}.

(c) $(f \circ f)(x) = f(f(x)) = f\left(x^2 + 1\right) = \left(x^2 + 1\right)^2 + 1 = x^4 + 2x^2 + 1 + 1 = x^4 + 2x^2 + 2$
Domain of $f \circ f$ is {Real Numbers}.

(d) $(g \circ g)(x) = g(g(x)) = g\left(\sqrt{x - 1}\right) = \sqrt{\sqrt{x - 1} - 1}$
Domain of $g \circ g$ is $\{x \mid x \geq 2\}$.

45. $f(x) = ax + b$ $g(x) = cx + d$ The domain of f is {Real Numbers}.
The domain of g is {Real Numbers}.

(a) $(f \circ g)(x) = f(g(x)) = f(cx + d) = a(cx + d) + b = acx + ad + b$
Domain of $f \circ g$ is {Real Numbers}.

(b) $(g \circ f)(x) = g(f(x)) = g(ax + b) = c(ax + b) + d = acx + bc + d$
Domain of $g \circ f$ is {Real Numbers}.

(c) $(f \circ f)(x) = f(f(x)) = f(ax + b) = a(ax + b) + b = a^2x + ab + b$
Domain of $f \circ f$ is {Real Numbers}.

(d) $(g \circ g)(x) = g(g(x)) = g(cx + d) = c(cx + d) + d = c^2x + cd + d$
Domain of $g \circ g$ is {Real Numbers}.

47. $(f \circ g)(x) = f(g(x)) = f\left(\dfrac{1}{2}x\right) = 2\left(\dfrac{1}{2}x\right) = x$

$(g \circ f)(x) = g(f(x)) = g(2x) = \dfrac{1}{2}(2x) = x$

49. $(f \circ g)(x) = f(g(x)) = f\left(\sqrt[3]{x}\right) = \left(\sqrt[3]{x}\right)^3 = x$

$(g \circ f)(x) = g(f(x)) = g\left(x^3\right) = \sqrt[3]{x^3} = x$

51. $(f \circ g)(x) = f(g(x)) = f\left(\frac{1}{2}(x+6)\right) = 2\left(\frac{1}{2}(x+6)\right) - 6 = x + 6 - 6 = x$

$(g \circ f)(x) = g(f(x)) = g(2x-6) = \frac{1}{2}((2x-6)+6) = \frac{1}{2}(2x) = x$

53. $(f \circ g)(x) = f(g(x)) = f\left(\frac{1}{a}(x-b)\right) = a\left(\frac{1}{a}(x-b)\right) + b = x - b + b = x$

$(g \circ f)(x) = g(f(x)) = g(ax+b) = \frac{1}{a}((ax+b)-b) = \frac{1}{a}(ax) = x$

55. $H(x) = (2x+3)^4 \qquad\qquad f(x) = x^4, \quad g(x) = 2x+3$

57. $H(x) = \sqrt{x^2+1} \qquad\qquad f(x) = \sqrt{x}, \quad g(x) = x^2+1$

59. $H(x) = |2x+1| \qquad\qquad f(x) = |x|, \quad g(x) = 2x+1$

61. $f(x) = 2x^3 - 3x^2 + 4x - 1 \qquad g(x) = 2$
$(f \circ g)(x) = f(g(x)) = f(2) = 2(2)^3 - 3(2)^2 + 4(2) - 1 = 16 - 12 + 8 - 1 = 11$
$(g \circ f)(x) = g(f(x)) = g(2x^3 - 3x^2 + 4x - 1) = 2$

63. $f(x) = 2x^2 + 5 \qquad g(x) = 3x + a$
$(f \circ g)(x) = f(g(x)) = f(3x+a) = 2(3x+a)^2 + 5$
When $x = 0$, $(f \circ g)(0) = 23$
Solving: $\quad 2(3 \cdot 0 + a)^2 + 5 = 23 \rightarrow 2a^2 + 5 = 23 \rightarrow 2a^2 = 18 \rightarrow a^2 = 9 \rightarrow a = -3$ or 3

65. $S(r) = 4\pi r^2 \qquad r(t) = \frac{2}{3}t^3, \ t \geq 0$

$S(r(t)) = S\left(\frac{2}{3}t^3\right) = 4\pi\left(\frac{2}{3}t^3\right)^2 = 4\pi\left(\frac{4}{9}t^6\right) = \frac{16}{9}\pi t^6$

67. $N(t) = 100t - 5t^2, \ 0 \leq t \leq 10 \qquad C(N) = 15000 + 8000N$
$C(N(t)) = C(100t - 5t^2) = 15000 + 8000(100t - 5t^2) = 15{,}000 + 800{,}000t - 40{,}000t^2$

69. $p = -\frac{1}{4}x + 100 \qquad 0 \leq x \leq 400$

$\frac{1}{4}x = 100 - p \rightarrow x = 4(100 - p)$

$C = \frac{\sqrt{x}}{25} + 600 = \frac{\sqrt{4(100-p)}}{25} + 600 = \frac{2\sqrt{100-p}}{25} + 600$

71. $V = \pi r^2 h \qquad h = 2r \qquad \rightarrow V(r) = \pi r^2 (2r) = 2\pi r^3$

73. Given that f and g are odd functions, we know that

$f(-x) = -f(x)$ and $g(-x) = -g(x)$ for all x in the domain of f and g respectively.

The composite function $f \circ g = f(g(x))$ has the following property:

$f(g(-x)) = f(-g(x))$ since g is odd

$= -f(g(x))$ since f is odd, $\therefore f \circ g$ is odd

Functions and Their Graphs

2.6   Mathematical Models; Constructing Functions

1.   If $V = \pi r^2 h$ and $h = 2r$, then $V(r) = \pi r^2 (2r) = 2\pi r^3$.

3.   (a)   If $P = -\dfrac{1}{6}x + 100$ and $R = x\,p$, then $R(x) = x\left(-\dfrac{1}{6}x + 100\right) = -\dfrac{1}{6}x^2 + 100x$.

   (b) $R(200) = -\dfrac{1}{6}(200)^2 + 100(200) = \$13,333$

   (c)   Graphing:

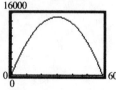

   (d) 300; \$15,000

   (e) $p = -\dfrac{1}{6}(300) + 100 = -50 + 100 = \$50$

5.   (a)   If $x = -5p + 100$ and $R = x\,p$, then $p = \dfrac{100 - x}{5}$ and

$$R(x) = x\left(\dfrac{100 - x}{5}\right) = -\dfrac{1}{5}x^2 + 20x.$$

   (b) $R(15) = -\dfrac{1}{5}(15)^2 + 20(15) = \$255$

   (c)   Graphing:

(d)   50; \$500

(e) $p = \dfrac{100 - 50}{5} = \dfrac{50}{5} = \$10$

7.   (a)   Let x be the width of the rectangle and let y be the length of the rectangle.
            Then, the perimeter is: $P = 2y + 2x = 400$.
            Solving for y: $y = \dfrac{400 - 2x}{2} = 200 - x$.
            The area function is: $A(x) = y(x) = (200 - x)x = -x^2 + 200x$.

      (b)   The domain is: $\{x \mid 0 < x < 200\}$

      (c)   Graphing:

            The area is largest when $x = 100$ yards.

9.   (a)   The distance d from P to the origin is $d = \sqrt{x^2 + y^2}$. Since P is a point on the graph
            of $y = x^2 - 8$, we have:
            $$d(x) = \sqrt{x^2 + (x^2 - 8)^2} = \sqrt{x^4 - 15x^2 + 64}$$

      (b) $d(0) = \sqrt{0^4 - 15(0)^2 + 64} = \sqrt{64} = 8$

      (c) $d(1) = \sqrt{(1)^4 - 15(1)^2 + 64} = \sqrt{1 - 15 + 64} = \sqrt{50} = 5\sqrt{2} \approx 7.07$

      (d)   Graphing:

      (e) d is smallest when x is 2.74.

11.   (a)   The distance d from P to the point $(1, 0)$ is $d = \sqrt{(x - 1)^2 + y^2}$. Since P is a point on
            the graph of $y = \sqrt{x}$, we have:
            $$d(x) = \sqrt{(x - 1)^2 + \left(\sqrt{x}\right)^2} = \sqrt{x^2 - x + 1}$$

      (b)   Graphing:

      (c) d is smallest when x is 0.50.

13.   By definition, a triangle has area $A = \dfrac{1}{2}bh, b =$ base, $h =$ height.
      Because a vertex of the triangle is at the origin, we know that $b = x$ and $h = y$.

Expressing the area of the triangle as a function of x, we have:

$$A(x) = \frac{1}{2}xy = \frac{1}{2}x(x^3) = \frac{1}{2}x^4.$$

15. (a) $A(x) = xy = x(16 - x^2) = -x^3 + 16x$

    (b)   Domain: $\{x | 0 < x < 4\}$

    (c)   Graphing: The area is largest when x is approximately 2.31.

17. (a) $A(x) = (2x)(2y) = 4x(4 - x^2)^{1/2}$

    (b) $p(x) = 2(2x) + 2(2y) = 4x + 4(4 - x^2)^{1/2}$

    (c)   Graphing: (d)   Graphing:

The area is largest when x is   The perimeter is largest when x is
approximately 1.41.   approximately 1.41.

19. (a) $C =$ circumference, $A =$ area, $r =$ radius, $x =$ side of square

    $C = 2\pi r = 10 - 4x \quad \rightarrow \quad r = \dfrac{5 - 2x}{\pi}$

    $A(x) = x^2 + \pi r^2 = x^2 + \pi\left(\dfrac{5 - 2x}{\pi}\right)^2 = x^2 + \dfrac{25 - 20x + 4x^2}{\pi}$

    (b)   Since the lengths must be positive, we have:
    $$10 - 4x > 0 \quad \text{and } x > 0$$
    $$-4x > -10 \rightarrow x < 2.5 \text{ and } x > 0$$
    Domain: $\{x | 0 < x < 2.5\}$

    (c)   Graphing: The area is smallest when x is approximately 1.40 meters.

21. (a) Since the wire of length x is bent into a circle, the circumference is x.
          Therefore, $C(x) = x$.

     (b) Since $C = x = 2\pi r$, $r = \dfrac{x}{2\pi}$.

          $$A(x) = \pi r^2 = \pi\left(\frac{x}{2\pi}\right)^2 = \frac{x^2}{4\pi}.$$

23. (a) A = area, r = radius; diameter = $2r$ (b) p = perimeter
          $A(r) = (2r)(r) = 2r^2$ $p(r) = 2(2r) + 2r = 6r$

25. Area of the equilateral triangle $= \dfrac{1}{2}x \cdot \dfrac{\sqrt{3}}{2}x = \dfrac{\sqrt{3}}{4}x^2$

     Area of $\frac{1}{3}$ of the equilateral triangle $= \dfrac{1}{2}x\sqrt{r^2 - \left(\dfrac{x}{2}\right)^2} = \dfrac{1}{2}x\sqrt{r^2 - \dfrac{x^2}{4}} = \dfrac{1}{3} \cdot \dfrac{\sqrt{3}}{4}x^2$

     Solving for r^2:

     $$\frac{1}{2}x\sqrt{r^2 - \frac{x^2}{4}} = \frac{1}{3} \cdot \frac{\sqrt{3}}{4}x^2$$

     $$\sqrt{r^2 - \frac{x^2}{4}} = \frac{2}{x} \cdot \frac{\sqrt{3}}{12}x^2$$

     $$\sqrt{r^2 - \frac{x^2}{4}} = \frac{\sqrt{3}}{6}x \rightarrow r^2 - \frac{x^2}{4} = \frac{3}{36}x^2 \rightarrow r^2 = \frac{x^2}{3}$$

     Area inside the circle, but outside the triangle:

     $$A(x) = \pi r^2 - \frac{\sqrt{3}}{4}x^2 = \pi\frac{x^2}{3} - \frac{\sqrt{3}}{4}x^2 = \left(\frac{\pi}{3} - \frac{\sqrt{3}}{4}\right)x^2$$

27. $$C = \begin{cases} 95 & \text{if } x = 7 \\ 119 & \text{if } 7 < x \le 8 \\ 143 & \text{if } 8 < x \le 9 \\ 167 & \text{if } 9 < x \le 10 \\ 190 & \text{if } 10 < x \le 14 \end{cases}$$

29.

     $d_2 = 40t$
     $d_1 = 30t$
     d

     $d^2 = d_1^{\,2} + d_2^{\,2}$
     $d^2 = (30t)^2 + (40t)^2$
     $d(t) = \sqrt{900t^2 + 1600t^2}$
     $d(t) = \sqrt{2500t^2} = 50t$

31. (a) length $= 24 - 2x$
          width $= 24 - 2x$
          height $= x$
          $V(x) = x(24 - 2x)(24 - 2x) = x(24 - 2x)^2$

(b) $V(3) = 3(24 - 2(3))^2 = 3(18)^2 = 3(324) = 972$ cu. in.

(c) $V(10) = 3(24 - 2(10))^2 = 3(4)^2 = 3(16) = 48$ cu. in.

(d)

The volume is largest when $x = 4$ inches.

33. r = radius of cylinder, h = height of cylinder, V = volume of cylinder

$$r^2 + \left(\frac{h}{2}\right)^2 = R^2$$

$$r^2 + \frac{h^2}{4} = R^2$$

$$r^2 = R^2 - \frac{h^2}{4}$$

$$r^2 = \frac{4R^2 - h^2}{4}$$

$$V = \pi r^2 h$$

$$V(h) = \pi \left(\frac{4R^2 - h^2}{4}\right)h = \frac{\pi}{4}\left(4R^2 h - h^3\right)$$

35. (a)   The total cost of installing the cable along the road is $10x$. If cable is installed x miles along the road, there are $5 - x$ miles left from the road to the house and where the cable ends.

$$d = \sqrt{(5 - x)^2 + 2^2} = \sqrt{25 - 10x + x^2 + 4}$$
$$= \sqrt{x^2 - 10x + 29}$$

The total cost of installing the cable is:
$$C(x) = 10x + 14\sqrt{x^2 - 10x + 29}$$
Domain: $\left\{x \mid 0 < x < 5\right\}$

(b) $C(1) = 10(1) + 14\sqrt{1^2 - 10(1) + 29} = 10 + 14\sqrt{20} \approx 10 + 62.61 = \$72.61$

(c) $C(3) = 10(3) + 14\sqrt{3^2 - 10(3) + 29} = 30 + 14\sqrt{8} \approx 30 + 39.60 = \$69.60$

(d)

The table indicates that $x = 3$ results in the least cost.

(e) Using MINIMUM, the graph indicates that $x = 2.96$ results in the least cost.

37.   Consider the diagram shown below.

We can extract a pair of similar triangles from the diagram.
Since the smaller triangle is similar to the larger triangle, we have the proportion

$$\frac{r}{h} = \frac{4}{16} \rightarrow \frac{r}{h} = \frac{1}{4} \rightarrow r = \frac{1}{4}h$$

Substituting into the volume formula for the conical portion of water,

$$V = \frac{1}{3}\pi r^2 h = \frac{1}{3}\pi \left(\frac{1}{4}h\right)^2 h = \frac{1}{48}\pi h^3,$$ so we have the volume function $V(h) = \frac{1}{48}\pi h^3$.

Chapter 2

Functions and Their Graphs

2.R Chapter Review

1. $f(4) = -5$ gives the ordered pair $(4,-5)$. $f(0) = 3$ gives $(0,3)$.
    Finding the slope: $m = \dfrac{3-(-5)}{0-4} = \dfrac{8}{-4} = -2$
    Using slope-intercept form: $f(x) = -2x + 3$

3. $f(x) = \dfrac{Ax+5}{6x-2}$ and $f(1) = 4$
    Solving:
    $$\dfrac{A(1)+5}{6(1)-2} = 4$$
    $$\dfrac{A+5}{4} = 4$$
    $$A+5 = 16$$
    $$A = 11$$

5. (b), (c), and (d) pass the vertical line test and therefore are functions.

7. $f(x) = \dfrac{3x}{x^2-4}$
    (a) $f(-x) = \dfrac{3(-x)}{(-x)^2-4} = \dfrac{-3x}{x^2-4}$
    (b) $-f(x) = -\left(\dfrac{3x}{x^2-4}\right) = \dfrac{-3x}{x^2-4}$

    (c) $f(x+2) = \dfrac{3(x+2)}{(x+2)^2-4} = \dfrac{3x+6}{x^2+4x+4-4} = \dfrac{3x+6}{x^2+4x}$
    (d) $f(x-2) = \dfrac{3(x-2)}{(x-2)^2-4} = \dfrac{3x-6}{x^2-4x+4-4} = \dfrac{3x-6}{x^2-4x}$
    (e) $f(2x) = \dfrac{3(2x)}{(2x)^2-4} = \dfrac{6x}{4x^2-4} = \dfrac{3x}{2x^2-2}$

9. $f(x) = \sqrt{x^2-4}$
    (a) $f(-x) = \sqrt{(-x)^2-4} = \sqrt{x^2-4}$
    (b) $-f(x) = -\sqrt{x^2-4}$
    (c) $f(x+2) = \sqrt{(x+2)^2-4} = \sqrt{x^2+4x+4-4} = \sqrt{x^2+4x}$
    (d) $f(x-2) = \sqrt{(x-2)^2-4} = \sqrt{x^2-4x+4-4} = \sqrt{x^2-4x}$
    (e) $f(2x) = \sqrt{(2x)^2-4} = \sqrt{4x^2-4} = 2\sqrt{x^2-1}$

11. $f(x) = \dfrac{x^2 - 4}{x^2}$

    (a) $f(-x) = \dfrac{(-x)^2 - 4}{(-x)^2} = \dfrac{x^2 - 4}{x^2}$

    (b) $-f(x) = -\left(\dfrac{x^2 - 4}{x^2}\right) = \dfrac{4 - x^2}{x^2}$

    (c) $f(x + 2) = \dfrac{(x + 2)^2 - 4}{(x + 2)^2} = \dfrac{x^2 + 4x + 4 - 4}{x^2 + 4x + 4} = \dfrac{x^2 + 4x}{x^2 + 4x + 4}$

    (d) $f(x - 2) = \dfrac{(x - 2)^2 - 4}{(x - 2)^2} = \dfrac{x^2 - 4x + 4 - 4}{x^2 - 4x + 4} = \dfrac{x^2 - 4x}{x^2 - 4x + 4}$

    (e) $f(2x) = \dfrac{(2x)^2 - 4}{(2x)^2} = \dfrac{4x^2 - 4}{4x^2} = \dfrac{x^2 - 1}{x^2}$

13. $f(x) = \dfrac{x}{x^2 - 9}$

The denominator cannot be zero:
$$x^2 - 9 \neq 0$$
$$(x + 3)(x - 3) \neq 0$$
$$x \neq -3 \text{ or } 3$$
Domain: $\{x \mid x \neq -3, x \neq 3\}$

15. $f(x) = \sqrt{2 - x}$

The radicand must be positive:
$$2 - x \geq 0$$
$$x \leq 2$$
Domain: $\{x \mid x \leq 2\}$ or $(-\infty, 2]$

17. $f(x) = \dfrac{\sqrt{x}}{|x|}$

The radicand must be positive and the
    denominator cannot be zero: $x > 0$
Domain: $\{x \mid x > 0\}$ or $(0, +\infty)$

19. $f(x) = \dfrac{x}{x^2 + 2x - 3}$

The denominator cannot be zero:
$$x^2 + 2x - 3 \neq 0$$
$$(x + 3)(x - 1) \neq 0$$
$$x \neq -3 \text{ or } 1$$
Domain: $\{x \mid x \neq -3, x \neq 1\}$

21. $f(x) = \begin{cases} 3x - 2 & \text{if } x \leq 1 \\ x + 1 & \text{if } x > 1 \end{cases}$

    (a) Domain: {Real Numbers}

    (b) x-intercept: $\left(\dfrac{2}{3}, 0\right)$

         y-intercept: $(0, -2)$

    (c)

    (d) Range: $\{y > 2\} \cup \{y \leq 1\}$

23. $f(x) = \begin{cases} x & \text{if } -4 \leq x < 0 \\ 1 & \text{if } x = 0 \\ 3x & \text{if } x > 0 \end{cases}$

    (a) Domain: $\{x \mid x \geq -4\}$

    (b) x-intercept: none

         y-intercept: $(0, 1)$

    (c)

    (d) Range: $\{y \mid y \geq -4, y \neq 0\}$

25. $f(x) = 2 - 5x$

$$\frac{f(x) - f(2)}{x - 2} = \frac{2 - 5x - (-8)}{x - 2} = \frac{-5x + 10}{x - 2} = \frac{-5(x - 2)}{x - 2} = -5$$

27. $f(x) = 3x - 4x^2$

$$\frac{f(x) - f(2)}{x - 2} = \frac{3x - 4x^2 - (-10)}{x - 2} = \frac{-4x^2 + 3x + 10}{x - 2}$$
$$= \frac{-(4x^2 - 3x - 10)}{x - 2} = \frac{-(4x + 5)(x - 2)}{x - 2} = -4x - 5$$

29. $f(x) = x^3 - 4x$

$$f(-x) = (-x)^3 - 4(-x) = -x^3 + 4x = -\left(x^3 - 4x\right) = -f(x) \qquad f \text{ is odd.}$$

31. $h(x) = \dfrac{1}{x^4} + \dfrac{1}{x^2} + 1$

$$h(-x) = \frac{1}{(-x)^4} + \frac{1}{(-x)^2} + 1 = \frac{1}{x^4} + \frac{1}{x^2} + 1 = h(x) \qquad h \text{ is even.}$$

33. $G(x) = 1 - x + x^3$

$$G(-x) = 1 - (-x) + (-x)^3 = 1 + x - x^3 \neq -G(x) \neq G(x)$$

G is neither even nor odd.

35. $f(x) = \dfrac{x}{1 + x^2}$

$$f(-x) = \frac{-x}{1 + (-x)^2} = \frac{-x}{1 + x^2} = -f(x) \quad f \text{ is odd.}$$

37. $F(x) = |x| - 4$

Using the graph of $y = |x|$, vertically shift the graph downward 4 units.

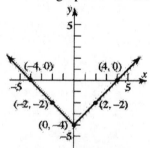

Intercepts: $(-4,0)$, $(4,0)$, $(0,-4)$
Domain: {Real Numbers}
Range: $\left\{y \mid y \geq -4\right\}$

39. $g(x) = -2|x|$

Reflect the graph of $y = |x|$ about the x-axis and vertically stretch the graph by a factor of 2.

Intercepts: $(0,0)$
Domain: {Real Numbers}
Range: $\left\{y \mid y \leq 0\right\}$

41. $h(x) = \sqrt{x-1}$

Using the graph of $y = \sqrt{x}$, horizontally shift the graph to the right 1 unit.

Intercepts: (1,0)
Domain: $\{x \mid x \geq 1\}$
Range: $\{y \mid y \geq 0\}$

43. $f(x) = \sqrt{1-x} = \sqrt{-1(x-1)}$

Reflect the graph of $y = \sqrt{x}$ about the y-axis and horizontally shift the graph to the right 1 unit..

Intercepts: (1,0), (0,1)
Domain: $\{x \mid x \leq 1\}$
Range: $\{y \mid y \geq 0\}$

45. $h(x) = (x-1)^2 + 2$

Using the graph of $y = x^2$, horizontally shift the graph to the right 1 unit and vertically shift the graph up 2 units.

Intercepts: (0,3)
Domain: {Real Numbers}
Range: $\{y \mid y \geq 2\}$

47. $g(x) = 3(x-1)^3 + 1$

Using the graph of $y = x^3$, horizontally shift the graph to the right 1 unit, vertically stretch the graph by a factor of 3, and vertically shift the graph up 1 unit.

Intercepts: $(0,-2), \left(1 - \dfrac{\sqrt[3]{3}}{3}, 0\right)$

Domain: {Real Numbers}
Range: {Real Numbers}

49. $f(x) = 3x - 5 \qquad g(x) = 1 - 2x^2$

(a) $(f \circ g)(2) = f(g(2)) = f\left(1 - 2(2)^2\right) = f(-7) = 3(-7) - 5 = -26$

(b) $(g \circ f)(-2) = g(f(-2)) = g(3(-2) - 5) = g(-11) = 1 - 2(-11)^2 = -241$

(c) $(f \circ f)(4) = f(f(4)) = f(3(4) - 5) = f(7) = 3(7) - 5 = 16$

(d) $(g \circ g)(-1) = g(g(-1)) = g\left(1 - 2(-1)^2\right) = g(-1) = 1 - 2(-1)^2 = -1$

51. $f(x) = \sqrt{x+2} \qquad g(x) = 2x^2 + 1$

(a) $(f \circ g)(2) = f(g(2)) = f\left(2(2)^2 + 1\right) = f(9) = \sqrt{9+2} = \sqrt{11}$

(b) $(g \circ f)(-2) = g(f(-2)) = g\left(\sqrt{-2+2}\right) = g(0) = 2(0)^2 + 1 = 1$

(c) $(f \circ f)(4) = f(f(4)) = f\left(\sqrt{4+2}\right) = f\left(\sqrt{6}\right) = \sqrt{\sqrt{6}+2}$

(d) $(g \circ g)(-1) = g(g(-1)) = g\left(2(-1)^2 + 1\right) = g(3) = 2(3)^2 + 1 = 19$

53. $f(x) = \dfrac{1}{x^2+4}$ $g(x) = 3x - 2$

(a) $(f \circ g)(2) = f(g(2)) = f(3(2)-2) = f(4) = \dfrac{1}{4^2+4} = \dfrac{1}{20}$

(b) $(g \circ f)(-2) = g(f(-2)) = g\left(\dfrac{1}{(-2)^2+4}\right) = g\left(\dfrac{1}{8}\right) = 3\left(\dfrac{1}{8}\right) - 2 = \dfrac{-13}{8}$

(c) $(f \circ f)(4) = f(f(4)) = f\left(\dfrac{1}{4^2+4}\right) = f\left(\dfrac{1}{20}\right) = \dfrac{1}{\left(\dfrac{1}{20}\right)^2+4} = \dfrac{1}{\left(\dfrac{1601}{400}\right)} = \dfrac{400}{1601}$

(d) $(g \circ g)(-1) = g(g(-1)) = g(3(-1)-2) = g(-5) = 3(-5) - 2 = -17$

55. $f(x) = 2 - x$ $g(x) = 3x + 1$
The domain of f is all real numbers. The domain of g is all real numbers.

(a) $(f \circ g)(x) = f(g(x)) = f(3x+1) = 2 - (3x+1) = 2 - 3x - 1 = 1 - 3x$
            Domain: All real numbers.

(b) $(g \circ f)(x) = g(f(x)) = g(2-x) = 3(2-x) + 1 = 6 - 3x + 1 = 7 - 3x$
            Domain: All real numbers.

(c) $(f \circ f)(x) = f(f(x)) = f(2-x) = 2 - (2-x) = 2 - 2 + x = x$
            Domain: All real numbers.

(d) $(g \circ g)(x) = g(g(x)) = g(3x+1) = 3(3x+1) + 1 = 9x + 3 + 1 = 9x + 4$
            Domain: All real numbers.

57. $f(x) = 3x^2 + x + 1$ $g(x) = |3x|$
The domain of f is all real numbers. The domain of g is all real numbers.

(a) $(f \circ g)(x) = f(g(x)) = f(|3x|) = 3(|3x|)^2 + (|3x|) + 1 = 27x^2 + 3|x| + 1$
            Domain: All real numbers.

(b) $(g \circ f)(x) = g(f(x)) = g(3x^2 + x + 1) = |3(3x^2 + x + 1)| = |9x^2 + 3x + 3|$
            Domain: All real numbers.

(c) $(f \circ f)(x) = f(f(x)) = f(3x^2 + x + 1) = 3(3x^2 + x + 1)^2 + (3x^2 + x + 1) + 1$
            $= 3(9x^4 + 6x^3 + 7x^2 + 2x + 1) + 3x^2 + x + 1 + 1$
            $= 27x^4 + 18x^3 + 24x^2 + 7x + 5$
            Domain: All real numbers.

(d) $(g \circ g)(x) = g(g(x)) = g(|3x|) = |3|3x|| = 9|x|$
            Domain: All real numbers.

59. $f(x) = \dfrac{x+1}{x-1}$ $g(x) = \dfrac{1}{x}$
The domain of f is $\{x \mid x \neq 1\}$. The domain of g is $\{x \mid x \neq 0\}$.

(a) $(f \circ g)(x) = f(g(x)) = f\left(\dfrac{1}{x}\right) = \dfrac{\left(\dfrac{1}{x}+1\right)}{\left(\dfrac{1}{x}-1\right)} = \dfrac{\left(\dfrac{1+x}{x}\right)}{\left(\dfrac{1-x}{x}\right)} = \dfrac{1+x}{1-x}$

Domain of $f \circ g$ is $\{x \mid x \neq 0, \, x \neq 1\}$.

(b) $(g \circ f)(x) = g(f(x)) = g\left(\dfrac{x+1}{x-1}\right) = \dfrac{1}{\left(\dfrac{x+1}{x-1}\right)} = \dfrac{x-1}{x+1}$

Domain of $g \circ f$ is $\{x \mid x \neq -1, \, x \neq 1\}$.

(c) $(f \circ f)(x) = f(f(x)) = f\left(\dfrac{x+1}{x-1}\right) = \dfrac{\left(\dfrac{x+1}{x-1}+1\right)}{\left(\dfrac{x+1}{x-1}-1\right)} = \dfrac{\left(\dfrac{x+1+x-1}{x-1}\right)}{\left(\dfrac{x+1-(x-1)}{x-1}\right)} = \dfrac{2x}{2} = x$

Domain of $f \circ f$ is $\{x \mid x \neq 1\}$.

(d) $(g \circ g)(x) = g(g(x)) = g\left(\dfrac{1}{x}\right) = \dfrac{1}{\left(\dfrac{1}{x}\right)} = x$ Domain of $g \circ g$ is $\{x \mid x \neq 0\}$.

61. (a) $y = f(-x)$
Reflect about the y-axis.

(b) $y = -f(x)$
Reflect about the x-axis.

(c) $y = f(x + 2)$
Horizontally shift left 2 units.

(d) $y = f(x) + 2$
Vertically shift up 2 units.

(e) $y = 2f(x)$

Vertical stretch by a factor of 2.

(f) $y = f(3x)$

Horizontal compression by $\frac{1}{3}$.

63.   we have the points $(h_1, T_1) = (0, 30)$ and $(h_2, T_2) = (10000, 5)$

$$slope = \frac{\Delta T}{\Delta h} = \frac{5 - 30}{10000 - 0} = \frac{-25}{10000} = -0.0025$$

using the point-slope formula yields

$$T - T_1 = m(h - h_1) \rightarrow T - 30 = -0.0025(h - 0)$$

$$T - 30 = -0.0025h \rightarrow T = -0.0025h + 30 \quad \therefore T(h) = -0.0025h + 30$$

65. $S = kxd^3$, $x = $ width; $d = $ depth

in the diagram, depth = diameter of the log = 6

$$S(x) = kx(6)^3 = 216kx \qquad \text{domain is } \{x \mid 0 \le x \le 6\}$$

67. (a), (b). (e)

(c)   average rate of change $= \dfrac{112.5 - 0}{5 - 0} = \dfrac{112.5}{5} = 22.5$ feet per second

(d)   for each 1 second increase in time, the distance fallen increases by 22.5 feet

(f)   average rate of change $= \dfrac{1960 - 1102.5}{20 - 15} = \dfrac{857.5}{5} = 171.5$ feet per second

(g)   for each 1 second increase in time, the distance fallen increases by 171.5 feet.

(h)   the average rate of change of distance is increasing as time passes

69. (a) We are given that the volume = 500 cubic feet, so we have

$$V = \pi r^2 h = 100 \rightarrow h = \frac{500}{\pi r^2}$$

Total Cost = cost of top + cost of bottom + cost of body

$= 2(\text{cost of top}) + \text{cost of body}$

$= 2(\text{area of top})(\text{cost per area of top}) + (\text{area of body})(\text{cost per area of body})$

$= 2(\pi r^2)(.06) + (2\pi rh)(.04) = 0.12\pi r^2 + .08\pi rh = .12\pi r^2 + .08\pi r\left(\dfrac{500}{\pi r^2}\right)$

$= .12\pi r^2 + \dfrac{40}{r}, \quad \therefore C(r) = .12\pi r^2 + \dfrac{40}{r}$

(b) $C(4) = .12\pi(4)^2 + \dfrac{40}{4} = 1.92\pi + 10 \approx 16.03$ dollars

(c) $C(8) = .12\pi(8)^2 + \dfrac{40}{8} = 7.68\pi + 5 \approx 29.13$ dollars

(d)

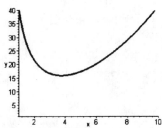

The minimum value occurs at $r \approx 3.79 \rightarrow C(3.79) \approx \$15.97$

71. $S = 4\pi r^2 \rightarrow r = \sqrt{\dfrac{S}{4\pi}} \qquad V(S) = \dfrac{4}{3}\pi r^3 = \dfrac{4\pi}{3}\left(\sqrt{\dfrac{S}{4\pi}}\right)^3 = \dfrac{4\pi}{3} \cdot \dfrac{S}{4\pi}\sqrt{\dfrac{S}{4\pi}} = \dfrac{S}{6}\sqrt{\dfrac{S}{\pi}}$

$V(2S) = \dfrac{2S}{6}\sqrt{\dfrac{2S}{\pi}} = 2\sqrt{2}\left(\dfrac{S}{6}\sqrt{\dfrac{S}{\pi}}\right)$ The volume is $2\sqrt{2}$ times as large.

Polynomial and Rational Functions

3.1   Quadratic Functions and Models

1.   D   3.   A   5.   B   7.   E

9. $f(x) = \dfrac{1}{4}x^2$

Using the function $y = x^2$, compress vertically by a factor of $\dfrac{1}{4}$.

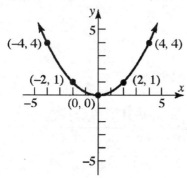

11. $f(x) = \dfrac{1}{4}x^2 - 2$

Using the function $y = x^2$, compress vertically by a factor of $\dfrac{1}{4}$, and shift downward 2 units.

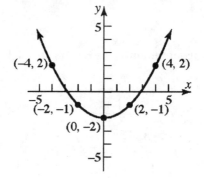

13. $f(x) = \dfrac{1}{4}x^2 + 2$

Using the function $y = x^2$, compress vertically by a factor of $\dfrac{1}{4}$, and shift upward 2 units.

15. $f(x) = \dfrac{1}{4}x^2 + 1$

Using the function $y = x^2$, compress vertically by a factor of $\dfrac{1}{4}$, and shift upward 1 unit.

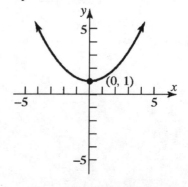

17. $f(x) = x^2 + 4x + 2$
Completing the square:
$f(x) = (x^2 + 4x + 4) + 2 - 4$
$\quad = (x + 2)^2 - 2$

Using the function $y = x^2$, shift the graph to the left 2 units, and shift downward 2 units.

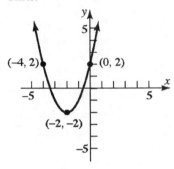

19. $f(x) = 2x^2 - 4x + 1$
Completing the square:
$f(x) = 2(x^2 - 2x + 1) + 1 - 2$
$\quad = 2(x - 1)^2 - 1$

Using the function $y = x^2$, shift the graph to the right 1 unit, stretch the graph vertically by a factor of 2, and shift downward 1 unit.

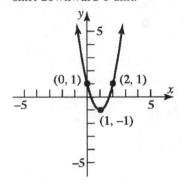

21. $f(x) = -x^2 - 2x$
Completing the square:
$f(x) = -(x^2 + 2x + 1) + 1$
$\quad = -(x + 1)^2 + 1$

Using the function $y = x^2$, shift the graph to the left 1 unit, reflect the graph on the x-axis, and shift upward 1 unit.

23. $f(x) = \frac{1}{2}x^2 + x - 1$
Completing the square:
$f(x) = \frac{1}{2}(x^2 + 2x + 1) - 1 - \frac{1}{2}$
$\quad = \frac{1}{2}(x + 1)^2 - \frac{3}{2}$

Using the function $y = x^2$, shift the graph to the left 1 unit, compress the graph vertically by a factor of $\frac{1}{2}$, and shift downward $\frac{3}{2}$ units.

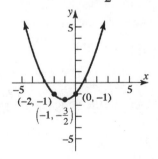

25. $f(x) = -x^2 - 6x$

$a = -1, b = -6, c = 0$. Since $a = -1 < 0$, the graph opens down.

The x-coordinate of the vertex is $x = \dfrac{-b}{2a} = \dfrac{-(-6)}{2(-1)} = \dfrac{6}{-2} = -3$.

The y-coordinate of the vertex is $f\left(\dfrac{-b}{2a}\right) = f(-3) = -(-3)^2 - 6(-3) = -9 + 18 = 9$.

Thus, the vertex is $(-3, 9)$.
The axis of symmetry is the line $x = -3$.
The discriminant is:

$$b^2 - 4ac = (-6)^2 - 4(-1)(0) = 36 > 0,$$

so the graph has two x-intercepts.
The x-intercepts are found by solving:

$$-x^2 - 6x = 0 \rightarrow -x(x+6) = 0$$

$$x = 0 \text{ or } x = -6$$

The x-intercepts are –6 and 0.
The y-intercept is $f(0) = 0$.

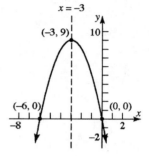

27. $f(x) = 2x^2 - 8x$

$a = 2, b = -8, c = 0$. Since $a = 2 > 0$, the graph opens up.

The x-coordinate of the vertex is $x = \dfrac{-b}{2a} = \dfrac{-(-8)}{2(2)} = \dfrac{8}{4} = 2$.

The y-coordinate of the vertex is $f\left(\dfrac{-b}{2a}\right) = f(2) = 2(2)^2 - 8(2) = 8 - 16 = -8$.

Thus, the vertex is $(2, -8)$.
The axis of symmetry is the line $x = 2$.
The discriminant is:

$$b^2 - 4ac = (-8)^2 - 4(2)(0) = 64 > 0,$$

so the graph has two x-intercepts.
The x-intercepts are found by solving:

$$2x^2 - 8x = 0 \rightarrow 2x(x-4) = 0$$

$$x = 0 \text{ or } x = 4$$

The x-intercepts are 0 and 4.
The y-intercept is $f(0) = 0$.

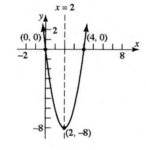

29. $f(x) = x^2 + 2x - 8$

$a = 1, b = 2, c = -8$. Since $a = 1 > 0$, the graph opens up.

The x-coordinate of the vertex is $x = \dfrac{-b}{2a} = \dfrac{-2}{2(1)} = \dfrac{-2}{2} = -1$.

The y-coordinate of the vertex is $f\left(\dfrac{-b}{2a}\right) = f(-1) = (-1)^2 + 2(-1) - 8 = 1 - 2 - 8 = -9$.

Thus, the vertex is $(-1, -9)$.
The axis of symmetry is the line $x = -1$.
The discriminant is:
$$b^2 - 4ac = 2^2 - 4(1)(-8) = 4 + 32 = 36 > 0,$$
so the graph has two x-intercepts.
The x-intercepts are found by solving:
$$x^2 + 2x - 8 = 0 \to (x + 4)(x - 2) = 0$$
$$x = -4 \text{ or } x = 2$$
The x-intercepts are -4 and 2.
The y-intercept is $f(0) = -8$.

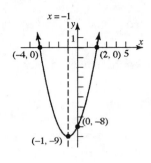

31. $f(x) = x^2 + 2x + 1$
$a = 1, b = 2, c = 1$. Since $a = 1 > 0$, the graph opens up.

The x-coordinate of the vertex is $x = \dfrac{-b}{2a} = \dfrac{-2}{2(1)} = \dfrac{-2}{2} = -1$.

The y-coordinate of the vertex is $f\left(\dfrac{-b}{2a}\right) = f(-1) = (-1)^2 + 2(-1) + 1 = 1 - 2 + 1 = 0$.

Thus, the vertex is $(-1, 0)$.
The axis of symmetry is the line $x = -1$.
The discriminant is:
$$b^2 - 4ac = 2^2 - 4(1)(1) = 4 - 4 = 0,$$
so the graph has one x-intercept.
The x-intercept is found by solving:
$$x^2 + 2x + 1 = 0 \to (x + 1)^2 = 0$$
$$x = -1$$
The x-intercept is -1.
The y-intercept is $f(0) = 1$.

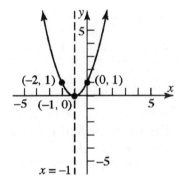

33. $f(x) = 2x^2 - x + 2$
$a = 2, b = -1, c = 2$. Since $a = 2 > 0$, the graph opens up.

The x-coordinate of the vertex is $x = \dfrac{-b}{2a} = \dfrac{-(-1)}{2(2)} = \dfrac{1}{4}$.

The y-coordinate of the vertex is $f\left(\dfrac{-b}{2a}\right) = f\left(\dfrac{1}{4}\right) = 2\left(\dfrac{1}{4}\right)^2 - \dfrac{1}{4} + 2 = \dfrac{1}{8} - \dfrac{1}{4} + 2 = \dfrac{15}{8}$.

Thus, the vertex is $\left(\dfrac{1}{4}, \dfrac{15}{8}\right)$.

The axis of symmetry is the line $x = \dfrac{1}{4}$.

The discriminant is:
$$b^2 - 4ac = (-1)^2 - 4(2)(2) = 1 - 16 = -15,$$
so the graph has no x-intercepts.
The y-intercept is $f(0) = 2$.

35. $f(x) = -2x^2 + 2x - 3$
$a = -2, b = 2, c = -3$. Since $a = -2 < 0$, the graph opens down.

The x-coordinate of the vertex is $x = \dfrac{-b}{2a} = \dfrac{-(2)}{2(-2)} = \dfrac{-2}{-4} = \dfrac{1}{2}$.

The y-coordinate of the vertex is $f\left(\dfrac{-b}{2a}\right) = f\left(\dfrac{1}{2}\right) = -2\left(\dfrac{1}{2}\right)^2 + 2\left(\dfrac{1}{2}\right) - 3 = -\dfrac{1}{2} + 1 - 3 = -\dfrac{5}{2}$.

Thus, the vertex is $\left(\dfrac{1}{2}, -\dfrac{5}{2}\right)$.

The axis of symmetry is the line $x = \dfrac{1}{2}$.

The discriminant is:
$$b^2 - 4ac = 2^2 - 4(-2)(-3) = 4 - 24 = -20,$$
so the graph has no x-intercepts.
The y-intercept is $f(0) = -3$.

37. $f(x) = 3x^2 + 6x + 2$
$a = 3, b = 6, c = 2$. Since $a = 3 > 0$, the graph opens up.

The x-coordinate of the vertex is $x = \dfrac{-b}{2a} = \dfrac{-6}{2(3)} = \dfrac{-6}{6} = -1$. The y-coordinate of the

vertex is $f\left(\dfrac{-b}{2a}\right) = f(-1) = 3(-1)^2 + 6(-1) + 2 = 3 - 6 + 2 = -1$.

Thus, the vertex is $(-1, -1)$.
The axis of symmetry is the line $x = -1$.
The discriminant is:
$$b^2 - 4ac = 6^2 - 4(3)(2) = 36 - 24 = 12,$$
so the graph has two x-intercepts.
The x-intercepts are found by solving:

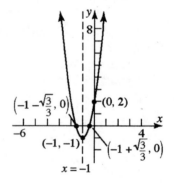

$$x = \dfrac{-b \pm \sqrt{b^2 - 4ac}}{2a} = \dfrac{-6 \pm \sqrt{12}}{2(3)}$$

$$= \dfrac{-6 \pm 2\sqrt{3}}{6} = \dfrac{-3 \pm \sqrt{3}}{3} \approx \dfrac{-3 \pm 1.732}{3}$$

The x-intercepts are approximately –0.42 and –1.58.
The y-intercept is $f(0) = 2$.

39. $f(x) = -4x^2 - 6x + 2$
$a = -4, b = -6, c = 2$. Since $a = -4 < 0$, the graph opens down.

The x-coordinate of the vertex is $x = \dfrac{-b}{2a} = \dfrac{-(-6)}{2(-4)} = \dfrac{6}{-8} = -\dfrac{3}{4}$.

The y-coordinate of the vertex is

$$f\left(\dfrac{-b}{2a}\right) = f\left(-\dfrac{3}{4}\right) = -4\left(-\dfrac{3}{4}\right)^2 - 6\left(-\dfrac{3}{4}\right) + 2 = -\dfrac{9}{4} + \dfrac{9}{2} + 2 = \dfrac{17}{4}.$$

Thus, the vertex is $\left(-\dfrac{3}{4}, \dfrac{17}{4}\right)$.

The axis of symmetry is the line $x = -\dfrac{3}{4}$.

The discriminant is:

$$b^2 - 4ac = (-6)^2 - 4(-4)(2) = 36 + 32 = 68,$$

so the graph has two x-intercepts.

The x-intercepts are found by solving:

$$x = \frac{-b \pm \sqrt{b^2 - 4ac}}{2a} = \frac{-(-6) \pm \sqrt{68}}{2(-4)}$$

$$= \frac{6 \pm 2\sqrt{17}}{-8} = \frac{-3 \pm \sqrt{17}}{4} \approx \frac{-3 \pm 4.123}{4}$$

The x-intercepts are approximately -1.78 and 0.28.
The y-intercept is $f(0) = 2$.

41. $f(x) = 2x^2 + 12x$, $a = 2, b = 12, c = 0$. Since $a = 2 > 0$, the graph opens up, so
the vertex is a minimum point. The minimum occurs at

$x = \dfrac{-b}{2a} = \dfrac{-12}{2(2)} = \dfrac{-12}{4} = -3$. The minimum value is

$f\left(\dfrac{-b}{2a}\right) = f(-3) = 2(-3)^2 + 12(-3) = 18 - 36 = -18$.

43. $f(x) = 2x^2 + 12x - 3$, $a = 2, b = 12, c = -3$. Since $a = 2 > 0$, the graph opens up,
so the vertex is a minimum point. The minimum occurs at

$x = \dfrac{-b}{2a} = \dfrac{-12}{2(2)} = \dfrac{-12}{4} = -3$. The minimum value is

$f\left(\dfrac{-b}{2a}\right) = f(-3) = 2(-3)^2 + 12(-3) - 3 = 18 - 36 - 3 = -21$.

45. $f(x) = -x^2 + 10x - 4$

$a = -1, b = 10, c = -4$. Since $a = -1 < 0$, the graph opens down, so the vertex is a

maximum point. The maximum occurs at $x = \dfrac{-b}{2a} = \dfrac{-10}{2(-1)} = \dfrac{-10}{-2} = 5$. The maximum

value is $f\left(\dfrac{-b}{2a}\right) = f(5) = -(5)^2 + 10(5) - 4 = -25 + 50 - 4 = 21$.

47. $f(x) = -3x^2 + 12x + 1$

$a = -3, b = 12, c = 1$. Since $a = -3 < 0$, the graph opens down, so the vertex is a

maximum point. The maximum occurs at $x = \dfrac{-b}{2a} = \dfrac{-12}{2(-3)} = \dfrac{-12}{-6} = 2$. The maximum

value is $f\left(\dfrac{-b}{2a}\right) = f(2) = -3(2)^2 + 12(2) + 1 = -12 + 24 + 1 = 13$.

49. (a) $f(x) = 1(x - (-3))(x - 1) = 1(x + 3)(x - 1) = 1(x^2 + 2x - 3) = x^2 + 2x - 3$

$f(x) = 2(x - (-3))(x - 1) = 2(x + 3)(x - 1) = 2(x^2 + 2x - 3) = 2x^2 + 4x - 6$

$f(x) = -2(x - (-3))(x - 1) = -2(x + 3)(x - 1)$

$$= -2(x^2 + 2x - 3) = -2x^2 - 4x + 6$$

$f(x) = 5(x - (-3))(x - 1) = 5(x + 3)(x - 1) = 5(x^2 + 2x - 3) = 5x^2 + 10x - 15$

(b) The value of a multiplies the value of the y-intercept by the value of a. The values of the x-intercepts are not changed.

(c) The axis of symmetry is unaffected by the value of a.

(d) The y-coordinate of the vertex is multiplied by the value of a.

(e) The x-coordinate of the vertex is the midpoint of the x-intercepts.

51. $R(p) = -4p^2 + 4000p$

$a = -4$, $b = 4000$, $c = 0$. Since $a = -4 < 0$, the graph is a parabola that opens down, so the vertex is a maximum point. The maximum occurs at

$$p = \frac{-b}{2a} = \frac{-4000}{2(-4)} = \frac{-4000}{-8} = 500.$$

$R(500) = -4(500)^2 + 4000(500) = -1000000 + 2000000 = 1,000,000$.

Thus, the unit price should be \$500 for maximum revenue. The maximum revenue is \$1,000,000.

53. (a) $R(x) = x\left(-\frac{1}{6}x + 100\right) = -\frac{1}{6}x^2 + 100x$

(b) $R(200) = \frac{-1}{6}(200)^2 + 100(200) = \frac{-20000}{3} + 20000 = \frac{40000}{3} \approx \$13,333$

(c) $x = \frac{-b}{2a} = \frac{-100}{2\left(-\frac{1}{6}\right)} = \frac{-100}{\left(-\frac{1}{3}\right)} = \frac{300}{1} = 300$

$R(300) = -\frac{1}{6}(300)^2 + 100(300) = -15000 + 30000 = \$15,000$

(d) $p = -\frac{1}{6}(300) + 100 = -50 + 100 = \$50$

55. (a)   If $x = -5p + 100$, then $p = \frac{100 - x}{5}$. $R(x) = x\left(\frac{100 - x}{5}\right) = -\frac{1}{5}x^2 + 20x$

(b) $R(15) = -\frac{1}{5}(15)^2 + 20(15) = -45 + 300 = \$255$

(c) $x = \frac{-b}{2a} = \frac{-20}{2\left(-\frac{1}{5}\right)} = \frac{-20}{\left(-\frac{2}{5}\right)} = \frac{100}{2} = 50$

$R(50) = -\frac{1}{5}(50)^2 + 20(50) = -500 + 1000 = \$500$

(d) $p = \frac{100 - 50}{5} = \frac{50}{5} = \$10$

57. (a) Let x = width and y = length of the rectangular area.
$$P = 2x + 2y = 400$$
$$y = \frac{400 - 2x}{2} = 200 - x$$
Then $A(x) = (200 - x)x = 200x - x^2 = -x^2 + 200x$

(b) $x = \dfrac{-b}{2a} = \dfrac{-200}{2(-1)} = \dfrac{-200}{-2} = 100$ yards

(c) $A(100) = -100^2 + 200(100) = -10000 + 20000 = 10{,}000$ sq yds.

59. Let x = width and y = length of the rectangular area.
$$2x + y = 4000 \quad \rightarrow \quad y = 4000 - 2x$$
Then $A(x) = (4000 - 2x)x = 4000x - 2x^2 = -2x^2 + 4000x$
$$x = \frac{-b}{2a} = \frac{-4000}{2(-2)} = \frac{-4000}{-4} = 1000$$
$$A(1000) = -2(1000)^2 + 4000(1000) = -2000000 + 4000000 = 2{,}000{,}000$$
The largest area that can be enclosed is 2,000,000 square meters.

61. (a) $a = -\dfrac{32}{2500}, b = 1, c = 200$. The maximum height occurs when
$$x = \frac{-b}{2a} = \frac{-1}{2\left(-\dfrac{32}{2500}\right)} = \frac{2500}{64} = 39.0625 \text{ feet from base of the cliff.}$$

(b) The maximum height is
$$h(39.0625) = \frac{-32(39.0625)^2}{2500} + 39.0625 + 200 = 219.53 \text{ feet.}$$

(c) Solving when $h(x) = 0$:
$$-\frac{32}{2500}x^2 + x + 200 = 0$$
$$x = \frac{-1 \pm \sqrt{1^2 - 4\left(-\dfrac{32}{2500}\right)(200)}}{2\left(-\dfrac{32}{2500}\right)} = \frac{-1 \pm \sqrt{11.24}}{-0.0256} \rightarrow x \approx -91.90 \text{ or } x \approx 170.02$$

Since the distance cannot be negative, the projectile strikes the water 170.02 feet from the base of the cliff.

(d) Graphing:

(e) Solving when $h(x) = 100$:
$$-\frac{32}{2500}x^2 + x + 200 = 100 \rightarrow -\frac{32}{2500}x^2 + x + 100 = 0$$

$$x = \dfrac{-1 \pm \sqrt{1^2 - 4\left(-\dfrac{32}{2500}\right)(100)}}{2\left(-\dfrac{32}{2500}\right)} = \dfrac{-1 \pm \sqrt{6.12}}{-0.0256}; \quad x \approx -57.57 \text{ or } x \approx 135.70$$

Since the distance cannot be negative, the projectile is 100 feet above the water 135.70 feet from the base of the cliff.

63. Locate the origin at the point where the cable touches the road. Then the equation of the parabola is of the form: $y = ax^2$, where $a > 0$. Since the point $(200, 75)$ is on the parabola, we can find the constant a:

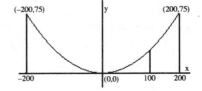

$$75 = a(200)^2 \quad \rightarrow \quad a = \dfrac{75}{200^2} = 0.001875$$

When $x = 100$, we have:
$$y = 0.001875(100)^2 = 18.75 \text{ meters}.$$

65. Let $x =$ the depth of the gutter and $y =$ the width of the gutter.
Then $A = xy$ is the cross-sectional area of the gutter.
Since the aluminum sheets for the gutter are 12 inches wide, we have
$$2x + y = 12 \text{ or } y = 12 - 2x.$$
The area is to be maximized, so: $A = xy = x(12 - 2x) = -2x^2 + 12x$.
This equation is a parabola opening down; thus, it has a maximum when
$$x = \dfrac{-b}{2a} = \dfrac{-12}{2(-2)} = \dfrac{-12}{-4} = 3.$$

Thus, a depth of 3 inches produces a maximum cross-sectional area.

67. Let $x =$ the width of the rectangle or the diameter of the semicircle.
Let $y =$ the length of the rectangle.

The perimeter of each semicircle is $\dfrac{\pi x}{2}$.

The perimeter of the track is given by: $\dfrac{\pi x}{2} + \dfrac{\pi x}{2} + y + y = 1500$.

Solving for x:

$$\dfrac{\pi x}{2} + \dfrac{\pi x}{2} + y + y = 1500 \rightarrow \pi x + 2y = 1500 \rightarrow \pi x = 1500 - 2y \rightarrow x = \dfrac{1500 - 2y}{\pi}$$

The area of the rectangle is: $A = xy = \left(\dfrac{1500 - 2y}{\pi}\right)y = \dfrac{-2}{\pi}y^2 + \dfrac{1500}{\pi}y$

This equation is a parabola opening down; thus, it has a maximum when

$$y = \dfrac{-b}{2a} = \dfrac{\dfrac{-1500}{\pi}}{2\left(\dfrac{-2}{\pi}\right)} = \dfrac{-1500}{-4} = 375. \quad \text{Thus, } x = \dfrac{1500 - 2(375)}{\pi} = \dfrac{750}{\pi} \approx 238.73.$$

The dimensions for the rectangle with maximum area are $\dfrac{750}{\pi} \approx 238.73$ meters by 375 meters.

Problems 69 – 73. The equations for the curves that are graphed on the screens use many more decimal places in order to get the desired accuracy.

69. (a) Graphing: The data appear to be quadratic with $a < 0$.

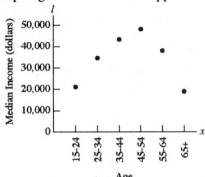

(b) $I(x) = -42.6x^2 + 3806x - 38526$

$$x = \frac{-b}{2a} = \frac{-3806}{2(-42.6)} = 44.695$$

An individual will earn the most income at an age of 44.7 years.

(c) The maximum income will be:

$$I(44.7) = -42.6(44.7)^2 + 3805(44.7) - 38526 = \$46,438.87$$

(d) and (e)   Graphing the quadratic function of best fit:

71. (a) Graphing: The data appears to be quadratic with $a > 0$.

(b) Note: The coefficients given in the text are incorrect. The correct function is:

$$I(x) = 17.199x^2 - 68147.78x + 67507955$$

$$x = \frac{-b}{2a} = \frac{-(-68147.78)}{2(17.199)} \approx 1981.16 \qquad \text{Imports were lowest in 1981.}$$

(c) The predicted number of barrels imported in 1998 (using the equation to many more decimal places) is: $I(1998) \approx 3165$ thousand barrels

(d) and (e)   Graphing the quadratic function of best fit:

73. (a)   Graphing: The data appears to be quadratic with $a < 0$.

(b) $h(x) = -0.0037x^2 + 1.03x + 5.7$

$$x = \frac{-b}{2a} = \frac{-1.03}{2(-0.0037)} \approx 139.19 \text{ feet}$$

The ball travels about 139 feet before reaching its maximum height.

(c)   The maximum height will be: (using the equation to many more decimal places)

$$h(139) = -0.0037(139)^2 + 1.03(139) + 5.7 \approx 77.38 \text{ feet}$$

(d) and (e)   Graphing the quadratic function of best fit:

75.   We have:

$$a(-h)^2 + b(-h) + c = ah^2 - bh + c = y_0$$
$$a(0)^2 + b(0) + c = c = y_1$$
$$a(h)^2 + b(h) + c = ah^2 + bh + c = y_2$$

Equating the two equations for the area, we have:

$$y_0 + 4y_1 + y_2 = ah^2 - bh + c + 4c + ah^2 + bh + c = 2ah^2 + 6c$$

Therefore,   Area $= \frac{h}{3}(2ah^2 + 6c) = \frac{h}{3}(y_0 + 4y_1 + y_2)$.

77. $f(x) = 2x^2 + 8, \quad h = 2$

Area $= \frac{2}{3}\left(2(2)(2)^2 + 6(8)\right) = \frac{2}{3}(16 + 48) = \frac{2}{3}(64) = \frac{128}{3}$

79. $f(x) = -x^2 + x + 4, \quad h = 1$

Area $= \frac{1}{3}\left(2(-1)(1)^2 + 6(4)\right) = \frac{1}{3}(-2 + 24) = \frac{1}{3}(22) = \frac{22}{3}$

81. If x is even, then ax^2 and bx are even. When two even numbers are added to an odd number the result is odd. Thus, $f(x)$ is odd.

If x is odd, then ax^2 and bx are odd. The sum of three odd numbers results in an odd number. Thus, $f(x)$ is odd.

83. $\qquad f(x) = x^2 + 2x - 3; \qquad f(x) = x^2 + 2x + 1; \qquad f(x) = x^2 + 2x$

each member of this family will be a parabola with the following characteristics:
- opens upwards since $a > 0$
- vertex occurs at $x = \dfrac{-b}{2a} = \dfrac{-2}{2(1)} = -1$
- there is at least one x-intercept since $b^2 - 4ac \geq 0$

Chapter 3

Polynomial and Rational Functions

3.2 Polynomial Functions

1. $f(x) = 4x + x^3$ is a polynomial function of degree 3.

3. $g(x) = \dfrac{1-x^2}{2} = \dfrac{1}{2} - \dfrac{1}{2}x^2$ is a polynomial function of degree 2.

5. $f(x) = 1 - \dfrac{1}{x} = 1 - x^{-1}$ is not a polynomial function because it contains a negative exponent.

7. $g(x) = x^{3/2} - x^2 + 2$ is not a polynomial function because it contains a fractional exponent.

9. $F(x) = 5x^4 - \pi x^3 + \dfrac{1}{2}$ is a polynomial function of degree 4.

11. $G(x) = 2(x-1)^2(x^2+1)$ is a polynomial function of degree 4.

13. $f(x) = (x+1)^4$
    Using the graph of $y = x^4$, shift the graph horizontally, 1 unit to the left.

    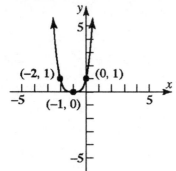

15. $f(x) = x^5 - 3$
    Using the graph of $y = x^5$, shift the graph vertically, 3 units down.

    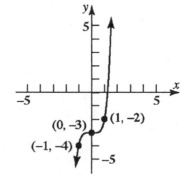

17. $f(x) = \dfrac{1}{2}x^4$

Using the graph of $y = x^4$, compress the

graph vertically by a factor of $\dfrac{1}{2}$.

19. $f(x) = -x^5$

Using the graph of $y = x^5$, reflect the
graph about the x-axis.

21. $f(x) = (x-1)^5 + 2$

Using the graph of $y = x^5$, shift the graph
horizontally, 1 unit to the right, and shift
vertically 2 units up.

23. $f(x) = 2(x+1)^4 + 1$

Using the graph of $y = x^4$, shift the
graph horizontally, 1 unit to the left,
stretch vertically by a factor of 2, and
shift vertically 1 unit up.

25. $f(x) = 4 - (x-2)^5 = -(x-2)^5 + 4$

Using the graph of $y = x^5$, shift the graph
horizontally, 2 units to the right, reflect
about the x-axis, and shift vertically 4
units up.

27. $f(x) = a(x-(-1))(x-1)(x-3)$

For $a = 1$: $f(x) = (x+1)(x-1)(x-3)$

$$= (x^2 - 1)(x-3)$$

$$= x^3 - 3x^2 - x + 3$$

29. $f(x) = a(x-(-3))(x-0)(x-4)$

For $a = 1$: $f(x) = (x+3)(x)(x-4)$

$$= (x^2 + 3x)(x-4)$$

$$= x^3 - 4x^2 + 3x^2 - 12x$$

$$= x^3 - x^2 - 12x$$

31. $f(x) = a(x-(-4))(x-(-1))(x-2)(x-3)$
     For $a = 1$: $f(x) = (x+4)(x+1)(x-2)(x-3)$
     $$f(x) = (x^2+5x+4)(x^2-5x+6)$$
     $$f(x) = x^4-5x^3+6x^2+5x^3-25x^2+30x+4x^2-20x+24$$
     $$f(x) = x^4-15x^2+10x+24$$

33. The real zeros of $f(x) = 3(x-7)(x+3)^2$ are: 7, with multiplicity one; and –3, with multiplicity two. The graph crosses the x-axis at 7 and touches it at –3.
     The function resembles $y = 3x^3$ for large values of $|x|$.

35. The real zeros of $f(x) = 4(x^2+1)(x-2)^3$ are: 2, with multiplicity three. $x^2+1 = 0$ has no real solution. The graph crosses the x-axis at 2.
     The function resembles $y = 4x^5$ for large values of $|x|$.

37. The real zeros of $f(x) = -2\left(x+\frac{1}{2}\right)^2(x^2+4)^2$ are: $-\frac{1}{2}$, with multiplicity two. $x^2+4 = 0$ has no real solution. The graph touches the x-axis at $\frac{-1}{2}$.
     The function resembles $y = 2x^6$ for large values of $|x|$.

39. The real zeros of $f(x) = (x-5)^3(x+4)^2$ are: 5, with multiplicity three; and –4, with multiplicity two. The graph crosses the x-axis at 5 and touches it at –4.
     The function resembles $y = x^5$ for large values of $|x|$.

41. $f(x) = 3(x^2+8)(x^2+9)^2$ has no real zeros. $x^2+8 = 0$ and $x^2+9 = 0$ have no real solutions. The graph neither touches nor crosses the x-axis. The function resembles $y = 3x^6$ for large values of $|x|$.

43. The real zeros of $f(x) = -2x^2(x^2-2)$ are: $-\sqrt{2}$ and $\sqrt{2}$ with multiplicity one; and 0, with multiplicity two. The graph touches the x-axis at $-\sqrt{2}$ and $\sqrt{2}$ and crosses the x-axis at 2. The function resembles $y = -2x^4$ for large values of $|x|$.

45. $f(x) = (x-1)^2$

    (a)   x-intercept: 1; y-intercept: 1   (b)   touches x-axis at x = 1

    (c) $y = x^2$ (d)   1

    (e)

Interval	$(-\infty, 1)$	$(1, \infty)$
Number Chosen	-1	2
Value of f	$f(-1) = 4$	$f(2) = 1$
Location of Graph	Above x-axis	Above x-axis
Point on Graph	(-1, 4)	(2, 1)

f is above the x-axis for $(-\infty,1) \cup (1,\infty)$

    (f)

47. $f(x) = x^2(x-3)$

    (a)   x-intercepts: 0, 3; y-intercept: 0

    (b)   touches x-axis at x = 0 ; crosses x-axis at x = 3

    (c) $y = x^3$ (d) 2

    (e)

Interval	$(-\infty, 0)$	$(0, 3)$	$(3, \infty)$
Number Chosen	-1	2	4
Value of f	$f(-1) = -4$	$f(2) = -4$	$f(4) = 16$
Location of Graph	Below x-axis	Below x-axis	Above x-axis
Point on Graph	(-1, -4)	(2, -4)	(4, 16)

f is below the x-axis for $(-\infty,0) \cup (0,3)$; f is above the x-axis for $(3,\infty)$

    (f)

49. $f(x) = 6x^3(x+4)$

(a)   x-intercepts: –4, 0; y-intercept: 0

(b)   crosses x-axis at x = –4 and x = 0

(c) $y = 6x^4$ (d)   3

(e)

	-4		0
Interval	$(-\infty, -4)$	$(-4, 0)$	$(0, \infty)$
Number Chosen	-5	-2	1
Value of f	$f(-5) = 750$	$f(-2) = -96$	$f(1) = 30$
Location of Graph	Above x-axis	Below x-axis	Above x-axis
Point on Graph	(-5, 750)	(-2, -96)	(1, 30)

f is above the x-axis for $(-\infty, -4) \cup (0, \infty)$; f is below the x-axis for $(-4, 0)$

(f)

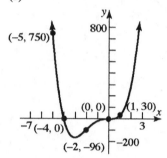

51. $f(x) = -4x^2(x+2)$

(a)   x-intercepts: 0, - 2; y-intercept: 0

(b)   crosses x-axis at x = - 2 ; touches x-axis at x = 0

(c) $y = -4x^3$ (d)   2

(e)

	-2		0
Interval	$(-\infty, -2)$	$(-2, 0)$	$(0, \infty)$
Number Chosen	-3	-1	1
Value of f	$f(-3) = 36$	$f(-1) = -4$	$f(1) = -12$
Location of Graph	Above x-axis	Below x-axis	Below x-axis
Point on Graph	(-3, 36)	(-1, -4)	(1, -12)

f is above the x-axis for $(-\infty, -2)$; f is below the x-axis for $(-2, 0) \cup (0, \infty)$

(f)

53. $f(x) = x(x-2)(x+4)$

(a)   x-intercepts: 0, - 4, 2; y-intercept: 0

(b)   crosses x-axis at x = 0, x = - 4 and x = 2

(c) $y = x^3$ (d)   2

(e)

Interval	$(-\infty, -4)$	$(-4, 0)$	$(0, 2)$	$(2, \infty)$
Number Chosen	-5	-2	1	3
Value of f	$f(-5) = -35$	$f(-2) = 16$	$f(1) = -5$	$f(3) = 21$
Location of Graph	Below x-axis	Above x-axis	Below x-axis	Above x-axis
Point on Graph	(-5, -35)	(-2, 16)	(1, -5)	(3, 21)

f is above the x-axis for $(-4,0)\cup(2,\infty)$; f is below the x-axis for $(-\infty,-4)\cup(0,2)$

(f)

55. $f(x) = 4x - x^3 = x(4 - x^2) = x(2 + x)(2 - x)$

(a)   x-intercepts: 0, - 2, 2; y-intercept: 0

(b)   crosses x-axis at x = 0, x = - 2 and x = 2

(c) $y = -x^3$ (d)   2

(e)

Interval	$(-\infty, -2)$	$(-2, 0)$	$(0, 2)$	$(2, \infty)$
Number Chosen	-3	-1	1	3
Value of f	$f(-3) = 15$	$f(-1) = -3$	$f(1) = 3$	$f(3) = -15$
Location of Graph	Above x-axis	Below x-axis	Above x-axis	Below x-axis
Point on Graph	(-3, 15)	(-1, -3)	(1, 3)	(3, -15)

f is above the x-axis for $(-\infty,-2)\cup(0,2)$; f is below the x-axis for $(-2,0)\cup(2,\infty)$

(f)

57. $f(x) = x^2(x-2)(x+2)$

(a)   x-intercepts: 0, -2, 2; y-intercept: 0

(b)   crosses x-axis at x = -2 and x = 2; touches x-axis at x = 0

(c) $y = x^4$ (d)   3

(e)

Interval	$(-\infty, -2)$	$(-2, 0)$	$(0, 2)$	$(2, \infty)$
Number Chosen	-3	-1	1	3
Value of f	$f(-3) = 45$	$f(-1) = -3$	$f(1) = -3$	$f(3) = 45$
Location of Graph	Above x-axis	Below x-axis	Below x-axis	Above x-axis
Point on Graph	(-3, 45)	(-1, -3)	(1, -3)	(3, 45)

f is above the x-axis for $(-\infty, -2) \cup (2, \infty)$; f is below the x-axis for $(-2, 0) \cup (0, 2)$

(f)

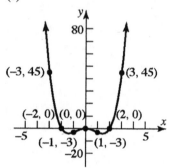

59. $f(x) = x^2(x-2)^2$

(a)   x-intercepts: 0, 2; y-intercept: 0   (b)   touches x-axis at x = 0 and x = 2

(c) $y = x^4$ (d)   3

(e)

Interval	$(-\infty, 0)$	$(0, 2)$	$(2, \infty)$
Number Chosen	-1	1	3
Value of f	$f(-1) = 9$	$f(1) = 1$	$f(3) = 9$
Location of Graph	Above x-axis	Above x-axis	Above x-axis
Point on Graph	(-1, 9)	(1, 1)	(3, 9)

f is above the x-axis for $(-\infty, 0) \cup (0, 2) \cup (2, \infty)$

(f)

61. $f(x) = x^2(x-3)(x+1)$

(a)   x-intercepts: 0, -1, 3; y-intercept: 0

(b)   crosses x-axis at x = -1 and x = 3; touches x-axis at x = 0

(c) $y = x^4$ (d)   3

(e)

	-1	0	3

Interval	$(-\infty, -1)$	$(-1, 0)$	$(0, 3)$	$(3, \infty)$
Number Chosen	-2	$-\frac{1}{2}$	2	4
Value of f	$f(-2) = 20$	$f(-\frac{1}{2}) = -\frac{7}{16}$	$f(2) = -12$	$f(4) = 80$
Location of Graph	Above x-axis	Below x-axis	Below x-axis	Above x-axis
Point on Graph	(-2, 20)	$(-\frac{1}{2}, -\frac{7}{16})$	(2, -12)	(4, 80)

f is above the x-axis for $(-\infty, -1) \cup (3, \infty)$; f is below the x-axis for $(-1, 0) \cup (0, 3)$

(f)

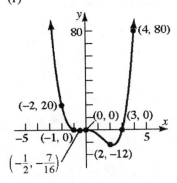

63. $f(x) = (x+2)^2(x-4)^2$

(a)   x-intercepts: -2, 4; y-intercept: 64   (b)   touches x-axis at x = -2 and x = 4

(c) $y = x^4$ (d)   3

(e)

	-2	4

Interval	$(-\infty, -2)$	$(-2, 4)$	$(4, \infty)$
Number Chosen	-3	0	5
Value of f	$f(-3) = 49$	$f(0) = 64$	$f(5) = 49$
Location of Graph	Above x-axis	Above x-axis	Above x-axis
Point on Graph	(-3, 49)	(0, 64)	(5, 49)

f is above the x-axis for $(-\infty, -2) \cup (-2, 4) \cup (4, \infty)$

(f)

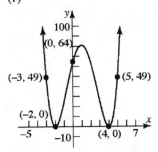

65. $f(x) = x^2(x-2)(x^2+3)$

(a)   x-intercepts: 0, 2; y-intercept: 0

(b)   crosses x-axis at x = 2 ; touches x-axis at x = 0

(c) $y = x^5$ (d)   4

(e)

| | 0 | | 2 |

Interval	$(-\infty, 0)$	$(0, 2)$	$(2, \infty)$
Number Chosen	-1	1	3
Value of f	$f(-1) = -12$	$f(1) = -4$	$f(3) = 108$
Location of Graph	Below x-axis	Below x-axis	Above x-axis
Point on Graph	(-1, -12)	(1, -4)	(3, 108)

f is above the x-axis for $(2,\infty)$; f is below the x-axis for $(-\infty,0)\cup(0,2)$

(f)

67. $f(x) = -x^2(x^2-1)(x+1) = -x^2(x-1)(x+1)(x+1) = -x^2(x-1)(x+1)^2$

(a)   x-intercepts: 0, - 1, 1; y-intercept: 0

(b)   crosses x-axis at x = 1; touches x-axis at x = 0 and x = - 1

(c) $y = -x^5$ (d)   4

(e)

| | -1 | | 0 | | 1 |

Interval	$(-\infty, -1)$	$(-1, 0)$	$(0, 1)$	$(1, \infty)$
Number Chosen	-2	$-\frac{1}{2}$	$\frac{1}{2}$	2
Value of f	$f(-2) = 12$	$f(-\frac{1}{2}) = \frac{3}{32}$	$f(\frac{1}{2}) = \frac{9}{32}$	$f(2) = -36$
Location of Graph	Above x-axis	Above x-axis	Above x-axis	Below x-axis
Point on Graph	(-2, 12)	$(-\frac{1}{2}, \frac{3}{32})$	$(\frac{1}{2}, \frac{9}{32})$	(2, -36)

f is above the x-axis for $(-\infty,-1)\cup(-1,0)\cup(0,1)$; f is below the x-axis for $(1,\infty)$

(f)

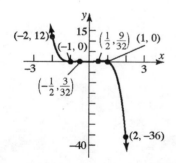

69.   c, e and f

71. c and e

73.
$$f(x) = x^3 + 0.2x^2 - 1.5876x - 0.31752$$

x-intercepts: −1.26, −0.2, 1.26

turning points: (−0.80, 0.57);
                        (0.66, −0.99)

75.
$$f(x) = x^3 + 2.56x^2 - 3.31x + 0.89$$

x-intercepts: −3.56, 0.50

turning points: (−2.21, 9.91);
                        (0.50, 0)

77. $f(x) = x^4 - 2.5x^2 + 0.5625$

x-intercepts: −1.50, −0.50, 0.50, 1.50

turning points: (0, 0.5625);
                        (−1.12, −1);
                        (1.12, −1)

79. $f(x) = x^4 + 0.65x^3 - 16.6319x^2$
                $+ 14.209335x - 3.1264785$

x-intercepts: −4.78, 0.45, 3.23

turning points: (0.45, 0)
                        (−3.32, −135.92);
                        (2.38, −22.67)

81. $f(x) = \pi x^3 + \sqrt{2}x^2 - x - 2$

x-intercept: 0.84
turning points: (−0.51, −1.54);
                        (0.21, −2.12)

103

83. $f(x) = 2x^4 - \pi x^3 + \sqrt{5}x - 4$

x-intercepts: –1.07, 1.62
turning point: (–0.42, –4.64)

85. $f(x) = -2x^5 - \sqrt{2}x^2 - x - \sqrt{2}$

x-intercept: –0.98
no turning points

87.   (a)   Graphing: The graph may be a cubic relation.

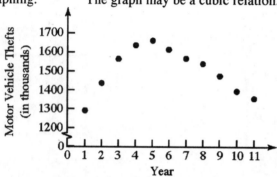

(b) $M(x) = 1.52x^3 - 39.81x^2 + 282.29x + 1035.5$
      $M(8) = 1.52(8)^3 - 39.81(8)^2 + 282.29(8) + 1035.5 = 1580.22$
      According to the function there would be approximately 1,580,220 motor vehicle thefts in 1994.

(c) and (d)   Graphing the cubic function of best fit:

(e) answers will vary

89.   (a)   Graphing: The graph may be a cubic relation.

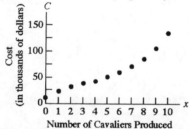

(b)   Average rate of change $= \dfrac{50 - 43}{5 - 4} = \dfrac{7}{1} = 7$

(c)   Average rate of change $= \dfrac{105 - 85}{9 - 8} = \dfrac{20}{1} = 20$

(d) $C(x) = 0.2x^3 - 2.3x^2 + 14.3x + 10.2$
    $C(11) = 0.2(11)^3 - 2.3(11)^2 + 14.3(11) + 10.2 \approx 155.4$
    The cost of manufacturing 11 Cavaliers in 1 hour would be approximately $155,400.

(e) and (f)   Graphing the cubic function of best fit:

(g)   The y-intercept would indicate the fixed costs before any cars are made.

91.   The graph of a polynomial function will always have a y-intercept since the domain
    of every polynomial function is the set of real numbers. Therefore $f(0)$ will always
    produce a y-coordinate on the graph.
    A polynomial function might have no x-intercepts. For example $f(x) = x^2 + 1$ has
    no x-intercepts since the equation $x^2 + 1 = 0$ has no real solutions.

93.   Answers will vary, one such polynomial is $f(x) = x^2(x + 1)(4 - x)(x - 2)^2$

95. $f(x) = \dfrac{1}{x}$ is piecewise smooth and not continuous; $g(x) = |x|$ is continuous but not smooth.

Chapter 3

Polynomial and Rational Functions

3.3   Rational Functions I

1.   In $R(x) = \dfrac{4x}{x-3}$, the denominator, $q(x) = x-3$, has a zero at 3. Thus, the domain of $R(x)$ is all real numbers except 3.

3.   In $H(x) = \dfrac{-4x^2}{(x-2)(x+4)}$, the denominator, $q(x) = (x-2)(x+4)$, has zeros at 2 and –4. Thus, the domain of $H(x)$ is all real numbers except 2 and –4.

5.   In $F(x) = \dfrac{3x(x-1)}{2x^2-5x-3}$, the denominator, $q(x) = 2x^2-5x-3 = (2x+1)(x-3)$, has zeros at $-\dfrac{1}{2}$ and 3. Thus, the domain of $F(x)$ is all real numbers except $-\dfrac{1}{2}$ and 3.

7.   In $R(x) = \dfrac{x}{x^3-8}$, the denominator, $q(x) = x^3-8 = (x-2)(x^2+2x+4)$, has a zero at 2. (x^2+2x+4 has no real zeros.) Thus, the domain of $R(x)$ is all real numbers except 2.

9.   In $H(x) = \dfrac{3x^2+x}{x^2+4}$, the denominator, $q(x) = x^2+4$, has no real zeros. Thus, the domain of $H(x)$ is all real numbers

11.   In $R(x) = \dfrac{3(x^2-x-6)}{4(x^2-9)}$, the denominator, $q(x) = 4(x^2-9) = 4(x-3)(x+3)$, has zeros at 3 and –3. Thus, the domain of $R(x)$ is all real numbers except 3 and –3.

13.   (a)   Domain: $\{x \mid x \neq 2\}$; Range: $\{y \mid y \neq 1\}$
      (b)   Intercept: $(0,0)$ (c)   Horizontal Asymptote: $y = 1$
      (d)   Vertical Asymptote: $x = 2$ (e)   Oblique Asymptote: none

15.   (a)   Domain: $\{x \mid x \neq 0\}$; Range: all real numbers
      (b)   Intercepts: $(-1,0),(1,0)$ (c)   Horizontal Asymptote: none
      (d)   Vertical Asymptote: $x = 0$ (e)   Oblique Asymptote: $y = 2x$

17.   (a)   Domain: $\{x \mid x \neq -2, x \neq 2\}$; Range: $\{y \mid y \leq 0 \text{ or } y > 1\}$
      (b)   Intercept: $(0,0)$ (c)   Horizontal Asymptote: $y = 1$
      (d)   Vertical Asymptotes: $x = -2, x = 2$ (e)   Oblique Asymptote: none

19. $F(x) = 2 + \dfrac{1}{x}$

Using the function, $y = \dfrac{1}{x}$, shift the graph vertically 2 units to up.

21. $R(x) = \dfrac{1}{(x-1)^2}$

Using the function, $y = \dfrac{1}{x^2}$, shift the graph horizontally 1 unit to the right.

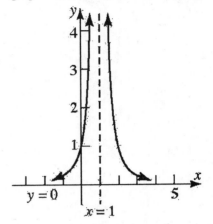

23. $H(x) = \dfrac{-2}{x+1}$

Using the function $y = \dfrac{1}{x}$, shift the graph horizontally 1 unit to the left, reflect about the x-axis, and stretch vertically by a factor of 2.

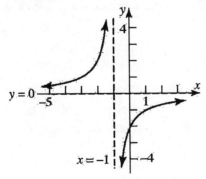

25. $R(x) = \dfrac{-1}{x^2 + 4x + 4} = \dfrac{-1}{(x+2)^2}$

Using the function $y = \dfrac{1}{x^2}$, shift the graph horizontally 2 units to the left, then reflect across the x-axis

27. $G(x) = 1 + \dfrac{2}{(x-3)^2} = \dfrac{2}{(x-3)^2} + 1$

Using the function $y = \dfrac{1}{x^2}$, shift the graph 3 units right, stretch vertically by a factor of 2, and shift vertically 1 unit up.

29. $R(x) = \dfrac{x^2 - 4}{x^2} = 1 - \dfrac{4}{x^2}$

Using the function $y = \dfrac{1}{x^2}$, reflect about the x-axis, stretch vertically by a factor of 4 and shift vertically 1 unit up.

31. $R(x) = \dfrac{3x}{x+4}$

The degree of the numerator, $p(x) = 3x$, is $n = 1$. The degree of the denominator, $q(x) = x + 4$, is $m = 1$. Since $n = m$, the line $y = \dfrac{3}{1} = 3$ is a horizontal asymptote. The denominator is zero at $x = -4$, so $x = -4$ is a vertical asymptote.

33. $H(x) = \dfrac{x^4 + 2x^2 + 1}{x^2 - x + 1}$

The degree of the numerator, $p(x) = x^4 + 2x^2 + 1$, is $n = 4$. The degree of the denominator, $q(x) = x^2 - x + 1$, is $m = 2$. Since $n > m + 1$, there is no horizontal asymptote or oblique asymptote. The denominator has no real zeros, so there is no vertical asymptote.

35. $T(x) = \dfrac{x^3}{x^4 - 1}$

The degree of the numerator, $p(x) = x^3$, is $n = 3$. The degree of the denominator, $q(x) = x^4 - 1$ is $m = 4$. Since $n < m$, the line $y = 0$ is a horizontal asymptote. The denominator is zero at $x = -1$ and $x = 1$, so $x = -1$ and $x = 1$ are vertical asymptotes.

37. $Q(x) = \dfrac{5 - x^2}{3x^4}$

The degree of the numerator, $p(x) = 5 - x^2$, is $n = 2$. The degree of the denominator, $q(x) = 3x^4$ is $m = 4$. Since $n < m$, the line $y = 0$ is a horizontal asymptote. The denominator is zero at $x = 0$, so $x = 0$ is a vertical asymptote.

39. $R(x) = \dfrac{3x^4 + 4}{x^3 + 3x}$

The degree of the numerator, $p(x) = 3x^4 + 4$, is $n = 4$. The degree of the denominator, $q(x) = x^3 + 3x$ is $m = 3$. Since $n = m + 1$, there is an oblique asymptote.

Dividing:

$$\begin{array}{r} 3x \\ x^3+3x\overline{)3x^4+0x^3+0x^2+0x+4} \\ \underline{3x^4+9x^2} \\ -9x^2+0x+4 \end{array} \qquad R(x)=3x+\dfrac{-9x^2+4}{x^3+3x}$$

Thus, the oblique asymptote is $y=3x$.

The denominator is zero at $x=0$, so $x=0$ is a vertical asymptote.

41. $G(x)=\dfrac{x^3-1}{x-x^2},\ x\neq 1$

The degree of the numerator, $p(x)=x^3-1$, is $n=3$. The degree of the denominator, $q(x)=x-x^2$ is $m=2$. Since $n=m+1$, there is an oblique asymptote.

Dividing:

$$\begin{array}{r} -x-1 \\ -x^2+x\overline{)x^3+0x^2+0x-1} \\ \underline{x^3-\ x^2} \\ x^2+0x \\ \underline{x^2-\ x} \\ x-1 \end{array} \qquad G(x)=-x-1+\dfrac{x-1}{x-x^2}=-x-1-\dfrac{1}{x},\ x\neq 1$$

Thus, the oblique asymptote is $y=-x-1$.

$G(x)$ must be in lowest terms to find the vertical asymptote:

$$G(x)=\dfrac{x^3-1}{x-x^2}=\dfrac{(x-1)(x^2+x+1)}{-x(x-1)}=\dfrac{x^2+x+1}{-x}$$

The denominator is zero at $x=0$, so $x=0$ is a vertical asymptotes.

43. $g(h)=\dfrac{3.99\times 10^{14}}{\left(6.374\times 10^6+h\right)^2}$

(a) $g(0)=\dfrac{3.99\times 10^{14}}{\left(6.374\times 10^6+0\right)^2}\approx 9.821\ m/s^2$

(b) $g(443)=\dfrac{3.99\times 10^{14}}{\left(6.374\times 10^6+443\right)^2}\approx 9.8195\ m/s^2$

(c) $g(8448)=\dfrac{3.99\times 10^{14}}{\left(6.374\times 10^6+8448\right)^2}\approx 9.795\ m/s^2$

(d) $g(h)=\dfrac{3.99\times 10^{14}}{\left(6.374\times 10^6+h\right)^2}\approx\dfrac{3.99\times 10^{14}}{h^2}\to 0$ as $h\to\infty$

$\therefore y=0$ is the horizontal asymptote.

(e)

(f) $g(h) = \dfrac{3.99 \times 10^{14}}{\left(6.374 \times 10^6 + h\right)^2} = 0$, to solve this equation would require that

$3.99 \times 10^{14} = 0$, which is impossible. Therefore, there is no height above sea level at which $g = 0$. In other words, there is no point in the entire universe that is unaffected by the Earth's gravity!

45. A rational function $R(x) = \dfrac{p(x)}{q(x)}$ has a vertical asymptote at x = c in each of these cases:

Case 1: $R(c) = \dfrac{nonzero}{zero}$

That is, whenever x = c yields a zero in the denominator of the function formula. And the denominator will equal zero only if it contains the factor $(x-c)^n$, for some $n > 0$.

Case 2: $R(c) = \dfrac{(x-c)^m}{(x-c)^n}$, where $n > 0$, $m > 0$ and $n > m$.

That is, whenever x = c yields a zero in the numerator and denominator of the function formula such that the multiplicity is greater in the denominator.

47. No, $R(x) = \dfrac{p(x)}{q(x)}$ has an oblique asymptote only when the degree of $p(x)$ is exceeds the degree of $q(x)$ by exactly 1. Moreover, $R(x) = \dfrac{p(x)}{q(x)}$ has a horizontal asymptote only when the degree of $p(x)$ is less than or equal to the degree of $q(x)$. These conditions are mutually exclusive.

Chapter 3

Polynomial and Rational Functions

3.4 Rational Functions II: Analyzing Graphs

In problems 1-37, we will use the terminology: $R(x) = \dfrac{p(x)}{q(x)}$, *where the degree of* $p(x) = n$ *and the degree of* $q(x) = m$.

1. $R(x) = \dfrac{x+1}{x(x+4)}$ $p(x) = x+1$; $q(x) = x(x+4) = x^2+4x$; $n = 1$; $m = 2$

Step 1: Domain: $\{x \mid x \neq -4, x \neq 0\}$

Step 2: (a) The x-intercept is the zero of $p(x)$: -1

   (b) There is no y-intercept; $R(0)$ is not defined, since $q(0) = 0$.

Step 3: $R(-x) = \dfrac{-x+1}{-x(-x+4)} = \dfrac{-x+1}{x^2-4x}$; this is neither $R(x)$ nor $-R(x)$, so there is no symmetry.

Step 4: The vertical asymptotes are the zeros of $q(x)$: $x = -4$ and $x = 0$

Step 5: Since $n < m$, the line $y = 0$ is the horizontal asymptote.

   $R(x)$ intersects $y = 0$ at $(-1, 0)$.

Step 6:

Interval	$(-\infty, -4)$	$(-4, -1)$	$(-1, 0)$	$(0, \infty)$
Number Chosen	-5	-2	$-\frac{1}{2}$	1
Value of R	$R(-5) = -\frac{4}{5}$	$R(-2) = \frac{1}{4}$	$R(-\frac{1}{2}) = -\frac{2}{7}$	$R(1) = \frac{2}{5}$
Location of Graph	Below x-axis	Above x-axis	Below x-axis	Above x-axis
Point on Graph	$(-5, -\frac{4}{5})$	$(-2, \frac{1}{4})$	$(-\frac{1}{2}, -\frac{2}{7})$	$(1, \frac{2}{5})$

Step 7: Graphing:

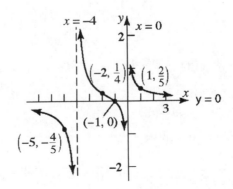

3. $R(x) = \dfrac{3x+3}{2x+4}$ $p(x) = 3x + 3$; $q(x) = 2x + 4$; $n = 1$; $m = 1$

Step 1:   Domain: $\{x \mid x \neq -2\}$

Step 2:   (a)   The x-intercept is the zero of $p(x)$: -1

           (b)   The y-intercept is $R(0) = \dfrac{3(0)+3}{2(0)+4} = \dfrac{3}{4}$.

Step 3: $R(-x) = \dfrac{3(-x)+3}{2(-x)+4} = \dfrac{-3x+3}{-2x+4} = \dfrac{3x-3}{2x-4}$; this is neither $R(x)$ nor $-R(x)$, so there is no symmetry.

Step 4:   The vertical asymptote is the zero of $q(x)$: $x = -2$

Step 5:   Since $n = m$, the line $y = \dfrac{3}{2}$ is the horizontal asymptote.

        $R(x)$ does not intersect $y = \dfrac{3}{2}$.

Step 6:

Interval	$(-\infty, -2)$	$(-2, -1)$	$(-1, \infty)$
Number Chosen	-3	$-\frac{3}{2}$	0
Value of R	$R(-3) = 3$	$R(-\frac{3}{2}) = -\frac{3}{2}$	$R(0) = \frac{3}{4}$
Location of Graph	Above x-axis	Below x-axis	Above x-axis
Point on Graph	$(-3, 3)$	$(-\frac{3}{2}, -\frac{3}{2})$	$(0, \frac{3}{4})$

Step 7:   Graphing:

5. $R(x) = \dfrac{3}{x^2 - 4}$ $p(x) = 3$; $q(x) = x^2 - 4$; $n = 0$; $m = 2$

Step 1:   Domain: $\{x \mid x \neq -2, x \neq 2\}$

Step 2:   (a)   There is no x-intercept.

           (b)   The y-intercept is $R(0) = \dfrac{3}{0^2 - 4} = \dfrac{3}{-4} = -\dfrac{3}{4}$.

Step 3: $R(-x) = \dfrac{3}{(-x)^2 - 4} = \dfrac{3}{x^2 - 4} = R(x)$; $R(x)$ is symmetric to the y-axis.

Step 4:   The vertical asymptotes are the zeros of $q(x)$: $x = -2$ and $x = 2$

Step 5:   Since $n < m$, the line $y = 0$ is the horizontal asymptote.

$R(x)$ does not intersect $y = 0$.

Step 6:

	-2	2	
Interval	$(-\infty, -2)$	$(-2, 2)$	$(2, \infty)$
Number Chosen	-3	0	3
Value of R	$R(-3) = \frac{3}{5}$	$R(0) = -\frac{3}{4}$	$R(3) = \frac{3}{5}$
Location of Graph	Above x-axis	Below x-axis	Above x-axis
Point on Graph	$(-3, \frac{3}{5})$	$(0, -\frac{3}{4})$	$(3, \frac{3}{5})$

Step 7:   Graphing:

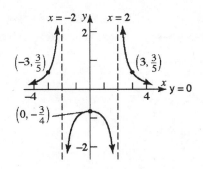

7. $P(x) = \dfrac{x^4 + x^2 + 1}{x^2 - 1}$ $p(x) = x^4 + x^2 + 1;\ q(x) = x^2 - 1;\ n = 4;\ m = 2$

Step 1:   Domain: $\left\{ x \mid x \neq -1,\ x \neq 1 \right\}$

Step 2:   (a)   There is no x-intercept.

(b)   The y-intercept is $P(0) = \dfrac{0^4 + 0^2 + 1}{0^2 - 1} = \dfrac{1}{-1} = -1$.

Step 3: $P(-x) = \dfrac{(-x)^4 + (-x)^2 + 1}{(-x)^2 - 1} = \dfrac{x^4 + x^2 + 1}{x^2 - 1} = P(x)$; $P(x)$ is symmetric to the y-axis.

Step 4:   The vertical asymptotes are the zeros of $q(x)$: $x = -1$ and $x = 1$

Step 5:   Since $n > m + 1$, there is no horizontal asymptote and no oblique asymptote.

Step 6:

	-1	1	
Interval	$(-\infty, -1)$	$(-1, 1)$	$(1, \infty)$
Number Chosen	-2	0	2
Value of P	$P(-2) = 7$	$P(0) = -1$	$P(2) = 7$
Location of Graph	Above x-axis	Below x-axis	Above x-axis
Point on Graph	(-2, 7)	(0, -1)	(2, 7)

Step 7:   Graphing:

9. $H(x) = \dfrac{x^3 - 1}{x^2 - 9}$ $p(x) = x^3 - 1;$ $q(x) = x^2 - 9;$ $n = 3;$ $m = 2$

Step 1:   Domain: $\left\{x \mid x \neq -3, x \neq 3\right\}$

Step 2:   (a)   The x-intercept is the zero of $p(x)$: 1.

   (b)   The y-intercept is $H(0) = \dfrac{0^3 - 1}{0^2 - 9} = \dfrac{-1}{-9} = \dfrac{1}{9}$.

Step 3: $H(-x) = \dfrac{(-x)^3 - 1}{(-x)^2 - 9} = \dfrac{-x^3 - 1}{x^2 - 9}$; this is neither $H(x)$ nor $-H(x)$, so there is no

   symmetry.

Step 4:   The vertical asymptotes are the zeros of $q(x)$: $x = -3$ and $x = 3$

Step 5:   Since $n = m + 1$, there is an oblique asymptote. Dividing:

$$x^2 - 9 \overline{\smash{)}\, x^3 + 0x^2 + 0x - 1} \qquad\qquad \begin{array}{c} x \\ \hline \end{array}$$
$$\underline{x^3 \qquad\ -9x} \qquad\qquad H(x) = x + \dfrac{9x - 1}{x^2 - 9}$$
$$9x - 1$$

The oblique asymptote is $y = x$.

Solve to find intersection points:

$$\dfrac{x^3 - 1}{x^2 - 9} = x \rightarrow x^3 - 1 = x^3 - 9x$$

$$-1 = -9x \rightarrow x = \dfrac{1}{9}$$

The oblique asymptote intersects $H(x)$ at $\left(\dfrac{1}{9}, \dfrac{1}{9}\right)$.

Step 6:

Interval	$(-\infty, -3)$	$(23, 1)$	$(1, 3)$	$(3, \infty)$
Number Chosen	-4	0	2	4
Value of H	$H(-4) \approx -9.3$	$H(0) = \frac{1}{9}$	$H(2) = -1.4$	$H(4) = 9$
Location of Graph	Below x-axis	Above x-axis	Below x-axis	Above x-axis
Point on Graph	(-4, -9.3)	$(0, \frac{1}{9})$	(2, -1.4)	(4, 9)

Step 7:   Graphing:

11. $R(x) = \dfrac{x^2}{x^2 + x - 6} = \dfrac{x^2}{(x+3)(x-2)}$ $p(x) = x^2$; $q(x) = x^2 + x - 6$; $n = 2$; $m = 2$

Step 1: Domain: $\{x \mid x \neq -3, x \neq 2\}$

Step 2: (a) The x-intercept is the zero of $p(x)$: 0

   (b) The y-intercept is $R(0) = \dfrac{0^2}{0^2 + 0 - 6} = \dfrac{0}{-6} = 0$.

Step 3: $R(-x) = \dfrac{(-x)^2}{(-x)^2 + (-x) - 6} = \dfrac{x^2}{x^2 - x - 6}$; this is neither $R(x)$ nor $-R(x)$, so there

   is no symmetry.

Step 4: The vertical asymptotes are the zeros of $q(x)$: $x = -3$ and $x = 2$

Step 5: Since $n = m$, the line $y = 1$ is the horizontal asymptote.

   $R(x)$ intersects $y = 1$ at $(6, 1)$, since:

$$\frac{x^2}{x^2 + x - 6} = 1 \rightarrow x^2 = x^2 + x - 6 \rightarrow 0 = x - 6 \rightarrow x = 6$$

Step 6:

Interval	$(-\infty, -3)$	$(-3, 0)$	$(0, 2)$	$(2, \infty)$
Number Chosen	-6	-1	1	3
Value of R	$R(-6) = 1.5$	$R(-1) = -\frac{1}{6}$	$R(1) = -0.25$	$R(3) = 1.5$
Location of Graph	Above x-axis	Below x-axis	Below x-axis	Above x-axis
Point on Graph	$(-6, 1.5)$	$(-1, -\frac{1}{6})$	$(1, -0.25)$	$(3, 1.5)$

Step 7: Graphing:

13. $G(x) = \dfrac{x}{x^2 - 4} = \dfrac{x}{(x+2)(x-2)}$ $p(x) = x$; $q(x) = x^2 - 4$; $n = 1$; $m = 2$

Step 1: Domain: $\{x \mid x \neq -2, x \neq 2\}$

Step 2: (a) The x-intercept is the zero of $p(x)$: 0

   (b) The y-intercept is $G(0) = \dfrac{0}{0^2 - 4} = \dfrac{0}{-4} = 0$.

Step 3: $G(-x) = \dfrac{-x}{(-x)^2 - 4} = \dfrac{-x}{x^2 - 4} = -G(x)$; $G(x)$ is symmetric to the origin.

Step 4: The vertical asymptotes are the zeros of $q(x)$: $x = -2$ and $x = 2$

Step 5: Since $n < m$, the line $y = 0$ is the horizontal asymptote.

   $G(x)$ intersects $y = 0$ at $(0, 0)$.

Step 6:

Interval	$(-\infty, -2)$	$(-2, 0)$	$(0, 2)$	$(2, \infty)$
Number Chosen	-3	-1	1	3
Value of G	$G(-3) = -\frac{3}{5}$	$G(-1) = \frac{1}{3}$	$G(1) = -\frac{1}{3}$	$G(3) = \frac{3}{5}$
Location of Graph	Below x-axis	Above x-axis	Below x-axis	Above x-axis
Point on Graph	$(-3, -\frac{3}{5})$	$(-1, \frac{1}{3})$	$(1, -\frac{1}{3})$	$(3, \frac{3}{5})$

Step 7:   Graphing:

15. $R(x) = \dfrac{3}{(x-1)(x^2-4)} = \dfrac{3}{(x-1)(x+2)(x-2)}$ $p(x) = 3;\ q(x) = (x-1)(x^2-4);$
      $n = 0;\ m = 3$

Step 1:   Domain: $\{x \mid x \neq -2,\ x \neq 1,\ x \neq 2\}$

Step 2:   (a)   There is no x-intercept.

          (b)   The y-intercept is $R(0) = \dfrac{3}{(0-1)(0^2-4)} = \dfrac{3}{4}$.

Step 3: $R(-x) = \dfrac{3}{(-x-1)\left((-x)^2-4\right)} = \dfrac{3}{(-x-1)(x^2-4)}$; this is neither $R(x)$ nor $-R(x)$,

          so there is no symmetry.

Step 4:   The vertical asymptotes are the zeros of $q(x)$: $x = -2,\ x = 1,$ and $x = 2$

Step 5:   Since $n < m$, the line $y = 0$ is the horizontal asymptote.

          $R(x)$ does not intersect $y = 0$.

Step 6:

Interval	$(-\infty, -2)$	$(-2, 1)$	$(1, 2)$	$(2, \infty)$
Number Chosen	-3	0	1.5	3
Value of R	$R(-3) = -\frac{3}{20}$	$R(0) = \frac{3}{4}$	$R(1.5) = -\frac{24}{7}$	$R(3) = \frac{3}{10}$
Location of Graph	Below x-axis	Above x-axis	Below x-axis	Above x-axis
Point on Graph	$(-3, -\frac{3}{20})$	$(0, \frac{3}{4})$	$(1.5, -\frac{24}{7})$	$(3, \frac{3}{10})$

Step 7:   Graphing:

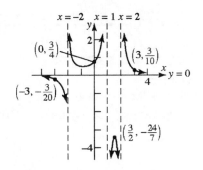

17. $H(x) = \dfrac{4(x^2-1)}{x^4-16} = \dfrac{4(x-1)(x+1)}{(x^2+4)(x+2)(x-2)}$ $p(x) = 4(x^2-1);\ q(x) = x^4-16;$
$n = 2;\ m = 4$

Step 1:   Domain: $\{x\,|\,x \neq -2,\ x \neq 2\}$

Step 2:   (a)   The x-intercepts are the zeros of $p(x)$: -1 and 1

(b)   The y-intercept is $H(0) = \dfrac{4(0^2-1)}{0^4-16} = \dfrac{-4}{-16} = \dfrac{1}{4}.$

Step 3: $H(-x) = \dfrac{4\big((-x)^2-1\big)}{(-x)^4-16} = \dfrac{4\big(x^2-1\big)}{x^4-16} = H(x);\ H(x)$ is symmetric to the y-axis.

Step 4:   The vertical asymptotes are the zeros of $q(x)$: $x = -2,$ and $x = 2$

Step 5:   Since $n < m$, the line $y = 0$ is the horizontal asymptote.
$H(x)$ intersects $y = 0$ at $(-1, 0)$ and $(1, 0)$.

Step 6:

Interval	$(-\infty, -2)$	$(-2, -1)$	$(-1, 1)$	$(1, 2)$	$(2, \infty)$
Number Chosen	-3	-1.5	0	1.5	3
Value of H	$H(-3) \approx 0.49$	$H(-1.5) \approx -0.46$	$H(0) = \frac{1}{4}$	$H(1.5) \approx -0.46$	$H(3) \approx 0.49$
Location of Graph	Above x-axis	Below x-axis	Above x-axis	Below x-axis	Above x-axis
Point on Graph	(-3, 0.49)	(-1.5, -0.46)	$(0, \frac{1}{4})$	(1.5, -0.46)	(3, 0.49)

Step 7:   Graphing:

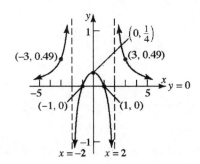

19. $F(x) = \dfrac{x^2-3x-4}{x+2} = \dfrac{(x+1)(x-4)}{x+2}$ $p(x) = x^2-3x-4;\ q(x) = x+2;\ n = 2;\ m = 1$

Step 1:   Domain: $\{x\,|\,x \neq -2\}$

Step 2:   (a)   The x-intercepts are the zeros of $p(x)$: -1 and 4.

117

(b)   The y-intercept is $F(0) = \dfrac{0^2 - 3(0) - 4}{0 + 2} = \dfrac{-4}{2} = -2$.

Step 3: $F(-x) = \dfrac{(-x)^2 - 3(-x) - 4}{-x + 2} = \dfrac{x^2 + 3x - 4}{-x + 2}$; this is neither $F(x)$ nor $-F(x)$, so

there is no symmetry.

Step 4:   The vertical asymptote is the zero of $q(x)$: $x = -2$

Step 5:   Since $n = m + 1$, there is an oblique asymptote. Dividing:

$$\begin{array}{r} x - 5 \\ x + 2 \overline{)\, x^2 - 3x - 4\,} \\ \underline{x^2 + 2x} \\ -5x - 4 \\ \underline{-5x - 10} \\ 6 \end{array} \qquad F(x) = x - 5 + \dfrac{6}{x + 2}$$

The oblique asymptote is $y = x - 5$.

Solve to find intersection points:

$$\dfrac{x^2 - 3x - 4}{x + 2} = x - 5 \rightarrow x^2 - 3x - 4 = x^2 - 3x - 10 \rightarrow -4 = -10$$

Since there is no solution, the oblique asymptote does not intersect $F(x)$.

Step 6:

Interval	$(-\infty, -2)$	$(-2, -1)$	$(-1, 4)$	$(4, \infty)$
Number Chosen	-3	-1.5	0	5
Value of F	$F(-3) = -14$	$F(-1.5) = 5.5$	$F(0) = -2$	$F(5) \approx 0.86$
Location of Graph	Below x-axis	Above x-axis	Below x-axis	Above x-axis
Point on Graph	$(-3, -14)$	$(-1.5, 5.5)$	$(0, -2)$	$(5, 0.86)$

Step 7:   Graphing:

21. $R(x) = \dfrac{x^2 + x - 12}{x - 4} = \dfrac{(x + 4)(x - 3)}{x - 4} \qquad p(x) = x^2 + x - 12; \; q(x) = x - 4; \; n = 2; \; m = 1$

Step 1:   Domain: $\{x \mid x \neq 4\}$

Step 2:   (a)   The x-intercepts are the zeros of $p(x)$: -4 and 3.

(b)   The y-intercept is $R(0) = \dfrac{0^2 + 0 - 12}{0 - 4} = \dfrac{-12}{-4} = 3$.

Step 3: $R(-x) = \dfrac{(-x)^2 + (-x) - 12}{-x - 4} = \dfrac{x^2 - x - 12}{-x - 4}$; this is neither $R(x)$ nor $-R(x)$, so

there is no symmetry.

Step 4:   The vertical asymptote is the zero of $q(x)$: $x = 4$

Step 5:   Since $n = m + 1$, there is an oblique asymptote. Dividing:

$$x - 4 \overline{)\begin{array}{r} x + 5 \\ x^2 + x - 12 \\ \underline{x^2 - 4x } \\ 5x - 12 \\ \underline{5x - 20} \\ 8 \end{array}} \qquad R(x) = x + 5 + \frac{8}{x - 4}$$

The oblique asymptote is $y = x + 5$.

Solve to find intersection points:

$$\frac{x^2 + x - 12}{x - 4} = x + 5$$
$$x^2 + x - 12 = x^2 + x - 20$$
$$-12 = -20$$

Since there is no solution, the oblique asymptote does not intersect $R(x)$.

Step 6:

Interval	$(-\infty, -4)$	$(-4, 3)$	$(3, 4)$	$(4, \infty)$
Number Chosen	-5	0	3.5	5
Value of R	$R(-5) = -\frac{8}{9}$	$R(0) = 3$	$R(3.5) = -7.5$	$R(5) = 18$
Location of Graph	Below x-axis	Above x-axis	Below x-axis	Above x-axis
Point on Graph	$(-5, -\frac{8}{9})$	$(0, 3)$	$(3.5, -7.5)$	$(5, 18)$

Step 7:   Graphing:

23. $F(x) = \dfrac{x^2 + x - 12}{x + 2} = \dfrac{(x + 4)(x - 3)}{x + 2}$ $p(x) = x^2 + x - 12$; $q(x) = x + 2$; $n = 2$; $m = 1$

Step 1:   Domain: $\{x \mid x \neq -2\}$

Step 2:   (a)   The x-intercepts are the zeros of $p(x)$: –4 and 3.

(b)   The y-intercept is $F(0) = \dfrac{0^2 + 0 - 12}{0 + 2} = \dfrac{-12}{2} = -6$.

Step 3: $F(-x) = \dfrac{(-x)^2 + (-x) - 12}{-x + 2} = \dfrac{x^2 - x - 12}{-x + 2}$; this is neither $F(x)$ nor $-F(x)$, so there is no symmetry.

Step 4:   The vertical asymptote is the zero of $q(x)$: $x = -2$

Step 5:   Since $n = m + 1$, there is an oblique asymptote. Dividing:

$$\begin{array}{r} x - 1 \\ x + 2 \overline{\smash{\big)}\ x^2 +\ x - 12} \\ \underline{x^2 + 2x} \\ -x - 12 \\ \underline{-x - 2} \\ -10 \end{array} \qquad F(x) = x - 1 + \dfrac{-10}{x + 2}$$

The oblique asymptote is $y = x - 1$.

Solve to find intersection points:

$$\dfrac{x^2 + x - 12}{x + 2} = x - 1 \rightarrow x^2 + x - 12 = x^2 + x - 2 \rightarrow -12 = -2$$

Since there is no solution, the oblique asymptote does not intersect $F(x)$.

Step 6:

Interval	$(-\infty, -4)$	$(-4, -2)$	$(-2, 3)$	$(3, \infty)$
Number Chosen	-5	-3	0	4
Value of F	$F(-5) = -\frac{8}{3}$	$F(-3) = 6$	$F(0) = -6$	$F(4) = \frac{4}{3}$
Location of Graph	Below x-axis	Above x-axis	Below x-axis	Above x-axis
Point on Graph	$(-5, -\frac{8}{3})$	$(-3, 6)$	$(0, -6)$	$(4, \frac{4}{3})$

Step 7:   Graphing:

25. $R(x) = \dfrac{x(x-1)^2}{(x+3)^3}$ $p(x) = x(x-1)^2;\ q(x) = (x+3)^3;\ n = 3;\ m = 3$

Step 1:   Domain: $\left\{ x \mid x \neq -3 \right\}$

Step 2:   (a)   The x-intercepts are the zeros of $p(x)$: 0 and 1

   (b)   The y-intercept is $R(0) = \dfrac{0(0-1)^2}{(0+3)^3} = \dfrac{0}{27} = 0$.

Step 3: $R(-x) = \dfrac{-x(-x-1)^2}{(-x+3)^3}$; this is neither $R(x)$ nor $-R(x)$, so there is no symmetry.

Step 4:   The vertical asymptote is the zero of $q(x)$: $x = -3$

Step 5:   Since $n = m$, the line $y = 1$ is the horizontal asymptote.

   Solve to find intersection points:

$$\dfrac{x(x-1)^2}{(x+3)^3} = 1$$

$$x^3 - 2x^2 + x = x^3 + 9x^2 + 27x + 27$$

$$0 = 11x^2 + 26x + 27$$

Since there is no real solution, $R(x)$ does not intersect $y = 1$.

Step 6:

	-3	0	1

Interval	$(-\infty, -3)$	$(-3, 0)$	$(0, 1)$	$(1, \infty)$
Number Chosen	-4	-1	$\frac{1}{2}$	2
Value of R	$R(-4) = 100$	$R(-1) = -0.5$	$R(\frac{1}{2}) \approx 0.003$	$R(2) = 0.016$
Location of Graph	Above x-axis	Below x-axis	Above x-axis	Above x-axis
Point on Graph	$(-4, 100)$	$(-1, -0.5)$	$(\frac{1}{2}, 0.003)$	$(2, 0.016)$

Step 7:   Graphing:

Enlarged view

27. $R(x) = \dfrac{x^2 + x - 12}{x^2 - x - 6} = \dfrac{(x+4)(x-3)}{(x-3)(x+2)} = \dfrac{x+4}{x+2}$ $p(x) = x^2 + x - 12$; $q(x) = x^2 - x - 6$; $n = 2$; $m = 2$

Step 1:   Domain: $\{x \mid x \neq -2, x \neq 3\}$

Step 2:   (a)   The x-intercept is the zero of $p(x)$: -4 (3 is not a zero because reduced form must be used to find the zeros.)

   (b)   The y-intercept is $R(0) = \dfrac{0^2 + 0 - 12}{0^2 - 0 - 6} = \dfrac{-12}{-6} = 2$.

Step 3: $R(-x) = \dfrac{(-x)^2 + (-x) - 12}{(-x)^2 - (-x) - 6} = \dfrac{x^2 - x - 12}{x^2 + x - 6}$; this is neither $R(x)$ nor $-R(x)$, so there is no symmetry.

Step 4:   The vertical asymptote is the zero of $q(x)$: $x = -2$ ($x = 3$ is not a vertical asymptote because reduced form must be used to find the them.)

Step 5:   Since $n = m$, the line $y = 1$ is the horizontal asymptote.

   $R(x)$ does not intersect $y = 1$ because $R(x)$ is not defined at $x = 3$.

   $$\dfrac{x^2 + x - 12}{x^2 - x - 6} = 1 \rightarrow x^2 + x - 12 = x^2 - x - 6$$

   $$2x = 6 \rightarrow x = 3$$

Step 6:

	-4	-2	3

Interval	$(-\infty, -4)$	$(-4, -2)$	$(-2, 3)$	$(3, \infty)$
Number Chosen	-5	-3	0	4
Value of R	$R(-5) = \frac{1}{3}$	$R(-3) = -1$	$R(0) = 2$	$R(4) = \frac{4}{3}$
Location of Graph	Above x-axis	Below x-axis	Above x-axis	Above x-axis
Point on Graph	$(-5, \frac{1}{3})$	$(-3, -1)$	$(0, 2)$	$(4, \frac{4}{3})$

Step 7:   Graphing:

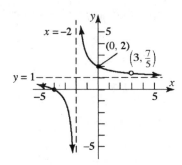

29. $R(x) = \dfrac{6x^2 - 7x - 3}{2x^2 - 7x + 6} = \dfrac{(3x+1)(2x-3)}{(2x-3)(x-2)} = \dfrac{3x+1}{x-2}$ $p(x) = 6x^2 - 7x - 3;$

$q(x) = 2x^2 - 7x + 6;$ $n = 2;$ $m = 2$

Step 1:   Domain: $\left\{ x \,\middle|\, x \neq \dfrac{3}{2}, \, x \neq 2 \right\}$

Step 2:   (a)   The x-intercept is the zero of $p(x)$: $-\dfrac{1}{3}$ ($\dfrac{3}{2}$ is not a zero because reduced

form must be used to find the zeros.)

(b)   The y-intercept is $R(0) = \dfrac{6(0)^2 - 7(0) - 3}{2(0)^2 - 7(0) + 6} = \dfrac{-3}{6} = \dfrac{-1}{2}.$

Step 3: $R(-x) = \dfrac{6(-x)^2 - 7(-x) - 3}{2(-x)^2 - 7(-x) + 6} = \dfrac{6x^2 + 7x - 3}{2x^2 + 7x + 6}$; this is neither $R(x)$ nor $-R(x)$, so

there is no symmetry.

Step 4:   The vertical asymptote is the zero of $q(x)$: $x = 2$ ($x = \dfrac{3}{2}$ is not a vertical

asymptote because reduced form must be used to find the them.)

Step 5:   Since $n = m$, the line $y = 3$ is the horizontal asymptote.

$R(x)$ does not intersect $y = 3$ because $R(x)$ is not defined at $x = \dfrac{3}{2}.$

$\dfrac{6x^2 - 7x - 3}{2x^2 - 7x + 6} = 3 \rightarrow 6x^2 - 7x - 3 = 6x^2 - 21x + 18$

$14x = 21 \rightarrow x = \dfrac{3}{2}$

Step 6:

Interval	$(-\infty, -\tfrac{1}{3})$	$(-\tfrac{1}{3}, \tfrac{3}{2})$	$(\tfrac{3}{2}, 2)$	$(2, \infty)$
Number Chosen	-1	0	1.7	6
Value of R	$R(-1) = \tfrac{2}{3}$	$R(0) = -\tfrac{1}{2}$	$R(1.7) \approx -20.3$	$R(6) = 4.75$
Location of Graph	Above x-axis	Below x-axis	Below x-axis	Above x-axis
Point on Graph	$(-1, \tfrac{2}{3})$	$(0, -\tfrac{1}{2})$	$(1.7, -20.3)$	$(6, 4.75)$

Step 7:   Graphing:

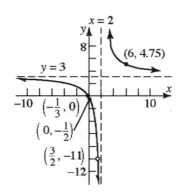

31. $R(x) = \dfrac{x^2 + 5x + 6}{x + 3} = \dfrac{(x+2)(x+3)}{x+3} = x + 2$ $p(x) = x^2 + 5x + 6;$ $q(x) = x + 3;$

$n = 2;\ m = 1$

Step 1: Domain: $\{x \mid x \neq -3\}$

Step 2: (a) The x-intercept is the zero of $p(x)$: -2 (-3 is not a zero because reduced form must be used to find the zeros.)

(b) The y-intercept is $R(0) = \dfrac{0^2 + 5(0) + 6}{0 + 3} = \dfrac{6}{3} = 2.$

Step 3: $R(-x) = \dfrac{(-x)^2 + 5(-x) + 6}{-x + 3} = \dfrac{x^2 - 5x + 6}{-x + 3}$; this is neither $R(x)$ nor $-R(x)$, so there is no symmetry.

Step 4: There are no vertical asymptotes. ($x = -3$ is not a vertical asymptote because reduced form must be used to find the them.)

Step 5: Since $n = m + 1$ there is a oblique asymptote. The line $y = x + 2$ is the oblique asymptote.

The oblique asymptote does not intersect $R(x)$.

Step 6:

Interval	$(-\infty, -3)$	$(-3, -2)$	$(-2, \infty)$
Number Chosen	-4	-2.5	0
Value of R	$R(-4) = -2$	$R(-2.5) = -\frac{1}{2}$	$R(0) = 2$
Location of Graph	Below x-axis	Below x-axis	Above x-axis
Point on Graph	$(-4, -2)$	$(-2.5, -\frac{1}{2})$	$(0, 2)$

Step 7: Graphing:

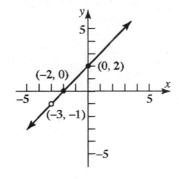

33. $f(x) = x + \dfrac{1}{x} = \dfrac{x^2 + 1}{x}$ $p(x) = x^2 + 1$; $q(x) = x$; $n = 2$; $m = 1$

Step 1:   Domain: $\{x \,|\, x \neq 0\}$

Step 2:   (a)   There are no x-intercepts.

(b)   There is no y-intercept because 0 is not in the domain.

Step 3: $f(-x) = \dfrac{(-x)^2 + 1}{-x} = \dfrac{x^2 + 1}{-x} = -f(x)$; The graph of $f(x)$ is symmetric to the origin.

Step 4:   The vertical asymptote is the zero of $q(x)$: $x = 0$

Step 5:   Since $n = m + 1$, there is an oblique asymptote. Dividing:

$$\begin{array}{r} x \\ x\overline{)\,x^2 + 1} \\ \underline{x^2} \\ 1 \end{array}$$ $f(x) = x + \dfrac{1}{x}$

The oblique asymptote is $y = x$.

Solve to find intersection points:

$$\dfrac{x^2 + 1}{x} = x \rightarrow x^2 + 1 = x^2 \rightarrow 1 = 0$$

Since there is no solution, the oblique asymptote does not intersect $f(x)$.

Step 6:

Interval	$(-\infty, 0)$	$(0, \infty)$
Number Chosen	-1	1
Value of f	$f(-1) = -2$	$f(1) = 2$
Location of Graph	Below x-axis	Above x-axis
Point on Graph	(-1, -2)	(1, 2)

Step 7:   Graphing:

35. $f(x) = x^2 + \dfrac{1}{x} = \dfrac{x^3+1}{x}$ $p(x) = x^3 + 1$; $q(x) = x$; $n = 3$; $m = 1$

Step 1:   Domain: $\{x \mid x \neq 0\}$

Step 2:   (a)   The x-intercept is the zero of $p(x)$: -1

(b)   There is no y-intercept because 0 is not in the domain.

Step 3: $f(-x) = \dfrac{(-x)^3 + 1}{-x} = \dfrac{-x^3 + 1}{-x}$; this is neither $f(x)$ nor $-f(x)$, so there is no symmetry.

Step 4:   The vertical asymptote is the zero of $q(x)$: $x = 0$

Step 5:   Since $n > m + 1$, there is no horizontal or oblique asymptote.

Step 6:

		-1		0	

Interval	$(-\infty, -1)$	$(-1, 0)$	$(0, \infty)$
Number Chosen	-2	$-\frac{1}{2}$	1
Value of f	$f(-2) = 3.5$	$f(-\frac{1}{2}) = -1.75$	$f(1) = 2$
Location of Graph	Above x-axis	Below x-axis	Above x-axis
Point on Graph	$(-2, 3.5)$	$(-\frac{1}{2}, -1.75)$	$(1, 2)$

Step 7:   Graphing :

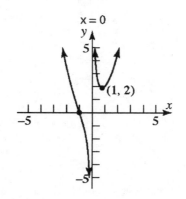

37. $f(x) = x + \dfrac{1}{x^3} = \dfrac{x^4+1}{x^3}$ $p(x) = x^4 + 1$; $q(x) = x^3$; $n = 4$; $m = 3$

Step 1:   Domain: $\{x \mid x \neq 0\}$

Step 2:   (a)   There are no x-intercepts.

(b)   There is no y-intercept because 0 is not in the domain.

Step 3: $f(-x) = \dfrac{(-x)^4 + 1}{(-x)^3} = \dfrac{x^4 + 1}{-x^3} = -f(x)$; The graph of $f(x)$ is symmetric to the origin.

Step 4:   The vertical asymptote is the zero of $q(x)$: $x = 0$

Step 5:   Since $n = m + 1$, there is an oblique asymptote. Dividing:

$$f(x) = x + \dfrac{1}{x^3}$$

The oblique asymptote is $y = x$.

Solve to find intersection points:

$$\frac{x^4+1}{x^3} = x \rightarrow x^4+1 = x^4 \rightarrow 1 = 0$$

Since there is no solution, the oblique asymptote does not intersect $f(x)$.

Step 6:

	0	

Interval	$(-\infty, 0)$	$(0, \infty)$
Number Chosen	-1	1
Value of f	$f(-1) = -2$	$f(1) = 2$
Location of Graph	Below x-axis	Above x-axis
Point on Graph	$(-1, -2)$	$(1, 2)$

Step 7:   Graphing:

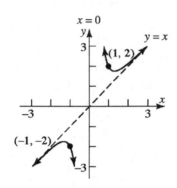

39. $f(x) = \dfrac{x^2}{x^2-4}$ 41. $f(x) = \dfrac{(x-1)^3(x-3)}{(x+1)^2(x-2)^2}$

43.   (a) $C(t) = \dfrac{t}{2t^2+1} \approx \dfrac{t}{2t^2} = \dfrac{1}{2t} \rightarrow 0$ as $t \rightarrow \pm\infty$

therefore the horizontal asymptote is $y = 0$.

The concentration of the drug decreases to 0 as time increases.

(b)   Graphing:

(c)   Using MAXIMUM, the concentration is highest when $t = 0.71$ hours.

45.   (a)   The average cost function is: $\overline{C}(x) = \dfrac{0.2x^3 - 2.3x^2 + 14.3x + 10.2}{x}$

(b) $\overline{C}(6) = \dfrac{0.2(6)^3 - 2.3(6)^2 + 14.3(6) + 10.2}{6} = \dfrac{56.4}{6} = 9.4$

The average cost of producing 6 Cavaliers per hour is $9400.

(c) $\overline{C}(9) = \dfrac{0.2(9)^3 - 2.3(9)^2 + 14.3(9) + 10.2}{9} = \dfrac{98.4}{9} = 10.933$

The average cost of producing 9 Cavaliers per hour is $10,933.

(d) Graphing:

(e) Using MINIMUM, the number of Cavaliers that should be produced per hour to minimize cost is 6.38.

(f) The minimum average cost is $9,366.

47. (a) The surface area is the sum of the areas of the six sides.
$$S = xy + xy + xy + xy + x^2 + x^2 = 4xy + 2x^2$$
The volume is $x \cdot x \cdot y = x^2 y = 10,000 \quad \rightarrow \quad y = \dfrac{10000}{x^2}$

Thus, $S(x) = 4x\left(\dfrac{10000}{x^2}\right) + 2x^2 = 2x^2 + \dfrac{40000}{x} = \dfrac{2x^3 + 40000}{x}$

(b) Graphing:

(c) The minimum surface area (amount of cardboard) is 2,785 square inches.

(d) The surface area is a minimum when $x = 21.544$.
$$y = \dfrac{10000}{21.544^2} = 21.545$$
The dimensions of the box are: 21.544 in. by 21.544 in. by 21.545 in.

49. (a) $500 = \pi r^2 h \quad \rightarrow \quad h = \dfrac{500}{\pi r^2}$

$$C(r) = 6(2\pi r^2) + 4(2\pi rh) = 12\pi r^2 + 8\pi r\left(\dfrac{500}{\pi r^2}\right) = 12\pi r^2 + \dfrac{4000}{r}$$

(b) Graphing:

The cost is least for $r = 3.76$ cm.

51.

x = 1 is not a vertical asymptote because of the following behavior:

$$y = \frac{x^2 - 1}{x - 1} = \frac{(x+1)(x-1)}{x-1} = x + 1 \text{ when } x \neq 1$$

$$y = \frac{x^3 - 1}{x - 1} = \frac{(x-1)(x^2 + x + 1)}{x-1} = x^2 + x + 1 \text{ when } x \neq 1$$

$$y = \frac{x^4 - 1}{x - 1} = \frac{(x^2+1)(x^2-1)}{x-1} = \frac{(x^2+1)(x-1)(x+1)}{x-1} = x^3 + x^2 + x + 1 \text{ when } x \neq 1$$

$$y = \frac{x^5 - 1}{x - 1} = \frac{(x^4 + x^3 + x^2 + x + 1)(x-1)}{x-1} = x^4 + x^3 + x^2 + x + 1 \text{ when } x \neq 1$$

In general, the graph of $y = \frac{x^n - 1}{x - 1}, n \geq 1$ an integer will have a "hole" with coordinates $(1, n)$.

53.

minimum value: $f(1) = 2$

55.

minimum value: $f\left(\sqrt[3]{\dfrac{1}{2}}\right) \approx 1.89$

57.

minimum value: $f\left(\sqrt[4]{3}\right) \approx 1.76$

59.   Answers will vary.

61.   Answers will vary, one example is $R(x) = \dfrac{2(x-3)(x+2)^2}{(x-1)^3}$

Polynomial and Rational Functions

3.5   Polynomial and Rational Inequalities

1. $(x-5)(x+2) < 0$ $f(x) = (x-5)(x+2)$
    $x = 5, x = -2$ are the zeros.

Interval	Test Number	$f(x)$	Positive/Negative
$-\infty < x < -2$	-3	8	Positive
$-2 < x < 5$	0	-10	Negative
$5 < x < \infty$	6	8	Positive

    The solution set is $\left\{ x \mid -2 < x < 5 \right\}$

3. $x^2 - 4x \geq 0$ $f(x) = x^2 - 4x$
    $x(x-4) \geq 0$
    $x = 0, x = 4$ are the zeros.

Interval	Test Number	$f(x)$	Positive/Negative
$-\infty < x < 0$	-1	5	Positive
$0 < x < 4$	1	-3	Negative
$4 < x < \infty$	5	5	Positive

    The solution set is $\left\{ x \mid x \leq 0 \text{ or } x \geq 4 \right\}$.

5. $x^2 - 9 < 0$ $f(x) = x^2 - 9$
    $(x+3)(x-3) < 0;$ $x = -3, x = 3$ are the zeros.

Interval	Test Number	$f(x)$	Positive/Negative
$-\infty < x < -3$	-4	7	Positive
$-3 < x < 3$	0	-9	Negative
$3 < x < \infty$	4	7	Positive

    The solution set is $\left\{ x \mid -3 < x < 3 \right\}$.

7. $x^2 + x \geq 2$ $f(x) = x^2 + x - 2$
    $x^2 + x - 2 \geq 0 \rightarrow (x+2)(x-1) \geq 0$
    $x = -2, x = 1$ are the zeros.

Interval	Test Number	$f(x)$	Positive/Negative
$-\infty < x < -2$	-5	18	Positive
$-2 < x < 1$	0	-2	Negative
$1 < x < \infty$	4	18	Positive

    The solution set is $\left\{ x \mid x \leq -2 \text{ or } x \geq 1 \right\}$.

9. $$2x^2 \le 5x + 3 \qquad f(x) = 2x^2 - 5x - 3$$
    $$2x^2 - 5x - 3 \le 0 \rightarrow (2x + 1)(x - 3) \le 0$$
    $$x = -\frac{1}{2},\ x = 3 \text{ are the zeros.}$$

Interval	Test Number	$f(x)$	Positive/Negative
$-\infty < x < -1/2$	-1	4	Positive
$-1/2 < x < 3$	0	-3	Negative
$3 < x < \infty$	4	9	Positive

The solution set is $\left\{ x \middle| -\frac{1}{2} \le x \le 3 \right\}$.

11. $$x(x - 7) > 8 \qquad f(x) = x^2 - 7x - 8$$
    $$x^2 - 7x > 8 \rightarrow x^2 - 7x - 8 > 0 \rightarrow (x + 1)(x - 8) > 0$$
    $$x = -1,\ x = 8 \text{ are the zeros.}$$

Interval	Test Number	$f(x)$	Positive/Negative
$-\infty < x < -1$	-2	10	Positive
$-1 < x < 8$	0	-8	Negative
$8 < x < \infty$	9	10	Positive

The solution set is $\left\{ x \middle| x < -1 \text{ or } x > 8 \right\}$.

13. $$4x^2 + 9 < 6x \qquad f(x) = 4x^2 - 6x + 9$$
    $$4x^2 - 6x + 9 < 0$$
    $$b^2 - 4ac = (-6)^2 - 4(4)(9) = 36 - 144 = -108$$

    Since the discriminant is negative, there are no real zeros.
    There is only one interval, the entire number line; choose any value and test.
    For $x = 0$, $4x^2 - 6x + 9 = 9 > 0$. Thus, there are no real zeros.

15. $$6\left(x^2 - 1\right) > 5x \qquad f(x) = 6x^2 - 5x - 6$$
    $$6x^2 - 6 > 5x \rightarrow 6x^2 - 5x - 6 > 0 \rightarrow (3x + 2)(2x - 3) > 0$$
    $$x = -\frac{2}{3},\ x = \frac{3}{2} \text{ are the zeros.}$$

Interval	Test Number	$f(x)$	Positive/Negative
$-\infty < x < -2/3$	-1	5	Positive
$-2/3 < x < 3/2$	0	-6	Negative
$3/2 < x < \infty$	2	8	Positive

The solution set is $\left\{ x \middle| x < -\frac{2}{3} \text{ or } x > \frac{3}{2} \right\}$.

17. $$(x - 1)\left(x^2 + x + 4\right) \ge 0 \qquad f(x) = (x - 1)\left(x^2 + x + 4\right)$$
    $x = 1$ is the zero. $x^2 + x + 4 = 0$ has no real zeros.

Interval	Test Number	$f(x)$	Positive/Negative
$-\infty < x < 1$	0	-4	Negative
$1 < x < \infty$	2	10	Positive

The solution set is $\left\{ x \middle| x \ge 1 \right\}$.

19.
$$(x-1)(x-2)(x-3) \le 0 \qquad f(x) = (x-1)(x-2)(x-3)$$
$x = 1$, $x = 2$, $x = 3$ are the zeros.

Interval	Test Number	$f(x)$	Positive/Negative
$-\infty < x < 1$	0	-6	Negative
$1 < x < 2$	1.5	0.375	Positive
$2 < x < 3$	2.5	-0.375	Negative
$3 < x < \infty$	4	6	Positive

The solution set is $\left\{ x \middle| x \le 1 \text{ or } 2 \le x \le 3 \right\}$.

21.
$$x^3 - 2x^2 - 3x > 0 \qquad f(x) = x^3 - 2x^2 - 3x$$
$$x\left(x^2 - 2x - 3\right) > 0 \rightarrow x(x+1)(x-3) > 0$$
$x = -1$, $x = 0$, $x = 3$ are the zeros.

Interval	Test Number	$f(x)$	Positive/Negative
$-\infty < x < -1$	-2	-10	Negative
$-1 < x < 0$	-0.5	0.875	Positive
$0 < x < 3$	1	-4	Negative
$3 < x < \infty$	4	20	Positive

The solution set is $\left\{ x \middle| -1 < x < 0 \text{ or } x > 3 \right\}$.

23.
$$x^4 > x^2 \qquad f(x) = x^4 - x^2$$
$$x^4 - x^2 > 0 \rightarrow x^2\left(x^2 - 1\right) > 0 \rightarrow x^2(x+1)(x-1) > 0$$
$x = -1$, $x = 0$, $x = 1$ are the zeros.

Interval	Test Number	$f(x)$	Positive/Negative
$-\infty < x < -1$	-2	12	Positive
$-1 < x < 0$	-0.5	-0.1875	Negative
$0 < x < 1$	0.5	-0.1875	Negative
$1 < x < \infty$	2	12	Positive

The solution set is $\left\{ x \middle| x < -1 \text{ or } x > 1 \right\}$.

25.
$$x^3 \ge 4x^2 \qquad f(x) = x^3 - 4x^2$$
$$x^3 - 4x^2 \ge 0 \rightarrow x^2(x-4) \ge 0$$
$x = 0$, $x = 4$ are the zeros.

Interval	Test Number	$f(x)$	Positive/Negative
$-\infty < x < 0$	-1	-3	Negative
$0 < x < 4$	0.5	-0.875	Negative
	5	25	Positive
$4 < x < \infty$			

The solution set is $\left\{ x \middle| x \ge 4 \right\}$.

27.
$$x^4 > 1 \qquad f(x) = x^4 - 1$$

$$x^4 - 1 > 0 \rightarrow \left(x^2 + 1\right)\left(x^2 - 1\right) > 0 \rightarrow \left(x^2 + 1\right)(x+1)(x-1) > 0$$
$x = -1$, $x = 1$ are the zeros.

Interval	Test Number	$f(x)$	Positive/Negative
$-\infty < x < -1$	-2	15	Positive
$-1 < x < 1$	0	-1	Negative
$1 < x < \infty$	2	15	Positive

The solution set is $\left\{ x \mid x < -1 \text{ or } x > 1 \right\}$.

29. $\dfrac{x+1}{x-1} > 0$ $f(x) = \dfrac{x+1}{x-1}$

The zeros and values where the expression is undefined are $x = -1$, and $x = 1$.

Interval	Test Number	$f(x)$	Positive/Negative
$-\infty < x < -1$	-2	$1/3$	Positive
$-1 < x < 1$	0	-1	Negative
$1 < x < \infty$	2	3	Positive

The solution set is $\left\{ x \mid x < -1 \text{ or } x > 1 \right\}$.

31. $\dfrac{(x-1)(x+1)}{x} < 0$ $f(x) = \dfrac{(x-1)(x+1)}{x}$

The zeros and values where the expression is undefined are $x = -1$, $x = 0$, and $x = 1$.

Interval	Test Number	$f(x)$	Positive/Negative
$-\infty < x < -1$	-2	-1.5	Negative
$-1 < x < 0$	-0.5	1.5	Positive
$0 < x < 1$	0.5	-1.5	Negative
$1 < x < \infty$	2	1.5	Positive

The solution set is $\left\{ x \mid x < -1 \text{ or } 0 < x < 1 \right\}$.

33. $\dfrac{(x-2)^2}{x^2-1} \geq 0$ $f(x) = \dfrac{(x-2)^2}{x^2-1}$

$\dfrac{(x-2)^2}{(x+1)(x-1)} \geq 0$

The zeros and values where the expression is undefined are $x = -1$, $x = 1$, and $x = 2$.

Interval	Test Number	$f(x)$	Positive/Negative
$-\infty < x < -1$	-2	$16/3$	Positive
$-1 < x < 1$	0	-4	Negative
$1 < x < 2$	1.5	0.2	Positive
$2 < x < \infty$	3	0.125	Positive

The solution set is $\left\{ x \mid x < -1 \text{ or } x > 1 \right\}$.

35. $6x - 5 < \dfrac{6}{x}$ $f(x) = 6x - 5 - \dfrac{6}{x}$

$6x - 5 - \dfrac{6}{x} < 0 \rightarrow \dfrac{6x^2 - 5x - 6}{x} < 0 \rightarrow \dfrac{(2x-3)(3x+2)}{x} < 0$

The zeros and values where the expression is undefined are $x = -\dfrac{2}{3}$, $x = 0$, and $x = \dfrac{3}{2}$.

Interval	Test Number	$f(x)$	Positive/Negative
$-\infty < x < -2/3$	-1	-5	Negative
$-2/3 < x < 0$	-0.5	4	Positive
$0 < x < 3/2$	1	-5	Negative
$3/2 < x < \infty$	2	4	Positive

The solution set is $\left\{ x \middle| x < -\dfrac{2}{3} \text{ or } 0 < x < \dfrac{3}{2} \right\}$.

37. $\qquad \dfrac{x+4}{x-2} \le 1 \qquad f(x) = \dfrac{x+4}{x-2} - 1$

$$\dfrac{x+4}{x-2} - 1 \le 0 \rightarrow \dfrac{x+4-(x-2)}{x-2} \le 0 \rightarrow \dfrac{6}{x-2} \le 0$$

The value where the expression is undefined is $x = 2$.

Interval	Test Number	$f(x)$	Positive/Negative
$-\infty < x < 2$	0	-3	Negative
$2 < x < \infty$	3	6	Positive

The solution set is $\left\{ x \middle| x < 2 \right\}$.

39. $\qquad \dfrac{3x-5}{x+2} \le 2 \qquad f(x) = \dfrac{3x-5}{x+2} - 2$

$$\dfrac{3x-5}{x+2} - 2 \le 0 \rightarrow \dfrac{3x-5-2(x+2)}{x+2} \le 0 \rightarrow \dfrac{x-9}{x+2} \le 0$$

The zeros and values where the expression is undefined are $x = -2$, and $x = 9$.

Interval	Test Number	$f(x)$	Positive/Negative
$-\infty < x < -2$	-3	12	Positive
$-2 < x < 9$	0	-4.5	Negative
$9 < x < \infty$	10	$1/12$	Positive

The solution set is $\left\{ x \middle| -2 < x \le 9 \right\}$.

41. $\qquad \dfrac{1}{x-2} < \dfrac{2}{3x-9} \qquad f(x) = \dfrac{1}{x+2} - \dfrac{2}{3x-9}$

$$\dfrac{1}{x-2} - \dfrac{2}{3x-9} < 0 \rightarrow \dfrac{3x-9-2(x-2)}{(x-2)(3x-9)} < 0 \rightarrow \dfrac{x-5}{(x-2)(3x-9)} < 0$$

The zeros and values where the expression is undefined are $x = 2$, $x = 3$, and $x = 5$.

Interval	Test Number	$f(x)$	Positive/Negative
$-\infty < x < 2$	0	$-5/18$	Negative
$2 < x < 3$	2.5	$10/3$	Positive
$3 < x < 5$	4	$-1/6$	Negative
$5 < x < \infty$	6	$1/36$	Positive

The solution set is $\left\{ x \middle| x < 2 \text{ or } 3 < x < 5 \right\}$.

43. $\qquad \dfrac{2x+5}{x+1} > \dfrac{x+1}{x-1} \qquad\qquad f(x) = \dfrac{2x+5}{x+1} - \dfrac{x+1}{x-1}$

$$\frac{2x+5}{x+1} - \frac{x+1}{x-1} > 0 \rightarrow \frac{(2x+5)(x-1)-(x+1)(x+1)}{(x+1)(x-1)} > 0$$

$$\frac{2x^2 + 3x - 5 - \left(x^2 + 2x + 1\right)}{(x+1)(x-1)} > 0 \rightarrow \frac{x^2 + x - 6}{(x+1)(x-1)} > 0 \rightarrow \frac{(x+3)(x-2)}{(x+1)(x-1)} > 0$$

The zeros and values where the expression is undefined are $x = -3, x = -1, x = 1, x = 2$.

Interval	Test Number	$f(x)$	Positive/Negative
$-\infty < x < -3$	-4	2/5	Positive
$-3 < x < -1$	-2	$-4/3$	Negative
$-1 < x < 1$	0	6	Positive
$1 < x < 2$	1.5	$-9/5$	Negative
$2 < x < \infty$	3	3/4	Positive

The solution set is $\left\{ x \mid x < -3, \ -1 < x < 1, \ x > 2 \right\}$.

45. $\quad \dfrac{x^2(3+x)(x+4)}{(x+5)(x-1)} \geq 0 \qquad f(x) = \dfrac{x^2(3+x)(x+4)}{(x+5)(x-1)}$

The zeros and values where the expression is undefined are
$x = -5, x = -4, x = -3, x = 0$ and $x = 1$.

Interval	Test Number	$f(x)$	Positive/Negative
$-\infty < x < -5$	-6	216/7	Positive
$-5 < x < -4$	-4.5	$-243/44$	Negative
$-4 < x < -3$	-3.5	49/108	Positive
$-3 < x < 0$	-1	$-3/4$	Negative
$0 < x < 1$	0.5	$-63/44$	Negative
$1 < x < \infty$	2	120/7	Positive

The solution set is $\left\{ x \mid x < -5, -4 \leq x \leq -3, x > 1 \right\}$.

47. Let x be the positive number. Then

$$x^3 > 4x^2 \rightarrow x^3 - 4x^2 > 0 \rightarrow x^2(x-4) > 0$$

The zeros are $x = 0$ and $x = 4$. $\qquad f(x) = x^3 - 4x^2$

Interval	Test Number	$f(x)$	Positive/Negative
$-\infty < x < 0$	-1	-5	Negative
$0 < x < 4$	1	-3	Negative
$4 < x < \infty$	5	25	Positive

The solution set is $\left\{ x \mid x > 4 \right\}$. All real numbers larger than 4 satisfy the condition.

49. The domain of the expression includes all values for which

$$x^2 - 16 \geq 0 \rightarrow (x+4)(x-4) \geq 0$$

The zeros are $x = -4$ and $x = 4$. $\qquad f(x) = x^2 - 16$

Interval	Test Number	$f(x)$	Positive/Negative
$-\infty < x < -4$	-5	9	Positive
$-4 < x < 4$	0	-16	Negative
$4 < x < \infty$	5	9	Positive

The domain is $\left\{ x \mid x \leq -4 \text{ or } x \geq 4 \right\}$.

51. The domain of the expression includes all values for which
$$\frac{x-2}{x+4} \geq 0$$
The zeros and values where the expression is undefined are $x = -4$ and $x = 2$.
$$f(x) = \frac{x-2}{x+4}$$

Interval	Test Number	$f(x)$	Positive/Negative
$-\infty < x < -4$	-5	7	Positive
$-4 < x < 2$	0	$-1/2$	Negative
$2 < x < \infty$	3	$1/7$	Positive

The domain is $\left\{ x \mid x < -4 \text{ or } x \geq 2 \right\}$.

53. Find the values of t for which
$$80t - 16t^2 > 96$$

$$-16t^2 + 80t - 96 > 0 \rightarrow 16t^2 - 80t + 96 < 0 \rightarrow 16(t^2 - 5t + 6) < 0 \rightarrow 16(t-2)(t-3) < 0$$
The zeros are $t = 2$ and $t = 3$. $s(t) = 16t^2 - 80t + 96$

Interval	Test Number	$s(t)$	Positive/Negative
$-\infty < t < 2$	1	32	Positive
$2 < t < 3$	2.5	-4	Negative
$3 < t < \infty$	4	32	Positive

The solution set is $\left\{ t \mid 2 < t < 3 \right\}$. The ball is more than 96 feet above the ground for times between 2 and 3 seconds.

55. Profit = Revenue – Cost
The zeros are approximately $x = 7.75$ and $x = 32.25$.
$$f(x) = x^2 - 40x + 250$$

Interval	Test Number	$f(x)$	Positive/Negative
$0 < x < 7.75$	7	19	Positive
$7.75 < x < 32.25$	10	-50	Negative
$32.25 < x < \infty$	40	250	Positive

The profit is at least $50 when at least 8 and no more than 32 watches are sold.

57. The equation $x^2 + kx + 1 = 0$ has no real solutions whenever the discriminant is less than zero.
Solving $b^2 - 4ac = k^2 - 4 < 0 \rightarrow (k+4)(k-4) < 0$
The zeros are $k = -4$ and $k = 4$.
$$f(k) = k^2 - 4$$

Interval	Test Number	$f(x)$	Positive/Negative
$-\infty < k < -4$	-5	21	Positive
$-4 < k < 4$	0	-4	Negative
$4 < k < \infty$	5	21	Positive

Therefore the equation $x^2 + kx + 1 = 0$ has no real solutions whenever $-4 < k < 4$.

59. Answers will vary, for example,
$x^2 < 0$ has no real solution $x^2 \leq 0$ has exactly one real solution

Chapter 3

Polynomial and Rational Functions

3.6 The Real Zeros of a Polynomial Function

1. $f(x) = 4x^3 - 3x^2 - 8x + 4; \quad c = 2$
   $f(2) = 4(2)^3 - 3(2)^2 - 8(2) + 4 = 32 - 12 - 16 + 4 = 8 \neq 0$
   Thus, 2 is not a zero of $f \therefore x - 2$ is not a factor of f.

3. $f(x) = 3x^4 - 6x^3 - 5x + 10; \quad c = 2$
   $f(2) = 3(2)^4 - 6(2)^3 - 5(2) + 10 = 48 - 48 - 10 + 10 = 0$
   Thus, 2 is a zero of $f \therefore x - 2$ is a factor of f.

5. $f(x) = 3x^6 + 82x^3 + 27; \quad c = -3$
   $f(-3) = 3(-3)^6 + 82(-3)^3 + 27 = 2187 - 2214 - 27 = 0$
   Thus, -3 is a zero of $f \therefore x + 3$ is a factor of f. Use synthetic division to find the factors.

   $$
   \begin{array}{r|rrrrrrr}
   -3 & 3 & 0 & 0 & 82 & 0 & 0 & 27 \\
      &   & -9 & 27 & -81 & -3 & 9 & -27 \\
   \hline
      & 3 & -9 & 27 & 1 & -3 & 9 & 0
   \end{array}
   $$

   The factored form is: $f(x) = (x + 3)(3x^5 - 9x^4 + 27x^3 + x^2 - 3x + 9)$.

7. $f(x) = 4x^6 - 64x^4 + x^2 - 15; \quad c = -4$
   Use synthetic division to determine whether -4 is a zero.

   $$
   \begin{array}{r|rrrrrrr}
   -4 & 4 & 0 & -64 & 0 & 1 & 0 & -15 \\
      &   & -16 & 64 & 0 & 0 & -4 & 16 \\
   \hline
      & 4 & -16 & 0 & 0 & 1 & -4 & 1
   \end{array}
   $$

   Thus, -4 is not a zero of $f \therefore x + 4$ is not a factor of f.

9. $f(x) = 2x^4 - x^3 + 2x - 1; \quad c = \dfrac{1}{2}$

   $f\left(\dfrac{1}{2}\right) = 2\left(\dfrac{1}{2}\right)^4 - \left(\dfrac{1}{2}\right)^3 + 2\left(\dfrac{1}{2}\right) - 1 = \dfrac{1}{8} - \dfrac{1}{8} + 1 - 1 = 0$

   Thus, $\dfrac{1}{2}$ is a zero of $f \therefore x - \dfrac{1}{2}$ is a factor of f.

Factoring:

$$f(x) = 2x^4 - x^3 + 2x - 1 = 2x^3\left(x - \frac{1}{2}\right) + 2\left(x - \frac{1}{2}\right) = \left(x - \frac{1}{2}\right)\left(2x^3 + 2\right)$$

$$= 2\left(x - \frac{1}{2}\right)\left(x^3 + 1\right) = 2\left(x - \frac{1}{2}\right)\left(x + 1\right)\left(x^2 - x + 1\right)$$

11. $f(x) = -4x^7 + x^3 - x^2 + 2$

The maximum number of zeros is the degree of the polynomial which is 7.
Examining $f(x) = -4x^7 + x^3 - x^2 + 2$, there are 3 variations in sign; thus, there are 3 or 1 positive real zeros.
Examining $f(-x) = -4(-x)^7 + (-x)^3 - (-x)^2 + 2 = 4x^7 - x^3 - x^2 + 2$, there are 2 variations in sign; thus, there are 2 or 0 negative real zeros.

13. $f(x) = 2x^6 - 3x^2 - x + 1$

The maximum number of zeros is the degree of the polynomial which is 6.
Examining $f(x) = 2x^6 - 3x^2 - x + 1$, there are 2 variations in sign; thus, there are 2 or 0 positive real zeros.
Examining $f(-x) = 2(-x)^6 - 3(-x)^2 - (-x) + 1 = 2x^6 - 3x^2 + x + 1$, there are 2 variations in sign; thus, there are 2 or 0 negative real zeros.

15. $f(x) = 3x^3 - 2x^2 + x + 2$

The maximum number of zeros is the degree of the polynomial which is 3.
Examining $f(x) = 3x^3 - 2x^2 + x + 2$, there are 2 variations in sign; thus, there are 2 or 0 positive real zeros.
Examining $f(-x) = 3(-x)^3 - 2(-x)^2 + (-x) + 2 = -3x^3 - 2x^2 - x + 2$, there is 1 variation in sign; thus, there is 1 negative real zero.

17. $f(x) = -x^4 + x^2 - 1$

The maximum number of zeros is the degree of the polynomial which is 4.
Examining $f(x) = -x^4 + x^2 - 1$, there are 2 variations in sign; thus, there are 2 or 0 positive real zeros.
Examining $f(-x) = -(-x)^4 + (-x)^2 - 1 = -x^4 + x^2 - 1$, there are 2 variations in sign; thus, there are 2 or 0 negative real zeros.

19. $f(x) = x^5 + x^4 + x^2 + x + 1$

The maximum number of zeros is the degree of the polynomial which is 5.
Examining $f(x) = x^5 + x^4 + x^2 + x + 1$, there are no variations in sign; thus, there are 0 positive real zeros.
Examining $f(-x) = (-x)^5 + (-x)^4 + (-x)^2 + (-x) + 1 = -x^5 + x^4 + x^2 - x + 1$, there are 3 variations in sign; thus, there are 3 or 1 negative real zeros.

21. $f(x) = x^6 - 1$

The maximum number of zeros is the degree of the polynomial which is 6.
Examining $f(x) = x^6 - 1$, there is 1 variation in sign; thus, there is 1 positive real zero.

Examining $f(-x) = (-x)^6 - 1 = x^6 - 1$, there is 1 variation in sign; thus, there is 1 negative real zero.

23. $f(x) = 3x^4 - 3x^3 + x^2 - x + 1$
     p must be a factor of 1: $p = \pm 1$
     q must be a factor of 3: $q = \pm 1, \pm 3$
     The possible rational zeros are: $\dfrac{p}{q} = \pm 1, \pm \dfrac{1}{3}$

25. $f(x) = x^5 - 6x^2 + 9x - 3$
     p must be a factor of -3: $p = \pm 1, \pm 3$
     q must be a factor of 1: $q = \pm 1$
     The possible rational zeros are: $\dfrac{p}{q} = \pm 1, \pm 3$

27. $f(x) = -4x^3 - x^2 + x + 2$
     p must be a factor of 2: $p = \pm 1, \pm 2$
     q must be a factor of -4: $q = \pm 1, \pm 2, \pm 4$
     The possible rational zeros are: $\dfrac{p}{q} = \pm 1, \pm 2, \pm \dfrac{1}{2}, \pm \dfrac{1}{4}$

29. $f(x) = 6x^4 - x^2 + 9$
     p must be a factor of 9: $p = \pm 1, \pm 3, \pm 9$
     q must be a factor of 6: $q = \pm 1, \pm 2, \pm 3, \pm 6$
     The possible rational zeros are: $\dfrac{p}{q} = \pm 1, \pm \dfrac{1}{2}, \pm \dfrac{1}{3}, \pm \dfrac{1}{6}, \pm 3, \pm \dfrac{3}{2}, \pm 9, \pm \dfrac{9}{2}$

31. $f(x) = 2x^5 - x^3 + 2x^2 + 12$
     p must be a factor of 12: $p = \pm 1, \pm 2, \pm 3, \pm 4, \pm 6, \pm 12$
     q must be a factor of 2: $q = \pm 1, \pm 2$
     The possible rational zeros are: $\dfrac{p}{q} = \pm 1, \pm 2, \pm 4, \pm \dfrac{1}{2}, \pm 3, \pm \dfrac{3}{2}, \pm 6, \pm 12$

33. $f(x) = 6x^4 + 2x^3 - x^2 + 20$
     p must be a factor of 20: $p = \pm 1, \pm 2, \pm 4, \pm 5, \pm 10, \pm 20$
     q must be a factor of 6: $q = \pm 1, \pm 2, \pm 3, \pm 6$
     The possible rational zeros are:
     $\dfrac{p}{q} = \pm 1, \pm 2, \pm \dfrac{1}{2}, \pm \dfrac{1}{3}, \pm \dfrac{2}{3}, \pm \dfrac{1}{6}, \pm 4, \pm \dfrac{4}{3}, \pm 5, \pm \dfrac{5}{2}, \pm \dfrac{5}{3}, \pm \dfrac{5}{6}, \pm 10, \pm \dfrac{10}{3}, \pm 20, \pm \dfrac{20}{3}$

35. $f(x) = x^3 + 2x^2 - 5x - 6$
     Step 1: $f(x)$ has at most 3 real zeros.
     Step 2: By Descartes Rule of Signs, there is 1 positive real zero.
              Also because $f(-x) = (-x)^3 + 2(-x)^2 - 5(-x) - 6 = -x^3 + 2x^2 + 5x - 6$, there are 2 or 0 negative real zeros.

Step 3: Possible rational zeros:

$$p = \pm 1, \pm 2, \pm 3, \pm 6; \quad q = \pm 1; \quad \frac{p}{q} = \pm 1, \pm 2, \pm 3, \pm 6$$

Step 4: Using synthetic division:

$$
\begin{array}{r|rrrr}
-3 & 1 & 2 & -5 & -6 \\
   &   & -3 & 3 & 6 \\
\hline
   & 1 & -1 & -2 & 0
\end{array}
$$

Since the remainder is 0, $x - (-3) = x + 3$ is a factor. The other factor is the quotient: $x^2 - x - 2$.

Thus, $f(x) = (x + 3)(x^2 - x - 2) = (x + 3)(x + 1)(x - 2)$.

The zeros are –3, –1, and 2.

37. $f(x) = 2x^3 - x^2 + 2x - 1; \; f(-x) = 2(-x)^3 - (-x)^2 + 2(-x) - 1 = -2x^3 - x^2 - 2x - 1$

Step 1: $f(x)$ has at most 3 real zeros.

Step 2: By Descartes Rule of Signs, there are 3 or 1 positive real zeros; thus, there are no negative real zeros.

Step 3: Possible rational zeros:

$$p = \pm 1 \quad q = \pm 1, \pm 2; \quad \frac{p}{q} = \pm 1, \pm \frac{1}{2}$$

Step 4: Using synthetic division:

$$
\begin{array}{r|rrrr}
-1 & 2 & -1 & 2 & -1 \\
   &   & -2 & 3 & -5 \\
\hline
   & 2 & -3 & 5 & \{-6\} \rightarrow
\end{array}
$$

$x + 1$ is **not** a factor

So we try $x - 1$

$$
\begin{array}{r|rrrr}
1 & 2 & -1 & 2 & -1 \\
  &   & 2 & 1 & 3 \\
\hline
  & 2 & 1 & 3 & \{2\} \rightarrow
\end{array}
$$

$x - 1$ is **not** a factor

Let's try $x - \dfrac{1}{2}$

$$
\begin{array}{r|rrrr}
\frac{1}{2} & 2 & -1 & 2 & -1 \\
            &   & 1 & 0 & 1 \\
\hline
            & 2 & 0 & 2 & 0 \rightarrow
\end{array}
$$

$x - \dfrac{1}{2}$ is a factor $\therefore$ the quotient is $2x^2 + 2$.

Thus, $f(x) = 2x^3 - x^2 + 2x - 1 = \left(x - \dfrac{1}{2}\right)(2x^2 + 2)$.

Since $2x^2 + 2 = 0$ has no real solutions, $x = \dfrac{1}{2}$ is the only real zero.

39. $f(x) = x^4 + x^2 - 2$

Step 1: $f(x)$ has at most 4 real zeros.

Step 2: By Descartes Rule of Signs, this is one positive real zero.

$f(-x) = (-x)^4 + (-x)^2 - 2 = x^4 + x^2 - 2$; thus, there is 1 negative real zero.

Step 3:   Possible rational zeros:

$$p = \pm 1, \pm 2; \quad q = \pm 1; \quad \frac{p}{q} = \pm 1, \pm 2$$

Step 4:   Using synthetic division:

$$-1)\overline{\begin{array}{rrrrr} 1 & 0 & 1 & 0 & 2 \\ & -1 & 1 & -2 & 2 \\ \hline 1 & -1 & 2 & -2 & 0 \end{array}} \rightarrow$$

Since the remainder is 0, $x - (-1) = x + 1$ is a factor. The other factor is the quotient: $x^3 - x^2 + 2x - 2$.

Thus, $f(x) = (x + 1)(x^3 - x^2 + 2x - 2)$. We can factor $x^3 - x^2 + 2x - 2$ by grouping terms: $x^3 - x^2 + 2x - 2 = x^2(x - 1) + 2(x - 1) = (x - 1)(x^2 + 2)$

Thus, $f(x) = (x + 1)(x - 1)(x^2 + 2)$. Since $x^2 + 2 = 0$ has no real solutions, we have two real zeros for f, namely -1 and 1.

41. $f(x) = 4x^4 + 7x^2 - 2$
Step 1: $f(x)$ has at most 4 real zeros.
Step 2:   By Descartes Rule of Signs, there is 1 positive real zero.
$f(-x) = 4(-x)^4 + 7(-x)^2 - 2 = 4x^4 + 7x^2 - 2$; thus, there is one negative real zero.

Step 3:   Possible rational zeros: $p = \pm 1, \pm 2; \quad q = \pm 1, \pm 2, \pm 4$

$$\frac{p}{q} = \pm 1, \pm \frac{1}{2}, \pm \frac{1}{4} \pm 2$$

We can factor f as follows:

$$f(x) = 4x^4 + 7x^2 - 2 = (4x^2 - 1)(x^2 + 2) = (2x + 1)(2x - 1)(x^2 + 2).$$

Thus, we have two real zeros, $-\dfrac{1}{2}$ and $\dfrac{1}{2}$.

43. $f(x) = x^4 + x^3 - 3x^2 - x + 2$
Step 1: $f(x)$ has at most 4 real zeros.
Step 2:   By Descartes Rule of Signs, there are 2 or 0 positive real zeros.
$f(-x) = (-x)^4 + (-x)^3 - 3(-x)^2 - (-x) + 2 = x^4 - x^3 - 3x^2 + x + 2$; thus, there are 2 or 0 negative real zeros.
Step 3:   Possible rational zeros:

$$p = \pm 1, \pm 2; \quad q = \pm 1; \quad \frac{p}{q} = \pm 1, \pm 2$$

Step 4:   Using synthetic division:

$$-2)\overline{\begin{array}{rrrrr} 1 & 1 & -3 & -1 & 2 \\ & -2 & 2 & 2 & -2 \\ \hline 1 & -1 & -1 & 1 & 0 \end{array}} \qquad -1)\overline{\begin{array}{rrrr} 1 & -1 & -1 & 1 \\ & -1 & 2 & -1 \\ \hline 1 & -2 & 1 & 0 \end{array}}$$

Since the remainder is 0, $x + 2$ and $x + 1$ are factors. The other factor is the quotient: $x^2 - 2x + 1$.

Thus, $f(x) = (x + 2)(x + 1)(x - 1)^2$.

The zeros are -2, -1, and 1 (multiplicity 2).

45. $f(x) = 4x^5 - 8x^4 - x + 2$

Step 1: $f(x)$ has at most 5 real zeros.

Step 2:   By Descartes Rule of Signs, there are 2 or 0 positive real zeros.

$f(-x) = 4(-x)^5 - 8(-x)^4 - (-x) + 2 = -4x^5 - 8x^4 + x + 2$;

thus, there is 1 negative real zero.

Step 3:   Possible rational zeros:

$$p = \pm 1, \pm 2; \quad q = \pm 1, \pm 2, \pm 4; \quad \frac{p}{q} = \pm 1, \pm 2, \pm \frac{1}{2}, \pm \frac{1}{4}$$

Step 4:   Using synthetic division:

$$\begin{array}{r}
2\overline{)\,4 \quad -8 \quad 0 \quad 0 \quad -1 \quad 2} \\
\underline{8 \quad 0 \quad 0 \quad 0 \quad -2} \\
4 \quad 0 \quad 0 \quad 0 \quad -1 \quad 0
\end{array}$$

Since the remainder is 0, $x - 2$ is a factor. The other factor is the quotient: $4x^4 - 1$.

Factoring,

$$f(x) = (x - 2)\left(4x^4 - 1\right) = (x - 2)(2x^2 - 1)(2x^2 + 1)$$

$$= (x - 2)\left(\sqrt{2}x - 1\right)\left(\sqrt{2}x + 1\right)\left(2x^2 + 1\right)$$

The zeros are $-\dfrac{\sqrt{2}}{2}, \dfrac{\sqrt{2}}{2}$, and 2 or –0.71, –0.71, and 2.

47. $x^4 - x^3 + 2x^2 - 4x - 8 = 0$

The solutions of the equation are the zeros of $f(x) = x^4 - x^3 + 2x^2 - 4x - 8$.

Step 1: $f(x)$ has at most 4 real zeros.

Step 2:   By Descartes Rule of Signs, there are 3 or 1 positive real zeros.

$f(-x) = (-x)^4 - (-x)^3 + 2(-x)^2 - 4(-x) - 8 = x^4 + x^3 + 2x^2 + 4x - 8$;

thus, there is 1 negative real zero.

Step 3:   Possible rational zeros:

$$p = \pm 1, \pm 2, \pm 4, \pm 8; \quad q = \pm 1; \quad \frac{p}{q} = \pm 1, \pm 2, \pm 4, \pm 8$$

Step 4:   Using synthetic division:

$$\begin{array}{r}
-1\overline{)\,1 \quad -1 \quad 2 \quad -4 \quad -8} \\
\underline{-1 \quad 2 \quad -4 \quad 8} \\
1 \quad -2 \quad 4 \quad -8 \quad 0
\end{array}
\qquad
\begin{array}{r}
2\overline{)\,1 \quad -2 \quad 4 \quad -8} \\
\underline{2 \quad 0 \quad 8} \\
1 \quad 0 \quad 4 \quad 0
\end{array}$$

Since the remainder is 0, $x + 1$ and $x - 2$ are factors. The other factor is the quotient: $x^2 + 4$.

The zeros are –1 and 2. ($x^2 + 4 = 0$ has no real solutions.)

49. $3x^3 + 4x^2 - 7x + 2 = 0$

The solutions of the equation are the zeros of $f(x) = 3x^3 + 4x^2 - 7x + 2$.

Step 1: $f(x)$ has at most 3 real zeros.

Step 2:   By Descartes Rule of Signs, there are 2 or 0 positive real zeros.

$f(-x) = 3(-x)^3 + 4(-x)^2 - 7(-x) + 2 = -3x^3 + 4x^2 + 7x + 2$;

thus, there is 1 negative real zero.

Step 3:   Possible rational zeros:

$$p = \pm 1, \pm 2; \quad q = \pm 1, \pm 3; \quad \frac{p}{q} = \pm 1, \pm 2, \pm \frac{1}{3}, \pm \frac{2}{3}$$

Step 4:   Using synthetic division:

$$\frac{2}{3}{\overline{)}}\begin{array}{rrrr} 3 & 4 & -7 & 2 \\ & 2 & 4 & -2 \\ \hline 3 & 6 & -3 & 0 \end{array}$$

Since the remainder is 0, $x - \dfrac{2}{3}$ is a factor. The other factor is the quotient:

$3x^2 + 6x - 3$.

$$f(x) = \left(x - \frac{2}{3}\right)\left(3x^2 + 6x - 3\right) = 3\left(x - \frac{2}{3}\right)\left(x^2 + 2x - 1\right)$$

Using the quadratic formula to solve $x^2 + 2x - 1 = 0$:

$$x = \frac{-2 \pm \sqrt{4 - 4(1)(-1)}}{2(1)} = \frac{-2 \pm \sqrt{8}}{2} = \frac{-2 \pm 2\sqrt{2}}{2} = -1 \pm \sqrt{2}$$

The zeros are $\dfrac{2}{3}, -1 + \sqrt{2}$, and $-1 - \sqrt{2}$ or $0.67, 0.41$, and -2.41.

51. $3x^3 - x^2 - 15x + 5 = 0$
Solving by factoring:

$$x^2(3x - 1) - 5(3x - 1) = 0 \rightarrow (3x - 1)\left(x^2 - 5\right) = 0 \rightarrow (3x - 1)\left(x - \sqrt{5}\right)\left(x + \sqrt{5}\right) = 0$$

The solutions of the equation are $\dfrac{1}{3}, \sqrt{5}$, and $-\sqrt{5}$ or $0.33, 2.24$, and -2.24.

53. $x^4 + 4x^3 + 2x^2 - x + 6 = 0$
The solutions of the equation are the zeros of $f(x) = x^4 + 4x^3 + 2x^2 - x + 6$.
Step 1: $f(x)$ has at most 4 real zeros.
Step 2:   By Descartes Rule of Signs, there are 2 or 0 positive real zeros.
$f(-x) = (-x)^4 + 4(-x)^3 + 2(-x)^2 - (-x) + 6 = x^4 - 4x^3 + 2x^2 + x + 6$;
thus, there are 2 or 0 negative real zeros.
Step 3:   Possible rational zeros:

$$p = \pm 1, \pm 2, \pm 3, \pm 6; \quad q = \pm 1; \quad \frac{p}{q} = \pm 1, \pm 2, \pm 3, \pm 6$$

Step 4:   Using synthetic division:

$$-3{\overline{)}}\begin{array}{rrrrr} 1 & 4 & 2 & -1 & 6 \\ & -3 & -3 & 3 & -6 \\ \hline 1 & 1 & -1 & 2 & 0 \end{array} \qquad -2{\overline{)}}\begin{array}{rrrr} 1 & 1 & -1 & 2 \\ & -2 & 2 & -2 \\ \hline 1 & -1 & 1 & 0 \end{array}$$

Since the remainder is 0, $x + 3$ and $x + 2$ are factors. The other factor is the
quotient: $x^2 - x + 1$.
The zeros are -3 and -2. ($x^2 - x + 1 = 0$ has no real solutions.)

55. $x^3 - \dfrac{2}{3}x^2 + \dfrac{8}{3}x + 1 = 0$

The solutions of the equation are the zeros of $f(x) = x^3 - \dfrac{2}{3}x^2 + \dfrac{8}{3}x + 1$.

Step 1: $f(x)$ has at most 3 real zeros.

Step 2:   By Descartes Rule of Signs, there are 2 or 0 positive real zeros.

$f(-x) = (-x)^3 - \dfrac{2}{3}(-x)^2 + \dfrac{8}{3}(-x) + 1 = -x^3 - \dfrac{2}{3}x^2 - \dfrac{8}{3}x + 1$;

thus, there is 1 negative real zero.

Step 3:   Use the equivalent equation $3x^3 - 2x^2 + 8x + 3 = 0$ to find the possible rational zeros:

$$p = \pm 1, \pm 3; \quad q = \pm 1, \pm 3; \quad \dfrac{p}{q} = \pm 1, \pm 3, \pm \dfrac{1}{3}$$

Step 4:   Using synthetic division:

$$
\begin{array}{r|rrrr}
-\dfrac{1}{3} & 1 & -\dfrac{2}{3} & \dfrac{8}{3} & 1 \\
 & & -\dfrac{1}{3} & \dfrac{1}{3} & -1 \\
\hline
 & 1 & -1 & 3 & 0
\end{array}
$$

Since the remainder is 0, $x + \dfrac{1}{3}$ is a factor. The other factor is the quotient: $x^2 - x + 3$.

The real zero is $-\dfrac{1}{3}$. ($x^2 - x + 3 = 0$ has no real solutions.)

57. $2x^4 - 19x^3 + 57x^2 - 64x + 20 = 0$

The solutions of the equation are the zeros of $f(x) = 2x^4 - 19x^3 + 57x^2 - 64x + 20$.

Step 1: $f(x)$ has at most 4 real zeros.

Step 2:   By Descartes Rule of Signs, there are 4, 2 or 0 positive real zeros.

$f(-x) = 2(-x)^4 - 19(-x)^3 + 57(-x)^2 - 64(-x) + 20$

$= 2x^4 + 19x^3 + 57x^2 + 64x + 20$

thus, there are no negative real zeros.

Step 3:   To find the possible rational zeros:

$p = \pm 1, \pm 2, \pm 4, \pm 5, \pm 10, \pm 20; \quad q = \pm 1, \pm 2;$

$$\dfrac{p}{q} = \pm 1, \pm \dfrac{1}{2}, \pm 2, \pm 4, \pm 5, \pm \dfrac{5}{2}, \pm 10, \pm 20$$

Step 4:   Using synthetic division:

$$
\begin{array}{r|rrrrr}
1 & 2 & -19 & 57 & -64 & 20 \\
 & & 2 & -17 & 40 & -24 \\
\hline
 & 2 & -17 & 40 & -24 & \{-4\}
\end{array}
\qquad
\begin{array}{r|rrrrr}
\dfrac{1}{2} & 2 & -19 & 57 & -64 & 20 \\
 & & 1 & -9 & 24 & -20 \\
\hline
 & 2 & -18 & 48 & -40 & 0
\end{array}
$$

Thus, $x - \dfrac{1}{2}$ is a factor.

So :

$$f(x) = 2x^4 - 19x^3 + 57x^2 - 64x + 20 = \left(x - \frac{1}{2}\right)\left(2x^3 - 18x^2 + 48x - 40\right)$$

$$= 2\left(x - \frac{1}{2}\right)\left(x^3 - 9x^2 + 24x - 20\right)$$

Now try x = 2 as a solution to the equation $x^3 - 9x^2 + 24x - 20 = 0$.

$$
\begin{array}{r}
2\overline{)\;1 \quad -9 \quad 24 \quad -20} \\
\underline{2 \quad -14 \quad 20} \\
1 \quad -7 \quad 10 \quad 0
\end{array}
$$

Thus, $x^3 - 9x^2 + 24x - 20 = (x - 2)(x^2 - 7x + 10) = (x - 2)(x - 2)(x - 5)$

So we have

$$f(x) = 2x^4 - 19x^3 + 57x^2 - 64x + 20 = 2\left(x - \frac{1}{2}\right)(x - 2)^2(x - 5).$$

Therefore, f has real zeros $\frac{1}{2}, 2, 5$, and 2 is a zero of multiplicity 2.

59. $f(x) = x^3 + 2x^2 + 5x - 6 = (x + 3)(x + 1)(x - 2)\cdot$

x-intercepts: –3, –1, 2;   y-intercept: –6;   crosses x axis at x = –3, –1, 2

Interval	$x < -3$	$-3 < x < -1$	$-1 < x < 2$	$x > 2$
Sign of f	-	+	-	+
Above or below x-axis	below	above	below	above

Graph of f is above the x-axis for $(-3, -1) \cup (2, \infty)$

Graph of f is below the x-axis for $(-\infty, -3) \cup (-1, -2)$

61.　$f(x) = 2x^3 - x^2 + 2x - 1 = \left(x - \dfrac{1}{2}\right)\left(2x^2 + 2\right).$

x-intercepts: $\dfrac{1}{2}$;　　　　y-intercept: -1;　　　　crosses x axis at x = $\dfrac{1}{2}$

Interval	$x < \dfrac{1}{2}$	$x > \dfrac{1}{2}$
Sign of f	-	+
Above or below x-axis	below	above

Graph of f is above the x-axis for $\left(\dfrac{1}{2}, \infty\right)$

Graph of f is below the x-axis for $\left(-\infty, \dfrac{1}{2}\right)$

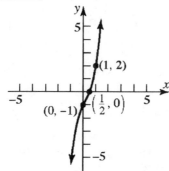

63.　$f(x) = x^4 + x^2 - 2 = (x + 1)(x - 1)(x^2 + 2).$

x-intercepts: –1, 1;　　　y-intercept: -2;　　　crosses x axis at x = –1, 1

Interval	$x < -1$	$-1 < x < 1$	$x > 1$
Sign of f	+	-	+
Above or below x-axis	above	below	above

Graph of f is above the x-axis for $(-\infty, -1) \cup (1, \infty)$

Graph of f is below the x-axis for $(-1, -1)$

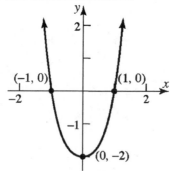

65. $f(x) = 4x^4 + 7x^2 - 2 = (2x+1)(2x-1)(x^2+2)$.

x-intercepts: $-\dfrac{1}{2}, \dfrac{1}{2}$;   y-intercept: -2;   crosses x axis at x = $-\dfrac{1}{2}, \dfrac{1}{2}$

Interval	$x < -\dfrac{1}{2}$	$-\dfrac{1}{2} < x < \dfrac{1}{2}$	$x > \dfrac{1}{2}$
Sign of f	+	-	+
Above or below x-axis	above	below	above

Graph of f is above the x-axis for $\left(-\infty, -\dfrac{1}{2}\right) \cup \left(\dfrac{1}{2}, \infty\right)$

Graph of f is below the x-axis for $\left(-\dfrac{1}{2}, \dfrac{1}{2}\right)$

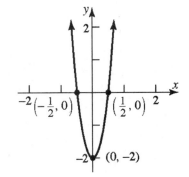

67. $f(x) = x^4 + x^3 - 3x^2 - x + 2 = (x+2)(x+1)(x-1)^2$.

x-intercepts: –2, –1, 1;   y-intercept: 2

crosses x axis at x = –2, –1;   touches x axis at x = 1

Interval	$x < -2$	$-2 < x < -1$	$-1 < x < 1$	$x > 1$
Sign of f	+	-	+	+
Above or below x-axis	above	below	above	above

Graph of f is above the x-axis for $(-\infty, -2) \cup (-1,1) \cup (1,\infty)$

Graph of f is below the x-axis for $(-2,-1)$

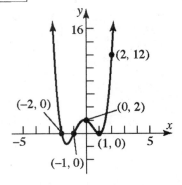

69. $f(x) = 4x^5 - 8x^4 - x + 2 = (x-2)(\sqrt{2}x - 1)(\sqrt{2}x + 1)(2x^2 + 1)$

x-intercepts: $-\dfrac{1}{\sqrt{2}}, \dfrac{1}{\sqrt{2}}, 2$;   y-intercept: 2

crosses x axis at x = $-\dfrac{1}{\sqrt{2}}, \dfrac{1}{\sqrt{2}}, 2$

Interval	$x < -\dfrac{1}{\sqrt{2}}$	$-\dfrac{1}{\sqrt{2}} < x < \dfrac{1}{\sqrt{2}}$	$\dfrac{1}{\sqrt{2}} < x < 2$	$x > 2$
Sign of f	-	+	-	+
Above or below x-axis	below	above	below	above

Graph of f is above the x-axis for $\left(-\dfrac{1}{\sqrt{2}},\dfrac{1}{\sqrt{2}}\right)\cup(2,\infty)$

Graph of f is below the x-axis for $\left(-\infty,-\dfrac{1}{\sqrt{2}}\right)\cup\left(\dfrac{1}{\sqrt{2}},2\right)$

71. $f(x)=x^4-3x^2-4$

$a_3=0, a_2=-3, a_1=0, a_0=-4$

$Max\{1,|-4|+|0|+|-3|+|0|\}=Max\{1,4+0+3+0\}=Max\{1,7\}=7$

$1+Max\{|-4|,|0|,|-3|,|0|\}=1+Max\{4,0,3,0\}=1+4=5$

The smaller of the two numbers is 5. Thus, every zero of f lies between -5 and 5.

73. $f(x)=x^4+x^3-x-1$

$a_3=1, a_2=0, a_1=-1, a_0=-1$

$Max\{1,|-1|+|-1|+|0|+|1|\}=Max\{1,1+1+0+1\}=Max\{1,3\}=3$

$1+Max\{|-1|,|-1|,|0|,|1|\}=1+Max\{1,1,0,1\}=1+1=2$

The smaller of the two numbers is 2. Thus, every zero of f lies between -2 and 2.

75. $f(x)=3x^4+3x^3-x^2-12x-12=3\left(x^4+x^3-\dfrac{1}{3}x^2-4x-4\right)$

Note: The leading coefficient must be 1.

$a_3=1, a_2=-\dfrac{1}{3}, a_1=-4, a_0=-4$

$Max\left\{1,|-4|+|-4|+\left|-\dfrac{1}{3}\right|+|1|\right\}=Max\left\{1,4+4+\dfrac{1}{3}+1\right\}=Max\left\{1,\dfrac{28}{3}\right\}=\dfrac{28}{3}$

$1+Max\left\{|-4|,|-4|,\left|-\dfrac{1}{3}\right|,|1|\right\}=1+Max\left\{4,4,\dfrac{1}{3},1\right\}=1+4=5$

The smaller of the two numbers is 5. Thus, every zero of f lies between -5 and 5.

77. $f(x)=4x^5-x^4+2x^3-2x^2+x-1=4\left(x^5-\dfrac{1}{4}x^4+\dfrac{1}{2}x^3-\dfrac{1}{2}x^2+\dfrac{1}{4}x-\dfrac{1}{4}\right)$

Note: The leading coefficient must be 1. $a_4=-\dfrac{1}{4}, a_3=\dfrac{1}{2}, a_2=-\dfrac{1}{2}, a_1=\dfrac{1}{4}, a_0=-\dfrac{1}{4}$

$Max\left\{1,\left|-\dfrac{1}{4}\right|+\left|\dfrac{1}{4}\right|+\left|-\dfrac{1}{2}\right|+\left|\dfrac{1}{2}\right|+\left|-\dfrac{1}{4}\right|\right\}=Max\left\{1,\dfrac{1}{4}+\dfrac{1}{4}+\dfrac{1}{2}+\dfrac{1}{2}+\dfrac{1}{4}\right\}=Max\left\{1,\dfrac{7}{4}\right\}=\dfrac{7}{4}$

$$1 + Max\left\{\left|-\frac{1}{4}\right|\left|\frac{1}{4}\right|\left|-\frac{1}{2}\right|\left|\frac{1}{2}\right|\left|-\frac{1}{4}\right|\right\} = 1 + Max\left\{\frac{1}{4},\frac{1}{4},\frac{1}{2},\frac{1}{2},\frac{1}{4}\right\} = 1 + \frac{1}{2} = \frac{3}{2}$$

The smaller of the two numbers is $\frac{3}{2}$. Thus, every zero of f lies between $-\frac{3}{2}$ and $\frac{3}{2}$.

79. $f(x) = 8x^4 - 2x^2 + 5x - 1$; $[0,1]$
$f(0) = -1 < 0$ and $f(1) = 10 > 0$
Since one is positive and one is negative, there is a zero in the interval.

81. $f(x) = 2x^3 + 6x^2 - 8x + 2$; $[-5,-4]$
$f(-5) = -58 < 0$ and $f(-4) = 2 > 0$
Since one is positive and one is negative, there is a zero in the interval.

83. $f(x) = x^5 - x^4 + 7x^3 - 7x^2 - 18x + 18$; $[1.4, 1.5]$
$f(1.4) = -0.1754 < 0$ and $f(1.5) = 1.4063 > 0$
Since one is positive and one is negative, there is a zero in the interval.

85. $8x^4 - 2x^2 + 5x - 1 = 0$; $0 \le r \le 1$

Consider the function $f(x) = 8x^4 - 2x^2 + 5x - 1$
Subdivide the interval [0,1] into 10 equal subintervals:

[0,0.1]; [0.1,0.2]; [0.2,0.3]; [0.3,0.4]; [0.4,0.5]; [0.5,0.6]; [0.6,0.7]; [0.7,0.8]; [0.8,0.9]; [0.9,1]

$f(0) = -1; f(0.1) = -0.5192$
$f(0.1) = -0.5192; f(0.2) = -0.0672$
$f(0.2) = -0.0672; f(0.3) = 0.3848$ so f has a real zero on the interval [0.2,0.3].
Subdivide the interval [0.2,0.3] into 10 equal subintervals:

[0.2,0.21]; [0.21,0.22]; [0.22,0.23]; [0.23,0.24]; [0.24,0.25]; [0.25,0.26];[0.26,0.27]; [0.27,0.28]; [0.28,0.29]; [0.29,0.3]

$f(0.2) = -0.0672; f(0.21) = -0.02264$
$f(0.21) = -0.02264; f(0.22) = 0.0219$ so f has a real zero on the interval [0.21,0.22], therefore $r = 0.21$, correct to 2 decimal places.

87. $2x^3 + 6x^2 - 8x + 2 = 0$; $-5 \le r \le -4$

Consider the function $f(x) = 2x^3 + 6x^2 - 8x + 2$
Subdivide the interval [-5,-4] into 10 equal subintervals:

[-5,-4.9]; [-4.9,-4.8]; [-4.8,-4.7]; [-4.7,-4.6]; [-4.6,-4.5]; [-4.5,-4.4]; [-4.4,-4.3]; [-4.3,-4.2]; [-4.2,-4.1]; [-4.1,-4]

$f(-5) = -58; f(-4.9) = -50.038$

$f(-4.9) = -50.038; f(-4.8) = -42.544$

$f(-4.8) = -42.544; f(-4.7) = -35.506$

$f(-4.7) = -35.506; f(-4.6) = -28.912$

$f(-4.6) = -28.912; f(-4.5) = -22.75$

$f(-4.5) = -22.75; f(-4.4) = -17$

$f(-4.4) = -17; f(-4.3) = -11.674$

$f(-4.3) = -11.674; f(-4.2) = -6.736$

$f(-4.2) = -6.736; f(-4.1) = -2.182$

$f(-4.1) = -2.182; f(-4) = 2$ so f has a real zero on the interval [-4.1,-4].

Subdivide the interval [-4.1,-4] into 10 equal subintervals:

[-4.1,-4.09]; [-4.09,-4.08]; [-4.08,-4.07]; [-4.07,-4.06]; [-4.06,-4.05]; [-4.05,-4.04];
[-4.04,-4.03]; [-4.03,-4.02]; [-4.02,-4.01]; [-4.01,-4]

$f(-4.1) = -2.182; f(-4.09) = -1.7473$

$f(-4.09) = -1.7473; f(-4.08) = -1.3162$

$f(-4.08) = -1.3162; f(-4.07) = -0.8889$

$f(-4.07) = -0.8889; f(-4.06) = -0.4652$

$f(-4.06) = -0.4652; f(-4.05) = -0.0452$

$f(-4.05) = -0.4652; f(-4.04) = 0.3711$ so f has a real zero on the interval [-4.05,-4.04], therefore $r = -4.05$, correct to 2 decimal places.

89. $f(x) = x^3 + x^2 + x - 4$

$f(1) = -1; f(2) = 10$ so f has a real zero on the interval [1,2],

Subdivide the interval [1,2] into 10 equal subintervals:

[1,1.1]; [1.1,1.2]; [1.2,1.3]; [1.3,1.4]; [1.4,1.5]; [1.5,1.6]; [1.6,1.7]; [1.7,1.8];
[1.8,1.9]; [1.9,2]

$f(1) = -1; f(1.1) = -0.359$

$f(1.1) = -0.359; f(1.2) = 0.368$ so f has a real zero on the interval [1.1,1.2].

Subdivide the interval [1,1.2] into 10 equal subintervals:

[1,1.11]; [1.11,1.12]; [1.12,1.13]; [1.13,1.14]; [1.14,1.15]; [1.15,1.16];[1.16,1.17];
[1.17,1.18]; [1.18,1.19]; [1.19,1.2]

$f(1) = -1; f(1.11) = -0.2903$ so f has a real zero on the interval

$f(1.11) = -0.2903; f(1.12) = -0.2207$ [1.15,1.16], therefore $r = 1.15$, correct

$f(1.12) = -0.2207; f(1.13) = -0.1502$ to 2 decimal places.

$f(1.13) = -0.1502; f(1.14) = -0.0789$

$f(1.14) = -0.0789; f(1.15) = -0.0066$

$f(1.15) = -0.0066; f(1.16) = 0.0665$

91. $f(x) = 2x^4 - 3x^3 - 4x^2 - 8$
$f(2) = -16; f(3) = 37$ so f has a real zero on the interval $[2,3]$,
Subdivide the interval $[2,3]$ into 10 equal subintervals:

$[2,2.1]; [2.1,2.2]; [2.2,2.3]; [2.3,2.4]; [2.4,2.5]; [2.5,2.6]; [2.6,2.7]; [2.7,2.8]; [2.8,2.9]; [2.9,3]$

$f(2) = -16; f(2.1) = -14.5268$
$f(2.1) = -14.5268; f(2.2) = -12.4528$
$f(2.2) = -12.4528; f(2.3) = -9.6928$
$f(2.3) = -9.6928; f(2.4) = -6.1568$
$f(2.4) = -6.1568; f(2.5) = -1.75$
$f(2.5) = -1.75; f(2.6) = 3.6272$ so f has a real zero on the interval $[2.5,2.6]$.
Subdivide the interval $[2.5,2.6]$ into 10 equal subintervals:

$[2.5,2.51]; [2.51,2.52]; [2.52,2.53]; [2.53,2.54]; [2.54,2.55]; [2.55,2.56]; [2.56,2.57]; [2.57,2.58]; [2.58,2.59]; [2.59,2.6]$

$f(2.5) = -1.75; f(2.51) = -1.2576$
$f(2.51) = -1.2576; f(2.52) = -0.7555$
$f(2.52) = -0.7555; f(2.53) = -0.2434$
$f(2.53) = -0.2434; f(2.54) = 0.2787$ so f has a real zero on the interval $[2.53,2.54]$, therefore $r = 2.53$, correct to 2 decimal places.

93. $x - 2$ is a factor of $f(x) = x^3 - kx^2 + kx + 2$ only if the remainder that results when $f(x)$ is divided by $x - 2$ is 0. Dividing, we have:

$$\begin{array}{r|rrrr} 2) & 1 & -k & k & 2 \\ & & 2 & -2k+4 & -2k+8 \\ \hline & 1 & -k+2 & -k+4 & -2k+10 \end{array}$$

Since we want the remainder to equal 0, set the remainder equal to zero and solve:
$-2k + 10 = 0 \rightarrow -2k = -10 \rightarrow k = 5$

95. By the Remainder Theorem we know that the remainder from synthetic division by c is equal to $f(c)$. Thus the easiest way to find the remainder is to evaluate:
$f(1) = 2(1)^{20} - 8(1)^{10} + 1 - 2 = 2 - 8 + 1 - 2 = -7$
The remainder is -7.

97. We want to prove that $x - c$ is a factor of $x^n - c^n$, for any positive integer n. By the Factor Theorem, $x - c$ will be a factor of $f(x)$ provided $f(c) = 0$. Here, $f(x) = x^n - c^n$, so that $f(c) = c^n - c^n = 0$. Therefore, $x - c$ is a factor of $x^n - c^n$.

99. $x^3 - 8x^2 + 16x - 3 = 0$ has solution $x = 3$, so $x - 3$ is a factor of
$f(x) = x^3 - 8x^2 + 16x - 3$.

Using synthetic division

$$
\begin{array}{r|rrrr}
3 & 1 & -8 & 16 & -3 \\
  &   & 3 & -15 & 3 \\
\hline
  & 1 & -5 & 1 & 0 \\
\end{array}
$$

$\therefore f(x) = x^3 - 8x^2 + 16x - 3 = (x - 3)(x^2 - 5x + 1)$.

Solving $x^2 - 5x + 1 = 0$

$$x = \frac{5 \pm \sqrt{25 - 4}}{2} = \frac{5 \pm \sqrt{21}}{2}$$

The sum of these two roots is $\dfrac{5 + \sqrt{21}}{2} + \dfrac{5 - \sqrt{21}}{2} = \dfrac{10}{2} = 5$.

101. $f(x) = 2x^3 + 3x^2 - 6x + 7$

By the Rational Zero Theorem, the only possible rational zeros are:

$\dfrac{p}{q} = \pm 1, \pm 7, \pm \dfrac{1}{2}, \pm \dfrac{7}{2}$ Since $\dfrac{1}{3}$ is not in the list of possible rational zeros, it is not a
zero of $f(x)$.

103. $f(x) = 2x^6 - 5x^4 + x^3 - x + 1$

By the Rational Zero Theorem, the only possible rational zeros are:

$\dfrac{p}{q} = \pm 1, \pm \dfrac{1}{2}$ Since $\dfrac{3}{5}$ is not in the list of possible rational zeros, it is not a zero of $f(x)$.

105.   Let x be the length of a side of the original cube.
After removing the 1 inch slice, one dimension will be $x - 1$.
The volume of the new solid will be:

$$(x - 1) \cdot x \cdot x = 294 \rightarrow x^3 - x^2 = 294 \rightarrow x^3 - x^2 - 294 = 0$$

By Descartes Rule of Signs, we know that there is one positive real solution.
The possible rational zeros are:

$p = \pm 1, \pm 2, \pm 3, \pm 6, \pm 7, \pm 14, \pm 21, \pm 42, \pm 49, \pm 98, \pm 147, \pm 294; \quad q = \pm 1$

The rational zeros are the same as the values for p.

Using synthetic division:

$$
\begin{array}{r|rrrr}
7 & 1 & -1 & 0 & -294 \\
  &   & 7 & 42 & 294 \\
\hline
  & 1 & 6 & 42 & 0 \\
\end{array}
$$

7 is a zero, so the length of the original edge of the cube was 7 inches.

107. $f(x) = x^n + a_{n-1}x^{n-1} + a_{n-2}x^{n-2} + ... + a_1x + a_0$; where $a_{n-1}, a_{n-2}, ...a_1, a_0$ are integers

If r is a real zero of f, then r is either rational or irrational. We know that the rational roots of f must be of the form $\frac{p}{q}$ where p is a divisor of a_0 and q is a divisor of 1. This means that $q = \pm 1$. So if r is rational, then $r = \frac{p}{q} = \pm p$. Therefore, r is an integer or r is irrational.

109. (a) $f(x) = 8x^4 - 2x^2 + 5x - 1 \quad 0 \le r \le 1$

At Step 0 we have the interval $[0,1]$.

$f(0) = -1; \quad f(1) = 10$

Let $m = $ the midpoint of the interval being considered.

So $m_0 = 0.5$

n	m_{n-1}	$f(m_{n-1})$	New interval
1	0.5	$f(0.5) = 1.5 > 0$	$[0,0.5]$
2	0.25	$f(0.25) = 0.15625 > 0$	$[0,0.25]$
3	0.125	$f(0.125) = -0.4043 < 0$	$[0.125,0.25]$
4	0.1875	$f(0.1875) = -0.1229 < 0$	$[0.1875,0.25]$
5	0.21875	$f(0.21875) = 0.0164 > 0$	$[0.1875,0.21875]$
6	0.203125	$f(0.203125) = -0.0533 < 0$	$[0.203125,0.21875]$
7	0.2109375	$f(0.2109375) = -0.0185 < 0$	$[0.2109375,0.21875]$
8	0.21484375		

Since the midpoint value at Step 8 agrees with the midpoint value at Step 7 to two decimal places, $r = 0.21$, correct to 2 decimal places.

(b) $f(x) = x^4 + 8x^3 - x^2 + 2; \quad -1 \le r \le 0$

At Step 0 we have the interval $[-1,0]$.

$f(-1) = -6; \quad f(0) = 2$

Let $m = $ the midpoint of the interval being considered.

So $m_0 = -0.5$

n	m_{n-1}	$f(m_{n-1})$	New interval
1	- 0.5	$f(-0.5) = 0.8125 > 0$	$[-1,- 0.5]$
2	- 0.75	$f(-0.75) = -1.6211 < 0$	$[-0.75,- 0.5]$
3	-0.625	$f(-0.625) = -0.1912 < 0$	$[-0.625,-0.5]$
4	-0.5625	$f(-0.5625) = 0.3599 > 0$	$[-0.625,-0.5625]$
5	-0.59375	$f(-0.59375) = 0.0972 > 0$	$[-0.625,-0.59375]$
6	-0.609375	$f(-0.609375) = -0.04372 < 0$	$[-0.609375,-0.59375]$
7	-0.6015625		

Since the midpoint value at Step 7 agrees with the midpoint value at Step 6 to two decimal places, $r = - 0.60$, correct to 2 decimal places.

(c) $f(x) = 2x^3 + 6x^2 - 8x + 2;$ $-5 \le r \le -4$
At Step 0 we have the interval [-5,-4].
$f(-5) = -58;$ $f(-4) = 2$
Let m = the midpoint of the interval being considered.
So $m_0 = -4.5$

n	m_{n-1}	$f(m_{n-1})$	New interval
1	- 4.5	$f(-4.5) = -22.75 < 0$	[-4.5,- 4]
2	- 4.25	$f(-4.25) = -9.1562 < 0$	[-4.25,- 4]
3	-4.125	$f(-4.125) = -3.2852 < 0$	[-4.125,-4]
4	-4.0625	$f(-4.0625) = -0.5708 < 0$	[-4.0625,-4]
5	-4.03125	$f(-4.03125) = 0.7324 > 0$	[-4.0625, -4.03125]
6	-4.046875	$f(-4.046875) = 0.0852 > 0$	[-4.0625, -4.046875]
7	-4.0546875	$f(-4.0546875) = -0.2417 < 0$	[-4.0546875, -4.046875]
8	-4.05078125		

Since the midpoint value at Step 8 agrees with the midpoint value at Step 7 to
two decimal places, $r = -4.05$, correct to 2 decimal places.

(d) $f(x) = 3x^3 - 10x + 9;$ $-3 \le r \le -2$
At Step 0 we have the interval [-3,-2].
$f(-3) = -42;$ $f(-2) = 5$
Let m = the midpoint of the interval being considered.
So $m_0 = -2.5$

n	m_{n-1}	$f(m_{n-1})$	New interval
1	- 2.5	$f(-2.5) = -12.875 < 0$	[-2.5,- 2]
2	- 2.25	$f(-2.25) = -2.6719 < 0$	[-2.25,- 2]
3	-2.125	$f(-2.125) = 1.4629 > 0$	[-2.25,- 2.125]
4	-2.1875	$f(-2.1875) = -0.5276 < 0$	[-2.1875,- 2.125]
5	-2.15625	$f(-2.15625) = 0.4866 > 0$	[-2.1875,-2.15625]
6	-2.171875	$f(-2.171875) = -0.0157 < 0$	[-2.171875,-2.15625]
7	-2.1640625	$f(-2.1640625) = 0.2366 > 0$	[-2.171875,- 2.1640625]
8	-2.16796875		

Since the midpoint value at Step 8 agrees with the midpoint value at Step 7 to
two decimal places, $r = -2.16$, correct to 2 decimal places.

(e) $f(x) = x^3 + x^2 + x - 4;$ $1 \le r \le 2$
At Step 0 we have the interval [1,2].
$f(1) = -1;$ $f(2) = 10$
Let m = the midpoint of the interval being considered.
So $m_0 = 1.5$

n	m_{n-1}	$f(m_{n-1})$	New interval
1	1.5	$f(1.5) = 3.125 > 0$	[1,1.5]
2	1.25	$f(1.25) = 0.7656 > 0$	[1,1.25]
3	1.125	$f(1.125) = -0.1855 < 0$	[1.125,1.25]

4	1.1875	$f(1.1875) = 0.2722 > 0$	[1.125,1.1875]
5	1.15625	$f(1.15625) = 0.0390 > 0$	[1.125,1.15625]
6	1.140625	$f(1.140625) = -0.0744 < 0$	[1.140625,1.15625]
7	1.1484375		

Since the midpoint value at Step 7 agrees with the midpoint value at Step 6 to two decimal places, $r = 1.14$, correct to 2 decimal places.

(f) $f(x) = 2x^4 + x^2 - 1;$ $0 \leq r \leq 1$
At Step 0 we have the interval [0,1].
$f(0) = -1;$ $f(1) = 2$
Let m = the midpoint of the interval being considered.
So $m_0 = 0.5$

n	m_{n-1}	$f(m_{n-1})$	New interval
1	0.5	$f(0.5) = -0.625 < 0$	[0.5,1]
2	0.75	$f(0.75) = 0.1593 > 0$	[0.5,0.75]
3	0.625	$f(0.625) = -0.3042 < 0$	[0.625,0.75]
4	0.6875	$f(0.6875) = -0.0805 < 0$	[0.6875,0.75]
5	0.71875	$f(0.71875) = 0.0504 > 0$	[0.6875,0.71875]
6	0.703125	$f(0.703125) = -0.0168 < 0$	[0.703125,0.71875]
7	0.7109375	$f(0.7109375) = 0.0164 > 0$	[0.703125, 0.7109375]
8	0.70703125	$f(0.70703125) = -0.0032 < 0$	[0.70703125, 0.7109375]
9	0.708984375		

Since the midpoint value at Step 9 agrees with the midpoint value at Step 8 to two decimal places, $r = 0.70$, correct to 2 decimal places.

(g) $f(x) = 2x^4 - 3x^3 - 4x^2 - 8;$ $2 \leq r \leq 3$
At Step 0 we have the interval [2,3]
$f(2) = -16;$ $f(3) = 37$
Let m = the midpoint of the interval being considered.
So $m_0 = 2.5$

n	m_{n-1}	$f(m_{n-1})$	New interval
1	2.5	$f(2.5) = -1.75 < 0$	[2.5,3]
2	2.75	$f(2.75) = 13.7422 > 0$	[2.5,2.75]
3	2.625	$f(2.625) = 5.1352 > 0$	[2.5,2.625]
4	2.5625	$f(2.5625) = 1.4905 > 0$	[2.5,2.5625]
5	2.53125	$f(2.53125) = -0.1787 < 0$	[2.53125,2.5625]
6	2.546875	$f(2.546875) = 0.6435 > 0$	[2.53125, 2.546875]
7	2.5390625	$f(2.5390625) = 0.2293 > 0$	[2.53125, 2.5390625]
8	2.53515625		

Since the midpoint value at Step 8 agrees with the midpoint value at Step 7 to two decimal places, $r = 2.53$, correct to 2 decimal places.

(h) $f(x) = 3x^3 - 2x^2 - 20;$ $2 \le r \le 3$

At Step 0 we have the interval [2,3].

$f(2) = -4;$ $f(3) = 43$

Let m = the midpoint of the interval being considered.

So $m_0 = 2.5$

n	m_{n-1}	$f(m_{n-1})$	New interval
1	2.5	$f(2.5) = 14.375 > 0$	[2,2.5]
2	2.25	$f(2.25) = 4.0469 > 0$	[2,2.25]
3	2.125	$f(2.125) = -0.2441 < 0$	[2.125,2.25]
4	2.1875	$f(2.1875) = 1.8323 > 0$	[2.125,2.1875]
5	2.15625	$f(2.15625) = 0.7771 > 0$	[2.125,2.15625]
6	2.140625	$f(2.140625) = 0.2622 > 0$	[2.125, 2.140625]
7	2.1328125	$f(2.1328125) = 0.0080 > 0$	[2.125, 2.1328125]
8	2.1315625		

Since the midpoint value at Step 8 agrees with the midpoint value at Step 7 to two decimal places, $r = 2.13$, correct to 2 decimal places.

Chapter 3

Polynomial and Rational Functions

3.7 Complex Zeros; Fundamental Theorem of Algebra

1. Since complex zeros appear in conjugate pairs, $4 + i$, the conjugate of $4 - i$, is the remaining zero of f.

3. Since complex zeros appear in conjugate pairs, $-i$, the conjugate of i, and $1 - i$, the conjugate of $1 + i$, are the remaining zeros of f.

5. Since complex zeros appear in conjugate pairs, $-i$, the conjugate of i, and $-2i$, the conjugate of $2i$, are the remaining zeros of f.

7. Since complex zeros appear in conjugate pairs, $-i$, the conjugate of i, is the remaining zero of f.

9. Since complex zeros appear in conjugate pairs, $2 - i$, the conjugate of $2 + i$, and $-3 + i$, the conjugate of $-3 - i$, are the remaining zeros of f.

11. Since $3 + 2i$ is a zero, its conjugate $3 - 2i$ is also a zero of f. Finding the function:
$$
\begin{aligned}
f(x) &= (x - 4)(x - 4)(x - (3 + 2i))(x - (3 - 2i)) \\
&= \left(x^2 - 8x + 16\right)((x - 3) - 2i)((x - 3) + 2i) \\
&= \left(x^2 - 8x + 16\right)\left(x^2 - 6x + 9 - 4i^2\right) \\
&= \left(x^2 - 8x + 16\right)\left(x^2 - 6x + 13\right) \\
&= x^4 - 6x^3 + 13x^2 - 8x^3 + 48x^2 - 104x + 16x^2 - 96x + 208 \\
&= x^4 - 14x^3 + 77x^2 - 200x + 208
\end{aligned}
$$

13. Since $-i$ is a zero, its conjugate i is also a zero, and since $1 + i$ is a zero, its conjugate $1 - i$ is also a zero of f. Finding the function:
$$
\begin{aligned}
f(x) &= (x - 2)(x + i)(x - i)(x - (1 + i))(x - (1 - i)) \\
&= (x - 2)\left(x^2 - i^2\right)((x - 1) - i)((x - 1) + i) \\
&= (x - 2)\left(x^2 + 1\right)\left(x^2 - 2x + 1 - i^2\right) \\
&= \left(x^3 - 2x^2 + x - 2\right)\left(x^2 - 2x + 2\right) \\
&= x^5 - 2x^4 + 2x^3 - 2x^4 + 4x^3 - 4x^2 + x^3 - 2x^2 + 2x - 2x^2 + 4x - 4 \\
&= x^5 - 4x^4 + 7x^3 - 8x^2 + 6x - 4
\end{aligned}
$$

15.　Since $-i$ is a zero, its conjugate i is also a zero of f. Finding the function:

$$f(x) = (x-3)(x-3)(x+i)(x-i)$$

$$= \left(x^2 - 6x + 9\right)\left(x^2 - i^2\right) = \left(x^2 - 6x + 9\right)\left(x^2 + 1\right)$$

$$= x^4 + x^2 - 6x^3 - 6x + 9x^2 + 9 = x^4 - 6x^3 + 10x^2 - 6x + 9$$

17.　Since $2i$ is a zero, its conjugate $-2i$ is also a zero of f. $x - 2i$ and $x + 2i$ are factors of f.
Thus, $(x - 2i)(x + 2i) = x^2 + 4$ is a factor of f. Using division to find the other factor:

$$
\begin{array}{r}
x - 4 \\
x^2 + 4\,\overline{)\,x^3 - 4x^2 + 4x - 16} \\
\underline{x^3 \qquad\;\; + 4x} \\
-4x^2 \qquad - 16 \\
\underline{-4x^2 \qquad - 16}
\end{array}
$$

$x - 4$ is a factor and the remaining zero is 4. The zeros of f are $4, 2i, -2i$.

19.　Since $-2i$ is a zero, its conjugate $2i$ is also a zero of f. $x - 2i$ and $x + 2i$ are factors of f.
Thus, $(x - 2i)(x + 2i) = x^2 + 4$ is a factor of f. Using division to find the other factor:

$$
\begin{array}{r}
2x^2 + 5x - 3 \\
x^2 + 4\,\overline{)\,2x^4 + 5x^3 + 5x^2 + 20x - 12} \\
\underline{2x^4 \qquad\;\; + 8x^2} \\
5x^3 - 3x^2 + 20x \\
\underline{5x^3 \qquad + 20x} \\
-3x^2 \qquad - 12 \\
\underline{-3x^2 \qquad - 12}
\end{array}
$$

$2x^2 + 5x - 3 = (2x - 1)(x + 3)$ are factors and the remaining zeros are $\dfrac{1}{2}$ and -3. The zeros of f are $2i, -2i, -3, \dfrac{1}{2}$.

21.　Since $3 - 2i$ is a zero, its conjugate $3 + 2i$ is also a zero of h. $x - (3 - 2i)$ and $x - (3 + 2i)$ are factors of h. Thus,

$$(x - (3 - 2i))(x - (3 + 2i)) = ((x - 3) + 2i)((x - 3) - 2i) = x^2 - 6x + 9 - 4i^2 = x^2 - 6x + 13 \text{ is}$$

a factor of h. Using division to find the other factor:

$$
\begin{array}{r}
x^2 - 3x - 10 \\
x^2 - 6x + 13\,\overline{)\,x^4 - 9x^3 + 21x^2 + 21x - 130} \\
\underline{x^4 - 6x^3 + 13x^2} \\
-3x^3 + 8x^2 + 21x \\
\underline{-3x^3 + 18x^2 - 39x} \\
-10x^2 + 60x - 130 \\
\underline{-10x^2 + 60x - 130}
\end{array}
$$

$x^2 - 3x - 10 = (x + 2)(x - 5)$ are factors and the remaining zeros are -2 and 5. The zeros of h are $3 - 2i, 3 + 2i, -2, 5$.

23.   Since $-4i$ is a zero, its conjugate $4i$ is also a zero of h. $x - 4i$ and $x + 4i$ are factors of h.

Thus, $(x - 4i)(x + 4i) = x^2 + 16$ is a factor of h. Using division to find the other factor:

$$
\begin{array}{r}
3x^3 + 2x^2 - 33x - 22 \\
x^2 + 16 \overline{)\,3x^5 + 2x^4 + 15x^3 + 10x^2 - 528x - 352} \\
\underline{3x^5 \qquad\quad + 48x^3} \\
2x^4 - 33x^3 + 10x^2 \\
\underline{2x^4 \qquad\quad + 32x^2} \\
-33x^3 - 22x^2 - 528x \\
\underline{-33x^3 \qquad\qquad - 528x} \\
-22x^2 \qquad\quad - 352 \\
\underline{-22x^2 \qquad\quad - 352}
\end{array}
$$

So, $3x^3 + 2x^2 - 33x - 22 = x^2(3x + 2) - 11(3x + 2) = (3x + 2)(x^2 - 11)$

$= (3x + 2)\left(x - \sqrt{11}\right)\left(x + \sqrt{11}\right)$ are factors and the remaining zeros are $-\dfrac{2}{3}, \sqrt{11}$, and $-\sqrt{11}$.

The zeros of h are $4i, -4i, -\sqrt{11}, \sqrt{11}, -\dfrac{2}{3}$.

25. $f(x) = x^3 - 1 = (x - 1)\left(x^2 + x + 1\right)$ The zeros of $x^2 + x + 1 = 0$ are:

$$x = \frac{-1 \pm \sqrt{1^2 - 4(1)(1)}}{2(1)} = \frac{-1 \pm \sqrt{-3}}{2} = -\frac{1}{2} + \frac{\sqrt{3}}{2}i \text{ or } -\frac{1}{2} - \frac{\sqrt{3}}{2}i$$

The zeros are: $1, -\dfrac{1}{2} + \dfrac{\sqrt{3}}{2}i, -\dfrac{1}{2} - \dfrac{\sqrt{3}}{2}i$.

27. $f(x) = x^3 - 8x^2 + 25x - 26$

Step 1: $f(x)$ has 3 complex zeros.

Step 2:   By Descartes Rule of Signs, there are 3 or 1 positive real zeros.

$f(-x) = (-x)^3 - 8(-x)^2 + 25(-x) - 26 = -x^3 - 8x^2 - 25x - 26$; thus, there are no negative real zeros.

Step 3:   Possible rational zeros:

$$p = \pm 1, \pm 2, \pm 13, \pm 26; \quad q = \pm 1; \quad \frac{p}{q} = \pm 1, \pm 2, \pm 13, \pm 26$$

Step 4:   Using synthetic division:

$$
\begin{array}{r}
2 \,\overline{)\,1 \quad -8 \quad\; 25 \quad -26} \\
\underline{2 \quad -12 \quad\;\; 26} \\
1 \quad -6 \quad\;\; 13 \qquad 0
\end{array}
$$

Since the remainder is 0, $x - 2$ is a factor. The other factor is the quotient: $x^2 - 6x + 13$.

Using the quadratic formula to find the zeros of $x^2 - 6x + 13 = 0$:

$$x = \frac{-(-6) \pm \sqrt{(-6)^2 - 4(1)(13)}}{2(1)} = \frac{6 \pm \sqrt{-16}}{2} = \frac{6 \pm 4i}{2} = 3 \pm 2i.$$

The complex zeros are $2, \ 3 - 2i, \ 3 + 2i$.

29. $f(x) = x^4 + 5x^2 + 4 = \left(x^2 + 4\right)\left(x^2 + 1\right) = (x + 2i)(x - 2i)(x + i)(x - i)$

The zeros are: $-2i, \ -i, \ i, \ 2i$.

31. $f(x) = x^4 + 2x^3 + 22x^2 + 50x - 75$

Step 1: $f(x)$ has 4 complex zeros.

Step 2: By Descartes Rule of Signs, there is 1 positive real zero.

$$f(-x) = (-x)^4 + 2(-x)^3 + 22(-x)^2 + 50(-x) - 75$$
$$= x^4 - 2x^3 + 22x^2 - 50x - 75$$

thus, there are 3 or 1 negative real zeros.

Step 3: Possible rational zeros:

$$p = \pm 1, \pm 3, \pm 5, \pm 15, \pm 25, \pm 75; \quad q = \pm 1;$$

$$\frac{p}{q} = \pm 1, \pm 3, \pm 5, \pm 15, \pm 25, \pm 75$$

Step 4: Using synthetic division:

$$
\begin{array}{r|rrrrr}
-3 & 1 & 2 & 22 & 50 & -75 \\
   &   & -3 & 3 & -75 & 75 \\
\hline
   & 1 & -1 & 25 & -25 & 0
\end{array}
$$

Since the remainder is 0, $x + 3$ is a factor. The other factor is the quotient:

$$x^3 - x^2 + 25x - 25 = x^2(x - 1) + 25(x - 1) = (x - 1)\left(x^2 + 25\right)$$
$$= (x - 1)(x + 5i)(x - 5i)$$

The complex zeros are $-3, \ 1, \ -5i, \ 5i$.

33. $f(x) = 3x^4 - x^3 - 9x^2 + 159x - 52$

Step 1: $f(x)$ has 4 complex zeros.

Step 2: By Descartes Rule of Signs, there are 3 or 1 positive real zeros.

$$f(-x) = 3(-x)^4 - (-x)^3 - 9(-x)^2 + 159(-x) - 52$$
$$= 3x^4 + x^3 - 9x^2 - 159x - 52$$

thus, there is 1 negative real zero.

Step 3: Possible rational zeros:

$$p = \pm 1, \pm 2, \pm 4, \pm 13, \pm 26, \pm 52; \quad q = \pm 1, \pm 3;$$

$$\frac{p}{q} = \pm 1, \pm 2, \pm 4, \pm 13, \pm 26, \pm 52, \pm \frac{1}{3}, \pm \frac{2}{3}, \pm \frac{4}{3}, \pm \frac{13}{3}, \pm \frac{26}{3}, \pm \frac{52}{3}$$

Step 4: Using synthetic division:

$$
\begin{array}{r|rrrrr}
-4 & 3 & -1 & -9 & 159 & -52 \\
   &   & -12 & 52 & -172 & 52 \\
\hline
   & 3 & -13 & 43 & -13 & 0
\end{array}
\qquad
\begin{array}{r|rrrr}
\frac{1}{3} & 3 & -13 & 43 & -13 \\
   &   & 1 & -4 & 13 \\
\hline
   & 3 & -12 & 39 & 0
\end{array}
$$

Since the remainder is 0, $x + 4$ and $x - \dfrac{1}{3}$ are factors. The other factor is the

quotient: $3x^2 - 12x + 39 = 3\left(x^2 - 4x + 13\right)$.

Using the quadratic formula to find the zeros of $x^2 - 4x + 13 = 0$:

$$x = \frac{-(-4) \pm \sqrt{(-4)^2 - 4(1)(13)}}{2(1)} = \frac{4 \pm \sqrt{-36}}{2} = \frac{4 \pm 6i}{2} = 2 \pm 3i.$$

The complex zeros are $-4, \ \frac{1}{3}, \ 2 - 3i, \ 2 + 3i$.

35. If the coefficients are real numbers and $2 + i$ is a zero, then $2 - i$ would also be a zero. This would then require a polynomial of degree 4.

37. If the coefficients are real numbers, then complex zeros must appear in conjugate pairs. We have a conjugate pair and one real zero. Thus, there is only one remaining zero and it must be real because a complex zero would require a pair and the polynomial would then have to be of degree 5.

Polynomial and Rational Functions

3.R   Chapter Review

1. $f(x) = (x-2)^2 + 2 = x^2 - 4x + 4 + 2 = x^2 - 4x + 6$

$a = 1, b = -4, c = 6.$ Since $a = 1 > 0,$ the graph opens up.

The x-coordinate of the vertex is $x = \dfrac{-b}{2a} = \dfrac{-(-4)}{2(1)} = \dfrac{4}{2} = 2.$

The y-coordinate of the vertex is $f\left(\dfrac{-b}{2a}\right) = f(2) = (2)^2 - 4(2) + 6 = 2.$

Thus, the vertex is (2, 2).
The axis of symmetry is the line $x = 2.$
The discriminant is:

$\qquad b^2 - 4ac = (-4)^2 - 4(1)(6) = -8 < 0,$

so the graph has no x-intercepts.
The y-intercept is $f(0) = 6.$

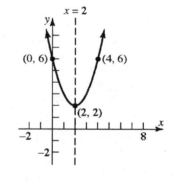

3. $f(x) = \dfrac{1}{4}x^2 - 16$

$a = \dfrac{1}{4}, b = 0, c = -16.$ Since $a = \dfrac{1}{4} > 0,$ the graph opens up.

The x-coordinate of the vertex is $x = \dfrac{-b}{2a} = \dfrac{-0}{2\left(\dfrac{1}{4}\right)} = \dfrac{0}{\left(\dfrac{1}{2}\right)} = 0.$

The y-coordinate of the vertex is $f\left(\dfrac{-b}{2a}\right) = f(0) = \dfrac{1}{4}(0)^2 - 16 = -16.$

Thus, the vertex is (0, –16).
The axis of symmetry is the line $x = 0.$
The discriminant is:

$\qquad b^2 - 4ac = (0)^2 - 4\left(\dfrac{1}{4}\right)(-16) = 16 > 0,$

so the graph has two x-intercepts.

The x-intercepts are found by solving:
$$\frac{1}{4}x^2 - 16 = 0$$
$$x^2 - 64 = 0 \rightarrow x^2 = 64 \rightarrow x = 8 \text{ or } x = -8$$
The x-intercepts are –8 and 8.
The y-intercept is $f(0) = -16$.

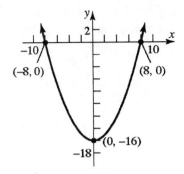

5.　$f(x) = -4x^2 + 4x$
$a = -4, b = 4, c = 0$. Since $a = -4 < 0$, the graph opens down.

The x-coordinate of the vertex is $x = \dfrac{-b}{2a} = \dfrac{-4}{2(-4)} = \dfrac{-4}{-8} = \dfrac{1}{2}$.

The y-coordinate of the vertex is $f\left(\dfrac{-b}{2a}\right) = f\left(\dfrac{1}{2}\right) = -4\left(\dfrac{1}{2}\right)^2 + 4\left(\dfrac{1}{2}\right) = -1 + 2 = 1$.

Thus, the vertex is $\left(\dfrac{1}{2}, 1\right)$.

The axis of symmetry is the line $x = \dfrac{1}{2}$.
The discriminant is:
$$b^2 - 4ac = 4^2 - 4(-4)(0) = 16 > 0,$$
so the graph has two x-intercepts.
The x-intercepts are found by solving:
$$-4x^2 + 4x = 0 \rightarrow -4x(x-1) = 0 \rightarrow x = 0 \text{ or } x = 1$$
The x-intercepts are 0 and 1.
The y-intercept is $f(0) = -4(0)^2 + 4(0) = 0$.

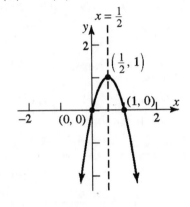

7.　$f(x) = \dfrac{9}{2}x^2 + 3x + 1$

$a = \dfrac{9}{2}, b = 3, c = 1$. Since $a = \dfrac{9}{2} > 0$, the graph opens up.

The x-coordinate of the vertex is $x = \dfrac{-b}{2a} = \dfrac{-3}{2\left(\dfrac{9}{2}\right)} = \dfrac{-3}{9} = -\dfrac{1}{3}$.

The y-coordinate of the vertex is $f\left(\dfrac{-b}{2a}\right) = f\left(-\dfrac{1}{3}\right) = \dfrac{9}{2}\left(-\dfrac{1}{3}\right)^2 + 3\left(-\dfrac{1}{3}\right) + 1 = \dfrac{1}{2} - 1 + 1 = \dfrac{1}{2}$.

Thus, the vertex is $\left(-\dfrac{1}{3}, \dfrac{1}{2}\right)$.

The axis of symmetry is the line $x = -\dfrac{1}{3}$.
The discriminant is:
$$b^2 - 4ac = 3^2 - 4\left(\dfrac{9}{2}\right)(1) = 9 - 18 = -9 < 0,$$

so the graph has no x-intercepts.

The y- intercept is $f(0) = \dfrac{9}{2}(0)^2 + 3(0) + 1 = 1$.

9. $f(x) = 3x^2 + 4x - 1$

$a = 3, b = 4, c = -1.$ Since $a = 3 > 0$, the graph opens up.

The x-coordinate of the vertex is $x = \dfrac{-b}{2a} = \dfrac{-4}{2(3)} = \dfrac{-4}{6} = -\dfrac{2}{3}$.

The y-coordinate of the vertex is $f\!\left(\dfrac{-b}{2a}\right) = f\!\left(-\dfrac{2}{3}\right) = 3\!\left(-\dfrac{2}{3}\right)^2 + 4\!\left(-\dfrac{2}{3}\right) - 1 = \dfrac{4}{3} - \dfrac{8}{3} - 1 = -\dfrac{7}{3}$.

Thus, the vertex is $\left(-\dfrac{2}{3}, -\dfrac{7}{3}\right)$.

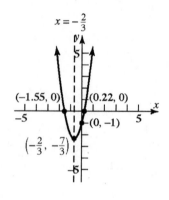

The axis of symmetry is the line $x = -\dfrac{2}{3}$.

The discriminant is:

$b^2 - 4ac = (4)^2 - 4(3)(-1) = 16 + 12 = 28 > 0,$

so the graph has two x-intercepts.

The x-intercepts are found by solving:

$x = \dfrac{-b \pm \sqrt{b^2 - 4ac}}{2a} = \dfrac{-4 \pm \sqrt{28}}{2(3)}$

$= \dfrac{-4 \pm 2\sqrt{7}}{6} = \dfrac{-2 \pm \sqrt{7}}{3} \approx \dfrac{-2 \pm 2.646}{3}$

The x-intercepts are approximately 0.22 and -1.55.

The y-intercept is $f(0) = 3(0)^2 + 4(0) - 1 = -1$.

11. $f(x) = (x + 2)^3$

Using the graph of $y = x^3$, shift the graph horizontally, 2 units to the left.

13. $f(x) = -(x-1)^4$

Using the graph of $y = x^4$, shift the graph horizontally, 1 unit right, and reflect about the x-axis.

15. $f(x) = (x-1)^4 + 2$

Using the graph of $y = x^4$, shift the graph horizontally, 1 unit to the right, and shift vertically 2 units up.

17. $f(x) = 3x^2 - 6x + 4$

$a = 3, b = -6, c = 4$. Since $a = 3 > 0$, the graph opens up, so the vertex is a minimum point. The minimum occurs at $x = \dfrac{-b}{2a} = \dfrac{-(-6)}{2(3)} = \dfrac{6}{6} = 1$.

The minimum value is $f\left(\dfrac{-b}{2a}\right) = f(1) = 3(1)^2 - 6(1) + 4 = 3 - 6 + 4 = 1$.

19. $f(x) = -x^2 + 8x - 4$

$a = -1, b = 8, c = -4$. Since $a = -1 < 0$, the graph opens down, so the vertex is a maximum point. The maximum occurs at $x = \dfrac{-b}{2a} = \dfrac{-8}{2(-1)} = \dfrac{-8}{-2} = 4$.

The maximum value is $f\left(\dfrac{-b}{2a}\right) = f(4) = -(4)^2 + 8(4) - 4 = -16 + 32 - 4 = 12$.

21. $f(x) = -3x^2 + 12x + 4$

$a = -3, b = 12, c = 4$. Since $a = -3 < 0$, the graph opens down, so the vertex is a maximum point. The maximum occurs at $x = \dfrac{-b}{2a} = \dfrac{-12}{2(-3)} = \dfrac{-12}{-6} = 2$.

The maximum value is $f\left(\dfrac{-b}{2a}\right) = f(2) = -3(2)^2 + 12(2) + 4 = -12 + 24 + 4 = 16$.

23. $f(x) = x(x+2)(x+4)$

(a)   x-intercepts: $-4, -2, 0$;   y-intercept: 0

(b)   crosses x axis at $x = -4, -2, 0$

(c) $y = x^3$

(d)   2

(e)

Interval	$(-\infty, -4)$	$(-4, -2)$	$(-2, 0)$	$(0, \infty)$
Number Chosen	-5	-3	-1	1
Value of f	$f(-5) = -15$	$f(-3) = 3$	$f(-1) = -3$	$f(1) = 15$
Location of Graph	Below x-axis	Above x-axis	Below x-axis	Above x-axis
Point on Graph	$(-5, -15)$	$(-3, 3)$	$(-1, -3)$	$(1, 15)$

Graph of f is above the x-axis for $(-4, -2) \cup (0, \infty)$

Graph of f is below the x-axis for $(-\infty, -4) \cup (-2, 0)$

(f)

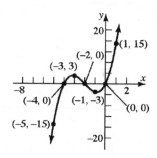

25. $f(x) = (x - 2)^2 (x + 4)$

(a)   x-intercepts: - 4, 2;   y-intercept: 16

(b)   crosses x axis at x = - 4 and touches the x axis at x = 2

(c) $y = x^3$ (d)   2

(e)

Interval	$(-\infty, -4)$	$(-4, 2)$	$(2, \infty)$
Number Chosen	-5	-2	3
Value of f	$f(-5) = -49$	$f(-2) = 32$	$f(3) = 7$
Location of Graph	Below x-axis	Above x-axis	Above x-axis
Point on Graph	$(-5, -49)$	$(-2, 32)$	$(3, 7)$

Graph of f is above the x-axis for $(-4, 2) \cup (2, \infty)$; Graph of f is below the x-axis for $(-\infty, -4)$

(f)

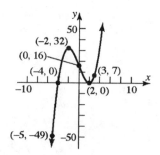

27. $f(x) = -2x^3 + 4x^2 = -2x^2(x-2)$

(a)   x-intercepts: 0, 2;   y-intercept: 0

(b)   crosses x axis at x = 2 and touches the x axis at x = 0

(c) $y = -2x^3$ (d)   2

(e)

Interval	$(-\infty, 0)$	$(0, 2)$	$(2, \infty)$
Number Chosen	-1	1	3
Value of f	$f(-1) = 6$	$f(1) = 2$	$f(3) = -18$
Location of Graph	Above x-axis	Above x-axis	Below x-axis
Point on Graph	(-1, 6)	(1, 2)	(3, -18)

Graph of f is above the x-axis for $(-\infty,0) \cup (0,2)$; Graph of f is below the x-axis for $(2,\infty)$

(f)

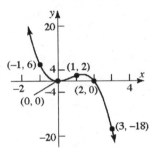

29. $f(x) = (x-1)^2(x+3)(x+1)$

(a)   x-intercepts: - 3, - 1, 1;   y-intercept: 3

(b)   crosses x axis at x = - 3, - 1 and touches x-axis at x = 1

(c) $y = x^4$ (d)   3

(e)

Interval	$(-\infty, -3)$	$(-3, -1)$	$(-1, 1)$	$(1, \infty)$
Number Chosen	-4	-2	0	2
Value of f	$f(-4) = 75$	$f(-2) = -9$	$f(0) = 3$	$f(2) = 15$
Location of Graph	Above x-axis	Below x-axis	Above x-axis	Above x-axis
Point on Graph	(-4, 75)	(-2, -9)	(0, 3)	(2, 15)

Graph of f is above the x-axis for $(-\infty,-3) \cup (-1,1) \cup (1,\infty)$

Graph of f is below the x-axis for $(-3,-1)$

(f)

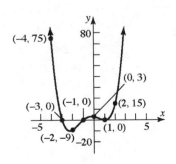

31. $R(x) = \dfrac{2x-6}{x}$ $p(x) = 2x - 6;$ $q(x) = x;$ $n = 1;$ $m = 1$

Step 1:   Domain: $\{x \mid x \neq 0\}$

Step 2:   (a)   The x-intercept is the zero of $p(x)$:   3

(b)   There is no y-intercept because 0 is not in the domain.

Step 3: $R(-x) = \dfrac{2(-x)-6}{-x} = \dfrac{-2x-6}{-x} = \dfrac{2x+6}{x}$; this is neither $R(x)$ nor $-R(x)$, so there

is no symmetry.

Step 4:   The vertical asymptote is the zero of $q(x)$: $x = 0$

Step 5:   Since $n = m$, the line $y = 2$ is the horizontal asymptote.

$R(x)$ does not intersect $y = 2$.

Step 6:

Interval	$(-\infty, 0)$	$(0, 3)$	$(3, \infty)$
Number Chosen	-2	1	4
Value of R	$R(-2) = 5$	$R(1) = -4$	$R(4) = \frac{1}{2}$
Location of Graph	Above x-axis	Below x-axis	Above x-axis
Point on Graph	$(-2, 5)$	$(1, -4)$	$(4, \frac{1}{2})$

Step 7:   Graphing:

33. $H(x) = \dfrac{x+2}{x(x-2)}$ $p(x) = x + 2;$ $q(x) = x(x-2) = x^2 - 2x;$ $n = 1;$ $m = 2$

Step 1:   Domain: $\{x \mid x \neq 0, x \neq 2\}$

Step 2:   (a)   The x-intercept is the zero of $p(x)$: -2

(b)   There is no y-intercept because 0 is not in the domain.

Step 3: $H(-x) = \dfrac{-x+2}{-x(-x-2)} = \dfrac{-x+2}{x^2+2x}$; this is neither $H(x)$ nor $-H(x)$, so there is no

symmetry.

Step 4:   The vertical asymptotes are the zeros of $q(x)$: $x = 0$ and $x = 2$

Step 5:   Since $n < m$, the line $y = 0$ is the horizontal asymptote.

$R(x)$ intersects $y = 0$ at $(-2, 0)$.

Step 6:

Interval	$(-\infty, -2)$	$(-2, 0)$	$(0, 2)$	$(2, \infty)$
Number Chosen	-3	-1	1	3
Value of H	$H(-3) = -\frac{1}{15}$	$H(-1) = \frac{1}{3}$	$H(1) = -3$	$H(3) = \frac{5}{3}$
Location of Graph	Below x-axis	Above x-axis	Below x-axis	Above x-axis
Point on Graph	$(-3, -\frac{1}{15})$	$(-1, \frac{1}{3})$	$(1, -3)$	$(3, \frac{5}{3})$

Step 7:   Graphing:

35. $R(x) = \dfrac{x^2 + x - 6}{x^2 - x - 6} = \dfrac{(x+3)(x-2)}{(x-3)(x+2)}$ $p(x) = x^2 + x - 6;\ q(x) = x^2 - x - 6;\ n = 2;\ m = 2$

Step 1:   Domain: $\{x \mid x \neq -2,\ x \neq 3\}$

Step 2:   (a)   The x-intercepts are the zeros of $p(x)$: -3 and 2

          (b)   The y-intercept is $R(0) = \dfrac{0^2 + 0 - 6}{0^2 - 0 - 6} = \dfrac{-6}{-6} = 1$.

Step 3: $R(-x) = \dfrac{(-x)^2 + (-x) - 6}{(-x)^2 - (-x) - 6} = \dfrac{x^2 - x - 6}{x^2 + x - 6}$; this is neither $R(x)$ nor $-R(x)$, so there

          is no symmetry.

Step 4:   The vertical asymptotes are the zeros of $q(x)$: $x = -2$ and $x = 3$

Step 5:   Since $n = m$, the line $y = 1$ is the horizontal asymptote.

          $R(x)$ intersects $y = 1$ at $(0, 1)$, since:

$$\frac{x^2 + x - 6}{x^2 - x - 6} = 1 \rightarrow x^2 + x - 6 = x^2 - x - 6 \rightarrow 2x = 0 \rightarrow x = 0$$

Step 6:

Interval	$(-\infty, -3)$	$(-3, -2)$	$(-2, 2)$	$(2, 3)$	$(3, \infty)$
Number Chosen	-4	-2.5	0	2.5	4
Value of R	$R(-4) \approx 0.43$	$R(-2.5) \approx -0.82$	$R(0) = 1$	$R(2.5) \approx -1.22$	$R(4) = \frac{7}{3}$
Location of Graph	Above x-axis	Below x-axis	Above x-axis	Below x-axis	Above x-axis
Point on Graph	$(-4, 0.43)$	$(-2.5, -0.82)$	$(0, 1)$	$(2.5, -1.22)$	$(4, \frac{7}{3})$

Step 7:   Graphing:

37. $F(x) = \dfrac{x^3}{x^2 - 4}$ $p(x) = x^3$; $q(x) = x^2 - 4$; $n = 3$; $m = 2$

Step 1: Domain: $\{x \mid x \neq -2, x \neq 2\}$

Step 2: (a) The x-intercept is the zero of $p(x)$: 0.

(b) The y-intercept is $F(0) = \dfrac{0^3}{0^2 - 4} = \dfrac{0}{-4} = 0$.

Step 3: $F(-x) = \dfrac{(-x)^3}{(-x)^2 - 4} = \dfrac{-x^3}{x^2 - 4} = -F(x)$; $F(x)$ is symmetric to the origin.

Step 4: The vertical asymptotes are the zeros of $q(x)$: $x = -2$ and $x = 2$

Step 5: Since $n = m + 1$, there is an oblique asymptote. Dividing:

$$x^2 - 4 \overline{)\, x^3 + 0x^2 + 0x + 0\,}$$

quotient x, with $x^3 - 4x$, remainder $4x$

$$F(x) = x + \dfrac{4x}{x^2 - 4}$$

The oblique asymptote is $y = x$.

Solve to find intersection points:

$$\dfrac{x^3}{x^2 - 4} = x \to x^3 = x^3 - 4x \to 0 = -4x \to x = 0$$

The oblique asymptote intersects $F(x)$ at $(0, 0)$.

Step 6:

Interval	$(-\infty, -2)$	$(-2, 0)$	$(0, 2)$	$(2, \infty)$
Number Chosen	-3	-1	1	3
Value of F	$F(-3) = -\frac{27}{5}$	$F(-1) = \frac{1}{3}$	$F(1) = -\frac{1}{3}$	$F(3) = \frac{27}{5}$
Location of Graph	Below x-axis	Above x-axis	Below x-axis	Above x-axis
Point on Graph	$(-3, -\frac{27}{5})$	$(-1, \frac{1}{3})$	$(1, -\frac{1}{3})$	$(3, \frac{27}{5})$

Step 7: Graphing:

39. $R(x) = \dfrac{2x^4}{(x-1)^2}$ $p(x) = 2x^4$; $q(x) = (x-1)^2$; $n = 4$; $m = 2$

Step 1: Domain: $\{x \mid x \neq 1\}$

Step 2: (a) The x-intercept is the zero of $p(x)$: 0

(b) The y-intercept is $R(0) = \dfrac{2(0)^4}{(0-1)^2} = \dfrac{0}{1} = 0$.

Step 3: $R(-x) = \dfrac{2(-x)^4}{(-x-1)^2} = \dfrac{2x^4}{(x+1)^2}$; this is neither $R(x)$ nor $-R(x)$, so there is no symmetry.

Step 4:   The vertical asymptote is the zero of $q(x)$: $x = 1$

Step 5:   Since $n > m + 1$, there is no horizontal asymptote and no oblique asymptote.

Step 6:

	0	1

Interval	$(-\infty, 0)$	$(0, 1)$	$(1, \infty)$
Number Chosen	-2	$\frac{1}{2}$	2
Value of R	$R(-2) \approx \frac{32}{9}$	$R(\frac{1}{2}) = \frac{1}{2}$	$R(2) = 32$
Location of Graph	Above x-axis	Above x-axis	Above x-axis
Point on Graph	$(-2, \frac{32}{9})$	$(\frac{1}{2}, \frac{1}{2})$	$(2, 32)$

Step 7:   Graphing:

41. $G(x) = \dfrac{x^2 - 4}{x^2 - x - 2} = \dfrac{(x + 2)(x - 2)}{(x - 2)(x + 1)} = \dfrac{x + 2}{x + 1}$ $p(x) = x^2 - 4;$ $q(x) = x^2 - x - 2;$

$n = 2;$ $m = 2$

Step 1:   Domain: $\{x \mid x \neq -1, \, x \neq 2\}$

Step 2:   (a)   The x-intercept is the zero of $p(x)$: -2 (2 is not a zero because reduced form must be used to find the zeros.)

(b)   The y-intercept is $G(0) = \dfrac{0^2 - 4}{0^2 - 0 - 2} = \dfrac{-4}{-2} = 2$.

Step 3: $G(-x) = \dfrac{(-x)^2 - 4}{(-x)^2 - (-x) - 2} = \dfrac{x^2 - 4}{x^2 + x - 2}$; this is neither $G(x)$ nor $-G(x)$, so there is no symmetry.

Step 4:   The vertical asymptote is the zero of $q(x)$: $x = -1$ ($x = 2$ is not a vertical asymptote because reduced form must be used to find the them.)

Step 5:   Since $n = m$, the line $y = 1$ is the horizontal asymptote.

$G(x)$ does not intersect $y = 1$ because $G(x)$ is not defined at $x = 2$.

$$\dfrac{x^2 - 4}{x^2 - x - 2} = 1 \rightarrow x^2 - 4 = x^2 - x - 2 \rightarrow -2 = -x \rightarrow x = 2$$

Step 6:

	-2	-1	2

Interval	$(-\infty, -2)$	$(-2, -1)$	$(-1, 2)$	$(2, \infty)$
Number Chosen	-3	-1.5	0	3
Value of G	$G(-3) = \frac{1}{2}$	$G(-1.5) = -1$	$G(0) = 2$	$G(3) = 1.25$
Location of Graph	Above x-axis	Below x-axis	Above x-axis	Above x-axis
Point on Graph	$(-3, \frac{1}{2})$	$(-1.5, -1)$	$(0, 2)$	$(3, 1.25)$

Step 7:   Graphing:

43. $2x^2 + 5x - 12 < 0$ $f(x) = 2x^2 + 5x - 12$
             $(x+4)(2x-3) < 0$

$x = -4, x = \dfrac{3}{2}$ are the zeros.

Interval	Test Number	$f(x)$	Positive/Negative
$-\infty < x < -4$	-5	13	Positive
$-4 < x < 3/2$	0	-12	Negative
$3/2 < x < \infty$	2	6	Positive

The solution set is $\left\{ x \,\middle|\, -4 < x < \dfrac{3}{2} \right\}$.

45. $\dfrac{6}{x+3} \geq 1$ $f(x) = \dfrac{6}{x+3} - 1$

$\dfrac{6}{x+3} - 1 \geq 0 \rightarrow \dfrac{6 - 1(x+3)}{x+3} \geq 0 \rightarrow \dfrac{-x+3}{x+3} \geq 0$

The zeros and values where the expression is undefined are $x = -3$, and $x = 3$.

Interval	Test Number	$f(x)$	Positive/Negative
$-\infty < x < -3$	-4	-7	Negative
$-3 < x < 3$	0	1	Positive
$3 < x < \infty$	4	$-1/7$	Negative

The solution set is $\left\{ x \,\middle|\, -3 < x \leq 3 \right\}$.

47. $\dfrac{2x-6}{1-x} < 2$ $f(x) = \dfrac{2x-6}{1-x} - 2$

$\dfrac{2x-6}{1-x} - 2 < 0 \rightarrow \dfrac{2x - 6 - 2(1-x)}{1-x} < 0 \rightarrow \dfrac{4x-8}{1-x} < 0$

The zeros and values where the expression is undefined are $x = 1$, and $x = 2$.

Interval	Test Number	$f(x)$	Positive/Negative
$-\infty < x < 1$	0	-8	Negative
$1 < x < 2$	1.5	4	Positive
$2 < x < \infty$	3	-2	Negative

The solution set is $\left\{ x \,\middle|\, x < 1 \text{ or } x > 2 \right\}$.

172

49. $\dfrac{(x-2)(x-1)}{x-3} > 0$ $f(x) = \dfrac{(x-2)(x-1)}{x-3}$

The zeros and values where the expression is undefined are $x = 1$, $x = 2$, and $x = 3$.

Interval	Test Number	$f(x)$	Positive/Negative
$-\infty < x < 1$	0	$-2/3$	Negative
$1 < x < 2$	1.5	$1/6$	Positive
$2 < x < 3$	2.5	$-3/2$	Negative
$3 < x < \infty$	4	6	Positive

The solution set is $\left\{ x \mid 1 < x < 2 \ \text{ or } \ x > 3 \right\}$.

51. $\dfrac{x^2 - 8x + 12}{x^2 - 16} > 0$ $f(x) = \dfrac{x^2 - 8x + 12}{x^2 - 16}$

$\dfrac{(x-2)(x-6)}{(x+4)(x-4)} > 0$

The zeros and values where the expression is undefined are $x = -4$, $x = 2$, $x = 4$, $x = 6$.

Interval	Test Number	$f(x)$	Positive/Negative
$-\infty < x < -4$	-5	$77/9$	Positive
$-4 < x < 2$	0	$-3/4$	Negative
$2 < x < 4$	3	$3/7$	Positive
$4 < x < 6$	5	$-1/3$	Negative
$6 < x < \infty$	7	$5/33$	Positive

The solution set is $\left\{ x \mid x < -4,\ 2 < x < 4,\ x > 6 \right\}$.

53. $f(x) = 12x^8 - x^7 + 8x^4 - 2x^3 + x + 3$

Examining $f(x)$, there are 4 variations in sign; thus, there are 4 or 2 or 0 positive real zeros.

Examining $f(-x) = 12(-x)^8 - (-x)^7 + 8(-x)^4 - 2(-x)^3 + (-x) + 3$

$= 12x^8 + x^7 + 8x^4 + 2x^3 - x + 3$, there are 2 variations in sign; thus, there are 2 or 0 negative real zeros.

55. $f(x) = 12x^8 - x^7 + 6x^4 - x^3 + x - 3$

p must be a factor of -3: $p = \pm 1,\ \pm 3$

q must be a factor of 12: $q = \pm 1,\ \pm 2,\ \pm 3,\ \pm 4,\ \pm 6,\ \pm 12$

The possible rational zeros are: $\dfrac{p}{q} = \pm 1,\ \pm 3,\ \pm \dfrac{1}{2},\ \pm \dfrac{3}{2},\ \pm \dfrac{1}{3},\ \pm \dfrac{1}{4},\ \pm \dfrac{3}{4},\ \pm \dfrac{1}{6},\ \pm \dfrac{1}{12}$

57. $f(x) = x^3 - 3x^2 - 6x + 8$

Step 1: $f(x)$ has at most 3 real zeros.

Step 2:   By Descartes Rule of Signs, there are 2 or 0 positive real zeros.
Also because $f(-x) = (-x)^3 - 3(-x)^2 - 6(-x) + 8 = -x^3 - 3x^2 + 6x + 8$, there is 1 negative real zero.

Step 3:   Possible rational zeros: $p = \pm 1,\ \pm 2,\ \pm 4,\ \pm 8;\quad q = \pm 1;\quad \dfrac{p}{q} = \pm 1,\ \pm 2,\ \pm 4,\ \pm 8$

Step 4:   Using the Bounds on Zeros Theorem:

$$a_2 = -3, \quad a_1 = -6, \quad a_0 = 8$$

$$\text{Max}\left\{1, |8| + |-6| + |-3|\right\} = \text{Max}\left\{1, 17\right\} = 17$$

$$1 + \text{Max}\left\{|8|, |-6|, |-3|\right\} = 1 + 8 = 9$$

The smaller of the two numbers is 9. Thus, every zero of f lies between -9 and 9.

Step 5:   Using synthetic division:

$$
\begin{array}{r|rrrr}
-2) & 1 & -3 & -6 & 8 \\
    &   & -2 & 10 & -8 \\
\hline
    & 1 & -5 & 4 & 0 \\
\end{array}
$$

Since the remainder is 0, $x - (-2) = x + 2$ is a factor. The other factor is the quotient: $x^2 - 5x + 4$.

Thus, $f(x) = (x + 2)(x^2 - 5x + 4) = (x + 2)(x - 1)(x - 4)$.

The zeros are -2, 1, and 4.

59. $f(x) = 4x^3 + 4x^2 - 7x + 2$

Step 1: $f(x)$ has at most 3 real zeros.

Step 2:   By Descartes Rule of Signs, there are 2 or 0 positive real zeros.

$f(-x) = 4(-x)^3 + 4(-x)^2 - 7(-x) + 2 = -4x^3 + 4x^2 + 7x + 2$; thus, there is 1 negative real zero.

Step 3:   Possible rational zeros: $p = \pm 1, \pm 2; \quad q = \pm 1, \pm 2, \pm 4; \quad \dfrac{p}{q} = \pm 1, \pm 2, \pm \dfrac{1}{2}, \pm \dfrac{1}{4}$

Step 4:   Using the Bounds on Zeros Theorem:

$$f(x) = 4\left(x^3 + x^2 - \frac{7}{4}x + \frac{1}{2}\right) \rightarrow a_2 = 1, \quad a_1 = -\frac{7}{4}, \quad a_0 = \frac{1}{2}$$

$$\text{Max}\left\{1, \left|\frac{1}{2}\right| + \left|-\frac{7}{4}\right| + |1|\right\} = \text{Max}\left\{1, \frac{13}{4}\right\} = \frac{13}{4} = 3.25$$

$$1 + \text{Max}\left\{\left|\frac{1}{2}\right|, \left|-\frac{7}{4}\right|, |1|\right\} = 1 + \frac{7}{4} = \frac{11}{4} = 2.75$$

The smaller of the two numbers is 2.75. Thus, every zero of f lies between -2.75 and 2.75.

Step 5:   Using synthetic division:

$$
\begin{array}{r|rrrr}
-2) & 4 & 4 & -7 & 2 \\
    &   & -8 & 8 & -2 \\
\hline
    & 4 & -4 & 1 & 0 \\
\end{array}
$$

Since the remainder is 0, $x - (-2) = x + 2$ is a factor. The other factor is the quotient: $4x^2 - 4x + 1$. Thus,

$$f(x) = (x + 2)(4x^2 - 4x + 1) = (x + 2)(2x - 1)(2x - 1).$$

The zeros are -2 and $\dfrac{1}{2}$ (multiplicity 2).

61. $f(x) = x^4 - 4x^3 + 9x^2 - 20x + 20$

Step 1: $f(x)$ has at most 4 real zeros.

Step 2:   By Descartes Rule of Signs, there are 4 or 2 or 0 positive real zeros.

$$f(-x) = (-x)^4 - 4(-x)^3 + 9(-x)^2 - 20(-x) + 20$$

$$= x^4 + 4x^3 + 9x^2 + 20x + 20;$$

thus, there are no negative real zeros.

Step 3:   Possible rational zeros:
$$p = \pm 1, \pm 2, \pm 4, \pm 5, \pm 10, \pm 20; \quad q = \pm 1;$$
$$\frac{p}{q} = \pm 1, \pm 2, \pm 4, \pm 5, \pm 10, \pm 20$$

Step 4:   Using the Bounds on Zeros Theorem:
$$a_3 = -4, \quad a_2 = 9, \quad a_1 = -20, \quad a_0 = 20$$
$$\text{Max} \left\{1, |20| + |-20| + |9| + |-4|\right\} = \text{Max}\left\{1, 53\right\} = 53$$
$$1 + \text{Max}\left\{|20|, |-20|, |9|, |-4|\right\} = 1 + 20 = 21$$

The smaller of the two numbers is 21. Thus, every zero of f lies between -21 and 21.

Step 5:   Using synthetic division:

```
2)1   -4    9   -20   20        2)1   -2   5   -10
        2   -4    10   -20             2   0    10
     ─────────────────────          ──────────────────
      1   -2    5   -10    0         1    0   5     0
```

Since the remainder is 0, $x - 2$ is a factor twice. The other factor is the quotient: $x^2 + 5$.

Thus, $f(x) = (x - 2)(x - 2)\left(x^2 + 5\right) = (x-2)^2\left(x^2 + 5\right)$.

The zero is 2 (multiplicity 2). ($x^2 + 5 = 0$ has no real solutions.)

63. $2x^4 + 2x^3 - 11x^2 + x - 6 = 0$

The solutions of the equation are the zeros of $f(x) = 2x^4 + 2x^3 - 11x^2 + x - 6$.

Step 1: $f(x)$ has at most 4 real zeros.

Step 2:   By Descartes Rule of Signs, there are 3 or 1 positive real zeros.
$$f(-x) = 2(-x)^4 + 2(-x)^3 - 11(-x)^2 + (-x) - 6 = 2x^4 - 2x^3 - 11x^2 - x - 6;$$
thus, there is 1 negative real zero.

Step 3:   Possible rational zeros:
$$p = \pm 1, \pm 2, \pm 3, \pm 6; \quad q = \pm 1, \pm 2; \quad \frac{p}{q} = \pm 1, \pm 2, \pm 3, \pm 6, \pm \frac{1}{2}, \pm \frac{3}{2}$$

Step 4:   Using the Bounds on Zeros Theorem:
$$f(x) = 2\left(x^4 + x^3 - \frac{11}{2}x^2 + \frac{1}{2}x - 3\right) \rightarrow a_3 = 1, \quad a_2 = -\frac{11}{2}, \quad a_1 = \frac{1}{2}, \quad a_0 = -3$$
$$\text{Max}\left\{1, |-3| + \left|\frac{1}{2}\right| + \left|-\frac{11}{2}\right| + |1|\right\} = \text{Max}\left\{1, 10\right\} = 10$$
$$1 + \text{Max}\left\{|-3|, \left|\frac{1}{2}\right|, \left|-\frac{11}{2}\right|, |1|\right\} = 1 + \frac{11}{2} = \frac{13}{2} = 6.5$$

The smaller of the two numbers is 6.5. Thus, every zero of f lies between -6.5 and 6.5.

Step 5:   Using synthetic division:

```
-3)2    2   -11    1   -6        2)2   -4   1   -2
        -6   12    -3    6            4    0    2
     ──────────────────────        ──────────────────
     2   -4     1   -2    0         2    0    1    0
```

Since the remainder is 0, $x + 3$ and $x - 2$ are factors. The other factor is the quotient: $2x^2 + 1$. The zeros are -3 and 2. ($2x^2 + 1 = 0$ has no real solutions.)

65.　$2x^4 + 7x^3 + x^2 - 7x - 3 = 0$

The solutions of the equation are the zeros of $f(x) = 2x^4 + 7x^3 + x^2 - 7x - 3$.

Step 1:　$f(x)$ has at most 4 real zeros.

Step 2:　By Descartes Rule of Signs, there is 1 positive real zero.
$$f(-x) = 2(-x)^4 + 7(-x)^3 + (-x)^2 - 7(-x) - 3 = 2x^4 - 7x^3 + x^2 + 7x - 3 ;$$
thus, there are 3 or 1 negative real zeros.

Step 3:　Possible rational zeros: $p = \pm 1, \pm 3; \quad q = \pm 1, \pm 2; \quad \dfrac{p}{q} = \pm 1, \pm 3, \pm \dfrac{1}{2}, \pm \dfrac{3}{2}$

Step 4:　Using the Bounds on Zeros Theorem:
$$f(x) = 2\left(x^4 + \frac{7}{2}x^3 + \frac{1}{2}x^2 - \frac{7}{2}x - \frac{3}{2}\right) \rightarrow a_3 = \frac{7}{2}, \ a_2 = \frac{1}{2}, \ a_1 = -\frac{7}{2}, \ a_0 = -\frac{3}{2}$$

$$\text{Max}\left\{1, \left|-\frac{3}{2}\right| + \left|-\frac{7}{2}\right| + \left|\frac{1}{2}\right| + \left|\frac{7}{2}\right|\right\} = \text{Max}\{1, 9\} = 9$$

$$1 + \text{Max}\left\{\left|-\frac{3}{2}\right|, \left|-\frac{7}{2}\right|, \left|\frac{1}{2}\right|, \left|\frac{7}{2}\right|\right\} = 1 + \frac{7}{2} = \frac{9}{2} = 4.5$$

The smaller of the two numbers is 4.5. Thus, every zero of f lies between –4.5 and 4.5.

Step 5:　Using synthetic division:

```
 -3)2   7   1  -7  -3          -1)2    1  -2  -1
      -6  -3   6   3                  -2   1   1
    ─────────────────            ─────────────────
    2   1  -2  -1   0            2   -1  -1   0
```

Since the remainder is 0, $x + 3$ and $x + 1$ are factors. The other factor is the quotient: $2x^2 - x - 1$.

Thus, $f(x) = (x+3)(x+1)\left(2x^2 - x - 1\right) = (x+3)(x+1)(2x+1)(x-1)$.

The zeros are $-3, \ -1, \ -\dfrac{1}{2}$, and 1.

67.　$f(x) = x^3 - 3x^2 - 6x + 8$.

Step 1:　$f(x)$ has at most 3 real zeros.

Step 2:　By Descartes Rule of Signs, there are 2 or no positive real zeros.
$$f(-x) = (-x)^3 - 3(-x)^2 - 6(-x) + 8 = -x^3 - 3x^2 + 6x + 8 ;$$
thus, there is 1 negative real zero.

Step 3:　Possible rational zeros:
$$p = \pm 1, \pm 2, \pm 4, \pm 8; \quad q = \pm 1;$$

$$\frac{p}{q} = \pm 1, \pm 2, \pm 4, \pm 8$$

Step 4:　Using synthetic division:

```
 -1)1  -3  -6   8            → x + 1 is not a factor
      -1   4   2
    ─────────────────
    1  -4  -2  {10}
```

$$1\overline{)\,1\quad -3\quad -6\quad 8\,}\qquad \rightarrow x-1 \text{ is a factor}$$
$$\underline{1\quad -2\quad -8}$$
$$1\quad -2\quad -8\quad 0$$

Thus, $f(x) = (x-1)\left(x^2 - 2x - 8\right) = (x-1)(x-4)(x+2)$.

The zeros are 1, 4, and - 2.

69. $f(x) = 4x^3 + 4x^2 - 7x + 2$.

Step 1: $f(x)$ has at most 3 real zeros.

Step 2:   By Descartes Rule of Signs, there are 2 or no positive real zeros.

$$f(-x) = 4(-x)^3 + 4(-x)^2 - 7(-x) + 2 = -4x^3 + 4x^2 + 7x + 2;$$

thus, there is 1 negative real zero.

Step 3:   Possible rational zeros:
$$p = \pm 1, \pm 2; \quad q = \pm 1, \pm 2, \pm 4;$$

$$\frac{p}{q} = \pm 1, \pm \frac{1}{2}, \pm \frac{1}{4}, \pm 2$$

Step 4:   Using synthetic division:

$$-1\overline{)\,4\quad 4\quad -7\quad 2\,}\qquad \rightarrow x+1 \text{ is not a factor}$$
$$\underline{-4\quad 0\quad 7}$$
$$4\quad 0\quad -7\quad \{9\}$$

$$1\overline{)\,4\quad 4\quad -7\quad 2\,}\qquad \rightarrow x-1 \text{ is not a factor}$$
$$\underline{4\quad 8\quad 1}$$
$$4\quad 8\quad 1\quad \{3\}$$

$$-2\overline{)\,4\quad 4\quad -7\quad 2\,}\qquad \rightarrow x+2 \text{ is a factor}$$
$$\underline{-8\quad 8\quad -2}$$
$$4\quad -4\quad 1\quad 0$$

Thus, $f(x) = (x+2)\left(4x^2 - 4x + 1\right) = (x+2)(4x-2)\left(x - \frac{1}{2}\right) = 4(x+2)\left(x - \frac{1}{2}\right)^2$.

The zeros are -2, and $\frac{1}{2}$ (with multiplicity 2).

71. $f(x) = x^4 - 4x^3 + 9x^2 - 20x + 20$.

Step 1: $f(x)$ has at most 4 real zeros.

Step 2:   By Descartes Rule of Signs, there are 4, 2 or no positive real zeros.

$$f(-x) = (-x)^4 - 4(-x)^3 + 9(-x)^2 - 20(-x) + 20 = x^4 + 4x^3 + 9x^2 + 20x + 20;$$

thus, there are no negative real zeros.

Step 3:   Possible rational zeros:
$$p = \pm 1, \pm 2, \pm 4, \pm 5, \pm 10, \pm 20; \quad q = \pm 1;$$

$$\frac{p}{q} = \pm 1, \pm 2, \pm 4, \pm 5, \pm 10, \pm 20$$

Step 4:   Using synthetic division:

$$
\begin{array}{r|rrrrr}
1) & 1 & -4 & 9 & -20 & 20 \\
   &   & 1 & -3 & 6 & -14 \\
\hline
   & 1 & -3 & 6 & -14 & \{6\}
\end{array}
$$
$\rightarrow x - 1$ is not a factor

$$
\begin{array}{r|rrrrr}
2) & 1 & -4 & 9 & -20 & 20 \\
   &   & 2 & -4 & 10 & -20 \\
\hline
   & 2 & -2 & 5 & -10 & 0
\end{array}
$$
$\rightarrow x - 2$ is a factor

Thus, $f(x) = (x - 2)\left(x^3 - 2x^2 + 5x - 10\right)$. We can factor $x^3 - 2x^2 + 5x - 10$ by grouping
$$x^3 - 2x^2 + 5x - 10 = x^2(x - 2) + 5(x - 2) = (x - 2)(x^2 + 5)$$

$$= (x - 2)\left(x + \sqrt{5}i\right)\left(x - \sqrt{5}i\right) \rightarrow f(x) = (x - 2)^2\left(x + \sqrt{5}i\right)\left(x - \sqrt{5}i\right)$$

The zeros are 2 (multiplicity 2), $\sqrt{5}i$, and $-\sqrt{5}i$.

73. $f(x) = 2x^4 + 2x^3 - 11x^2 + x - 6$.
Step 1: $f(x)$ has at most 4 real zeros.
Step 2:   By Descartes Rule of Signs, there are 3 or 1 positive real zeros.
$$f(-x) = 2(-x)^4 + 2(-x)^3 - 11(-x)^2 + (-x) - 6 = 2x^4 - 2x^3 - 11x^2 - x - 6;$$
thus, there is 1 negative real zero.
Step 3:   Possible rational zeros:
$$p = \pm 1, \pm 2, \pm 3, \pm 6; \quad q = \pm 1, \pm 2;$$

$$\frac{p}{q} = \pm 1, \pm \frac{1}{2}, \pm 2, \pm 3, \pm \frac{3}{2} \pm 6$$

Step 4:   Using synthetic division:

$$
\begin{array}{r|rrrrr}
-1) & 2 & 2 & -11 & 1 & -6 \\
    &   & -2 & 0 & 11 & -12 \\
\hline
    & 2 & 0 & -11 & 12 & \{-18\}
\end{array}
$$
$\rightarrow x + 1$ is not a factor

$$
\begin{array}{r|rrrrr}
1) & 2 & 2 & -11 & 1 & -6 \\
   &   & 2 & 4 & -7 & -6 \\
\hline
   & 2 & 4 & -7 & -6 & \{-12\}
\end{array}
$$
$\rightarrow x - 1$ is not a factor

$$
\begin{array}{r|rrrrr}
-2) & 2 & 2 & -11 & 1 & -6 \\
    &   & -4 & 4 & 14 & -30 \\
\hline
    & 2 & -2 & -7 & 15 & \{-36\}
\end{array}
$$
$\rightarrow x + 2$ is not a factor

$$\begin{array}{r|rrrrr} 2) & 2 & 2 & -11 & 1 & -6 \\ & & 4 & 12 & 2 & 6 \\ \hline & 2 & 6 & 1 & 3 & 0 \end{array} \quad \rightarrow x-2 \text{ is a factor}$$

Thus, $f(x) = (x-2)(2x^3 + 6x^2 + x + 3)$.

We can factor $2x^3 + 6x^2 + x + 3$ by grouping

$$2x^3 + 6x^2 + x + 3 = 2x^2(x+3) + (x+3) = (x+3)(2x^2+1)$$

$$= (x+3)(\sqrt{2}x + i)(\sqrt{2}x - i) \rightarrow f(x) = (x-2)(x+3)(\sqrt{2}x+i)(\sqrt{2}x-i)$$

The zeros are 2, -3, $-\dfrac{\sqrt{2}}{2}i$, and $\dfrac{\sqrt{2}}{2}i$.

75. $f(x) = 2x^4 + 7x^3 + x^2 - 7x - 3$.

   Step 1: $f(x)$ has at most 4 real zeros.

   Step 2:   By Descartes Rule of Signs, there is 1 positive real zero.

   $$f(-x) = 2(-x)^4 + 7(-x)^3 + (-x)^2 - 7(-x) - 3 = 2x^4 - 7x^3 + x^2 + 7x - 3;$$

   thus, there are 3 or 1 negative real zeros.

   Step 3:   Possible rational zeros:

   $$p = \pm 1, \pm 3; \quad q = \pm 1, \pm 2;$$

   $$\frac{p}{q} = \pm 1, \pm \frac{1}{2}, \pm 3, \pm \frac{3}{2}$$

   Step 4:   Using synthetic division:

   $$\begin{array}{r|rrrrr} 1) & 2 & 7 & 1 & -7 & -3 \\ & & 2 & 9 & 10 & 3 \\ \hline & 2 & 9 & 10 & 3 & 0 \end{array} \quad \rightarrow x-1 \text{ is a factor}$$

   Thus, $f(x) = (x-1)(2x^3 + 9x^2 + 10x + 3)$.

   Note: $g(x) = 2x^3 + 9x^2 + 10x + 3$ has the same possible rational roots as f.
   However, since we have already found the only positive real zero for f, so
   we only need to look at the possible negative zeros, $\dfrac{p}{q} = -1, -\dfrac{1}{2}, -3, -\dfrac{3}{2}$.

   $$\begin{array}{r|rrrr} -1) & 2 & 9 & 10 & 3 \\ & & -2 & -7 & -3 \\ \hline & 2 & 7 & 3 & 0 \end{array} \quad \rightarrow x+1 \text{ is a factor}$$

   $$2x^3 + 9x^2 + 10x + 3 = (x+1)(2x^2 + 7x + 3)$$

   $$= (x+1)(2x+1)(x+3) \rightarrow f(x) = (x-1)(x+1)(2x+1)(x+3)$$

   The zeros are 1, -1, $-\dfrac{1}{2}$, and -3.

77. $f(x) = x^3 - x^2 - 4x + 2$
$a_2 = -1, \quad a_1 = -4, \quad a_0 = 2$
Max $\{1, |2| + |-4| + |-1|\}$ = Max $\{1, 7\}$ = 7
$1 +$ Max $\{|2|, |-4|, |-1|\} = 1 + 4 = 5$
The smaller of the two numbers is 5, $\therefore$ every zero of f lies between -5 and 5.

79. $f(x) = 2x^3 - 7x^2 - 10x + 35 = 2\left(x^3 - \frac{7}{2}x^2 - 5x + \frac{35}{2}\right)$

$a_2 = -\frac{7}{2}, \quad a_1 = -5, \quad a_0 = \frac{35}{2}$

Max $\left\{1, \left|\frac{35}{2}\right| + |-5| + \left|-\frac{7}{2}\right|\right\}$ = Max $\{1, 26\}$ = 26

$1 +$ Max $\left\{\left|\frac{35}{2}\right|, |-5|, \left|-\frac{7}{2}\right|\right\} = 1 + \frac{35}{2} = \frac{37}{2} = 18.5$

The smaller of the two numbers is 18.5, $\therefore$ every zero of f lies between -18.5 and 18.5.

81. $f(x) = 3x^3 - x - 1; \quad [0, 1]$
$f(0) = -1 < 0$ and $f(1) = 1 > 0$
Since one is positive and one is negative, there is a zero in the interval.

83. $f(x) = 8x^4 - 4x^3 - 2x - 1; \quad [0, 1]$
$f(0) = -1 < 0$ and $f(1) = 1 > 0$
Since one is positive and one is negative, there is a zero in the interval.

85. $f(x) = x^3 - x - 2$
$f(1) = -2; \quad f(2) = 4$,   so by the Intermediate Value Theorem, f has a zero on the interval [1,2].
Subdivide the interval [1,2] into 10 equal subintervals:

[1,1.1]; [1.1,1.2]; [1.2,1.3]; [1.3,1.4]; [1.4,1.5]; [1.5,1.6]; [1.6,1.7]; [1.7,1.8]; [1.8,1.9]; [1.9,2]

$f(1) = -2; f(1.1) = -1.769$
$f(1.1) = -1.769; f(1.2) = -1.472$
$f(1.2) = -1.472; f(1.3) = -1.103$
$f(1.3) = -1.103; f(1.4) = -0.656$
$f(1.4) = -0.656; f(1.5) = -0.125$
$f(1.5) = -0.125; f(1.6) = 0.496$ so f has a real zero on the interval [1.5,1.6].
Subdivide the interval [1.5,1.6] into 10 equal subintervals:

[1.5,1.51]; [1.51,1.52]; [1.52,1.53]; [1.53,1.54]; [1.54,1.55]; [1.55,1.56];[1.56,1.57]; [1.57,1.58]; [1.58,1.59]; [1.59,1.6]

$f(1.5) = -0.125; f(1.51) = -0.0670$
$f(1.51) = -0.0670; f(1.52) = -0.0082$
$f(1.52) = -0.0082; f(1.53) = 0.0516$ so f has a real zero on the interval $[1.52, 1.53]$, therefore $r = 1.52$, correct to 2 decimal places.

87. $f(x) = 8x^4 - 4x^3 - 2x - 1$

$f(0) = -1; \; f(1) = 1,$ so by the Intermediate Value Theorem, f has a zero on the interval $[0,1]$.

Subdivide the interval $[0,1]$ into 10 equal subintervals:
$[0, 0.1]$; $[0.1, 0.2]$; $[0.2, 0.3]$; $[0.3, 0.4]$; $[0.4, 0.5]$; $[0.5, 0.6]$; $[0.6, 0.7]$; $[0.7, 0.8]$; $[0.8, 0.9]$; $[0.9, 1]$

$f(0) = -1; f(0.1) = -1.2032$
$f(0.1) = -1.2032; f(0.2) = -1.4192$
$f(0.2) = -1.4192; f(0.3) = -1.6432$
$f(0.3) = -1.6432; f(0.4) = -1.8512$
$f(0.4) = -1.8512; f(0.5) = -2$
$f(0.5) = -2; f(0.6) = -2.0272$
$f(0.6) = -2.0272; f(0.7) = -1.8512$
$f(0.7) = -1.8512; f(0.8) = -1.3712$
$f(0.8) = -1.3712; f(0.9) = -0.4672$
$f(0.9) = -0.4672; f(1) = 1$ so f has a real zero on the interval $[0.9, 1]$.

Subdivide the interval $[0.9, 1]$ into 10 equal subintervals:
$[0.9, 0.91]$; $[0.91, 0.92]$; $[0.92, 0.93]$; $[0.93, 0.94]$; $[0.94, 0.95]$; $[0.95, 0.96]$; $[0.96, 0.97]$; $[0.97, 0.98]$; $[0.98, 0.99]$; $[0.99, 1]$

$f(0.9) = -0.4672; f(0.91) = -0.3483$ so f has a real zero on the
$f(0.91) = -0.3483; f(0.92) = -0.2236$ interval $[0.93, 0.94]$, therefore
$f(0.92) = -0.2236; f(0.93) = -0.0930$ $r = 0.93$, correct to 2 decimal
$f(0.93) = -0.0930; f(0.94) = 0.0437$ places.

89.   Since complex zeros appear in conjugate pairs, $4 - i$, the conjugate of $4 + i$, is the remaining zero of f.

91.   Since complex zeros appear in conjugate pairs, $-i$, the conjugate of i, and $1 - i$, the conjugate of $1 + i$, are the remaining zeros of f.

93. $x^2 + x + 1 = 0$

$a = 1, b = 1, c = 1, \quad b^2 - 4ac = 1^2 - 4(1)(1) = 1 - 4 = -3$

$x = \dfrac{-1 \pm \sqrt{-3}}{2(1)} = \dfrac{-1 \pm \sqrt{3}\,i}{2} = -\dfrac{1}{2} \pm \dfrac{\sqrt{3}}{2}i$

The solution set is $\left\{ -\dfrac{1}{2} - \dfrac{\sqrt{3}}{2}i, \ -\dfrac{1}{2} + \dfrac{\sqrt{3}}{2}i \right\}$.

95. $2x^2 + x - 2 = 0$

$a = 2, b = 1, c = -2, \quad b^2 - 4ac = 1^2 - 4(2)(-2) = 1 + 16 = 17$

$x = \dfrac{-1 \pm \sqrt{17}}{2(2)} = \dfrac{-1 \pm \sqrt{17}}{4} \quad \rightarrow \quad$ The solution set is $\left\{ \dfrac{-1 - \sqrt{17}}{4}, \ \dfrac{-1 + \sqrt{17}}{4} \right\}$.

97. $x^2 + 3 = x$

$x^2 - x + 3 = 0$

$a = 1, b = -1, c = 3, \quad b^2 - 4ac = (-1)^2 - 4(1)(3) = 1 - 12 = -11$

$x = \dfrac{-(-1) \pm \sqrt{-11}}{2(1)} = \dfrac{1 \pm \sqrt{11}\,i}{2} = \dfrac{1}{2} \pm \dfrac{\sqrt{11}}{2}i \rightarrow$ The solution set is $\left\{ \dfrac{1}{2} - \dfrac{\sqrt{11}}{2}i, \ \dfrac{1}{2} + \dfrac{\sqrt{11}}{2}i \right\}$.

99. $x(1 - x) = 6$

$-x^2 + x - 6 = 0$

$a = -1, b = 1, c = -6, \quad b^2 - 4ac = 1^2 - 4(-1)(-6) = 1 - 24 = -23$

$x = \dfrac{-1 \pm \sqrt{-23}}{2(-1)} = \dfrac{-1 \pm \sqrt{23}\,i}{-2} = \dfrac{1}{2} \pm \dfrac{\sqrt{23}}{2}i \rightarrow$ The solution set is $\left\{ \dfrac{1}{2} - \dfrac{\sqrt{23}}{2}i, \ \dfrac{1}{2} + \dfrac{\sqrt{23}}{2}i \right\}$.

101. $x^4 + 2x^2 - 8 = 0$

$\left(x^2 + 4 \right)\left(x^2 - 2 \right) = 0 \rightarrow x^2 + 4 = 0 \ $ or $ \ x^2 - 2 = 0$

$\qquad\qquad x^2 = -4 \ \rightarrow x = \pm 2i$

$\qquad\qquad x^2 = 2 \rightarrow x = \pm \sqrt{2}$

The solution set is $\left\{ -2i, \ 2i, \ -\sqrt{2}, \ \sqrt{2} \right\}$.

103. $x^3 - x^2 - 8x + 12 = 0$

The solutions of the equation are the zeros of the function $f(x) = x^3 - x^2 - 8x + 12$.

Step 1: $f(x)$ has 3 complex zeros.

Step 2: By Descartes Rule of Signs, there are 2 or 0 positive real zeros.

$\qquad f(-x) = (-x)^3 - (-x)^2 - 8(-x) + 12 = -x^3 - x^2 + 8x + 12$; thus, there is 1 negative real zero.

Step 3: Possible rational zeros:

$\qquad p = \pm 1, \pm 2, \pm 3, \pm 4, \pm 6, \pm 12; \quad q = \pm 1; \quad \dfrac{p}{q} = \pm 1, \pm 2, \pm 3, \pm 4, \pm 6, \pm 12$

Step 4: Using synthetic division:

$$
\begin{array}{r}
2\overline{)\,1 \quad -1 \quad -8 \quad 12} \\
 2 \quad 2 \quad -12 \\
\hline
1 \quad 1 \quad -6 \quad 0
\end{array}
$$

Since the remainder is 0, $x - 2$ is a factor. The other factor is the quotient:
$x^2 + x - 6 = (x + 3)(x - 2)$.
The complex zeros are $-3, 2$ (multiplicity 2).

105. $3x^4 - 4x^3 + 4x^2 - 4x + 1 = 0$

The solutions of the equation are the zeros of the function $f(x) = 3x^4 - 4x^3 + 4x^2 - 4x + 1$

Step 1: $f(x)$ has 4 complex zeros.

Step 2: By Descartes Rule of Signs, there are 4 or 2 or 0 positive real zeros.
$$f(-x) = 3(-x)^4 - 4(-x)^3 + 4(-x)^2 - 4(-x) + 1 = 3x^4 + 4x^3 + 4x^2 + 4x + 1;$$
thus, there are no negative real zeros.

Step 3: Possible rational zeros: $p = \pm 1$; $q = \pm 1, \pm 3$; $\dfrac{p}{q} = \pm 1, \pm \dfrac{1}{3}$

Step 4: Using synthetic division:

$$
\begin{array}{r}
1\overline{)\,3 \quad -4 \quad 4 \quad -4 \quad 1} \\
 3 \quad -1 \quad 3 \quad -1 \\
\hline
3 \quad -1 \quad 3 \quad -1 \quad 0
\end{array}
\qquad
\begin{array}{r}
\frac{1}{3}\overline{)\,3 \quad -1 \quad 3 \quad -1} \\
\phantom{\frac{1}{3})\,3} 1 \quad 0 \quad 1 \\
\hline
3 \quad 0 \quad 3 \quad 0
\end{array}
$$

Since the remainder is 0, $x - 1$ and $x - \dfrac{1}{3}$ are factors. The other factor is the
quotient: $3x^2 + 3 = 3\left(x^2 + 1\right)$.

Solving $x^2 + 1 = 0 \rightarrow x^2 = -1 \rightarrow x = \pm i$

The complex zeros are $1, \dfrac{1}{3}, -i, i$.

107.

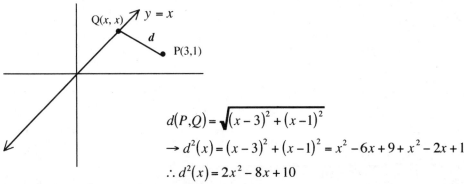

$$d(P,Q) = \sqrt{(x-3)^2 + (x-1)^2}$$
$$\rightarrow d^2(x) = (x-3)^2 + (x-1)^2 = x^2 - 6x + 9 + x^2 - 2x + 1$$
$$\therefore d^2(x) = 2x^2 - 8x + 10$$

Since $d^2(x) = 2x^2 - 8x + 10$ is a quadratic function with $a = 2 > 0$, the vertex corresponds
to the minimum value for the function.

The vertex occurs at $x = -\dfrac{b}{2a} = -\dfrac{-8}{2(2)} = 2$. Therefore the point Q on the line $y = x$ will be
closest to the point $P = (3,1)$ when $Q = (2,2)$.

109. Since there are 200 feet of border, we know that $2x + 2y = 200$.
The area is to be maximized, so $A = x \cdot y$.
Solving the perimeter formula for y:

$$2x + 2y = 200 \rightarrow 2y = 200 - 2x \rightarrow y = 100 - x$$

The area function is: $A(x) = x(100 - x) = -x^2 + 100x$

The maximum value occurs at the vertex: $x = \dfrac{-b}{2a} = \dfrac{-100}{2(-1)} = \dfrac{-100}{-2} = 50$

The pond should be 50 feet by 50 feet for maximum area.

111. Consider the diagram

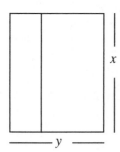

Total amount of fence $= 3x + 2y = 10000$

$$\therefore y = \frac{10000 - 3x}{2} = 5000 - \frac{3}{2}x$$

Total enclosed area $= (x)(y) = (x)\left(5000 - \frac{3}{2}x\right)$

$$\therefore A(x) = 5000x - \frac{3}{2}x^2 = -\frac{3}{2}x^2 + 5000x \text{ is a quadratic function with } a = -\frac{3}{2} < 0.$$

So the vertex corresponds to the maximum value for this function.
The vertex occurs when

$$x = -\frac{b}{2a} = -\frac{5000}{2\left(-\frac{3}{2}\right)} = \frac{5000}{3} \rightarrow \text{the maximum area is } A\left(\frac{5000}{3}\right) = -\frac{3}{2}\left(\frac{5000}{3}\right)^2 + 5000\left(\frac{5000}{3}\right)$$

$$= -\frac{3}{2}\left(\frac{25000000}{9}\right) + \frac{25000000}{3} = -\frac{12500000}{3} + \frac{25000000}{3}$$

$$= \frac{12500000}{3} \approx 4166666.67 \text{ square meters}$$

113. Consider the diagram

d = diameter of the semicircles = width of the rectangle
x = length of the rectangle
$\therefore$ outside dimension length $= 2x + 2(\text{circumference of a semicircle})$

$$= 2x + \text{circumference of a circle} = 2x + \pi d = 100 \rightarrow x = \frac{100 - \pi d}{2} = 50 - \frac{1}{2}\pi d$$

Total enclosed area = (area of the rectangle) + 2(area of a semicircle)

= (area of the rectangle) + area of a circle

$$= (x)(d) + \pi r^2 = (x)(d) + \pi\left(\frac{d}{2}\right)^2 = \left(50 - \frac{1}{2}\pi d\right)(d) + \pi\left(\frac{d}{2}\right)^2$$

$$= 50d - \frac{1}{2}\pi d^2 + \frac{1}{4}\pi d^2 = 50d - \frac{1}{4}\pi d^2 = -\frac{1}{4}\pi d^2 + 50d$$

$\therefore A(d) = -\frac{1}{4}\pi d^2 + 50d$ is a quadratic function with $a = -\frac{1}{4}\pi < 0$. Therefore the vertex corresponds to the maximum value for the function.

The vertex occurs when $x = -\dfrac{b}{2a} = -\dfrac{50}{2\left(-\dfrac{1}{4}\pi\right)} = \dfrac{100}{\pi}$

→ the maximum area is $A\left(\dfrac{100}{\pi}\right) = -\dfrac{1}{4}\pi\left(\dfrac{100}{\pi}\right)^2 + 50\left(\dfrac{100}{\pi}\right) \approx 795.78$ square meters

115. (a)   Graphing:

$A(t) = -212t^3 + 2429t^2 + 59569t + 130003$

(b) $A(11) = -212(11)^3 + 2429(11)^2 + 59569(11) + 130003 = 796999$ cases

(c) and (d)   Graphing the cubic function of best fit:

(e)   answers will vary

117. Answers will vary, one example is $p(x) = -5\left(x^2 + 1\right)(x + 1)^3\left(x - \dfrac{3}{5}\right)$

119. (a)   The degree is even.   (b)   The leading coefficient is positive.
   (c)   The function is even - it is symmetric to the y-axis.
   (d) x^2 is a factor because the curve touches the x-axis at the origin.
   (e)   The minimum degree is 8.
   (f)   Answers will vary, 5 possibilities are:

$$p_1(x) = x^2(x + 3)(x + 2)(x + 1)(x - 1)(x - 2)(x - 3)$$
$$p_2(x) = x^4(x + 3)(x + 2)(x + 1)(x - 1)(x - 2)(x - 3)$$
$$p_3(x) = x^6(x + 3)(x + 2)(x + 1)(x - 1)(x - 2)(x - 3)$$
$$p_4(x) = x^8(x + 3)(x + 2)(x + 1)(x - 1)(x - 2)(x - 3)$$
$$p_5(x) = x^{10}(x + 3)(x + 2)(x + 1)(x - 1)(x - 2)(x - 3)$$

Exponential and Logarithmic Functions

4.1   One-to-One Functions; Inverse Functions

1.   (a)
Domain   Range

(b)   Inverse is a function.

3.   (a)
Domain   Range

(b)   Inverse is not a function since $200 corresponds to two elements in the range.

5.   (a) $\{(6,2), (6,-3), (9,4), (10,1)\}$

(b)   Inverse is not a function since 6 corresponds to 2 and –3.

7.   (a) $\{(0,0), (1,1), (16,2), (81,3)\}$

(b)   Inverse is a function.

9.   Every horizontal line intersects the graph of f at exactly one point. One-to-One.

11.   There are horizontal lines that intersect the graph of f at more than one point. Not One-to-One.

13.   Every horizontal line intersects the graph of f at exactly one point. One-to-One.

15.   Graphing the inverse:

17.   Graphing the inverse:

19. Graphing the inverse:

21. $f(x) = 3x + 4, \qquad g(x) = \dfrac{1}{3}(x - 4)$

$f(g(x)) = f\left(\dfrac{1}{3}(x - 4)\right) = 3\left(\dfrac{1}{3}(x - 4)\right) + 4 = (x - 4) + 4 = x$

$g(f(x)) = g(3x + 4) = \dfrac{1}{3}\left((3x + 4) - 4\right) = \dfrac{1}{3}(3x) = x$

23. $f(x) = 4x - 8, \qquad g(x) = \dfrac{x}{4} + 2$

$f(g(x)) = f\left(\dfrac{x}{4} + 2\right) = 4\left(\dfrac{x}{4} + 2\right) - 8 = (x + 8) - 8 = x$

$g(f(x)) = g(4x - 8) = \dfrac{4x - 8}{4} + 2 = x - 2 + 2 = x$

25. $f(x) = x^3 - 8, \qquad g(x) = \sqrt[3]{x + 8}$

$f(g(x)) = f\left(\sqrt[3]{x + 8}\right) = \left(\sqrt[3]{x + 8}\right)^3 - 8 = (x + 8) - 8 = x$

$g(f(x)) = g(x^3 - 8) = \sqrt[3]{(x^3 - 8) + 8} = \sqrt[3]{x^3} = x$

27. $f(x) = \dfrac{1}{x}, \qquad g(x) = \dfrac{1}{x}$

$f(g(x)) = f\left(\dfrac{1}{x}\right) = \dfrac{1}{\left(\dfrac{1}{x}\right)} = x$

$g(f(x)) = g\left(\dfrac{1}{x}\right) = \dfrac{1}{\left(\dfrac{1}{x}\right)} = x$

29. $f(x) = \dfrac{2x+3}{x+4}$, $g(x) = \dfrac{4x-3}{2-x}$

$$f(g(x)) = f\left(\dfrac{4x-3}{2-x}\right) = \dfrac{2\left(\dfrac{4x-3}{2-x}\right)+3}{\left(\dfrac{4x-3}{2-x}\right)+4} = \dfrac{\left(\dfrac{8x-6+6-3x}{2-x}\right)}{\left(\dfrac{4x-3+8-4x}{2-x}\right)} = \dfrac{\left(\dfrac{5x}{2-x}\right)}{\left(\dfrac{5}{2-x}\right)}$$

$$= \dfrac{5x}{2-x} \cdot \dfrac{2-x}{5} = x$$

$$g(f(x)) = g\left(\dfrac{2x+3}{x+4}\right) = \dfrac{4\left(\dfrac{2x+3}{x+4}\right)-3}{2-\left(\dfrac{2x+3}{x+4}\right)} = \dfrac{\left(\dfrac{8x+12-3x-12}{x+4}\right)}{\left(\dfrac{2x+8-2x-3}{x+4}\right)} = \dfrac{\left(\dfrac{5x}{x+4}\right)}{\left(\dfrac{5}{x+4}\right)}$$

$$= \dfrac{5x}{x+4} \cdot \dfrac{x+4}{5} = x$$

31. $f(x) = 3x$

$y = 3x$

$x = 3y$ Inverse

$y = \dfrac{x}{3} \rightarrow f^{-1}(x) = \dfrac{x}{3}$

Verify: $f\left(f^{-1}(x)\right) = f\left(\dfrac{x}{3}\right) = 3\left(\dfrac{x}{3}\right) = x$

$f^{-1}\left(f(x)\right) = f^{-1}(3x) = \dfrac{3x}{3} = x$

Domain of f = range of f^{-1} = $(-\infty, \infty)$

Range of f = domain of f^{-1} = $(-\infty, \infty)$

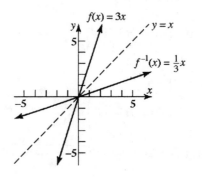

33. $f(x) = 4x + 2$

$y = 4x + 2$

$x = 4y + 2$ Inverse

$4y = x - 2 \rightarrow y = \dfrac{x-2}{4} \rightarrow f^{-1}(x) = \dfrac{x-2}{4}$

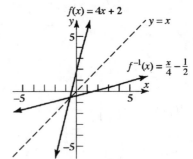

Verify: $f\left(f^{-1}(x)\right) = f\left(\dfrac{x-2}{4}\right) = 4\left(\dfrac{x-2}{4}\right) + 2 = x - 2 + 2 = x$

$f^{-1}\left(f(x)\right) = f^{-1}(4x+2) = \dfrac{(4x+2)-2}{4} = \dfrac{4x}{4} = x$

Domain of f = range of f^{-1} = $(-\infty, \infty)$

Range of f = domain of f^{-1} = $(-\infty, \infty)$

35. $f(x) = x^3 - 1$

$y = x^3 - 1$

$x = y^3 - 1$ Inverse

$y^3 = x + 1 \rightarrow y = \sqrt[3]{x+1} \rightarrow f^{-1}(x) = \sqrt[3]{x+1}$

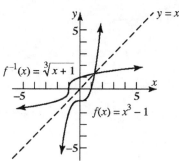

Verify: $f(f^{-1}(x)) = f(\sqrt[3]{x+1}) = (\sqrt[3]{x+1})^3 - 1 = x + 1 - 1 = x$

$f^{-1}(f(x)) = f^{-1}(x^3 - 1) = \sqrt[3]{(x^3-1)+1} = \sqrt[3]{x^3} = x$

Domain of f = range of $f^{-1} = (-\infty, \infty)$

Range of f = domain of $f^{-1} = (-\infty, \infty)$

37. $f(x) = x^2 + 4, \ x \geq 0$

$y = x^2 + 4 \ \ x \geq 0$

$x = y^2 + 4 \ \ y \geq 0$ Inverse

$y^2 = x - 4 \ \ y \geq 0$

$y = \sqrt{x-4} \rightarrow f^{-1}(x) = \sqrt{x-4}$

Verify:

$f(f^{-1}(x)) = f(\sqrt{x-4}) =$

$(\sqrt{x-4})^2 + 4 = x - 4 + 4 = x$

$f^{-1}(f(x)) = f^{-1}(x^2 + 4) = \sqrt{(x^2+4)-4}$

$= \sqrt{x^2} = |x| = x, x \geq 0$

Domain of f =

range of $f^{-1} = [0, \infty)$

Range of f =

domain of $f^{-1} = [4, \infty)$

39. $f(x) = \dfrac{4}{x}$

$y = \dfrac{4}{x}$

$x = \dfrac{4}{y}$ Inverse

$xy = 4 \rightarrow y = \dfrac{4}{x} \rightarrow f^{-1}(x) = \dfrac{4}{x}$

Verify: $f(f^{-1}(x)) = f\left(\dfrac{4}{x}\right) = \dfrac{4}{\left(\dfrac{4}{x}\right)} = 4 \cdot \dfrac{x}{4} = x$; $f^{-1}(f(x)) = f^{-1}\left(\dfrac{4}{x}\right) = \dfrac{4}{\left(\dfrac{4}{x}\right)} = 4 \cdot \dfrac{x}{4} = x$

Domain of f = range of f^{-1} = all real numbers except 0

Range of f = domain of f^{-1} = all real numbers except 0

41. $f(x) = \dfrac{1}{x-2}$

$y = \dfrac{1}{x-2} \rightarrow x = \dfrac{1}{y-2}$ Inverse

$x(y-2) = 1 \rightarrow xy - 2x = 1 \rightarrow xy = 2x + 1$

$y = \dfrac{2x+1}{x} \rightarrow f^{-1}(x) = \dfrac{2x+1}{x}$

Verify: $f\left(f^{-1}(x)\right) = f\left(\dfrac{2x+1}{x}\right) = \dfrac{1}{\left(\dfrac{2x+1}{x} - 2\right)} = \dfrac{1}{\left(\dfrac{2x+1-2x}{x}\right)} = \dfrac{1}{\left(\dfrac{1}{x}\right)} = x$

$f^{-1}(f(x)) = f^{-1}\left(\dfrac{1}{x-2}\right) = \dfrac{2\left(\dfrac{1}{x-2}\right)+1}{\left(\dfrac{1}{x-2}\right)} = \dfrac{\left(\dfrac{2+x-2}{x-2}\right)}{\left(\dfrac{1}{x-2}\right)} = \dfrac{x}{x-2} \cdot \dfrac{x-2}{1} = x$

Domain of f = range of f^{-1} = all real numbers except 2
Range of f = domain of f^{-1} = all real numbers except 0

43. $f(x) = \dfrac{2}{3+x}$

$y = \dfrac{2}{3+x}$

$x = \dfrac{2}{3+y}$ Inverse

$x(3+y) = 2$

$3x + xy = 2 \rightarrow xy = 2 - 3x$

$y = \dfrac{2-3x}{x} \rightarrow f^{-1}(x) = \dfrac{2-3x}{x}$

Domain of f =
range of f^{-1} = all real numbers except -3

Range of f =
domain of f^{-1} = all real numbers except 0

Verify: $f\left(f^{-1}(x)\right) = f\left(\dfrac{2-3x}{x}\right) = \dfrac{2}{3+\left(\dfrac{2-3x}{x}\right)} = \dfrac{2}{\left(\dfrac{3x+2-3x}{x}\right)} = \dfrac{2}{\left(\dfrac{2}{x}\right)} = 2 \cdot \dfrac{x}{2} = x$

$f^{-1}(f(x)) = f^{-1}\left(\dfrac{2}{3+x}\right) = \dfrac{2-3\left(\dfrac{2}{3+x}\right)}{\left(\dfrac{2}{3+x}\right)} = \dfrac{\left(\dfrac{6+2x-6}{3+x}\right)}{\left(\dfrac{2}{3+x}\right)} = \dfrac{2x}{3+x} \cdot \dfrac{3+x}{2} = x$

45. $f(x) = \dfrac{3x}{x+2}$

Domain of f =
range of f^{-1} = all real numbers except -2

$y = \dfrac{3x}{x+2}$

$x = \dfrac{3y}{y+2}$ Inverse

Range of f =
domain of f^{-1} = all real numbers except 3

$x(y+2) = 3y$

$xy + 2x = 3y$

$2x = 3y - xy$

$2x = y(3-x)$

$\dfrac{2x}{3-x} = y \rightarrow f^{-1}(x) = \dfrac{2x}{3-x}$

Verify: $f\left(f^{-1}(x)\right) = f\left(\dfrac{2x}{3-x}\right) = \dfrac{3\left(\dfrac{2x}{3-x}\right)}{\left(\dfrac{2x}{3-x}\right)+2} = \dfrac{\left(\dfrac{6x}{3-x}\right)}{\left(\dfrac{2x+2(3-x)}{3-x}\right)} = \dfrac{\left(\dfrac{6x}{3-x}\right)}{\left(\dfrac{2x+6-2x}{3-x}\right)}$

$= \dfrac{\left(\dfrac{6x}{3-x}\right)}{\left(\dfrac{6}{3-x}\right)} = \left(\dfrac{6x}{3-x}\right) \cdot \left(\dfrac{3-x}{6}\right) = x$

$f^{-1}(f(x)) = f^{-1}\left(\dfrac{3x}{x+2}\right) = \dfrac{2\left(\dfrac{3x}{x+2}\right)}{3-\left(\dfrac{3x}{x+2}\right)} = \dfrac{\left(\dfrac{6x}{x+2}\right)}{\left(\dfrac{3(x+2)-3x}{x+2}\right)} = \dfrac{\left(\dfrac{6x}{x+2}\right)}{\left(\dfrac{3x+6-3x}{x+2}\right)}$

$= \dfrac{\left(\dfrac{6x}{x+2}\right)}{\left(\dfrac{6}{x+2}\right)} = \left(\dfrac{6x}{x+2}\right) \cdot \left(\dfrac{x+2}{6}\right) = x$

47. $f(x) = \dfrac{2x}{3x-1}$

   $y = \dfrac{2x}{3x-1}$

   $x = \dfrac{2y}{3y-1}$ Inverse

   $x(3y-1) = 2y$

   $3xy - x = 2y$

   $3xy - 2y = x$

   $y(3x-2) = x \rightarrow y = \dfrac{x}{3x-2}$

   $f^{-1}(x) = \dfrac{x}{3x-2}$

Domain of f =

range of f^{-1} = all real numbers except $\dfrac{1}{3}$

Range of f =

domain of f^{-1} = all real numbers except $\dfrac{2}{3}$

Verify:

$$f\left(f^{-1}(x)\right) = f\left(\dfrac{x}{3x-2}\right) = \dfrac{2\left(\dfrac{x}{3x-2}\right)}{3\left(\dfrac{x}{3x-2}\right)-1} = \dfrac{\left(\dfrac{2x}{3x-2}\right)}{\left(\dfrac{3x-1(3x-2)}{3x-2}\right)} = \dfrac{\left(\dfrac{2x}{3x-2}\right)}{\left(\dfrac{3x-3x+2}{3x-2}\right)} = \dfrac{2x}{3x-2}\cdot\dfrac{3x-2}{2} = x$$

$$f^{-1}\left(f(x)\right) = f^{-1}\left(\dfrac{2x}{3x-1}\right) = \dfrac{\left(\dfrac{2x}{3x-1}\right)}{3\left(\dfrac{2x}{3x-1}\right)-2} = \dfrac{\left(\dfrac{2x}{3x-1}\right)}{\left(\dfrac{6x-2(3x-1)}{3x-1}\right)} = \dfrac{\left(\dfrac{2x}{3x-1}\right)}{\left(\dfrac{6x-6x+2}{3x-1}\right)} = \dfrac{2x}{3x-1}\cdot\dfrac{3x-1}{2} = x$$

49. $f(x) = \dfrac{3x+4}{2x-3}$

   $y = \dfrac{3x+4}{2x-3}$

   $x = \dfrac{3y+4}{2y-3}$ Inverse

   $x(2y-3) = 3y+4$

   $2xy - 3x = 3y+4$

   $2xy - 3y = 3x+4$

   $y(2x-3) = 3x+4$

   $y = \dfrac{3x+4}{2x-3}$

   $f^{-1}(x) = \dfrac{3x+4}{2x-3}$

Domain of f =

range of f^{-1} = all real numbers except $\dfrac{3}{2}$

Range of f =

domain of f^{-1} = all real numbers except $\dfrac{3}{2}$

$$f\left(f^{-1}(x)\right) = f\left(\dfrac{3x+4}{2x-3}\right) = \dfrac{3\left(\dfrac{3x+4}{2x-3}\right)+4}{2\left(\dfrac{3x+4}{2x-3}\right)-3} = \dfrac{\left(\dfrac{9x+12+8x-12}{2x-3}\right)}{\left(\dfrac{6x+8-6x+9}{2x-3}\right)} = \dfrac{\left(\dfrac{17x}{2x-3}\right)}{\left(\dfrac{17}{2x-3}\right)}$$

Verify:

$$= \dfrac{17x}{2x-3}\cdot\dfrac{2x-3}{17} = x$$

$$f^{-1}(f(x)) = f^{-1}\left(\frac{3x+4}{2x-3}\right) = \frac{3\left(\frac{3x+4}{2x-3}\right)+4}{2\left(\frac{3x+4}{2x-3}\right)-3} = \frac{\left(\frac{9x+12+8x-12}{2x-3}\right)}{\left(\frac{6x+8-6x+9}{2x-3}\right)} = \frac{\left(\frac{17x}{2x-3}\right)}{\left(\frac{17}{2x-3}\right)}$$

$$= \frac{17x}{2x-3} \cdot \frac{2x-3}{17} = x$$

51. $f(x) = \dfrac{2x+3}{x+2}$

$y = \dfrac{2x+3}{x+2}$

$x = \dfrac{2y+3}{y+2}$ Inverse

$x(y+2) = 2y+3$

$xy + 2x = 2y + 3$

$xy - 2y = -2x + 3$

$y(x-2) = -2x+3$

$y = \dfrac{-2x+3}{x-2}$

$f^{-1}(x) = \dfrac{-2x+3}{x-2}$

Domain of f =
range of f^{-1} = all real numbers except -2

Range of f =
domain of f^{-1} = all real numbers except 2

Verify:

$$f\left(f^{-1}(x)\right) = f\left(\frac{-2x+3}{x-2}\right) = \frac{2\left(\frac{-2x+3}{x-2}\right)+3}{\left(\frac{-2x+3}{x-2}\right)+2} = \frac{\left(\frac{-4x+6+3x-6}{x-2}\right)}{\left(\frac{-2x+3+2x-4}{x-2}\right)} = \frac{\left(\frac{-x}{x-2}\right)}{\left(\frac{-1}{x-2}\right)}$$

$$= \frac{-x}{x-2} \cdot \frac{x-2}{-1} = x$$

$$f^{-1}(f(x)) = f^{-1}\left(\frac{2x+3}{x+2}\right) = \frac{-2\left(\frac{2x+3}{x+2}\right)+3}{\left(\frac{2x+3}{x+2}\right)-2} = \frac{\left(\frac{-4x-6+3x+6}{x+2}\right)}{\left(\frac{2x+3-2x-4}{x+2}\right)} = \frac{\left(\frac{-x}{x+2}\right)}{\left(\frac{-1}{x+2}\right)}$$

$$= \frac{-x}{x+2} \cdot \frac{x+2}{-1} = x$$

53. $f(x) = \dfrac{x^2 - 4}{2x^2}, x > 0$

$y = \dfrac{x^2 - 4}{2x^2}, x > 0$

$x = \dfrac{y^2 - 4}{2y^2}, y > 0$ Inverse

$2xy^2 = y^2 - 4, y > 0$

$2xy^2 - y^2 = -4, y > 0$

$y^2(2x - 1) = -4, y > 0$

$y^2 = \dfrac{-4}{2x-1} \rightarrow y = \sqrt{\dfrac{-4}{2x-1}}$

$f^{-1}(x) = \sqrt{\dfrac{-4}{2x-1}}, x > 0$

Domain of f =
range of f^{-1} = $(0, \infty)$

Range of f =
domain of = $f^{-1} = \left(0, \dfrac{1}{2}\right)$

Verify:

$$f\left(f^{-1}(x)\right) = f\left(\sqrt{\dfrac{-4}{2x-1}}\right) = \dfrac{\left(\sqrt{\dfrac{-4}{2x-1}}\right)^2 - 4}{2\left(\sqrt{\dfrac{-4}{2x-1}}\right)^2} = \dfrac{\left(\dfrac{-4}{2x-1} - 4\right)}{2\left(\dfrac{-4}{2x-1}\right)} = \dfrac{\left(\dfrac{-4 - 4(2x-1)}{2x-1}\right)}{\left(\dfrac{-8}{2x-1}\right)} = \dfrac{\left(\dfrac{-4 - 8x + 4}{2x-1}\right)}{\left(\dfrac{-8}{2x-1}\right)}$$

$$= \dfrac{\left(\dfrac{-8x}{2x-1}\right)}{\left(\dfrac{-8}{2x-1}\right)} = \left(\dfrac{-8x}{2x-1}\right) \cdot \left(\dfrac{2x-1}{-8}\right) = x$$

$$f^{-1}\left(f(x)\right) = f^{-1}\left(\dfrac{x^2-4}{2x^2}\right) = \sqrt{\dfrac{-4}{2\left(\dfrac{x^2-4}{2x^2}\right) - 1}} = \sqrt{\dfrac{-4}{\left(\dfrac{x^2-4}{x^2}\right) - 1}} = \sqrt{\dfrac{-4}{\left(\dfrac{x^2-4-x^2}{x^2}\right)}} = \sqrt{\dfrac{-4}{\left(\dfrac{-4}{x^2}\right)}}$$

$$= \sqrt{\dfrac{-4}{1} \cdot \dfrac{x^2}{-4}} = \sqrt{x^2} = |x| = x \text{ when } x > 0$$

55. $f(x) = mx + b, \quad m \neq 0$

$y = mx + b$

$x = my + b$ Inverse

$x - b = my$

$y = \dfrac{x - b}{m}$

$f^{-1}(x) = \dfrac{x - b}{m}, \quad m \neq 0$

57. f^{-1} lies in quadrant I. Whenever (a,b) is on f, then (b,a) is on f^{-1}. Since both coordinates of (a,b) are positive, both coordinates of (b,a) are positive and it is in quadrant I.

59. $f(x) = |x|, x \geq 0$ is one-to-one. Thus, $f(x) = x, x \geq 0$ and $f^{-1}(x) = x, x \geq 0$.

61. $f(x) = \dfrac{9}{5}x + 32 \qquad g(x) = \dfrac{5}{9}(x - 32)$

$$f(g(x)) = f\left(\frac{5}{9}(x - 32)\right) = \frac{9}{5}\left(\frac{5}{9}(x - 32)\right) + 32 = x - 32 + 32 = x$$

$$g(f(x)) = g\left(\frac{9}{5}x + 32\right) = \frac{5}{9}\left(\frac{9}{5}x + 32 - 32\right) = \frac{5}{9}\left(\frac{9}{5}x\right) = x$$

63. $T(l) = 2\pi\sqrt{\dfrac{l}{g}}, \quad g \approx 32.2$

$$T = 2\pi\sqrt{\frac{l}{g}} \quad \rightarrow \quad \frac{T}{2\pi} = \sqrt{\frac{l}{g}}$$

$$\rightarrow \quad \frac{T^2}{4\pi^2} = \frac{l}{g} \quad \rightarrow \quad l = \frac{gT^2}{4\pi^2}$$

$$l(T) = \frac{gT^2}{4\pi^2}$$

65.   An even function cannot be one-to-one. When a function is even, $f(-x) = f(x)$. Thus, both x and $-x$ produce the same y value.

67.   If the graph of a function and its inverse intersect, they must intersect at a point on the line $y = x$. However, the graphs do not have to intersect.

69. Answers will vary.

195

Chapter 4

Exponential and Logarithmic Functions

4.2   Exponential Functions

1. (a) $3^{2.2} \approx 11.212$ (b) $3^{2.23} \approx 11.587$ (c) $3^{2.236} \approx 11.664$ (d) $3^{\sqrt{5}} \approx 11.665$

3. (a) $2^{3.14} \approx 8.815$ (b) $2^{3.141} \approx 8.821$ (c) $2^{3.1415} \approx 8.824$ (d) $2^{\pi} \approx 8.825$

5. (a) $3.1^{2.7} \approx 21.217$ (b) $3.14^{2.71} \approx 22.217$
   (c) $3.141^{2.718} \approx 22.440$ (d) $\pi^{e} \approx 22.459$

7. $e^{1.2} \approx 3.320$ 9. $e^{-0.85} \approx 0.427$

11. B   13. D   15. A   17. E

19. $f(x) = 2^x + 1$
    Using the graph of $y = 2^x$, shift the graph
    up 1 unit.
    Domain: $(-\infty, \infty)$
    Range: $(1, \infty)$
    Horizontal Asymptote: $y = 1$

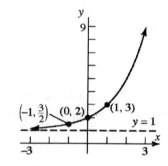

21. $f(x) = 3^{-x} - 2$
    Using the graph of $y = 3^x$, reflect the
    graph about the y-axis, and shift down 2
    units.
    Domain: $(-\infty, \infty)$
    Range: $(-2, \infty)$
    Horizontal Asymptote: $y = -2$

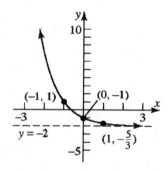

23. $f(x) = 2 + 3\left(4^x\right)$
Using the graph of $y = 4^x$, stretch the graph vertically by a factor of 3, and shift up 2 units.
Domain: $(-\infty, \infty)$
Range: $(2, \infty)$
Horizontal Asymptote: $y = 2$

25. $f(x) = 2 + 3^{x/2}$
Using the graph of $y = 3^x$, stretch the graph horizontally by a factor of 2, and shift up 2 units.
Domain: $(-\infty, \infty)$
Range: $(2, \infty)$
Horizontal Asymptote: $y = 2$

27. $f(x) = e^{-x}$
Using the graph of $y = e^x$, reflect the graph about the y-axis.
Domain: $(-\infty, \infty)$
Range: $(0, \infty)$
Horizontal Asymptote: $y = 0$

29. $f(x) = e^{x+2}$
Using the graph of $y = e^x$, shift the graph 2 units to the left.
Domain: $(-\infty, \infty)$
Range: $(0, \infty)$
Horizontal Asymptote: $y = 0$

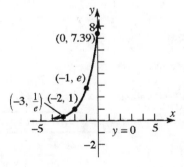

31. $f(x) = 5 - e^{-x}$

Using the graph of $y = e^x$, reflect the
graph about the y-axis, reflect about the
x-axis, and shift up 5 units.
Domain: $(-\infty, \infty)$
Range: $(-\infty, 5)$
Horizontal Asymptote: $y = 5$

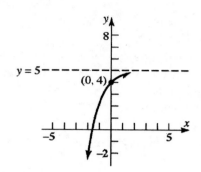

33. $f(x) = 2 - e^{-x/2}$

Using the graph of $y = e^x$, reflect the
graph about the y-axis, stretch
horizontally by a factor of 2, reflect about
the x-axis, and shift up 2 units.
Domain: $(-\infty, \infty)$
Range: $(-\infty, 2)$
Horizontal Asymptote: $y = 2$

35. $2^{2x+1} = 4$
$2^{2x+1} = 2^2$

$2x + 1 = 2 \rightarrow 2x = 1 \rightarrow x = \dfrac{1}{2}$

The solution set is $\left\{ \dfrac{1}{2} \right\}$.

37. $3^{x^3} = 9^x$

$3^{x^3} = \left(3^2\right)^x$

$3^{x^3} = 3^{2x}$

$x^3 = 2x \rightarrow x^3 - 2x = 0 \rightarrow x(x^2 - 2) = 0$

$x = 0$ or $x^2 = 2$

$x = 0$ or $x = \pm\sqrt{2}$

The solution set is $\left\{ -\sqrt{2}, \ 0, \ \sqrt{2} \right\}$.

39. $8^{x^2 - 2x} = \dfrac{1}{2}$

$\left(2^3\right)^{x^2 - 2x} = 2^{-1}$

$2^{3x^2 - 6x} = 2^{-1}$

$3x^2 - 6x = -1 \rightarrow 3x^2 - 6x + 1 = 0$

$x = \dfrac{-(-6) \pm \sqrt{(-6)^2 - 4(3)(1)}}{2(3)} = \dfrac{6 \pm \sqrt{24}}{6} = \dfrac{6 \pm 2\sqrt{6}}{6} = \dfrac{3 \pm \sqrt{6}}{3}$

The solution set is $\left\{ \dfrac{3 - \sqrt{6}}{3}, \ \dfrac{3 + \sqrt{6}}{3} \right\}$.

41. $2^x \cdot 8^{-x} = 4^x$

$2^x \cdot \left(2^3\right)^{-x} = \left(2^2\right)^x$

$2^x \cdot 2^{-3x} = 2^{2x}$

$2^{-2x} = 2^{2x}$

$-2x = 2x \rightarrow -4x = 0 \rightarrow x = 0$

The solution set is $\{0\}$.

43. $\left(\dfrac{1}{5}\right)^{2-x} = 25$

$\left(5^{-1}\right)^{2-x} = 5^2$

$5^{x-2} = 5^2$

$x - 2 = 2 \rightarrow x = 4$

The solution set is $\{4\}$.

45. $4^x = 8$

$\left(2^2\right)^x = 2^3$

$2^{2x} = 2^3$

$2x = 3 \rightarrow x = \dfrac{3}{2}$

The solution set is $\left\{\dfrac{3}{2}\right\}$.

47. $e^{x^2} = e^{3x} \cdot \dfrac{1}{e^2}$

$e^{x^2} = e^{3x-2}$

$x^2 = 3x - 2$

$x^2 - 3x + 2 = 0$

$(x-1)(x-2) = 0 \rightarrow x = 1$ or $x = 2$

The solution set is $\{1, 2\}$.

49. $4^x = 7$

$\left(4^x\right)^{-2} = 7^{-2} \quad \rightarrow \quad 4^{-2x} = \dfrac{1}{7^2} = \dfrac{1}{49}$

51. $3^{-x} = 2$

$(3^{-x})^{-2} = 2^{-2} \quad \rightarrow \quad 3^{2x} = \dfrac{1}{2^2} = \dfrac{1}{4}$

53. $f(x) = \begin{cases} e^{-x} & \text{if } x < 0 \\ e^x & \text{if } x \geq 0 \end{cases}$

domain $= (-\infty, \infty)$
range $= [1, \infty)$
y-intercept $(0, 1)$

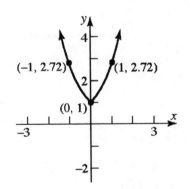

55. $f(x) = \begin{cases} -e^x & \text{if } x < 0 \\ -e^{-x} & \text{if } x \geq 0 \end{cases}$

domain $= (-\infty, \infty)$
range $= [-1, 0)$
y-intercept $(0, -1)$
horizontal asymptote $y = 0$

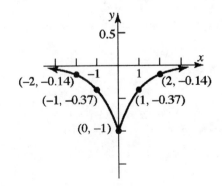

57. $p = 100e^{-0.03\,n}$

    (a) $p = 100e^{-0.03(10)} = 100e^{-03} \approx 100(0.741) = 74.1\%$ of light

    (b) $p = 100e^{-0.03(25)} = 100e^{-0.75} \approx 100(0.472) = 47.2\%$ of light

59. $w(d) = 50e^{-0.004\,d}$

    (a) $w(30) = 50e^{-0.004(30)} = 50e^{-0.12} \approx 50(0.887) = 44.35$ watts

    (b) $w(365) = 50e^{-0.004(365)} = 50e^{-1.46} \approx 50(0.232) = 11.61$ watts

61. $D(h) = 5e^{-0.4\,h}$

$$D(1) = 5e^{-04(1)} = 5e^{-0.4} \approx 5(0.670) = 3.35 \text{ milligrams}$$
$$D(6) = 5e^{-04(6)} = 5e^{-24} \approx 5(0.091) = 0.45 \text{ milligrams}$$

63. $F(t) = 1 - e^{-0.1t}$

    (a) $F(10) = 1 - e^{-0.1(10)} = 1 - e^{-1} \approx 1 - 0.368 = 0.632 = 63.2\%$

    (b) $F(40) = 1 - e^{-0.1(40)} = 1 - e^{-4} \approx 1 - 0.018 = 0.982 = 98.2\%$

    (c)   as $t \to +\infty,\ F(t) = 1 - e^{-0.1t} \to 1 - 0 = 1$

    (d)   Graphing the function:

    (e) $F(7) \approx 50$, so 7 minutes are needed for the probability to reach 50%.

65. $P(x) = \dfrac{20^x e^{-20}}{x!}$

    (a) $P(15) = \dfrac{20^{15} e^{-20}}{15!} \approx 0.0516 = 5.16\%$ The probability that 15 cars will arrive between

        5:00 p.m. and 6:00 p.m. is 5.16%.

    (b) $P(20) = \dfrac{20^{20} e^{-20}}{20!} \approx 0.0888 = 8.88\%$ The probability that 20 cars will arrive between

        5:00 p.m. and 6:00 p.m. is 8.88%.

67. $R = 10^{\left(\frac{2345}{T} - \frac{2345}{D} + 2\right)}$

    (a) $R = 10^{\left(\frac{2345}{283} - \frac{2345}{278} + 2\right)} \approx 10^{1.851} \approx 70.96\%$

    (b) $R = 10^{\left(\frac{2345}{293} - \frac{2345}{288} + 2\right)} \approx 10^{1.861} \approx 72.61\%$

    (c) $R = 10^{\left(\frac{2345}{x} - \frac{2345}{x} + 2\right)} = 10^2 = 100\%$

69. $I = \dfrac{E}{R}\left[1 - e^{-\left(\frac{R}{L}\right)t}\right]$

(a) $I = \dfrac{120}{10}\left[1 - e^{-\left(\frac{10}{5}\right)0.3}\right] = 12\left[1 - e^{-0.6}\right] \approx 5.414$ amperes after 0.3 second

$I = \dfrac{120}{10}\left[1 - e^{-\left(\frac{10}{5}\right)0.5}\right] = 12\left[1 - e^{-1}\right] \approx 7.585$ amperes after 0.5 second

$I = \dfrac{120}{10}\left[1 - e^{-\left(\frac{10}{5}\right)1}\right] = 12\left[1 - e^{-2}\right] \approx 10.376$ amperes after 1 second

(b) As $t \to \infty$, $e^{-\left(\frac{10}{5}\right)t} \to 0$. Therefore, the maximum current is 12 amperes.

(c), (f) Graphing the function:

(d) $I = \dfrac{120}{5}\left[1 - e^{-\left(\frac{5}{10}\right)0.3}\right] = 24\left[1 - e^{-0.15}\right] \approx 3.343$ amperes after 0.3 second

$I = \dfrac{120}{5}\left[1 - e^{-\left(\frac{5}{10}\right)0.5}\right] = 24\left[1 - e^{-0.25}\right] \approx 5.309$ amperes after 0.5 second

$I = \dfrac{120}{5}\left[1 - e^{-\left(\frac{5}{10}\right)1}\right] = 24\left[1 - e^{-0.5}\right] \approx 9.443$ amperes after 1 second

(e) As $t \to \infty$, $e^{-\left(\frac{5}{10}\right)t} \to 0$. Therefore, the maximum current is 24 amperes.

71. $2 + \dfrac{1}{2!} + \dfrac{1}{3!} + \dfrac{1}{4!} + \ldots + \dfrac{1}{n!}$

$n = 4; \quad 2 + \dfrac{1}{2!} + \dfrac{1}{3!} + \dfrac{1}{4!} = 2.7083$

$n = 6; \quad 2 + \dfrac{1}{2!} + \dfrac{1}{3!} + \dfrac{1}{4!} + \dfrac{1}{5!} + \dfrac{1}{6!} = 2.7181$

$n = 8; \quad 2 + \dfrac{1}{2!} + \dfrac{1}{3!} + \dfrac{1}{4!} + \dfrac{1}{5!} + \dfrac{1}{6!} + \dfrac{1}{7!} + \dfrac{1}{8!} = 2.7182788$

$n = 10; \quad 2 + \dfrac{1}{2!} + \dfrac{1}{3!} + \dfrac{1}{4!} + \dfrac{1}{5!} + \dfrac{1}{6!} + \dfrac{1}{7!} + \dfrac{1}{8!} + \dfrac{1}{9!} + \dfrac{1}{10!} = 2.7182818$

$e = 2.718281828$

73. $f(x) = a^x$

$$\frac{f(x+h) - f(x)}{h} = \frac{a^{x+h} - a^x}{h} = \frac{a^x a^h - a^x}{h} = \frac{a^x(a^h - 1)}{h} = a^x\left(\frac{a^h - 1}{h}\right)$$

75. $f(x) = a^x$

$$f(-x) = a^{-x} = \frac{1}{a^x} = \frac{1}{f(x)}$$

77.   (a) $y = \dfrac{6}{1 + e^{-(5.085 - 0.1156(100))}} \approx 0.0092$ O - rings

(b) $y = \dfrac{6}{1 + e^{-(5.085 - 0.1156(60))}} \approx 0.8145$ O - rings

(c) $y = \dfrac{6}{1 + e^{-(5.085 - 0.1156(30))}} \approx 5.0063$ O - rings

(d)   Graphing:

At 58°F, there would be 1 leaky O-ring.
At 44°F, there would be 3 leaky O-rings.
At 30°F, there would be 5 leaky O-rings.

79.   We can use the function $f(t) = f(0)e^{kt}$
the number of bacteria doubles
every minute means

$f(1) = 2f(0)$

$f(0)e^{k\,(1)} = 2f(0)$

$e^k = 2$

the container is full after
60 minutes means

$f(60) = 4$

$f(0)e^{k\,(60)} = 4$

$f(0)\left(e^k\right)^{(60)} = 4$

$f(0)(2)^{(60)} = 4$

$\rightarrow f(0) = \dfrac{4}{2^{60}} = \dfrac{2^2}{2^{60}} = \dfrac{1}{2^{58}}$

We want to find t so that $f(t) = 2$.

$$f(t) = f(0)e^{kt} = 2 \rightarrow f(0)\left(e^k\right)^t = 2$$

$$f(0)(2)^t = 2 \rightarrow \left(\frac{1}{2^{58}}\right)(2)^t = 2 \rightarrow 2^t = 2 \cdot 2^{58} = 2^{59} \rightarrow t = 59 \text{ minutes}$$

81.   Answers will vary.

Chapter 4

Exponential and Logarithmic Functions

4.3   Logarithmic Functions

1. $9 = 3^2$ is equivalent to $2 = \log_3 9$

3. $a^2 = 1.6$ is equivalent to $2 = \log_a 1.6$

5. $1.1^2 = M$ is equivalent to $2 = \log_{1.1} M$

7. $2^x = 7.2$ is equivalent to $x = \log_2 7.2$

9. $x^{\sqrt{2}} = \pi$ is equivalent to $\sqrt{2} = \log_x \pi$

11. $e^x = 8$ is equivalent to $x = \ln 8$

13. $\log_2 8 = 3$ is equivalent to $2^3 = 8$

15. $\log_a 3 = 6$ is equivalent to $a^6 = 3$

17. $\log_3 2 = x$ is equivalent to $3^x = 2$

19. $\log_2 M = 1.3$ is equivalent to $2^{1.3} = M$

21. $\log_{\sqrt{2}} \pi = x$ is equivalent to $\left(\sqrt{2}\right)^x = \pi$

23. $\ln 4 = x$ is equivalent to $e^x = 4$

25. $\log_2 1 = 0$ since $2^0 = 1$

27. $\log_5 25 = 2$ since $5^2 = 25$

29. $\log_{\frac{1}{2}} 16 = -4$ since $\left(\dfrac{1}{2}\right)^{-4} = 2^4 = 16$

31. $\log_{10} \sqrt{10} = \dfrac{1}{2}$ since $10^{1/2} = \sqrt{10}$

33. $\log_{\sqrt{2}} 4 = 4$ since $\left(\sqrt{2}\right)^4 = 4$

35. $\ln \sqrt{e} = \dfrac{1}{2}$ since $e^{1/2} = \sqrt{e}$

37. The domain of $f(x) = \ln(x - 3)$ is:
$$x - 3 > 0 \to x > 3$$
$$\{x \mid x > 3\}$$

39. The domain of $F(x) = \log_2 x^2$ is:
$$x^2 > 0$$
$$\{x \mid x \neq 0\}$$

41. The domain of $h(x) = \log_{\frac{1}{2}}\left(x^2 - 2x + 1\right)$ is:
$$x^2 - 2x + 1 > 0 \to (x - 1)^2 > 0$$
$$\{x \mid x \neq 1\}$$

43. The domain of $f(x) = \ln\left(\dfrac{1}{x+1}\right)$ is:
$$\dfrac{1}{x+1} > 0 \to x + 1 > 0$$
$$x > -1$$
$$\{x \mid x > -1\}$$

45. The domain of $g(x) = \log_5\left(\dfrac{x+1}{x}\right)$ requires that $\dfrac{x+1}{x} > 0$.

The expression is zero or undefined when $x = -1$ or $x = 0$.

$$f(x) = \frac{x+1}{x}$$

Interval	Test Number	$f(x) = \frac{x+1}{x}$	Positive/Negative
$-\infty < x < -1$	-2	$1/2$	Positive
$-1 < x < 0$	-0.5	-1	Negative
$0 < x < \infty$	1	2	Positive

The domain is $\{x \mid x < -1 \text{ or } x > 0\}$

47. $\ln\left(\dfrac{5}{3}\right) \approx 0.511$

49. $\dfrac{\ln(10/3)}{0.04} \approx 30.099$

51. For $f(x) = \log_a x$, find a so that $f(2) = \log_a 2 = 2$ or $a^2 = 2$ or $a = \sqrt{2}$.
(The base a must be positive by definition.)

53. B   55. D   57. A   59. E

61. $f(x) = \ln(x+4)$
Using the graph of $y = \ln x$, shift the
graph 4 units to the left.
Domain: $(-4, \infty)$
Range: $(-\infty, \infty)$
Vertical Asymptote: $x = -4$

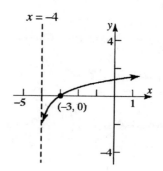

63. $f(x) = \ln(-x)$
Using the graph of $y = \ln x$, reflect the
graph about the y-axis.
Domain: $(-\infty, 0)$
Range: $(-\infty, \infty)$
Vertical Asymptote: $x = 0$

65. $g(x) = \ln(2x)$
Using the graph of $y = \ln x$, compress the
graph horizontally by a factor of $\dfrac{1}{2}$.
Domain: $(0, \infty)$
Range: $(-\infty, \infty)$
Vertical Asymptote: $x = 0$

67. $f(x) = 3\ln x$
Using the graph of $y = \ln x$, stretch the
graph vertically by a factor of 3.
Domain: $(0, \infty)$
Range: $(-\infty, \infty)$
Vertical Asymptote: $x = 0$

69. $g(x) = \ln(3 - x) = \ln\left(-(x - 3)\right)$
Using the graph of $y = \ln x$, reflect the
graph about the y-axis, and shift 3 units to
the right.
Domain: $(-\infty, 3)$
Range: $(-\infty, \infty)$
Vertical Asymptote: $x = 3$

71. $f(x) = -\ln(x - 1)$
Using the graph of $y = \ln x$, shift the
graph 1 unit to the right, and reflect about
the x-axis.
Domain: $(1, \infty)$
Range: $(-\infty, \infty)$
Vertical Asymptote: $x = 1$

73. $\log_3 x = 2$
$$x = 3^2 \rightarrow x = 9$$

75. $\log_2(2x + 1) = 3$
$$2x + 1 = 2^3 \rightarrow 2x + 1 = 8$$
$$2x = 7 \rightarrow x = \dfrac{7}{2}$$

77. $\log_x 4 = 2$
$x^2 = 4$
$x = 2$ $(x \neq -2$, base is positive$)$

79. $\ln e^x = 5$
$e^x = e^5 \rightarrow x = 5$

81. $\log_4 64 = x$
$4^x = 64 \rightarrow 4^x = 4^3 \rightarrow x = 3$

83. $\log_3 243 = 2x + 1$
$3^{2x+1} = 243$
$3^{2x+1} = 3^5$
$2x + 1 = 5 \rightarrow 2x = 4 \rightarrow x = 2$

85. $e^{3x} = 10$
$3x = \ln 10 \rightarrow x = \dfrac{\ln 10}{3}$

87. $e^{2x+5} = 8$
$2x + 5 = \ln 8$
$2x = -5 + \ln 8 \rightarrow x = \dfrac{-5 + \ln 8}{2}$

89. $\log_3\left(x^2 + 1\right) = 2$
$x^2 + 1 = 3^2$
$x^2 + 1 = 9 \rightarrow x^2 = 8$
$x = \pm\sqrt{8} = \pm 2\sqrt{2}$

91. $\log_2 8^x = -3$
$8^x = 2^{-3}$
$8^x = \dfrac{1}{8} \rightarrow 8^x = 8^{-1} \rightarrow x = -1$

93. $f(x) = \begin{cases} \ln(-x) & \text{if } x < 0 \\ \ln x & \text{if } x > 0 \end{cases}$

Domain: $(-\infty, 0) \cup (0, \infty)$
Range: $(-\infty, \infty)$
x-intercept: $(-1, 0), (1, 0)$
vertical asymptote: $x = 0$

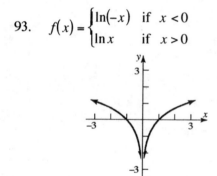

95. $f(x) = \begin{cases} -\ln x & \text{if } 0 < x < 1 \\ \ln x & \text{if } x \geq 1 \end{cases}$

Domain: $(0, \infty)$
Range: $[0, \infty)$
x-intercept: $(1, 0)$
vertical asymptote: $x = 0$

97. $P = 100e^{-0.1\,n}$

(a) $50 = 100e^{-0.1\,n}$

$0.5 = e^{-0.1n}$

$\ln(0.5) = -0.1n$

$n = \dfrac{\ln(0.5)}{-0.1} \to n \approx 6.93$

7 panes of glass are needed.

(b) $25 = 100e^{-0.1n}$

$0.25 = e^{-0.1n}$

$\ln(0.25) = -0.1n$

$n = \dfrac{\ln(0.25)}{-0.1} \to n \approx 13.86$

14 panes of glass are needed.

99. $w = 50e^{-0.004\,d}$

(a) $30 = 50e^{-0.004\,d}$

$0.6 = e^{-0.004d}$

$\ln(0.6) = -0.004\,d$

$d = \dfrac{\ln(0.6)}{-0.004} \approx 127.7$

Approximately 128 days.

(b) $5 = 50e^{-0.004\,d}$

$0.1 = e^{-0.004d}$

$\ln(0.1) = -0.004\,d$

$d = \dfrac{\ln(0.1)}{-0.004} \approx 575.6$

Approximately 576 days.

101. $F(t) = 1 - e^{-0.1\,t}$

(a) $0.5 = 1 - e^{-0.1t}$

$-0.5 = -e^{-0.1t} \to 0.5 = e^{-0.1t} \to \ln(0.5) = -0.1t$

$t = \dfrac{\ln(0.5)}{-0.1} \approx 6.93$

Approximately 7 minutes.

(b) $0.8 = 1 - e^{-0.1t}$

$-0.2 = -e^{-0.1t} \to 0.2 = e^{-0.1t} \to \ln(0.2) = -0.1t$

$t = \dfrac{\ln(0.2)}{-0.1} \approx 16.09$

Approximately 16 minutes.

(c)   It is impossible for the probability to reach 100% because $e^{-0.1t}$ will never equal zero.

103. $D = 5e^{-0.4h}$

$2 = 5e^{-0.4h} \to 0.4 = e^{-0.4h}$

$\ln(0.4) = -0.4h \to h = \dfrac{\ln(0.4)}{-0.4} \approx 2.29$ hours

105. $I = \dfrac{E}{R}\left[1 - e^{-\left(\frac{R}{L}\right)t}\right]$

0.5 ampere:

$0.5 = \dfrac{12}{10}\left[1 - e^{-\left(\frac{10}{5}\right)t}\right]$

$0.4167 = 1 - e^{-2t}$

$e^{-2t} = 0.5833$

$-2t = \ln(0.5833)$

$t = \dfrac{\ln(0.5833)}{-2} \approx 0.2695$ seconds

1.0 ampere:

$1.0 = \dfrac{12}{10}\left[1 - e^{-\left(\frac{10}{5}\right)t}\right]$

$0.8333 = 1 - e^{-2t}$

$e^{-2t} = 0.1667$

$-2t = \ln(0.1667)$

$t = \dfrac{\ln(0.1667)}{-2} \approx 0.8958$ seconds

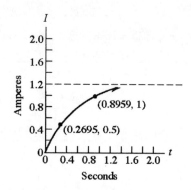

107.　$R = 3e^{kx}$

(a)
$$10 = 3e^{k(0.06)}$$

$$\frac{10}{3} = e^{k(0.06)}$$

$$\ln\left(\frac{10}{3}\right) = k(0.06)$$

$$k = \frac{\ln\left(\dfrac{10}{3}\right)}{0.06} \approx 20.07$$

(b)
$$R = 3e^{(20.07)(0.17)}$$

$$R = 3e^{3.4119} \approx 90.97\%$$

(c)
$$100 = 3e^{(20.07)x}$$

$$\frac{100}{3} = e^{(20.07)x} \rightarrow \ln\left(\frac{100}{3}\right) = (20.07)x$$

$$x = \frac{\ln\left(\dfrac{100}{3}\right)}{(20.07)} \approx 0.1747$$

(d)
$$15 = 3e^{(20.07)(x)}$$

$$5 = e^{(20.07)(x)} \rightarrow \ln(5) = (20.07)(x)$$

$$x = \frac{\ln(5)}{20.07} \approx 0.080$$

(e)　Answers will vary.

109.　New $= \text{Old}\left(e^{Rt}\right)$

Age	Depreciation rate	Age	Depreciation rate
1	$38000 = 36600e^{R\,(1)}$ $$\frac{38000}{36600} = e^{R}$$ $$\ln\left(\frac{38000}{36600}\right) = R$$ $$\rightarrow R \approx 0.03754 = 3.8\%$$	2	$38000 = 32400e^{R\,(2)}$ $$\frac{38000}{32400} = e^{2R}$$ $$\ln\left(\frac{38000}{32400}\right) = 2R$$ $$\frac{\ln\left(\dfrac{38000}{32400}\right)}{2} = R$$ $$\rightarrow R \approx 0.07971 = 8\%$$

Age	Depreciation rate	Age	Depreciation rate
3	$38000 = 28750e^{R(3)}$ $$\frac{38000}{28750} = e^{3R}$$ $$\ln\left(\frac{38000}{28750}\right) = 3R$$ $$\frac{\ln\left(\frac{38000}{28750}\right)}{3} = R$$ $\rightarrow R \approx 0.0930 = 9.3\%$	4	$38000 = 25400e^{R(4)}$ $$\frac{38000}{25400} = e^{4R}$$ $$\ln\left(\frac{38000}{25400}\right) = 4R$$ $$\frac{\ln\left(\frac{38000}{24500}\right)}{4} = R$$ $\rightarrow R \approx 0.1007 = 10.1\%$

Age	Depreciation rate
5	$38000 = 21200e^{R(5)}$ $$\frac{38000}{21200} = e^{5R} \rightarrow \ln\left(\frac{38000}{21200}\right) = 5R$$ $$\frac{\ln\left(\frac{38000}{21200}\right)}{5} = R \rightarrow R \approx 0.1167 = 11.7\%$$

Exponential and Logarithmic Functions

4.4 Properties of Logarithms; Exponential and Logarithmic Models

1. $\log_3 3^{71} = 71$

3. $\ln e^{-4} = -4$

5. $2^{\log_2 7} = 7$

7. $\log_8 2 + \log_8 4 = \log_8(4 \cdot 2) = \log_8(8) = 1$

9. $\log_6 18 - \log_6 3 = \log_6\left(\dfrac{18}{3}\right) = \log_6(6) = 1$

11. $\log_2 6 \cdot \log_6 4$

$$= \log_6\left(4^{\log_2 6}\right) = \log_6\left(\left(2^2\right)^{\log_2 6}\right)$$

$$= \log_6\left((2)^{2\log_2 6}\right) = \log_6\left((2)^{\log_2 6^2}\right) = \log_6\left(6^2\right) = 2$$

13. $3^{\log_3 5 - \log_3 4} = 3^{\log_3\left(\frac{5}{4}\right)} = \dfrac{5}{4}$

15. $e^{\log_{e^2} 16}$

Simplify the exponent:

$$\text{Let} \qquad a = \log_{e^2} 16$$

$$\left(e^2\right)^a = 16$$

$$e^{2a} = 16 = 4^2$$

$$e^a = 4 \rightarrow a = \ln 4$$

$$\text{Thus, } e^{\log_{e^2} 16} = e^{\ln 4} = 4$$

17. $\ln 6 = \ln(3 \cdot 2) = \ln 3 + \ln 2 = b + a$

19. $\ln(1.5) = \ln\left(\dfrac{3}{2}\right) = \ln 3 - \ln 2 = b - a$

21. $\ln 8 = \ln 2^3 = 3 \cdot \ln 2 = 3a$

23. $\ln\left(\sqrt[5]{6}\right) = \ln\left(6^{1/5}\right) = \dfrac{1}{5} \cdot \ln 6 = \dfrac{1}{5} \cdot \ln(2 \cdot 3) = \dfrac{1}{5} \cdot (\ln 2 + \ln 3) = \dfrac{1}{5} \cdot (a + b)$

25. $\log_a(u^2v^3) = \log_a u^2 + \log_a v^3 = 2\log_a u + 3\log_a v$

27. $\log\left(\dfrac{1}{M^3}\right) = \log(M^{-3}) = -3\log M$

29. $\log_5\sqrt{\dfrac{a^3}{b}} = \log_5\left(\dfrac{a^3}{b}\right)^{1/2} = \log_5\left(\dfrac{a^{3/2}}{b^{1/2}}\right) = \log_5 a^{3/2} - \log_5 b^{1/2} = \dfrac{3}{2}\log_5 a - \dfrac{1}{2}\log_5 b$

31. $\ln\left(x^2\sqrt{1-x}\right) = \ln x^2 + \ln\sqrt{1-x} = \ln x^2 + \ln(1-x)^{1/2} = 2\ln x + \dfrac{1}{2}\ln(1-x)$

33. $\log_2\left(\dfrac{x^3}{x-3}\right) = \log_2 x^3 - \log_2(x-3) = 3\log_2 x - \log_2(x-3)$

35. $\log\left[\dfrac{x(x+2)}{(x+3)^2}\right] = \log(x(x+2)) - \log(x+3)^2 = \log x + \log(x+2) - 2\log(x+3)$

37. $\ln\left[\dfrac{x^2-x-2}{(x+4)^2}\right]^{1/3} = \dfrac{1}{3}\ln\left[\dfrac{(x-2)(x+1)}{(x+4)^2}\right] = \dfrac{1}{3}\left[\ln(x-2)(x+1) - \ln(x+4)^2\right]$

$= \dfrac{1}{3}\left[\ln(x-2) + \ln(x+1) - 2\ln(x+4)\right] = \dfrac{1}{3}\ln(x-2) + \dfrac{1}{3}\ln(x+1) - \dfrac{2}{3}\ln(x+4)$

39. $\ln\left(\dfrac{5x\sqrt{1-3x}}{(x-4)^3}\right) = \ln\left(5x\sqrt{1-3x}\right) - \ln(x-4)^3 = \ln 5 + \ln x + \ln\sqrt{1-3x} - 3\ln(x-4)$

$= \ln 5 + \ln x + \ln(1-3x)^{1/2} - 3\ln(x-4) = \ln 5 + \ln x + \dfrac{1}{2}\ln(1-3x) - 3\ln(x-4)$

41. $3\log_5 u + 4\log_5 v = \log_5 u^3 + \log_5 v^4 = \log_5(u^3v^4)$

43. $\log_{\frac{1}{2}}\sqrt{x} - \log_{\frac{1}{2}} x^3 = \log_{\frac{1}{2}}\left(\dfrac{\sqrt{x}}{x^3}\right) = \log_{\frac{1}{2}}\left(\dfrac{x^{1/2}}{x^3}\right) = \log_{\frac{1}{2}}\left(x^{-5/2}\right) = -\dfrac{5}{2}\log_{\frac{1}{2}} x$

45. $\ln\left(\dfrac{x}{x-1}\right) + \ln\left(\dfrac{x+1}{x}\right) - \ln(x^2-1) = \ln\left[\dfrac{x}{x-1}\cdot\dfrac{x+1}{x}\right] - \ln(x^2-1) = \ln\left[\dfrac{x+1}{x-1}\div(x^2-1)\right]$

$= \ln\left[\dfrac{x+1}{(x-1)(x-1)(x+1)}\right] = \ln\left(\dfrac{1}{(x-1)^2}\right) = \ln(x-1)^{-2} = -2\ln(x-1)$

47. $8\log_2\sqrt{3x-2} - \log_2\left(\dfrac{4}{x}\right) + \log_2 4 = \log_2\left(\sqrt{3x-2}\right)^8 - (\log_2 4 - \log_2 x) + \log_2 4$

$= \log_2(3x-2)^4 - \log_2 4 + \log_2 x + \log_2 4 = \log_2\left[x(3x-2)^4\right]$

49. $2\log_a\left(5x^3\right) - \dfrac{1}{2}\log_a(2x+3) = \log_a\left(5x^3\right)^2 - \log_a(2x-3)^{1/2} = \log_a\left[\dfrac{25x^6}{(2x-3)^{1/2}}\right]$

51. $\log_3 21 = \dfrac{\log 21}{\log 3} \approx \dfrac{1.32222}{0.47712} \approx 2.771$

53. $\log_{\frac{1}{3}} 71 = \dfrac{\log 71}{\log\left(\dfrac{1}{3}\right)} = \dfrac{\log 71}{-\log 3} \approx \dfrac{1.85126}{-0.47712} \approx -3.880$

55. $\log_{\sqrt{2}} 7 = \dfrac{\log 7}{\log\sqrt{2}} = \dfrac{\log 7}{\log 2^{1/2}} = \dfrac{\log 7}{\left(\dfrac{1}{2}\log 2\right)} \approx \dfrac{0.84510}{0.5(0.30103)} \approx 5.615$

57. $\log_{\pi} e = \dfrac{\ln e}{\ln \pi} \approx \dfrac{1}{1.14473} \approx 0.874$

59. $\log_2 3 \cdot \log_3 4 \cdot \log_4 5 \cdot \log_5 6 \cdot \log_6 7 \cdot \log_7 8$

     $= \dfrac{\log 3}{\log 2} \cdot \dfrac{\log 4}{\log 3} \cdot \dfrac{\log 5}{\log 4} \cdot \dfrac{\log 6}{\log 5} \cdot \dfrac{\log 7}{\log 6} \cdot \dfrac{\log 8}{\log 7} = \dfrac{\log 8}{\log 2} = \dfrac{\log 2^3}{\log 2} = \dfrac{3\log 2}{\log 2} = 3$

61. $\log_2 3 \cdot \log_3 4 \cdot \ldots \cdot \log_n (n+1) \cdot \log_{n+1} 2$

     $= \dfrac{\log 3}{\log 2} \cdot \dfrac{\log 4}{\log 3} \cdot \ldots \cdot \dfrac{\log(n+1)}{\log n} \cdot \dfrac{\log 2}{\log(n+1)} = \dfrac{\log 2}{\log 2} = 1$

63. $y = \log_4 x = \dfrac{\ln x}{\ln 4}$ or $y = \dfrac{\log x}{\log 4}$

65. $y = \log_2(x+2) = \dfrac{\ln(x+2)}{\ln 2}$

          or $y = \dfrac{\log(x+2)}{\log 2}$

67. $y = \log_{x-1}(x+1) = \dfrac{\ln(x+1)}{\ln(x-1)}$

$\text{or } y = \dfrac{\log(x+1)}{\log(x-1)}$

69. $\ln y = \ln x + \ln C$
$\ln y = \ln(xC)$
$y = Cx$

71. $\ln y = \ln x + \ln(x+1) + \ln C$
$\ln y = \ln(x(x+1)C)$
$y = Cx(x+1)$

73. $\ln y = 3x + \ln C$
$\ln y = \ln e^{3x} + \ln C$
$\ln y = \ln(Ce^{3x})$
$y = Ce^{3x}$

75. $\ln(y-3) = -4x + \ln C$
$\ln(y-3) = \ln e^{-4x} + \ln C$
$\ln(y-3) = \ln(Ce^{-4x})$
$y - 3 = Ce^{-4x} \rightarrow y = Ce^{-4x} + 3$

77. $3\ln y = \dfrac{1}{2}\ln(2x+1) - \dfrac{1}{3}\ln(x+4) + \ln C$

$\ln y^3 = \ln(2x+1)^{1/2} - \ln(x+4)^{1/3} + \ln C$

$\ln y^3 = \ln\left[\dfrac{C(2x+1)^{1/2}}{(x+4)^{1/3}}\right] \rightarrow y^3 = \dfrac{C(2x+1)^{1/2}}{(x+4)^{1/3}}$

$y = \left[\dfrac{C(2x+1)^{1/2}}{(x+4)^{\frac{1}{3}}}\right]^{1/3} \rightarrow y = \dfrac{\sqrt[3]{C}(2x+1)^{1/6}}{(x+4)^{1/9}}$

79. Verifying:
$$\log_a\left(x + \sqrt{x^2-1}\right) + \log_a\left(x - \sqrt{x^2-1}\right) = \log_a\left[\left(x + \sqrt{x^2-1}\right)\left(x - \sqrt{x^2-1}\right)\right]$$
$$= \log_a\left[x^2 - \left(x^2-1\right)\right] = \log_a\left[x^2 - x^2 + 1\right] = \log_a 1 = 0$$

81. Verifying:
$$2x + \ln\left(1 + e^{-2x}\right) = \ln e^{2x} + \ln\left(1 + e^{-2x}\right) = \ln\left(e^{2x}\left(1 + e^{-2x}\right)\right) = \ln\left(e^{2x} + e^0\right) = \ln\left(e^{2x} + 1\right)$$

83. $f(x) = \log_a x$

$x = a^{f(x)} \rightarrow x^{-1} = a^{-f(x)} = \left(a^{-1}\right)^{f(x)} = \left(\dfrac{1}{a}\right)^{f(x)}$

$\log_{\frac{1}{a}} x^{-1} = f(x) \rightarrow -\log_{\frac{1}{a}} x = f(x) \rightarrow -f(x) = \log_{\frac{1}{a}} x$

85. $f(x) = \log_a x$

$a^{f(x)} = x \rightarrow \dfrac{1}{a^{f(x)}} = \dfrac{1}{x}$

$a^{-f(x)} = \dfrac{1}{x} \rightarrow -f(x) = \log_a \dfrac{1}{x} = f\left(\dfrac{1}{x}\right)$

87.   If $A = \log_a M$ and $B = \log_a N$, then $a^A = M$ and $a^B = N$.

$$\log_a\left(\frac{M}{N}\right) = \log_a\left(\frac{a^A}{a^B}\right) = \log_a a^{A-B} = A - B = \log_a M - \log_a N$$

89.   (a)   Graphing:

   (d) $A = 100e^{-0.1278(50)}$
         ≈ 0.1678 grams

   (e) and (f) Graphing:

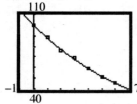

   (b) $y = 100(0.88)^x$
         $0.88 = e^{\ln(0.88)}$
         $\rightarrow y = 100\left(e^{\ln(0.88)}\right)^x = 100e^{\ln(0.88)x}$

         $A = A_0 e^{-0.1278t}, A_0 = 100$

   (c)

         $100e^{-0.1278t} = 50$

         $e^{-0.1278t} = \dfrac{50}{100}$

         $e^{-0.1278t} = 0.5$

         $\ln\left(e^{-0.1278t}\right) = \ln(0.5)$

         $-0.1278t = \ln(0.5)$

         $t = \dfrac{\ln(0.5)}{-0.1278} \approx 5.42$ weeks

91. (a)   Graphing:

(d) and (e) Graphing

(b) $y = 96(0.98)^x$

$0.98 = e^{\ln(0.98)}$

$\rightarrow y = 96\left(e^{\ln(0.98)}\right)^x = 96e^{\ln(0.98)x}$

$A = A_0 e^{-0.0202\,t},\ A_0 = 96$

(c)

$96e^{-0.0202t} = 60$

$e^{-0.0202t} = \dfrac{60}{96}$

$\ln(e^{-0.0202t}) = \ln\left(\dfrac{60}{96}\right)$

$-0.0202t = \ln\left(\dfrac{60}{96}\right)$

$t = \dfrac{\ln\left(\dfrac{60}{96}\right)}{-0.0202} \approx 23$ shoes

93. (a)   Graphing:

(c) and (d) Graphing:

(b)

$y = 32741 - 6071 \cdot \ln x$

$1650 = 32741 - 6071 \cdot \ln x$

$-31091 = -6071 \cdot \ln x$

$\dfrac{-31091}{-6071} = \ln x$

$\dfrac{31091}{6071} = \ln x$

$e^{\left(\frac{31091}{6071}\right)} = e^{\ln x} = x$

$x \approx 168$ computers

Chapter 4

Exponential and Logarithmic Functions

4.5   Logarithmic and Exponential Equations

1. $\log_4(x+2) = \log_4 8$
    $$x+2 = 8 \rightarrow x = 6$$

3. $\dfrac{1}{2}\log_3 x = 2\log_3 2$
    $$\log_3 x^{1/2} = \log_3 2^2$$
    $$x^{1/2} = 4 \rightarrow x = 16$$

5. $2\log_5 x = 3\log_5 4$
    $$\log_5 x^2 = \log_5 4^3$$
    $$x^2 = 64$$
    $$x = \pm 8$$
    Since $\log_5(-8)$ is undefined, the only solution is $x = 8$.

7. $3\log_2(x-1) + \log_2 4 = 5$
    $$\log_2(x-1)^3 + \log_2 4 = 5$$
    $$\log_2\left(4(x-1)^3\right) = 5$$
    $$4(x-1)^3 = 2^5$$
    $$(x-1)^3 = \frac{32}{4}$$
    $$(x-1)^3 = 8$$
    $$x-1 = 2$$
    $$x = 3$$

9. $\log x + \log(x+15) = 2$
    $$\log\big(x(x+15)\big) = 2$$
    $$x(x+15) = 10^2$$
    $$x^2 + 15x - 100 = 0 \rightarrow (x+20)(x-5) = 0$$
    $$x = -20 \text{ or } x = 5$$
    Since $\log(-20)$ is undefined, the only solution is $x = 5$.

11. $\ln x + \ln(x+2) = 4$
    $$\ln\big(x(x+2)\big) = 4$$
    $$x(x+2) = e^4 \rightarrow x^2 + 2x - e^4 = 0$$
    $$x = \frac{-2 \pm \sqrt{2^2 - 4(1)(-e^4)}}{2(1)} = \frac{-2 \pm \sqrt{4 + 4e^4}}{2} = \frac{-2 \pm 2\sqrt{1+e^4}}{2} = -1 \pm \sqrt{1+e^4}$$
    Since $\ln\left(-1 - \sqrt{1+e^4}\right)$ is undefined, the only solution is $x = -1 + \sqrt{1+e^4} \approx 6.456$.

13. $2^{2x}+2^x-12=0$
$\left(2^x\right)^2+2^x-12=0$
$\left(2^x-3\right)\left(2^x+4\right)=0$

$2^x-3=0$ or $2^x+4=0$
$2^x=3$ or $2^x=-4$
$x=\log_2 3$ No solution
$x\approx1.585$

15. $3^{2x}+3^{x+1}-4=0$
$\left(3^x\right)^2+3\cdot3^x-4=0$
$\left(3^x-1\right)\left(3^x+4\right)=0$

$3^x-1=0$ or $3^x+4=0$
$3^x=1$ or $3^x=-4$
$x=0$ No solution

17. $2^x=10$
$\log\left(2^x\right)=\log10$
$x\log2=1$
$x=\dfrac{1}{\log2}\approx3.322$

19. $8^{-x}=1.2$
$\log\left(8^{-x}\right)=\log(1.2)$
$-x\log8=\log(1.2)$
$x=\dfrac{\log(1.2)}{-\log8}\approx-0.088$

21. $3^{1-2x}=4^x$
$\log\left(3^{1-2x}\right)=\log\left(4^x\right)$
$(1-2x)\log3=x\log4$
$\log3-2x\log3=x\log4\rightarrow\log3=x\log4+2x\log3\rightarrow\log3=x(\log4+2\log3)$
$x=\dfrac{\log3}{\log4+2\log3}\approx0.307$

23. $\left(\dfrac{3}{5}\right)^x=7^{1-x}$
$\log\left(\left(\dfrac{3}{5}\right)^x\right)=\log\left(7^{1-x}\right)$
$x\log\left(\dfrac{3}{5}\right)=(1-x)\log7\rightarrow x(\log3-\log5)=\log7-x\log7$
$x\log3-x\log5+x\log7=\log7\rightarrow x(\log3-\log5+\log7)=\log7$
$x=\dfrac{\log7}{\log3-\log5+\log7}\approx1.356$

25.
$$1.2^x = (0.5)^{-x}$$
$$\log 1.2^x = \log(0.5)^{-x}$$
$$x\log(1.2) = -x\log(0.5)$$
$$x\log(1.2) + x\log(0.5) = 0$$
$$x(\log(1.2) + \log(0.5)) = 0$$
$$x = 0$$

27.
$$\pi^{1-x} = e^x$$
$$\ln \pi^{1-x} = \ln e^x$$
$$(1-x)\ln\pi = x$$
$$\ln\pi - x\ln\pi = x$$
$$\ln\pi = x + x\ln\pi$$
$$\ln\pi = x(1 + \ln\pi)$$
$$x = \frac{\ln\pi}{1 + \ln\pi} \approx 0.534$$

29. $\quad 5\left(2^{3x}\right) = 8$
$$2^{3x} = \frac{8}{5} \rightarrow \log 2^{3x} = \log\left(\frac{8}{5}\right)$$
$$3x\log 2 = \log 8 - \log 5 \rightarrow x = \frac{\log 8 - \log 5}{3\log 2} \approx 0.226$$

31. $\quad \log_a(x-1) - \log_a(x+6) = \log_a(x-2) - \log_a(x+3)$

$$\log_a\left(\frac{x-1}{x+6}\right) = \log_a\left(\frac{x-2}{x+3}\right) \rightarrow a^{\left(\log_a\left(\frac{x-1}{x+6}\right)\right)} = a^{\left(\log_a\left(\frac{x-2}{x+3}\right)\right)}$$

so

$$\frac{x-1}{x+6} = \frac{x-2}{x+3} \rightarrow (x-1)(x+3) = (x-2)(x+6)$$

$$x^2 + 2x - 3 = x^2 + 4x - 12 \rightarrow 2x - 3 = 4x - 12 \rightarrow 9 = 2x \rightarrow x = \frac{9}{2}$$

Since each of the original logarithms is defined for $x = \frac{9}{2}$, the solution is $x = \frac{9}{2}$.

33. $\quad \log_{\frac{1}{3}}(x^2 + x) - \log_{\frac{1}{3}}(x^2 - x) = -1$

$$\log_{\frac{1}{3}}\left(\frac{x^2 + x}{x^2 - x}\right) = -1$$

$$\frac{x^2 + x}{x^2 - x} = \left(\frac{1}{3}\right)^{-1}$$

$$\frac{x(x+1)}{x(x-1)} = 3 \rightarrow x + 1 = 3(x-1)$$

$$x + 1 = 3x - 3 \rightarrow -2x = -4 \rightarrow x = 2$$

35. $\quad \log_2(x+1) - \log_4 x = 1$

$$\log_2(x+1) - \frac{\log_2 x}{\log_2 4} = 1$$

$$\log_2(x+1) - \frac{\log_2 x}{2} = 1$$

$$2\log_2(x+1) - \log_2 x = 2$$

$$\log_2(x+1)^2 - \log_2 x = 2$$

$$\log_2\left(\frac{(x+1)^2}{x}\right) = 2 \rightarrow \frac{(x+1)^2}{x} = 2^2$$

$$x^2 + 2x + 1 = 4x$$

$$x^2 - 2x + 1 = 0$$

$$(x-1)^2 = 0 \rightarrow x - 1 = 0$$

$$x = 1$$

37.
$$\log_{16} x + \log_4 x + \log_2 x = 7$$
$$\frac{\log_2 x}{\log_2 16} + \frac{\log_2 x}{\log_2 4} + \log_2 x = 7$$
$$\frac{\log_2 x}{4} + \frac{\log_2 x}{2} + \log_2 x = 7$$
$$\log_2 x + 2\log_2 x + 4\log_2 x = 28$$
$$7\log_2 x = 28$$
$$\log_2 x = 4$$
$$x = 2^4 = 16$$

39.
$$\left(\sqrt[3]{2}\right)^{2-x} = 2^{x^2}$$
$$\left(2^{1/3}\right)^{2-x} = 2^{x^2}$$
$$2^{\frac{1}{3}(2-x)} = 2^{x^2}$$
$$\frac{1}{3}(2-x) = x^2 \rightarrow 2 - x = 3x^2$$
$$3x^2 + x - 2 = 0 \rightarrow (3x-2)(x+1) = 0$$
$$x = \frac{2}{3} \text{ or } x = -1$$

41.
$$\frac{e^x + e^{-x}}{2} = 1$$
$$e^x + e^{-x} = 2$$
$$e^x\left(e^x + e^{-x}\right) = 2e^x$$
$$e^{2x} + 1 = 2e^x$$
$$(e^x)^2 - 2e^x + 1 = 0$$
$$\left(e^x - 1\right)^2 = 0$$
$$e^x - 1 = 0$$
$$e^x = 1$$
$$x = 0$$

43.

$$\frac{e^x - e^{-x}}{2} = 2$$
$$e^x - e^{-x} = 4$$
$$e^x\left(e^x - e^{-x}\right) = 4e^x$$
$$e^{2x} - 1 = 4e^x$$
$$(e^x)^2 - 4e^x - 1 = 0$$
$$e^x = \frac{-(-4) \pm \sqrt{(-4)^2 - 4(1)(-1)}}{2(1)}$$
$$= \frac{4 \pm \sqrt{20}}{2}$$
$$= \frac{4 \pm 2\sqrt{5}}{2} = 2 \pm \sqrt{5}$$
$$x = \ln\left(2 + \sqrt{5}\right)$$
$\ln\left(2 - \sqrt{5}\right)$ is undefined; it is not a solution.

45. Using INTERSECT to solve:
$$y_1 = \ln(x) / \ln(5) + \ln(x) / \ln(3)$$
$$y_2 = 1$$

The solution is 1.92.

47. Using INTERSECT to solve:
$$y_1 = \ln(x+1) / \ln(5) - \ln(x-2) / \ln(4)$$
$$y_2 = 1$$

The solution is 2.79.

49. Using INTERSECT to solve:

$y_1 = e^x$; $y_2 = -x$

The solution is –0.57.

51. Using INTERSECT to solve:

$y_1 = e^x$; $y_2 = x^2$

The solution is –0.70.

53. Using INTERSECT to solve:

$y_1 = \ln x$; $y_2 = -x$

The solution is 0.57.

55. Using INTERSECT to solve:

$y_1 = \ln x$; $y_2 = x^3 - 1$

The solutions are 0.39, 1.00.

57. Using INTERSECT to solve:

$y_1 = e^x + \ln x$; $y_2 = 4$

The solution is 1.32.

59. Using INTERSECT to solve:

$y_1 = e^{-x}$; $y_2 = \ln x$

The solution is 1.31.

Chapter 4

Exponential and Logarithmic Functions

4.6 Compound Interest

1. $P = \$100,\ r = 0.04,\ n = 4,\ t = 2$

$$A = P\left[1+\frac{r}{n}\right]^{nt} = 100\left[1+\frac{0.04}{4}\right]^{(4)(2)} \approx \$108.29$$

3. $P = \$500,\ r = 0.08,\ n = 4,\ t = 2.5$

$$A = P\left[1+\frac{r}{n}\right]^{nt} = 500\left[1+\frac{0.08}{4}\right]^{(4)(2.5)} \approx \$609.50$$

5. $P = \$600,\ r = 0.05,\ n = 365,\ t = 3$

$$A = P\left[1+\frac{r}{n}\right]^{nt} = 600\left[1+\frac{0.05}{365}\right]^{(365)(3)} \approx \$697.09$$

7. $P = \$10,\ r = 0.11,\ t = 2$

$$A = Pe^{rt} = 10e^{(0.11)(2)} \approx \$12.46$$

9. $P = \$100,\ r = 0.10,\ t = 2.25$

$$A = Pe^{rt} = 100e^{(0.10)(2.25)} \approx \$125.23$$

11. $A = \$100,\ r = 0.06,\ n = 12,\ t = 2$

$$P = A\left[1+\frac{r}{n}\right]^{-nt} = 100\left[1+\frac{0.06}{12}\right]^{(-12)(2)} \approx \$88.72$$

13. $A = \$1000,\ r = 0.06,\ n = 365,\ t = 2.5$

$$P = A\left[1+\frac{r}{n}\right]^{-nt} = 1000\left[1+\frac{0.06}{365}\right]^{(-365)(2.5)} \approx \$860.72$$

15. $A = \$600,\ r = 0.04,\ n = 4,\ t = 2$

$$P = A\left[1+\frac{r}{n}\right]^{-nt} = 600\left[1+\frac{0.04}{4}\right]^{(-4)(2)} \approx \$554.09$$

17. $A = \$80,\ r = 0.09,\ t = 3.25$

$$P = Ae^{-rt} = 80e^{(-0.09)(3.25)} \approx \$59.71$$

19. $A = \$400,\ r = 0.10,\ t = 1$

$$P = Ae^{-rt} = 400e^{(-0.10)(1)} \approx \$361.93$$

21. $\quad r_e = \left[1 + \dfrac{r}{n}\right]^n - 1 = \left[1 + \dfrac{0.0525}{4}\right]^4 - 1 \approx 1.0535 - 1 = 0.0535 = 5.35\%$

23. $\quad 2P = P(1+r)^3$

$\qquad 2 = (1+r)^3$

$\qquad \sqrt[3]{2} = 1 + r$

$\qquad r = \sqrt[3]{2} - 1 \approx 1.26 - 1 = 0.26 = 26\%$

25. 6% compounded quarterly:

$$A = 10{,}000\left[1 + \frac{0.06}{4}\right]^{(4)(1)} \approx \$10{,}613.64$$

$6\frac{1}{4}\%$ compounded annually:

$$A = 10{,}000\left[1 + 0.0625\right]^1 \approx \$10{,}625$$

$6\frac{1}{4}\%$ compounded annually yields the larger amount.

27. 9% compounded monthly:

$$A = 10{,}000\left[1 + \frac{0.09}{12}\right]^{(12)(1)} \approx \$10{,}938.07$$

8.8% compounded daily:

$$A = 10{,}000\left[1 + \frac{0.088}{365}\right]^{365} \approx \$10{,}919.77$$

9% compounded monthly yields the larger amount.

29. Compounded monthly: Compounded continuously:

$$2P = P\left[1 + \frac{0.08}{12}\right]^{12t}$$

$$2 \approx (1.00667)^{12t}$$

$$\ln 2 \approx 12t\,\ln(1.00667)$$

$$t \approx \frac{\ln 2}{12\,\ln(1.00667)} \approx 8.69 \text{ years}$$

$$2P = Pe^{0.08\,t}$$

$$2 = e^{0.08\,t}$$

$$\ln 2 = 0.08t$$

$$t = \frac{\ln 2}{0.08} \approx 8.66 \text{ years}$$

31. Compounded monthly: Compounded continuously:

$$150 = 100\left[1 + \frac{0.08}{12}\right]^{12t}$$

$$1.5 \approx (1.00667)^{12t}$$

$$\ln(1.5) \approx 12t\,\ln(1.00667)$$

$$t \approx \frac{\ln(1.5)}{12\,\ln(1.00667)} \approx 5.083 \text{ years}$$

$$150 = 100e^{0.08t}$$

$$1.5 = e^{0.08\,t}$$

$$\ln(1.5) = 0.08t$$

$$t = \frac{\ln(1.5)}{0.08} \approx 5.068 \text{ years}$$

33. $25,000 = 10,000e^{0.06t}$

   $2.5 = e^{0.06t}$

   $\ln(2.5) = 0.06t$

   $t = \dfrac{\ln(2.5)}{0.06} \approx 15.27$ years

35. $A = 90,000(1 + 0.03)^5 \approx \$104,335$

37. $P = 15,000e^{(-0.05)(3)} \approx \$12,910.62$

39. $A = 1500(1 + 0.15)^5 = 1500(1.15)^5 \approx \$3017$

41. $850,000 = 650,000(1 + r)^3$

   $\dfrac{85}{65} = (1 + r)^3$

   $\sqrt[3]{\dfrac{85}{65}} = 1 + r \rightarrow r \approx \sqrt[3]{1.3077} - 1 \approx 0.0935 = 9.35\%$

43. 5.6% compounded continuously:

   $A = 1000e^{(0.056)(1)} \approx \$1057.60$

   Jim does not have enough money to buy the computer.

   5.9% compounded monthly:

   $A = 1000\left[1 + \dfrac{0.059}{12}\right]^{12} \approx \$1060.62$

   The second bank offers the better deal.

45. Will - 9% compounded semiannually:

   $A = 2000\left[1 + \dfrac{0.09}{2}\right]^{(2)(20)} \approx \$11,632.73$

   Henry - 8.5% compounded continuously:

   $A = 2000e^{(0.085)(20)} \approx \$10,947.89$

   Will has more money after 20 years.

47. $P = 50,000;\ t = 5$

   (a) Simple interest at 12% per annum:

   $A = 50,000 + 50,000(0.12)(5) = \$80,000$

   (b) 11.5% compounded monthly:

   $A = 50,000\left[1 + \dfrac{0.115}{12}\right]^{(12)(5)} \approx \$88,613.59$

   (c) 11.25% compounded continuously:

   $A = 50,000e^{(0.1125)(5)} \approx \$87,752.73$

   Subtract $50,000 from each to get the amount of interest:

   (a) \$30,000   (b) \$38,613.59   (c) \$37.752.73

   Option (a) results in the least interest.

49. (a) $A = \$10,000$, $r = 0.10$, $n = 12$, $t = 20$ (compounded monthly)

$$P = 10,000\left[1 + \frac{0.10}{12}\right]^{(-12)(20)} \approx \$1364.62$$

(b) $A = \$10,000$, $r = 0.10$, $t = 20$ (compounded continuously)

$$P = 10,000e^{(-0.10)(20)} \approx \$1353.35$$

51. $A = \$10,000$, $r = 0.08$, $n = 1$, $t = 10$ (compounded annually)

$$P = 10,000\left[1 + \frac{0.08}{1}\right]^{(-1)(10)} \approx \$4631.93$$

53. (a) $y = \dfrac{\ln 2}{1 \cdot \ln\left(1 + \dfrac{0.12}{1}\right)} = \dfrac{\ln 2}{\ln(1.12)} \approx 6.12$ years

(b) $y = \dfrac{\ln 3}{4 \cdot \ln\left(1 + \dfrac{0.06}{4}\right)} = \dfrac{\ln 3}{4\ln(1.015)} \approx 18.45$ years

(c) $mP = P\left[1 + \dfrac{r}{n}\right]^{nt}$

$m = \left[1 + \dfrac{r}{n}\right]^{nt} \rightarrow \ln m = nt \cdot \ln\left[1 + \dfrac{r}{n}\right] \rightarrow t = \dfrac{\ln m}{n \cdot \ln\left[1 + \dfrac{r}{n}\right]}$

55. Answers will vary.   57. Answers will vary.

Chapter 4

Exponential and Logarithmic Functions

4.7   Growth and Decay; Newton's Law; Logistic Models

1. $P(t) = 500e^{0.02t}$

   Find t when $P = 1000$:

   $$1000 = 500e^{0.02t}$$
   $$2 = e^{0.02t}$$
   $$\ln 2 = 0.02t$$
   $$t = \frac{\ln 2}{0.02} \approx 34.7 \text{ days}$$

   Find t when $P = 2000$:

   $$2000 = 500e^{0.02t}$$
   $$4 = e^{0.02t}$$
   $$\ln 4 = 0.02t$$
   $$t = \frac{\ln 4}{0.02} \approx 69.3 \text{ days}$$

3. Find t when $A(t) = \frac{1}{2}A_0$:

   $$\frac{1}{2}A_0 = A_0 e^{-0.0244t}$$
   $$\frac{1}{2} = e^{-0.0244t}$$
   $$\ln\left(\frac{1}{2}\right) = -0.0244t$$
   $$t = \frac{\ln\left(\frac{1}{2}\right)}{-0.0244} \approx 28.4 \text{ years}$$

5. Use $N(t) = N_0 e^{kt}$ and solve for k:

   $$1800 = 1000e^{k(1)}$$
   $$1.8 = e^k$$
   $$k = \ln(1.8) \approx 0.5878$$

   When $t = 3$:

   $$N(3) = 1000e^{0.5878(3)} = 5832 \text{ mosquitos}$$

   Find t when $N(t) = 10,000$

   $$10,000 = 1000e^{0.5878t}$$
   $$10 = e^{0.5878t}$$
   $$\ln 10 = 0.5878t \rightarrow t = \frac{\ln 10}{0.5878} \approx 3.9 \text{ days}$$

7. Use $P(t) = P_0 e^{kt}$ and solve for k:

   $$2P_0 = P_0 e^{k(1.5)}$$
   $$2 = e^{1.5k}$$
   $$\ln 2 = 1.5k$$
   $$k = \frac{\ln 2}{1.5} \approx 0.4621$$

   When $t = 2$:

   $$P(2) = 10,000e^{0.4621(2)} = 25,199$$

   is the population 2 years from now.

9. Use $A = A_0 e^{kt}$ and solve for k:

   $$\frac{1}{2}A_0 = A_0 e^{k(1690)} \rightarrow \frac{1}{2} = e^{1690k}$$
   $$\ln\left(\frac{1}{2}\right) = 1690k$$
   $$k = \frac{\ln(0.5)}{1690} \approx -0.00041$$

   When $A_0 = 10$ and $t = 50$:

   $$A = 10e^{-0.00041(50)} = 9.797 \text{ grams}$$

11. Use $A = A_0 e^{kt}$ and solve for k:

$$\frac{1}{2} A_0 = A_0 e^{k(5600)}$$

$$\frac{1}{2} = e^{5600k}$$

$$\ln\left(\frac{1}{2}\right) = 5600k \rightarrow k = \frac{\ln(0.5)}{5600} \approx -0.000124$$

Solve for t when $A = 0.3A_0$:

$$0.3A_0 = A_0 e^{-0.000124\,t}$$

$$0.3 = e^{-0.000124\,t}$$

$$\ln(0.3) = -0.000124\,t$$

$$t = \frac{\ln(0.3)}{-0.000124}$$

$$\approx 9709 \text{ years ago}$$

13. (a) Using $u = T + (u_0 - T)e^{kt}$ where $t = 5$,

$T = 70,\ u_0 = 450,\ u = 300$:

$$300 = 70 + (450 - 70)e^{k(5)}$$
$$230 = 380e^{5k}$$
$$0.6053 = e^{5k}$$
$$5k = \ln(0.6053)$$
$$k = \frac{\ln(0.6053)}{5} \approx -0.1004$$

$T = 70,\ u_0 = 450,\ u = 135$:

$$135 = 70 + (450 - 70)e^{-0.1004\,t}$$
$$65 = 380e^{-0.1004\,t}$$
$$0.17105 = e^{-0.1004\,t}$$
$$-0.1004\,t = \ln(0.17105)$$
$$t = \frac{\ln(0.17105)}{-0.1004} \approx 17.6 \text{ minutes}$$

The pizza will be cool enough to eat at 5:18 p.m

(b) $T = 70,\ u_0 = 450,\ u = 160$:

$$160 = 70 + (450 - 70)e^{-0.1004\,t}$$
$$90 = 380e^{-0.1004\,t}$$
$$\frac{90}{380} = e^{-0.1004\,t}$$
$$\ln\left(\frac{90}{380}\right) = \ln e^{-0.1004\,t} = -0.1004\,t$$
$$t = \frac{\ln\left(\frac{90}{380}\right)}{-0.1004} \approx 14.35 \text{ minutes}$$

The pizza will be 160°F after about 14.3 minutes.

(c) As time passes the temperature gets closer to 70°F.

15.   Using $u = T + (u_0 - T)e^{kt}$ where $t = 3$,
      $T = 35$, $u_0 = 8$, $u = 15$:

$$15 = 35 + (8 - 35)e^{k(3)} \rightarrow -20 = -27e^{3k} \rightarrow 0.74074 = e^{3k}$$

$$3k = \ln(0.74074) \rightarrow k = \frac{\ln(0.74074)}{3} \approx -0.100035$$

At $t = 5$:
$$u = 35 + (8 - 35)e^{-0.100035\,(5)} = 18.63°C$$

At $t = 10$:
$$u = 35 + (8 - 35)e^{-0.100035\,(10)} = 25.1°C$$

17.   Use $A = A_0 e^{kt}$ and solve for k:
$$15 = 25e^{k(10)}$$

$$0.6 = e^{10k} \rightarrow \ln(0.6) = 10k \rightarrow k = \frac{\ln(0.6)}{10} \approx -0.0511$$

When $A_0 = 25$ and $t = 24$:
$$A = 25e^{-0.0511(24)} \approx 7.33 \text{ kilograms}$$

Find t when $A = \frac{1}{2}A_0$:
$$0.5 = 25\,e^{-0.0511\,t}$$

$$0.02 = e^{-0.0511\,t} \rightarrow \ln(0.02) = -0.0511t \rightarrow t = \frac{\ln(0.02)}{-0.0511} \approx 76.6 \text{ hours}$$

19.   Use $A = A_0 e^{kt}$ and solve for k:

$$\frac{1}{2}A_0 = A_0 e^{k(8)}$$

$$0.5 = e^{8k} \rightarrow \ln(0.5) = 8k$$

$$k = \frac{\ln(0.5)}{8} \approx -0.0866$$

Find t when $A = 0.1A_0$:

$$0.1A_0 = A_0 e^{-0.0866\,t}$$

$$0.1 = e^{-0.0866\,t} \rightarrow \ln(0.1) = -0.0866t$$

$$t = \frac{\ln(0.1)}{-0.0866} \approx 26.6 \text{ days}$$

The farmers need to wait about 27 days before using the hay.

21.   (a) $P(0) = \dfrac{0.9}{1 + 6e^{-0.32\,(0)}} = \dfrac{0.9}{1 + 6 \cdot 1} = \dfrac{0.9}{7} = 0.1286$

      (b)   The maximum proportion is the carrying capacity, 0.9.

      (c)
$$0.8 = \frac{0.9}{1 + 6e^{-0.32\,t}}$$

$$0.8(1 + 6e^{-0.32t}) = 0.9 \rightarrow 1 + 6e^{-0.32t} = 1.125 \rightarrow 6e^{-0.32t} = 0.125$$

$$e^{-0.32t} = 0.020833 \rightarrow -0.32t = \ln(0.020833)$$

$$t = \frac{\ln(0.020833)}{-0.32} \approx 12.1$$

80% of households will own VCR's in 1996 (t = 12).

23. (a)   As $t \to \infty$, $e^{-0.439\,t} \to 0$. Thus, $P(t) \to 1000$. The carrying capacity is 1000.

(b) $P(0) = \dfrac{1000}{1 + 32.33e^{-0.439\,(0)}} = \dfrac{1000}{33.33} = 30$

(c) $$800 = \dfrac{1000}{1 + 32.33e^{-0.439\,t}}$$

$$800\left(1 + 32.33e^{-0.439\,t}\right) = 1000 \to 1 + 32.33e^{-0.439\,t} = 1.25$$

$$32.33e^{-0.439\,t} = 0.25 \to e^{-0.439\,t} = 0.007733$$

$$-0.439t = \ln(0.007733) \to t = \dfrac{\ln(0.007733)}{-0.439} \approx 11.076 \text{ hours}$$

Chapter 4

Exponential and Logarithmic Functions

4.8 Logarithmic Scales

1. $L(10^{-5}) = 10 \log\left(\dfrac{10^{-5}}{10^{-12}}\right) = 10 \log 10^7 = 10 \cdot 7 = 70$ decibels

3. $L(0.15) = 10 \log\left(\dfrac{1.5 \times 10^{-1}}{10^{-12}}\right) = 10 \log\left(1.5 \times 10^{11}\right)$

$= 10\left(\log(1.5) + \log(10^{11})\right) = 10\left(\log(1.5) + 11\right) \approx 111.76$ decibels

5. $L(x) = 10 \log\left(\dfrac{x}{10^{-12}}\right) = 130$ decibels

$10 \log\left(\dfrac{x}{10^{-12}}\right) = 130 \to \log\left(\dfrac{x}{10^{-12}}\right) = 13$

$\log\left(x \cdot 10^{12}\right) = 13 \to \log(x) + \log(10^{12}) = 13$

$\log(x) + 12 = 13 \to \log(x) = 1 \to 10^{\log(x)} = 10^1$

$x = 10^1 = 10$ watts per square meter

7. $M(10) = \log\dfrac{10}{10^{-3}} = \log 10^4 = 4$

9. Mexico City:
$\log\left(x \cdot 10^3\right) = 8.1$

$\log(x) + \log(10^3) = 8.1$

$\log(x) + 3 = 8.1$

$\log(x) = 5.1$

$10^{\log x} = 10^{5.1}$

$x = 10^{5.1} \approx 125892.54$

San Francisco: $M(x) = \log\dfrac{x}{10^{-3}} = 6.9$

$\log\left(x \cdot 10^3\right) = 6.9$

$\log(x) + \log(10^3) = 6.9$

$\log(x) + 3 = 6.9$

$\log(x) = 3.9$

$10^{\log x} = 10^{3.9}$

$x = 10^{3.9} \approx 7943.28$

11.   Delta Center   NBA guidelines

$$L(x_1) = 10 \log\left(\frac{x_1}{10^{-12}}\right) = 110 \text{ decibels} \qquad L(x_2) = 10 \log\left(\frac{x_2}{10^{-12}}\right) = 95 \text{ decibels}$$

$$10 \log\left(\frac{x_1}{10^{-12}}\right) = 110 \qquad\qquad 10 \log\left(\frac{x_2}{10^{-12}}\right) = 95$$

$$\log\left(x_1 \cdot 10^{12}\right) = 11 \qquad\qquad \log\left(x_2 \cdot 10^{12}\right) = 9.5$$

$$\log(x_1) + \log(10^{12}) = 11 \qquad\qquad \log(x_2) + \log(10^{12}) = 9.5$$

$$\log(x_1) + 12 = 11 \qquad\qquad \log(x_2) + 12 = 9.5$$

$$\log(x_1) = -1 \qquad\qquad \log(x_2) = -2.5$$

$$10^{\log(x_1)} = 10^{-1} \qquad\qquad 10^{\log(x_2)} = 10^{-2.5}$$

$$x_1 = 10^{-1} = 0.1 \qquad\qquad x_2 = 10^{-2.5} \approx 0.0032$$

Therefore $\dfrac{x_1}{x_2} = \dfrac{0.1}{0.0032} = 31.25$, which means that the crowd noise was approximately 31 times louder than NBA guidelines allow.

Exponential and Logarithmic Functions

4.R Chapter Review

1. $f(x) = \dfrac{2x+3}{5x-2}$

    $y = \dfrac{2x+3}{5x-2}$

    $x = \dfrac{2y+3}{5y-2}$ Inverse

    $x(5y-2) = 2y+3$

    $5xy-2x = 2y+3$

    $5xy-2y = 2x+3$

    $y(5x-2) = 2x+3$

    $y = \dfrac{2x+3}{5x-2}$

    $f^{-1}(x) = \dfrac{2x+3}{5x-2}$

    Domain of f =

    range of f^{-1} = all real numbers except $\dfrac{2}{5}$

    Range of f =

    domain of f^{-1} = all real numbers except $\dfrac{2}{5}$

3. $f(x) = \dfrac{1}{x-1}$

    $y = \dfrac{1}{x-1}$

    $x = \dfrac{1}{y-1}$ Inverse

    $x(y-1) = 1$

    $xy - x = 1$

    $xy = x+1$

    $y = \dfrac{x+1}{x}$

    $f^{-1}(x) = \dfrac{x+1}{x}$

    Domain of f =

    range of f^{-1} = all real numbers except 1

    Range of f =

    domain of f^{-1} = all real numbers except 0

5. $f(x) = \dfrac{3}{x^{1/3}}$

    $y = \dfrac{3}{x^{1/3}}$

    $x = \dfrac{3}{y^{1/3}}$ Inverse

    $xy^{1/3} = 3$

    $y^{1/3} = \dfrac{3}{x}$

    $y = \dfrac{27}{x^3}$

    $f^{-1}(x) = \dfrac{27}{x^3}$

    Domain of f =
    range of f^{-1} = all real numbers except 0

    Range of f =
    domain of f^{-1} = all real numbers except 0

7. $\log_2\left(\dfrac{1}{8}\right) = \log_2 2^{-3} = -3\log_2 2 = -3$

9. $\ln e^{\sqrt{2}} = \sqrt{2}$

11. $2^{\log_2 0.4} = 0.4$

13. $\log_3\left(\dfrac{uv^2}{w}\right) = \log_3 uv^2 - \log_3 w = \log_3 u + \log_3 v^2 - \log_3 w = \log_3 u + 2\log_3 v - \log_3 w$

15. $\log\left(x^2\sqrt{x^3+1}\right) = \log x^2 + \log\left(x^3+1\right)^{1/2} = 2\log x + \dfrac{1}{2}\log\left(x^3+1\right)$

17. $\ln\left(\dfrac{x\sqrt[3]{x^2+1}}{x-3}\right) = \ln\left(x\sqrt[3]{x^2+1}\right) - \ln(x-3) = \ln x + \ln\left(x^2+1\right)^{1/3} - \ln(x-3)$

    $= \ln x + \dfrac{1}{3}\ln\left(x^2+1\right) - \ln(x-3)$

19. $3\log_4 x^2 + \dfrac{1}{2}\log_4\sqrt{x} = \log_4\left(x^2\right)^3 + \log_4\left(x^{1/2}\right)^{1/2} = \log_4 x^6 + \log_4 x^{1/4} = \log_4 x^6 \cdot x^{1/4}$

    $= \log_4 x^{25/4} = \dfrac{25}{4}\log_4 x$

21. $\ln\left(\dfrac{x-1}{x}\right) + \ln\left(\dfrac{x}{x+1}\right) - \ln\left(x^2-1\right) = \ln\left(\dfrac{x-1}{x}\cdot\dfrac{x}{x+1}\right) - \ln\left(x^2-1\right) = \ln\left[\dfrac{\left(\dfrac{x-1}{x+1}\right)}{x^2-1}\right]$

    $= \ln\left(\dfrac{x-1}{x+1}\cdot\dfrac{1}{(x-1)(x+1)}\right) = \ln\dfrac{1}{(x+1)^2} = \ln(x+1)^{-2} = -2\ln(x+1)$

23. $2\log 2 + 3\log x - \dfrac{1}{2}\left[\log(x+3) + \log(x-2)\right] = \log 2^2 + \log x^3 - \dfrac{1}{2}\log\left[(x+3)(x-2)\right]$

$$= \log 4x^3 - \log\left((x+3)(x-2)\right)^{1/2} = \log\left[\dfrac{4x^3}{\left((x+3)(x-2)\right)^{1/2}}\right]$$

25. $\log_4 19 = \dfrac{\log 19}{\log 4} \approx 2.124$

27. $\ln y = 2x^2 + \ln C$

$\ln y = \ln e^{2x^2} + \ln C$

$\ln y = \ln\left(Ce^{2x^2}\right) \rightarrow y = Ce^{2x^2}$

29. $\ln(y-3) + \ln(y+3) = x + C$

$\ln(y-3)(y+3) = x + C$

$(y-3)(y+3) = e^{x+C}$

$y^2 - 9 = e^{x+C}$

$y^2 = 9 + e^{x+C}$

$y = \sqrt{9 + e^{x+C}}$

31. $e^{y+C} = x^2 + 4$

$\ln e^{y+C} = \ln(x^2 + 4)$

$y + C = \ln(x^2 + 4)$

$y = \ln(x^2 + 4) - C$

33. $f(x) = 2^{x-3}$

Using the graph of $y = 2^x$, shift the graph
3 units to the right.
Domain: $(-\infty, \infty)$
Range: $(0, \infty)$
Horizontal Asymptote: $y = 0$

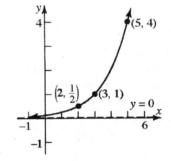

35. $f(x) = \dfrac{1}{2} \cdot 3^{-x}$

Using the graph of $y = 3^x$, reflect the
graph about the y-axis, and shrink

vertically by a factor of $\dfrac{1}{2}$.
Domain: $(-\infty, \infty)$
Range: $(0, \infty)$
Horizontal Asymptote: $y = 0$

37. $f(x) = 1 - e^x$
Using the graph of $y = e^x$, reflect about
the x-axis, and shift up 1 unit.
Domain: $(-\infty, \infty)$
Range: $(-\infty, 1)$
Horizontal Asymptote: $y = 1$

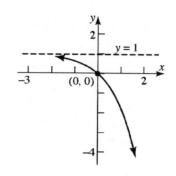

39. $f(x) = 3 + \ln x$
Using the graph of $y = \ln x$, shift the
graph up 3 units.
Domain: $(0, \infty)$
Range: $(-\infty, \infty)$
Vertical Asymptote: $x = 0$

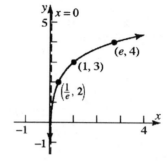

41. $f(x) = 3 - e^{-x}$
Using the graph of $y = e^x$, reflect the
graph about the y-axis, reflect about the
x-axis, and shift up 3 units.
Domain: $(-\infty, \infty)$
Range: $(-\infty, 3)$
Horizontal Asymptote: $y = 3$

43. $4^{1-2x} = 2$
$\left(2^2\right)^{1-2x} = 2$
$2^{2-4x} = 2^1$
$2 - 4x = 1 \rightarrow -4x = -1 \rightarrow x = \dfrac{1}{4}$

45. $3^{x^2 + x} = \sqrt{3}$
$3^{x^2 + x} = 3^{1/2}$

$x^2 + x = \dfrac{1}{2} \rightarrow 2x^2 + 2x - 1 = 0 \rightarrow x = \dfrac{-2 \pm \sqrt{4 - 4(2)(-1)}}{2(2)} = \dfrac{-2 \pm \sqrt{12}}{4} = \dfrac{-2 \pm 2\sqrt{3}}{4} = \dfrac{-1 \pm \sqrt{3}}{2}$

$x = \dfrac{-1 - \sqrt{3}}{2}$ or $x = \dfrac{-1 + \sqrt{3}}{2}$

47. $\log_x 64 = -3$

$x^{-3} = 64$

$\left(x^{-3}\right)^{-1/3} = 64^{-1/3} \rightarrow x = \dfrac{1}{\sqrt[3]{64}} = \dfrac{1}{4}$

49.

$5^x = 3^{x+2}$

$\log\left(5^x\right) = \log\left(3^{x+2}\right)$

$x\log 5 = (x+2)\log 3$

$x\log 5 = x\log 3 + 2\log 3$

$x\log 5 - x\log 3 = 2\log 3$

$x(\log 5 - \log 3) = 2\log 3$

$x = \dfrac{2\log 3}{\log 5 - \log 3}$

$x \approx 4.301$

51. $9^{2x} = 27^{3x-4}$

$\left(3^2\right)^{2x} = \left(3^3\right)^{3x-4}$

$3^{4x} = 3^{9x-12}$

$4x = 9x - 12 \rightarrow -5x = -12$

$x = \dfrac{12}{5}$

53. $\log_3 \sqrt{x-2} = 2$

$\sqrt{x-2} = 3^2$

$x-2 = 9^2 \rightarrow x-2 = 81 \rightarrow x = 83$

55. $8 = 4^{x^2} \cdot 2^{5x}$

$2^3 = \left(2^2\right)^{x^2} \cdot 2^{5x}$

$2^3 = 2^{2x^2+5x}$

$3 = 2x^2 + 5x \rightarrow 0 = 2x^2 + 5x - 3$

$0 = (2x-1)(x+3) \rightarrow x = \dfrac{1}{2}$ or $x = -3$

57. $\log_6(x+3) + \log_6(x+4) = 1$

$\log_6\big((x+3)(x+4)\big) = 1$

$(x+3)(x+4) = 6^1$

$x^2 + 7x + 12 = 6$

$x^2 + 7x + 6 = 0$

$(x+6)(x+1) = 0$

$x = -6$ or $x = -1$

The logarithms are undefined when $x = -6$, so $x = -1$ is the only solution.

59. $e^{1-x} = 5$

$1 - x = \ln 5$

$-x = -1 + \ln 5$

$x = 1 - \ln 5 \approx -0.609$

61.

$2^{3x} = 3^{2x+1}$

$\ln 2^{3x} = \ln 3^{2x+1}$

$3x\ln 2 = (2x+1)\ln 3$

$3x\ln 2 = 2x\ln 3 + \ln 3$

$3x\ln 2 - 2x\ln 3 = \ln 3$

$x(3\ln 2 - 2\ln 3) = \ln 3$

$x = \dfrac{\ln 3}{3\ln 2 - 2\ln 3}$

$x \approx -9.327$

63. $h(300) = \big(30(0) + 8000\big)\log\left(\dfrac{760}{300}\right) \approx 8000\log(2.53333) \approx 3229.5$ meters

65. $h(x) = 10000$

$$10000 = (30(-100) + 8000)\log\left(\frac{760}{x}\right)$$

$$10000 = (5000)\log\left(\frac{760}{x}\right)$$

$$2 = \log\left(\frac{760}{x}\right)$$

$$10^2 = 10^{\log\left(\frac{760}{x}\right)}$$

$$100 = \frac{760}{x} \to x = \frac{760}{100} = 7.6 \text{ mm}$$

67. $P = 25e^{0.1d}$

(a) $P = 25e^{0.1(4)}$
$$= 25e^{0.4}$$
$$\approx 37.3 \text{ watts}$$

(b) $50 = 25e^{0.1d}$
$$2 = e^{0.1d}$$
$$\ln 2 = 0.1d$$
$$d = \frac{\ln 2}{0.1} \approx 6.9 \text{ decibels}$$

69. (a) $P = 90 - 80\left(\frac{3}{4}\right)^5 \approx 71.02\%$ (b) $P = 90 - 80\left(\frac{3}{4}\right)^{10} \approx 85.5\%$

(c) as $t \to \infty$, $P = 90 - 80\left(\frac{3}{4}\right)^t \to 90 - 0 = 90\%$

(d) $40 = 90 - 80\left(\frac{3}{4}\right)^t \to -50 = -80\left(\frac{3}{4}\right)^t \to \frac{5}{8} = \left(\frac{3}{4}\right)^t \to \ln\left(\frac{5}{8}\right) = \ln\left(\frac{3}{4}\right)^t$

$$\to \ln\left(\frac{5}{8}\right) = t \cdot \ln\left(\frac{3}{4}\right) \to t = \frac{\ln\left(\frac{5}{8}\right)}{\ln\left(\frac{3}{4}\right)} \approx 1.63 \text{ months}$$

(e) $70 = 90 - 80\left(\frac{3}{4}\right)^t$

$$-20 = -80\left(\frac{3}{4}\right)^t \to 0.25 = \left(\frac{3}{4}\right)^t \to \ln(0.25) = \ln\left(\frac{3}{4}\right)^t$$

$$\to \ln(0.25) = t \cdot \ln\left(\frac{3}{4}\right) \to t = \frac{\ln(0.25)}{\ln\left(\frac{3}{4}\right)} \approx 4.82 \text{ months}$$

71. (a) $n = \dfrac{\log 10000 - \log 90000}{\log(1 - 0.20)} = 9.85 \text{ years}$

(b) $n = \dfrac{\log(0.5i) - \log(i)}{\log(1 - 0.15)} = \dfrac{\log\left(\frac{0.5i}{i}\right)}{\log(0.85)} = \dfrac{\log(0.5)}{\log(0.85)} \approx 4.27 \text{ years}$

73. $P = A\left(1 + \dfrac{r}{n}\right)^{-nt} = 85000\left(1 + \dfrac{0.04}{2}\right)^{-2(18)} = \$41,668.97$

75. $L(10^{-4}) = 10\log\left(\dfrac{10^{-4}}{10^{-12}}\right) = 10\log(10^8) = 10 \cdot 8 = 80 \text{ decibels}$

77. $A = A_0 e^{kt}$

$$\frac{1}{2} A_0 = A_0 e^{k(5600)} \rightarrow 0.5 = e^{5600\,k} \rightarrow \ln(0.5) = 5600k \rightarrow k = \frac{\ln(0.5)}{5600} \approx -0.000124$$

$$0.05 A_0 = A_0 e^{-0.000124\,t}$$

$$0.05 = e^{-0.000124\,t} \rightarrow \ln(0.05) = -0.000124t \rightarrow t = \frac{\ln(0.05)}{-0.000124} \approx 24{,}159 \text{ years ago}$$

79. (a) Graphing:

(b) $y = 1000\left(\sqrt{2}\right)^x$

$\sqrt{2} = e^{\ln\sqrt{2}}$

$\rightarrow y = 1000\left(e^{\ln\sqrt{2}}\right)^x = 1000 e^{(\ln\sqrt{2})x}$

$N = N_0 e^{0.3466t}, N_0 = 1000, k = 0.3466$

(d) and (e) Graphing:

(c) $N = 1000 e^{0.3466(7)} \approx 11314$ bacteria

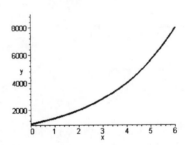

81. $P = P_0 e^{kt} = 5{,}840{,}445{,}216 e^{0.0133(3)} = 6{,}078{,}190{,}457$

83. (a) $P(0) = \dfrac{0.8}{1 + 1.67 e^{-0.16(0)}} = \dfrac{0.8}{1 + 1.67} \approx 0.2996$

(b) 0.8

(c) Graphing:

(d) Using INTERSECT we have:

75% use Windows 98 in 2018.

Chapter 5

Trigonometric Functions

5.1 Angles and Their Measure

1.

3.

5.

7.

9.

11.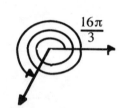

13. $30° = 30 \cdot \dfrac{\pi}{180}$ radian $= \dfrac{\pi}{6}$ radians

15. $240° = 240 \cdot \dfrac{\pi}{180}$ radian $= \dfrac{4\pi}{3}$ radians

17. $-60° = -60 \cdot \dfrac{\pi}{180}$ radian $= -\dfrac{\pi}{3}$ radians

19. $180° = 180 \cdot \dfrac{\pi}{180}$ radian $= \pi$ radians

21. $-135° = -135 \cdot \dfrac{\pi}{180}$ radian $= -\dfrac{3\pi}{4}$ radians

23. $-90° = -90 \cdot \dfrac{\pi}{180}$ radian $= -\dfrac{\pi}{2}$ radians

25. $\dfrac{\pi}{3} = \dfrac{\pi}{3} \cdot \dfrac{180}{\pi}$ degrees $= 60°$

27. $-\dfrac{5\pi}{4} = -\dfrac{5\pi}{4} \cdot \dfrac{180}{\pi}$ degrees $= -225°$

29. $\dfrac{\pi}{2} = \dfrac{\pi}{2} \cdot \dfrac{180}{\pi}$ degrees $= 90°$

31. $\dfrac{\pi}{12} = \dfrac{\pi}{12} \cdot \dfrac{180}{\pi}$ degrees $= 15°$

33. $-\dfrac{\pi}{2} = -\dfrac{\pi}{2} \cdot \dfrac{180}{\pi}$ degrees $= -90°$

35. $-\dfrac{\pi}{6} = -\dfrac{\pi}{6} \cdot \dfrac{180}{\pi}$ degrees $= -30°$

238

37. $r = 10$ meters; $\theta = \dfrac{1}{2}$ radian; $s = r\theta = 10 \cdot \dfrac{1}{2} = 5$ meters

39. $\theta = \dfrac{1}{3}$ radian; $s = 2$ feet; $s = r\theta \;\rightarrow\; r = \dfrac{s}{\theta} = \dfrac{2}{\left(\dfrac{1}{3}\right)} = 6$ feet

41. $r = 5$ miles; $s = 3$ miles; $s = r\theta$ or $\theta = \dfrac{s}{r} = \dfrac{3}{5} = 0.6$ radians

43. $r = 2$ inches; $\theta = 30°$;   Convert to radians: $30° = 30 \cdot \dfrac{\pi}{180} = \dfrac{\pi}{6}$ radians

$\qquad s = r\theta = 2 \cdot \dfrac{\pi}{6} = \dfrac{\pi}{3}$ inches

45. $r = 10$ meters; $\theta = \dfrac{1}{2}$ radian

$\qquad A = \dfrac{1}{2}r^2\theta = \dfrac{1}{2}(10)^2\left(\dfrac{1}{2}\right) = \dfrac{100}{4} = 25$ square meters

47. $\theta = \dfrac{1}{3}$ radian; $A = 2$ square feet

$\qquad A = \dfrac{1}{2}r^2\theta \rightarrow 2 = \dfrac{1}{2}r^2\left(\dfrac{1}{3}\right) = \dfrac{1}{6}r^2$

$\qquad 2 = \dfrac{1}{6}r^2 \rightarrow 12 = r^2 \rightarrow r = \sqrt{12} \approx 3.464$ feet

49. $r = 5$ miles; $A = 3$ square miles

$\qquad A = \dfrac{1}{2}r^2\theta \rightarrow 3 = \dfrac{1}{2}(5)^2\theta = \dfrac{25}{2}\theta$

$\qquad 3 = \dfrac{25}{2}\theta \rightarrow \dfrac{6}{25} = \theta \rightarrow \theta \approx 0.24$ radians

51. $r = 2$ inches; $\theta = 30°$;   Convert to radians: $30° = 30 \cdot \dfrac{\pi}{180} = \dfrac{\pi}{6}$ radians

$\qquad A = \dfrac{1}{2}r^2\theta = \dfrac{1}{2}(2)^2\left(\dfrac{\pi}{6}\right) = \dfrac{1}{2}\cdot 4\left(\dfrac{\pi}{6}\right) = \dfrac{\pi}{3} \approx 1.047$ square inches

53. $r = 2$ feet; $\theta = \dfrac{\pi}{3}$ radians

$\qquad s = r\theta = 2 \cdot \dfrac{\pi}{3} = \dfrac{2\pi}{3} \approx 2.094$ feet

$\qquad A = \dfrac{1}{2}r^2\theta = \dfrac{1}{2}(2)^2\left(\dfrac{\pi}{3}\right) = \dfrac{1}{2}\cdot 4\left(\dfrac{\pi}{3}\right) = \dfrac{2\pi}{3} \approx 2.094$ square feet

55. $r = 12$ yards; $\theta = 70°$;   Convert to radians: $70° = 70 \cdot \dfrac{\pi}{180} = \dfrac{7\pi}{18}$ radians

$s = r\theta = 12 \cdot \dfrac{7\pi}{18} \approx 14.661$ yards

$$A = \frac{1}{2}r^2\theta = \frac{1}{2}(12)^2\left(\frac{7\pi}{18}\right) = \frac{1}{2} \cdot 144\left(\frac{7\pi}{18}\right) = 72\left(\frac{7\pi}{18}\right) \approx 87.965 \text{ square yards}$$

57. $17° = 17 \cdot \dfrac{\pi}{180}$ radian $= \dfrac{17\pi}{180}$ radians ≈ 0.30 radians

59. $-40° = -40 \cdot \dfrac{\pi}{180}$ radian $= -\dfrac{2\pi}{9}$ radians ≈ -0.70 radians

61. $125° = 125 \cdot \dfrac{\pi}{180}$ radian $= \dfrac{25\pi}{36}$ radians ≈ 2.18 radians

63. 3.14 radians $= 3.14 \cdot \dfrac{180}{\pi}$ degrees $\approx 179.91°$

65. 2 radians $= 2 \cdot \dfrac{180}{\pi}$ degrees $\approx 114.59°$

67. 6.32 radians $= 6.32 \cdot \dfrac{180}{\pi}$ degrees $\approx 362.11°$

69. $40°10'25" = \left(40 + 10 \cdot \dfrac{1}{60} + 25 \cdot \dfrac{1}{60} \cdot \dfrac{1}{60}\right)^° \approx (40 + 0.1667 + 0.00694)° \approx 40.17°$

71. $1°2'3" = \left(1 + 2 \cdot \dfrac{1}{60} + 3 \cdot \dfrac{1}{60} \cdot \dfrac{1}{60}\right)^° \approx (1 + 0.0333 + 0.00083)° \approx 1.03°$

73. $9°9'9" = \left(9 + 9 \cdot \dfrac{1}{60} + 9 \cdot \dfrac{1}{60} \cdot \dfrac{1}{60}\right)^° = (9 + 0.15 + 0.0025)° \approx 9.15°$

75. $40.32° = ?$
   $0.32° = 0.32(1°) = 0.32(60') = 19.2'$
   $0.2' = 0.2(1') = 0.2(60") = 12"$
   $40.32° = 40° + 0.32° = 40° + 19.2' = 40° + 19' + 0.2' = 40° + 19' + 12" = 40°19'12"$

77. $18.255° = ?$
   $0.255° = 0.255(1°) = 0.255(60') = 15.3'$
   $0.3' = 0.3(1') = 0.3(60") = 18"$
   $18.255° = 18° + 0.255° = 18° + 15.3' = 18° + 15' + 0.3' = 18° + 15' + 18" = 18°15'18"$

79. $19.99° = ?$
   $0.99° = 0.99(1°) = 0.99(60') = 59.4'$
   $0.4' = 0.4(1') = 0.4(60") = 24"$
   $19.99° = 19° + 0.99° = 19° + 59.4' = 19° + 59' + 0.4' = 19° + 59' + 24" = 19°59'24"$

81. $r = 6$ inches; $\theta = 90° = \dfrac{\pi}{2}$ radians
   $s = r\theta = 6 \cdot \dfrac{\pi}{2} = 3\pi$ inches ≈ 9.42 inches
   $r = 6$ inches; $\theta = \dfrac{25}{60}$ rev $= \dfrac{5}{12} \cdot 360° = 150° = \dfrac{5\pi}{6}$ radians
   $s = r\theta = 6 \cdot \dfrac{5\pi}{6} = 5\pi$ inches ≈ 15.71 inches

83. $r = 4$ m; $\theta = 45°$;   Convert to radians : $45° = 45 \cdot \dfrac{\pi}{180} = \dfrac{\pi}{4}$ radians
   $A = \dfrac{1}{2}r^2\theta = \dfrac{1}{2}(4)^2\left(\dfrac{\pi}{4}\right) = \dfrac{1}{2} \cdot 16\left(\dfrac{\pi}{4}\right) = 2\pi \approx 6.283$ square meters

85. $r = 30$ feet; $\theta = 135°$;   Convert to radians : $135° = 135 \cdot \dfrac{\pi}{180} = \dfrac{3\pi}{4}$ radians
   $A = \dfrac{1}{2}r^2\theta = \dfrac{1}{2}(30)^2\left(\dfrac{3\pi}{4}\right) = \dfrac{1}{2} \cdot (900)\left(\dfrac{3\pi}{4}\right) = \dfrac{2700\pi}{8} \approx 1060.29$ square feet

87. $r = 5$ cm.; $t = 20$ seconds; $\theta = \dfrac{1}{3}$ radian
   $\omega = \dfrac{\theta}{t} = \dfrac{\left(\frac{1}{3}\right)}{20} = \dfrac{1}{3} \cdot \dfrac{1}{20} = \dfrac{1}{60}$ radian/sec; $v = \dfrac{s}{t} = \dfrac{r\theta}{t} = \dfrac{5 \cdot \left(\frac{1}{3}\right)}{20} = \dfrac{5}{3} \cdot \dfrac{1}{20} = \dfrac{1}{12}$ cm/sec

89. $d = 26$ inches; $r = 13$ inches; $v = 35$ mi / hr
   $v = \dfrac{35 \text{ mi}}{\text{hr}} \cdot \dfrac{5280 \text{ ft}}{\text{mi}} \cdot \dfrac{12 \text{ in}}{\text{ft}} \cdot \dfrac{1 \text{ hr}}{60 \text{ min}} = 36960$ in/min
   $\omega = \dfrac{v}{r} = \dfrac{36960 \text{ in/min}}{13 \text{ in}} = 2843.08$ radians/min
   $= \dfrac{2843.08 \text{ rad}}{\text{min}} \cdot \dfrac{1 \text{ rev}}{2\pi \text{ rad}} \approx 452.5$ rev/min

91. $r = 3960$ miles; $\theta = 35°9' - 29°57' = 5°12' = 5.2° = 5.2 \cdot \dfrac{\pi}{180} \approx 0.09076$ radian
   $s = r\theta = 3960 \cdot 0.09076 \approx 359.4$ miles

93. $r = 3429.5$ miles; $\omega = 1$ rev / day $= 2\pi$ radians / day $= \dfrac{\pi}{12}$ radians / hr
   $v = r\omega = 3429.5 \cdot \dfrac{\pi}{12} \approx 897.8$ miles/hr

95. $r = 2.39 \times 10^5$ miles;

$$\omega = 1 \text{ rev}/27.3 \text{ days} = 2\pi \text{ radians}/27.3 \text{ day} = \frac{\pi}{12 \cdot 27.3} \text{ radians/hr}$$

$$v = r\omega = \left(2.39 \times 10^5\right) \cdot \frac{\pi}{327.6} \approx 2292 \text{ miles/hr}$$

97. $r_1 = 2$ inches; $r_2 = 8$ inches; $\omega_1 = 3$ rev / min = 6π radians / min
Find ω_2:

$$v_1 = v_2$$

$$r_1\omega_1 = r_2\omega_2 \rightarrow 2(6\pi) = 8\omega_2$$

$$\omega_2 = \frac{12\pi}{8} = 1.5\pi \text{ radians/min} = \frac{1.5\pi}{2\pi} \text{ rev/min} = \frac{3}{4} \text{ rev/min}$$

99. $r = 4$ feet; $\omega = 10$ rev / min = 20π radians / min

$$v = r\omega = 4 \cdot 20\pi = 80\pi \ \frac{\text{ft}}{\text{min}} = \frac{80\pi \text{ ft}}{\text{min}} \cdot \frac{1 \text{ mi}}{5280 \text{ ft}} \cdot \frac{60 \text{ min}}{\text{hr}} \approx 2.86 \text{ mi/hr}$$

101. $d = 8.5$ feet; $r = 4.25$ feet; $v = 9.55$ mi/hr

$$\omega = \frac{v}{r} = \frac{9.55 \text{ mi/hr}}{4.25 \text{ ft}} = \frac{9.55 \text{ mi}}{\text{hr}} \cdot \frac{1}{4.25 \text{ ft}} \cdot \frac{5280 \text{ ft}}{\text{mi}} \cdot \frac{1 \text{ hr}}{60 \text{ min}} \cdot \frac{1 \text{ rev}}{2\pi} \approx 31.47 \text{ rev/min}$$

103. The earth makes one full rotation in 24 hours. The distance traveled in 24 hours is the circumference of the earth. At the equator the circumference is $2\pi(3960)$ miles. Therefore, the linear velocity a person must travel to keep up with the sun is:

$$v = \frac{s}{t} = \frac{2\pi(3960)}{24} \approx 1037 \text{ miles / hr}$$

105. r_1 rotates at ω_1 rev / min; r_2 rotates at ω_2 rev / min;
$v = r_1\omega_1 = r_2\omega_2$

So, $\dfrac{r_1}{r_2} = \dfrac{\omega_2}{\omega_1}$

107. Answers will vary.

Trigonometric Functions

5.2 Trigonometric Functions: Unit Circle Approach

1. $P = \left(\dfrac{\sqrt{2}}{2}, \dfrac{1}{2} \right) \Rightarrow x = \dfrac{\sqrt{2}}{2}, y = \dfrac{1}{2}$

$\sin t = \dfrac{1}{2};$
$\qquad \csc t = \dfrac{1}{\left(\dfrac{1}{2} \right)} = 1 \cdot \left(\dfrac{2}{1} \right) = 2$

$\cos t = \dfrac{\sqrt{2}}{2};$
$\qquad \sec t = \dfrac{1}{\left(\dfrac{\sqrt{2}}{2} \right)} = 1 \cdot \left(\dfrac{2}{\sqrt{2}} \right) = \dfrac{2}{\sqrt{2}} \cdot \dfrac{\sqrt{2}}{\sqrt{2}} = \dfrac{2\sqrt{2}}{2} = \sqrt{2}$

$\tan t = \dfrac{\left(\dfrac{1}{2} \right)}{\left(\dfrac{\sqrt{2}}{2} \right)} = \left(\dfrac{1}{2} \right) \cdot \left(\dfrac{2}{\sqrt{2}} \right)$
$\qquad \cot t = \dfrac{\left(\dfrac{\sqrt{2}}{2} \right)}{\left(\dfrac{1}{2} \right)} = \left(\dfrac{\sqrt{2}}{2} \right) \cdot \left(\dfrac{2}{1} \right) = \sqrt{2}$

$\qquad\qquad = \dfrac{1}{\sqrt{2}} \cdot \dfrac{\sqrt{2}}{\sqrt{2}} = \dfrac{\sqrt{2}}{2}$

3. $\left(-\dfrac{2}{5}, \dfrac{\sqrt{21}}{5} \right) \Rightarrow x = -\dfrac{2}{5}, y = \dfrac{\sqrt{21}}{5}$

$\sin t = \dfrac{\sqrt{21}}{5}$
$\qquad \tan t = \dfrac{\left(\dfrac{\sqrt{21}}{5} \right)}{\left(-\dfrac{2}{5} \right)} = \dfrac{\sqrt{21}}{5} \cdot \left(-\dfrac{5}{2} \right) = -\dfrac{\sqrt{21}}{2}$

$\cos t = -\dfrac{2}{5}$
$\qquad \sec t = \dfrac{1}{\left(-\dfrac{2}{5} \right)} = 1 \cdot \left(-\dfrac{5}{2} \right) = -\dfrac{5}{2}$

$\csc t = \dfrac{1}{\left(\dfrac{\sqrt{21}}{5} \right)} = 1 \cdot \dfrac{5}{\sqrt{21}} \cdot \dfrac{\sqrt{21}}{\sqrt{21}} = \dfrac{5\sqrt{21}}{21}$
$\qquad \cot t = \dfrac{\left(-\dfrac{2}{5} \right)}{\left(\dfrac{\sqrt{21}}{5} \right)} = -\dfrac{2}{5} \cdot \dfrac{5}{\sqrt{21}} \cdot \dfrac{\sqrt{21}}{\sqrt{21}} = -\dfrac{2\sqrt{21}}{21}$

5. $P = \left(-\dfrac{\sqrt{2}}{2}, \dfrac{\sqrt{2}}{2}\right) \Rightarrow x = -\dfrac{\sqrt{2}}{2}, y = \dfrac{\sqrt{2}}{2}$

$\sin t = \dfrac{\sqrt{2}}{2}$ $\csc t = \dfrac{1}{\left(\dfrac{\sqrt{2}}{2}\right)} = 1 \cdot \left(\dfrac{2}{\sqrt{2}}\right) \cdot \dfrac{\sqrt{2}}{\sqrt{2}} = \dfrac{2\sqrt{2}}{2} = \sqrt{2}$

$\cos t = -\dfrac{\sqrt{2}}{2}$ $\sec t = \dfrac{1}{\left(-\dfrac{\sqrt{2}}{2}\right)} = 1 \cdot \left(-\dfrac{2}{\sqrt{2}}\right) \cdot \dfrac{\sqrt{2}}{\sqrt{2}} = -\dfrac{2\sqrt{2}}{2} = -\sqrt{2}$

$\tan t = \dfrac{\left(\dfrac{\sqrt{2}}{2}\right)}{\left(-\dfrac{\sqrt{2}}{2}\right)} = \left(\dfrac{\sqrt{2}}{2}\right) \cdot \left(-\dfrac{2}{\sqrt{2}}\right) = -1$ $\cot t = \dfrac{\left(-\dfrac{\sqrt{2}}{2}\right)}{\left(\dfrac{\sqrt{2}}{2}\right)} = \left(-\dfrac{\sqrt{2}}{2}\right) \cdot \left(\dfrac{2}{\sqrt{2}}\right) = -1$

7. $\left(\dfrac{2\sqrt{2}}{3}, -\dfrac{1}{3}\right) \Rightarrow x = \dfrac{2\sqrt{2}}{3}, y = -\dfrac{1}{3}$

$\sin t = -\dfrac{1}{3}$ $\csc t = \dfrac{1}{\left(-\dfrac{1}{3}\right)} = 1 \cdot \left(-\dfrac{3}{1}\right) = -3$

$\cos t = \dfrac{2\sqrt{2}}{3}$ $\sec t = \dfrac{1}{\left(\dfrac{2\sqrt{2}}{3}\right)} = 1 \cdot \dfrac{3}{2\sqrt{2}} \cdot \dfrac{\sqrt{2}}{\sqrt{2}} = \dfrac{3\sqrt{2}}{4}$

$\tan t = \dfrac{\left(-\dfrac{1}{3}\right)}{\left(\dfrac{2\sqrt{2}}{3}\right)} = -\dfrac{1}{3} \cdot \dfrac{3}{2\sqrt{2}} \cdot \dfrac{\sqrt{2}}{\sqrt{2}} = -\dfrac{\sqrt{2}}{4}$ $\cot t = \dfrac{\left(\dfrac{2\sqrt{2}}{3}\right)}{\left(-\dfrac{1}{3}\right)} = \dfrac{2\sqrt{2}}{3} \cdot \left(-\dfrac{3}{1}\right) = -2\sqrt{2}$

9. $\sin\left(\dfrac{11\pi}{2}\right) = \sin\left(\dfrac{3\pi}{2} + \dfrac{8\pi}{2}\right) = \sin\left(\dfrac{3\pi}{2} + 4\pi\right) = \sin\left(\dfrac{3\pi}{2} + 2 \cdot 2\pi\right) = \sin\left(\dfrac{3\pi}{2}\right) = -1$

11. $\tan(6\pi) = \tan(0 + 6\pi) = \tan(0) = 0$

13. $\csc\left(\dfrac{11\pi}{2}\right) = \csc\left(\dfrac{3\pi}{2} + \dfrac{8\pi}{2}\right) = \csc\left(\dfrac{3\pi}{2} + 4\pi\right) = \csc\left(\dfrac{3\pi}{2} + 2 \cdot 2\pi\right) = \csc\left(\dfrac{3\pi}{2}\right) = -1$

15. $\cos\left(-\dfrac{3\pi}{2}\right) = \cos\left(\dfrac{3\pi}{2}\right) = \cos\left(-\dfrac{\pi}{2} + \dfrac{4\pi}{2}\right) = \cos\left(-\dfrac{\pi}{2} + 2\pi\right) = \cos\left(-\dfrac{\pi}{2}\right) = \cos\left(\dfrac{\pi}{2}\right) = 0$

17. $\sec(-\pi) = \sec(\pi) = -1$

19. $\sin(45°) + \cos(60°) = \dfrac{\sqrt{2}}{2} + \dfrac{1}{2} = \dfrac{1+\sqrt{2}}{2}$

21. $\sin(90°) + \tan(45°) = 1 + 1 = 2$

23. $\sin(45°)\cos(45°) = \dfrac{\sqrt{2}}{2} \cdot \dfrac{\sqrt{2}}{2} = \dfrac{2}{4} = \dfrac{1}{2}$

25. $\csc(45°)\tan(60°) = \sqrt{2} \cdot \sqrt{3} = \sqrt{6}$

27. $4\sin(90°) - 3\tan(180°) = 4 \cdot 1 - 3 \cdot 0 = 4$

29. $2\sin\left(\dfrac{\pi}{3}\right) - 3\tan\left(\dfrac{\pi}{6}\right) = 2 \cdot \dfrac{\sqrt{3}}{2} - 3 \cdot \dfrac{\sqrt{3}}{3} = \sqrt{3} - \sqrt{3} = 0$

31. $\sin\left(\dfrac{\pi}{4}\right) - \cos\left(\dfrac{\pi}{4}\right) = \dfrac{\sqrt{2}}{2} - \dfrac{\sqrt{2}}{2} = 0$

33. $2\sec\left(\dfrac{\pi}{4}\right) + 4\cot\left(\dfrac{\pi}{3}\right) = 2 \cdot \sqrt{2} + 4 \cdot \dfrac{\sqrt{3}}{3} = 2\sqrt{2} + \dfrac{4\sqrt{3}}{3}$

35. $\tan(\pi) - \cos(0) = 0 - 1 = -1$

37. $\csc\left(\dfrac{\pi}{2}\right) + \cot\left(\dfrac{\pi}{2}\right) = 1 + 0 = 1$

39. The point on the unit circle that corresponds to $\theta = \dfrac{2\pi}{3} = 120°$ is $\left(-\dfrac{1}{2}, \dfrac{\sqrt{3}}{2}\right)$.

$\sin\theta = \dfrac{\sqrt{3}}{2}$
$\qquad \csc\theta = \dfrac{1}{\left(\dfrac{\sqrt{3}}{2}\right)} = 1 \cdot \dfrac{2}{\sqrt{3}} \cdot \dfrac{\sqrt{3}}{\sqrt{3}} = \dfrac{2\sqrt{3}}{3}$

$\cos\theta = -\dfrac{1}{2}$
$\qquad \sec\theta = \dfrac{1}{\left(-\dfrac{1}{2}\right)} = 1 \cdot \left(-\dfrac{2}{1}\right) = -2$

$\tan\theta = \dfrac{\left(\dfrac{\sqrt{3}}{2}\right)}{\left(-\dfrac{1}{2}\right)} = \dfrac{\sqrt{3}}{2} \cdot \left(-\dfrac{2}{1}\right) = -\sqrt{3}$
$\qquad \cot\theta = \dfrac{\left(-\dfrac{1}{2}\right)}{\left(\dfrac{\sqrt{3}}{2}\right)} = -\dfrac{1}{2} \cdot \dfrac{2}{\sqrt{3}} \cdot \dfrac{\sqrt{3}}{\sqrt{3}} = -\dfrac{\sqrt{3}}{3}$

Chapter 5 Trigonometric Functions

41. The point on the unit circle that corresponds to $\theta = 210° = \dfrac{7\pi}{6}$ is $\left(-\dfrac{\sqrt{3}}{2}, -\dfrac{1}{2}\right)$.

$\sin\theta = -\dfrac{1}{2}$

$\csc\theta = \dfrac{1}{\left(-\dfrac{1}{2}\right)} = 1\cdot\left(-\dfrac{2}{1}\right) = -2$

$\cos\theta = -\dfrac{\sqrt{3}}{2}$

$\sec\theta = \dfrac{1}{\left(-\dfrac{\sqrt{3}}{2}\right)} = 1\cdot\left(-\dfrac{2}{\sqrt{3}}\right)\cdot\dfrac{\sqrt{3}}{\sqrt{3}} = -\dfrac{2\sqrt{3}}{3}$

$\tan\theta = \dfrac{\left(-\dfrac{1}{2}\right)}{\left(-\dfrac{\sqrt{3}}{2}\right)} = -\dfrac{1}{2}\cdot\left(-\dfrac{2}{\sqrt{3}}\right)\cdot\dfrac{\sqrt{3}}{\sqrt{3}} = \dfrac{\sqrt{3}}{3}$

$\cot\theta = \dfrac{\left(-\dfrac{\sqrt{3}}{2}\right)}{\left(-\dfrac{1}{2}\right)} = -\dfrac{\sqrt{3}}{2}\cdot\left(-\dfrac{2}{1}\right) = \sqrt{3}$

43. The point on the unit circle that corresponds to $\theta = \dfrac{3\pi}{4} = 135°$ is $\left(-\dfrac{\sqrt{2}}{2}, \dfrac{\sqrt{2}}{2}\right)$.

$\sin\theta = \dfrac{\sqrt{2}}{2}$

$\csc\theta = \dfrac{1}{\left(\dfrac{\sqrt{2}}{2}\right)} = 1\cdot\dfrac{2}{\sqrt{2}}\cdot\dfrac{\sqrt{2}}{\sqrt{2}} = \dfrac{2\sqrt{2}}{2} = \sqrt{2}$

$\cos\theta = -\dfrac{\sqrt{2}}{2}$

$\sec\theta = \dfrac{1}{\left(-\dfrac{\sqrt{2}}{2}\right)} = 1\cdot\left(-\dfrac{2}{\sqrt{2}}\right)\cdot\dfrac{\sqrt{2}}{\sqrt{2}} = -\dfrac{2\sqrt{2}}{2} = -\sqrt{2}$

$\tan\theta = \dfrac{\left(\dfrac{\sqrt{2}}{2}\right)}{\left(-\dfrac{\sqrt{2}}{2}\right)} = \dfrac{\sqrt{2}}{2}\cdot\left(-\dfrac{2}{\sqrt{2}}\right) = -1$

$\cot\theta = \dfrac{\left(-\dfrac{\sqrt{2}}{2}\right)}{\left(\dfrac{\sqrt{2}}{2}\right)} = -\dfrac{\sqrt{2}}{2}\cdot\dfrac{2}{\sqrt{2}} = -1$

45. The point on the unit circle that corresponds to $\theta = \dfrac{8\pi}{3} = 480°$ is $\left(-\dfrac{1}{2}, \dfrac{\sqrt{3}}{2}\right)$.

$\sin\theta = \dfrac{\sqrt{3}}{2}$

$\csc\theta = \dfrac{1}{\left(\dfrac{\sqrt{3}}{2}\right)} = 1\cdot\dfrac{2}{\sqrt{3}}\cdot\dfrac{\sqrt{3}}{\sqrt{3}} = \dfrac{2\sqrt{3}}{3}$

$\cos\theta = -\dfrac{1}{2}$

$\sec\theta = \dfrac{1}{\left(-\dfrac{1}{2}\right)} = 1\cdot\left(-\dfrac{2}{1}\right) = -2$

$$\tan\theta = \dfrac{\left(\dfrac{\sqrt{3}}{2}\right)}{\left(-\dfrac{1}{2}\right)} = \dfrac{\sqrt{3}}{2}\cdot\left(-\dfrac{2}{1}\right) = -\sqrt{3} \qquad\qquad \cot\theta = \dfrac{\left(-\dfrac{1}{2}\right)}{\left(\dfrac{\sqrt{3}}{2}\right)} = -\dfrac{1}{2}\cdot\dfrac{2}{\sqrt{3}}\cdot\dfrac{\sqrt{3}}{\sqrt{3}} = -\dfrac{\sqrt{3}}{3}$$

47. The point on the unit circle that corresponds to $\theta = 405° = \dfrac{9\pi}{4}$ is $\left(\dfrac{\sqrt{2}}{2},\dfrac{\sqrt{2}}{2}\right)$.

$$\sin\theta = \dfrac{\sqrt{2}}{2} \qquad\qquad \csc\theta = \dfrac{1}{\left(\dfrac{\sqrt{2}}{2}\right)} = 1\cdot\dfrac{2}{\sqrt{2}}\cdot\dfrac{\sqrt{2}}{\sqrt{2}} = \sqrt{2}$$

$$\cos\theta = \dfrac{\sqrt{2}}{2} \qquad\qquad \sec\theta = \dfrac{1}{\left(\dfrac{\sqrt{2}}{2}\right)} = 1\cdot\dfrac{2}{\sqrt{2}}\cdot\dfrac{\sqrt{2}}{\sqrt{2}} = \sqrt{2}$$

$$\tan\theta = \dfrac{\left(\dfrac{\sqrt{2}}{2}\right)}{\left(\dfrac{\sqrt{2}}{2}\right)} = \dfrac{\sqrt{2}}{2}\cdot\dfrac{2}{\sqrt{2}} = 1 \qquad\qquad \cot\theta = \dfrac{\left(\dfrac{\sqrt{2}}{2}\right)}{\left(\dfrac{\sqrt{2}}{2}\right)} = 1$$

49. The point on the unit circle that corresponds to $\theta = -\dfrac{\pi}{6} = -30°$ is $\left(\dfrac{\sqrt{3}}{2},-\dfrac{1}{2}\right)$.

$$\sin\theta = -\dfrac{1}{2} \qquad\qquad \csc\theta = \dfrac{1}{\left(-\dfrac{1}{2}\right)} = 1\cdot\left(-\dfrac{2}{1}\right) = -2$$

$$\cos\theta = \dfrac{\sqrt{3}}{2} \qquad\qquad \sec\theta = \dfrac{1}{\left(\dfrac{\sqrt{3}}{2}\right)} = 1\cdot\dfrac{2}{\sqrt{3}}\cdot\dfrac{\sqrt{3}}{\sqrt{3}} = \dfrac{2\sqrt{3}}{3}$$

$$\tan\theta = \dfrac{\left(-\dfrac{1}{2}\right)}{\left(\dfrac{\sqrt{3}}{2}\right)} = -\dfrac{1}{2}\cdot\dfrac{2}{\sqrt{3}}\cdot\dfrac{\sqrt{3}}{\sqrt{3}} = -\dfrac{\sqrt{3}}{3} \qquad \cot\theta = \dfrac{\left(\dfrac{\sqrt{3}}{2}\right)}{\left(-\dfrac{1}{2}\right)} = \dfrac{\sqrt{3}}{2}\cdot\left(-\dfrac{2}{1}\right) = -\sqrt{3}$$

51. The point on the unit circle that corresponds to $\theta = -45° = -\dfrac{\pi}{4}$ is $\left(\dfrac{\sqrt{2}}{2},-\dfrac{\sqrt{2}}{2}\right)$.

$$\sin\theta = -\dfrac{\sqrt{2}}{2} \qquad\qquad \csc\theta = \dfrac{1}{\left(-\dfrac{\sqrt{2}}{2}\right)} = 1\cdot\left(-\dfrac{2}{\sqrt{2}}\right)\cdot\dfrac{\sqrt{2}}{\sqrt{2}} = -\sqrt{2}$$

$$\cos\theta = \frac{\sqrt{2}}{2}$$

$$\sec\theta = \frac{1}{\left(\frac{\sqrt{2}}{2}\right)} = 1\cdot\frac{2}{\sqrt{2}}\cdot\frac{\sqrt{2}}{\sqrt{2}} = \sqrt{2}$$

$$\tan\theta = \frac{\left(-\frac{\sqrt{2}}{2}\right)}{\left(\frac{\sqrt{2}}{2}\right)} = -\frac{\sqrt{2}}{2}\cdot\frac{2}{\sqrt{2}} = -1$$

$$\cot\theta = \frac{\left(\frac{\sqrt{2}}{2}\right)}{\left(-\frac{\sqrt{2}}{2}\right)} = -1$$

53.   The point on the unit circle that corresponds to $\theta = \dfrac{5\pi}{2} = 450°$ is $(0,1)$.

$$\sin\theta = 1 \qquad\qquad\qquad \csc\theta = \frac{1}{1} = 1$$

$$\cos\theta = 0 \qquad\qquad\qquad \sec\theta = \frac{1}{0} = \text{not defined}$$

$$\tan\theta = \frac{1}{0} = \text{not defined} \qquad\qquad \cot\theta = \frac{0}{1} = 0$$

55.   The point on the unit circle that corresponds to $\theta = 720° = 4\pi$ is $(1,0)$.

$$\sin\theta = 0 \qquad\qquad\qquad \csc\theta = \frac{1}{0} = \text{not defined}$$

$$\cos\theta = 1 \qquad\qquad\qquad \sec\theta = \frac{1}{1} = 1$$

$$\tan\theta = \frac{0}{1} = 0 \qquad\qquad\qquad \cot\theta = \frac{1}{0} = \text{not defined}$$

57.   Set the calculator to degree mode : $\sin(28°) \approx 0.47$

59.   Set the calculator to degree mode : $\tan(21°) \approx 0.38$

61.   Set the calculator to degree mode : $\sec(41°) = \dfrac{1}{\cos(41°)} \approx 1.33$

63.   Set the calculator to radian mode : $\sin\left(\dfrac{\pi}{10}\right) \approx 0.31$

65.   Set the calculator to radian mode : $\tan\left(\dfrac{5\pi}{12}\right) \approx 3.73$

67.   Set the calculator to radian mode : $\sec\left(\dfrac{\pi}{12}\right) = \dfrac{1}{\cos\left(\dfrac{\pi}{12}\right)} \approx 1.04$

69. Set the calculator to radian mode: $\sin(1) \approx 0.84$.

71. Set the calculator to degree mode: $\sin(1°) \approx 0.02$.

73. For the point $(-3, 4)$, $x = -3$, $y = 4$, $r = \sqrt{x^2 + y^2} = \sqrt{9 + 16} = \sqrt{25} = 5$

$$\sin\theta = \frac{4}{5} \qquad \cos\theta = -\frac{3}{5} \qquad \tan\theta = -\frac{4}{3}$$

$$\csc\theta = \frac{5}{4} \qquad \sec\theta = -\frac{5}{3} \qquad \cot\theta = -\frac{3}{4}$$

75. For the point $(2, -3)$, $x = 2$, $y = -3$, $r = \sqrt{x^2 + y^2} = \sqrt{4 + 9} = \sqrt{13}$

$$\sin\theta = \frac{-3}{\sqrt{13}} \cdot \frac{\sqrt{13}}{\sqrt{13}} = -\frac{3\sqrt{13}}{13} \qquad \cos\theta = \frac{2}{\sqrt{13}} \cdot \frac{\sqrt{13}}{\sqrt{13}} = \frac{2\sqrt{13}}{13} \qquad \tan\theta = -\frac{3}{2}$$

$$\csc\theta = -\frac{\sqrt{13}}{3} \qquad\qquad \sec\theta = \frac{\sqrt{13}}{2} \qquad\qquad \cot\theta = -\frac{2}{3}$$

77. For the point $(-2, -2)$, $x = -2$, $y = -2$, $r = \sqrt{x^2 + y^2} = \sqrt{4 + 4} = \sqrt{8} = 2\sqrt{2}$

$$\sin\theta = \frac{-2}{2\sqrt{2}} \cdot \frac{\sqrt{2}}{\sqrt{2}} = -\frac{\sqrt{2}}{2} \qquad \cos\theta = \frac{-2}{2\sqrt{2}} \cdot \frac{\sqrt{2}}{\sqrt{2}} = \frac{-\sqrt{2}}{2} \qquad \tan\theta = \frac{-2}{-2} = 1$$

$$\csc\theta = \frac{2\sqrt{2}}{-2} = -\sqrt{2} \qquad\qquad \sec\theta = \frac{2\sqrt{2}}{-2} = -\sqrt{2} \qquad\qquad \cot\theta = \frac{-2}{-2} = 1$$

79. For the point $(-3, -2)$, $x = -3$, $y = -2$, $r = \sqrt{x^2 + y^2} = \sqrt{9 + 4} = \sqrt{13}$

$$\sin\theta = \frac{-2}{\sqrt{13}} \cdot \frac{\sqrt{13}}{\sqrt{13}} = -\frac{2\sqrt{13}}{13} \qquad \cos\theta = \frac{-3}{\sqrt{13}} \cdot \frac{\sqrt{13}}{\sqrt{13}} = -\frac{3\sqrt{13}}{13} \qquad \tan\theta = \frac{-2}{-3} = \frac{2}{3}$$

$$\csc\theta = \frac{\sqrt{13}}{-2} = -\frac{\sqrt{13}}{2} \qquad\quad \sec\theta = \frac{\sqrt{13}}{-3} = -\frac{\sqrt{13}}{3} \qquad\quad \cot\theta = \frac{-3}{-2} = \frac{3}{2}$$

81. For the point $\left(\frac{1}{3}, -\frac{1}{4}\right)$, $x = \frac{1}{3}$, $y = -\frac{1}{4}$, $r = \sqrt{x^2 + y^2} = \sqrt{\frac{1}{9} + \frac{1}{16}} = \sqrt{\frac{25}{144}} = \frac{5}{12}$

$$\sin\theta = \frac{\left(-\frac{1}{4}\right)}{\left(\frac{5}{12}\right)} = -\frac{1}{4} \cdot \frac{12}{5} = -\frac{3}{5} \qquad \csc\theta = \frac{\left(\frac{5}{12}\right)}{\left(-\frac{1}{4}\right)} = \frac{5}{12} \cdot -\frac{4}{1} = -\frac{5}{3}$$

$$\cos\theta = \frac{\left(\frac{1}{3}\right)}{\left(\frac{5}{12}\right)} = \frac{1}{3} \cdot \frac{12}{5} = \frac{4}{5} \qquad \sec\theta = \frac{\left(\frac{5}{12}\right)}{\left(\frac{1}{3}\right)} = \frac{5}{12} \cdot \frac{3}{1} = \frac{5}{4}$$

$$\tan\theta = \frac{\left(-\frac{1}{4}\right)}{\left(\frac{1}{3}\right)} = -\frac{1}{4}\cdot\frac{3}{1} = -\frac{3}{4} \qquad \cot\theta = \frac{\left(\frac{1}{3}\right)}{\left(-\frac{1}{4}\right)} = \frac{1}{3}\cdot-\frac{4}{1} = -\frac{4}{3}$$

83. $\sin(45°) + \sin(135°) + \sin(225°) + \sin(315°) = \dfrac{\sqrt{2}}{2} + \dfrac{\sqrt{2}}{2} - \dfrac{\sqrt{2}}{2} - \dfrac{\sqrt{2}}{2} = 0$

85. If $\sin\theta = 0.1$, then $\sin(\theta + \pi) = -0.1$

87. If $\tan\theta = 3$, then $\tan(\theta + \pi) = 3$

89. If $\sin\theta = \dfrac{1}{5}$, then $\csc\theta = \dfrac{1}{\left(\frac{1}{5}\right)} = 1\cdot\dfrac{5}{1} = 5$

91. $\sin(60°) = \dfrac{\sqrt{3}}{2}$
93. $\sin\left(\dfrac{60°}{2}\right) = \sin(30°) = \dfrac{1}{2}$

95. $(\sin 60°)^2 = \left(\dfrac{\sqrt{3}}{2}\right)^2 = \dfrac{3}{4}$
97. $\sin(2\cdot 60°) = \sin(120°) = \dfrac{\sqrt{3}}{2}$

99. $2\sin(60°) = 2\cdot\dfrac{\sqrt{3}}{2} = \sqrt{3}$
101. $\sin(-60°) = \sin(300°) = -\dfrac{\sqrt{3}}{2}$

103. Complete the table:

θ	0.5	0.4	0.2	0.1	0.01	0.001	0.0001	0.00001
$\sin\theta$	0.4794	0.3894	0.1987	0.0998	0.0100	0.0010	0.0001	0.00001
$\dfrac{\sin\theta}{\theta}$	0.9589	0.9735	0.9933	0.9983	1.0000	1.0000	1.0000	1.0000

As θ approaches 0, the ratio $\dfrac{\sin\theta}{\theta}$ is 1.

105. Use the formula $R = \dfrac{v_0^{\,2}\sin(2\theta)}{g}$ with $g = 32.2\text{ft / sec}^2$; $\theta = 45°$; $v_0 = 100\text{ ft / sec}$:

$$R = \frac{100^2\sin(2(45°))}{32.2} = \frac{10000\sin(90°)}{32.2} = \frac{10000}{32.2} \approx 310.56 \text{ feet}$$

Use the formula $H = \dfrac{v_0^{\,2}\sin^2\theta}{2g}$ with $g = 32.2\text{ft / sec}^2$; $\theta = 45°$; $v_0 = 100\text{ ft / sec}$:

$$H = \frac{100^2\sin^2(45°)}{2(32.2)} \approx \frac{10000(0.7071)^2}{64.4} \approx 77.64 \text{ feet}$$

107. Use the formula $R = \dfrac{v_0^2 \sin(2\theta)}{g}$ with $g = 9.8$ m / sec^2; $\theta = 25°$; $v_0 = 500$ m / sec :

$$R = \frac{500^2 \sin(2(25°))}{9.8} = \frac{250{,}000 \sin 50°}{9.8} \approx 19{,}542 \text{ meters}$$

Use the formula $H = \dfrac{v_0^2 \sin^2 \theta}{2g}$ with $g = 9.8$ m / sec^2; $\theta = 25°$; $v_0 = 500$ m / sec :

$$H = \frac{500^2 \sin^2 (25°)}{2(9.8)} \approx \frac{250{,}000(0.4226)^2}{19.6} \approx 2278 \text{ meters}$$

109. Use the formula $t = \sqrt{\dfrac{2a}{g \sin \theta \cos \theta}}$ with $g = 32$ ft / sec^2 and $a = 10$ feet :

(a) $t = \sqrt{\dfrac{2(10)}{32 \sin(30°) \cos(30°)}} = \sqrt{\dfrac{20}{\left(32 \cdot \dfrac{1}{2} \cdot \dfrac{\sqrt{3}}{2}\right)}} = \sqrt{\dfrac{20}{8\sqrt{3}}} = \sqrt{\dfrac{5}{2\sqrt{3}}} \approx 1.20$ seconds

(b) $t = \sqrt{\dfrac{2(10)}{32 \sin(45°) \cos(45°)}} = \sqrt{\dfrac{20}{\left(32 \cdot \dfrac{\sqrt{2}}{2} \cdot \dfrac{\sqrt{2}}{2}\right)}} = \sqrt{\dfrac{20}{16}} = \sqrt{\dfrac{5}{4}} \approx 1.12$ seconds

(c) $t = \sqrt{\dfrac{2(10)}{32 \sin(60°) \cos(60°)}} = \sqrt{\dfrac{20}{\left(32 \cdot \dfrac{\sqrt{3}}{2} \cdot \dfrac{1}{2}\right)}} = \sqrt{\dfrac{20}{8\sqrt{3}}} = \sqrt{\dfrac{5}{2\sqrt{3}}} \approx 1.20$ seconds

111. (a) $T(30°) = 1 + \dfrac{2}{3\sin(30°)} - \dfrac{1}{4\tan(30°)} = 1 + \dfrac{2}{\left(3 \cdot \dfrac{1}{2}\right)} - \dfrac{1}{\left(4 \cdot \dfrac{1}{\sqrt{3}}\right)} = 1 + \dfrac{4}{3} - \dfrac{\sqrt{3}}{4} \approx 1.9$ hrs

$\dfrac{1}{x} = \tan \theta \;\rightarrow\; x = \dfrac{1}{\tan \theta}$

Distance traveled on road is : $8 - 2x = 8 - \dfrac{2}{\tan \theta}$

Time on road $= \dfrac{\text{distance on road}}{\text{rate on road}} = \dfrac{\left(8 - \dfrac{2}{\tan \theta}\right)}{8}$

Time on road $= \dfrac{\left(8 - \dfrac{2}{\tan(30°)}\right)}{8} \approx 0.57$ hours

(b) $T(45°) = 1 + \dfrac{2}{3\sin(45°)} - \dfrac{1}{4\tan(45°)} = 1 + \dfrac{2}{\left(3 \cdot \dfrac{1}{\sqrt{2}}\right)} - \dfrac{1}{4 \cdot 1} = 1 + \dfrac{2\sqrt{2}}{3} - \dfrac{1}{4} \approx 1.69$ hrs

Time on road $= \dfrac{\left(8 - \dfrac{2}{\tan(45°)}\right)}{8} \approx 0.75$ hours

(c) $T(60°) = 1 + \dfrac{2}{3\sin(60°)} - \dfrac{1}{4\tan(60°)} = 1 + \dfrac{2}{\left(3 \cdot \dfrac{\sqrt{3}}{2}\right)} - \dfrac{1}{4 \cdot \sqrt{3}}$

$= 1 + \dfrac{4}{3\sqrt{3}} - \dfrac{1}{4\sqrt{3}} \approx 1.63 \text{ hrs}$

Time on road $= \dfrac{\left(8 - \dfrac{2}{\tan 60°}\right)}{8} \approx 0.86 \text{ hours}$

(d) $T(90°) = 1 + \dfrac{2}{3\sin(90°)} - \dfrac{1}{4\tan(90°)} = (\tan(90°) \text{ is undefined})$

The distance would be 2 miles in the sand and 8 miles on the road. The total time would be: $\dfrac{2}{3} + 1 = \dfrac{5}{3} \approx 1.67$ hours.

113. (a) $R = \dfrac{\left(32^2\right)\sqrt{2}}{32} \cdot \left[\sin(2(60°)) - \cos(2(60°)) - 1\right] \approx 32\sqrt{2}\left(0.866 - (-0.5) - 1\right) \approx 16.6 \text{ ft}$

(b)   Graph:

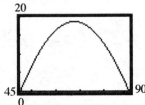

(c)   Using MAXIMUM, R is largest when $\theta \approx 67.5°$.

115. – 1119.   Answers will vary.

Chapter 5

Trigonometric Functions

5.3 Properties of the Trigonometric Functions

1. $\sin(405°) = \sin(360° + 45°) = \sin(45°) = \dfrac{\sqrt{2}}{2}$

3. $\tan(405°) = \tan(180° + 180° + 45°) = \tan(45°) = 1$

5. $\csc(450°) = \csc(360° + 90°) = \csc(90°) = 1$

7. $\cot(390°) = \cot(180° + 180° + 30°) = \cot(30°) = \sqrt{3}$

9. $\cos\left(\dfrac{33\pi}{4}\right) = \cos\left(\dfrac{\pi}{4} + \dfrac{32\pi}{4}\right) = \cos\left(\dfrac{\pi}{4} + 8\pi\right) = \cos\left(\dfrac{\pi}{4} + 4\cdot 2\pi\right) = \cos\left(\dfrac{\pi}{4}\right) = \dfrac{\sqrt{2}}{2}$

11. $\tan(21\pi) = \tan(0 + 21\pi) = \tan(0) = 0$

13. $\sec\left(\dfrac{17\pi}{4}\right) = \sec\left(\dfrac{\pi}{4} + \dfrac{16\pi}{4}\right) = \sec\left(\dfrac{\pi}{4} + 4\pi\right) = \sec\left(\dfrac{\pi}{4} + 2\cdot 2\pi\right) = \sec\left(\dfrac{\pi}{4}\right) = \sqrt{2}$

15. $\tan\left(\dfrac{19\pi}{6}\right) = \tan\left(\dfrac{\pi}{6} + \dfrac{18\pi}{6}\right) = \tan\left(\dfrac{\pi}{6} + 3\pi\right) = \tan\left(\dfrac{\pi}{6}\right) = \dfrac{\sqrt{3}}{3}$

17. Since $\sin\theta > 0$ for points in quadrants I and II, and $\cos\theta < 0$ for points in quadrants II and III, the angle θ lies in quadrant II.

19. Since $\sin\theta < 0$ for points in quadrants III and IV, and $\tan\theta < 0$ for points in quadrants II and IV, the angle θ lies in quadrant IV.

21. Since $\cos\theta > 0$ for points in quadrants I and IV, and $\tan\theta < 0$ for points in quadrants II and IV, the angle θ lies in quadrant IV.

23. Since $\sec\theta < 0$ for points in quadrants II and III, and $\sin\theta > 0$ for points in quadrants I and II, the angle θ lies in quadrant II.

25. $\sin\theta = -\dfrac{3}{5}, \quad \cos\theta = \dfrac{4}{5}$

$$\tan\theta = \frac{\sin\theta}{\cos\theta} = \frac{\left(-\dfrac{3}{5}\right)}{\left(\dfrac{4}{5}\right)} = -\frac{3}{5}\cdot\frac{5}{4} = -\frac{3}{4} \qquad\qquad \sec\theta = \frac{1}{\cos\theta} = \frac{1}{\left(\dfrac{4}{5}\right)} = \frac{5}{4}$$

$$\csc\theta = \frac{1}{\sin\theta} = \frac{1}{\left(-\dfrac{3}{5}\right)} = -\frac{5}{3} \qquad\qquad \cot\theta = \frac{1}{\tan\theta} = -\frac{4}{3}$$

27. $\sin\theta = \dfrac{2\sqrt{5}}{5}, \quad \cos\theta = \dfrac{\sqrt{5}}{5}$

$$\tan\theta = \frac{\sin\theta}{\cos\theta} = \frac{\left(\dfrac{2\sqrt{5}}{5}\right)}{\left(\dfrac{\sqrt{5}}{5}\right)} = \frac{2\sqrt{5}}{5}\cdot\frac{5}{\sqrt{5}} = 2$$

$$\sec\theta = \frac{1}{\cos\theta} = \frac{1}{\left(\dfrac{\sqrt{5}}{5}\right)}$$
$$= \frac{5}{\sqrt{5}}\cdot\frac{\sqrt{5}}{\sqrt{5}} = \sqrt{5}$$

$$\csc\theta = \frac{1}{\sin\theta} = \frac{1}{\left(\dfrac{2\sqrt{5}}{5}\right)} = 1\cdot\frac{5}{2\sqrt{5}}\cdot\frac{\sqrt{5}}{\sqrt{5}} = \frac{\sqrt{5}}{2} \qquad\qquad \cot\theta = \frac{1}{\tan\theta} = \frac{1}{2}$$

29. $\sin\theta = \dfrac{1}{2}, \quad \cos\theta = \dfrac{\sqrt{3}}{2}$

$$\tan\theta = \frac{\sin\theta}{\cos\theta} = \frac{\left(\dfrac{1}{2}\right)}{\left(\dfrac{\sqrt{3}}{2}\right)}$$
$$= \frac{1}{2}\cdot\frac{2}{\sqrt{3}}\cdot\frac{\sqrt{3}}{\sqrt{3}} = \frac{\sqrt{3}}{3}$$

$$\sec\theta = \frac{1}{\cos\theta} = \frac{1}{\left(\dfrac{\sqrt{3}}{2}\right)}$$
$$= \frac{2}{\sqrt{3}}\cdot\frac{\sqrt{3}}{\sqrt{3}} = \frac{2\sqrt{3}}{3}$$

$$\csc\theta = \frac{1}{\sin\theta} = \frac{1}{\left(\dfrac{1}{2}\right)}$$
$$= 1\cdot\frac{2}{1} = 2$$

$$\cot\theta = \frac{1}{\tan\theta} = \frac{1}{\left(\dfrac{\sqrt{3}}{3}\right)}$$
$$= \frac{3}{\sqrt{3}}\cdot\frac{\sqrt{3}}{\sqrt{3}} = \sqrt{3}$$

31. $\sin \theta = -\dfrac{1}{3}, \quad \cos \theta = \dfrac{2\sqrt{2}}{3}$

$$\tan \theta = \frac{\sin \theta}{\cos \theta} = \frac{\left(-\dfrac{1}{3}\right)}{\left(\dfrac{2\sqrt{2}}{3}\right)}$$

$$\sec \theta = \frac{1}{\cos \theta} = \frac{1}{\left(\dfrac{2\sqrt{2}}{3}\right)}$$

$$= -\frac{1}{3} \cdot \frac{3}{2\sqrt{2}} \cdot \frac{\sqrt{2}}{\sqrt{2}} = -\frac{\sqrt{2}}{4}$$

$$= \frac{3}{2\sqrt{2}} \cdot \frac{\sqrt{2}}{\sqrt{2}} = \frac{3\sqrt{2}}{4}$$

$$\csc \theta = \frac{1}{\sin \theta} = \frac{1}{\left(-\dfrac{1}{3}\right)}$$

$$\cot \theta = \frac{1}{\tan \theta} = \frac{1}{\left(-\dfrac{\sqrt{2}}{4}\right)}$$

$$= 1 \cdot \left(-\frac{3}{1}\right) = -3$$

$$= -\frac{4}{\sqrt{2}} \cdot \frac{\sqrt{2}}{\sqrt{2}} = -2\sqrt{2}$$

33. $\sin \theta = \dfrac{12}{13}, \quad \theta$ in quadrant II

Solve for $\cos \theta$:

$$\sin^2 \theta + \cos^2 \theta = 1$$
$$\cos^2 \theta = 1 - \sin^2 \theta$$
$$\cos \theta = \pm \sqrt{1 - \sin^2 \theta}$$

Since θ is in quadrant II, $\cos \theta < 0$.

$$\cos \theta = -\sqrt{1 - \sin^2 \theta} = -\sqrt{1 - \left(\frac{12}{13}\right)^2} = -\sqrt{1 - \frac{144}{169}} = -\sqrt{\frac{25}{169}} = -\frac{5}{13}$$

$$\tan \theta = \frac{\sin \theta}{\cos \theta} = \frac{\left(\dfrac{12}{13}\right)}{\left(-\dfrac{5}{13}\right)} = \frac{12}{13} \cdot \left(-\frac{13}{5}\right) = -\frac{12}{5}$$

$$\sec \theta = \frac{1}{\cos \theta} = \frac{1}{\left(-\dfrac{5}{13}\right)} = -\frac{13}{5}$$

$$\csc \theta = \frac{1}{\sin \theta} = \frac{1}{\left(\dfrac{12}{13}\right)} = \frac{13}{12}$$

$$\cot \theta = \frac{1}{\tan \theta} = \frac{1}{\left(-\dfrac{12}{5}\right)} = -\frac{5}{12}$$

35. $\cos \theta = -\dfrac{4}{5}, \quad \theta$ in quadrant III

Solve for $\sin \theta$:

$$\sin^2 \theta + \cos^2 \theta = 1$$
$$\sin^2 \theta = 1 - \cos^2 \theta$$
$$\sin \theta = \pm \sqrt{1 - \cos^2 \theta}$$

Since θ is in quadrant III, $\sin \theta < 0$.

$$\sin\theta = -\sqrt{1-\cos^2\theta} = -\sqrt{1-\left(-\frac{4}{5}\right)^2} = -\sqrt{1-\frac{16}{25}} = -\sqrt{\frac{9}{25}} = -\frac{3}{5}$$

$$\tan\theta = \frac{\sin\theta}{\cos\theta} = \frac{\left(-\frac{3}{5}\right)}{\left(-\frac{4}{5}\right)} = -\frac{3}{5}\cdot\left(-\frac{5}{4}\right) = \frac{3}{4} \qquad \sec\theta = \frac{1}{\cos\theta} = \frac{1}{\left(-\frac{4}{5}\right)} = -\frac{5}{4}$$

$$\csc\theta = \frac{1}{\sin\theta} = \frac{1}{\left(-\frac{3}{5}\right)} = -\frac{5}{3} \qquad\qquad\qquad \cot\theta = \frac{1}{\tan\theta} = \frac{1}{\left(\frac{3}{4}\right)} = \frac{4}{3}$$

37. $\sin\theta = \dfrac{5}{13}$, $90° < \theta < 180°$, θ in quadrant II

Solve for $\cos\theta$:

$$\sin^2\theta + \cos^2\theta = 1$$
$$\cos^2\theta = 1 - \sin^2\theta$$
$$\cos\theta = \pm\sqrt{1 - \sin^2\theta}$$

Since θ is in quadrant II, $\cos\theta < 0$.

$$\cos\theta = -\sqrt{1-\sin^2\theta} = -\sqrt{1-\left(\frac{5}{13}\right)^2} = -\sqrt{1-\frac{25}{169}} = -\sqrt{\frac{144}{169}} = -\frac{12}{13}$$

$$\tan\theta = \frac{\sin\theta}{\cos\theta} = \frac{\left(\frac{5}{13}\right)}{\left(-\frac{12}{13}\right)} = \frac{5}{13}\cdot\left(-\frac{13}{12}\right) = -\frac{5}{12} \qquad \sec\theta = \frac{1}{\cos\theta} = \frac{1}{\left(-\frac{12}{13}\right)} = -\frac{13}{12}$$

$$\csc\theta = \frac{1}{\sin\theta} = \frac{1}{\left(\frac{5}{13}\right)} = \frac{13}{5} \qquad\qquad\qquad \cot\theta = \frac{1}{\tan\theta} = \frac{1}{\left(-\frac{5}{12}\right)} = -\frac{12}{5}$$

39. $\cos\theta = -\dfrac{1}{3}$, $\dfrac{\pi}{2} < \theta < \pi$, θ in quadrant II

Solve for $\sin\theta$:

$$\sin^2\theta + \cos^2\theta = 1$$
$$\sin^2\theta = 1 - \cos^2\theta$$
$$\sin\theta = \pm\sqrt{1 - \cos^2\theta}$$

Since θ is in quadrant II, $\sin\theta > 0$.

$$\sin\theta = \sqrt{1-\cos^2\theta} = \sqrt{1-\left(-\frac{1}{3}\right)^2} = \sqrt{1-\frac{1}{9}} = \sqrt{\frac{8}{9}} = \frac{2\sqrt{2}}{3}$$

$$\tan\theta = \frac{\sin\theta}{\cos\theta} = \frac{\left(\frac{2\sqrt{2}}{3}\right)}{\left(-\frac{1}{3}\right)} = \frac{2\sqrt{2}}{3}\cdot\left(-\frac{3}{1}\right) = -2\sqrt{2} \qquad \sec\theta = \frac{1}{\cos\theta} = \frac{1}{\left(-\frac{1}{3}\right)} = -3$$

$$\csc\theta = \frac{1}{\sin\theta} = \frac{1}{\left(\dfrac{2\sqrt{2}}{3}\right)} = \frac{3}{2\sqrt{2}} \cdot \frac{\sqrt{2}}{\sqrt{2}} = \frac{3\sqrt{2}}{4} \qquad \cot\theta = \frac{1}{\tan\theta} = \frac{1}{-2\sqrt{2}} \cdot \frac{\sqrt{2}}{\sqrt{2}} = -\frac{\sqrt{2}}{4}$$

41. $\sin\theta = \dfrac{2}{3}, \quad \tan\theta < 0, \quad \Rightarrow \quad \theta$ in quadrant II

Solve for $\cos\theta$:

$$\sin^2\theta + \cos^2\theta = 1$$
$$\cos^2\theta = 1 - \sin^2\theta$$
$$\cos\theta = \pm\sqrt{1 - \sin^2\theta}$$

Since θ is in quadrant II, $\cos\theta < 0$.

$$\cos\theta = -\sqrt{1 - \sin^2\theta} = -\sqrt{1 - \left(\frac{2}{3}\right)^2} = -\sqrt{1 - \frac{4}{9}} = -\sqrt{\frac{5}{9}} = -\frac{\sqrt{5}}{3}$$

$$\tan\theta = \frac{\sin\theta}{\cos\theta} = \frac{\left(\dfrac{2}{3}\right)}{\left(-\dfrac{\sqrt{5}}{3}\right)} = \frac{2}{3} \cdot \left(-\frac{3}{\sqrt{5}}\right) = -\frac{2\sqrt{5}}{5} \qquad \sec\theta = \frac{1}{\cos\theta} = \frac{1}{\left(-\dfrac{\sqrt{5}}{3}\right)} = -\frac{3}{\sqrt{5}} = -\frac{3\sqrt{5}}{5}$$

$$\csc\theta = \frac{1}{\sin\theta} = \frac{1}{\left(\dfrac{2}{3}\right)} = \frac{3}{2} \qquad\qquad \cot\theta = \frac{1}{\tan\theta} = \frac{1}{\left(-\dfrac{2\sqrt{5}}{5}\right)} = -\frac{5}{2\sqrt{5}} = -\frac{\sqrt{5}}{2}$$

43. $\sec\theta = 2, \quad \sin\theta < 0, \quad \Rightarrow \quad \theta$ in quadrant IV

Solve for $\cos\theta$:

$$\cos\theta = \frac{1}{\sec\theta} = \frac{1}{2}$$

Solve for $\sin\theta$:

$$\sin^2\theta + \cos^2\theta = 1$$
$$\sin^2\theta = 1 - \cos^2\theta$$
$$\sin\theta = \pm\sqrt{1 - \cos^2\theta}$$

Since θ is in quadrant IV, $\sin\theta < 0$.

$$\sin\theta = -\sqrt{1 - \cos^2\theta} = -\sqrt{1 - \left(\frac{1}{2}\right)^2} = -\sqrt{1 - \frac{1}{4}} = -\sqrt{\frac{3}{4}} = -\frac{\sqrt{3}}{2}$$

$$\tan\theta = \frac{\sin\theta}{\cos\theta} = \frac{\left(-\dfrac{\sqrt{3}}{2}\right)}{\left(\dfrac{1}{2}\right)} = -\frac{\sqrt{3}}{2} \cdot \frac{2}{1} = -\sqrt{3}$$

$$\csc\theta = \frac{1}{\sin\theta} = \frac{1}{\left(-\dfrac{\sqrt{3}}{2}\right)} = -\frac{2}{\sqrt{3}} = -\frac{2\sqrt{3}}{3} \qquad \cot\theta = \frac{1}{\tan\theta} = \frac{1}{-\sqrt{3}} = -\frac{\sqrt{3}}{3}$$

45. $\tan\theta = \dfrac{3}{4}, \quad \sin\theta < 0, \quad \Rightarrow \quad \theta$ in quadrant III

Solve for $\sec\theta$:

$$\sec^2\theta = 1 + \tan^2\theta$$

$$\sec\theta = \pm\sqrt{1 + \tan^2\theta}$$

Since θ is in quadrant III, $\sec\theta < 0$.

$$\sec\theta = -\sqrt{1 + \tan^2\theta} = -\sqrt{1 + \left(\dfrac{3}{4}\right)^2} = -\sqrt{1 + \dfrac{9}{16}} = -\sqrt{\dfrac{25}{16}} = -\dfrac{5}{4}$$

$$\cos\theta = \dfrac{1}{\sec\theta} = -\dfrac{4}{5}$$

$$\sin\theta = -\sqrt{1 - \cos^2\theta} = -\sqrt{1 - \left(-\dfrac{4}{5}\right)^2} = -\sqrt{1 - \dfrac{16}{25}} = -\sqrt{\dfrac{9}{25}} = -\dfrac{3}{5}$$

$$\csc\theta = \dfrac{1}{\sin\theta} = \dfrac{1}{\left(-\dfrac{3}{5}\right)} = -\dfrac{5}{3} \qquad\qquad \cot\theta = \dfrac{1}{\tan\theta} = \dfrac{1}{\left(\dfrac{3}{4}\right)} = \dfrac{4}{3}$$

47. $\tan\theta = -\dfrac{1}{3}, \quad \sin\theta > 0, \quad \Rightarrow \quad \theta$ in quadrant II

Solve for $\sec\theta$:

$$\sec^2\theta = 1 + \tan^2\theta$$

$$\sec\theta = \pm\sqrt{1 + \tan^2\theta}$$

Since θ is in quadrant II, $\sec\theta < 0$.

$$\sec\theta = -\sqrt{1 + \tan^2\theta} = -\sqrt{1 + \left(-\dfrac{1}{3}\right)^2} = -\sqrt{1 + \dfrac{1}{9}} = -\sqrt{\dfrac{10}{9}} = -\dfrac{\sqrt{10}}{3}$$

$$\cos\theta = \dfrac{1}{\sec\theta} = \dfrac{1}{\left(-\dfrac{\sqrt{10}}{3}\right)} = -\dfrac{3}{\sqrt{10}} = -\dfrac{3\sqrt{10}}{10}$$

$$\sin\theta = \sqrt{1 - \cos^2\theta} = \sqrt{1 - \left(-\dfrac{3\sqrt{10}}{10}\right)^2} = \sqrt{1 - \dfrac{90}{100}} = \sqrt{\dfrac{10}{100}} = \dfrac{\sqrt{10}}{10}$$

$$\csc\theta = \dfrac{1}{\sin\theta} = \dfrac{1}{\left(\dfrac{\sqrt{10}}{10}\right)} = \sqrt{10} \qquad\qquad \cot\theta = \dfrac{1}{\tan\theta} = \dfrac{1}{\left(-\dfrac{1}{3}\right)} = -3$$

49. $\sin(-60°) = -\sin(60°) = -\dfrac{\sqrt{3}}{2}$

51. $\tan(-30°) = -\tan(30°) = -\dfrac{\sqrt{3}}{3}$

53. $\sec(-60°) = \sec(60°) = 2$

55. $\sin(-90°) = -\sin(90°) = -1$

57. $\tan\left(-\dfrac{\pi}{4}\right) = -\tan\left(\dfrac{\pi}{4}\right) = -1$

59. $\cos\left(-\dfrac{\pi}{4}\right) = \cos\left(\dfrac{\pi}{4}\right) = \dfrac{\sqrt{2}}{2}$

61. $\tan(-\pi) = -\tan(\pi) = 0$

63. $\csc\left(-\dfrac{\pi}{4}\right) = -\csc\left(\dfrac{\pi}{4}\right) = -\sqrt{2}$

65. $\sec\left(-\dfrac{\pi}{6}\right) = \sec\left(\dfrac{\pi}{6}\right) = \dfrac{2\sqrt{3}}{3}$

67. $\sin^2(40°) + \cos^2(40°) = 1$

69. $\sin(80°)\csc(80°) = \sin(80°) \cdot \dfrac{1}{\sin(80°)} = 1$

71. $\tan(40°) - \dfrac{\sin(40°)}{\cos(40°)} = \tan(40°) - \tan(40°) = 0$

73. $\cos(400°) \cdot \sec(40°) = \cos(40°+360°) \cdot \sec(40°) = \cos(40°) \cdot \sec(40°) = \cos(40°) \cdot \dfrac{1}{\cos(40°)} = 1$

75. $\sin\left(-\dfrac{\pi}{12}\right)\csc\left(\dfrac{25\pi}{12}\right) = -\sin\left(\dfrac{\pi}{12}\right)\csc\left(\dfrac{25\pi}{12}\right) = -\sin\left(\dfrac{\pi}{12}\right)\csc\left(\dfrac{\pi}{12} + \dfrac{24\pi}{12}\right) = -\sin\left(\dfrac{\pi}{12}\right)\csc\left(\dfrac{\pi}{12} + 2\pi\right)$

$= -\sin\left(\dfrac{\pi}{12}\right)\csc\left(\dfrac{\pi}{12}\right) = -\sin\left(\dfrac{\pi}{12}\right) \cdot \dfrac{1}{\sin\left(\dfrac{\pi}{12}\right)} = -1$

77. $\dfrac{\sin(-20°)}{\cos(380°)} + \tan(200°) = \dfrac{-\sin(20°)}{\cos(20°+360°)} + \tan(20°+180°)$

$= \dfrac{-\sin(20°)}{\cos(20°)} + \tan(20°) = -\tan(20°) + \tan(20°) = 0$

79. If $\sin\theta = 0.3$, then $\sin(\theta + \pi) = -0.3$

81. If $\tan\theta = 3$, then $\tan(\theta + \pi) = 3$

83. Find the value:

$\sin(1°) + \sin(2°) + \sin(3°) + \dots + \sin(357°) + \sin(358°) + \sin(359°)$

$= \sin(1°) + \sin(2°) + \sin(3°) + \dots + \sin(360°-3°) + \sin(360°-2°) + \sin(360°-1°)$

$= \sin(1°) + \sin(2°) + \sin(3°) + \dots + \sin(-3°) + \sin(-2°) + \sin(-1°)$

$= \sin(1°) + \sin(2°) + \sin(3°) + \dots - \sin(3°) - \sin(2°) - \sin(1°)$

$= \sin(180°)$

$= 0$

85. The domain of the sine function is the set of all real numbers.

87. $f(\theta) = \tan\theta$ is not defined for numbers that are odd multiples of $\frac{\pi}{2}$.

89. $f(\theta) = \sec\theta$ is not defined for numbers that are odd multiples of $\frac{\pi}{2}$.

91. The range of the sine function is the set of all real numbers between -1 and 1, inclusive.

93. The range of the tangent function is the set of all real numbers.

95. The range of the secant function is the set of all real number greater than or equal to 1 and all real numbers less than or equal to -1.

97. The sine function is odd because $\sin(-\theta) = -\sin\theta$. Its graph is symmetric to the origin.

99. The tangent function is odd because $\tan(-\theta) = -\tan\theta$. Its graph is symmetric to the origin.

101. The secant function is even because $\sec(-\theta) = \sec\theta$. Its graph is symmetric to the y-axis.

103. (a) $f(-a) = -f(a) = -\frac{1}{3}$

       (b) $f(a) + f(a+2\pi) + f(a+4\pi) = f(a) + f(a) + f(a) = \frac{1}{3} + \frac{1}{3} + \frac{1}{3} = 1$

105. (a) $f(-a) = -f(a) = -2$
       (b) $f(a) + f(a+\pi) + f(a+2\pi) = f(a) + f(a) + f(a) = 2 + 2 + 2 = 6$

107. (a) $f(-a) = f(a) = -4$
       (b) $f(a) + f(a+2\pi) + f(a+4\pi) = f(a) + f(a) + f(a) = (-4) + (-4) + (-4) = -12$

109. Since $\tan\theta = \frac{500}{1500} = \frac{1}{3}$, then $\sin\theta = \frac{1}{\sqrt{1+9}} = \frac{1}{\sqrt{10}}$.

       $T = 5 - \dfrac{5}{\left(3\cdot\frac{1}{3}\right)} + \dfrac{5}{\left(\frac{1}{\sqrt{10}}\right)} = 5 - 5 + 5\sqrt{10} = 5\sqrt{10} \approx 15.8$ minutes

111. Let $P = (x, y)$ be the point on the unit circle that corresponds to an angle θ.

       Consider the equation $\tan\theta = \frac{y}{x} = a$. Then $y = ax$. Now $x^2 + y^2 = 1$, so $x^2 + a^2x^2 = 1$.

       Thus, $x = \pm\dfrac{1}{\sqrt{1+a^2}}$ and $y = \pm\dfrac{a}{\sqrt{1+a^2}}$; that is, for any real number a, there is a point

       $P = (x, y)$ on the unit circle for which $\tan\theta = a$. In other words, $-\infty < \tan\theta < +\infty$, and the range of the tangent function is the set of all real numbers.

113. Suppose there is a number $p, 0 < p < 2\pi$, for which $\sin(\theta + p) = \sin\theta$ for all θ. If

       $\theta = 0$, then $\sin(0 + p) = \sin p = \sin 0 = 0$; so that $p = \pi$. If $\theta = \frac{\pi}{2}$, then

$\sin\left(\dfrac{\pi}{2}+p\right)=\sin\left(\dfrac{\pi}{2}\right)$. But $p=\pi$. Thus, $\sin\left(\dfrac{3\pi}{2}\right)=-1=\sin\left(\dfrac{\pi}{2}\right)=1$, or $-1=1$. This is impossible. The smallest positive number p for which $\sin(\theta+p)=\sin\theta$ for all θ is therefore $p=2\pi$.

115. $\sec\theta=\dfrac{1}{\cos\theta}$: since $\cos\theta$ has period 2π, so does $\sec\theta$.

117. If $P=(a,b)$ is the point on the unit circle corresponding to θ, then $Q=(-a,-b)$ is the point on the unit circle corresponding to $\theta+\pi$.

   Thus, $\tan(\theta+\pi)=\dfrac{-b}{-a}=\dfrac{b}{a}=\tan\theta$. If there exists a number p, $0<p<\pi$, for which $\tan(\theta+p)=\tan\theta$ for all θ, then if $\theta=0$, then $\tan(p)=\tan(0)=0$.

   But this means that p is a multiple of π. Since no multiple of π exists in the interval $(0,\pi)$, this is impossible. Therefore, the fundamental period of $f(\theta)=\tan\theta$ is π.

119. Let $P=(a,b)$ be the point on the unit circle corresponding to θ.

   Then $\csc\theta=\dfrac{1}{b}=\dfrac{1}{\sin\theta}$; $\sec\theta=\dfrac{1}{a}=\dfrac{1}{\cos\theta}$; $\cot\theta=\dfrac{a}{b}=\dfrac{1}{\left(\dfrac{b}{a}\right)}=\dfrac{1}{\tan\theta}$

121. $(\sin\theta\cos\phi)^2+(\sin\theta\sin\phi)^2+\cos^2\theta=\sin^2\theta\cos^2\phi+\sin^2\theta\sin^2\phi+\cos^2\theta$
   $$=\sin^2\theta(\cos^2\phi+\sin^2\phi)+\cos^2\theta=\sin^2\theta+\cos^2\theta=1$$

123. Answers will vary.

Chapter 5

Trigonometric Functions

5.4 Graphs of the Sine and Cosine Functions

1. 0

3. The graph of $y = \sin x$ is increasing for $-\dfrac{\pi}{2} < x < \dfrac{\pi}{2}$.

5. The greatest value of $y = \sin x$ is 1.

7. $\sin x = 0$ when $x = 0, \pi, 2\pi$

9. $\sin x = 1$ when $x = -\dfrac{3\pi}{2}, \dfrac{\pi}{2}$; $\sin x = -1$ when $x = -\dfrac{\pi}{2}, \dfrac{3\pi}{2}$

11. B, C, F

13. $y = 3\sin x$; The graph of $y = \sin x$ is stretched vertically by a factor of 3.

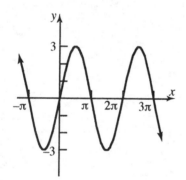

15. $y = \cos\left(x + \dfrac{\pi}{4}\right)$; The graph of $y = \cos x$ is shifted $\dfrac{\pi}{4}$ units to the left.

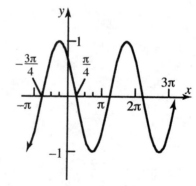

17. $y = \sin x - 1$; The graph of $y = \sin x$ is shifted 1 unit down.

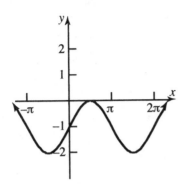

19. $y = -2\sin x$; The graph of $y = \sin x$ is reflected across the x-axis and then vertically stretched by a factor of 2.

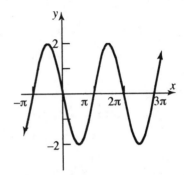

21. $y = \sin(\pi x)$; The graph of $y = \sin x$ is compressed horizontally by a factor of π.

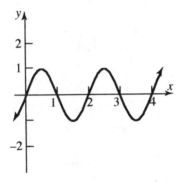

23. $y = 2\sin x + 2$; The graph of $y = \sin x$ is stretched vertically by a factor of 2, and then shifted up 2 units.

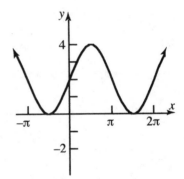

25. $y = -2\cos\left(x - \dfrac{\pi}{2}\right)$; The graph of $y = \cos x$

is shifted right $\dfrac{\pi}{2}$ units, then reflected

across the x-axis and stretched vertically by a factor of 2.

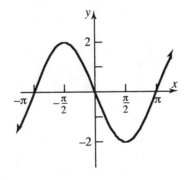

27. $y = 3\sin(\pi - x) = 3\sin\left(-(-\pi + x)\right)$
$= -3\sin(-\pi + x)$, since sine is an odd function.
The graph of $y = \sin x$ is shifted right π units, then reflected across the x-axis and stretched vertically by a factor of 3.

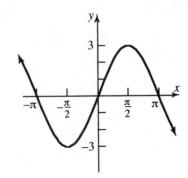

29. $y = 2\sin x$

This is in the form $y = A\sin(\omega x)$ where $A = 2$ and $\omega = 1$.

Thus, the amplitude is $|A| = |2| = 2$ and the period is $T = \dfrac{2\pi}{\omega} = \dfrac{2\pi}{1} = 2\pi$.

31. $y = -4\cos(2x)$

This is in the form $y = A\cos(\omega x)$ where $A = -4$ and $\omega = 2$.

Thus, the amplitude is $|A| = |-4| = 4$ and the period is $T = \dfrac{2\pi}{\omega} = \dfrac{2\pi}{2} = \pi$.

33. $y = 6\sin(\pi x)$

This is in the form $y = A\sin(\omega x)$ where $A = 6$ and $\omega = \pi$.

Thus, the amplitude is $|A| = |6| = 6$ and the period is $T = \dfrac{2\pi}{\omega} = \dfrac{2\pi}{\pi} = 2$.

35. $y = -\dfrac{1}{2}\cos\left(\dfrac{3}{2}x\right)$

This is in the form $y = A\cos(\omega x)$ where $A = -\dfrac{1}{2}$ and $\omega = \dfrac{3}{2}$.

Thus, the amplitude is $|A| = \left|-\dfrac{1}{2}\right| = \dfrac{1}{2}$ and the period is $T = \dfrac{2\pi}{\omega} = \dfrac{2\pi}{\left(\dfrac{3}{2}\right)} = \dfrac{4\pi}{3}$.

37. $y = \dfrac{5}{3}\sin\left(-\dfrac{2\pi}{3}x\right) = -\dfrac{5}{3}\sin\left(\dfrac{2\pi}{3}x\right)$

This is in the form $y = A\sin(\omega x)$ where $A = -\dfrac{5}{3}$ and $\omega = \dfrac{2\pi}{3}$.

Thus, the amplitude is $|A| = \left|-\dfrac{5}{3}\right| = \dfrac{5}{3}$ and the period is $T = \dfrac{2\pi}{\omega} = \dfrac{2\pi}{\left(\dfrac{2\pi}{3}\right)} = 3$.

39. F   41. A   43. H   45. C

47. J   49. A   51. B

53. $y = 5\sin(4x)$ $A = 5;\ \ T = \dfrac{\pi}{2}$ 55. $y = 5\cos(\pi x)$ $A = 5;\ \ T = 2$

          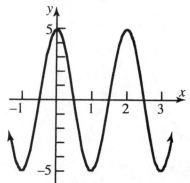

57. $y = -2\cos(2\pi x)$ $A = -2;\;\; T = 1$

59. $y = -4\sin\left(\dfrac{1}{2}x\right)$ $A = -4;\;\; T = 4\pi$

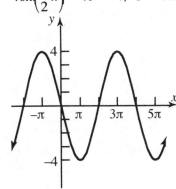

61. $y = \dfrac{3}{2}\sin\left(-\dfrac{2}{3}x\right) = -\dfrac{3}{2}\sin\left(\dfrac{2}{3}x\right)$

$$A = -\dfrac{3}{2};\;\; T = 3\pi$$

63. $|A| = 3;\;\; T = \pi;\;\;\; \omega = \dfrac{2\pi}{T} = \dfrac{2\pi}{\pi} = 2;\;\;\;\; y = 3\sin(2x)$

65. $|A| = 3;\;\; T = 2;\;\;\; \omega = \dfrac{2\pi}{T} = \dfrac{2\pi}{2} = \pi;\;\;\;\; y = 3\sin(\pi x)$

67. The graph is a cosine graph with an amplitude of 5 and a period of 8. Find ω:

$$8 = \dfrac{2\pi}{\omega} \;\rightarrow\; 8\omega = 2\pi \;\rightarrow\; \omega = \dfrac{2\pi}{8} = \dfrac{\pi}{4}$$

The equation is: $y = 5\cos\left(\dfrac{\pi}{4}x\right)$.

69. The graph is a reflected cosine graph with an amplitude of 3 and a period of 4π.
Find ω:

$$4\pi = \dfrac{2\pi}{\omega} \;\rightarrow\; 4\pi\omega = 2\pi \;\rightarrow\; \omega = \dfrac{2\pi}{4\pi} = \dfrac{1}{2}$$

The equation is: $y = -3\cos\left(\dfrac{1}{2}x\right)$.

71. The graph is a sine graph with an amplitude of $\frac{3}{4}$ and a period of 1. Find ω:

$$1 = \frac{2\pi}{\omega} \;\rightarrow\; \omega = 2\pi$$

The equation is: $y = \frac{3}{4}\sin(2\pi x)$.

73. The graph is a reflected sine graph with an amplitude of 1 and a period of $\frac{4\pi}{3}$. Find ω:

$$\frac{4\pi}{3} = \frac{2\pi}{\omega} \;\rightarrow\; 4\pi\omega = 6\pi \;\rightarrow\; \omega = \frac{6\pi}{4\pi} = \frac{3}{2}$$

The equation is: $y = -\sin\left(\frac{3}{2}x\right)$.

75. The graph is a reflected cosine graph with an amplitude of 2 and a period of $\frac{4}{3}$. Find ω:

$$\frac{4}{3} = \frac{2\pi}{\omega} \;\rightarrow\; 4\omega = 6\pi \;\rightarrow\; \omega = \frac{6\pi}{4} = \frac{3\pi}{2}$$

The equation is: $y = -2\cos\left(\frac{3\pi}{2}x\right)$.

77. The graph is a sine graph with an amplitude of 3 and a period of 4. Find ω:

$$4 = \frac{2\pi}{\omega} \;\rightarrow\; 4\omega = 2\pi \;\rightarrow\; \omega = \frac{2\pi}{4} = \frac{\pi}{2}$$

The equation is: $y = 3\sin\left(\frac{\pi}{2}x\right)$.

79. The graph is a reflected cosine graph with an amplitude of 4 and a period of $\frac{2\pi}{3}$. Find ω:

$$\frac{2\pi}{3} = \frac{2\pi}{\omega} \;\rightarrow\; 2\pi\omega = 6\pi \;\rightarrow\; \omega = \frac{6\pi}{2\pi} = 3$$

The equation is: $y = -4\cos(3x)$.

81. $I = 220\sin(60\pi t),\; t \geq 0$

Period: $\quad T = \frac{2\pi}{\omega} = \frac{2\pi}{60\pi} = \frac{1}{30}$

Amplitude: $|A| = |220| = 220$

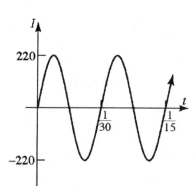

83. $V = 220\sin(120\pi t)$

    (a) Amplitude: $|A| = |220| = 220$

        Period: $T = \dfrac{2\pi}{\omega} = \dfrac{2\pi}{120\pi} = \dfrac{1}{60}$

    (b), (e)

    (c) $V = IR$

        $220\sin(120\pi t) = 10I$

        $22\sin(120\pi t) = I$

    (d) Amplitude: $|A| = |22| = 22$

        Period: $T = \dfrac{2\pi}{\omega} = \dfrac{2\pi}{120\pi} = \dfrac{1}{60}$

85. (a) $P = \dfrac{V^2}{R} = \dfrac{\left(V_0\sin(2\pi f t)\right)^2}{R} = \dfrac{V_0^2\sin^2(2\pi f t)}{R}$

    (b) The graph is the reflected cosine graph translated up a distance equivalent to the amplitude. The period is $\dfrac{1}{2f}$, so $\omega = 4\pi f$. The amplitude is $\dfrac{1}{2}\cdot\dfrac{V_0^2}{R} = \dfrac{V_0^2}{2R}$.

        The equation is: $P = -\dfrac{V_0^2}{2R}\cos(4\pi f t) + \dfrac{V_0^2}{2R} = \dfrac{V_0^2}{R}\cdot\dfrac{1}{2}\left(1 - \cos(4\pi f t)\right)$

    (c) Comparing the formulas:

        $\sin^2(2\pi f t) = \dfrac{1}{2}\left(1 - \cos(4\pi f t)\right)$

87. $y = |\cos x|,\quad -2\pi \le x \le 2\pi$

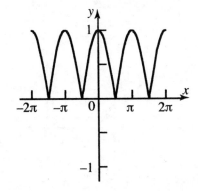

Chapter 5

Trigonometric Functions

5.5 Graphs of the Tangent, Cotangent, Cosecant and Secant Functions

1. y-intercept: 0

3. y-intercept: 1

5. $\sec x = 1$ for $x = -2\pi, 0, 2\pi$; $\sec x = -1$ for $x = -\pi, \pi$

7. $y = \sec x$ has vertical asymptotes for $x = -\dfrac{3\pi}{2}, -\dfrac{\pi}{2}, \dfrac{\pi}{2}, \dfrac{3\pi}{2}$

9. $y = \tan x$ has vertical asymptotes for $x = -\dfrac{3\pi}{2}, -\dfrac{\pi}{2}, \dfrac{\pi}{2}, \dfrac{3\pi}{2}$

11. B

13. A

15. $y = -\sec x$; The graph of $y = \sec x$ is reflected across the x-axis.

17. $y = \sec\left(x - \dfrac{\pi}{2}\right)$; The graph of $y = \sec x$ is shifted right $\dfrac{\pi}{2}$ units.

19. $y = \tan(x - \pi)$; The graph of $y = \tan x$ is
shifted right π units.

21. $y = 3\tan(2x)$; The graph of $y = \tan x$ is
compressed horizontally by a factor of 2
and stretched vertically by a factor of 3.

23. $y = \sec(2x)$; The graph of $y = \sec x$ is
compressed horizontally by a factor of 2.

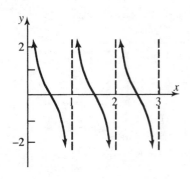

25. $y = \cot(\pi x)$; The graph of $y = \cot x$ is
compressed horizontally by a factor
of π.

27. $y = -3\tan(4x)$; The graph of $y = \tan x$
is compressed horizontally by a factor
of 4, then reflected across the x-axis and
vertically stretched by a factor of 3.

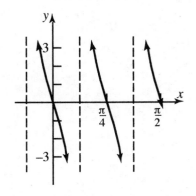

29. $y = 2\sec\left(\dfrac{1}{2}x\right)$; The graph of $y = \sec(x)$ is
compressed horizontally by a factor of 0.5,
then vertically stretched by a factor of 2.

31. $y = -3\csc\left(x + \dfrac{\pi}{4}\right)$; The graph of $y = \csc(x)$ is shifted left $\dfrac{\pi}{4}$ units, reflected across the x-axis, then vertically stretched by a factor of 3.

33. $y = \dfrac{1}{2}\cot\left(x + \dfrac{\pi}{4}\right)$; The graph of $y = \cot(x)$ is shifted left $\dfrac{\pi}{4}$ units, then vertically compressed by a factor of 2.

35. (a) $L = \dfrac{3}{\cos\theta} + \dfrac{4}{\sin\theta} = 3\sec\theta + 4\csc\theta$

(b) Graph:

(c) Use MINIMUM to find the least value: L is least when $\theta = 0.83$.

(d) $L = \dfrac{3}{\cos(0.83)} + \dfrac{4}{\sin(0.83)} \approx 9.86$ feet

Trigonometric Functions

5.6 Phase Shift; Sinusoidal Curve Fitting

1. $y = 4\sin(2x - \pi)$

    Amplitude: $|A| = |4| = 4$

    Period: $T = \dfrac{2\pi}{\omega} = \dfrac{2\pi}{2} = \pi$

    Phase Shift: $\dfrac{\phi}{\omega} = \dfrac{\pi}{2}$

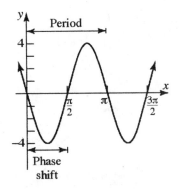

3. $y = 2\cos\left(3x + \dfrac{\pi}{2}\right)$

    Amplitude: $|A| = |2| = 2$

    Period: $T = \dfrac{2\pi}{\omega} = \dfrac{2\pi}{3}$

    Phase Shift: $\dfrac{\phi}{\omega} = \dfrac{\left(-\dfrac{\pi}{2}\right)}{3} = -\dfrac{\pi}{6}$

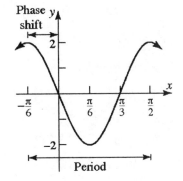

5. $y = -3\sin\left(2x + \dfrac{\pi}{2}\right)$

    Amplitude: $|A| = |-3| = 3$

    Period: $T = \dfrac{2\pi}{\omega} = \dfrac{2\pi}{2} = \pi$

    Phase Shift: $\dfrac{\phi}{\omega} = \dfrac{\left(-\dfrac{\pi}{2}\right)}{2} = -\dfrac{\pi}{4}$

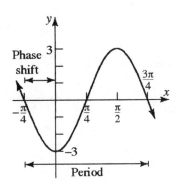

7. $y = 4\sin(\pi x + 2)$

Amplitude : $|A| = |4| = 4$

Period : $T = \dfrac{2\pi}{\omega} = \dfrac{2\pi}{\pi} = 2$

Phase Shift : $\dfrac{\phi}{\omega} = -\dfrac{2}{\pi}$

9. $y = 3\cos(\pi x - 2)$

Amplitude : $|A| = |3| = 3$

Period : $T = \dfrac{2\pi}{\omega} = \dfrac{2\pi}{\pi} = 2$

Phase Shift : $\dfrac{\phi}{\omega} = \dfrac{2}{\pi}$

11. $y = 3\sin\left(-2x + \dfrac{\pi}{2}\right) = -3\sin\left(2x - \dfrac{\pi}{2}\right)$

Amplitude : $|A| = |-3| = 3$

Period : $T = \dfrac{2\pi}{\omega} = \dfrac{2\pi}{2} = \pi$

Phase Shift : $\dfrac{\phi}{\omega} = \dfrac{\left(\dfrac{\pi}{2}\right)}{2} = \dfrac{\pi}{4}$

13. $|A| = 2;\quad T = \pi;\quad \dfrac{\phi}{\omega} = \dfrac{1}{2};\quad \omega = \dfrac{2\pi}{T} = \dfrac{2\pi}{\pi} = 2;\quad \dfrac{\phi}{\omega} = \dfrac{\phi}{2} = \dfrac{1}{2} \;\rightarrow\; \phi = 1$

$y = 2\sin(2x - 1) = 2\sin\left[2\left(x - \dfrac{1}{2}\right)\right]$

15. $|A| = 3;\quad T = 3\pi;\quad \dfrac{\phi}{\omega} = -\dfrac{1}{3};\quad \omega = \dfrac{2\pi}{T} = \dfrac{2\pi}{3\pi} = \dfrac{2}{3};$

$\dfrac{\phi}{\omega} = \dfrac{\phi}{\frac{2}{3}} = -\dfrac{1}{3} \;\rightarrow\; \phi = -\dfrac{1}{3}\cdot\dfrac{2}{3} = -\dfrac{2}{9}$

$y = 3\sin\left(\dfrac{2}{3}x + \dfrac{2}{9}\right) = 3\sin\left[\dfrac{2}{3}\left(x + \dfrac{1}{3}\right)\right]$

17. $I = 120\sin\left(30\pi t - \dfrac{\pi}{3}\right), \quad t \geq 0$

Period : $T = \dfrac{2\pi}{\omega} = \dfrac{2\pi}{30\pi} = \dfrac{1}{15}$

Amplitude : $|A| = |120| = 120$

Phase Shift : $\dfrac{\phi}{\omega} = \dfrac{\left(\dfrac{\pi}{3}\right)}{30\pi} = \dfrac{1}{90}$

19. (a)   Draw a scatter diagram:

(b)   Amplitude: $A = \dfrac{56.0 - 24.2}{2} = \dfrac{31.8}{2} = 15.9$

Vertical Shift: $\dfrac{56.0 + 24.2}{2} = \dfrac{80.2}{2} = 40.1$

$\omega = \dfrac{2\pi}{12} = \dfrac{\pi}{6}$

Phase shift (use $y = 24.2$, $x = 1$):

$$24.2 = 15.9\sin\left(\dfrac{\pi}{6} \cdot 1 - \phi\right) + 40.1$$

$$-15.9 = 15.9\sin\left(\dfrac{\pi}{6} - \phi\right) \rightarrow -1 = \sin\left(\dfrac{\pi}{6} - \phi\right) \rightarrow -\dfrac{\pi}{2} = \dfrac{\pi}{6} - \phi$$

$$\phi = \dfrac{2\pi}{3}$$

Thus, $y = 15.9\sin\left(\dfrac{\pi}{6}x - \dfrac{2\pi}{3}\right) + 40.1$

(c)

(d) $y = 15.62\sin(0.517x - 2.096) + 40.377$

(e)

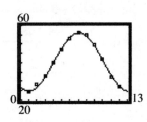

273

21. (a)   Draw a scatter diagram:

(b)   Amplitude: $A = \dfrac{75.4 - 25.5}{2} = \dfrac{49.9}{2} = 24.95$

Vertical Shift: $\dfrac{75.4 + 25.5}{2} = \dfrac{100.9}{2} = 50.45$

$\omega = \dfrac{2\pi}{12} = \dfrac{\pi}{6}$

Phase shift (use $y = 25.5$, $x = 1$):

$$25.5 = 24.95\sin\left(\dfrac{\pi}{6} \cdot 1 - \phi\right) + 50.45$$

$$-24.95 = 24.95\sin\left(\dfrac{\pi}{6} - \phi\right) \rightarrow -1 = \sin\left(\dfrac{\pi}{6} - \phi\right) \rightarrow -\dfrac{\pi}{2} = \dfrac{\pi}{6} - \phi$$

$$\phi = \dfrac{2\pi}{3}$$

Thus, $y = 24.95\sin\left(\dfrac{\pi}{6}x - \dfrac{2\pi}{3}\right) + 50.45$

(c)

(e)

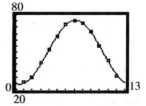

(d) $y = 25.693\sin(0.476x - 1.814) + 49.854$

23. (a) $3.6333 + 12.5 = 16.1333$ hours which is at 4:08 p.m.

(b)   Amplitude: $A = \dfrac{8.2 - (-0.6)}{2} = \dfrac{8.8}{2} = 4.4$

Vertical Shift: $\dfrac{8.2 + (-0.6)}{2} = \dfrac{7.6}{2} = 3.8$

$\omega = \dfrac{2\pi}{12.5} = \dfrac{\pi}{6.25}$

Phase shift (use $y = -0.6$, $x = 10.1333$):

$$-0.6 = 4.4\sin\left(\frac{\pi}{6.25} \cdot 10.1333 - \phi\right) + 3.8$$

$$-4.4 = 4.4\sin\left(\frac{\pi}{6.25} \cdot 10.1333 - \phi\right) \rightarrow -1 = \sin\left(\frac{10.1333\pi}{6.25} - \phi\right)$$

$$-\frac{\pi}{2} = \frac{10.1333\pi}{6.25} - \phi \rightarrow \phi = 6.6643$$

Thus, $y = 4.4\sin\left(\frac{\pi}{6.25}x - 6.6643\right) + 3.8$

(c)

(d) $y = 4.4\sin\left(\frac{\pi}{6.25}(16.1333) - 6.6643\right) + 3.8 \approx 8.2$ feet

25. (a)   Amplitude: $A = \dfrac{12.75 - 10.583}{2} = \dfrac{2.167}{2} = 1.0835$

   Vertical Shift: $\dfrac{12.75 + 10.583}{2} = \dfrac{23.333}{2} = 11.6665$

   $\omega = \dfrac{2\pi}{365}$

   Phase shift (use $y = 10.583$, $x = 356$):

$$10.583 = 1.0835\sin\left(\frac{2\pi}{365} \cdot 356 - \phi\right) + 11.6665$$

$$-1.0835 = 1.0835\sin\left(\frac{2\pi}{365} \cdot 356 - \phi\right) \rightarrow -1 = \sin\left(\frac{712\pi}{365} - \phi\right)$$

$$-\frac{\pi}{2} = \frac{712\pi}{365} - \phi \rightarrow \phi = 7.6991$$

   Thus, $y = 1.0835\sin\left(\frac{2\pi}{365}x - 7.6991\right) + 11.6665$

(b)

(c) $y = 1.0835 \sin\left(\dfrac{2\pi}{365}(92) - 7.6991\right) + 11.6665 \approx 11.85$ hours

27. (a)   Amplitude: $A = \dfrac{16.233 - 5.45}{2} = \dfrac{10.783}{2} = 5.3915$

Vertical Shift: $\dfrac{16.233 + 5.45}{2} = \dfrac{21.683}{2} = 10.8415$

$\omega = \dfrac{2\pi}{365}$

Phase shift (use $y = 5.45$, $x = 356$):

$$5.45 = 5.3915 \sin\left(\dfrac{2\pi}{365} \cdot 356 - \phi\right) + 10.8415$$

$$-5.3915 = 5.3915 \sin\left(\dfrac{2\pi}{365} \cdot 356 - \phi\right) \rightarrow -1 = \sin\left(\dfrac{712\pi}{365} - \phi\right)$$

$$-\dfrac{\pi}{2} = \dfrac{712\pi}{365} - \phi \rightarrow \phi = 7.6991$$

Thus, $y = 5.3915 \sin\left(\dfrac{2\pi}{365}x - 7.6991\right) + 10.8415$

(b)

(c) $y = 5.3915 \sin\left(\dfrac{2\pi}{365}(92) - 7.6991\right) + 10.8415 \approx 11.74$ hours

29. Answers will vary.

Chapter 5

Trigonometric Functions

5.R   Chapter Review

1. $135° = 135 \cdot \dfrac{\pi}{180}$ radian $= \dfrac{3\pi}{4}$ radians

3. $18° = 18 \cdot \dfrac{\pi}{180}$ radian $= \dfrac{\pi}{10}$ radians

5. $\dfrac{3\pi}{4} = \dfrac{3\pi}{4} \cdot \dfrac{180}{\pi}$ degrees $= 135°$

7. $-\dfrac{5\pi}{2} = -\dfrac{5\pi}{2} \cdot \dfrac{180}{\pi}$ degrees $= -450°$

9. $\tan\left(\dfrac{\pi}{4}\right) - \sin\left(\dfrac{\pi}{6}\right) = 1 - \dfrac{1}{2} = \dfrac{1}{2}$

11. $3\sin(45°) - 4\tan\left(\dfrac{\pi}{6}\right) = 3 \cdot \dfrac{\sqrt{2}}{2} - 4 \cdot \dfrac{\sqrt{3}}{3} = \dfrac{3\sqrt{2}}{2} - \dfrac{4\sqrt{3}}{3}$

13. $6\cos\left(\dfrac{3\pi}{4}\right) + 2\tan\left(-\dfrac{\pi}{3}\right) = 6\left(-\dfrac{\sqrt{2}}{2}\right) + 2\left(-\sqrt{3}\right) = -3\sqrt{2} - 2\sqrt{3}$

15. $\sec\left(-\dfrac{\pi}{3}\right) - \cot\left(-\dfrac{5\pi}{4}\right) = \sec\left(\dfrac{\pi}{3}\right) + \cot\left(\dfrac{5\pi}{4}\right) = 2 + 1 = 3$

17. $\tan(\pi) + \sin(\pi) = 0 + 0 = 0$

19. $\cos(180°) - \tan(-45°) = -1 - (-1) = -1 + 1 = 0$

21. $\sin^2(20°) + \dfrac{1}{\sec^2(20°)} = \sin^2(20°) + \cos^2(20°) = 1$

23. $\sec(50°)\cos(50°) = \dfrac{1}{\cos(50°)} \cdot \cos(50°) = 1$

25. $\dfrac{\cos(400°)}{\cos(-40°)} = \dfrac{\cos(40° + 360°)}{\cos(40°)} = = \dfrac{\cos(40°)}{\cos(40°)} = 1$

27. $\sin\theta = -\dfrac{4}{5}, \quad \cos\theta > 0, \quad \Rightarrow \quad \theta$ in quadrant IV

Solve for $\cos\theta$:

$$\sin^2\theta + \cos^2\theta = 1 \rightarrow \cos^2\theta = 1 - \sin^2\theta \rightarrow \cos\theta = \pm\sqrt{1-\sin^2\theta}$$

Since θ is in quadrant IV, $\cos\theta > 0$.

$$\cos\theta = \sqrt{1-\sin^2\theta} = \sqrt{1-\left(-\dfrac{4}{5}\right)^2} = \sqrt{1-\dfrac{16}{25}} = \sqrt{\dfrac{9}{25}} = \dfrac{3}{5}$$

$$\tan\theta = \dfrac{\sin\theta}{\cos\theta} = \dfrac{\left(-\dfrac{4}{5}\right)}{\left(\dfrac{3}{5}\right)} = -\dfrac{4}{5}\cdot\dfrac{5}{3} = -\dfrac{4}{3} \qquad \sec\theta = \dfrac{1}{\cos\theta} = \dfrac{1}{\left(\dfrac{3}{5}\right)} = \dfrac{5}{3}$$

$$\csc\theta = \dfrac{1}{\sin\theta} = \dfrac{1}{\left(-\dfrac{4}{5}\right)} = -\dfrac{5}{4} \qquad \cot\theta = \dfrac{1}{\tan\theta} = \dfrac{1}{\left(-\dfrac{4}{3}\right)} = -\dfrac{3}{4}$$

29. $\tan\theta = \dfrac{12}{5}, \quad \sin\theta < 0, \quad \Rightarrow \quad \theta$ in quadrant III

Solve for $\sec\theta$:

$$\sec^2\theta = 1 + \tan^2\theta \rightarrow \sec\theta = \pm\sqrt{1+\tan^2\theta}$$

Since θ is in quadrant III, $\sec\theta < 0$.

$$\sec\theta = -\sqrt{1+\tan^2\theta} = -\sqrt{1+\left(\dfrac{12}{5}\right)^2} = -\sqrt{1+\dfrac{144}{25}} = -\sqrt{\dfrac{169}{25}} = -\dfrac{13}{5}$$

$$\cos\theta = -\dfrac{5}{13}$$

$$\sin\theta = -\sqrt{1-\cos^2\theta} = -\sqrt{1-\left(-\dfrac{5}{13}\right)^2} = -\sqrt{1-\dfrac{25}{169}} = -\sqrt{\dfrac{144}{169}} = -\dfrac{12}{13}$$

$$\csc\theta = \dfrac{1}{\sin\theta} = \dfrac{1}{\left(-\dfrac{12}{13}\right)} = -\dfrac{13}{12} \qquad \cot\theta = \dfrac{1}{\tan\theta} = \dfrac{1}{\left(\dfrac{12}{5}\right)} = \dfrac{5}{12}$$

31. $\sec\theta = -\dfrac{5}{4}, \quad \tan\theta < 0, \quad \Rightarrow \quad \theta$ in quadrant II

Solve for $\cos\theta$:

$$\cos\theta = \dfrac{1}{\sec\theta} = \dfrac{1}{\left(-\dfrac{5}{4}\right)} = -\dfrac{4}{5}$$

Solve for $\sin\theta$:

$$\sin^2\theta + \cos^2\theta = 1 \rightarrow \sin^2\theta = 1 - \cos^2\theta \rightarrow \sin\theta = \pm\sqrt{1-\cos^2\theta}$$

Since θ is in quadrant II, $\sin\theta > 0$.

$$\sin\theta = \sqrt{1-\cos^2\theta} = \sqrt{1-\left(-\dfrac{4}{5}\right)^2} = \sqrt{1-\dfrac{16}{25}} = \sqrt{\dfrac{9}{25}} = \dfrac{3}{5}$$

$$\tan\theta = \frac{\sin\theta}{\cos\theta} = \frac{\left(\dfrac{3}{5}\right)}{\left(-\dfrac{4}{5}\right)} = \frac{3}{5} \cdot -\frac{5}{4} = -\frac{3}{4}$$

$$\csc\theta = \frac{1}{\sin\theta} = \frac{1}{\left(\dfrac{3}{5}\right)} = \frac{5}{3} \qquad\qquad \cot\theta = \frac{1}{\tan\theta} = \frac{1}{\left(-\dfrac{3}{4}\right)} = -\frac{4}{3}$$

33. $\sin\theta = \dfrac{12}{13}, \quad \theta$ in quadrant II

Solve for $\cos\theta$:

$$\sin^2\theta + \cos^2\theta = 1 \rightarrow \cos^2\theta = 1 - \sin^2\theta \rightarrow \cos\theta = \pm\sqrt{1-\sin^2\theta}$$

Since θ is in quadrant II, $\cos\theta < 0.$

$$\cos\theta = -\sqrt{1-\sin^2\theta} = -\sqrt{1-\left(\frac{12}{13}\right)^2} = -\sqrt{1-\frac{144}{169}} = -\sqrt{\frac{25}{169}} = -\frac{5}{13}$$

$$\tan\theta = \frac{\sin\theta}{\cos\theta} = \frac{\left(\dfrac{12}{13}\right)}{\left(-\dfrac{5}{13}\right)} = \frac{12}{13} \cdot -\frac{13}{5} = -\frac{12}{5} \qquad\qquad \sec\theta = \frac{1}{\cos\theta} = \frac{1}{\left(-\dfrac{5}{13}\right)} = -\frac{13}{5}$$

$$\csc\theta = \frac{1}{\sin\theta} = \frac{1}{\left(\dfrac{12}{13}\right)} = \frac{13}{12} \qquad\qquad \cot\theta = \frac{1}{\tan\theta} = \frac{1}{\left(-\dfrac{12}{5}\right)} = -\frac{5}{12}$$

35. $\sin\theta = -\dfrac{5}{13}, \quad \dfrac{3\pi}{2} < \theta < 2\pi, \Rightarrow \theta$ in quadrant IV

Solve for $\cos\theta$:

$$\sin^2\theta + \cos^2\theta = 1 \rightarrow \cos^2\theta = 1 - \sin^2\theta \rightarrow \cos\theta = \pm\sqrt{1-\sin^2\theta}$$

Since θ is in quadrant IV, $\cos\theta > 0.$

$$\cos\theta = \sqrt{1-\sin^2\theta} = \sqrt{1-\left(-\frac{5}{13}\right)^2} = \sqrt{1-\frac{25}{169}} = \sqrt{\frac{144}{169}} = \frac{12}{13}$$

$$\tan\theta = \frac{\sin\theta}{\cos\theta} = \frac{\left(-\dfrac{5}{13}\right)}{\left(\dfrac{12}{13}\right)} = -\frac{5}{13} \cdot \frac{13}{12} = -\frac{5}{12} \qquad\qquad \sec\theta = \frac{1}{\cos\theta} = \frac{1}{\left(\dfrac{12}{13}\right)} = \frac{13}{12}$$

$$\csc\theta = \frac{1}{\sin\theta} = \frac{1}{\left(-\dfrac{5}{13}\right)} = -\frac{13}{5} \qquad\qquad \cot\theta = \frac{1}{\tan\theta} = \frac{1}{\left(-\dfrac{5}{12}\right)} = -\frac{12}{5}$$

37. $\tan\theta = \dfrac{1}{3},\ \ 180°< \theta < 270°,\ \ \Rightarrow\ \ \theta$ in quadrant III

Solve for $\sec\theta$:

$$\sec^2\theta = 1+\tan^2\theta \rightarrow \sec\theta = \pm\sqrt{1+\tan^2\theta}$$

Since θ is in quadrant III, $\sec\theta < 0$.

$$\sec\theta = -\sqrt{1+\tan^2\theta} = -\sqrt{1+\left(\dfrac{1}{3}\right)^2} = -\sqrt{1+\dfrac{1}{9}} = -\sqrt{\dfrac{10}{9}} = -\dfrac{\sqrt{10}}{3}$$

$$\cos\theta = \dfrac{1}{\sec\theta} = \dfrac{1}{\left(-\dfrac{\sqrt{10}}{3}\right)} = -\dfrac{3}{\sqrt{10}} \cdot \dfrac{\sqrt{10}}{\sqrt{10}} = -\dfrac{3\sqrt{10}}{10}$$

$$\sin\theta = -\sqrt{1-\cos^2\theta} = -\sqrt{1-\left(-\dfrac{3\sqrt{10}}{3}\right)^2} = -\sqrt{1-\dfrac{90}{100}} = -\sqrt{\dfrac{10}{100}} = -\dfrac{\sqrt{10}}{10}$$

$$\csc\theta = \dfrac{1}{\sin\theta} = \dfrac{1}{\left(-\dfrac{\sqrt{10}}{10}\right)} = -\dfrac{10}{\sqrt{10}} = -\sqrt{10} \qquad\qquad \cot\theta = \dfrac{1}{\tan\theta} = \dfrac{1}{\left(\dfrac{1}{3}\right)} = 3$$

39. $\sec\theta = 3,\ \ \dfrac{3\pi}{2} < \theta < 2\pi,\ \ \Rightarrow\ \ \theta$ in quadrant IV

$$\cos\theta = \dfrac{1}{\sec\theta} = \dfrac{1}{3}$$

Solve for $\sin\theta$:

$$\sin^2\theta + \cos^2\theta = 1\ \ \rightarrow\ \ \sin^2\theta = 1-\cos^2\theta\ \ \rightarrow\ \ \sin\theta = \pm\sqrt{1-\cos^2\theta}$$

Since θ is in quadrant IV, $\sin\theta < 0$.

$$\sin\theta = -\sqrt{1-\cos^2\theta} = -\sqrt{1-\left(\dfrac{1}{3}\right)^2} = -\sqrt{1-\dfrac{1}{9}} = -\sqrt{\dfrac{8}{9}} = -\dfrac{2\sqrt{2}}{3}$$

$$\tan\theta = \dfrac{\sin\theta}{\cos\theta} = \dfrac{\left(-\dfrac{2\sqrt{2}}{3}\right)}{\left(\dfrac{1}{3}\right)} = -\dfrac{2\sqrt{2}}{3} \cdot \dfrac{3}{1} = -2\sqrt{2}$$

$$\csc\theta = \dfrac{1}{\sin\theta} = \dfrac{1}{\left(-\dfrac{2\sqrt{2}}{3}\right)} = -\dfrac{3}{2\sqrt{2}} = -\dfrac{3\sqrt{2}}{4} \qquad\qquad \cot\theta = \dfrac{1}{\tan\theta} = \dfrac{1}{-2\sqrt{2}} = -\dfrac{\sqrt{2}}{4}$$

41. $\cot\theta = -2,\ \ \dfrac{\pi}{2} < \theta < \pi,\ \ \Rightarrow\ \ \theta$ in quadrant II

$$\tan\theta = \dfrac{1}{\cot\theta} = \dfrac{1}{-2} = -\dfrac{1}{2}$$

Solve for $\sec\theta$:

$$\sec^2\theta = 1+\tan^2\theta \rightarrow \sec\theta = \pm\sqrt{1+\tan^2\theta}$$

Since θ is in quadrant II, $\sec\theta < 0$.

$$\sec\theta = -\sqrt{1+\tan^2\theta} = -\sqrt{1+\left(-\frac{1}{2}\right)^2} = -\sqrt{1+\frac{1}{4}} = -\sqrt{\frac{5}{4}} = -\frac{\sqrt{5}}{2}$$

$$\cos\theta = \frac{1}{\sec\theta} = \frac{1}{\left(-\frac{\sqrt{5}}{2}\right)} = -\frac{2}{\sqrt{5}}\cdot\frac{\sqrt{5}}{\sqrt{5}} = -\frac{2\sqrt{5}}{5}$$

$$\sin\theta = \sqrt{1-\cos^2\theta} = \sqrt{1-\left(-\frac{2\sqrt{5}}{2}\right)^2} = \sqrt{1-\frac{20}{25}} = \sqrt{\frac{5}{25}} = \frac{\sqrt{5}}{5}$$

$$\csc\theta = \frac{1}{\sin\theta} = \frac{1}{\left(\frac{\sqrt{5}}{5}\right)} = \frac{5}{\sqrt{5}} = \sqrt{5}$$

43. $y = 2\sin(4x)$

45. $y = -2\cos\left(x+\frac{\pi}{2}\right)$

47. $y = \tan(x+\pi)$

49. $y = -2\tan(3x)$

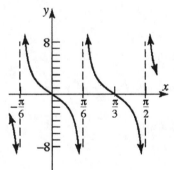

51. $y = \cot\left(x + \dfrac{\pi}{4}\right)$

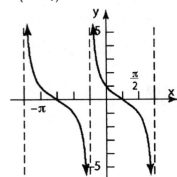

53. $y = \sec\left(x - \dfrac{\pi}{4}\right)$

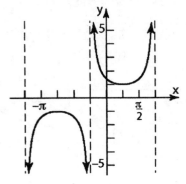

55. $y = 4\cos x$

Amplitude $= 4$
Period $= 2\pi$

57. $y = -8\sin\left(\dfrac{\pi}{2}x\right)$

Amplitude $= 8$
Period $= 4$

59. $y = 4\sin(3x)$
Amplitude: $|A| = |4| = 4$

Period: $\quad T = \dfrac{2\pi}{\omega} = \dfrac{2\pi}{3}$

Phase Shift: $\dfrac{\phi}{\omega} = \dfrac{0}{3} = 0$

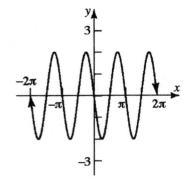

61.
$y = 2\sin(2x - \pi)$
Amplitude: $|A| = |2| = 2$

Period: $\quad T = \dfrac{2\pi}{\omega} = \dfrac{2\pi}{2} = \pi$

Phase Shift: $\dfrac{\phi}{\omega} = \dfrac{-\pi}{2} = -\dfrac{\pi}{2}$

63. $y = \dfrac{1}{2}\sin\left(\dfrac{3}{2}x - \pi\right)$

Amplitude: $|A| = \left|\dfrac{1}{2}\right| = \dfrac{1}{2}$

Period: $T = \dfrac{2\pi}{\omega} = \dfrac{2\pi}{\left(\dfrac{3}{2}\right)} = \dfrac{4\pi}{3}$

Phase Shift: $\dfrac{\phi}{\omega} = \dfrac{\pi}{\left(\dfrac{3}{2}\right)} = \dfrac{2\pi}{3}$

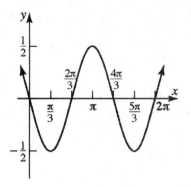

65. $y = -\dfrac{2}{3}\cos(\pi x - 6)$

Amplitude: $|A| = \left|-\dfrac{2}{3}\right| = \dfrac{2}{3}$

Period: $T = \dfrac{2\pi}{\omega} = \dfrac{2\pi}{\pi} = 2$

Phase Shift: $\dfrac{\phi}{\omega} = \dfrac{6}{\pi}$

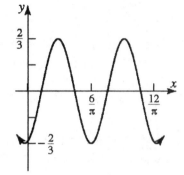

67. The graph is a cosine graph with an amplitude of 5 and a period of 8π. Find ω:

$8\pi = \dfrac{2\pi}{\omega} \;\rightarrow\; 8\pi\omega = 2\pi \;\rightarrow\; \omega = \dfrac{2\pi}{8\pi} = \dfrac{1}{4}$

The equation is: $y = 5\cos\left(\dfrac{1}{4}x\right)$.

69. The graph is a reflected cosine graph with an amplitude of 6 and a period of 8.
Find ω:

$8 = \dfrac{2\pi}{\omega} \;\rightarrow\; 8\omega = 2\pi \;\rightarrow\; \omega = \dfrac{2\pi}{8} = \dfrac{\pi}{4}$

The equation is: $y = -6\cos\left(\dfrac{\pi}{4}x\right)$.

71. $r = 2$ feet, $\theta = 30°$ or $\theta = \dfrac{\pi}{6}$

$s = r\theta = 2 \cdot \dfrac{\pi}{6} = \dfrac{\pi}{3}$ feet; $A = \dfrac{1}{2} \cdot r^2\theta = \dfrac{1}{2} \cdot (2)^2 \cdot \dfrac{\pi}{6} = \dfrac{\pi}{3}$ square feet

73. $v = 180$ mi/hr, $d = \dfrac{1}{2}$ mile $\rightarrow$ $r = \dfrac{1}{4} = 0.25$ mile

$\omega = \dfrac{v}{r} = \dfrac{180 \text{ mi/hr}}{0.25 \text{ mi}} = 720 \text{ rad/hr} = \dfrac{720 \text{ rad}}{\text{hr}} \cdot \dfrac{1 \text{ rev}}{2\pi \text{ rad}} = \dfrac{360 \text{ rev}}{\pi \text{ hr}} \approx 114.6 \text{ rev/hr}$

283

75. Since there are two lights on opposite sides and the light is seen every 5 seconds, the beacon makes 1 revolution every 10 seconds.

$$\omega = \frac{1 \text{ rev}}{10 \text{ sec}} \cdot \frac{2\pi}{1 \text{ rev}} = \frac{\pi}{5} \text{ radians / second}$$

77. $E(t) = 120\sin(120\pi t), \quad t \geq 0$

 (a) The maximum value of E is the amplitude which is 120.

 (b) Period $= \dfrac{2\pi}{120\pi} = \dfrac{1}{60}$ seconds

 (c) Graphing:

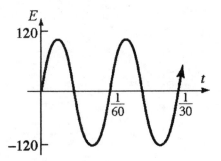

79. (a) Draw a scatter diagram:

 (b) Amplitude: $A = \dfrac{90 - 51}{2} = \dfrac{39}{2} = 19.5$

   Vertical Shift: $\dfrac{90 + 51}{2} = \dfrac{141}{2} = 70.5$

   $\omega = \dfrac{2\pi}{12} = \dfrac{\pi}{6}$

   Phase shift (use $y = 51$, $x = 1$):

   $$51 = 19.5\sin\left(\frac{\pi}{6} \cdot 1 - \phi\right) + 70.5$$

   $$-19.5 = 19.5\sin\left(\frac{\pi}{6} - \phi\right) \rightarrow -1 = \sin\left(\frac{\pi}{6} - \phi\right)$$

   $$-\frac{\pi}{2} = \frac{\pi}{6} - \phi \rightarrow \phi = \frac{2\pi}{3}$$

   Thus, $y = 19.5\sin\left(\dfrac{\pi}{6}x - \dfrac{2\pi}{3}\right) + 70.5$

(c)

(d) $y = 19.518\sin(0.541x - 2.283) + 71.01$

(e)

Chapter 6

Analytic Trigonometry

6.1 The Inverse Sine, Cosine and Tangent Functions

1. $\sin^{-1}(0)$

    We are finding the angle θ, $-\dfrac{\pi}{2} \le \theta \le \dfrac{\pi}{2}$, whose sine equals 0.

    $$\sin\theta = 0 \qquad -\dfrac{\pi}{2} \le \theta \le \dfrac{\pi}{2}$$

    $$\theta = 0 \to \sin^{-1}(0) = 0$$

3. $\sin^{-1}(-1)$

    We are finding the angle θ, $-\dfrac{\pi}{2} \le \theta \le \dfrac{\pi}{2}$, whose sine equals -1.

    $$\sin\theta = -1 \qquad -\dfrac{\pi}{2} \le \theta \le \dfrac{\pi}{2}$$

    $$\theta = -\dfrac{\pi}{2} \to \sin^{-1}(-1) = -\dfrac{\pi}{2}$$

5. $\tan^{-1}(0)$

    We are finding the angle θ, $-\dfrac{\pi}{2} < \theta < \dfrac{\pi}{2}$, whose tangent equals 0.

    $$\tan\theta = 0 \qquad -\dfrac{\pi}{2} < \theta < \dfrac{\pi}{2}$$

    $$\theta = 0 \to \tan^{-1}(0) = 0$$

7. $\sin^{-1}\left(\dfrac{\sqrt{2}}{2}\right)$

    We are finding the angle θ, $-\dfrac{\pi}{2} \le \theta \le \dfrac{\pi}{2}$, whose sine equals $\dfrac{\sqrt{2}}{2}$.

    $$\sin\theta = \dfrac{\sqrt{2}}{2} \qquad -\dfrac{\pi}{2} \le \theta \le \dfrac{\pi}{2}$$

    $$\theta = \dfrac{\pi}{4} \to \sin^{-1}\left(\dfrac{\sqrt{2}}{2}\right) = \dfrac{\pi}{4}$$

9. $\tan^{-1}\left(\sqrt{3}\right)$

We are finding the angle $\theta, -\dfrac{\pi}{2} < \theta < \dfrac{\pi}{2}$, whose tangent equals $\sqrt{3}$.

$$\tan\theta = \sqrt{3} \qquad -\dfrac{\pi}{2} < \theta < \dfrac{\pi}{2}$$

$$\theta = \dfrac{\pi}{3} \rightarrow \tan^{-1}\left(\sqrt{3}\right) = \dfrac{\pi}{3}$$

11. $\cos^{-1}\left(-\dfrac{\sqrt{3}}{2}\right)$

We are finding the angle θ, $0 \le \theta \le \pi$, whose cosine equals $-\dfrac{\sqrt{3}}{2}$.

$$\cos\theta = -\dfrac{\sqrt{3}}{2} \qquad 0 \le \theta \le \pi$$

$$\theta = \dfrac{5\pi}{6} \rightarrow \cos^{-1}\left(-\dfrac{\sqrt{3}}{2}\right) = \dfrac{5\pi}{6}$$

13. $\sin^{-1}(0.1) \approx 0.10$

15. $\tan^{-1}(5) \approx 1.37$

17. $\cos^{-1}\left(\dfrac{7}{8}\right) \approx 0.51$

19. $\tan^{-1}(-0.4) \approx -0.38$

21. $\sin^{-1}(-0.12) \approx -0.12$

23. $\cos^{-1}\left(\dfrac{\sqrt{2}}{3}\right) \approx 1.08$

25. $\sin\left[\sin^{-1}(0.54)\right] = 0.54$

27. $\cos^{-1}\left[\cos\left(\dfrac{4\pi}{5}\right)\right] = \dfrac{4\pi}{5}$

29. $\tan\left[\tan^{-1}(-3.5)\right] = -3.5$

31. $\sin^{-1}\left[\sin\left(-\dfrac{3\pi}{7}\right)\right] = -\dfrac{3\pi}{7}$

33.   yes, $\sin^{-1}\left[\sin\left(-\dfrac{\pi}{6}\right)\right] = -\dfrac{\pi}{6}$ since $\sin^{-1}\left[\sin(x)\right] = x$ where $-\dfrac{\pi}{2} \le x \le \dfrac{\pi}{2}$

and $-\dfrac{\pi}{6}$ is in the restricted domain of $f(x) = \sin(x)$.

35.   no, $\sin\left[\sin^{-1}(2)\right] \ne 2$ since $\sin\left[\sin^{-1}(x)\right] = x$ where $-1 \le x \le 1$
and 2 is not in the domain of $f(x) = \sin^{-1}(x)$.

37. no, $\cos^{-1}\left[\cos\left(-\dfrac{\pi}{6}\right)\right] \neq -\dfrac{\pi}{6}$ since $\cos^{-1}[\cos(x)] = x$ where $0 \leq x \leq \pi$

and $-\dfrac{\pi}{6}$ is not in the restricted domain of $f(x) = \cos(x)$.

39. yes, $\cos\left[\cos^{-1}\left(-\dfrac{1}{2}\right)\right] = -\dfrac{1}{2}$ since $\cos\left[\cos^{-1}(x)\right] = x$ where $-1 \leq x \leq 1$

and $-\dfrac{1}{2}$ is in the domain of $f(x) = \cos^{-1}(x)$.

41. yes, $\tan^{-1}\left[\tan\left(-\dfrac{\pi}{3}\right)\right] = -\dfrac{\pi}{3}$ since $\tan^{-1}[\tan(x)] = x$ where $-\dfrac{\pi}{2} < x < \dfrac{\pi}{2}$

and $-\dfrac{\pi}{3}$ is in the restricted domain of $f(x) = \tan(x)$.

43. yes, $\tan\left[\tan^{-1}(2)\right] = 2$ since $\tan\left[\tan^{-1}(x)\right] = x$ where $-\infty < x < \infty$

45. (a) $D = 24 \cdot \left[1 - \dfrac{\cos^{-1}\left(\tan\left(23.5 \cdot \dfrac{\pi}{180}\right)\tan\left(29.75 \cdot \dfrac{\pi}{180}\right)\right)}{\pi}\right] \approx 13.92$ hours

(b) $D = 24 \cdot \left[1 - \dfrac{\cos^{-1}\left(\tan\left(0 \cdot \dfrac{\pi}{180}\right)\tan\left(29.75 \cdot \dfrac{\pi}{180}\right)\right)}{\pi}\right] \approx 12$ hours

(c) $D = 24 \cdot \left[1 - \dfrac{\cos^{-1}\left(\tan\left(22.8 \cdot \dfrac{\pi}{180}\right)\tan\left(29.75 \cdot \dfrac{\pi}{180}\right)\right)}{\pi}\right] \approx 13.85$ hours

47. (a) $D = 24 \cdot \left(1 - \dfrac{\cos^{-1}\left(\tan\left(23.5 \cdot \dfrac{\pi}{180}\right)\tan\left(21.3 \cdot \dfrac{\pi}{180}\right)\right)}{\pi}\right) \approx 13.30$ hours

(b) $D = 24 \cdot \left(1 - \dfrac{\cos^{-1}\left(\tan\left(0 \cdot \dfrac{\pi}{180}\right)\tan\left(21.3 \cdot \dfrac{\pi}{180}\right)\right)}{\pi}\right) \approx 12$ hours

(c) $D = 24 \cdot \left(1 - \dfrac{\cos^{-1}\left(\tan\left(22.8 \cdot \dfrac{\pi}{180} \right) \tan\left(21.3 \cdot \dfrac{\pi}{180} \right) \right)}{\pi} \right) \approx 13.26$ hours

49. (a) $D = 24 \cdot \left(1 - \dfrac{\cos^{-1}\left(\tan\left(23.5 \cdot \dfrac{\pi}{180} \right) \tan\left(0 \cdot \dfrac{\pi}{180} \right) \right)}{\pi} \right) \approx 12$ hours

(b) $D = 24 \cdot \left(1 - \dfrac{\cos^{-1}\left(\tan\left(0 \cdot \dfrac{\pi}{180} \right) \tan\left(0 \cdot \dfrac{\pi}{180} \right) \right)}{\pi} \right) \approx 12$ hours

(c) $D = 24 \cdot \left(1 - \dfrac{\cos^{-1}\left(\tan\left(22.8 \cdot \dfrac{\pi}{180} \right) \tan\left(0 \cdot \dfrac{\pi}{180} \right) \right)}{\pi} \right) \approx 12$ hours

(d)   The number of hours of daylight per day is approximately 12 hours throughout the year.

51. $1530 \text{ ft} \cdot \dfrac{1 \text{ mile}}{5280 \text{ feet}} = 0.29$ mile

$\cos\theta = \dfrac{3960}{3960.29}$

$\theta \approx 0.0121$ radians

$s = r\theta = 3960(0.0121) = 47.92$ miles

$\dfrac{2\pi(2710)}{24} = \dfrac{47.92}{t}$

$t \approx 0.0675$ hour $= 4.05$ minutes

Chapter 6

Analytic Trigonometry

6.2 The Inverse Trigonometric Functions (Continued)

1. $\cos\left(\sin^{-1}\left(\frac{\sqrt{2}}{2}\right)\right)$

Find the angle θ, $-\frac{\pi}{2} \le \theta \le \frac{\pi}{2}$, whose sine equals $\frac{\sqrt{2}}{2}$.

$$\sin\theta = \frac{\sqrt{2}}{2} \qquad -\frac{\pi}{2} \le \theta \le \frac{\pi}{2}$$

$$\theta = \frac{\pi}{4} \rightarrow \cos\left(\sin^{-1}\left(\frac{\sqrt{2}}{2}\right)\right) = \cos\left(\frac{\pi}{4}\right) = \frac{\sqrt{2}}{2}$$

3. $\tan\left(\cos^{-1}\left(-\frac{\sqrt{3}}{2}\right)\right)$

Find the angle θ, $0 \le \theta \le \pi$, whose cosine equals $-\frac{\sqrt{3}}{2}$.

$$\cos\theta = -\frac{\sqrt{3}}{2} \qquad 0 \le \theta \le \pi$$

$$\theta = \frac{5\pi}{6} \rightarrow \tan\left(\cos^{-1}\left(-\frac{\sqrt{3}}{2}\right)\right) = \tan\left(\frac{5\pi}{6}\right) = -\frac{\sqrt{3}}{3}$$

5. $\sec\left(\cos^{-1}\left(\frac{1}{2}\right)\right)$

Find the angle θ, $0 \le \theta \le \pi$, whose cosine equals $\frac{1}{2}$.

$$\cos\theta = \frac{1}{2} \qquad 0 \le \theta \le \pi$$

$$\theta = \frac{\pi}{3} \rightarrow \sec\left(\cos^{-1}\left(\frac{1}{2}\right)\right) = \sec\left(\frac{\pi}{3}\right) = 2$$

7. $\csc\left(\tan^{-1}(1)\right)$

Find the angle θ, $-\dfrac{\pi}{2} < \theta < \dfrac{\pi}{2}$, whose tangent equals 1.

$\tan\theta = 1 \qquad -\dfrac{\pi}{2} < \theta < \dfrac{\pi}{2}$

$\theta = \dfrac{\pi}{4} \rightarrow \csc\left(\tan^{-1}(1)\right) = \csc\left(\dfrac{\pi}{4}\right) = \sqrt{2}$

9. $\sin\left(\tan^{-1}(-1)\right)$

Find the angle θ, $-\dfrac{\pi}{2} < \theta < \dfrac{\pi}{2}$, whose tangent equals -1.

$\tan\theta = -1 \qquad -\dfrac{\pi}{2} < \theta < \dfrac{\pi}{2}$

$\theta = -\dfrac{\pi}{4} \rightarrow \sin\left(\tan^{-1}(-1)\right) = \sin\left(-\dfrac{\pi}{4}\right) = -\dfrac{\sqrt{2}}{2}$

11. $\sec\left(\sin^{-1}\left(-\dfrac{1}{2}\right)\right)$

Find the angle θ, $-\dfrac{\pi}{2} \le \theta \le \dfrac{\pi}{2}$, whose sine equals $-\dfrac{1}{2}$.

$\sin\theta = -\dfrac{1}{2} \qquad -\dfrac{\pi}{2} \le \theta \le \dfrac{\pi}{2}$

$\theta = -\dfrac{\pi}{6} \rightarrow \sec\left(\sin^{-1}\left(-\dfrac{1}{2}\right)\right) = \sec\left(-\dfrac{\pi}{6}\right) = \dfrac{2\sqrt{3}}{3}$

13. $\cos^{-1}\left(\cos\dfrac{5\pi}{4}\right) = \cos^{-1}\left(-\dfrac{\sqrt{2}}{2}\right)$

Find the angle θ, $0 \le \theta \le \pi$, whose cosine equals $-\dfrac{\sqrt{2}}{2}$.

$\cos\theta = -\dfrac{\sqrt{2}}{2} \qquad 0 \le \theta \le \pi$

$\theta = \dfrac{3\pi}{4}$

15. $\sin^{-1}\left(\sin\left(-\dfrac{7\pi}{6}\right)\right) = \sin^{-1}\left(\dfrac{1}{2}\right)$

Find the angle θ, $-\dfrac{\pi}{2} \le \theta \le \dfrac{\pi}{2}$, whose sine equals $\dfrac{1}{2}$.

$\sin\theta = \dfrac{1}{2} \qquad -\dfrac{\pi}{2} \le \theta \le \dfrac{\pi}{2}$

$\theta = \dfrac{\pi}{6}$

17. $\tan\left(\sin^{-1}\left(\dfrac{1}{3}\right)\right)$

Since $\sin\theta = \dfrac{1}{3}$, $-\dfrac{\pi}{2} \leq \theta \leq \dfrac{\pi}{2}$, let $y = 1$ and $r = 3$. Solve for x:

$$x^2 + 1 = 9 \rightarrow x^2 = 8 \rightarrow x = \pm\sqrt{8} = \pm 2\sqrt{2}$$

Since θ is in quadrant I, $x = 2\sqrt{2}$.

$$\tan\left(\sin^{-1}\left(\frac{1}{3}\right)\right) = \tan\theta = \frac{y}{x} = \frac{1}{2\sqrt{2}} \cdot \frac{\sqrt{2}}{\sqrt{2}} = \frac{\sqrt{2}}{4}$$

19. $\sec\left(\tan^{-1}\left(\dfrac{1}{2}\right)\right)$

Since $\tan\theta = \dfrac{1}{2}$, $-\dfrac{\pi}{2} < \theta < \dfrac{\pi}{2}$, let $x = 2$ and $y = 1$. Solve for r:

$$2^2 + 1 = r^2 \rightarrow r^2 = 5 \rightarrow r = \sqrt{5}$$

θ is in quadrant I.

$$\sec\left(\tan^{-1}\left(\frac{1}{2}\right)\right) = \sec\theta = \frac{r}{x} = \frac{\sqrt{5}}{2}$$

21. $\cot\left(\sin^{-1}\left(-\dfrac{\sqrt{2}}{3}\right)\right)$

Since $\sin\theta = -\dfrac{\sqrt{2}}{3}$, $-\dfrac{\pi}{2} \leq \theta \leq \dfrac{\pi}{2}$, let $y = -\sqrt{2}$ and $r = 3$. Solve for x:

$$x^2 + 2 = 9 \rightarrow x^2 = 7 \rightarrow x = \pm\sqrt{7}$$

Since θ is in quadrant IV, $x = \sqrt{7}$.

$$\cot\left(\sin^{-1}\left(-\frac{\sqrt{2}}{3}\right)\right) = \cot\theta = \frac{x}{y} = \frac{\sqrt{7}}{-\sqrt{2}} \cdot \frac{\sqrt{2}}{\sqrt{2}} = -\frac{\sqrt{14}}{2}$$

23. $\sin\left(\tan^{-1}(-3)\right)$

Since $\tan\theta = -3$, $-\dfrac{\pi}{2} < \theta < \dfrac{\pi}{2}$, let $x = 1$ and $y = -3$. Solve for r:

$$1 + 9 = r^2 \rightarrow r^2 = 10 \rightarrow r = \sqrt{10}; \ \theta \text{ is in quadrant IV.}$$

$$\sin\left(\tan^{-1}(-3)\right) = \sin\theta = \frac{y}{r} = \frac{-3}{\sqrt{10}} \cdot \frac{\sqrt{10}}{\sqrt{10}} = -\frac{3\sqrt{10}}{10}$$

25. $\sec\left(\sin^{-1}\left(\dfrac{2\sqrt{5}}{5}\right)\right)$

Since $\sin\theta = \dfrac{2\sqrt{5}}{5}$, $-\dfrac{\pi}{2} \leq \theta \leq \dfrac{\pi}{2}$, let $y = 2\sqrt{5}$ and $r = 5$. Solve for x:

$$x^2 + 20 = 25 \rightarrow x^2 = 5 \rightarrow x = \pm\sqrt{5}$$

Since θ is in quadrant I, $x = \sqrt{5}$.

$$\sec\left(\sin^{-1}\left(\frac{2\sqrt{5}}{5}\right)\right) = \sec\theta = \frac{r}{x} = \frac{5}{\sqrt{5}} \cdot \frac{\sqrt{5}}{\sqrt{5}} = \sqrt{5}$$

27. $\sin^{-1}\left(\cos\left(\dfrac{3\pi}{4}\right)\right) = \sin^{-1}\left(-\dfrac{\sqrt{2}}{2}\right) = -\dfrac{\pi}{4}$

29. $\cot^{-1}\left(\sqrt{3}\right)$

We are finding the angle θ, $0 < \theta < \pi$, whose cotangent equals $\sqrt{3}$.

$$\cot\theta = \sqrt{3} \quad 0 < \theta < \pi$$

$$\theta = \dfrac{\pi}{6} \rightarrow \cot^{-1}\left(\sqrt{3}\right) = \dfrac{\pi}{6}$$

31. $\csc^{-1}(-1)$

We are finding the angle θ, $-\dfrac{\pi}{2} \le \theta \le \dfrac{\pi}{2}$, $\theta \ne 0$, whose cosecant equals -1.

$$\csc\theta = -1 \quad -\dfrac{\pi}{2} \le \theta \le \dfrac{\pi}{2},\ \theta \ne 0$$

$$\theta = -\dfrac{\pi}{2} \rightarrow \csc^{-1}(-1) = -\dfrac{\pi}{2}$$

33. $\sec^{-1}\left(\dfrac{2\sqrt{3}}{3}\right)$

We are finding the angle θ, $0 \le \theta \le \pi$, $\theta \ne \dfrac{\pi}{2}$, whose secant equals $\dfrac{2\sqrt{3}}{3}$.

$$\sec\theta = \dfrac{2\sqrt{3}}{3} \quad 0 \le \theta \le \pi,\ \theta \ne \dfrac{\pi}{2}$$

$$\theta = \dfrac{\pi}{6} \rightarrow \sec^{-1}\left(\dfrac{2\sqrt{3}}{3}\right) = \dfrac{\pi}{6}$$

35. $\cot^{-1}\left(-\dfrac{\sqrt{3}}{3}\right)$

We are finding the angle θ, $0 < \theta < \pi$, whose cotangent equals $-\dfrac{\sqrt{3}}{3}$.

$$\cot\theta = -\dfrac{\sqrt{3}}{3} \quad 0 < \theta < \pi$$

$$\theta = \dfrac{2\pi}{3} \rightarrow \cot^{-1}\left(-\dfrac{\sqrt{3}}{3}\right) = \dfrac{2\pi}{3}$$

37. $\sec^{-1}(4) = \cos^{-1}\left(\dfrac{1}{4}\right)$

We are finding the angle θ, $0 \le \theta \le \pi$, whose cosine equals $\dfrac{1}{4}$.

$$\cos\theta = \frac{1}{4} \Rightarrow \theta \text{ in quadrant I}$$

The calculator yields $\theta = \cos^{-1}\left(\frac{1}{4}\right) \approx 1.32$, which is an angle in quadrant I.

$$\therefore \quad \sec^{-1}(4) \approx 1.32$$

39. $\cot^{-1}(2) = \tan^{-1}\left(\frac{1}{2}\right)$

We are finding the angle $\theta, \ 0 \le \theta \le \pi,$ whose tangent equals $\frac{1}{2}.$

$$\tan\theta = \frac{1}{2} \Rightarrow \theta \text{ in quadrant I}$$

The calculator yields $\theta = \tan^{-1}\left(\frac{1}{2}\right) \approx 0.46,$ which is an angle in quadrant I.

$$\therefore \quad \cot^{-1}(2) \approx 0.46$$

41. $\csc^{-1}(-3) = \sin^{-1}\left(-\frac{1}{3}\right) \approx -0.34$

We are finding the angle $\theta, \ -\frac{\pi}{2} \le \theta \le \frac{\pi}{2},$ whose sine equals $-\frac{1}{3}.$

$$\sin\theta = -\frac{1}{3} \Rightarrow \theta \text{ in quadrant IV}$$

The calculator yields $\theta = \sin^{-1}\left(-\frac{1}{3}\right) \approx -0.34,$ which is an angle in quadrant IV.

$$\therefore \quad \csc^{-1}(-3) \approx -0.34$$

43. $\cot^{-1}\left(-\sqrt{5}\right) = \tan^{-1}\left(-\frac{1}{\sqrt{5}}\right)$

We are finding the angle $\theta, \ 0 \le \theta \le \pi,$ whose tangent equals $-\frac{1}{\sqrt{5}}.$

$$\tan\theta = -\frac{1}{\sqrt{5}} \Rightarrow \theta \text{ in quadrant II}$$

The calculator yields $\tan^{-1}\left(-\frac{1}{\sqrt{5}}\right) \approx -0.42$, which is an angle in quadrant IV.

Since θ is in quadrant II, $\theta \approx -0.42 + \pi \approx 2.72.$

$$\therefore \quad \cot^{-1}\left(-\sqrt{5}\right) \approx 2.72$$

45. $\csc^{-1}\left(-\frac{3}{2}\right) = \sin^{-1}\left(-\frac{2}{3}\right)$

We are finding the angle $\theta, \ -\frac{\pi}{2} \le \theta \le \frac{\pi}{2},$ whose sine equals $-\frac{2}{3}.$

$$\sin\theta = -\frac{2}{3} \Rightarrow \theta \text{ in quadrant IV}$$

The calculator yields $\sin^{-1}\left(-\dfrac{2}{3}\right) \approx -0.73$, which is an angle in quadrant IV.

$\therefore \quad \csc^{-1}\left(-\dfrac{3}{2}\right) \approx -0.73$

47. $\cot^{-1}\left(-\dfrac{3}{2}\right) = \tan^{-1}\left(-\dfrac{2}{3}\right)$

We are finding the angle $\theta,\ 0 \le \theta \le \pi,$ whose tangent equals $-\dfrac{2}{3}$.

$\tan\theta = -\dfrac{2}{3} \Rightarrow \theta$ in quadrant II

The calculator yields $\tan^{-1}\left(-\dfrac{2}{3}\right) \approx -0.59$, which is an angle in quadrant IV.

Since θ is in quadrant II, $\theta \approx -0.59 + \pi \approx 2.55$.

$\therefore \quad \cot^{-1}\left(-\dfrac{3}{2}\right) \approx 2.55$

49. $y = \cot^{-1} x$

51. $y = \csc^{-1} x$

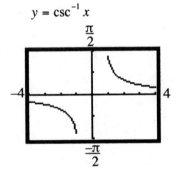

53.   Answers will vary.

Chapter 6

Analytic Trigonometry

6.3 Trigonometric Identities

1. $\csc\theta\cdot\cos\theta = \dfrac{1}{\sin\theta}\cdot\cos\theta = \dfrac{\cos\theta}{\sin\theta} = \cot\theta$

3. $1+\tan^2(-\theta) = 1+(-\tan\theta)^2 = 1+\tan^2\theta = \sec^2\theta$

5. $\cos\theta(\tan\theta+\cot\theta) = \cos\theta\left(\dfrac{\sin\theta}{\cos\theta}+\dfrac{\cos\theta}{\sin\theta}\right) = \cos\theta\left(\dfrac{\sin^2\theta+\cos^2\theta}{\cos\theta\sin\theta}\right) = \dfrac{1}{\sin\theta} = \csc\theta$

7. $\tan\theta\cot\theta - \cos^2\theta = \tan\theta\cdot\dfrac{1}{\tan\theta} - \cos^2\theta = 1-\cos^2\theta = \sin^2\theta$

9. $(\sec\theta-1)(\sec\theta+1) = \sec^2\theta - 1 = \tan^2\theta$

11. $(\sec\theta+\tan\theta)(\sec\theta-\tan\theta) = \sec^2\theta - \tan^2\theta = 1$

13. $\cos^2\theta(1+\tan^2\theta) = \cos^2\theta\cdot\sec^2\theta = \cos^2\theta\cdot\dfrac{1}{\cos^2\theta} = 1$

15. $(\sin\theta+\cos\theta)^2 + (\sin\theta-\cos\theta)^2$
$$= \sin^2\theta + 2\sin\theta\cos\theta + \cos^2\theta + \sin^2\theta - 2\sin\theta\cos\theta + \cos^2\theta$$
$$= 2\sin^2\theta + 2\cos^2\theta = 2(\sin^2\theta+\cos^2\theta) = 2\cdot1 = 2$$

17. $\sec^4\theta - \sec^2\theta = \sec^2\theta(\sec^2\theta-1) = (\tan^2\theta+1)\tan^2\theta = \tan^4\theta + \tan^2\theta$

19. $\sec\theta-\tan\theta = \dfrac{1}{\cos\theta} - \dfrac{\sin\theta}{\cos\theta} = \left(\dfrac{1-\sin\theta}{\cos\theta}\right)\cdot\left(\dfrac{1+\sin\theta}{1+\sin\theta}\right) = \dfrac{1-\sin^2\theta}{\cos\theta(1+\sin\theta)}$
$$= \dfrac{\cos^2\theta}{\cos\theta(1+\sin\theta)} = \dfrac{\cos\theta}{1+\sin\theta}$$

21. $3\sin^2\theta + 4\cos^2\theta = 3\sin^2\theta + 3\cos^2\theta + \cos^2\theta = 3(\sin^2\theta+\cos^2\theta) + \cos^2\theta$
$$= 3\cdot1 + \cos^2\theta = 3 + \cos^2\theta$$

23. $1-\dfrac{\cos^2\theta}{1+\sin\theta} = 1-\dfrac{1-\sin^2\theta}{1+\sin\theta} = 1-\dfrac{(1-\sin\theta)(1+\sin\theta)}{1+\sin\theta} = 1-1+\sin\theta = \sin\theta$

25. $\dfrac{1+\tan\theta}{1-\tan\theta}=\dfrac{\left(1+\dfrac{1}{\cot\theta}\right)}{\left(1-\dfrac{1}{\cot\theta}\right)}=\dfrac{\left(\dfrac{\cot\theta+1}{\cot\theta}\right)}{\left(\dfrac{\cot\theta-1}{\cot\theta}\right)}=\dfrac{\cot\theta+1}{\cot\theta}\cdot\dfrac{\cot\theta}{\cot\theta-1}=\dfrac{\cot\theta+1}{\cot\theta-1}$

27. $\dfrac{\sec\theta}{\csc\theta}+\dfrac{\sin\theta}{\cos\theta}=\dfrac{\left(\dfrac{1}{\cos\theta}\right)}{\left(\dfrac{1}{\sin\theta}\right)}+\dfrac{\sin\theta}{\cos\theta}=\dfrac{\sin\theta}{\cos\theta}+\dfrac{\sin\theta}{\cos\theta}=\tan\theta+\tan\theta=2\tan\theta$

29. $\dfrac{1+\sin\theta}{1-\sin\theta}=\dfrac{\left(1+\dfrac{1}{\csc\theta}\right)}{\left(1-\dfrac{1}{\csc\theta}\right)}=\dfrac{\left(\dfrac{\csc\theta+1}{\csc\theta}\right)}{\left(\dfrac{\csc\theta-1}{\csc\theta}\right)}=\dfrac{\csc\theta+1}{\csc\theta}\cdot\dfrac{\csc\theta}{\csc\theta-1}=\dfrac{\csc\theta+1}{\csc\theta-1}$

31. $\dfrac{1-\sin\theta}{\cos\theta}+\dfrac{\cos\theta}{1-\sin\theta}=\dfrac{(1-\sin\theta)^2+\cos^2\theta}{\cos\theta(1-\sin\theta)}=\dfrac{1-2\sin\theta+\sin^2\theta+\cos^2\theta}{\cos\theta(1-\sin\theta)}$

$=\dfrac{1-2\sin\theta+1}{\cos\theta(1-\sin\theta)}=\dfrac{2-2\sin\theta}{\cos\theta(1-\sin\theta)}=\dfrac{2(1-\sin\theta)}{\cos\theta(1-\sin\theta)}=\dfrac{2}{\cos\theta}=2\sec\theta$

33. $\dfrac{\sin\theta}{\sin\theta-\cos\theta}=\dfrac{\sin\theta}{\sin\theta-\cos\theta}\cdot\dfrac{\left(\dfrac{1}{\sin\theta}\right)}{\left(\dfrac{1}{\sin\theta}\right)}=\dfrac{1}{\left(1-\dfrac{\cos\theta}{\sin\theta}\right)}=\dfrac{1}{1-\cot\theta}$

35. $(\sec\theta-\tan\theta)^2=\sec^2\theta-2\sec\theta\tan\theta+\tan^2\theta=\dfrac{1}{\cos^2\theta}-2\cdot\dfrac{1}{\cos\theta}\cdot\dfrac{\sin\theta}{\cos\theta}+\dfrac{\sin^2\theta}{\cos^2\theta}$

$=\dfrac{1-2\sin\theta+\sin^2\theta}{\cos^2\theta}=\dfrac{(1-\sin\theta)(1-\sin\theta)}{1-\sin^2\theta}=\dfrac{(1-\sin\theta)(1-\sin\theta)}{(1-\sin\theta)(1+\sin\theta)}=\dfrac{1-\sin\theta}{1+\sin\theta}$

37. $\dfrac{\cos\theta}{1-\tan\theta}+\dfrac{\sin\theta}{1-\cot\theta}=\dfrac{\cos\theta}{\left(1-\dfrac{\sin\theta}{\cos\theta}\right)}+\dfrac{\sin\theta}{\left(1-\dfrac{\cos\theta}{\sin\theta}\right)}=\dfrac{\cos\theta}{\left(\dfrac{\cos\theta-\sin\theta}{\cos\theta}\right)}+\dfrac{\sin\theta}{\left(\dfrac{\sin\theta-\cos\theta}{\sin\theta}\right)}$

$=\dfrac{\cos^2\theta}{\cos\theta-\sin\theta}+\dfrac{\sin^2\theta}{\sin\theta-\cos\theta}=\dfrac{\cos^2\theta-\sin^2\theta}{\cos\theta-\sin\theta}$

$=\dfrac{(\cos\theta-\sin\theta)(\cos\theta+\sin\theta)}{\cos\theta-\sin\theta}=\cos\theta+\sin\theta=\sin\theta+\cos\theta$

39. $\tan\theta+\dfrac{\cos\theta}{1+\sin\theta}=\dfrac{\sin\theta}{\cos\theta}+\dfrac{\cos\theta}{1+\sin\theta}=\dfrac{\sin\theta(1+\sin\theta)+\cos^2\theta}{\cos\theta(1+\sin\theta)}$

$=\dfrac{\sin\theta+\sin^2\theta+\cos^2\theta}{\cos\theta(1+\sin\theta)}=\dfrac{\sin\theta+1}{\cos\theta(1+\sin\theta)}=\dfrac{1}{\cos\theta}=\sec\theta$

41. $\dfrac{\tan\theta+\sec\theta-1}{\tan\theta-\sec\theta+1}=\dfrac{\tan\theta+(\sec\theta-1)}{\tan\theta-(\sec\theta-1)}\cdot\dfrac{\tan\theta+(\sec\theta-1)}{\tan\theta+(\sec\theta-1)}$

$$= \frac{\tan^2\theta + 2\tan\theta(\sec\theta - 1) + \sec^2\theta - 2\sec\theta + 1}{\tan^2\theta - (\sec^2\theta - 2\sec\theta + 1)}$$

$$= \frac{\sec^2\theta - 1 + 2\tan\theta(\sec\theta - 1) + \sec^2\theta - 2\sec\theta + 1}{\sec^2\theta - 1 - \sec^2\theta + 2\sec\theta - 1}$$

$$= \frac{2\sec^2\theta - 2\sec\theta + 2\tan\theta(\sec\theta - 1)}{2\sec\theta - 2}$$

$$= \frac{2\sec\theta(\sec\theta - 1) + 2\tan\theta(\sec\theta - 1)}{2\sec\theta - 2}$$

$$= \frac{2(\sec\theta - 1)(\sec\theta + \tan\theta)}{2(\sec\theta - 1)} = \sec\theta + \tan\theta = \tan\theta + \sec\theta$$

43. $\displaystyle \frac{\tan\theta - \cot\theta}{\tan\theta + \cot\theta} = \frac{\left(\dfrac{\sin\theta}{\cos\theta} - \dfrac{\cos\theta}{\sin\theta}\right)}{\left(\dfrac{\sin\theta}{\cos\theta} + \dfrac{\cos\theta}{\sin\theta}\right)} = \frac{\left(\dfrac{\sin^2\theta - \cos^2\theta}{\cos\theta\sin\theta}\right)}{\left(\dfrac{\sin^2\theta + \cos^2\theta}{\cos\theta\sin\theta}\right)} = \frac{\sin^2\theta - \cos^2\theta}{1} = \sin^2\theta - \cos^2\theta$

45. $\displaystyle \frac{\tan\theta - \cot\theta}{\tan\theta + \cot\theta} + 1 = \frac{\left(\dfrac{\sin\theta}{\cos\theta} - \dfrac{\cos\theta}{\sin\theta}\right)}{\left(\dfrac{\sin\theta}{\cos\theta} + \dfrac{\cos\theta}{\sin\theta}\right)} + 1 = \frac{\left(\dfrac{\sin^2\theta - \cos^2\theta}{\cos\theta\sin\theta}\right)}{\left(\dfrac{\sin^2\theta + \cos^2\theta}{\cos\theta\sin\theta}\right)} + 1 = \frac{\sin^2\theta - \cos^2\theta}{1} + 1$

$$= \sin^2\theta - \cos^2\theta + 1 = \sin^2\theta + (1 - \cos^2\theta) = \sin^2\theta + \sin^2\theta = 2\sin^2\theta$$

47. $\displaystyle \frac{\sec\theta + \tan\theta}{\cot\theta + \cos\theta} = \frac{\left(\dfrac{1}{\cos\theta} + \dfrac{\sin\theta}{\cos\theta}\right)}{\left(\dfrac{\cos\theta}{\sin\theta} + \cos\theta\right)} = \frac{\left(\dfrac{1 + \sin\theta}{\cos\theta}\right)}{\left(\dfrac{\cos\theta + \cos\theta\sin\theta}{\sin\theta}\right)} = \frac{1 + \sin\theta}{\cos\theta} \cdot \frac{\sin\theta}{\cos\theta(1 + \sin\theta)}$

$$= \frac{\sin\theta}{\cos\theta} \cdot \frac{1}{\cos\theta} = \tan\theta\sec\theta$$

49. $\displaystyle \frac{1 - \tan^2\theta}{1 + \tan^2\theta} + 1 = \frac{1 - \tan^2\theta + 1 + \tan^2\theta}{1 + \tan^2\theta} = \frac{2}{\sec^2\theta} = 2 \cdot \frac{1}{\sec^2\theta} = 2\cos^2\theta$

51. $\displaystyle \frac{\sec\theta - \csc\theta}{\sec\theta\csc\theta} = \frac{\left(\dfrac{1}{\cos\theta} - \dfrac{1}{\sin\theta}\right)}{\left(\dfrac{1}{\cos\theta} \cdot \dfrac{1}{\sin\theta}\right)} = \frac{\left(\dfrac{\sin\theta - \cos\theta}{\cos\theta\sin\theta}\right)}{\left(\dfrac{1}{\cos\theta\sin\theta}\right)} = \sin\theta - \cos\theta$

53. $\displaystyle \sec\theta - \cos\theta - \sin\theta\tan\theta = \frac{1}{\cos\theta} - \cos\theta - \sin\theta \cdot \frac{\sin\theta}{\cos\theta} = \frac{1 - \cos^2\theta - \sin^2\theta}{\cos\theta}$

$$= \frac{\sin^2\theta - \sin^2\theta}{\cos\theta} = 0$$

55. $\displaystyle \frac{1}{1 - \sin\theta} + \frac{1}{1 + \sin\theta} = \frac{1 + \sin\theta + 1 - \sin\theta}{(1 - \sin\theta)(1 + \sin\theta)} = \frac{2}{1 - \sin^2\theta} = \frac{2}{\cos^2\theta} = 2\sec^2\theta$

57. $\dfrac{\sec\theta}{1-\sin\theta} = \left(\dfrac{\sec\theta}{1-\sin\theta}\right)\left(\dfrac{1+\sin\theta}{1+\sin\theta}\right) = \dfrac{\sec\theta(1+\sin\theta)}{1-\sin^2\theta} = \dfrac{\sec\theta(1+\sin\theta)}{\cos^2\theta}$

$\qquad = \dfrac{1}{\cos\theta}\cdot\dfrac{1+\sin\theta}{\cos^2\theta} = \dfrac{1+\sin\theta}{\cos^3\theta}$

59. $\dfrac{(\sec\theta-\tan\theta)^2+1}{\csc\theta(\sec\theta-\tan\theta)} = \dfrac{\sec^2\theta-2\sec\theta\tan\theta+\tan^2\theta+1}{\csc\theta(\sec\theta-\tan\theta)} = \dfrac{2\sec^2\theta-2\sec\theta\tan\theta}{\csc\theta(\sec\theta-\tan\theta)}$

$\qquad = \dfrac{2\sec\theta(\sec\theta-\tan\theta)}{\csc\theta(\sec\theta-\tan\theta)} = \dfrac{2\sec\theta}{\csc\theta} = \dfrac{\left(2\cdot\dfrac{1}{\cos\theta}\right)}{\left(\dfrac{1}{\sin\theta}\right)} = 2\cdot\dfrac{1}{\cos\theta}\cdot\dfrac{\sin\theta}{1} = 2\tan\theta$

61. $\dfrac{\sin\theta+\cos\theta}{\cos\theta} - \dfrac{\sin\theta-\cos\theta}{\sin\theta} = \dfrac{\sin\theta}{\cos\theta}+\dfrac{\cos\theta}{\cos\theta}-\dfrac{\sin\theta}{\sin\theta}+\dfrac{\cos\theta}{\sin\theta} = \dfrac{\sin\theta}{\cos\theta}+1-1+\dfrac{\cos\theta}{\sin\theta}$

$\qquad = \dfrac{\sin^2\theta+\cos^2\theta}{\cos\theta\sin\theta} = \dfrac{1}{\cos\theta\sin\theta} = \sec\theta\csc\theta$

63. $\dfrac{\sin^3\theta+\cos^3\theta}{\sin\theta+\cos\theta} = \dfrac{(\sin\theta+\cos\theta)(\sin^2\theta-\sin\theta\cos\theta+\cos^2\theta)}{\sin\theta+\cos\theta} = 1-\sin\theta\cos\theta$

65. $\dfrac{\cos^2\theta-\sin^2\theta}{1-\tan^2\theta} = \dfrac{\cos^2\theta-\sin^2\theta}{\left(1-\dfrac{\sin^2\theta}{\cos^2\theta}\right)} = \dfrac{\cos^2\theta-\sin^2\theta}{\left(\dfrac{\cos^2\theta-\sin^2\theta}{\cos^2\theta}\right)} = \cos^2\theta$

67. $\dfrac{(2\cos^2\theta-1)^2}{\cos^4\theta-\sin^4\theta} = \dfrac{\left[2\cos^2\theta-(\sin^2\theta+\cos^2\theta)\right]^2}{(\cos^2\theta-\sin^2\theta)(\cos^2\theta+\sin^2\theta)}$

$\qquad = \dfrac{(\cos^2\theta-\sin^2\theta)^2}{(\cos^2\theta-\sin^2\theta)(\cos^2\theta+\sin^2\theta)} = \dfrac{\cos^2\theta-\sin^2\theta}{\cos^2\theta+\sin^2\theta}$

$\qquad = \cos^2\theta-\sin^2\theta = 1-\sin^2\theta-\sin^2\theta = 1-2\sin^2\theta$

69. $\dfrac{1+\sin\theta+\cos\theta}{1+\sin\theta-\cos\theta} = \dfrac{(1+\sin\theta)+\cos\theta}{(1+\sin\theta)-\cos\theta}\cdot\dfrac{(1+\sin\theta)+\cos\theta}{(1+\sin\theta)+\cos\theta}$

$\qquad = \dfrac{1+2\sin\theta+\sin^2\theta+2\cos\theta(1+\sin\theta)+\cos^2\theta}{1+2\sin\theta+\sin^2\theta-\cos^2\theta}$

$\qquad = \dfrac{1+2\sin\theta+\sin^2\theta+2\cos\theta(1+\sin\theta)+(1-\sin^2\theta)}{1+2\sin\theta+\sin^2\theta-(1-\sin^2\theta)}$

$\qquad = \dfrac{2+2\sin\theta+2\cos\theta(1+\sin\theta)}{2\sin\theta+2\sin^2\theta} = \dfrac{2(1+\sin\theta)+2\cos\theta(1+\sin\theta)}{2\sin\theta(1+\sin\theta)}$

$\qquad = \dfrac{2(1+\sin\theta)(1+\cos\theta)}{2\sin\theta(1+\sin\theta)} = \dfrac{1+\cos\theta}{\sin\theta}$

71. $(a\sin\theta+b\cos\theta)^2+(a\cos\theta-b\sin\theta)^2$

$\qquad = a^2\sin^2\theta+2ab\sin\theta\cos\theta+b^2\cos^2\theta+a^2\cos^2\theta-2ab\sin\theta\cos\theta+b^2\sin^2\theta$

$\qquad = a^2(\sin^2\theta+\cos^2\theta)+b^2(\sin^2\theta+\cos^2\theta) = a^2+b^2$

73. $\dfrac{\tan\alpha + \tan\beta}{\cot\alpha + \cot\beta} = \dfrac{\tan\alpha + \tan\beta}{\left(\dfrac{1}{\tan\alpha} + \dfrac{1}{\tan\beta}\right)} = \dfrac{\tan\alpha + \tan\beta}{\left(\dfrac{\tan\beta + \tan\alpha}{\tan\alpha\tan\beta}\right)}$

$$= (\tan\alpha + \tan\beta)\cdot\left(\dfrac{\tan\alpha\tan\beta}{\tan\alpha + \tan\beta}\right) = \tan\alpha\tan\beta$$

75. $(\sin\alpha + \cos\beta)^2 + (\cos\beta + \sin\alpha)(\cos\beta - \sin\alpha)$
$$= \sin^2\alpha + 2\sin\alpha\cos\beta + \cos^2\beta + \cos^2\beta - \sin^2\alpha$$
$$= 2\sin\alpha\cos\beta + 2\cos^2\beta = 2\cos\beta(\sin\alpha + \cos\beta)$$

77. $\ln\left|\sec\theta\right| = \ln\left|\dfrac{1}{\cos\theta}\right| = \ln\left|\cos\theta\right|^{-1} = -\ln\left|\cos\theta\right|$

79. $\ln\left|1 + \cos\theta\right| + \ln\left|1 - \cos\theta\right| = \ln\left(\left|1 + \cos\theta\right|\cdot\left|1 - \cos\theta\right|\right) = \ln\left|1 - \cos^2\theta\right|$
$$= \ln\left|\sin^2\theta\right| = 2\ln\left|\sin\theta\right|$$

81. Show that $\sec\left(\tan^{-1}v\right) = \sqrt{1 + v^2}$.

Let $\alpha = \tan^{-1}v$. Then $\tan\alpha = v$, $-\dfrac{\pi}{2} < \alpha < \dfrac{\pi}{2}$.

$$\sec\left(\tan^{-1}v\right) = \sec\alpha = \sqrt{1 + \tan^2\alpha} = \sqrt{1 + v^2}$$

83. Show that $\tan\left(\cos^{-1}v\right) = \dfrac{\sqrt{1 - v^2}}{v}$.

Let $\alpha = \cos^{-1}v$. Then $\cos\alpha = v$, $0 \le \alpha \le \pi$.

$$\tan\left(\cos^{-1}v\right) = \tan\alpha = \dfrac{\sin\alpha}{\cos\alpha} = \dfrac{\sqrt{1 - \cos^2\alpha}}{\cos\alpha} = \dfrac{\sqrt{1 - v^2}}{v}$$

85. Show that $\cos\left(\sin^{-1}v\right) = \sqrt{1 - v^2}$.

Let $\alpha = \sin^{-1}v$. Then $\sin\alpha = v$, $-\dfrac{\pi}{2} \le \alpha \le \dfrac{\pi}{2}$.

$$\cos\left(\sin^{-1}v\right) = \cos\alpha = \sqrt{1 - \sin^2\alpha} = \sqrt{1 - v^2}$$

Analytic Trigonometry

6.4　Sum and Difference Formulas

1.　$\sin\left(\dfrac{5\pi}{12}\right) = \sin\left(\dfrac{3\pi}{12}+\dfrac{2\pi}{12}\right) = \sin\left(\dfrac{\pi}{4}\right)\cos\left(\dfrac{\pi}{6}\right) + \cos\left(\dfrac{\pi}{4}\right)\sin\left(\dfrac{\pi}{6}\right) = \dfrac{\sqrt{2}}{2}\cdot\dfrac{\sqrt{3}}{2} + \dfrac{\sqrt{2}}{2}\cdot\dfrac{1}{2}$

$\quad\quad = \dfrac{1}{4}\left(\sqrt{6}+\sqrt{2}\right)$

3.　$\cos\left(\dfrac{7\pi}{12}\right) = \cos\left(\dfrac{4\pi}{12}+\dfrac{3\pi}{12}\right) = \cos\left(\dfrac{\pi}{3}\right)\cos\left(\dfrac{\pi}{4}\right) - \sin\left(\dfrac{\pi}{3}\right)\sin\left(\dfrac{\pi}{4}\right) = \dfrac{1}{2}\cdot\dfrac{\sqrt{2}}{2} - \dfrac{\sqrt{3}}{2}\cdot\dfrac{\sqrt{2}}{2}$

$\quad\quad = \dfrac{1}{4}\left(\sqrt{2}-\sqrt{6}\right)$

5.　$\cos(165°) = \cos(120°+45°) = \cos(120°)\cos(45°) - \sin(120°)\sin(45°)$

$\quad\quad = -\dfrac{1}{2}\cdot\dfrac{\sqrt{2}}{2} - \dfrac{\sqrt{3}}{2}\cdot\dfrac{\sqrt{2}}{2} = -\dfrac{1}{4}\left(\sqrt{2}+\sqrt{6}\right)$

7.　$\tan(15°) = \tan(45°-30°) = \dfrac{\tan(45°)-\tan(30°)}{1+\tan(45°)\tan(30°)} = \dfrac{\left(1-\dfrac{\sqrt{3}}{3}\right)}{\left(1+1\cdot\dfrac{\sqrt{3}}{3}\right)} = \dfrac{\left(\dfrac{3-\sqrt{3}}{3}\right)}{\left(\dfrac{3+\sqrt{3}}{3}\right)}$

$\quad\quad = \left(\dfrac{3-\sqrt{3}}{3+\sqrt{3}}\right)\cdot\left(\dfrac{3-\sqrt{3}}{3-\sqrt{3}}\right) = \dfrac{9-6\sqrt{3}+3}{9-3} = \dfrac{12-6\sqrt{3}}{6} = \dfrac{6\left(2-\sqrt{3}\right)}{6} = 2-\sqrt{3}$

9.　$\sin\left(\dfrac{17\pi}{12}\right) = \sin\left(\dfrac{15\pi}{12}+\dfrac{2\pi}{12}\right) = \sin\left(\dfrac{5\pi}{4}\right)\cos\left(\dfrac{\pi}{6}\right) + \cos\left(\dfrac{5\pi}{4}\right)\sin\left(\dfrac{\pi}{6}\right) = -\dfrac{\sqrt{2}}{2}\cdot\dfrac{\sqrt{3}}{2} + -\dfrac{\sqrt{2}}{2}\cdot\dfrac{1}{2}$

$\quad\quad = -\dfrac{1}{4}\left(\sqrt{6}+\sqrt{2}\right)$

11.　$\sec\left(-\dfrac{\pi}{12}\right) = \dfrac{1}{\cos\left(-\dfrac{\pi}{12}\right)} = \dfrac{1}{\cos\left(\dfrac{3\pi}{12}-\dfrac{4\pi}{12}\right)} = \dfrac{1}{\cos\left(\dfrac{\pi}{4}\right)\cos\left(\dfrac{\pi}{3}\right) + \sin\left(\dfrac{\pi}{4}\right)\sin\left(\dfrac{\pi}{3}\right)}$

$$= \frac{1}{\left(\frac{\sqrt{2}}{2}\cdot\frac{1}{2}+\frac{\sqrt{2}}{2}\cdot\frac{\sqrt{3}}{2}\right)} = \frac{1}{\left(\frac{\sqrt{2}+\sqrt{6}}{4}\right)} = \left(\frac{4}{\sqrt{2}+\sqrt{6}}\right)\cdot\left(\frac{\sqrt{2}-\sqrt{6}}{\sqrt{2}-\sqrt{6}}\right)$$

$$= \frac{4\left(\sqrt{2}-\sqrt{6}\right)}{2-6} = \frac{4\left(\sqrt{2}-\sqrt{6}\right)}{-4} = -\left(\sqrt{2}-\sqrt{6}\right) = \sqrt{6}-\sqrt{2}$$

13. $\sin(20°)\cos(10°) + \cos(20°)\sin(10°) = \sin(20°+10°) = \sin(30°) = \dfrac{1}{2}$

15. $\cos(70°)\cos(20°) - \sin(70°)\sin(20°) = \cos(70°+20°) = \cos(90°) = 0$

17. $\dfrac{\tan(20°)+\tan(25°)}{1-\tan(20°)\tan(25°)} = \tan(20°+25°) = \tan(45°) = 1$

19. $\sin\left(\dfrac{\pi}{12}\right)\cos\left(\dfrac{7\pi}{12}\right) - \cos\left(\dfrac{\pi}{12}\right)\sin\left(\dfrac{7\pi}{12}\right) = \sin\left(\dfrac{\pi}{12}-\dfrac{7\pi}{12}\right) = \sin\left(-\dfrac{\pi}{2}\right) = -1$

21. $\cos\left(\dfrac{\pi}{12}\right)\cos\left(\dfrac{5\pi}{12}\right) + \sin\left(\dfrac{5\pi}{12}\right)\sin\left(\dfrac{\pi}{12}\right) = \cos\left(\dfrac{\pi}{12}-\dfrac{5\pi}{12}\right) = \cos\left(-\dfrac{\pi}{3}\right) = \cos\left(\dfrac{\pi}{3}\right) = \dfrac{1}{2}$

23. $\sin\alpha = \dfrac{3}{5},\ 0 < \alpha < \dfrac{\pi}{2};$ $\cos\beta = \dfrac{2\sqrt{5}}{5},\ -\dfrac{\pi}{2} < \beta < 0$

$x^2 + 3^2 = 5^2,\ x > 0$ $\left(2\sqrt{5}\right)^2 + y^2 = 5^2,\ y < 0$

$\qquad x^2 = 25 - 9 = 16,\ x > 0$ $\qquad y^2 = 25 - 20 = 5.\ y < 0$

$\qquad\qquad x = 4$ $\qquad\qquad y = -\sqrt{5}$

$\qquad \cos\alpha = \dfrac{4}{5},\ \tan\alpha = \dfrac{3}{4}$ $\qquad \sin\beta = -\dfrac{\sqrt{5}}{5},\ \tan\beta = \dfrac{-\sqrt{5}}{2\sqrt{5}} = -\dfrac{1}{2}$

(a) $\sin(\alpha+\beta) = \sin\alpha\cos\beta + \cos\alpha\sin\beta = \dfrac{3}{5}\cdot\dfrac{2\sqrt{5}}{5} + \dfrac{4}{5}\cdot-\dfrac{\sqrt{5}}{5} = \dfrac{6\sqrt{5}-4\sqrt{5}}{25} = \dfrac{2\sqrt{5}}{25}$

(b) $\cos(\alpha+\beta) = \cos\alpha\cos\beta - \sin\alpha\sin\beta = \dfrac{4}{5}\cdot\dfrac{2\sqrt{5}}{5} - \dfrac{3}{5}\cdot-\dfrac{\sqrt{5}}{5} = \dfrac{8\sqrt{5}+3\sqrt{5}}{25} = \dfrac{11\sqrt{5}}{25}$

(c) $\sin(\alpha - \beta) = \sin\alpha\cos\beta - \cos\alpha\sin\beta = \dfrac{3}{5}\cdot\dfrac{2\sqrt{5}}{5} - \dfrac{4}{5}\cdot-\dfrac{\sqrt{5}}{5} = \dfrac{6\sqrt{5}+4\sqrt{5}}{25}$

$= \dfrac{10\sqrt{5}}{25} = \dfrac{2\sqrt{5}}{5}$

(d) $\tan(\alpha - \beta) = \dfrac{\tan\alpha - \tan\beta}{1+\tan\alpha\tan\beta} = \dfrac{\left(\dfrac{3}{4}-\left(-\dfrac{1}{2}\right)\right)}{\left(1+\left(\dfrac{3}{4}\right)\cdot\left(-\dfrac{1}{2}\right)\right)} = \dfrac{\left(\dfrac{5}{4}\right)}{\left(\dfrac{5}{8}\right)} = 2$

25. $\tan\alpha = -\dfrac{4}{3},\ \dfrac{\pi}{2} < \alpha < \pi;$ $\cos\beta = \dfrac{1}{2},\ 0 < \beta < \dfrac{\pi}{2}$

$r^2 = (-3)^2 + 4^2 = 25$

$r = 5$

$\sin\alpha = \dfrac{4}{5},\ \cos\alpha = \dfrac{-3}{5} = -\dfrac{3}{5}$

$1^2 + y^2 = 2^2,\ y > 0$

$y^2 = 4 - 1 = 3.\ y > 0$

$y = \sqrt{3}$

$\sin\beta = \dfrac{\sqrt{3}}{2},\quad \tan\beta = \dfrac{\sqrt{3}}{1} = \sqrt{3}$

(a) $\sin(\alpha + \beta) = \sin\alpha\cos\beta + \cos\alpha\sin\beta = \left(\dfrac{4}{5}\right)\cdot\left(\dfrac{1}{2}\right) + \left(-\dfrac{3}{5}\right)\cdot\left(\dfrac{\sqrt{3}}{2}\right) = \dfrac{4-3\sqrt{3}}{10}$

(b) $\cos(\alpha + \beta) = \cos\alpha\cos\beta - \sin\alpha\sin\beta = \left(-\dfrac{3}{5}\right)\cdot\left(\dfrac{1}{2}\right) - \left(\dfrac{4}{5}\right)\cdot\left(\dfrac{\sqrt{3}}{2}\right) = \dfrac{-3-4\sqrt{3}}{10}$

(c) $\sin(\alpha - \beta) = \sin\alpha\cos\beta - \cos\alpha\sin\beta = \left(\dfrac{4}{5}\right)\cdot\left(\dfrac{1}{2}\right) - \left(-\dfrac{3}{5}\right)\cdot\left(\dfrac{\sqrt{3}}{2}\right) = \dfrac{4+3\sqrt{3}}{10}$

(d) $\tan(\alpha - \beta) = \dfrac{\tan\alpha - \tan\beta}{1+\tan\alpha\tan\beta} = \dfrac{\left(-\dfrac{4}{3}-\sqrt{3}\right)}{\left(1+\left(-\dfrac{4}{3}\right)\cdot\sqrt{3}\right)} = \dfrac{\left(\dfrac{-4-3\sqrt{3}}{3}\right)}{\left(\dfrac{3-4\sqrt{3}}{3}\right)} = \left(\dfrac{-4-3\sqrt{3}}{3-4\sqrt{3}}\right)\cdot\left(\dfrac{3+4\sqrt{3}}{3+4\sqrt{3}}\right)$

$= \dfrac{-48-25\sqrt{3}}{-39} = \dfrac{48+25\sqrt{3}}{39}$

27. $\sin\alpha = \dfrac{5}{13},\ -\dfrac{3\pi}{2} < \alpha < -\pi;\qquad \tan\beta = -\sqrt{3},\ \dfrac{\pi}{2} < \beta < \pi$

$x^2 + 5^2 = 13^2,\ x < 0$ $r^2 = (-1)^2 + \sqrt{3}^2 = 4$

$\quad x^2 = 169 - 25 = 144,\ x < 0$ $r = 2$

$\quad\quad x = -12$ $\sin\beta = \dfrac{\sqrt{3}}{2},\quad \cos\beta = \dfrac{-1}{2} = -\dfrac{1}{2}$

$\cos\alpha = \dfrac{-12}{13} = -\dfrac{12}{13},\ \tan\alpha = -\dfrac{5}{12}$

(a) $\sin(\alpha+\beta) = \sin\alpha\cos\beta + \cos\alpha\sin\beta = \left(\dfrac{5}{13}\right)\cdot\left(-\dfrac{1}{2}\right) + \left(-\dfrac{12}{13}\right)\cdot\left(\dfrac{\sqrt{3}}{2}\right) = \dfrac{-5 - 12\sqrt{3}}{26}$

(b) $\cos(\alpha+\beta) = \cos\alpha\cos\beta - \sin\alpha\sin\beta = \left(-\dfrac{12}{13}\right)\cdot\left(-\dfrac{1}{2}\right) - \left(\dfrac{5}{13}\right)\cdot\left(\dfrac{\sqrt{3}}{2}\right) = \dfrac{12 - 5\sqrt{3}}{26}$

(c) $\sin(\alpha-\beta) = \sin\alpha\cos\beta - \cos\alpha\sin\beta = \left(\dfrac{5}{13}\right)\cdot\left(-\dfrac{1}{2}\right) - \left(-\dfrac{12}{13}\right)\cdot\left(\dfrac{\sqrt{3}}{2}\right) = \dfrac{-5 + 12\sqrt{3}}{26}$

(d) $\tan(\alpha-\beta) = \dfrac{\tan\alpha - \tan\beta}{1 + \tan\alpha\tan\beta} = \dfrac{\left(-\dfrac{5}{12} - (-\sqrt{3})\right)}{\left(1 + \left(-\dfrac{5}{12}\right)\cdot(-\sqrt{3})\right)} = \dfrac{\left(\dfrac{-5 + 12\sqrt{3}}{12}\right)}{\left(\dfrac{12 + 5\sqrt{3}}{12}\right)}$

$\qquad\qquad = \left(\dfrac{-5 + 12\sqrt{3}}{12 + 5\sqrt{3}}\right)\cdot\left(\dfrac{12 - 5\sqrt{3}}{12 - 5\sqrt{3}}\right) = \dfrac{-240 + 169\sqrt{3}}{69}$

29. $\sin\theta = \dfrac{1}{3},\quad \theta$ in quadrant II

(a) $\cos\theta = -\sqrt{1 - \sin^2\theta} = -\sqrt{1 - \left(\dfrac{1}{3}\right)^2} = -\sqrt{1 - \dfrac{1}{9}} = -\sqrt{\dfrac{8}{9}} = -\dfrac{2\sqrt{2}}{3}$

(b) $\sin\left(\theta + \dfrac{\pi}{6}\right) = \sin\theta\cos\left(\dfrac{\pi}{6}\right) + \cos\theta\sin\left(\dfrac{\pi}{6}\right) = \left(\dfrac{1}{3}\right)\cdot\left(\dfrac{\sqrt{3}}{2}\right) + \left(-\dfrac{2\sqrt{2}}{3}\right)\cdot\left(\dfrac{1}{2}\right) = \dfrac{\sqrt{3} - 2\sqrt{2}}{6}$

(c) $\cos\left(\theta - \dfrac{\pi}{3}\right) = \cos\theta\cos\left(\dfrac{\pi}{3}\right) + \sin\theta\sin\left(\dfrac{\pi}{3}\right) = \left(-\dfrac{2\sqrt{2}}{3}\right)\cdot\left(\dfrac{1}{2}\right) + \left(\dfrac{1}{3}\right)\cdot\left(\dfrac{\sqrt{3}}{2}\right) = \dfrac{-2\sqrt{2} + \sqrt{3}}{6}$

(d) $\tan\left(\theta+\dfrac{\pi}{4}\right)=\dfrac{\tan\theta+\tan\left(\dfrac{\pi}{4}\right)}{1-\tan\theta\tan\left(\dfrac{\pi}{4}\right)}=\dfrac{\left(-\dfrac{1}{2\sqrt{2}}+1\right)}{\left(1-\left(-\dfrac{1}{2\sqrt{2}}\right)\cdot1\right)}=\dfrac{\left(\dfrac{-1+2\sqrt{2}}{2\sqrt{2}}\right)}{\left(\dfrac{2\sqrt{2}+1}{2\sqrt{2}}\right)}$

$$=\left(\dfrac{2\sqrt{2}-1}{2\sqrt{2}+1}\right)\cdot\left(\dfrac{2\sqrt{2}-1}{2\sqrt{2}-1}\right)=\dfrac{9-4\sqrt{2}}{7}$$

31. $\sin\left(\dfrac{\pi}{2}+\theta\right)=\sin\left(\dfrac{\pi}{2}\right)\cos\theta+\cos\left(\dfrac{\pi}{2}\right)\sin\theta=1\cdot\cos\theta+0\cdot\sin\theta=\cos\theta$

33. $\sin(\pi-\theta)=\sin(\pi)\cos\theta-\cos(\pi)\sin\theta=0\cdot\cos\theta-(-1)\sin\theta=\sin\theta$

35. $\sin(\pi+\theta)=\sin(\pi)\cos\theta+\cos(\pi)\sin\theta=0\cdot\cos\theta+(-1)\sin\theta=-\sin\theta$

37. $\tan(\pi-\theta)=\dfrac{\tan(\pi)-\tan\theta}{1+\tan(\pi)\tan\theta}=\dfrac{0-\tan\theta}{1+0\cdot\tan\theta}=\dfrac{-\tan\theta}{1}=-\tan\theta$

39. $\sin\left(\dfrac{3\pi}{2}+\theta\right)=\sin\left(\dfrac{3\pi}{2}\right)\cos\theta+\cos\left(\dfrac{3\pi}{2}\right)\sin\theta=-1\cdot\cos\theta+0\cdot\sin\theta=-\cos\theta$

41. $\sin(\alpha+\beta)+\sin(\alpha-\beta)=\sin\alpha\cos\beta+\cos\alpha\sin\beta+\sin\alpha\cos\beta-\cos\alpha\sin\beta$
$$=2\sin\alpha\cos\beta$$

43. $\dfrac{\sin(\alpha+\beta)}{\sin\alpha\cos\beta}=\dfrac{\sin\alpha\cos\beta+\cos\alpha\sin\beta}{\sin\alpha\cos\beta}=\dfrac{\sin\alpha\cos\beta}{\sin\alpha\cos\beta}+\dfrac{\cos\alpha\sin\beta}{\sin\alpha\cos\beta}=1+\cot\alpha\tan\beta$

45. $\dfrac{\cos(\alpha+\beta)}{\cos\alpha\cos\beta}=\dfrac{\cos\alpha\cos\beta-\sin\alpha\sin\beta}{\cos\alpha\cos\beta}=\dfrac{\cos\alpha\cos\beta}{\cos\alpha\cos\beta}-\dfrac{\sin\alpha\sin\beta}{\cos\alpha\cos\beta}=1-\tan\alpha\tan\beta$

47. $\dfrac{\sin(\alpha+\beta)}{\sin(\alpha-\beta)}=\dfrac{\sin\alpha\cos\beta+\cos\alpha\sin\beta}{\sin\alpha\cos\beta-\cos\alpha\sin\beta}=\dfrac{\left(\dfrac{\sin\alpha\cos\beta}{\cos\alpha\cos\beta}+\dfrac{\cos\alpha\sin\beta}{\cos\alpha\cos\beta}\right)}{\left(\dfrac{\sin\alpha\cos\beta}{\cos\alpha\cos\beta}-\dfrac{\cos\alpha\sin\beta}{\cos\alpha\cos\beta}\right)}=\dfrac{\tan\alpha+\tan\beta}{\tan\alpha-\tan\beta}$

49. $\cot(\alpha+\beta)=\dfrac{\cos(\alpha+\beta)}{\sin(\alpha+\beta)}=\dfrac{\cos\alpha\cos\beta-\sin\alpha\sin\beta}{\sin\alpha\cos\beta+\cos\alpha\sin\beta}$

$$=\dfrac{\left(\dfrac{\cos\alpha\cos\beta}{\sin\alpha\sin\beta}-\dfrac{\sin\alpha\sin\beta}{\sin\alpha\sin\beta}\right)}{\left(\dfrac{\sin\alpha\cos\beta}{\sin\alpha\sin\beta}+\dfrac{\cos\alpha\sin\beta}{\sin\alpha\sin\beta}\right)}=\dfrac{\cot\alpha\cot\beta-1}{\cot\beta+\cot\alpha}$$

51.　$\sec(\alpha + \beta) = \dfrac{1}{\cos(\alpha + \beta)} = \dfrac{1}{\cos\alpha\cos\beta - \sin\alpha\sin\beta}$

$$= \dfrac{\left(\dfrac{1}{\sin\alpha\sin\beta}\right)}{\left(\dfrac{\cos\alpha\cos\beta}{\sin\alpha\sin\beta} - \dfrac{\sin\alpha\sin\beta}{\sin\alpha\sin\beta}\right)} = \dfrac{\csc\alpha\csc\beta}{\cot\alpha\cot\beta - 1}$$

53.　$\sin(\alpha - \beta)\sin(\alpha + \beta) = (\sin\alpha\cos\beta - \cos\alpha\sin\beta)(\sin\alpha\cos\beta + \cos\alpha\sin\beta)$

$\qquad\qquad = \sin^2\alpha\cos^2\beta - \cos^2\alpha\sin^2\beta = \sin^2\alpha(1 - \sin^2\beta) - (1 - \sin^2\alpha)\sin^2\beta$

$\qquad\qquad \sin^2\alpha - \sin^2\alpha\sin^2\beta - \sin^2\beta + \sin^2\alpha\sin^2\beta = \sin^2\alpha - \sin^2\beta$

55.　$\sin(\theta + k\pi) = \sin\theta\cos(k\pi) + \cos\theta\sin(k\pi) = \sin\theta(-1)^k + \cos\theta\cdot 0$

$\qquad\qquad = (-1)^k\sin\theta,\ k$ any integer

57.　$\sin\left(\sin^{-1}\left(\dfrac{1}{2}\right) + \cos^{-1}(0)\right) = \sin\left(\dfrac{\pi}{6} + \dfrac{\pi}{2}\right) = \sin\left(\dfrac{\pi}{6}\right)\cos\left(\dfrac{\pi}{2}\right) + \cos\left(\dfrac{\pi}{6}\right)\sin\left(\dfrac{\pi}{2}\right) = \dfrac{1}{2}\cdot 0 + \dfrac{\sqrt{3}}{2}\cdot 1 = \dfrac{\sqrt{3}}{2}$

59.　$\sin\left[\sin^{-1}\left(\dfrac{3}{5}\right) - \cos^{-1}\left(-\dfrac{4}{5}\right)\right]$

Let $\alpha = \sin^{-1}\left(\dfrac{3}{5}\right)$ and $\beta = \cos^{-1}\left(-\dfrac{4}{5}\right)$. α is in quadrant I; β is in quadrant II.

Then $\sin\alpha = \dfrac{3}{5},\ -\dfrac{\pi}{2} \le \alpha \le \dfrac{\pi}{2}$, and $\cos\beta = -\dfrac{4}{5}, 0 \le \beta \le \pi$.

$$\cos\alpha = \sqrt{1 - \sin^2\alpha} = \sqrt{1 - \left(\dfrac{3}{5}\right)^2} = \sqrt{1 - \dfrac{9}{25}} = \sqrt{\dfrac{16}{25}} = \dfrac{4}{5}$$

$$\sin\beta = \sqrt{1 - \cos^2\beta} = \sqrt{1 - \left(-\dfrac{4}{5}\right)^2} = \sqrt{1 - \dfrac{16}{25}} = \sqrt{\dfrac{9}{25}} = \dfrac{3}{5}$$

$$\sin\left[\sin^{-1}\left(\dfrac{3}{5}\right) - \cos^{-1}\left(-\dfrac{4}{5}\right)\right] = \sin(\alpha - \beta) = \sin\alpha\cos\beta - \cos\alpha\sin\beta$$

$$= \left(\dfrac{3}{5}\right)\cdot\left(-\dfrac{4}{5}\right) - \left(\dfrac{4}{5}\right)\cdot\left(\dfrac{3}{5}\right) = -\dfrac{12}{25} - \dfrac{12}{25} = -\dfrac{24}{25}$$

61.　$\cos\left(\tan^{-1}\left(\dfrac{4}{3}\right) + \cos^{-1}\left(\dfrac{5}{13}\right)\right)$

Let $\alpha = \tan^{-1}\left(\dfrac{4}{3}\right)$ and $\beta = \cos^{-1}\left(\dfrac{5}{13}\right)$. α is in quadrant I; β is in quadrant I.

Then $\tan\alpha = \dfrac{4}{3},\ -\dfrac{\pi}{2} < \alpha < \dfrac{\pi}{2}$, and $\cos\beta = \dfrac{5}{13}, 0 \le \beta \le \pi$.

$$\sec\alpha = \sqrt{1+\tan^2\alpha} = \sqrt{1+\left(\frac{4}{3}\right)^2} = \sqrt{1+\frac{16}{9}} = \sqrt{\frac{25}{9}} = \frac{5}{3}; \quad \cos\alpha = \frac{3}{5}$$

$$\sin\alpha = \sqrt{1-\cos^2\alpha} = \sqrt{1-\left(\frac{3}{5}\right)^2} = \sqrt{1-\frac{9}{25}} = \sqrt{\frac{16}{25}} = \frac{4}{5}$$

$$\sin\beta = \sqrt{1-\cos^2\beta} = \sqrt{1-\left(\frac{5}{13}\right)^2} = \sqrt{1-\frac{25}{169}} = \sqrt{\frac{144}{169}} = \frac{12}{13}$$

$$\cos\left(\tan^{-1}\left(\frac{4}{3}\right) + \cos^{-1}\left(\frac{5}{13}\right)\right) = \cos(\alpha+\beta) = \cos\alpha\cos\beta - \sin\alpha\sin\beta$$

$$= \left(\frac{3}{5}\right)\cdot\left(\frac{5}{13}\right) - \left(\frac{4}{5}\right)\cdot\left(\frac{12}{13}\right) = \frac{15}{65} - \frac{48}{65} = -\frac{33}{65}$$

63. $\cos\left(\sin^{-1}\left(\frac{5}{13}\right) - \tan^{-1}\left(\frac{3}{4}\right)\right)$

Let $\alpha = \sin^{-1}\left(\frac{5}{13}\right)$ and $\beta = \tan^{-1}\left(\frac{3}{4}\right)$. α is in quadrant I; β is in quadrant I.

Then $\sin\alpha = \frac{5}{13}$, $-\frac{\pi}{2} \leq \alpha \leq \frac{\pi}{2}$, and $\tan\beta = \frac{3}{4}$, $-\frac{\pi}{2} < \beta < \frac{\pi}{2}$.

$$\cos\alpha = \sqrt{1-\sin^2\alpha} = \sqrt{1-\left(\frac{5}{13}\right)^2} = \sqrt{1-\frac{25}{169}} = \sqrt{\frac{144}{169}} = \frac{12}{13}$$

$$\sec\beta = \sqrt{1+\tan^2\beta} = \sqrt{1+\left(\frac{3}{4}\right)^2} = \sqrt{1+\frac{9}{16}} = \sqrt{\frac{25}{16}} = \frac{5}{4}; \quad \cos\beta = \frac{4}{5}$$

$$\sin\beta = \sqrt{1-\cos^2\beta} = \sqrt{1-\left(\frac{4}{5}\right)^2} = \sqrt{1-\frac{16}{25}} = \sqrt{\frac{9}{25}} = \frac{3}{5}$$

$$\cos\left(\sin^{-1}\left(\frac{5}{13}\right) - \tan^{-1}\left(\frac{3}{4}\right)\right) = \cos(\alpha-\beta)$$

$$= \cos\alpha\cos\beta + \sin\alpha\sin\beta = \frac{12}{13}\cdot\frac{4}{5} + \frac{5}{13}\cdot\frac{3}{5} = \frac{48}{65} + \frac{15}{65} = \frac{63}{65}$$

65. $\tan\left(\sin^{-1}\left(\frac{3}{5}\right) + \frac{\pi}{6}\right)$

Let $\alpha = \sin^{-1}\left(\frac{3}{5}\right)$. α is in quadrant I.

Then $\sin\alpha = \frac{3}{5}$, $-\frac{\pi}{2} \leq \alpha \leq \frac{\pi}{2}$.

$$\cos\alpha = \sqrt{1-\sin^2\alpha} = \sqrt{1-\left(\frac{3}{5}\right)^2} = \sqrt{1-\frac{9}{25}} = \sqrt{\frac{16}{25}} = \frac{4}{5}$$

$$\tan\alpha = \frac{\sin\alpha}{\cos\alpha} = \frac{\left(\frac{3}{5}\right)}{\left(\frac{4}{5}\right)} = \frac{3}{5}\cdot\frac{5}{4} = \frac{3}{4}$$

$$\tan\left(\sin^{-1}\left(\frac{3}{5}\right)+\frac{\pi}{6}\right) = \frac{\tan(\alpha)+\tan\left(\frac{\pi}{6}\right)}{1-\tan(\alpha)\tan\left(\frac{\pi}{6}\right)} = \frac{\left(\frac{3}{4}+\frac{\sqrt{3}}{3}\right)}{\left(1-\frac{3}{4}\cdot\frac{\sqrt{3}}{3}\right)} = \frac{\left(\frac{9+4\sqrt{3}}{12}\right)}{\left(1-\frac{\sqrt{3}}{4}\right)}$$

$$= \frac{\left(\frac{9+4\sqrt{3}}{12}\right)}{\left(\frac{4-\sqrt{3}}{4}\right)} = \left(\frac{9+4\sqrt{3}}{12}\right)\left(\frac{4}{4-\sqrt{3}}\right)\left(\frac{4+\sqrt{3}}{4+\sqrt{3}}\right)$$

$$= \left(\frac{9+4\sqrt{3}}{12}\right)\left(\frac{4\left(4+\sqrt{3}\right)}{16-3}\right) = \frac{4\left(9+4\sqrt{3}\right)\left(4+\sqrt{3}\right)}{12\cdot 13}$$

$$= \frac{\left(9+4\sqrt{3}\right)\left(4+\sqrt{3}\right)}{3\cdot 13} = \frac{48+25\sqrt{3}}{39}$$

67. $\tan\left(\sin^{-1}\left(\frac{4}{5}\right)+\cos^{-1}(1)\right)$

Let $\alpha = \sin^{-1}\left(\frac{4}{5}\right)$ and $\beta = \cos^{-1}(1)$; α is in quadrant I.

Then $\sin\alpha = \frac{4}{5}$, $-\frac{\pi}{2}\le\alpha\le\frac{\pi}{2}$ and $\cos\beta = 1$, $0\le\beta\le\pi$.

$\cos\beta = 1$, $0\le\beta\le\pi \Rightarrow \beta = 0$ $\therefore \cos^{-1}1 = 0$

$\cos\alpha = \sqrt{1-\sin^2\alpha} = \sqrt{1-\left(\frac{4}{5}\right)^2} = \sqrt{1-\frac{16}{25}} = \sqrt{\frac{9}{25}} = \frac{3}{5}$

$\tan\alpha = \frac{\sin\alpha}{\cos\alpha} = \frac{\left(\frac{4}{5}\right)}{\left(\frac{3}{5}\right)} = \frac{4}{5}\cdot\frac{5}{3} = \frac{4}{3}$

$$\tan\left(\sin^{-1}\left(\frac{4}{5}\right)+\cos^{-1}(1)\right) = \frac{\tan\left(\sin^{-1}\left(\frac{4}{5}\right)\right)+\tan\left(\cos^{-1}(1)\right)}{1-\tan\left(\sin^{-1}\left(\frac{4}{5}\right)\right)\tan\left(\cos^{-1}(1)\right)} = \frac{\left(\frac{4}{3}+0\right)}{\left(1-\frac{4}{3}\cdot 0\right)} = \frac{\left(\frac{4}{3}\right)}{(1)} = \frac{4}{3}$$

69. $\cos\left(\cos^{-1}u + \sin^{-1}v\right)$

Let $\alpha = \cos^{-1}u$ and $\beta = \sin^{-1}v$.

Then $\cos\alpha = u$, $0 \le \alpha \le \pi$, and $\sin\beta = v$, $-\dfrac{\pi}{2} \le \beta \le \dfrac{\pi}{2}$

$$\sin\alpha = \sqrt{1 - \cos^2\alpha} = \sqrt{1 - u^2}$$
$$\cos\beta = \sqrt{1 - \sin^2\beta} = \sqrt{1 - v^2}$$
$$\cos\left(\cos^{-1}u + \sin^{-1}v\right) = \cos(\alpha + \beta) = \cos\alpha\cos\beta - \sin\alpha\sin\beta = u\sqrt{1 - v^2} - v\sqrt{1 - u^2}$$

71. $\sin\left(\tan^{-1}u - \sin^{-1}v\right)$

Let $\alpha = \tan^{-1}u$ and $\beta = \sin^{-1}v$.

Then $\tan\alpha = u$, $-\dfrac{\pi}{2} < \alpha < \dfrac{\pi}{2}$, and $\sin\beta = v$, $-\dfrac{\pi}{2} \le \beta \le \dfrac{\pi}{2}$

$$\sec\alpha = \sqrt{\tan^2\alpha + 1} = \sqrt{u^2 + 1}; \quad \cos\alpha = \frac{1}{\sqrt{u^2 + 1}}$$

$$\sin\alpha = \sqrt{1 - \cos^2\alpha} = \sqrt{1 - \frac{1}{u^2 + 1}} = \sqrt{\frac{u^2 + 1 - 1}{u^2 + 1}} = \sqrt{\frac{u^2}{u^2 + 1}} = \frac{u}{\sqrt{u^2 + 1}}$$

$$\cos\beta = \sqrt{1 - \sin^2\beta} = \sqrt{1 - v^2}$$

$$\sin\left(\tan^{-1}u - \sin^{-1}v\right) = \sin(\alpha - \beta) = \sin\alpha\cos\beta - \cos\alpha\sin\beta$$

$$= \frac{u}{\sqrt{u^2 + 1}} \cdot \sqrt{1 - v^2} - \frac{1}{\sqrt{u^2 + 1}} \cdot v = \frac{u\sqrt{1 - v^2} - v}{\sqrt{u^2 + 1}}$$

73. $\tan\left(\sin^{-1}u - \cos^{-1}v\right)$

Let $\alpha = \sin^{-1}u$ and $\beta = \cos^{-1}v$.

Then $\sin\alpha = u$, $-\dfrac{\pi}{2} \le \alpha \le \dfrac{\pi}{2}$, and $\cos\beta = v$, $0 \le \beta \le \pi$

$$\cos\alpha = \sqrt{1 - \sin^2\alpha} = \sqrt{1 - u^2}; \quad \tan\alpha = \frac{\sin\alpha}{\cos\alpha} = \frac{u}{\sqrt{1 - u^2}}$$

$$\sin\beta = \sqrt{1 - \cos^2\beta} = \sqrt{1 - v^2}; \quad \tan\beta = \frac{\sin\beta}{\cos\beta} = \frac{\sqrt{1 - v^2}}{v}$$

$$\tan\left(\sin^{-1}u - \cos^{-1}v\right) = \tan(\alpha - \beta) = \frac{\tan\alpha - \tan\beta}{1 + \tan\alpha\tan\beta} = \frac{\left(\dfrac{u}{\sqrt{1 - u^2}} - \dfrac{\sqrt{1 - v^2}}{v}\right)}{\left(1 + \dfrac{u}{\sqrt{1 - u^2}} \cdot \dfrac{\sqrt{1 - v^2}}{v}\right)}$$

$$= \frac{\left(\dfrac{uv - \sqrt{1 - u^2}\sqrt{1 - v^2}}{v\sqrt{1 - u^2}}\right)}{\left(\dfrac{v\sqrt{1 - u^2} + u\sqrt{1 - v^2}}{v\sqrt{1 - u^2}}\right)} = \frac{uv - \sqrt{1 - u^2}\sqrt{1 - v^2}}{v\sqrt{1 - u^2} + u\sqrt{1 - v^2}}$$

75. Show that $\sin^{-1}v + \cos^{-1}v = \frac{\pi}{2}$.

Let $\alpha = \sin^{-1}v$ and $\beta = \cos^{-1}v$.

Then $\sin\alpha = v = \cos\beta$, and since $\sin\alpha = \cos\left(\frac{\pi}{2}-\alpha\right)$, $\cos\left(\frac{\pi}{2}-\alpha\right) = \cos\beta$.

If $v \geq 0$, then $0 \leq \alpha \leq \frac{\pi}{2}$, so that $\left(\frac{\pi}{2}-\alpha\right)$ and β both lie in the interval $\left[0,\frac{\pi}{2}\right]$.

If $v < 0$, then $-\frac{\pi}{2} \leq \alpha < 0$, so that $\left(\frac{\pi}{2}-\alpha\right)$ and β both lie in the interval $\left[\frac{\pi}{2},\pi\right]$. Either

way, $\cos\left(\frac{\pi}{2}-\alpha\right) = \cos\beta$ implies $\frac{\pi}{2}-\alpha = \beta$, or $\alpha+\beta = \frac{\pi}{2}$. Thus, $\sin^{-1}v + \cos^{-1}v = \frac{\pi}{2}$.

77. Show that $\tan^{-1}\left(\frac{1}{v}\right) = \frac{\pi}{2} - \tan^{-1}v$, if $v > 0$.

Let $\alpha = \tan^{-1}\left(\frac{1}{v}\right)$ and $\beta = \tan^{-1}v$. Because $\frac{1}{v}$ must be defined, $v \neq 0$ and so $\alpha,\beta \neq 0$.

Then $\tan\alpha = \frac{1}{v} = \frac{1}{\tan\beta} = \cot\beta$, and since $\tan\alpha = \cot\left(\frac{\pi}{2}-\alpha\right)$, $\cot\left(\frac{\pi}{2}-\alpha\right) = \cot\beta$.

Because $v > 0, 0 < \alpha < \frac{\pi}{2}$ and so $\frac{\pi}{2}-\alpha$ and β both lie in the interval

$\left(0,\frac{\pi}{2}\right)$. Then, $\cot\left(\frac{\pi}{2}-\alpha\right) = \cot\beta$ implies $\frac{\pi}{2}-\alpha = \beta$ or $\alpha = \frac{\pi}{2}-\beta$.

Thus, $\tan^{-1}\left(\frac{1}{v}\right) = \frac{\pi}{2} - \tan^{-1}v$, if $v > 0$.

79. $\sin\left(\sin^{-1}v + \cos^{-1}v\right) = \sin\left(\sin^{-1}v\right)\cos\left(\cos^{-1}v\right) + \cos\left(\sin^{-1}v\right)\sin\left(\cos^{-1}v\right)$

$= v \cdot v + \sqrt{1-v^2}\sqrt{1-v^2} = v^2 + 1 - v^2 = 1$

81. $\dfrac{\sin(x+h)-\sin x}{h} = \dfrac{\sin x\cos h + \cos x\sin h - \sin x}{h} = \dfrac{\cos x\sin h - \sin x + \sin x\cos h}{h}$

$= \cos x \cdot \left(\dfrac{\sin h}{h}\right) - \sin x \cdot \left(\dfrac{1-\cos h}{h}\right)$

83. $\tan\left(\frac{\pi}{2}-\theta\right) = \dfrac{\tan\left(\frac{\pi}{2}\right)-\tan\theta}{1+\tan\left(\frac{\pi}{2}\right)\tan\theta}$ This is impossible because $\tan\left(\frac{\pi}{2}\right)$ is undefined.

$\tan\left(\frac{\pi}{2}-\theta\right) = \dfrac{\sin\left(\frac{\pi}{2}-\theta\right)}{\cos\left(\frac{\pi}{2}-\theta\right)} = \dfrac{\cos\theta}{\sin\theta} = \cot\theta$

85. $\tan\theta = \tan(\theta_2 - \theta_1) = \dfrac{\tan\theta_2 - \tan\theta_1}{1+\tan\theta_2\tan\theta_1} = \dfrac{m_2 - m_1}{1+m_2 m_1}$

87. The first step in the derivation

$$\tan\left(\theta + \frac{\pi}{2}\right) = \frac{\tan\theta + \tan\left(\frac{\pi}{2}\right)}{1 - \tan\theta\tan\left(\frac{\pi}{2}\right)}$$

is impossible because $\tan\left(\frac{\pi}{2}\right)$ is undefined.

Analytic Trigonometry

6.5   Double-Angle and Half-Angle Formulas

1. $\sin\theta = \dfrac{3}{5}, \ \ 0 < \theta < \dfrac{\pi}{2};$ thus, $0 < \dfrac{\theta}{2} < \dfrac{\pi}{4} \Rightarrow \dfrac{\theta}{2}$ is in quadrant I.

$y = 3, \ r = 5$

$x^2 + 3^2 = 5^2, \ x > 0 \ \ \rightarrow \ \ x^2 = 25 - 9 = 16, \ x > 0 \ \ \rightarrow \ \ x = 4$

$\cos\theta = \dfrac{4}{5}$

(a) $\sin(2\theta) = 2\sin\theta\cos\theta = 2 \cdot \dfrac{3}{5} \cdot \dfrac{4}{5} = \dfrac{24}{25}$

(b) $\cos(2\theta) = \cos^2\theta - \sin^2\theta = \left(\dfrac{4}{5}\right)^2 - \left(\dfrac{3}{5}\right)^2 = \dfrac{16}{25} - \dfrac{9}{25} = \dfrac{7}{25}$

(c) $\sin\left(\dfrac{\theta}{2}\right) = \sqrt{\dfrac{1-\cos\theta}{2}} = \sqrt{\dfrac{\left(1-\dfrac{4}{5}\right)}{2}} = \sqrt{\dfrac{\left(\dfrac{1}{5}\right)}{2}} = \sqrt{\dfrac{1}{10}} = \dfrac{1}{\sqrt{10}} = \dfrac{\sqrt{10}}{10}$

3. $\tan\theta = \dfrac{4}{3}, \ \ \pi < \theta < \dfrac{3\pi}{2};$ thus, $\dfrac{\pi}{2} < \dfrac{\theta}{2} < \dfrac{3\pi}{4} \ \Rightarrow \ \dfrac{\theta}{2}$ is in quadrant II.

$x = -3, \ y = -4$

$r^2 = (-3)^2 + (-4)^2 = 9 + 16 = 25 \ \ \rightarrow \ \ r = 5$

$\sin\theta = -\dfrac{4}{5}, \ \ \cos\theta = -\dfrac{3}{5}$

(a) $\sin(2\theta) = 2\sin\theta\cos\theta = 2 \cdot \left(-\dfrac{4}{5}\right) \cdot \left(-\dfrac{3}{5}\right) = \dfrac{24}{25}$

(b) $\cos(2\theta) = \cos^2\theta - \sin^2\theta = \left(-\dfrac{3}{5}\right)^2 - \left(-\dfrac{4}{5}\right)^2 = \dfrac{9}{25} - \dfrac{16}{25} = -\dfrac{7}{25}$

(c) $\sin\left(\dfrac{\theta}{2}\right) = \sqrt{\dfrac{1-\cos\theta}{2}} = \sqrt{\dfrac{1-\left(-\dfrac{3}{5}\right)}{2}} = \sqrt{\dfrac{\left(\dfrac{8}{5}\right)}{2}} = \sqrt{\dfrac{4}{5}} = \dfrac{2}{\sqrt{5}} = \dfrac{2\sqrt{5}}{5}$

(d) $\cos\left(\dfrac{\theta}{2}\right) = -\sqrt{\dfrac{1+\cos\theta}{2}} = -\sqrt{\dfrac{1+\left(-\dfrac{3}{5}\right)}{2}} = -\sqrt{\dfrac{\left(\dfrac{2}{5}\right)}{2}} = -\sqrt{\dfrac{1}{5}} = -\dfrac{1}{\sqrt{5}} = -\dfrac{\sqrt{5}}{5}$

5. $\cos\theta = -\dfrac{\sqrt{6}}{3}$, $\dfrac{\pi}{2} < \theta < \pi$;   thus, $\dfrac{\pi}{4} < \dfrac{\theta}{2} < \dfrac{\pi}{2} \Rightarrow$ $\dfrac{\theta}{2}$ is in quadrant I.

$x = -\sqrt{6}$, $r = 3$

$(-\sqrt{6})^2 + y^2 = 3^2$ $\rightarrow$ $y^2 = 9 - 6 = 3$ $\rightarrow$ $y = \sqrt{3}$

$\sin\theta = \dfrac{\sqrt{3}}{3}$

(a) $\sin(2\theta) = 2\sin\theta\cos\theta = 2\cdot\left(\dfrac{\sqrt{3}}{3}\right)\cdot\left(-\dfrac{\sqrt{6}}{3}\right) = -\dfrac{2\sqrt{18}}{9} = -\dfrac{6\sqrt{2}}{9} = -\dfrac{2\sqrt{2}}{3}$

(b) $\cos(2\theta) = \cos^2\theta - \sin^2\theta = \left(-\dfrac{\sqrt{6}}{3}\right)^2 - \left(\dfrac{\sqrt{3}}{3}\right)^2 = \dfrac{6}{9} - \dfrac{3}{9} = \dfrac{3}{9} = \dfrac{1}{3}$

(c) $\sin\left(\dfrac{\theta}{2}\right) = \sqrt{\dfrac{1-\cos\theta}{2}} = \sqrt{\dfrac{1-\left(-\dfrac{\sqrt{6}}{3}\right)}{2}} = \sqrt{\dfrac{\dfrac{3+\sqrt{6}}{3}}{2}} = \sqrt{\dfrac{3+\sqrt{6}}{6}}$

(d) $\cos\left(\dfrac{\theta}{2}\right) = \sqrt{\dfrac{1+\cos\theta}{2}} = \sqrt{\dfrac{1+\left(-\dfrac{\sqrt{6}}{3}\right)}{2}} = \sqrt{\dfrac{\dfrac{3-\sqrt{6}}{3}}{2}} = \sqrt{\dfrac{3-\sqrt{6}}{6}}$

7. $\sec\theta = 3$, $\sin\theta > 0$; $0 < \theta < \dfrac{\pi}{2}$;   thus, $0 < \dfrac{\theta}{2} < \dfrac{\pi}{4} \Rightarrow$ $\dfrac{\theta}{2}$ is in quadrant I.

$x = 1$, $r = 3$

$1^2 + y^2 = 3^2$ $\rightarrow$ $y^2 = 9 - 1 = 8$ $\rightarrow$ $y = 2\sqrt{2}$

$\sin\theta = \dfrac{2\sqrt{2}}{3}$, $\cos\theta = \dfrac{1}{3}$

(a) $\sin(2\theta) = 2\sin\theta\cos\theta = 2\cdot\dfrac{2\sqrt{2}}{3}\cdot\dfrac{1}{3} = \dfrac{4\sqrt{2}}{9}$

(b) $\cos(2\theta) = \cos^2\theta - \sin^2\theta = \left(\dfrac{1}{3}\right)^2 - \left(\dfrac{2\sqrt{2}}{3}\right)^2 = \dfrac{1}{9} - \dfrac{8}{9} = -\dfrac{7}{9}$

(c) $\sin\left(\dfrac{\theta}{2}\right) = \sqrt{\dfrac{1-\cos\theta}{2}} = \sqrt{\dfrac{1-\dfrac{1}{3}}{2}} = \sqrt{\dfrac{\dfrac{2}{3}}{2}} = \sqrt{\dfrac{1}{3}} = \dfrac{1}{\sqrt{3}} = \dfrac{\sqrt{3}}{3}$

(d) $\cos\left(\dfrac{\theta}{2}\right) = \sqrt{\dfrac{1+\cos\theta}{2}} = \sqrt{\dfrac{1+\dfrac{1}{3}}{2}} = \sqrt{\dfrac{\dfrac{4}{3}}{2}} = \sqrt{\dfrac{2}{3}} = \dfrac{\sqrt{6}}{3}$

9. $\cot\theta = -2$, $\sec\theta < 0$; $\dfrac{\pi}{2} < \theta < \pi$; thus, $\dfrac{\pi}{4} < \dfrac{\theta}{2} < \dfrac{\pi}{2} \Rightarrow \dfrac{\theta}{2}$ is in quadrant I.

$x = -2$, $y = 1$

$r^2 = (-2)^2 + 1^2 = 4 + 1 = 5 \rightarrow r = \sqrt{5}$

$\sin\theta = \dfrac{1}{\sqrt{5}} = \dfrac{\sqrt{5}}{5}$, $\cos\theta = -\dfrac{2}{\sqrt{5}} = -\dfrac{2\sqrt{5}}{5}$

(a) $\sin(2\theta) = 2\sin\theta\cos\theta = 2\cdot\left(\dfrac{\sqrt{5}}{5}\right)\cdot\left(-\dfrac{2\sqrt{5}}{5}\right) = -\dfrac{20}{25} = -\dfrac{4}{5}$

(b) $\cos(2\theta) = \cos^2\theta - \sin^2\theta = \left(-\dfrac{2\sqrt{5}}{5}\right)^2 - \left(\dfrac{\sqrt{5}}{5}\right)^2 = \dfrac{20}{25} - \dfrac{5}{25} = \dfrac{15}{25} = \dfrac{3}{5}$

(c) $\sin\left(\dfrac{\theta}{2}\right) = \sqrt{\dfrac{1-\cos\theta}{2}} = \sqrt{\dfrac{1-\left(-\frac{2\sqrt{5}}{5}\right)}{2}} = \sqrt{\dfrac{\frac{5+2\sqrt{5}}{5}}{2}} = \sqrt{\dfrac{5+2\sqrt{5}}{10}}$

(d) $\cos\left(\dfrac{\theta}{2}\right) = \sqrt{\dfrac{1+\cos\theta}{2}} = \sqrt{\dfrac{1+\left(-\frac{2\sqrt{5}}{5}\right)}{2}} = \sqrt{\dfrac{\frac{5-2\sqrt{5}}{5}}{2}} = \sqrt{\dfrac{5-2\sqrt{5}}{10}}$

11. $\tan\theta = -3$, $\sin\theta < 0$; $\dfrac{3\pi}{2} < \theta < 2\pi$; thus, $\dfrac{3\pi}{4} < \dfrac{\theta}{2} < \pi$ or $\dfrac{\theta}{2}$ is in quadrant II.

$x = 1$, $y = -3$

$r^2 = 1^2 + (-3)^2 = 1 + 9 = 10 \rightarrow r = \sqrt{10}$

$\sin\theta = \dfrac{-3}{\sqrt{10}} = -\dfrac{3\sqrt{10}}{10}$, $\cos\theta = \dfrac{1}{\sqrt{10}} = \dfrac{\sqrt{10}}{10}$

(a) $\sin(2\theta) = 2\sin\theta\cos\theta = 2\cdot\left(-\dfrac{3\sqrt{10}}{10}\right)\cdot\left(\dfrac{\sqrt{10}}{10}\right) = -\dfrac{60}{100} = -\dfrac{3}{5}$

(b) $\cos(2\theta) = \cos^2\theta - \sin^2\theta = \left(\dfrac{\sqrt{10}}{10}\right)^2 - \left(-\dfrac{3\sqrt{10}}{10}\right)^2 = \dfrac{10}{100} - \dfrac{90}{100} = -\dfrac{80}{100} = -\dfrac{4}{5}$

(c) $\sin\left(\dfrac{\theta}{2}\right) = \sqrt{\dfrac{1-\cos\theta}{2}} = \sqrt{\dfrac{1-\frac{\sqrt{10}}{10}}{2}} = \sqrt{\dfrac{\frac{10-\sqrt{10}}{10}}{2}} = \sqrt{\dfrac{10-\sqrt{10}}{20}} = \dfrac{1}{2}\sqrt{\dfrac{10-\sqrt{10}}{5}}$

(d) $\cos\left(\dfrac{\theta}{2}\right) = -\sqrt{\dfrac{1+\cos\theta}{2}} = -\sqrt{\dfrac{1+\frac{\sqrt{10}}{10}}{2}} = -\sqrt{\dfrac{\frac{10+\sqrt{10}}{10}}{2}} = -\sqrt{\dfrac{10+\sqrt{10}}{20}}$

$= -\dfrac{1}{2}\sqrt{\dfrac{10+\sqrt{10}}{5}}$

13. $\sin(22.5°) = \sin\left(\dfrac{45°}{2}\right) = \sqrt{\dfrac{1-\cos(45°)}{2}} = \sqrt{\dfrac{\left(1-\dfrac{\sqrt{2}}{2}\right)}{2}} = \sqrt{\dfrac{2-\sqrt{2}}{4}} = \dfrac{\sqrt{2-\sqrt{2}}}{2}$

15. $\tan\left(\dfrac{7\pi}{8}\right) = \tan\left(\dfrac{\left(\dfrac{7\pi}{4}\right)}{2}\right) = -\sqrt{\dfrac{\left(1-\cos\left(\dfrac{7\pi}{4}\right)\right)}{\left(1+\cos\left(\dfrac{7\pi}{4}\right)\right)}} = -\sqrt{\dfrac{\left(1-\dfrac{\sqrt{2}}{2}\right)}{\left(1+\dfrac{\sqrt{2}}{2}\right)}} = -\sqrt{\left(\dfrac{2-\sqrt{2}}{2+\sqrt{2}}\right)\cdot\left(\dfrac{2-\sqrt{2}}{2-\sqrt{2}}\right)}$

$= -\sqrt{\dfrac{6-4\sqrt{2}}{2}} = -\sqrt{3-2\sqrt{2}}$

17. $\cos(165°) = \cos\left(\dfrac{330°}{2}\right) = -\sqrt{\dfrac{1+\cos(330°)}{2}} = -\sqrt{\dfrac{\left(1+\dfrac{\sqrt{3}}{2}\right)}{2}} = -\sqrt{\dfrac{2+\sqrt{3}}{4}} = -\dfrac{\sqrt{2+\sqrt{3}}}{2}$

19. $\sec\left(\dfrac{15\pi}{8}\right) = \dfrac{1}{\cos\left(\dfrac{15\pi}{8}\right)} = \dfrac{1}{\cos\left(\dfrac{\left(\dfrac{15\pi}{4}\right)}{2}\right)} = \dfrac{1}{\sqrt{\dfrac{\left(1+\cos\left(\dfrac{15\pi}{4}\right)\right)}{2}}} = \dfrac{1}{\sqrt{\dfrac{\left(1+\dfrac{\sqrt{2}}{2}\right)}{2}}} = \dfrac{1}{\sqrt{\dfrac{2+\sqrt{2}}{4}}}$

$= \left(\dfrac{2}{\sqrt{2+\sqrt{2}}}\right)\cdot\left(\dfrac{\sqrt{2+\sqrt{2}}}{\sqrt{2+\sqrt{2}}}\right) = \left(\dfrac{2\sqrt{2+\sqrt{2}}}{2+\sqrt{2}}\right)\cdot\left(\dfrac{2-\sqrt{2}}{2-\sqrt{2}}\right)$

$= \dfrac{2\left(2-\sqrt{2}\right)\sqrt{2+\sqrt{2}}}{2} = \left(2-\sqrt{2}\right)\sqrt{2+\sqrt{2}}$

21. $\sin\left(-\dfrac{\pi}{8}\right) = \sin\left(\dfrac{\left(-\dfrac{\pi}{4}\right)}{2}\right) = -\sqrt{\dfrac{1-\cos\left(-\dfrac{\pi}{4}\right)}{2}} = -\sqrt{\dfrac{\left(1-\dfrac{\sqrt{2}}{2}\right)}{2}} = -\sqrt{\dfrac{2-\sqrt{2}}{4}} = -\dfrac{\sqrt{2-\sqrt{2}}}{2}$

23. $\sin^4\theta = \left(\sin^2\theta\right)^2 = \left(\dfrac{1-\cos(2\theta)}{2}\right)^2 = \dfrac{1}{4}\left(1-2\cos(2\theta)+\cos^2(2\theta)\right)$

$= \dfrac{1}{4} - \dfrac{1}{2}\cos(2\theta) + \dfrac{1}{4}\cos^2(2\theta) = \dfrac{1}{4} - \dfrac{1}{2}\cos(2\theta) + \dfrac{1}{4}\left(\dfrac{1+\cos(4\theta)}{2}\right)$

$= \dfrac{1}{4} - \dfrac{1}{2}\cos(2\theta) + \dfrac{1}{8} + \dfrac{1}{8}\cos(4\theta) = \dfrac{3}{8} - \dfrac{1}{2}\cos(2\theta) + \dfrac{1}{8}\cos(4\theta)$

25. $\sin(4\theta) = \sin(2(2\theta)) = 2\sin(2\theta)\cos(2\theta) = 2(2\sin\theta\cos\theta)(1 - 2\sin^2\theta)$

$\qquad\qquad = \cos\theta(4\sin\theta - 8\sin^3\theta)$

27. Use the result of problem 25 to help solve the problem:

$\qquad \sin(5\theta) = \sin(4\theta + \theta) = \sin(4\theta)\cos\theta + \cos(4\theta)\sin\theta$

$\qquad\qquad = \cos\theta(4\sin\theta - 8\sin^3\theta)\cos\theta + \cos(2(2\theta))\sin\theta$

$\qquad\qquad = \cos^2\theta(4\sin\theta - 8\sin^3\theta) + (1 - 2\sin^2(2\theta))\sin\theta$

$\qquad\qquad = (1 - \sin^2\theta)(4\sin\theta - 8\sin^3\theta) + \sin\theta(1 - 2(2\sin\theta\cos\theta)^2)$

$\qquad\qquad = 4\sin\theta - 12\sin^3\theta + 8\sin^5\theta + \sin\theta(1 - 8\sin^2\theta\cos^2\theta)$

$\qquad\qquad = 4\sin\theta - 12\sin^3\theta + 8\sin^5\theta + \sin\theta - 8\sin^3\theta(1 - \sin^2\theta)$

$\qquad\qquad = 5\sin\theta - 12\sin^3\theta + 8\sin^5\theta - 8\sin^3\theta + 8\sin^5\theta$

$\qquad\qquad = 5\sin\theta - 20\sin^3\theta + 16\sin^5\theta$

29. $\cos^4\theta - \sin^4\theta = (\cos^2\theta + \sin^2\theta)(\cos^2\theta - \sin^2\theta) = 1 \cdot \cos(2\theta) = \cos(2\theta)$

31. $\cot(2\theta) = \dfrac{1}{\tan(2\theta)} = \dfrac{1}{\left(\dfrac{2\tan\theta}{1-\tan^2\theta}\right)} = \dfrac{1-\tan^2\theta}{2\tan\theta} = \dfrac{\left(1 - \dfrac{1}{\cot^2\theta}\right)}{\left(\dfrac{2}{\cot\theta}\right)} = \dfrac{\left(\dfrac{\cot^2\theta - 1}{\cot^2\theta}\right)}{\left(\dfrac{2}{\cot\theta}\right)} = \dfrac{\cot^2\theta - 1}{2\cot\theta}$

33. $\sec(2\theta) = \dfrac{1}{\cos(2\theta)} = \dfrac{1}{2\cos^2\theta - 1} = \dfrac{1}{\left(\dfrac{2}{\sec^2\theta} - 1\right)} = \dfrac{1}{\left(\dfrac{2 - \sec^2\theta}{\sec^2\theta}\right)} = \dfrac{\sec^2\theta}{2 - \sec^2\theta}$

35. $\cos^2(2\theta) - \sin^2(2\theta) = \cos(2(2\theta)) = \cos(4\theta)$

37. $\dfrac{\cos(2\theta)}{1 + \sin(2\theta)} = \dfrac{\cos^2\theta - \sin^2\theta}{1 + 2\sin\theta\cos\theta} = \dfrac{(\cos\theta - \sin\theta)(\cos\theta + \sin\theta)}{\cos^2\theta + \sin^2\theta + 2\sin\theta\cos\theta}$

$\qquad\qquad = \dfrac{(\cos\theta - \sin\theta)(\cos\theta + \sin\theta)}{(\cos\theta + \sin\theta)(\cos\theta + \sin\theta)} = \dfrac{\cos\theta - \sin\theta}{\cos\theta + \sin\theta} = \dfrac{\left(\dfrac{\cos\theta}{\sin\theta} - \dfrac{\sin\theta}{\sin\theta}\right)}{\left(\dfrac{\cos\theta}{\sin\theta} + \dfrac{\sin\theta}{\sin\theta}\right)} = \dfrac{\cot\theta - 1}{\cot\theta + 1}$

39. $\sec^2\left(\dfrac{\theta}{2}\right) = \dfrac{1}{\cos^2\left(\dfrac{\theta}{2}\right)} = \dfrac{1}{\left(\dfrac{1 + \cos\theta}{2}\right)} = \dfrac{2}{1 + \cos\theta}$

41. $\cot^2\left(\dfrac{\theta}{2}\right) = \dfrac{1}{\tan^2\left(\dfrac{\theta}{2}\right)} = \dfrac{1}{\left(\dfrac{1-\cos\theta}{1+\cos\theta}\right)} = \dfrac{1+\cos\theta}{1-\cos\theta} = \dfrac{\left(1+\dfrac{1}{\sec\theta}\right)}{\left(1-\dfrac{1}{\sec\theta}\right)} = \dfrac{\left(\dfrac{\sec\theta+1}{\sec\theta}\right)}{\left(\dfrac{\sec\theta-1}{\sec\theta}\right)} = \dfrac{\sec\theta+1}{\sec\theta-1}$

43. $\dfrac{\left(1-\tan^2\left(\dfrac{\theta}{2}\right)\right)}{\left(1+\tan^2\left(\dfrac{\theta}{2}\right)\right)} = \dfrac{\left(1-\dfrac{1-\cos\theta}{1+\cos\theta}\right)}{\left(1+\dfrac{1-\cos\theta}{1+\cos\theta}\right)} = \dfrac{\left(\dfrac{1+\cos\theta-(1-\cos\theta)}{1+\cos\theta}\right)}{\left(\dfrac{1+\cos\theta+1-\cos\theta}{1+\cos\theta}\right)} = \dfrac{2\cos\theta}{2} = \cos\theta$

45. $\dfrac{\sin(3\theta)}{\sin\theta} - \dfrac{\cos(3\theta)}{\cos\theta} = \dfrac{\sin(3\theta)\cos\theta - \cos(3\theta)\sin\theta}{\sin\theta\cos\theta} = \dfrac{\sin(3\theta-\theta)}{\sin\theta\cos\theta}$

$= \dfrac{\sin 2\theta}{\sin\theta\cos\theta} = \dfrac{2\sin\theta\cos\theta}{\sin\theta\cos\theta} = 2$

47. $\tan(3\theta) = \tan(2\theta + \theta) = \dfrac{\tan(2\theta) + \tan\theta}{1-\tan(2\theta)\tan\theta} = \dfrac{\left(\dfrac{2\tan\theta}{1-\tan^2\theta} + \tan\theta\right)}{\left(1 - \dfrac{2\tan\theta}{1-\tan^2\theta}\cdot\tan\theta\right)}$

$= \dfrac{\left(\dfrac{2\tan\theta + \tan\theta - \tan^3\theta}{1-\tan^2\theta}\right)}{\left(\dfrac{1-\tan^2\theta - 2\tan^2\theta}{1-\tan^2\theta}\right)} = \dfrac{3\tan\theta - \tan^3\theta}{1 - 3\tan^2\theta}$

49. $\dfrac{1}{2}\cdot\left(\ln|1-\cos(2\theta)| - \ln 2\right) = \dfrac{1}{2}\cdot\ln\left|\dfrac{1-\cos 2\theta}{2}\right| = \ln\left(\left|\dfrac{1-\cos(2\theta)}{2}\right|^{1/2}\right) = \ln\left(\left|\sin^2\theta\right|^{1/2}\right) = \ln|\sin\theta|$

51. $\sin\left(2\sin^{-1}\left(\dfrac{1}{2}\right)\right) = \sin\left(2\left(\dfrac{\pi}{6}\right)\right) = \sin\left(\dfrac{\pi}{3}\right) = \dfrac{\sqrt{3}}{2}$

53. $\cos\left(2\sin^{-1}\left(\dfrac{3}{5}\right)\right) = 1 - 2\sin^2\left(\sin^{-1}\left(\dfrac{3}{5}\right)\right) = 1 - 2\left(\dfrac{3}{5}\right)^2 = 1 - 2\left(\dfrac{9}{25}\right) = 1 - \dfrac{18}{25} = \dfrac{7}{25}$

55. $\tan\left[2\cos^{-1}\left(-\dfrac{3}{5}\right)\right]$

Let $\alpha = \cos^{-1}\left(-\dfrac{3}{5}\right)$. α is in quadrant II.

Then $\cos\alpha = -\dfrac{3}{5},\ 0 \le \alpha \le \pi$.

$\sec\alpha = -\dfrac{5}{3};\quad \tan\alpha = -\sqrt{\sec^2\alpha - 1} = -\sqrt{\left(-\dfrac{5}{3}\right)^2 - 1} = -\sqrt{\dfrac{25}{9} - 1} = -\sqrt{\dfrac{16}{9}} = -\dfrac{4}{3}$

$$\tan\left[2\cos^{-1}\left(-\frac{3}{5}\right)\right] = \tan 2\alpha = \frac{2\tan\alpha}{1-\tan^2\alpha} = \frac{2\left(-\frac{4}{3}\right)}{\left(1-\left(-\frac{4}{3}\right)^2\right)} = \frac{\left(-\frac{8}{3}\right)}{\left(1-\frac{16}{9}\right)} = \frac{\left(-\frac{8}{3}\right)}{\left(-\frac{7}{9}\right)} = \left(-\frac{8}{3}\right)\cdot\left(-\frac{9}{7}\right) = \frac{24}{7}$$

57. $\sin\left(2\cos^{-1}\left(\frac{4}{5}\right)\right)$

Let $\alpha = \cos^{-1}\left(\frac{4}{5}\right)$. α is in quadrant I.

Then $\cos\alpha = \frac{4}{5}, 0 \le \alpha \le \pi$.

$$\sin\alpha = \sqrt{1-\cos^2\alpha} = \sqrt{1-\left(\frac{4}{5}\right)^2} = \sqrt{1-\frac{16}{25}} = \sqrt{\frac{9}{25}} = \frac{3}{5}$$

$$\sin\left[2\cos^{-1}\left(\frac{4}{5}\right)\right] = \sin 2\alpha = 2\sin\alpha\cos\alpha = 2\cdot\frac{3}{5}\cdot\frac{4}{5} = \frac{24}{25}$$

59. $\sin^2\left[\frac{1}{2}\cdot\cos^{-1}\left(\frac{3}{5}\right)\right] = \frac{1-\cos\left(\cos^{-1}\left(\frac{3}{5}\right)\right)}{2} = \frac{\left(1-\frac{3}{5}\right)}{2} = \frac{\left(\frac{2}{5}\right)}{2} = \frac{1}{5}$

61. $\sec\left(2\tan^{-1}\left(\frac{3}{4}\right)\right)$

Let $\alpha = \tan^{-1}\left(\frac{3}{4}\right)$. α is in quadrant I.

Then $\tan\alpha = \frac{3}{4}, -\frac{\pi}{2} < \alpha < \frac{\pi}{2}$.

$$\sec\alpha = \sqrt{\tan^2\alpha+1} = \sqrt{\left(\frac{3}{4}\right)^2+1} = \sqrt{\frac{9}{16}+1} = \sqrt{\frac{25}{16}} = \frac{5}{4}; \quad \cos\alpha = \frac{4}{5}$$

$$\sec\left[2\tan^{-1}\left(\frac{3}{4}\right)\right] = \sec(2\alpha) = \frac{1}{\cos(2\alpha)} = \frac{1}{2\cos^2\alpha-1} = \frac{1}{\left(2\left(\frac{4}{5}\right)^2-1\right)} = \frac{1}{\left(2\cdot\frac{16}{25}-1\right)} = \frac{1}{\left(\frac{7}{25}\right)} = \frac{25}{7}$$

63.   If $x = 2\tan\theta$, then:

$$\sin(2\theta) = 2\sin\theta\cos\theta = \frac{2\sin\theta}{\cos\theta}\cdot\frac{\cos^2\theta}{1} = \frac{\left(2\cdot\frac{\sin\theta}{\cos\theta}\right)}{\left(\frac{1}{\cos^2\theta}\right)} = \frac{2\tan\theta}{\sec^2\theta} = \frac{2\tan\theta}{1+\tan^2\theta}\cdot\frac{4}{4}$$

$$= \frac{4(2\tan\theta)}{4+(2\tan\theta)^2} = \frac{4x}{4+x^2}$$

65.   Solve for C:

$$\frac{1}{2}\cdot \sin^2 x + C = -\frac{1}{4}\cdot \cos(2x)$$

$$C = -\frac{1}{4}\cdot \cos(2x) - \frac{1}{2}\cdot \sin^2 x = -\frac{1}{4}\cdot \left(\cos(2x) + 2\sin^2 x\right)$$

$$= -\frac{1}{4}\cdot \left(1 - 2\sin^2 x + 2\sin^2 x\right) = -\frac{1}{4}\cdot (1) = -\frac{1}{4}$$

67.

$$z = \tan\left(\frac{\alpha}{2}\right) = \frac{1 - \cos\alpha}{\sin\alpha}$$

$$z\sin\alpha = 1 - \cos\alpha$$

$$z\sin\alpha = 1 - \sqrt{1 - \sin^2\alpha}$$

$$z\sin\alpha - 1 = -\sqrt{1 - \sin^2\alpha}$$

$$z^2\sin^2\alpha - 2z\sin\alpha + 1 = 1 - \sin^2\alpha \rightarrow z^2\sin^2\alpha + \sin^2\alpha = 2z\sin\alpha$$

$$\sin^2\alpha(z^2 + 1) = 2z\sin\alpha \rightarrow \sin\alpha(z^2 + 1) = 2z \rightarrow \sin\alpha = \frac{2z}{z^2 + 1}$$

69.   Let b represent the base of the triangle.

$$\cos\left(\frac{\theta}{2}\right) = \frac{h}{s} \;\rightarrow\; h = s\cos\left(\frac{\theta}{2}\right) \qquad \sin\left(\frac{\theta}{2}\right) = \frac{\left(\frac{1}{2}b\right)}{s} \;\rightarrow\; b = 2s\sin\left(\frac{\theta}{2}\right)$$

$$A = \frac{1}{2}b\cdot h = \frac{1}{2}\cdot \left(2s\sin\left(\frac{\theta}{2}\right)\right)\left(s\cos\left(\frac{\theta}{2}\right)\right) = s^2\sin\left(\frac{\theta}{2}\right)\cos\left(\frac{\theta}{2}\right) = \frac{1}{2}\cdot s^2\sin\theta$$

71. $f(x) = \sin^2 x = \dfrac{1 - \cos(2x)}{2}$

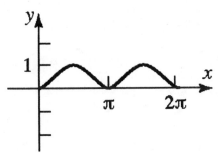

73. $\sin\left(\dfrac{\pi}{24}\right) = \sin\left(\dfrac{\left(\dfrac{\pi}{12}\right)}{2}\right) = \sqrt{\dfrac{1 - \cos\left(\dfrac{\pi}{12}\right)}{2}} = \sqrt{\dfrac{1 - \left(\dfrac{1}{4}\left(\sqrt{6} + \sqrt{2}\right)\right)}{2}} = \sqrt{\dfrac{1}{2} - \dfrac{1}{8}\left(\sqrt{6} + \sqrt{2}\right)}$

$$= \sqrt{\dfrac{8 - 2\left(\sqrt{6} + \sqrt{2}\right)}{16}} = \dfrac{\sqrt{8 - 2\left(\sqrt{6} + \sqrt{2}\right)}}{4}$$

$$\cos\left(\frac{\pi}{24}\right) = \cos\left(\frac{\left(\frac{\pi}{12}\right)}{2}\right) = \sqrt{\frac{1+\cos\left(\frac{\pi}{12}\right)}{2}} = \sqrt{\frac{1+\frac{1}{4}\left(\sqrt{6}+\sqrt{2}\right)}{2}} = \sqrt{\frac{1}{2}+\frac{1}{8}\left(\sqrt{6}+\sqrt{2}\right)}$$

$$= \sqrt{\frac{8+2\left(\sqrt{6}+\sqrt{2}\right)}{16}} = \frac{\sqrt{8+2\left(\sqrt{6}+\sqrt{2}\right)}}{4}$$

75. $\sin^3\theta + \sin^3(\theta+120°) + \sin^3(\theta+240°)$

$= \sin^3\theta + \left(\sin\theta\cos(120°)+\cos\theta\sin(120°)\right)^3 + \left(\sin\theta\cos(240°)+\cos\theta\sin(240°)\right)^3$

$= \sin^3\theta + \left(-\frac{1}{2}\cdot\sin\theta+\frac{\sqrt{3}}{2}\cdot\cos\theta\right)^3 + \left(-\frac{1}{2}\cdot\sin\theta-\frac{\sqrt{3}}{2}\cdot\cos\theta\right)^3$

$= \sin^3\theta + \frac{1}{8}\cdot\left(-\sin^3\theta+3\sqrt{3}\sin^2\theta\cos\theta-9\sin\theta\cos^2\theta+3\sqrt{3}\cos^3\theta\right)$

$\quad -\frac{1}{8}\left(\sin^3\theta+3\sqrt{3}\sin^2\theta\cos\theta+9\sin\theta\cos^2\theta+3\sqrt{3}\cos^3\theta\right)$

$= \sin^3\theta - \frac{1}{8}\cdot\sin^3\theta + \frac{3\sqrt{3}}{8}\cdot\sin^2\theta\cos\theta - \frac{9}{8}\cdot\sin\theta\cos^2\theta + \frac{3\sqrt{3}}{8}\cdot\cos^3\theta$

$\quad -\frac{1}{8}\cdot\sin^3\theta - \frac{3\sqrt{3}}{8}\cdot\sin^2\theta\cos\theta - \frac{9}{8}\cdot\sin\theta\cos^2\theta - \frac{3\sqrt{3}}{8}\cdot\cos^3\theta$

$= \frac{3}{4}\cdot\sin^3\theta - \frac{9}{4}\cdot\sin\theta\cos^2\theta = \frac{3}{4}\cdot\left(\sin^3\theta - 3\sin\theta\left(1-\sin^2\theta\right)\right)$

$= \frac{3}{4}\cdot\left(\sin^3\theta - 3\sin\theta + 3\sin^3\theta\right) = \frac{3}{4}\cdot\left(4\sin^3\theta - 3\sin\theta\right) = -\frac{3}{4}\cdot\sin(3\theta)$

(See the formula for $\sin(3\theta)$ on page 420 of the text.)

77. (a) $R(\theta) = \frac{v_0^2\sqrt{2}}{16}\cos\theta(\sin\theta-\cos\theta) = \frac{v_0^2\sqrt{2}}{16}(\cos\theta\sin\theta-\cos^2\theta)$

$= \frac{v_0^2\sqrt{2}}{16}\cdot\frac{1}{2}(2\cos\theta\sin\theta-2\cos^2\theta) = \frac{v_0^2\sqrt{2}}{32}\left(\sin(2\theta)-2\left(\frac{1+\cos2\theta}{2}\right)\right)$

$= \frac{v_0^2\sqrt{2}}{32}\left(\sin(2\theta)-1-\cos(2\theta)\right) = \frac{v_0^2\sqrt{2}}{32}\left(\sin(2\theta)-\cos(2\theta)-1\right)$

(b)

(c) Using the MAXIMUM feature on the calculator:
R has the largest value when $\theta \approx 67.5°$.

Analytic Trigonometry

6.6   Product-to-Sum and Sum-to-Product Formulas

For Problems 1-9, use the formulas:

$$\sin\alpha\sin\beta = \frac{1}{2}\cdot\left[\cos(\alpha-\beta)-\cos(\alpha+\beta)\right] \qquad \cos\alpha\cos\beta = \frac{1}{2}\cdot\left[\cos(\alpha-\beta)+\cos(\alpha+\beta)\right]$$

$$\sin\alpha\cos\beta = \frac{1}{2}\cdot\left[\sin(\alpha+\beta)+\sin(\alpha-\beta)\right]$$

---

1. $\sin(4\theta)\sin(2\theta) = \frac{1}{2}\cdot\left[\cos(4\theta-2\theta)-\cos(4\theta+2\theta)\right] = \frac{1}{2}\cdot\left[\cos(2\theta)-\cos(6\theta)\right]$

3. $\sin(4\theta)\cos(2\theta) = \frac{1}{2}\cdot\left[\sin(4\theta+2\theta)+\sin(4\theta-2\theta)\right] = \frac{1}{2}\cdot\left[\sin(6\theta)+\sin(2\theta)\right]$

5. $\cos(3\theta)\cos(5\theta) = \frac{1}{2}\cdot\left[\cos(3\theta-5\theta)+\cos(3\theta+5\theta)\right] = \frac{1}{2}\cdot\left[\cos(-2\theta)+\cos(8\theta)\right]$

$$= \frac{1}{2}\cdot\left[\cos(2\theta)+\cos(8\theta)\right]$$

7. $\sin\theta\sin(2\theta) = \frac{1}{2}\cdot\left[\cos(\theta-2\theta)-\cos(\theta+2\theta)\right] = \frac{1}{2}\cdot\left[\cos(-\theta)-\cos(3\theta)\right]$

$$= \frac{1}{2}\cdot\left[\cos\theta-\cos(3\theta)\right]$$

9. $\sin\left(\frac{3\theta}{2}\right)\cos\left(\frac{\theta}{2}\right) = \frac{1}{2}\cdot\left[\sin\left(\frac{3\theta}{2}+\frac{\theta}{2}\right)+\sin\left(\frac{3\theta}{2}-\frac{\theta}{2}\right)\right] = \frac{1}{2}\cdot\left[\sin(2\theta)+\sin\theta\right]$

For Problems 11-17, use the formulas:

$$\sin\alpha+\sin\beta = 2\sin\left(\frac{\alpha+\beta}{2}\right)\cos\left(\frac{\alpha-\beta}{2}\right) \qquad \sin\alpha-\sin\beta = 2\sin\left(\frac{\alpha-\beta}{2}\right)\cos\left(\frac{\alpha+\beta}{2}\right)$$

$$\cos\alpha+\cos\beta = 2\cos\left(\frac{\alpha+\beta}{2}\right)\cos\left(\frac{\alpha-\beta}{2}\right) \qquad \cos\alpha-\cos\beta = -2\sin\left(\frac{\alpha+\beta}{2}\right)\sin\left(\frac{\alpha-\beta}{2}\right)$$

---

11. $\sin(4\theta)-\sin(2\theta) = 2\sin\left(\frac{4\theta-2\theta}{2}\right)\cos\left(\frac{4\theta+2\theta}{2}\right) = 2\sin\theta\cos(3\theta)$

13. $\cos(2\theta)+\cos(4\theta)=2\cos\left(\dfrac{2\theta+4\theta}{2}\right)\cos\left(\dfrac{2\theta-4\theta}{2}\right)=2\cos(3\theta)\cos(-\theta)=2\cos(3\theta)\cos\theta$

15. $\sin\theta+\sin(3\theta)=2\sin\left(\dfrac{\theta+3\theta}{2}\right)\cos\left(\dfrac{\theta-3\theta}{2}\right)=2\sin(2\theta)\cos(-\theta)=2\sin(2\theta)\cos\theta$

17. $\cos\left(\dfrac{\theta}{2}\right)-\cos\left(\dfrac{3\theta}{2}\right)$

$=-2\sin\left(\dfrac{\left(\dfrac{\theta}{2}+\dfrac{3\theta}{2}\right)}{2}\right)\sin\left(\dfrac{\left(\dfrac{\theta}{2}-\dfrac{3\theta}{2}\right)}{2}\right)=-2\sin\theta\sin\left(-\dfrac{\theta}{2}\right)=-2\sin\theta\left(-\sin\left(\dfrac{\theta}{2}\right)\right)=2\sin\theta\sin\left(\dfrac{\theta}{2}\right)$

19. $\dfrac{\sin\theta+\sin(3\theta)}{2\sin(2\theta)}=\dfrac{2\sin(2\theta)\cos(-\theta)}{2\sin(2\theta)}=\cos(-\theta)=\cos\theta$

21. $\dfrac{\sin(4\theta)+\sin(2\theta)}{\cos(4\theta)+\cos(2\theta)}=\dfrac{2\sin(3\theta)\cos\theta}{2\cos(3\theta)\cos\theta}=\dfrac{\sin(3\theta)}{\cos(3\theta)}=\tan(3\theta)$

23. $\dfrac{\cos\theta-\cos(3\theta)}{\sin\theta+\sin(3\theta)}=\dfrac{-2\sin(2\theta)\sin(-\theta)}{2\sin(2\theta)\cos(-\theta)}=\dfrac{-(-\sin\theta)}{\cos\theta}=\tan\theta$

25. $\sin\theta[\sin\theta+\sin(3\theta)]=\sin\theta[2\sin(2\theta)\cos(-\theta)]=\cos\theta[2\sin(2\theta)\sin\theta]$

$=\cos\theta\left[2\cdot\dfrac{1}{2}(\cos\theta-\cos(3\theta))\right]=\cos\theta(\cos\theta-\cos(3\theta))$

27. $\dfrac{\sin(4\theta)+\sin(8\theta)}{\cos(4\theta)+\cos(8\theta)}=\dfrac{2\sin(6\theta)\cos(-2\theta)}{2\cos(6\theta)\cos(-2\theta)}=\dfrac{\sin(6\theta)}{\cos(6\theta)}=\tan(6\theta)$

29. $\dfrac{\sin(4\theta)+\sin(8\theta)}{\sin(4\theta)-\sin(8\theta)}=\dfrac{2\sin(6\theta)\cos(-2\theta)}{2\sin(-2\theta)\cos(6\theta)}=\dfrac{\sin(6\theta)\cos(2\theta)}{-\sin(2\theta)\cos(6\theta)}$

$=-\tan(6\theta)\cot(2\theta)=-\dfrac{\tan(6\theta)}{\tan(2\theta)}$

31. $\dfrac{\sin\alpha+\sin\beta}{\sin\alpha-\sin\beta}=\dfrac{2\sin\left(\dfrac{\alpha+\beta}{2}\right)\cos\left(\dfrac{\alpha-\beta}{2}\right)}{2\sin\left(\dfrac{\alpha-\beta}{2}\right)\cos\left(\dfrac{\alpha+\beta}{2}\right)}=\tan\left(\dfrac{\alpha+\beta}{2}\right)\cot\left(\dfrac{\alpha-\beta}{2}\right)$

33. $\dfrac{\sin\alpha+\sin\beta}{\cos\alpha+\cos\beta}=\dfrac{2\sin\left(\dfrac{\alpha+\beta}{2}\right)\cos\left(\dfrac{\alpha-\beta}{2}\right)}{2\cos\left(\dfrac{\alpha+\beta}{2}\right)\cos\left(\dfrac{\alpha-\beta}{2}\right)}=\tan\left(\dfrac{\alpha+\beta}{2}\right)$

35. $1 + \cos(2\theta) + \cos(4\theta) + \cos(6\theta) = \cos 0 + \cos(6\theta) + \cos(2\theta) + \cos(4\theta)$

$\qquad = 2\cos(3\theta)\cos(-3\theta) + 2\cos(3\theta)\cos(-\theta) = 2\cos^2(3\theta) + 2\cos(3\theta)\cos\theta$

$\qquad = 2\cos(3\theta)(\cos(3\theta) + \cos\theta) = 2\cos(3\theta)2\cos(2\theta)\cos\theta$

$\qquad = 4\cos\theta\cos(2\theta)\cos(3\theta)$

37. (a) $y = \sin\left[2\pi(852)t\right] + \sin\left[2\pi(1209)t\right]$

$\qquad = 2\sin\left(\dfrac{2\pi(852)t + 2\pi(1209)t}{2}\right)\cos\left(\dfrac{2\pi(852)t - 2\pi(1209)t}{2}\right)$

$\qquad = 2\sin(2061\pi t)\cos(357\pi t)$

   (b)   The maximum value of y is 2.

   (c)

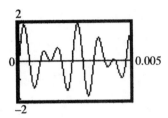

39. $\sin(2\alpha) + \sin(2\beta) + \sin(2\gamma)$

$\qquad = 2\sin\left(\dfrac{2\alpha + 2\beta}{2}\right)\cos\left(\dfrac{2\alpha - 2\beta}{2}\right) + \sin(2\gamma)$

$\qquad = 2\sin(\alpha + \beta)\cos(\alpha - \beta) + 2\sin\gamma\cos\gamma$

$\qquad = 2\sin(\pi - \gamma)\cos(\alpha - \beta) + 2\sin\gamma\cos\gamma$

$\qquad = 2\sin\gamma\cos(\alpha - \beta) + 2\sin\gamma\cos\gamma = 2\sin\gamma\left[\cos(\alpha - \beta) + \cos\gamma\right]$

$\qquad = 2\sin\gamma\left(2\cos\left(\dfrac{\alpha - \beta + \gamma}{2}\right)\cos\left(\dfrac{\alpha - \beta - \gamma}{2}\right)\right)$

$\qquad = 4\sin\gamma\cos\left(\dfrac{\pi}{2} - \beta\right)\cos\left(\alpha - \dfrac{\pi}{2}\right) = 4\sin\gamma\sin\beta\sin\alpha = 4\sin\alpha\sin\beta\sin\gamma$

41. Add the two sum formulas for $\sin(\alpha + \beta)$ and $\sin(\alpha - \beta)$ and solve:

$\qquad \sin(\alpha + \beta) = \sin\alpha\cos\beta + \cos\alpha\sin\beta$

$\qquad \sin(\alpha - \beta) = \sin\alpha\cos\beta - \cos\alpha\sin\beta$

$\qquad \sin(\alpha + \beta) + \sin(\alpha - \beta) = 2\sin\alpha\cos\beta$

$\qquad \sin\alpha\cos\beta = \dfrac{1}{2}\cdot\left[\sin(\alpha + \beta) + \sin(\alpha - \beta)\right]$

43. $2\cos\left(\dfrac{\alpha + \beta}{2}\right)\cos\left(\dfrac{\alpha - \beta}{2}\right) = 2\cdot\dfrac{1}{2}\left[\cos\left(\dfrac{\alpha + \beta}{2} - \dfrac{\alpha - \beta}{2}\right) + \cos\left(\dfrac{\alpha + \beta}{2} + \dfrac{\alpha - \beta}{2}\right)\right]$

$\qquad = \cos\left(\dfrac{2\beta}{2}\right) + \cos\left(\dfrac{2\alpha}{2}\right) = \cos\beta + \cos\alpha$

Therefore, $\cos\alpha + \cos\beta = 2\cos\left(\dfrac{\alpha + \beta}{2}\right)\cos\left(\dfrac{\alpha - \beta}{2}\right)$

Chapter 6

Analytic Trigonometry

6.7   Trigonometric Equations (I)

1. $\sin\theta = \dfrac{1}{2}$

$\theta = \dfrac{\pi}{6} + 2k\pi$ or $\theta = \dfrac{5\pi}{6} + 2k\pi$, where k is any integer

Six solutions are $\theta = \dfrac{\pi}{6}, \dfrac{5\pi}{6}, \dfrac{13\pi}{6}, \dfrac{17\pi}{6}, \dfrac{25\pi}{6}, \dfrac{29\pi}{6}$

3. $\tan\theta = -\dfrac{\sqrt{3}}{3}$

$\theta = \dfrac{5\pi}{6} + k\pi$, where k is any integer

Six solutions are $\theta = \dfrac{5\pi}{6}, \dfrac{11\pi}{6}, \dfrac{17\pi}{6}, \dfrac{23\pi}{6}, \dfrac{29\pi}{6}, \dfrac{35\pi}{6}$

5. $\cos\theta = 0$

$\theta = \dfrac{\pi}{2} + 2k\pi$ or $\theta = \dfrac{3\pi}{2} + 2k\pi$, where k is any integer

Six solutions are $\theta = \dfrac{\pi}{2}, \dfrac{3\pi}{2}, \dfrac{5\pi}{2}, \dfrac{7\pi}{2}, \dfrac{9\pi}{2}, \dfrac{11\pi}{2}$

7. $\cos(2\theta) = -\dfrac{1}{2}$

$2\theta = \dfrac{2\pi}{3} + 2k\pi \quad \rightarrow \quad \theta = \dfrac{\pi}{3} + k\pi$, where k is any integer

$2\theta = \dfrac{4\pi}{3} + 2k\pi \quad \rightarrow \quad \theta = \dfrac{2\pi}{3} + k\pi$, where k is any integer

Six solutions are $\theta = \dfrac{\pi}{3}, \dfrac{2\pi}{3}, \dfrac{4\pi}{3}, \dfrac{5\pi}{3}, \dfrac{7\pi}{3}, \dfrac{8\pi}{3}$

9. $\sin\left(\dfrac{\theta}{2}\right) = -\dfrac{\sqrt{3}}{2}$

$\dfrac{\theta}{2} = \dfrac{4\pi}{3} + 2k\pi \quad \rightarrow \quad \theta = \dfrac{8\pi}{3} + 4k\pi$, where k is any integer

$\dfrac{\theta}{2} = \dfrac{5\pi}{3} + 2k\pi \quad \rightarrow \quad \theta = \dfrac{10\pi}{3} + 4k\pi$, where k is any integer

Six solutions are $\theta = \dfrac{8\pi}{3}, \dfrac{10\pi}{3}, \dfrac{20\pi}{3}, \dfrac{22\pi}{3}, \dfrac{32\pi}{3}, \dfrac{34\pi}{3}$

11. $2\sin\theta + 3 = 2$

$2\sin\theta = -1 \rightarrow \sin\theta = -\dfrac{1}{2}$

$\theta = \dfrac{7\pi}{6} + 2k\pi$ or $\theta = \dfrac{11\pi}{6} + 2k\pi$, k is any integer

The solutions on the interval $[0, 2\pi)$ are $\theta = \dfrac{7\pi}{6}, \dfrac{11\pi}{6}$.

13. $4\cos^2\theta = 1$

$\cos^2\theta = \dfrac{1}{4} \rightarrow \cos\theta = \pm\dfrac{1}{2}$

$\theta = \dfrac{\pi}{3} + k\pi$ or $\theta = \dfrac{2\pi}{3} + k\pi$, k is any integer

The solutions on the interval $[0, 2\pi)$ are $\theta = \dfrac{\pi}{3}, \dfrac{2\pi}{3}, \dfrac{4\pi}{3}, \dfrac{5\pi}{3}$.

15. $2\sin^2\theta - 1 = 0$

$2\sin^2\theta = 1 \rightarrow \sin^2\theta = \dfrac{1}{2} \rightarrow \sin\theta = \pm\sqrt{\dfrac{1}{2}} = \pm\dfrac{\sqrt{2}}{2}$

$\theta = \dfrac{\pi}{4} + k\pi$ or $\theta = \dfrac{3\pi}{4} + k\pi$, k is any integer

The solutions on the interval $[0, 2\pi)$ are $\theta = \dfrac{\pi}{4}, \dfrac{3\pi}{4}, \dfrac{5\pi}{4}, \dfrac{7\pi}{4}$.

17. $\sin(3\theta) = -1$

$3\theta = \dfrac{3\pi}{2} + 2k\pi \quad \rightarrow \quad \theta = \dfrac{\pi}{2} + \dfrac{2k\pi}{3}$, where k is any integer

The solutions on the interval $[0, 2\pi)$ are $\theta = \dfrac{\pi}{2}, \dfrac{7\pi}{6}, \dfrac{11\pi}{6}$

19. $\cos(2\theta) = -\dfrac{1}{2}$

$2\theta = \dfrac{2\pi}{3} + 2k\pi \quad \rightarrow \quad \theta = \dfrac{\pi}{3} + k\pi$, where k is any integer

$2\theta = \dfrac{4\pi}{3} + 2k\pi \quad \rightarrow \quad \theta = \dfrac{2\pi}{3} + k\pi$, where k is any integer

The solutions on the interval $[0, 2\pi)$ are $\theta = \dfrac{\pi}{3}, \dfrac{2\pi}{3}, \dfrac{4\pi}{3}, \dfrac{5\pi}{3}$

21. $\sec\left(\dfrac{3\theta}{2}\right) = -2$

$\dfrac{3\theta}{2} = \dfrac{2\pi}{3} + 2k\pi \quad \rightarrow \quad \theta = \dfrac{4\pi}{9} + \dfrac{4k\pi}{3}$, where k is any integer

$\dfrac{3\theta}{2} = \dfrac{4\pi}{3} + 2k\pi \quad \rightarrow \quad \theta = \dfrac{8\pi}{9} + \dfrac{4k\pi}{3}$, where k is any integer

The solutions on the interval $[0, 2\pi)$ are $\theta = \dfrac{4\pi}{9}, \dfrac{8\pi}{9}, \dfrac{16\pi}{9}$

23. $\cos\left(2\theta - \dfrac{\pi}{2}\right) = -1$

$2\theta - \dfrac{\pi}{2} = \pi + 2k\pi \quad \rightarrow \quad 2\theta = \dfrac{3\pi}{2} + 2k\pi \quad \rightarrow \quad \theta = \dfrac{3\pi}{4} + k\pi, \ k$ is any integer

The solutions on the interval $[0, 2\pi)$ are $\theta = \dfrac{3\pi}{4}, \dfrac{7\pi}{4}$.

25. $\tan\left(\dfrac{\theta}{2} + \dfrac{\pi}{3}\right) = 1$

$\dfrac{\theta}{2} + \dfrac{\pi}{3} = \dfrac{\pi}{4} + k\pi \quad \rightarrow \quad \dfrac{\theta}{2} = -\dfrac{\pi}{12} + k\pi \quad \rightarrow \quad \theta = -\dfrac{\pi}{6} + 2k\pi, \ k$ is any integer

The solutions on the interval $[0, 2\pi)$ are $\theta = \dfrac{11\pi}{6}$.

27. $2\sin\theta + 1 = 0 \quad \rightarrow \quad 2\sin\theta = -1 \quad \rightarrow \quad \sin\theta = -\dfrac{1}{2}$

$\theta = \dfrac{7\pi}{6} + 2k\pi \quad$ or $\quad \theta = \dfrac{11\pi}{6} + 2k\pi, \ k$ is any integer

The solutions on the interval $[0, 2\pi)$ are $\theta = \dfrac{7\pi}{6}, \dfrac{11\pi}{6}$.

29. $\tan\theta + 1 = 0 \quad \rightarrow \quad \tan\theta = -1$

$\theta = \dfrac{3\pi}{4} + k\pi, \ k$ is any integer

The solutions on the interval $[0, 2\pi)$ are $\theta = \dfrac{3\pi}{4}, \dfrac{7\pi}{4}$.

31. $4\sec\theta + 6 = -2 \quad \rightarrow \quad 4\sec\theta = -8 \quad \rightarrow \quad \sec\theta = -2$

$\theta = \dfrac{2\pi}{3} + 2k\pi \quad$ or $\quad \theta = \dfrac{4\pi}{3} + 2k\pi, \ k$ is any integer

The solutions on the interval $[0, 2\pi)$ are $\theta = \dfrac{2\pi}{3}, \dfrac{4\pi}{3}$.

33. $3\sqrt{2}\cos\theta + 2 = -1 \quad \rightarrow \quad 3\sqrt{2}\cos\theta = -3 \quad \rightarrow \quad \cos\theta = -\dfrac{1}{\sqrt{2}} = -\dfrac{\sqrt{2}}{2}$

$\theta = \dfrac{3\pi}{4} + 2k\pi \quad$ or $\quad \theta = \dfrac{5\pi}{4} + 2k\pi, \ k$ is any integer

The solutions on the interval $[0, 2\pi)$ are $\theta = \dfrac{3\pi}{4}, \dfrac{5\pi}{4}$.

35. $\sin\theta = 0.4$

$\theta \approx 0.4115168 \ $ or $\ \theta \approx \pi - 0.4115168 \approx 2.7300758$
$\theta \approx 0.41, 2.73$

37. $\tan\theta = 5$

$\theta \approx 1.3734008 \ $ or $\ \theta \approx \pi + 1.3734008 \approx 4.5149934$
$\theta \approx 1.37, 4.51$

39. $\cos\theta = -0.9$

$\theta \approx 2.6905658 \ $ or $\ \theta \approx 2\pi - 2.6905658 \approx 3.5926195$
$\theta \approx 2.69, 3.59$

41. $\sec\theta = -4 \;\;\rightarrow\;\; \cos\theta = -\dfrac{1}{4}$

$\theta \approx 1.8234766$ or $\theta \approx 2\pi - 1.8234766 \approx 4.4597087$
$\theta \approx 1.82,\, 4.46$

43. Use Snell's Law to solve:

$$\frac{\sin(40°)}{\sin\theta_2} = 1.33 \rightarrow \sin(40°) = 1.33\sin\theta_2 \rightarrow \sin\theta_2 = \frac{\sin(40°)}{1.33} \approx 0.4833$$

$$\theta_2 = \sin^{-1}(0.4833) \approx 28.9°$$

45. Calculate the index of refraction for each:

$\theta_1 = 10°,\; \theta_2 = 7°45' = 7.75°$ $\qquad \dfrac{\sin\theta_1}{\sin\theta_2} = \dfrac{\sin(10°)}{\sin(7.75°)} \approx 1.2877$

$\theta_1 = 20°,\; \theta_2 = 15°30' = 15.5°$ $\qquad \dfrac{\sin\theta_1}{\sin\theta_2} = \dfrac{\sin(20°)}{\sin(15.5°)} \approx 1.2798$

$\theta_1 = 30°,\; \theta_2 = 22°30' = 22.5°$ $\qquad \dfrac{\sin\theta_1}{\sin\theta_2} = \dfrac{\sin(30°)}{\sin(22.5°)} \approx 1.3066$

$\theta_1 = 40°,\; \theta_2 = 29°0' = 29°$ $\qquad \dfrac{\sin\theta_1}{\sin\theta_2} = \dfrac{\sin(40°)}{\sin(29°)} \approx 1.3259$

$\theta_1 = 50°,\; \theta_2 = 35°0' = 35°$ $\qquad \dfrac{\sin\theta_1}{\sin\theta_2} = \dfrac{\sin(50°)}{\sin(35°)} \approx 1.3356$

$\theta_1 = 60°,\; \theta_2 = 40°30' = 40.5°$ $\qquad \dfrac{\sin\theta_1}{\sin\theta_2} = \dfrac{\sin(60°)}{\sin(40.5°)} \approx 1.3335$

$\theta_1 = 70°,\; \theta_2 = 45°30' = 45.5°$ $\qquad \dfrac{\sin\theta_1}{\sin\theta_2} = \dfrac{\sin(70°)}{\sin(45.5°)} \approx 1.3175$

$\theta_1 = 80°,\; \theta_2 = 50°0' = 50°$ $\qquad \dfrac{\sin\theta_1}{\sin\theta_2} = \dfrac{\sin(80°)}{\sin(50°)} \approx 1.2856$

The results range from 1.28 to 1.34 and are surprisingly close to Snell's Law.

47. Calculate the index of refraction:

$\theta_1 = 40°,\; \theta_2 = 26°$ $\qquad \dfrac{\sin\theta_1}{\sin\theta_2} = \dfrac{\sin(40°)}{\sin(26°)} \approx 1.47$

49. If θ is the original angle of incidence and ϕ is the angle of refraction, then $\dfrac{\sin\theta}{\sin\phi} = n_2$. The angle of incidence of the emerging beam is also ϕ, and the index of refraction is $\dfrac{1}{n_2}$. Thus, θ is the angle of refraction of the emerging beam. The two beams are parallel since the original angle of incidence and the angle of refraction of the emerging beam are equal.

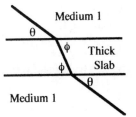

327

Chapter 6

Analytic Trigonometry

6.8 Trigonometric Equations (II)

1. $2\cos^2\theta + \cos\theta = 0 \rightarrow \cos\theta(2\cos\theta + 1) = 0$

   $\cos\theta = 0 \rightarrow \theta = \dfrac{\pi}{2}, \dfrac{3\pi}{2}$

   or $2\cos\theta + 1 = 0$

   $2\cos\theta = -1 \rightarrow \cos\theta = -\dfrac{1}{2} \rightarrow \theta = \dfrac{2\pi}{3}, \dfrac{4\pi}{3}$

3. $2\sin^2\theta - \sin\theta - 1 = 0 \rightarrow (2\sin\theta + 1)(\sin\theta - 1) = 0$

   $2\sin\theta + 1 = 0 \rightarrow 2\sin\theta = -1 \rightarrow \sin\theta = -\dfrac{1}{2} \rightarrow \theta = \dfrac{7\pi}{6}, \dfrac{11\pi}{6}$

   or $\sin\theta - 1 = 0 \rightarrow \sin\theta = 1 \rightarrow \theta = \dfrac{\pi}{2}$

5. $(\tan\theta - 1)(\sec\theta - 1) = 0$

   $\tan\theta - 1 = 0 \rightarrow \tan\theta = 1 \rightarrow \theta = \dfrac{\pi}{4}, \dfrac{5\pi}{4}$

   or $\sec\theta - 1 = 0 \rightarrow \sec\theta = 1 \rightarrow \theta = 0$

7. $\sin^2\theta - \cos^2\theta = 1 + \cos\theta$

   $(1 - \cos^2\theta) - \cos^2\theta = 1 + \cos\theta$

   $1 - 2\cos^2\theta = 1 + \cos\theta$

   $2\cos^2\theta + \cos\theta = 0$

   $(\cos\theta)(2\cos\theta + 1) = 0$

   $\cos\theta = 0 \rightarrow \theta = \dfrac{\pi}{2}, \dfrac{3\pi}{2}$

   or $2\cos\theta + 1 = 0$

   $\cos\theta = -\dfrac{1}{2} \rightarrow \theta = \dfrac{2\pi}{3}, \dfrac{4\pi}{3}$

9. $\sin^2\theta = 6(\cos\theta + 1)$

   $1 - \cos^2\theta = 6(\cos\theta + 1) \rightarrow 1 - \cos^2\theta = 6\cos\theta + 6$

   $\cos^2\theta + 6\cos\theta + 5 = 0 \rightarrow (\cos\theta + 5)(\cos\theta + 1) = 0$

   $\cos\theta + 5 = 0 \rightarrow \cos\theta = -5$, which is impossible

   or $\cos\theta + 1 \rightarrow \cos\theta = -1 \rightarrow \theta = \pi$

   the solution is $\theta = \pi$.

11. $\cos(2\theta)+6\sin^2\theta=4$

$1-2\sin^2\theta+6\sin^2\theta=4\rightarrow4\sin^2\theta=3\rightarrow\sin^2\theta=\dfrac{3}{4}\rightarrow\sin\theta=\pm\dfrac{\sqrt{3}}{2}$

$\theta=\dfrac{\pi}{3},\dfrac{2\pi}{3},\dfrac{4\pi}{3},\dfrac{5\pi}{3}$

13. $\cos\theta=\sin\theta$

$\dfrac{\sin\theta}{\cos\theta}=1\rightarrow\tan\theta=1$

$\theta=\dfrac{\pi}{4},\dfrac{5\pi}{4}$

15. $\tan\theta=2\sin\theta$

$\dfrac{\sin\theta}{\cos\theta}=2\sin\theta$

$\sin\theta=2\sin\theta\cos\theta$

$0=2\sin\theta\cos\theta-\sin\theta$

$0=\sin\theta(2\cos\theta-1)$

$2\cos\theta-1=0\rightarrow\cos\theta=\dfrac{1}{2}\rightarrow\theta=\dfrac{\pi}{3},\dfrac{5\pi}{3}$

or $\sin\theta=0\rightarrow\theta=0,\pi$

17. $\sin\theta=\csc\theta$

$\sin\theta=\dfrac{1}{\sin\theta}$

$\sin^2\theta=1\rightarrow\sin\theta=\pm1$

$\theta=\dfrac{\pi}{2},\dfrac{3\pi}{2}$

19. $\cos(2\theta)=\cos\theta$

$2\cos^2\theta-1=\cos\theta\rightarrow2\cos^2\theta-\cos\theta-1=0\rightarrow(2\cos\theta+1)(\cos\theta-1)=0$

$2\cos\theta+1=0\rightarrow\cos\theta=-\dfrac{1}{2}\rightarrow\theta=\dfrac{2\pi}{3},\dfrac{4\pi}{3}$

or $\cos\theta-1=0\rightarrow\cos\theta=1\rightarrow\theta=0$

21. $\sin(2\theta)+\sin(4\theta)=0$

$\sin(2\theta)+2\sin(2\theta)\cos(2\theta)=0$

$\sin(2\theta)(1+2\cos(2\theta))=0$

$1+2\cos(2\theta)=0\rightarrow\cos(2\theta)=-\dfrac{1}{2}\rightarrow2\theta=\dfrac{2\pi}{3}+2k\pi\rightarrow\theta=\dfrac{\pi}{3}+k\pi$

$2\theta=\dfrac{4\pi}{3}+2k\pi\rightarrow\theta=\dfrac{2\pi}{3}+k\pi$

or $\sin(2\theta)=0\rightarrow2\theta=0+2k\pi\rightarrow\theta=k\pi$

$2\theta=\pi+2k\pi\rightarrow\theta=\dfrac{\pi}{2}+k\pi$

$\theta=0,\dfrac{\pi}{3},\dfrac{\pi}{2},\dfrac{2\pi}{3},\pi,\dfrac{4\pi}{3},\dfrac{3\pi}{2},\dfrac{5\pi}{3}$

23. $\cos(4\theta) - \cos(6\theta) = 0$

$\cos(5\theta - \theta) - \cos(5\theta + \theta) = 0$

$-2\sin(5\theta)\sin(-\theta) = 0$

$2\sin(5\theta)\sin\theta = 0$

$\sin(5\theta) = 0 \rightarrow 5\theta = 0 + 2k\pi \quad \rightarrow \quad \theta = \dfrac{2k\pi}{5}$

$5\theta = \pi + 2k\pi \quad \rightarrow \quad \theta = \dfrac{\pi}{5} + \dfrac{2k\pi}{5}$

or $\sin\theta = 0 \rightarrow \theta = 0 + 2k\pi$

$\theta = \pi + 2k\pi$

$\theta = 0, \dfrac{\pi}{5}, \dfrac{2\pi}{5}, \dfrac{3\pi}{5}, \dfrac{4\pi}{5}, \pi, \dfrac{6\pi}{5}, \dfrac{7\pi}{5}, \dfrac{8\pi}{5}, \dfrac{9\pi}{5}$

25. $1 + \sin\theta = 2\cos^2\theta$

$1 + \sin\theta = 2(1 - \sin^2\theta) \rightarrow 1 + \sin\theta = 2 - 2\sin^2\theta$

$2\sin^2\theta + \sin\theta - 1 = 0 \rightarrow (2\sin\theta - 1)(\sin\theta + 1) = 0$

$2\sin\theta - 1 = 0 \rightarrow \sin\theta = \dfrac{1}{2} \rightarrow \theta = \dfrac{\pi}{6}, \dfrac{5\pi}{6}$

or $\sin\theta + 1 = 0 \rightarrow \sin\theta = -1 \rightarrow \theta = \dfrac{3\pi}{2}$

27. $\tan^2\theta = \dfrac{3}{2}\sec\theta$

$\sec^2\theta - 1 = \dfrac{3}{2}\sec\theta \rightarrow 2\sec^2\theta - 2 = 3\sec\theta$

$2\sec^2\theta - 3\sec\theta - 2 = 0 \rightarrow (2\sec\theta + 1)(\sec\theta - 2) = 0$

$2\sec\theta + 1 = 0 \quad \rightarrow \sec\theta = -\dfrac{1}{2}$, which is impossible

or $\sec\theta - 2 = 0 \rightarrow \sec\theta = 2 \rightarrow \theta = \dfrac{\pi}{3}, \dfrac{5\pi}{3}$

the solutions are $\theta = \dfrac{\pi}{3}, \dfrac{5\pi}{3}$

29. $3 - \sin\theta = \cos(2\theta)$

$3 - \sin\theta = 1 - 2\sin^2\theta$

$2\sin^2\theta - \sin\theta + 2 = 0$

This is a quadratic equation in $\sin\theta$. The discriminant is $b^2 - 4ac = 1 - 16 = -15 < 0$.
The equation has no real solutions.

31. $\sec^2\theta + \tan\theta = 0$

$\tan^2\theta + 1 + \tan\theta = 0$

This is a quadratic equation in $\tan\theta$. The discriminant is $b^2 - 4ac = 1 - 4 = -3 < 0$.
The equation has no real solutions.

33. $\sin\theta - \sqrt{3}\cos\theta = 1$
Divide each side by 2:
$$\frac{1}{2}\sin\theta - \frac{\sqrt{3}}{2}\cos\theta = \frac{1}{2}$$
Rewrite in the difference of two angles form where
$$\cos\phi = \frac{1}{2} \text{ and } \sin\phi = \frac{\sqrt{3}}{2} \text{ and } \phi = \frac{\pi}{3}:$$
$$\sin\theta\cos\phi - \cos\theta\sin\phi = \frac{1}{2} \to \sin(\theta - \phi) = \frac{1}{2}$$
$$\theta - \phi = \frac{\pi}{6} \quad \text{or} \quad \theta - \phi = \frac{5\pi}{6}$$
$$\theta - \frac{\pi}{3} = \frac{\pi}{6} \quad \text{or} \quad \theta - \frac{\pi}{3} = \frac{5\pi}{6}$$
$$\theta = \frac{\pi}{2} \quad \text{or} \quad \theta = \frac{7\pi}{6}$$

35. $\tan(2\theta) + 2\sin\theta = 0$
$$\frac{\sin(2\theta)}{\cos(2\theta)} + 2\sin\theta = 0$$
$$\frac{\sin 2\theta + 2\sin\theta\cos 2\theta}{\cos 2\theta} = 0$$
$$2\sin\theta\cos\theta + 2\sin\theta(2\cos^2\theta - 1) = 0$$
$$2\sin\theta(\cos\theta + 2\cos^2\theta - 1) = 0$$
$$2\sin\theta(2\cos^2\theta + \cos\theta - 1) = 0$$
$$2\sin\theta(2\cos\theta - 1)(\cos\theta + 1) = 0$$
$$2\cos\theta - 1 = 0 \to \cos\theta = \frac{1}{2} \to \theta = \frac{\pi}{3}, \frac{5\pi}{3}$$
$$\text{or} \quad 2\sin\theta = 0 \to \sin\theta = 0 \to \theta = 0, \pi$$
$$\text{or} \quad \cos\theta + 1 = 0 \to \cos\theta = -1 \to \theta = \pi$$

the solutions are $\theta = 0, \frac{\pi}{3}, \pi, \frac{5\pi}{3}$

37. $\sin\theta + \cos\theta = \sqrt{2}$
Divide each side by $\sqrt{2}$: $\qquad \frac{1}{\sqrt{2}}\sin\theta + \frac{1}{\sqrt{2}}\cos\theta = 1$

Rewrite in the sum of two angles form where $\cos\phi = \frac{1}{\sqrt{2}}$ and $\sin\phi = \frac{1}{\sqrt{2}}$ and $\phi = \frac{\pi}{4}:$
$$\sin\theta\cos\phi + \cos\theta\sin\phi = 1 \to \sin(\theta + \phi) = 1$$
$$\theta + \phi = \frac{\pi}{2}$$
$$\theta + \frac{\pi}{4} = \frac{\pi}{2} \to \theta = \frac{\pi}{4}$$

39. Use INTERSECT to solve:

$x \approx -1.29, 0$

41. Use INTERSECT to solve:

$x \approx -2.24, 0, 2.24$

43. Use INTERSECT to solve:

$x \approx -0.82, 0.82$

45. $x + 5\cos x = 0$
     Find the intersection of
     $y_1 = x + 5\cos x$ and $y_2 = 0$:

$x \approx -1.31, 1.98, 3.84$

47. $22x - 17\sin x = 3$
     Find the intersection of
     $y_1 = 22x - 17\sin x$ and $y_2 = 3$:

$x \approx 0.52$

49. $\sin x + \cos x = x$
     Find the intersection of
     $y_1 = \sin x + \cos x$ and $y_2 = x$:

$x \approx 1.26$

51. $x^2 - 2\cos x = 0$
     Find the intersection of
     $y_1 = x^2 - 2\cos x$ and $y_2 = 0$:

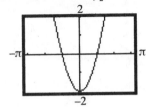

$x \approx -1.02, 1.02$

53. $x^2 - 2\sin 2x = 3x$
     Find the intersection of
     $y_1 = x^2 - 2\sin 2x$ and $y_2 = 3x$:

$x \approx 0, 2.15$

55. $6\sin x - e^x = 2,\ x > 0$
     Find the intersection of
     $y_1 = 6\sin x - e^x$ and $y_2 = 2$:

$x \approx 0.76, 1.35$

57. (a) Solve: $\cos(2\theta) + \cos\theta = 0,\ \ 0° < \theta < 90°$

$$2\cos^2\theta - 1 + \cos\theta = 0 \to 2\cos^2\theta + \cos\theta - 1 = 0$$

$$(2\cos\theta - 1)(\cos\theta + 1) = 0$$

$$2\cos\theta - 1 = 0 \ \to \cos\theta = \frac{1}{2} \to \theta = 60°, 300°$$

$$\text{or}\ \ \cos\theta + 1 = 0 \to \cos\theta = -1 \to \theta = 180°$$

The solution is $60°$.

     (b) Solve: $\cos(2\theta) + \cos\theta = 0,\ \ 0° < \theta < 90°$

$$2\cos\left(\frac{3\theta}{2}\right)\cos\left(\frac{\theta}{2}\right) = 0 \to \cos\left(\frac{3\theta}{2}\right) = 0 \ \text{ or } \ \cos\left(\frac{\theta}{2}\right) = 0$$

$$\frac{3\theta}{2} = 90° \ \ \to \ \ \theta = 60°$$

$$\frac{3\theta}{2} = 270° \ \ \to \ \ \theta = 180°$$

$$\frac{\theta}{2} = 90° \ \ \to \ \ \theta = 180°$$

$$\frac{\theta}{2} = 270° \ \ \to \ \ \theta = 540°$$

The solution is $60°$.

     (c) $A(60°) = 16\sin(60°)(\cos(60°) + 1) = 16 \cdot \dfrac{\sqrt{3}}{2}\left(\dfrac{1}{2} + 1\right) = 12\sqrt{3}\ \text{in}^2 \approx 20.78\ \text{in}^2$

(d) Graph and use the MAXIMUM feature:

The maximum area is approximately 20.78 in^2 when the angle is 60°.

59. Graph:

The first two positive solutions are 2.03 and 4.91.

61. (a) $107 = \dfrac{(34.8)^2 \sin(2\theta)}{9.8}$

$\sin(2\theta) = \dfrac{107(9.8)}{(34.8)^2} \approx 0.8659$

$2\theta = \sin^{-1}(0.8659) \approx 59.98°$ or $120.02°$

$\therefore \ \theta \approx 29.99°$ or $60.01°$

(b) Graph and use the MAXIMUM feature:

The maximum distance is 123.58 meters when the angle is 45°.

(c) Graph:

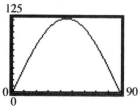

Chapter 6

Analytic Trigonometry

6.R   Chapter Review

1. $\sin^{-1}(1)$

We are finding the angle θ, $-\dfrac{\pi}{2} \le \theta \le \dfrac{\pi}{2}$, whose sine equals 1.

$$\sin\theta = 1 \quad -\dfrac{\pi}{2} \le \theta \le \dfrac{\pi}{2}$$

$$\theta = \dfrac{\pi}{2} \rightarrow \sin^{-1}(1) = \dfrac{\pi}{2}$$

3. $\tan^{-1}(1)$

We are finding the angle θ, $-\dfrac{\pi}{2} < \theta < \dfrac{\pi}{2}$, whose tangent equals 1.

$$\tan\theta = 1 \quad -\dfrac{\pi}{2} < \theta < \dfrac{\pi}{2}$$

$$\theta = \dfrac{\pi}{4} \rightarrow \tan^{-1}(1) = \dfrac{\pi}{4}$$

5. $\cos^{-1}\left(-\dfrac{\sqrt{3}}{2}\right)$

We are finding the angle θ, $0 \le \theta \le \pi$, whose cosine equals $-\dfrac{\sqrt{3}}{2}$.

$$\cos\theta = -\dfrac{\sqrt{3}}{2} \quad 0 \le \theta \le \pi$$

$$\theta = \dfrac{5\pi}{6} \rightarrow \cos^{-1}\left(-\dfrac{\sqrt{3}}{2}\right) = \dfrac{5\pi}{6}$$

7. $\sin\left(\cos^{-1}\left(\dfrac{\sqrt{2}}{2}\right)\right)$

Find the angle θ, $0 \le \theta \le \pi$, whose cosine equals $\dfrac{\sqrt{2}}{2}$.

$$\cos\theta = \dfrac{\sqrt{2}}{2} \quad 0 \le \theta \le \pi$$

$$\theta = \dfrac{\pi}{4} \rightarrow \sin\left(\cos^{-1}\left(\dfrac{\sqrt{2}}{2}\right)\right) = \sin\left(\dfrac{\pi}{4}\right) = \dfrac{\sqrt{2}}{2}$$

9. $\tan\left(\sin^{-1}\left(-\frac{\sqrt{3}}{2}\right)\right)$

Find the angle θ, $-\frac{\pi}{2} \le \theta \le \frac{\pi}{2}$, whose sine equals $-\frac{\sqrt{3}}{2}$.

$$\sin\theta = -\frac{\sqrt{3}}{2} \qquad -\frac{\pi}{2} \le \theta \le \frac{\pi}{2}$$

$$\theta = -\frac{\pi}{3} \to \tan\left(\sin^{-1}\left(-\frac{\sqrt{3}}{2}\right)\right) = \tan\left(-\frac{\pi}{3}\right) = -\sqrt{3}$$

11. $\sec\left(\tan^{-1}\left(\frac{\sqrt{3}}{3}\right)\right)$

Find the angle θ, $-\frac{\pi}{2} < \theta < \frac{\pi}{2}$, whose tangent is $\frac{\sqrt{3}}{3}$

$$\tan\theta = \frac{\sqrt{3}}{3}, \qquad -\frac{\pi}{2} < \theta < \frac{\pi}{2}$$

$$\theta = \frac{\pi}{6} \to \sec\left(\tan^{-1}\left(\frac{\sqrt{3}}{3}\right)\right) = \sec\left(\frac{\pi}{6}\right) = \frac{2\sqrt{3}}{3}$$

13. $\sin\left(\tan^{-1}\left(\frac{3}{4}\right)\right)$

Since $\tan\theta = \frac{3}{4}$, $-\frac{\pi}{2} < \theta < \frac{\pi}{2}$, let $x = 4$ and $y = 3$. Solve for r:

$$16 + 9 = r^2 \to r^2 = 25 \to r = 5$$
θ is in quadrant I.
$$\sin\left(\tan^{-1}\left(\frac{3}{4}\right)\right) = \sin\theta = \frac{y}{r} = \frac{3}{5}$$

15. $\tan\left(\sin^{-1}\left(-\frac{4}{5}\right)\right)$

Since $\sin\theta = -\frac{4}{5}$, $-\frac{\pi}{2} \le \theta \le \frac{\pi}{2}$, let $y = -4$ and $r = 5$. Solve for x:

$$x^2 + 16 = 25 \to x^2 = 9 \to x = \pm3$$
Since θ is in quadrant IV, $x = 3$.
$$\tan\left(\sin^{-1}\left(-\frac{4}{5}\right)\right) = \tan\theta = \frac{y}{x} = \frac{-4}{3} = -\frac{4}{3}$$

17. $\sin^{-1}\left(\cos\left(\frac{2\pi}{3}\right)\right) = \sin^{-1}\left(-\frac{1}{2}\right) = -\frac{\pi}{6}$

19. $\tan^{-1}\left(\tan\left(\frac{7\pi}{4}\right)\right) = \tan^{-1}(-1) = -\frac{\pi}{4}$

21. $\tan\theta\cot\theta - \sin^2\theta = \tan\theta\cdot\dfrac{1}{\tan\theta} - \sin^2\theta = 1 - \sin^2\theta = \cos^2\theta$

23. $\cos^2\theta(1+\tan^2\theta) = \cos^2\theta\cdot\sec^2\theta = \cos^2\theta\cdot\dfrac{1}{\cos^2\theta} = 1$

25. $4\cos^2\theta + 3\sin^2\theta = \cos^2\theta + 3\cos^2\theta + 3\sin^2\theta = \cos^2\theta + 3(\cos^2\theta + \sin^2\theta)$
$$= \cos^2\theta + 3\cdot 1 = \cos^2\theta + 3 = 3 + \cos^2\theta$$

27. $\dfrac{1-\cos\theta}{\sin\theta} + \dfrac{\sin\theta}{1-\cos\theta} = \dfrac{(1-\cos\theta)^2 + \sin^2\theta}{\sin\theta(1-\cos\theta)} = \dfrac{1 - 2\cos\theta + \cos^2\theta + \sin^2\theta}{\sin\theta(1-\cos\theta)}$
$$= \dfrac{1 - 2\cos\theta + 1}{\sin\theta(1-\cos\theta)} = \dfrac{2 - 2\cos\theta}{\sin\theta(1-\cos\theta)} = \dfrac{2(1-\cos\theta)}{\sin\theta(1-\cos\theta)} = \dfrac{2}{\sin\theta} = 2\csc\theta$$

29. $\dfrac{\cos\theta}{\cos\theta - \sin\theta} = \dfrac{\cos\theta}{\cos\theta - \sin\theta}\cdot\dfrac{\left(\dfrac{1}{\cos\theta}\right)}{\left(\dfrac{1}{\cos\theta}\right)} = \dfrac{1}{\left(1 - \dfrac{\sin\theta}{\cos\theta}\right)} = \dfrac{1}{1 - \tan\theta}$

31. $\dfrac{\csc\theta}{1+\csc\theta} = \dfrac{\left(\dfrac{1}{\sin\theta}\right)}{\left(1 + \dfrac{1}{\sin\theta}\right)} = \dfrac{\left(\dfrac{1}{\sin\theta}\right)}{\left(\dfrac{\sin\theta + 1}{\sin\theta}\right)} = \left(\dfrac{1}{1+\sin\theta}\right)\cdot\left(\dfrac{1-\sin\theta}{1-\sin\theta}\right) = \dfrac{1-\sin\theta}{1-\sin^2\theta} = \dfrac{1-\sin\theta}{\cos^2\theta}$

33. $\csc\theta - \sin\theta = \dfrac{1}{\sin\theta} - \sin\theta = \dfrac{1-\sin^2\theta}{\sin\theta} = \dfrac{\cos^2\theta}{\sin\theta} = \cos\theta\cdot\dfrac{\cos\theta}{\sin\theta} = \cos\theta\cot\theta$

35. $\dfrac{1-\sin\theta}{\sec\theta} = \cos\theta(1-\sin\theta) = \cos\theta(1-\sin\theta)\cdot\dfrac{1+\sin\theta}{1+\sin\theta} = \dfrac{\cos\theta(1-\sin^2\theta)}{1+\sin\theta}$
$$= \dfrac{\cos\theta(\cos^2\theta)}{1+\sin\theta} = \dfrac{\cos^3\theta}{1+\sin\theta}$$

37. $\cot\theta - \tan\theta = \dfrac{\cos\theta}{\sin\theta} - \dfrac{\sin\theta}{\cos\theta} = \dfrac{\cos^2\theta - \sin^2\theta}{\sin\theta\cos\theta} = \dfrac{1-\sin^2\theta - \sin^2\theta}{\sin\theta\cos\theta} = \dfrac{1 - 2\sin^2\theta}{\sin\theta\cos\theta}$

39. $\dfrac{\cos(\alpha+\beta)}{\cos\alpha\sin\beta} = \dfrac{\cos\alpha\cos\beta - \sin\alpha\sin\beta}{\cos\alpha\sin\beta} = \dfrac{\cos\alpha\cos\beta}{\cos\alpha\sin\beta} - \dfrac{\sin\alpha\sin\beta}{\cos\alpha\sin\beta} = \cot\beta - \tan\alpha$

41. $\dfrac{\cos(\alpha-\beta)}{\cos\alpha\cos\beta} = \dfrac{\cos\alpha\cos\beta + \sin\alpha\sin\beta}{\cos\alpha\cos\beta} = \dfrac{\cos\alpha\cos\beta}{\cos\alpha\cos\beta} + \dfrac{\sin\alpha\sin\beta}{\cos\alpha\cos\beta} = 1 + \tan\alpha\tan\beta$

43. $(1+\cos\theta)\left(\tan\left(\dfrac{\theta}{2}\right)\right) = (1+\cos\theta)\cdot\dfrac{\sin\theta}{1+\cos\theta} = \sin\theta$

45. $2\cot\theta\cot(2\theta) = 2\cdot\dfrac{\cos\theta}{\sin\theta}\cdot\dfrac{\cos 2\theta}{\sin 2\theta} = \dfrac{2\cos\theta(\cos^2\theta - \sin^2\theta)}{\sin\theta 2\sin\theta\cos\theta} = \dfrac{\cos^2\theta - \sin^2\theta}{\sin^2\theta}$
$$= \cot^2\theta - 1$$

47. $1 - 8\sin^2\theta\cos^2\theta = 1 - 2(2\sin\theta\cos\theta)^2 = 1 - 2\sin^2(2\theta) = \cos(4\theta)$

49. $\dfrac{\sin(2\theta) + \sin(4\theta)}{\cos(2\theta) + \cos(4\theta)} = \dfrac{2\sin(3\theta)\cos(-\theta)}{2\cos(3\theta)\cos(-\theta)} = \dfrac{\sin(3\theta)}{\cos(3\theta)} = \tan(3\theta)$

51. $\dfrac{\cos(2\theta) - \cos(4\theta)}{\cos(2\theta) + \cos(4\theta)} - \tan\theta\tan(3\theta) = \dfrac{-2\sin(3\theta)\sin(-\theta)}{2\cos(3\theta)\cos(-\theta)} - \tan\theta\tan(3\theta)$

$$= \dfrac{2\sin(3\theta)\sin\theta}{2\cos(3\theta)\cos\theta} - \tan\theta\tan(3\theta) = \tan(3\theta)\tan\theta - \tan\theta\tan(3\theta) = 0$$

53. $\sin(165°) = \sin(120° + 45°) = \sin(120°)\cos(45°) + \cos(120°)\sin(45°)$

$$= \left(\dfrac{\sqrt{3}}{2}\right)\cdot\left(\dfrac{\sqrt{2}}{2}\right) + \left(-\dfrac{1}{2}\right)\cdot\left(\dfrac{\sqrt{2}}{2}\right) = \dfrac{1}{4}\left(\sqrt{6} - \sqrt{2}\right)$$

55. $\cos\left(\dfrac{5\pi}{12}\right) = \cos\left(\dfrac{3\pi}{12} + \dfrac{2\pi}{12}\right) = \cos\left(\dfrac{\pi}{4}\right)\cos\left(\dfrac{\pi}{6}\right) - \sin\left(\dfrac{\pi}{4}\right)\sin\left(\dfrac{\pi}{6}\right) = \dfrac{\sqrt{2}}{2}\cdot\dfrac{\sqrt{3}}{2} - \dfrac{\sqrt{2}}{2}\cdot\dfrac{1}{2}$

$$= \dfrac{1}{4}\left(\sqrt{6} - \sqrt{2}\right)$$

57. $\cos(80°)\cos(20°) + \sin(80°)\sin(20°) = \cos(80° - 20°) = \cos(60°) = \dfrac{1}{2}$

59. $\tan\left(\dfrac{\pi}{8}\right) = \tan\left(\dfrac{\left(\dfrac{\pi}{4}\right)}{2}\right) = \sqrt{\dfrac{1 - \cos\left(\dfrac{\pi}{4}\right)}{1 + \cos\left(\dfrac{\pi}{4}\right)}} = \sqrt{\dfrac{1 - \dfrac{\sqrt{2}}{2}}{1 + \dfrac{\sqrt{2}}{2}}} = \sqrt{\left(\dfrac{2 - \sqrt{2}}{2 + \sqrt{2}}\right)\cdot\left(\dfrac{2 - \sqrt{2}}{2 - \sqrt{2}}\right)}$

$$= \sqrt{\dfrac{6 - 4\sqrt{2}}{2}} = \sqrt{3 - 2\sqrt{2}}$$

61. $\sin\alpha = \dfrac{4}{5},\ 0 < \alpha < \dfrac{\pi}{2};\qquad \sin\beta = \dfrac{5}{13},\ \dfrac{\pi}{2} < \beta < \pi$

$\cos\alpha = \dfrac{3}{5},\ \tan\alpha = \dfrac{4}{3},\ \cos\beta = -\dfrac{12}{13},\ \tan\beta = -\dfrac{5}{12},\ \ 0 < \dfrac{\alpha}{2} < \dfrac{\pi}{4},\ \ \dfrac{\pi}{4} < \dfrac{\beta}{2} < \dfrac{\pi}{2}$

(a) $\sin(\alpha + \beta) = \sin\alpha\cos\beta + \cos\alpha\sin\beta = \left(\dfrac{4}{5}\right)\cdot\left(-\dfrac{12}{13}\right) + \left(\dfrac{3}{5}\right)\cdot\left(\dfrac{5}{13}\right) = \dfrac{-48 + 15}{65} = -\dfrac{33}{65}$

(b) $\cos(\alpha + \beta) = \cos\alpha\cos\beta - \sin\alpha\sin\beta = \left(\dfrac{3}{5}\right)\cdot\left(-\dfrac{12}{13}\right) - \left(\dfrac{4}{5}\right)\cdot\left(\dfrac{5}{13}\right) = \dfrac{-36 - 20}{65} = -\dfrac{56}{65}$

(c) $\sin(\alpha - \beta) = \sin\alpha\cos\beta - \cos\alpha\sin\beta = \left(\dfrac{4}{5}\right)\cdot\left(-\dfrac{12}{13}\right) - \left(\dfrac{3}{5}\right)\cdot\left(\dfrac{5}{13}\right) = \dfrac{-48 - 15}{65} = -\dfrac{63}{65}$

(d) $\tan(\alpha + \beta) = \dfrac{\tan\alpha + \tan\beta}{1 - \tan\alpha\tan\beta} = \dfrac{\left(\dfrac{4}{3} + \left(-\dfrac{5}{12}\right)\right)}{\left(1 - \left(\dfrac{4}{3}\right)\cdot\left(-\dfrac{5}{12}\right)\right)} = \dfrac{\left(\dfrac{11}{12}\right)}{\left(\dfrac{14}{9}\right)} = \dfrac{11}{12}\cdot\dfrac{9}{14} = \dfrac{33}{56}$

(e) $\sin(2\alpha) = 2\sin\alpha\cos\alpha = 2\cdot\frac{4}{5}\cdot\frac{3}{5} = \frac{24}{25}$

(f) $\cos(2\beta) = \cos^2\beta - \sin^2\beta = \left(-\frac{12}{13}\right)^2 - \left(\frac{5}{13}\right)^2 = \frac{144}{169} - \frac{25}{169} = \frac{119}{169}$

(g) $\sin\left(\frac{\beta}{2}\right) = \sqrt{\dfrac{1-\cos\beta}{2}} = \sqrt{\dfrac{1-\left(-\frac{12}{13}\right)}{2}} = \sqrt{\dfrac{\left(\frac{25}{13}\right)}{2}} = \sqrt{\dfrac{25}{26}} = \dfrac{5}{\sqrt{26}} = \dfrac{5\sqrt{26}}{26}$

(h) $\cos\left(\frac{\alpha}{2}\right) = \sqrt{\dfrac{1+\cos\alpha}{2}} = \sqrt{\dfrac{\left(1+\frac{3}{5}\right)}{2}} = \sqrt{\dfrac{\left(\frac{8}{5}\right)}{2}} = \sqrt{\dfrac{4}{5}} = \dfrac{2}{\sqrt{5}} = \dfrac{2\sqrt{5}}{5}$

63. $\sin\alpha = -\dfrac{3}{5},\ \pi < \alpha < \dfrac{3\pi}{2};\qquad \cos\beta = \dfrac{12}{13},\ \dfrac{3\pi}{2} < \beta < 2\pi$

$\cos\alpha = -\dfrac{4}{5},\ \tan\alpha = \dfrac{3}{4},\ \sin\beta = -\dfrac{5}{13},\ \tan\beta = -\dfrac{5}{12},\ \dfrac{\pi}{2} < \dfrac{\alpha}{2} < \dfrac{3\pi}{4},\ \dfrac{3\pi}{4} < \dfrac{\beta}{2} < \pi$

(a) $\sin(\alpha+\beta) = \sin\alpha\cos\beta + \cos\alpha\sin\beta = \left(-\dfrac{3}{5}\right)\cdot\left(\dfrac{12}{13}\right) + \left(-\dfrac{4}{5}\right)\cdot\left(-\dfrac{5}{13}\right) = \dfrac{-36+20}{65} = -\dfrac{16}{65}$

(b) $\cos(\alpha+\beta) = \cos\alpha\cos\beta - \sin\alpha\sin\beta = \left(-\dfrac{4}{5}\right)\cdot\left(\dfrac{12}{13}\right) - \left(-\dfrac{3}{5}\right)\cdot\left(-\dfrac{5}{13}\right) = \dfrac{-48-15}{65} = -\dfrac{63}{65}$

(c) $\sin(\alpha-\beta) = \sin\alpha\cos\beta - \cos\alpha\sin\beta = \left(-\dfrac{3}{5}\right)\cdot\left(\dfrac{12}{13}\right) - \left(-\dfrac{4}{5}\right)\cdot\left(-\dfrac{5}{13}\right) = \dfrac{-36-20}{65} = -\dfrac{56}{65}$

(d) $\tan(\alpha+\beta) = \dfrac{\tan\alpha + \tan\beta}{1 - \tan\alpha\tan\beta} = \dfrac{\left(\frac{3}{4}+\left(-\frac{5}{12}\right)\right)}{\left(1-\left(\frac{3}{4}\right)\cdot\left(-\frac{5}{12}\right)\right)} = \dfrac{\left(\frac{1}{3}\right)}{\left(\frac{21}{16}\right)} = \dfrac{1}{3}\cdot\dfrac{16}{21} = \dfrac{16}{63}$

(e) $\sin(2\alpha) = 2\sin\alpha\cos\alpha = 2\cdot\left(-\dfrac{3}{5}\right)\cdot\left(-\dfrac{4}{5}\right) = \dfrac{24}{25}$

(f) $\cos(2\beta) = \cos^2\beta - \sin^2\beta = \left(\dfrac{12}{13}\right)^2 - \left(-\dfrac{5}{13}\right)^2 = \dfrac{144}{169} - \dfrac{25}{169} = \dfrac{119}{169}$

(g) $\sin\left(\dfrac{\beta}{2}\right) = \sqrt{\dfrac{1-\cos\beta}{2}} = \sqrt{\dfrac{\left(1-\left(\frac{12}{13}\right)\right)}{2}} = \sqrt{\dfrac{\left(\frac{1}{13}\right)}{2}} = \sqrt{\dfrac{1}{26}} = \dfrac{1}{\sqrt{26}} = \dfrac{\sqrt{26}}{26}$

(h) $\cos\left(\dfrac{\alpha}{2}\right) = -\sqrt{\dfrac{1+\cos\alpha}{2}} = -\sqrt{\dfrac{\left(1+\left(-\frac{4}{5}\right)\right)}{2}} = -\sqrt{\dfrac{\left(\frac{1}{5}\right)}{2}} = -\sqrt{\dfrac{1}{10}} = -\dfrac{1}{\sqrt{10}} = -\dfrac{\sqrt{10}}{10}$

65. $\tan\alpha = \dfrac{3}{4},\ \pi < \alpha < \dfrac{3\pi}{2};\qquad \tan\beta = \dfrac{12}{5},\ 0 < \beta < \dfrac{\pi}{2}$

$\sin\alpha = -\dfrac{3}{5},\ \cos\alpha = -\dfrac{4}{5},\ \sin\beta = \dfrac{12}{13},\ \cos\beta = \dfrac{5}{13},\ \dfrac{\pi}{2} < \dfrac{\alpha}{2} < \dfrac{3\pi}{4},\ 0 < \dfrac{\beta}{2} < \dfrac{\pi}{4}$

(a) $\sin(\alpha+\beta) = \sin\alpha\cos\beta + \cos\alpha\sin\beta = \left(-\dfrac{3}{5}\right)\cdot\left(\dfrac{5}{13}\right) + \left(-\dfrac{4}{5}\right)\cdot\left(\dfrac{12}{13}\right) = \dfrac{-15-48}{65} = -\dfrac{63}{65}$

(b) $\cos(\alpha+\beta) = \cos\alpha\cos\beta - \sin\alpha\sin\beta = \left(-\dfrac{4}{5}\right)\cdot\left(\dfrac{5}{13}\right) - \left(-\dfrac{3}{5}\right)\cdot\left(\dfrac{12}{13}\right) = \dfrac{-20+36}{65} = \dfrac{16}{65}$

(c) $\sin(\alpha - \beta) = \sin\alpha\cos\beta - \cos\alpha\sin\beta = \left(-\dfrac{3}{5}\right)\cdot\left(\dfrac{5}{13}\right) - \left(-\dfrac{4}{5}\right)\cdot\left(\dfrac{12}{13}\right) = \dfrac{-15+48}{65} = \dfrac{33}{65}$

(d) $\tan(\alpha + \beta) = \dfrac{\tan\alpha + \tan\beta}{1 - \tan\alpha\tan\beta} = \dfrac{\left(\dfrac{3}{4} + \left(\dfrac{12}{5}\right)\right)}{\left(1 - \left(\dfrac{3}{4}\right)\cdot\left(\dfrac{12}{5}\right)\right)} = \dfrac{\left(\dfrac{15+48}{20}\right)}{\left(-\dfrac{4}{5}\right)} = \left(\dfrac{63}{20}\right)\cdot\left(-\dfrac{5}{4}\right) = -\dfrac{63}{16}$

(e) $\sin(2\alpha) = 2\sin\alpha\cos\alpha = 2\cdot\left(-\dfrac{3}{5}\right)\cdot\left(-\dfrac{4}{5}\right) = \dfrac{24}{25}$

(f) $\cos(2\beta) = \cos^2\beta - \sin^2\beta = \left(\dfrac{5}{13}\right)^2 - \left(\dfrac{12}{13}\right)^2 = \dfrac{25}{169} - \dfrac{144}{169} = -\dfrac{119}{169}$

(g) $\sin\left(\dfrac{\beta}{2}\right) = \sqrt{\dfrac{1-\cos\beta}{2}} = \sqrt{\dfrac{\left(1-\left(\dfrac{5}{13}\right)\right)}{2}} = \sqrt{\dfrac{\left(\dfrac{8}{13}\right)}{2}} = \sqrt{\dfrac{4}{13}} = \dfrac{2}{\sqrt{13}} = \dfrac{2\sqrt{13}}{13}$

(h) $\cos\left(\dfrac{\alpha}{2}\right) = -\sqrt{\dfrac{1+\cos\alpha}{2}} = -\sqrt{\dfrac{\left(1+\left(-\dfrac{4}{5}\right)\right)}{2}} = -\sqrt{\dfrac{\left(\dfrac{1}{5}\right)}{2}} = -\sqrt{\dfrac{1}{10}} = -\dfrac{1}{\sqrt{10}} = -\dfrac{\sqrt{10}}{10}$

67. $\sec\alpha = 2, \ -\dfrac{\pi}{2} < \alpha < 0; \qquad \sec\beta = 3, \ \dfrac{3\pi}{2} < \beta < 2\pi$

$\sin\alpha = -\dfrac{\sqrt{3}}{2}, \cos\alpha = \dfrac{1}{2}, \tan\alpha = -\sqrt{3}, \sin\beta = -\dfrac{2\sqrt{2}}{3}, \cos\beta = \dfrac{1}{3}, \tan\beta = -2\sqrt{2},$

$-\dfrac{\pi}{4} < \dfrac{\alpha}{2} < 0, \quad \dfrac{3\pi}{4} < \dfrac{\beta}{2} < \pi$

(a) $\sin(\alpha + \beta) = \sin\alpha\cos\beta + \cos\alpha\sin\beta = \left(-\dfrac{\sqrt{3}}{2}\right)\cdot\left(\dfrac{1}{3}\right) + \left(\dfrac{1}{2}\right)\cdot\left(-\dfrac{2\sqrt{2}}{3}\right) = \dfrac{-\sqrt{3} - 2\sqrt{2}}{6}$

(b) $\cos(\alpha + \beta) = \cos\alpha\cos\beta - \sin\alpha\sin\beta = \left(\dfrac{1}{2}\right)\cdot\left(\dfrac{1}{3}\right) - \left(-\dfrac{\sqrt{3}}{2}\right)\cdot\left(-\dfrac{2\sqrt{2}}{3}\right) = \dfrac{1 - 2\sqrt{6}}{6}$

(c) $\sin(\alpha - \beta) = \sin\alpha\cos\beta - \cos\alpha\sin\beta = \left(-\dfrac{\sqrt{3}}{2}\right)\cdot\left(\dfrac{1}{3}\right) - \left(\dfrac{1}{2}\right)\cdot\left(-\dfrac{2\sqrt{2}}{3}\right) = \dfrac{-\sqrt{3} + 2\sqrt{2}}{6}$

(d) $\tan(\alpha + \beta) = \dfrac{\tan\alpha + \tan\beta}{1 - \tan\alpha\tan\beta} = \dfrac{\left(-\sqrt{3} + \left(-2\sqrt{2}\right)\right)}{\left(1 - \left(-\sqrt{3}\right)\left(-2\sqrt{2}\right)\right)} = \left(\dfrac{-\sqrt{3} - 2\sqrt{2}}{1 - 2\sqrt{6}}\right)\cdot\left(\dfrac{1 + 2\sqrt{6}}{1 + 2\sqrt{6}}\right)$

$= \dfrac{-9\sqrt{3} - 8\sqrt{2}}{-23} = \dfrac{9\sqrt{3} + 8\sqrt{2}}{23}$

(e) $\sin(2\alpha) = 2\sin\alpha\cos\alpha = 2\cdot\left(-\dfrac{\sqrt{3}}{2}\right)\cdot\left(\dfrac{1}{2}\right) = -\dfrac{\sqrt{3}}{2}$

(f) $\cos(2\beta) = \cos^2\beta - \sin^2\beta = \left(\dfrac{1}{3}\right)^2 - \left(\dfrac{-2\sqrt{2}}{3}\right)^2 = \dfrac{1}{9} - \dfrac{8}{9} = -\dfrac{7}{9}$

(g) $\sin\left(\dfrac{\beta}{2}\right) = \sqrt{\dfrac{1-\cos\beta}{2}} = \sqrt{\dfrac{\left(1-\left(\dfrac{1}{3}\right)\right)}{2}} = \sqrt{\dfrac{\left(\dfrac{2}{3}\right)}{2}} = \sqrt{\dfrac{1}{3}} = \dfrac{1}{\sqrt{3}} = \dfrac{\sqrt{3}}{3}$

(h) $\cos\left(\dfrac{\alpha}{2}\right) = \sqrt{\dfrac{1+\cos\alpha}{2}} = \sqrt{\dfrac{\left(1+\dfrac{1}{2}\right)}{2}} = \sqrt{\dfrac{\left(\dfrac{3}{2}\right)}{2}} = \sqrt{\dfrac{3}{4}} = \dfrac{\sqrt{3}}{2}$

69. $\sin\alpha = -\dfrac{2}{3},\ \pi < \alpha < \dfrac{3\pi}{2};\qquad \cos\beta = -\dfrac{2}{3},\ \pi < \beta < \dfrac{3\pi}{2}$

$\cos\alpha = -\dfrac{\sqrt{5}}{3},\ \tan\alpha = \dfrac{2\sqrt{5}}{5},\ \sin\beta = -\dfrac{\sqrt{5}}{3},\ \tan\beta = \dfrac{\sqrt{5}}{2},\ \dfrac{\pi}{2} < \dfrac{\alpha}{2} < \dfrac{3\pi}{4},\ \dfrac{\pi}{2} < \dfrac{\beta}{2} < \dfrac{3\pi}{4}$

(a) $\sin(\alpha+\beta) = \sin\alpha\cos\beta + \cos\alpha\sin\beta = \left(-\dfrac{2}{3}\right)\cdot\left(-\dfrac{2}{3}\right) + \left(-\dfrac{\sqrt{5}}{3}\right)\cdot\left(-\dfrac{\sqrt{5}}{3}\right) = \dfrac{4+5}{9} = 1$

(b) $\cos(\alpha+\beta) = \cos\alpha\cos\beta - \sin\alpha\sin\beta = \left(-\dfrac{\sqrt{5}}{3}\right)\cdot\left(-\dfrac{2}{3}\right) - \left(-\dfrac{2}{3}\right)\cdot\left(-\dfrac{\sqrt{5}}{3}\right) = \dfrac{2\sqrt{5}-2\sqrt{5}}{9} = 0$

(c) $\sin(\alpha-\beta) = \sin\alpha\cos\beta - \cos\alpha\sin\beta = \left(-\dfrac{2}{3}\right)\cdot\left(-\dfrac{2}{3}\right) - \left(-\dfrac{\sqrt{5}}{3}\right)\cdot\left(-\dfrac{\sqrt{5}}{3}\right) = \dfrac{4-5}{9} = -\dfrac{1}{9}$

(d) $\tan(\alpha+\beta) = \dfrac{\tan\alpha + \tan\beta}{1 - \tan\alpha\tan\beta} = \dfrac{\left(\dfrac{2\sqrt{5}}{5} + \dfrac{\sqrt{5}}{2}\right)}{\left(1 - \left(\dfrac{2\sqrt{5}}{5}\right)\cdot\left(\dfrac{\sqrt{5}}{2}\right)\right)} = \dfrac{\left(\dfrac{4\sqrt{5}+5\sqrt{5}}{10}\right)}{\left(\dfrac{10-10}{10}\right)} = \dfrac{\left(\dfrac{9\sqrt{5}}{10}\right)}{0}$ Undefined

(e) $\sin(2\alpha) = 2\sin\alpha\cos\alpha = 2\cdot\left(-\dfrac{2}{3}\right)\cdot\left(-\dfrac{\sqrt{5}}{3}\right) = \dfrac{4\sqrt{5}}{9}$

(f) $\cos(2\beta) = \cos^2\beta - \sin^2\beta = \left(-\dfrac{2}{3}\right)^2 - \left(-\dfrac{\sqrt{5}}{3}\right)^2 = \dfrac{4}{9} - \dfrac{5}{9} = -\dfrac{1}{9}$

(g) $\sin\left(\dfrac{\beta}{2}\right) = \sqrt{\dfrac{1-\cos\beta}{2}} = \sqrt{\dfrac{1-\left(-\dfrac{2}{3}\right)}{2}} = \sqrt{\dfrac{\left(\dfrac{5}{3}\right)}{2}} = \sqrt{\dfrac{5}{6}} = \dfrac{\sqrt{30}}{6}$

(h) $\cos\left(\dfrac{\alpha}{2}\right) = -\sqrt{\dfrac{1+\cos\alpha}{2}} = -\sqrt{\dfrac{1+\left(-\dfrac{\sqrt{5}}{3}\right)}{2}} = -\sqrt{\dfrac{\left(\dfrac{3-\sqrt{5}}{3}\right)}{2}} = -\sqrt{\dfrac{3-\sqrt{5}}{6}} = -\dfrac{\sqrt{18-6\sqrt{5}}}{6}$

71. $\cos\left(\sin^{-1}\left(\dfrac{3}{5}\right) - \cos^{-1}\left(\dfrac{1}{2}\right)\right)$

Let $\alpha = \sin^{-1}\left(\dfrac{3}{5}\right)$ and $\beta = \cos^{-1}\left(\dfrac{1}{2}\right)$. α is in quadrant I; β is in quadrant I.

Then $\sin\alpha = \dfrac{3}{5},\ -\dfrac{\pi}{2} \le \alpha \le \dfrac{\pi}{2}$, and $\cos\beta = \dfrac{1}{2},\ 0 \le \beta \le \pi$.

$\cos\alpha = \sqrt{1 - \sin^2\alpha} = \sqrt{1 - \left(\dfrac{3}{5}\right)^2} = \sqrt{1 - \dfrac{9}{25}} = \sqrt{\dfrac{16}{25}} = \dfrac{4}{5}$

$\sin\beta = \sqrt{1 - \cos^2\beta} = \sqrt{1 - \left(\dfrac{1}{2}\right)^2} = \sqrt{1 - \dfrac{1}{4}} = \sqrt{\dfrac{3}{4}} = \dfrac{\sqrt{3}}{2}$

$$\cos\left(\sin^{-1}\left(\frac{3}{5}\right)-\cos^{-1}\left(\frac{1}{2}\right)\right)=\cos(\alpha-\beta)=\cos\alpha\cos\beta+\sin\alpha\sin\beta$$

$$=\left(\frac{4}{5}\right)\cdot\left(\frac{1}{2}\right)+\left(\frac{3}{5}\right)\cdot\left(\frac{\sqrt{3}}{2}\right)=\frac{4+3\sqrt{3}}{10}$$

73. $\tan\left[\sin^{-1}\left(-\frac{1}{2}\right)-\tan^{-1}\left(\frac{3}{4}\right)\right]$

Let $\alpha=\sin^{-1}\left(-\frac{1}{2}\right)$ and $\beta=\tan^{-1}\left(\frac{3}{4}\right)$. α is in quadrant IV; β is in quadrant I.

Then $\sin\alpha=-\frac{1}{2}$, $-\frac{\pi}{2}\le\alpha\le\frac{\pi}{2}$, and $\tan\beta=\frac{3}{4}$, $-\frac{\pi}{2}<\beta<\frac{\pi}{2}$.

$$\cos\alpha=\sqrt{1-\sin^2\alpha}=\sqrt{1-\left(-\frac{1}{2}\right)^2}=\sqrt{1-\frac{1}{4}}=\sqrt{\frac{3}{4}}=\frac{\sqrt{3}}{2};\quad \tan\alpha=-\frac{1}{\sqrt{3}}=-\frac{\sqrt{3}}{3}$$

$$\tan\left[\sin^{-1}\left(-\frac{1}{2}\right)-\tan^{-1}\left(\frac{3}{4}\right)\right]=\tan(\alpha-\beta)=\frac{\tan\alpha-\tan\beta}{1+\tan\alpha\tan\beta}=\frac{\left(-\frac{\sqrt{3}}{3}-\frac{3}{4}\right)}{\left(1+\left(-\frac{\sqrt{3}}{3}\right)\cdot\left(\frac{3}{4}\right)\right)}$$

$$=\frac{\left(\frac{-4\sqrt{3}-9}{12}\right)}{\left(1-\frac{3\sqrt{3}}{12}\right)}=\left(\frac{-9-4\sqrt{3}}{12-3\sqrt{3}}\right)\cdot\left(\frac{12+3\sqrt{3}}{12+3\sqrt{3}}\right)$$

$$=\frac{-144-75\sqrt{3}}{117}=\frac{-48-25\sqrt{3}}{39}$$

75. $\sin\left[2\cos^{-1}\left(-\frac{3}{5}\right)\right]$

Let $\alpha=\cos^{-1}\left(-\frac{3}{5}\right)$. α is in quadrant II.

Then $\cos\alpha=-\frac{3}{5}$, $0\le\alpha\le\pi$.

$$\sin\alpha=\sqrt{1-\cos^2\alpha}=\sqrt{1-\left(-\frac{3}{5}\right)^2}=\sqrt{1-\frac{9}{25}}=\sqrt{\frac{16}{25}}=\frac{4}{5}$$

$$\sin\left[2\cos^{-1}\left(-\frac{3}{5}\right)\right]=\sin 2\alpha=2\sin\alpha\cos\alpha=2\cdot\left(\frac{4}{5}\right)\cdot\left(-\frac{3}{5}\right)=-\frac{24}{25}$$

77. $\cos\theta=\frac{1}{2}$

$\theta=\frac{\pi}{3}+2k\pi$ or $\theta=\frac{5\pi}{3}+2k\pi$, where k is any integer

The solutions on the interval $[0, 2\pi)$ are $\theta = \dfrac{\pi}{3}, \dfrac{5\pi}{3}$

79. $2\cos\theta + \sqrt{2} = 0 \quad \rightarrow \quad 2\cos\theta = -\sqrt{2} \quad \rightarrow \quad \cos\theta = -\dfrac{\sqrt{2}}{2}$

$\theta = \dfrac{3\pi}{4} + 2k\pi \quad$ or $\quad \theta = \dfrac{5\pi}{4} + 2k\pi, \ k$ is any integer

The solutions on the interval $[0, 2\pi)$ are $\theta = \dfrac{3\pi}{4}, \dfrac{5\pi}{4}$.

81. $\sin(2\theta) + 1 = 0 \quad \rightarrow \quad \sin(2\theta) = -1$

$2\theta = \dfrac{3\pi}{2} + 2k\pi \quad \rightarrow \quad \theta = \dfrac{3\pi}{4} + k\pi$, where k is any integer

The solutions on the interval $[0, 2\pi)$ are $\theta = \dfrac{3\pi}{4}, \dfrac{7\pi}{4}$

83. $\tan(2\theta) = 0$

$2\theta = 0 + k\pi \quad \rightarrow \quad \theta = \dfrac{k\pi}{2}$, where k is any integer

The solutions on the interval $[0, 2\pi)$ are $\theta = 0, \dfrac{\pi}{2}, \pi, \dfrac{3\pi}{2}$.

85. $\sec^2\theta = 4$

$\sec\theta = \pm 2 \rightarrow \cos\theta = \pm\dfrac{1}{2}$

$\theta = \dfrac{\pi}{3} + k\pi$, where k is any integer

$\theta = \dfrac{2\pi}{3} + k\pi$, where k is any integer

The solutions on the interval $[0, 2\pi)$ are $\theta = \dfrac{\pi}{3}, \dfrac{2\pi}{3}, \dfrac{4\pi}{3}, \dfrac{5\pi}{3}$

87. $\sin\theta = \tan\theta$

$\sin\theta = \dfrac{\sin\theta}{\cos\theta} \rightarrow \sin\theta\cos\theta = \sin\theta$

$\sin\theta\cos\theta - \sin\theta = 0 \rightarrow \sin\theta(\cos\theta - 1) = 0$

$\cos\theta - 1 = 0 \rightarrow \cos\theta = 1 \rightarrow \theta = \pi$

or $\sin\theta = 0 \rightarrow \theta = 0$

The solutions on the interval $[0, 2\pi)$ are $\theta = 0, \pi$.

89. $\sin\theta + \sin(2\theta) = 0$

$\sin\theta + 2\sin\theta\cos\theta = 0 \rightarrow \sin\theta(1 + 2\cos\theta) = 0$

$1 + 2\cos\theta = 0 \rightarrow \cos\theta = -\dfrac{1}{2} \quad \rightarrow \theta = \dfrac{2\pi}{3}, \dfrac{4\pi}{3}$

or $\sin\theta = 0 \rightarrow \theta = 0, \pi$

The solutions on the interval $[0, 2\pi)$ are $\theta = 0, \dfrac{2\pi}{3}, \pi, \dfrac{4\pi}{3}$.

91. $\sin(2\theta) - \cos\theta - 2\sin\theta + 1 = 0$
     $2\sin\theta\cos\theta - \cos\theta - 2\sin\theta + 1 = 0 \rightarrow \cos\theta(2\sin\theta - 1) - 1(2\sin\theta - 1) = 0$

$$(2\sin\theta - 1)(\cos\theta - 1) = 0$$

$$\sin\theta = \frac{1}{2} \quad \rightarrow \theta = \frac{\pi}{6}, \frac{5\pi}{6}$$

$$\text{or } \cos\theta = 1 \rightarrow \theta = 0$$

The solutions on the interval $[0, 2\pi)$ are $\theta = 0, \frac{\pi}{6}, \frac{5\pi}{6}$.

93. $2\sin^2\theta - 3\sin\theta + 1 = 0$
     $(2\sin\theta - 1)(\sin\theta - 1) = 0$

$$2\sin\theta - 1 = 0 \rightarrow \sin\theta = \frac{1}{2} \rightarrow \theta = \frac{\pi}{6}, \frac{5\pi}{6}$$

$$\text{or } \sin\theta - 1 = 0 \rightarrow \sin\theta = 1 \rightarrow \theta = \frac{\pi}{2}$$

The solutions on the interval $[0, 2\pi)$ are $\theta = \frac{\pi}{6}, \frac{\pi}{2}, \frac{5\pi}{6}$.

95. $4\sin^2\theta = 1 + 4\cos\theta$

$$4\left(1 - \cos^2\theta\right) = 1 + 4\cos\theta$$

$$4 - 4\cos^2\theta = 1 + 4\cos\theta$$

$$4\cos^2\theta + 4\cos\theta - 3 = 0$$

$$(2\cos\theta - 1)(2\cos\theta + 3) = 0$$

$$2\cos\theta - 1 = 0 \rightarrow \cos\theta = \frac{1}{2} \rightarrow \theta = \frac{\pi}{3}, \frac{5\pi}{3}$$

$$\text{or } 2\cos\theta + 3 = 0 \rightarrow \cos\theta = -\frac{3}{2}, \text{ which is impossible}$$

The solutions on the interval $[0, 2\pi)$ are $\theta = \frac{\pi}{3}, \frac{5\pi}{3}$.

97. $\sin(2\theta) = \sqrt{2}\cos\theta$
     $2\sin\theta\cos\theta = \sqrt{2}\cos\theta$
     $2\sin\theta\cos\theta - \sqrt{2}\cos\theta = 0$
     $\cos\theta\left(2\sin\theta - \sqrt{2}\right) = 0$

$$\cos\theta = 0 \rightarrow \theta = \frac{\pi}{2}, \frac{3\pi}{2}$$

$$\text{or } 2\sin\theta - \sqrt{2} = 0 \rightarrow \sin\theta = \frac{\sqrt{2}}{2} \rightarrow \theta = \frac{\pi}{4}, \frac{3\pi}{4}$$

The solutions on the interval $[0, 2\pi)$ are $\theta = \frac{\pi}{4}, \frac{\pi}{2}, \frac{3\pi}{4}, \frac{3\pi}{2}$.

99. $\sin\theta - \cos\theta = 1$
Divide each side by $\sqrt{2}$:
$$\frac{1}{\sqrt{2}}\sin\theta - \frac{1}{\sqrt{2}}\cos\theta = \frac{1}{\sqrt{2}}$$
Rewrite in the difference of two angles form where
$$\cos\phi = \frac{1}{\sqrt{2}} \text{ and } \sin\phi = \frac{1}{\sqrt{2}} \text{ and } \phi = \frac{\pi}{4}:$$
$$\sin\theta\cos\phi - \cos\theta\sin\phi = \frac{1}{\sqrt{2}}$$
$$\sin(\theta - \phi) = \frac{\sqrt{2}}{2}$$
$$\theta - \phi = \frac{\pi}{4} \text{ or } \theta - \phi = \frac{3\pi}{4}$$
$$\theta - \frac{\pi}{4} = \frac{\pi}{4} \text{ or } \theta - \frac{\pi}{4} = \frac{3\pi}{4}$$
$$\theta = \frac{\pi}{2} \text{ or } \theta = \pi$$

The solutions on the interval $[0, 2\pi)$ are $\theta = \frac{\pi}{2}, \pi.$

101. $2x = 5\cos x$
Find the intersection of
$y_1 = 2x$ and $y_2 = 5\cos x$:

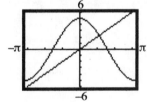

$x \approx 1.11$

103. $2\sin x + 3\cos x = 4x$
Find the intersection of
$y_1 = 2\sin x + 3\cos x$ and $y_2 = 4x$:

$x \approx 0.87$

105. $\sin x = \ln x$
Find the intersection of
$y_1 = \sin x$ and $y_2 = \ln x$:

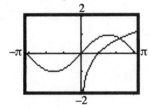

$x \approx 2.22$

Chapter 7

Applications of Trigonometric Functions

7.1   Right Triangle Trigonometry

1.   opposite = 5; adjacent = 12
Find the hypotenuse:
$$5^2 + 12^2 = (\text{hypotenuse})^2$$

$$(\text{hypotenuse})^2 = 25 + 144 = 169 \rightarrow \text{hypotenuse} = 13$$

$$\sin\theta = \frac{\text{opp}}{\text{hyp}} = \frac{5}{13} \qquad \cos\theta = \frac{\text{adj}}{\text{hyp}} = \frac{12}{13} \qquad \tan\theta = \frac{\text{opp}}{\text{adj}} = \frac{5}{12}$$

$$\csc\theta = \frac{\text{hyp}}{\text{opp}} = \frac{13}{5} \qquad \sec\theta = \frac{\text{hyp}}{\text{adj}} = \frac{13}{12} \qquad \cot\theta = \frac{\text{adj}}{\text{opp}} = \frac{12}{5}$$

3.   opposite = 2; adjacent = 3
Find the hypotenuse:
$$2^2 + 3^2 = (\text{hypotenuse})^2$$

$$(\text{hypotenuse})^2 = 4 + 9 = 13 \rightarrow \text{hypotenuse} = \sqrt{13}$$

$$\sin\theta = \frac{\text{opp}}{\text{hyp}} = \frac{2}{\sqrt{13}} = \frac{2\sqrt{13}}{13} \qquad \cos\theta = \frac{\text{adj}}{\text{hyp}} = \frac{3}{\sqrt{13}} = \frac{3\sqrt{13}}{13} \qquad \tan\theta = \frac{\text{opp}}{\text{adj}} = \frac{2}{3}$$

$$\csc\theta = \frac{\text{hyp}}{\text{opp}} = \frac{\sqrt{13}}{2} \qquad \sec\theta = \frac{\text{hyp}}{\text{adj}} = \frac{\sqrt{13}}{3} \qquad \cot\theta = \frac{\text{adj}}{\text{opp}} = \frac{3}{2}$$

5.   adjacent = 2; hypotenuse = 4
Find the opposite side:
$$(\text{opposite})^2 + 2^2 = 4^2$$

$$(\text{opposite})^2 = 16 - 4 = 12 \rightarrow \text{opposite} = \sqrt{12} = 2\sqrt{3}$$

$$\sin\theta = \frac{\text{opp}}{\text{hyp}} = \frac{2\sqrt{3}}{4} = \frac{\sqrt{3}}{2} \qquad \cos\theta = \frac{\text{adj}}{\text{hyp}} = \frac{2}{4} = \frac{1}{2} \qquad \tan\theta = \frac{\text{opp}}{\text{adj}} = \frac{2\sqrt{3}}{2} = \sqrt{3}$$

$$\csc\theta = \frac{\text{hyp}}{\text{opp}} = \frac{4}{2\sqrt{3}} = \frac{2\sqrt{3}}{3} \qquad \sec\theta = \frac{\text{hyp}}{\text{adj}} = \frac{4}{2} = 2 \qquad \cot\theta = \frac{\text{adj}}{\text{opp}} = \frac{2}{2\sqrt{3}} = \frac{\sqrt{3}}{3}$$

7.   opposite = $\sqrt{2}$; adjacent = 1
Find the hypotenuse:
$$\left(\sqrt{2}\right)^2 + 1^2 = (\text{hypotenuse})^2$$

$$(\text{hypotenuse})^2 = 2 + 1 = 3 \rightarrow \text{hypotenuse} = \sqrt{3}$$

$$\sin\theta = \frac{\text{opp}}{\text{hyp}} = \frac{\sqrt{2}}{\sqrt{3}} = \frac{\sqrt{6}}{3} \qquad \cos\theta = \frac{\text{adj}}{\text{hyp}} = \frac{1}{\sqrt{3}} = \frac{\sqrt{3}}{3} \qquad \tan\theta = \frac{\text{opp}}{\text{adj}} = \frac{\sqrt{2}}{1} = \sqrt{2}$$

$$\csc\theta = \frac{\text{hyp}}{\text{opp}} = \frac{\sqrt{3}}{\sqrt{2}} = \frac{\sqrt{6}}{2} \qquad \sec\theta = \frac{\text{hyp}}{\text{adj}} = \frac{\sqrt{3}}{1} = \sqrt{3} \qquad \cot\theta = \frac{\text{adj}}{\text{opp}} = \frac{1}{\sqrt{2}} = \frac{\sqrt{2}}{2}$$

9. opposite = 1; hypotenuse = $\sqrt{5}$
Find the adjacent side:

$$1^2 + (\text{adjacent})^2 = \left(\sqrt{5}\right)^2$$

$$(\text{adjacent})^2 = 5 - 1 = 4 \rightarrow \text{adjacent} = 2$$

$$\sin\theta = \frac{\text{opp}}{\text{hyp}} = \frac{1}{\sqrt{5}} = \frac{\sqrt{5}}{5} \qquad \cos\theta = \frac{\text{adj}}{\text{hyp}} = \frac{2}{\sqrt{5}} = \frac{2\sqrt{5}}{5} \qquad \tan\theta = \frac{\text{opp}}{\text{adj}} = \frac{1}{2}$$

$$\csc\theta = \frac{\text{hyp}}{\text{opp}} = \frac{\sqrt{5}}{1} = \sqrt{5} \qquad \sec\theta = \frac{\text{hyp}}{\text{adj}} = \frac{\sqrt{5}}{2} \qquad \cot\theta = \frac{\text{adj}}{\text{opp}} = \frac{2}{1} = 2$$

11. $\sin(38°) - \cos(52°) = \sin(38°) - \sin(90° - 52°) = \sin(38°) - \sin(38°) = 0$

13. $\dfrac{\cos(10°)}{\sin(80°)} = \dfrac{\sin(90° - 10°)}{\sin(80°)} = \dfrac{\sin(80°)}{\sin(80°)} = 1$

15. $1 - \cos^2(20°) - \cos^2(70°) = \sin^2(20°) - \sin^2(90° - 70°) = \sin^2(20°) - \sin^2(20°) = 0$

17. $\tan(20°) - \dfrac{\cos(70°)}{\cos(20°)} = \tan 20° - \dfrac{\sin(90° - 70°)}{\cos(20°)} = \tan(20°) - \dfrac{\sin(20°)}{\cos(20°)} = \tan(20°) - \tan(20°) = 0$

19. $\cos(35°)\sin(55°) + \sin(35°)\cos(55°) = \sin(55° + 35°) = \sin(90°) = 1$

21. $b = 5, \ \beta = 20°$

$$\sin\beta = \frac{b}{c} \ \rightarrow \ \sin(20°) = \frac{5}{c} \ \rightarrow \ c = \frac{5}{\sin(20°)} \approx \frac{5}{0.3420} \approx 14.62$$

$$\tan\beta = \frac{b}{a} \ \rightarrow \ \tan(20°) = \frac{5}{a} \ \rightarrow \ a = \frac{5}{\tan(20°)} \approx \frac{5}{0.3640} \approx 13.74$$

$$\alpha = 90° - \beta = 90° - 20° = 70°$$

23. $a = 6, \ \beta = 40°$

$$\cos\beta = \frac{a}{c} \ \rightarrow \ \cos(40°) = \frac{6}{c} \ \rightarrow \ c = \frac{6}{\cos(40°)} \approx \frac{6}{0.7660} \approx 7.83$$

$$\tan\beta = \frac{b}{a} \ \rightarrow \ \tan(40°) = \frac{b}{6} \ \rightarrow \ b = 6\tan(40°) \approx 6 \cdot (0.8391) \approx 5.03$$

$$\alpha = 90° - \beta = 90° - 40° = 50°$$

25. $b = 4, \ \alpha = 10°$

$\tan\alpha = \dfrac{a}{b} \ \rightarrow \ \tan(10°) = \dfrac{a}{4} \ \rightarrow \ a = 4\tan(10°) \approx 4 \cdot (0.1763) \approx 0.71$

$\cos\alpha = \dfrac{b}{c} \ \rightarrow \ \cos(10°) = \dfrac{4}{c} \ \rightarrow \ c = \dfrac{4}{\cos(10°)} \approx \dfrac{4}{0.9848} \approx 4.06$

$\beta = 90° - \alpha = 90° - 10° = 80°$

27. $a = 5, \ \alpha = 25°$

$\cot\alpha = \dfrac{b}{a} \ \rightarrow \ \cot(25°) = \dfrac{b}{5} \ \rightarrow \ b = 5\cot(25°) \approx 5 \cdot (2.1445) \approx 10.72$

$\csc\alpha = \dfrac{c}{a} \ \rightarrow \ \csc(25°) = \dfrac{c}{5} \ \rightarrow \ c = 5\csc(25°) \approx 5 \cdot (2.3662) \approx 11.83$

$\beta = 90° - \alpha = 90° - 25° = 65°$

29. $c = 9, \ \beta = 20°$

$\sin\beta = \dfrac{b}{c} \ \rightarrow \ \sin(20°) = \dfrac{b}{9} \ \rightarrow \ b = 9\sin(20°) \approx 9 \cdot (0.3420) \approx 3.08$

$\cos\beta = \dfrac{a}{c} \ \rightarrow \ \cos(20°) = \dfrac{a}{9} \ \rightarrow \ a = 9\cos(20°) \approx 9 \cdot (0.9397) \approx 8.46$

$\alpha = 90° - \alpha = 90° - 20° = 70°$

31. $a = 5, \ b = 3$

$c^2 = a^2 + b^2 = 5^2 + 3^2 = 25 + 9 = 34 \ \rightarrow \ c = \sqrt{34} \approx 5.83$

$\tan\alpha = \dfrac{a}{b} = \dfrac{5}{3} \approx 1.6667 \ \rightarrow \ \alpha \approx 59.0°$

$\beta = 90° - \alpha = 90° - 59.0° = 31.0°$

33. $a = 2, \ c = 5$

$c^2 = a^2 + b^2 \ \rightarrow \ b^2 = c^2 - a^2 = 5^2 - 2^2 = 25 - 4 = 21 \ \rightarrow \ b = \sqrt{21} \approx 4.58$

$\sin\alpha = \dfrac{a}{c} = \dfrac{2}{5} = 0.4000 \ \rightarrow \ \alpha \approx 23.6°$

$\beta = 90° - \alpha = 90° - 23.6° = 66.4°$

35. $c = 8, \ \alpha = 35°$

$\sin(35°) = \dfrac{a}{8} \ \rightarrow \ a = 8\sin(35°) \approx 8(0.5736) \approx 4.59 \text{ in.}$

$\cos(35°) = \dfrac{b}{8} \ \rightarrow \ b = 8\cos(35°) \approx 8(0.8192) \approx 6.55 \text{ in.}$

37. $\alpha = 25°, \ a = 5$

$\sin(25°) = \dfrac{5}{c} \ \rightarrow \ c = \dfrac{5}{\sin(25°)} \approx \dfrac{5}{0.4226} \approx 11.83 \text{ in.}$

$\alpha = 25°,\ b = 5$

$\cos(25°) = \dfrac{5}{c} \ \rightarrow\ c = \dfrac{5}{\cos(25°)} \approx \dfrac{5}{0.9063} \approx 5.52$ in.

39. $c = 5,\ a = 2$

$\sin\alpha = \dfrac{2}{5} = 0.4000 \ \rightarrow\ \alpha \approx 23.6 \ \rightarrow\ \beta = 90° - \alpha = 90° - 23.6° \approx 66.4°$

41. $\tan(35°) = \dfrac{b}{100} \ \rightarrow\ b = 100\tan(35°) \approx 100(0.7002) \approx 70.02$ feet

43. $\tan(85.361°) = \dfrac{a}{80} \ \rightarrow\ a = 80\tan(85.361°) \approx 80(12.3239) \approx 985.91$ feet

45.

$\tan(20°) = \dfrac{50}{x} \ \rightarrow\ x = \dfrac{50}{\tan(20°)} \approx \dfrac{50}{0.3640} \approx 137.37$ meters

47.

$\sin(70°) = \dfrac{x}{22} \ \rightarrow\ x = 22\sin(70°) \approx 22(0.9397) \approx 20.67$ feet

49. opposite side = 10 feet, adjacent side = 35 feet

$\tan\theta = \dfrac{10}{35} \approx 0.2857$

$\theta = \tan^{-1}\left(\dfrac{10}{35}\right) \approx 15.9°$

51. Let h represent the height of Lincoln's face.

$\tan(32°) = \dfrac{b}{800} \ \rightarrow\ b = 800\tan(32°) \approx 800(0.6249) \approx 499.9$

$\tan(35°) = \dfrac{b+h}{800} \ \rightarrow\ b + h = 800\tan(35°) \approx 800(0.7002) \approx 560.2$

$h = (b + h) - b = 560.2 - 499.9 \approx 60.27$ feet

53.

$\sin(21°) = \dfrac{190}{x} \ \rightarrow\ x = \dfrac{190}{\sin(21°)} \approx \dfrac{190}{0.3584} \approx 530.18$ ft.

55.

$$\tan(35.1°) = \frac{x}{789} \quad \rightarrow \quad x = 789\tan(35.1°) \approx 789(0.7028) \approx 554.52 \text{ ft}$$

57.   (a) $\tan(15°) = \dfrac{30}{x} \quad \rightarrow \quad x = \dfrac{30}{\tan(15°)} \approx \dfrac{30}{0.2679} \approx 111.96$ feet

The truck is traveling at 111.96 ft/sec.

$$\frac{111.96 \text{ ft}}{\text{sec}} \cdot \frac{1 \text{ mile}}{5280 \text{ ft}} \cdot \frac{3600 \text{ sec}}{\text{hr}} \approx 76.34 \text{ mi / hr}$$

(b) $\tan(20°) = \dfrac{30}{x} \quad \rightarrow \quad x = \dfrac{30}{\tan(20°)} \approx \dfrac{30}{0.3640} \approx 82.42$ feet

The truck is traveling at 82.42 ft/sec.

$$\frac{82.42 \text{ ft}}{\text{sec}} \cdot \frac{1 \text{ mile}}{5280 \text{ ft}} \cdot \frac{3600 \text{ sec}}{\text{hr}} \approx 56.20 \text{ mi / hr}$$

(c)   A ticket is issued for traveling at a speed of 60 mi/hr or more.

$$\frac{60 \text{ mi}}{\text{hr}} \cdot \frac{5280 \text{ ft}}{\text{mi}} \cdot \frac{1\text{hr}}{3600 \text{ sec}} = 88 \text{ ft / sec.}$$

If $\tan \theta < \dfrac{30}{88}$, the trooper should issue a ticket.

A ticket is issued if $\theta < 18.8°$.

59.   Find angle θ: (see the figure)

$$\tan \theta = \frac{1}{0.5} = 2 \quad \rightarrow \quad \theta \approx 63.4°$$

$$\angle DAC = 40° + 63.4° = 103.4°$$

$$\angle EAC = 103.4° - 90° = 13.4°$$

The bearing the control tower should use is S76.6°E.

61. $\tan \alpha = \dfrac{10 - 6}{15} = \dfrac{4}{15} \quad \rightarrow \quad \alpha = 14.9°$

63.   The length of the highway $= x + y + z$

$$\sin(40°) = \frac{1}{x} \quad \rightarrow \quad x = \frac{1}{\sin(40°)} \approx 1.56 \text{ mi}$$

$$\sin(50°) = \frac{1}{z} \quad \rightarrow \quad z = \frac{1}{\sin(50°)} \approx 1.31 \text{ mi}$$

$$\tan(40°) = \frac{1}{a} \quad \rightarrow \quad a = \frac{1}{\tan(40°)} \approx 1.19 \text{ mi}$$

$$\tan(50°) = \frac{1}{b} \quad \rightarrow \quad b = \frac{1}{\tan(50°)} \approx 0.84 \text{ mi}$$

$a + y + b = 3 \quad \rightarrow \quad y = 3 - a - b = 3 - 1.19 - 0.84 = 0.97$ mi

The length of the highway is: $1.56 + 0.97 + 1.31 = 3.83$ miles.

65. In order to see George's head and feet the camera must be x feet from George.
     Solve:

$$\tan(20°) = \frac{4}{x} \quad \rightarrow \quad x = \frac{4}{\tan(20°)} \approx 10.99 \text{ feet}$$

The camera will need to be moved back 1 foot to see George's feet.

67. $\sin\theta = \frac{y}{1} = y; \qquad \cos\theta = \frac{x}{1} = x$

(a) $A = 2xy = 2\cos\theta\sin\theta$

(b) $2\cos\theta\sin\theta = 2\sin\theta\cos\theta = \sin(2\theta)$

(c) The largest value of the sine function is 1. Solve:
$$\sin(2\theta) = 1$$

$$2\theta = \frac{\pi}{2}$$

$$\theta = \frac{\pi}{4}$$

(d) $x = \cos\left(\frac{\pi}{4}\right) = \frac{\sqrt{2}}{2} \qquad y = \sin\left(\frac{\pi}{4}\right) = \frac{\sqrt{2}}{2}$

The dimensions are $\sqrt{2}$ by $\frac{\sqrt{2}}{2}$.

69. Make the following observations related to the figure:

$$\tan\theta = \frac{1}{z} \quad \rightarrow \quad z = \frac{1}{\tan\theta}$$

$$\sin\theta = \frac{1}{x} \quad \rightarrow \quad x = \frac{1}{\sin\theta}$$

$$2z + y = 8 \quad \rightarrow \quad y = 8 - 2z = 8 - 2\left(\frac{1}{\tan\theta}\right)$$

(a) The time equation is:

$$T = \frac{x}{3} + \frac{y}{8} + \frac{x}{3} = \frac{2x}{3} + \frac{y}{8} = \frac{\left(2 \cdot \frac{1}{\sin\theta}\right)}{3} + \frac{\left(8 - \frac{2}{\tan\theta}\right)}{8} = \frac{2}{3\sin\theta} + 1 - \frac{1}{4\tan\theta}$$

(b) Graph:

The least time occurs when the angle is approximately 68°. The least time is approximately
1.62 hours. The distance and time on the paved road is:

$$8 - \frac{2}{\tan\theta} = 8 - \frac{2}{\tan(68°)} \approx 7.2 \text{ miles}$$

$$t = \frac{7.2}{8} = 0.9 \text{ hours}$$

71. (a) $T = \dfrac{1500}{300} + \dfrac{500}{100} = 5 + 5 = 10$ minutes

(b) $T = \dfrac{500}{100} + \dfrac{1500}{100} = 5 + 15 = 20$ minutes

(c) $\tan\theta = \dfrac{500}{x} \;\rightarrow\; x = \dfrac{500}{\tan\theta}$

$\sin\theta = \dfrac{500}{\text{distance in sand}} \;\rightarrow\; \text{distance in sand} = \dfrac{500}{\sin\theta}$

$T = \dfrac{1500 - x}{300} + \dfrac{\text{distance in sand}}{100} = \dfrac{\left(1500 - \dfrac{500}{\tan\theta}\right)}{300} + \dfrac{\left(\dfrac{500}{\sin\theta}\right)}{100} = 5 - \dfrac{5}{3\tan\theta} + \dfrac{5}{\sin\theta}$

(d) 1000 feet along the paved path leaves an additional 500 feet in the direction of the path, so the angle of the path across the sand is 45°.

$T = 5 - \dfrac{5}{3\tan(45°)} + \dfrac{5}{\sin(45°)} = 5 - \dfrac{5}{3\cdot 1} + \dfrac{5}{\left(\dfrac{\sqrt{2}}{2}\right)} = 5 - \dfrac{5}{3} + \dfrac{10}{\sqrt{2}} \approx 10.4$ minutes

(e) Graph:

The time is least when the angle is approximately 70.53°. The least time is approximately 9.71 minutes. The value of x is:

$x = \dfrac{500}{\tan(70.53°)} \approx 176.8$ feet

73. Let h be the height of the triangle and b be the base of the triangle.

$\sin\theta = \dfrac{h}{a}$

$h = a\sin\theta$

$\cos\theta = \dfrac{\left(\dfrac{1}{2}b\right)}{a}$

$b = 2a\cos\theta$

$A = \dfrac{1}{2}\cdot b\cdot h = \dfrac{1}{2}(2a\cos\theta)(a\sin\theta) = a^2\sin\theta\cos\theta$

75. $\sin\theta = \dfrac{a}{x+a} = \dfrac{b}{x+2a+b}$

$xb + ab = xa + 2a^2 + ab$

$x(b - a) = 2a^2$

$x = \dfrac{2a^2}{b - a}$

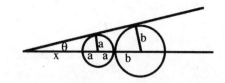

$$\sin\theta = \frac{a}{x+a} = \frac{a}{\left(\dfrac{2a^2}{b-a}+a\right)} = \frac{a}{\left(\dfrac{2a^2+ab-a^2}{b-a}\right)} = \frac{a(b-a)}{a^2+ab} = \frac{a(b-a)}{a(b+a)} = \frac{b-a}{b+a}$$

$$\cos\theta = \sqrt{1-\sin^2\theta} = \sqrt{1-\left(\frac{b-a}{b+a}\right)^2} = \sqrt{1-\frac{b^2-2ab+a^2}{b^2+2ab+a^2}}$$

$$= \sqrt{\frac{b^2+2ab+a^2-b^2+2ab-a^2}{b^2+2ab+a^2}} = \sqrt{\frac{4ab}{(a+b)^2}} = \frac{2\sqrt{ab}}{a+b} = \frac{\sqrt{ab}}{\left(\dfrac{a+b}{2}\right)}$$

77. $\sin\alpha = \dfrac{\sin\alpha}{\cos\alpha}\cdot\cos\alpha = \tan\alpha\cos\alpha = \cos\beta\cos\alpha = \cos\beta\tan\beta = \cos\beta\cdot\dfrac{\sin\beta}{\cos\beta} = \sin\beta$

$$\sin^2\alpha + \cos^2\alpha = 1$$

$$\sin^2\alpha + \tan^2\beta = 1$$

$$\sin^2\alpha + \frac{\sin^2\beta}{\cos^2\beta} = 1$$

$$\sin^2\alpha + \frac{\sin^2\alpha}{1-\sin^2\alpha} = 1$$

$$\left(1-\sin^2\alpha\right)\left(\sin^2\alpha + \frac{\sin^2\alpha}{1-\sin^2\alpha}\right) = 1\left(1-\sin^2\alpha\right)$$

$$\sin^2\alpha - \sin^4\alpha + \sin^2\alpha = 1 - \sin^2\alpha$$

$$\sin^4\alpha - 3\sin^2\alpha + 1 = 0, \text{ this is quadratic in } \sin^2\alpha$$

The quadratic fomula yields: $\sin^2\alpha = \dfrac{3\pm\sqrt{5}}{2}$

$\therefore \ \sin^2\alpha = \dfrac{3-\sqrt{5}}{2} \Rightarrow \sin\alpha = \sqrt{\dfrac{3-\sqrt{5}}{2}}$

Chapter 7

Applications of Trigonometric Functions

7.2 The Law of Sines

1. $c = 5,\ \beta = 45°,\ \gamma = 95°$
$\alpha = 180° - \beta - \gamma = 180° - 45° - 95° = 40°$

$$\frac{\sin\alpha}{a} = \frac{\sin\gamma}{c} \quad \rightarrow \quad \frac{\sin(40°)}{a} = \frac{\sin(95°)}{5} \quad \rightarrow \quad a = \frac{5\sin(40°)}{\sin(95°)} \approx 3.23$$

$$\frac{\sin\beta}{b} = \frac{\sin\gamma}{c} \quad \rightarrow \quad \frac{\sin(45°)}{b} = \frac{\sin(95°)}{5} \quad \rightarrow \quad b = \frac{5\sin(45°)}{\sin(95°)} \approx 3.55$$

3. $b = 3,\ \alpha = 50°,\ \gamma = 85°$
$\beta = 180° - \alpha - \gamma = 180° - 50° - 85° = 45°$

$$\frac{\sin\alpha}{a} = \frac{\sin\beta}{b} \quad \rightarrow \quad \frac{\sin(50°)}{a} = \frac{\sin(45°)}{3} \quad \rightarrow \quad a = \frac{3\sin(50°)}{\sin(45°)} \approx 3.25$$

$$\frac{\sin\gamma}{c} = \frac{\sin\beta}{b} \quad \rightarrow \quad \frac{\sin(85°)}{c} = \frac{\sin(45°)}{3} \quad \rightarrow \quad c = \frac{3\sin(85°)}{\sin(45°)} \approx 4.23$$

5. $b = 7,\ \alpha = 40°,\ \beta = 45°$
$\gamma = 180° - \alpha - \beta = 180° - 40° - 45° = 95°$

$$\frac{\sin\alpha}{a} = \frac{\sin\beta}{b} \quad \rightarrow \quad \frac{\sin(40°)}{a} = \frac{\sin(45°)}{7} \quad \rightarrow \quad a = \frac{7\sin(40°)}{\sin(45°)} \approx 6.36$$

$$\frac{\sin\gamma}{c} = \frac{\sin\beta}{b} \quad \rightarrow \quad \frac{\sin(95°)}{c} = \frac{\sin(45°)}{7} \quad \rightarrow \quad c = \frac{7\sin(95°)}{\sin(45°)} \approx 9.86$$

7. $b = 2,\ \beta = 40°,\ \gamma = 100°$
$\alpha = 180° - \beta - \gamma = 180° - 40° - 100° = 40°$

$$\frac{\sin\alpha}{a} = \frac{\sin\beta}{b} \quad \rightarrow \quad \frac{\sin(40°)}{a} = \frac{\sin(40°)}{2} \quad \rightarrow \quad a = \frac{2\sin(40°)}{\sin(40°)} = 2$$

$$\frac{\sin\gamma}{c} = \frac{\sin\beta}{b} \quad \rightarrow \quad \frac{\sin(100°)}{c} = \frac{\sin(40°)}{2} \quad \rightarrow \quad c = \frac{2\sin(100°)}{\sin(40°)} \approx 3.06$$

9. $\alpha = 40°,\ \beta = 20°,\ a = 2$
$\gamma = 180° - \alpha - \beta = 180° - 40° - 20° = 120°$

$$\frac{\sin\alpha}{a} = \frac{\sin\beta}{b} \rightarrow \frac{\sin(40°)}{2} = \frac{\sin(20°)}{b} \rightarrow b = \frac{2\sin(20°)}{\sin(40°)} \approx 1.06$$

$$\frac{\sin\gamma}{c} = \frac{\sin\alpha}{a} \rightarrow \frac{\sin(120°)}{c} = \frac{\sin(40°)}{2} \rightarrow c = \frac{2\sin(120°)}{\sin(40°)} \approx 2.69$$

11. $\beta = 70°,\ \gamma = 10°,\ b = 5$
$\alpha = 180° - \beta - \gamma = 180° - 70° - 10° = 100°$

$$\frac{\sin\alpha}{a} = \frac{\sin\beta}{b} \rightarrow \frac{\sin(100°)}{a} = \frac{\sin(70°)}{5} \rightarrow a = \frac{5\sin(100°)}{\sin(70°)} \approx 5.24$$

$$\frac{\sin\gamma}{c} = \frac{\sin\beta}{b} \rightarrow \frac{\sin(10°)}{c} = \frac{\sin(70°)}{5} \rightarrow c = \frac{5\sin(10°)}{\sin(70°)} \approx 0.92$$

13. $\alpha = 110°,\ \gamma = 30°,\ c = 3$
$\beta = 180° - \alpha - \gamma = 180° - 110° - 30° = 40°$

$$\frac{\sin\alpha}{a} = \frac{\sin\gamma}{c} \rightarrow \frac{\sin(110°)}{a} = \frac{\sin(30°)}{3} \rightarrow a = \frac{3\sin(110°)}{\sin(30°)} \approx 5.64$$

$$\frac{\sin\gamma}{c} = \frac{\sin\beta}{b} \rightarrow \frac{\sin(30°)}{3} = \frac{\sin(40°)}{b} \rightarrow b = \frac{3\sin(40°)}{\sin(30°)} \approx 3.86$$

15. $\alpha = 40°,\ \beta = 40°,\ c = 2$
$\gamma = 180° - \alpha - \beta = 180° - 40° - 40° = 100°$

$$\frac{\sin\alpha}{a} = \frac{\sin\gamma}{c} \rightarrow \frac{\sin(40°)}{a} = \frac{\sin(100°)}{2} \rightarrow a = \frac{2\sin(40°)}{\sin(100°)} \approx 1.31$$

$$\frac{\sin\beta}{b} = \frac{\sin\gamma}{c} \rightarrow \frac{\sin(40°)}{b} = \frac{\sin(100°)}{2} \rightarrow b = \frac{2\sin(40°)}{\sin(100°)} \approx 1.31$$

17. $a = 3,\ b = 2,\ \alpha = 50°$

$$\frac{\sin\beta}{b} = \frac{\sin\alpha}{a} \rightarrow \frac{\sin\beta}{2} = \frac{\sin(50°)}{3} \rightarrow \sin\beta = \frac{2\sin(50°)}{3} \approx 0.5107$$

$$\rightarrow \beta = 30.7°\ \text{ or }\ \beta = 149.3°$$

The second value is discarded because $\alpha + \beta > 180°$.
$\gamma = 180° - \alpha - \beta = 180° - 50° - 30.7° = 99.3°$

$$\frac{\sin\gamma}{c} = \frac{\sin\alpha}{a} \rightarrow \frac{\sin(99.3°)}{c} = \frac{\sin(50°)}{3} \rightarrow c = \frac{3\sin(99.3°)}{\sin(50°)} \approx 3.86$$

One triangle: $\beta \approx 30.7°,\ \gamma \approx 99.3°,\ c \approx 3.86$

19. $b = 5$, $c = 3$, $\beta = 100°$

$$\frac{\sin\beta}{b} = \frac{\sin\gamma}{c} \rightarrow \frac{\sin(100°)}{5} = \frac{\sin\gamma}{3} \rightarrow \sin\gamma = \frac{3\sin(100°)}{5} = 0.5909$$

$$\rightarrow \gamma = 36.2° \text{ or } \gamma = 143.8°$$

The second value is discarded because $\beta + \gamma > 180°$.

$\alpha = 180° - \beta - \gamma = 180° - 100° - 36.2° = 43.8°$

$$\frac{\sin\beta}{b} = \frac{\sin\alpha}{a} \rightarrow \frac{\sin(100°)}{5} = \frac{\sin(43.8°)}{a} \rightarrow a = \frac{5\sin(43.8°)}{\sin(100°)} \approx 3.51$$

One triangle: $\alpha \approx 43.8°$, $\gamma \approx 36.2°$, $a \approx 3.51$

21. $a = 4$, $b = 5$, $\alpha = 60°$

$$\frac{\sin\beta}{b} = \frac{\sin\alpha}{a} \rightarrow \frac{\sin\beta}{5} = \frac{\sin(60°)}{4} \rightarrow \sin\beta = \frac{5\sin(60°)}{4} = 1.0825$$

There is no angle β for which $\sin\beta > 1$. Therefore, there is no triangle with the given measurements.

23. $b = 4$, $c = 6$, $\beta = 20°$

$$\frac{\sin\beta}{b} = \frac{\sin\gamma}{c} \rightarrow \frac{\sin(20°)}{4} = \frac{\sin\gamma}{6} \rightarrow \sin\gamma = \frac{6\sin(20°)}{4} \approx 0.5130$$

$$\rightarrow \gamma_1 = 30.9° \text{ or } \gamma_2 = 149.1°$$

For both values, $\beta + \gamma < 180°$. Therefore, there are two triangles.

$\alpha_1 = 180° - \beta - \gamma_1 = 180° - 20° - 30.9° = 129.1°$

$$\frac{\sin\beta}{b} = \frac{\sin\alpha_1}{a_1} \rightarrow \frac{\sin(20°)}{4} = \frac{\sin(129.1°)}{a_1} \rightarrow a_1 = \frac{4\sin(129.1°)}{\sin(20°)} \approx 9.07$$

$\alpha_2 = 180° - \beta - \gamma_2 = 180° - 20° - 149.1° = 10.9°$

$$\frac{\sin\beta}{b} = \frac{\sin\alpha_2}{a_2} \rightarrow \frac{\sin(20°)}{4} = \frac{\sin(10.9°)}{a_2} \rightarrow a_2 = \frac{4\sin(10.9°)}{\sin(20°)} \approx 2.20$$

Two triangles: $\alpha_1 \approx 129.1°$, $\gamma_1 \approx 30.9°$, $a_1 \approx 9.08$
or $\alpha_2 \approx 10.9°$, $\gamma_2 \approx 149.1°$, $a_2 \approx 2.21$

25. $a = 2$, $c = 1$, $\gamma = 100°$

$$\frac{\sin\gamma}{c} = \frac{\sin\alpha}{a} \rightarrow \frac{\sin(100°)}{1} = \frac{\sin\alpha}{2} \rightarrow \sin\alpha = \frac{2\sin(100°)}{1} \approx 1.9696$$

There is no angle α for which $\sin\alpha > 1$. Therefore, there is no triangle with the given measurements.

27. $a = 2$, $c = 1$, $\gamma = 25°$

$$\frac{\sin\alpha}{a} = \frac{\sin\gamma}{c} \rightarrow \frac{\sin\alpha}{2} = \frac{\sin(25°)}{1} \rightarrow \sin\alpha = \frac{2\sin(25°)}{1} \approx 0.8452$$

$$\rightarrow \alpha_1 = 57.7° \text{ or } \alpha_2 = 122.3°$$

For both values, $\alpha + \gamma < 180°$. Therefore, there are two triangles.

$$\beta_1 = 180° - \alpha_1 - \gamma = 180° - 57.7° - 25° = 97.3°$$

$$\frac{\sin\beta_1}{b_1} = \frac{\sin\gamma}{c} \quad\rightarrow\quad \frac{\sin(97.3°)}{b_1} = \frac{\sin(25°)}{1} \quad\rightarrow\quad b_1 = \frac{1\sin(97.3°)}{\sin(25°)} \approx 2.35$$

$$\beta_2 = 180° - \alpha_2 - \gamma = 180° - 122.3° - 25° = 32.7°$$

$$\frac{\sin\beta_2}{b_2} = \frac{\sin\gamma}{c} \quad\rightarrow\quad \frac{\sin(32.7°)}{b_2} = \frac{\sin(25°)}{1} \quad\rightarrow\quad b_2 = \frac{1\sin(32.7°)}{\sin(25°)} \approx 1.28$$

Two triangles: $\alpha_1 \approx 57.7°$, $\beta_1 \approx 97.3°$, $b_1 \approx 2.35$
     or $\alpha_2 \approx 122.3°$, $\beta_2 \approx 32.7°$, $b_2 \approx 1.28$

29. (a)   Find γ ; then use the Law of Sines:

$$\gamma = 180° - 60° - 55° = 65°$$

$$\frac{\sin(55°)}{a} = \frac{\sin(65°)}{150} \quad\rightarrow\quad a = \frac{150\sin(55°)}{\sin(65°)} \approx 135.58 \text{ miles}$$

$$\frac{\sin(60°)}{b} = \frac{\sin(65°)}{150} \quad\rightarrow\quad b = \frac{150\sin(60°)}{\sin(65°)} \approx 143.3 \text{ miles}$$

(b) $t = \dfrac{a}{r} = \dfrac{135.6}{200} \approx 0.68$ hours or ≈ 41 minutes

31. $\angle CAB = 180° - 25° = 155°$ $\angle ABC = 180° - 155° - 15° = 10°$

Let c represent the distance from A to B.

$$\frac{\sin(15°)}{c} = \frac{\sin(10°)}{1000} \quad\rightarrow\quad c = \frac{1000\sin(15°)}{\sin(10°)} \approx 1490.48 \text{ feet}$$

The length of the proposed ski lift is approximately 1490 feet.

33. Find the distance from B to the plane:

$$\gamma = 180° - 40° - 35° = 105° \qquad (\gamma = \angle APB)$$

$$\frac{\sin(40°)}{x} = \frac{\sin(105°)}{1000} \quad\rightarrow\quad x = \frac{1000\sin(40°)}{\sin(105°)} \approx 665.5 \text{ feet}$$

Find the height:

$$\sin(35°) = \frac{h}{x} = \frac{h}{665.5} \quad\rightarrow\quad h = (665.5)\sin(35°) \approx 381.69 \text{ feet}$$

The plane is 381.69 feet high.

35. (a) $\angle ABC = 180° - 40° = 140°$

Find the angle at city C:

$$\frac{\sin C}{150} = \frac{\sin(140°)}{300} \quad\rightarrow\quad \sin C = \frac{150\sin(140°)}{300} \approx 0.3214 \quad\rightarrow\quad C \approx 18.7°$$

Find the angle at city A :

$$A = 180° - 140° - 18.7° = 21.3°$$

$$\frac{\sin(21.3°)}{y} = \frac{\sin(140°)}{300} \quad\rightarrow\quad y = \frac{300\sin(21.3°)}{\sin(140°)} \approx 169.18 \text{ miles}$$

The distance from city B to city C is approximately 169.18 miles.
(b)   To find the angle to turn, subtract angle C from 180°:
$$180° - 18.7° = 161.3°$$
The pilot needs to turn through an angle of 161.3° to return to city A.

37.   Find angle β ($\angle ACB$):
$$\frac{\sin\beta}{123} = \frac{\sin(60°)}{184.5} \ \rightarrow \ \sin\beta = \frac{123\sin(60°)}{184.5} \approx 0.5774$$
$$\beta \approx 35.3°$$
$$\angle CAB = 180° - 60° - 35.3° \approx 84.7°$$
Find the perpendicular distance:
$$\sin(84.7°) = \frac{h}{184.5} \rightarrow h = 184.5\sin(84.7°) = 183.72 \text{ feet}$$

39. $\alpha = 180° - 140° = 40°$ $\beta = 180° - 135° = 45°$
$\gamma = 180° - 40° - 45° = 95°$
$$\frac{\sin(40°)}{a} = \frac{\sin(95°)}{2} \ \rightarrow \ a = \frac{2\sin(40°)}{\sin95°} \approx 1.290 \text{ mi}$$
$$\frac{\sin(45°)}{b} = \frac{\sin(95°)}{2} \ \rightarrow \ b = \frac{2\sin(45°)}{\sin(95°)} \approx 1.420 \text{ mi}$$
$$\overline{BE} = 1.290 - 0.125 = 1.165 \text{ mi}$$
$$\overline{AD} = 1.420 - 0.125 = 1.295 \text{ mi}$$
$$\angle CDE = \angle CED = \frac{180° - 95°}{2} = 42.5°$$
For the isosceles triangle, $\frac{\sin(95°)}{DE} = \frac{\sin(42.5°)}{0.125} \ \rightarrow \ DE = \frac{0.125\sin(95°)}{\sin(42.5°)} \approx 0.184 \text{ miles}$
The length of the highway is $1.165 + 1.295 + 0.184 = 2.64$ miles.

41. $\angle ABD = 180° - 30° = 150°$ $\gamma = 180° - 150° - 20° = 10°$
$$\frac{\sin(150°)}{y} = \frac{\sin(10°)}{1} \ \rightarrow \ y = \frac{1\sin(150°)}{\sin(10°)} \approx 2.88 \text{ mi}$$
$$\frac{\sin\beta}{2.88} = \frac{\sin(20°)}{1} \ \rightarrow \ \sin\beta = \frac{2.88\sin(20°)}{1} \approx 0.9850$$
$$\beta \approx 80°$$
$$\alpha = 180° - 80° - 30° = 70°$$
$$\frac{\sin(70°)}{x} = \frac{\sin(30°)}{1} \ \rightarrow \ x = \frac{\sin(70°)}{\sin(30°)} \approx 1.88 \text{ mi}$$
The ship is about 1.88 miles from the harbor.

43. Using the Law of Sines:

$$\frac{\sin(46.27°)}{x} = \frac{\sin(90° - 46.27°)}{y + 100}$$

$$(y + 100)\sin(46.27°) = x\sin(43.73°)$$

$$y\sin(46.27°) + 100\sin(46.27°) = x\sin(43.73°)$$

$$y = \frac{x\sin(43.73°) - 100\sin(46.27°)}{\sin(46.27°)}$$

$$\frac{\sin(40.3°)}{x} = \frac{\sin(90° - 40.3°)}{y + 200}$$

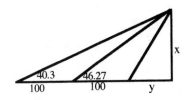

$$(y + 200)\sin(40.3°) = x\sin(49.7°)$$

$$y\sin(40.3°) + 200\sin(40.3°) = x\sin(49.7°)$$

$$y = \frac{x\sin(49.7°) - 200\sin(40.3°)}{\sin(40.3°)}$$

Set the two equations equal to each other and solve:

$$\frac{x\sin(43.73°) - 100\sin(46.27°)}{\sin(46.27°)} = \frac{x\sin(49.7°) - 200\sin(40.3°)}{\sin(40.3°)}$$

$$x\sin(43.73°) \cdot \sin(40.3°) - 100\sin(46.27°) \cdot \sin(40.3°)$$
$$= x\sin(49.7°) \cdot \sin(46.27°) - 200\sin(40.3°) \cdot \sin(46.27°)$$

$$x\sin(43.73°) \cdot \sin(40.3°) - x\sin(49.7°) \cdot \sin(46.27°)$$
$$= 100\sin(46.27°) \cdot \sin(40.3°) - 200\sin(40.3°) \cdot \sin(46.27°)$$

$$x = \frac{100\sin(46.27°) \cdot \sin(40.3°) - 200\sin(40.3°) \cdot \sin(46.27°)}{\sin(43.73°) \cdot \sin(40.3°) - \sin(49.7°) \cdot \sin(46.27°)} \approx 449.36 \text{ feet}$$

45. Using the Law of Sines:

$$\frac{\sin(30°)}{h} = \frac{\sin(60°)}{x} \quad \rightarrow \quad x = \frac{h\sin(60°)}{\sin(30°)}$$

$$\frac{\sin(20°)}{h} = \frac{\sin(70°)}{x + 40} \quad \rightarrow \quad x = \frac{h\sin(70°)}{\sin(20°)} - 40$$

$$\frac{h\sin(60°)}{\sin(30°)} = \frac{h\sin(70°)}{\sin(20°)} - 40$$

$$h\left(\frac{\sin(60°)}{\sin(30°)} - \frac{\sin(70°)}{\sin(20°)}\right) = -40 \rightarrow h = \frac{-40}{\left(\frac{\sin(60°)}{\sin(30°)} - \frac{\sin(70°)}{\sin(20°)}\right)} \approx 39.39 \text{ feet}$$

47.   Find the distance from B to the helicopter:

$$\gamma = 180° - 40° - 25° = 115°   \qquad (\gamma = \angle APB)$$

$$\frac{\sin(40°)}{x} = \frac{\sin(115°)}{100} \quad \rightarrow \quad x = \frac{100\sin(40°)}{\sin(115°)} \approx 70.9 \text{ feet}$$

Find the height:

$$\sin(25°) = \frac{h}{x} \approx \frac{h}{70.9} \quad \rightarrow \quad h \approx 70.9(\sin 25°) \approx 29.97 \text{ feet}$$

The helicopter is about 30 feet high.

49. $$\frac{a-b}{c} = \frac{a}{c} - \frac{b}{c} = \frac{\sin\alpha}{\sin\gamma} - \frac{\sin\beta}{\sin\gamma} = \frac{\sin\alpha - \sin\beta}{\sin\gamma} = \frac{2\sin\left(\frac{\alpha-\beta}{2}\right)\cos\left(\frac{\alpha+\beta}{2}\right)}{\sin\left(2\cdot\frac{\gamma}{2}\right)}$$

$$= \frac{2\sin\left(\frac{\alpha-\beta}{2}\right)\cos\left(\frac{\alpha+\beta}{2}\right)}{2\sin\left(\frac{\gamma}{2}\right)\cos\left(\frac{\gamma}{2}\right)} = \frac{\sin\left(\frac{\alpha-\beta}{2}\right)\cos\left(\frac{\pi}{2}-\frac{\gamma}{2}\right)}{\sin\left(\frac{\gamma}{2}\right)\cos\left(\frac{\gamma}{2}\right)} = \frac{\sin\left(\frac{\alpha-\beta}{2}\right)\sin\left(\frac{\gamma}{2}\right)}{\sin\left(\frac{\gamma}{2}\right)\cos\left(\frac{\gamma}{2}\right)}$$

$$= \frac{\sin\left(\frac{\alpha-\beta}{2}\right)}{\cos\left(\frac{\gamma}{2}\right)} = \frac{\sin\left(\frac{1}{2}(\alpha-\beta)\right)}{\cos\left(\frac{1}{2}\gamma\right)}$$

51.   Derive the Law of Tangents:

$$\frac{a-b}{a+b} = \frac{\left(\frac{a-b}{c}\right)}{\left(\frac{a+b}{c}\right)} = \frac{\left(\dfrac{\sin\left(\frac{1}{2}(\alpha-\beta)\right)}{\cos\left(\frac{1}{2}\gamma\right)}\right)}{\left(\dfrac{\cos\left(\frac{1}{2}(\alpha-\beta)\right)}{\sin\left(\frac{1}{2}\gamma\right)}\right)} = \frac{\sin\left(\frac{1}{2}(\alpha-\beta)\right)}{\cos\left(\frac{1}{2}\gamma\right)} \cdot \frac{\sin\left(\frac{1}{2}\gamma\right)}{\cos\left(\frac{1}{2}(\alpha-\beta)\right)} = \tan\left(\frac{1}{2}(\alpha-\beta)\right)\tan\left(\frac{1}{2}\gamma\right)$$

$$= \tan\left(\frac{1}{2}(\alpha-\beta)\right)\tan\left(\frac{1}{2}(\pi-(\alpha+\beta))\right) = \tan\left(\frac{1}{2}(\alpha-\beta)\right)\tan\left(\frac{\pi}{2}-\left(\frac{\alpha+\beta}{2}\right)\right)$$

$$= \tan\left(\frac{1}{2}(\alpha-\beta)\right)\cot\left(\frac{\alpha+\beta}{2}\right) = \frac{\tan\left(\frac{1}{2}(\alpha-\beta)\right)}{\tan\left(\frac{1}{2}(\alpha+\beta)\right)}$$

Applications of Trigonometric Functions

7.3 The Law of Cosines

1. $a = 2, \ c = 4, \ \beta = 45°$ $b^2 = a^2 + c^2 - 2ac\cos\beta$

$b^2 = 2^2 + 4^2 - 2 \cdot 2 \cdot 4\cos(45°) = 20 - 16 \cdot \dfrac{\sqrt{2}}{2} = 20 - 8\sqrt{2} \approx 8.6863$

$b \approx 2.95$

$a^2 = b^2 + c^2 - 2bc\cos\alpha \ \rightarrow \ 2bc\cos\alpha = b^2 + c^2 - a^2 \ \rightarrow \ \cos\alpha = \dfrac{b^2 + c^2 - a^2}{2bc}$

$\cos\alpha = \dfrac{2.95^2 + 4^2 - 2^2}{2(2.95)(4)} \approx \dfrac{20.6863}{23.6} \approx 0.8765 \rightarrow \alpha \approx 28.7°$

$c^2 = a^2 + b^2 - 2ab\cos\gamma \ \rightarrow \ \cos\gamma = \dfrac{a^2 + b^2 - c^2}{2ab} = \dfrac{2^2 + 2.95^2 - 4^2}{2(2)(2.95)} \approx -0.2794 \rightarrow \gamma \approx 106.3°$

3. $a = 2, \ b = 3, \ \gamma = 95°$ $c^2 = a^2 + b^2 - 2ab\cos\gamma$

$c^2 = 2^2 + 3^2 - 2 \cdot 2 \cdot 3\cos(95°) \approx 13 - 12 \cdot (-0.0872) \approx 14.0459$

$c \approx 3.75$

$a^2 = b^2 + c^2 - 2bc\cos\alpha \ \rightarrow \ \cos\alpha = \dfrac{b^2 + c^2 - a^2}{2bc} = \dfrac{3^2 + 3.75^2 - 2^2}{2(3)(3.75)} \approx 0.8472 \rightarrow \alpha \approx 32.1°$

$b^2 = a^2 + c^2 - 2ac\cos\beta \ \rightarrow \ \cos\beta = \dfrac{a^2 + c^2 - b^2}{2ac} = \dfrac{2^2 + 3.75^2 - 3^2}{2(2)(3.75)} \approx 0.6042 \rightarrow \beta \approx 52.9°$

5. $a = 6, \ b = 5, \ c = 8$

$a^2 = b^2 + c^2 - 2bc\cos\alpha \ \rightarrow \ \cos\alpha = \dfrac{b^2 + c^2 - a^2}{2bc} = \dfrac{5^2 + 8^2 - 6^2}{2(5)(8)} \approx 0.6625 \rightarrow \alpha \approx 48.5°$

$b^2 = a^2 + c^2 - 2ac\cos\beta \ \rightarrow \ \cos\beta = \dfrac{a^2 + c^2 - b^2}{2ac} = \dfrac{6^2 + 8^2 - 5^2}{2(6)(8)} \approx 0.7813 \rightarrow \beta = 38.6°$

$c^2 = a^2 + b^2 - 2ab\cos\gamma \ \rightarrow \ \cos\gamma = \dfrac{a^2 + b^2 - c^2}{2ab} = \dfrac{6^2 + 5^2 - 8^2}{2(6)(5)} \approx -0.0500 \rightarrow \gamma = 92.9°$

7. $a = 9, \; b = 6, \; c = 4$

$$a^2 = b^2 + c^2 - 2bc \cos\alpha \;\; \rightarrow \;\; \cos\alpha = \frac{b^2 + c^2 - a^2}{2bc} = \frac{6^2 + 4^2 - 9^2}{2(6)(4)} \approx -0.6042 \rightarrow \alpha \approx 127.2°$$

$$b^2 = a^2 + c^2 - 2ac \cos\beta \;\; \rightarrow \;\; \cos\beta = \frac{a^2 + c^2 - b^2}{2ac} = \frac{9^2 + 4^2 - 6^2}{2(9)(4)} \approx 0.8472 \rightarrow \beta = 32.1°$$

$$c^2 = a^2 + b^2 - 2ab \cos\gamma \;\; \rightarrow \;\; \cos\gamma = \frac{a^2 + b^2 - c^2}{2ab} = \frac{9^2 + 6^2 - 4^2}{2(9)(6)} \approx 0.9352 \rightarrow \gamma = 20.7°$$

9. $a = 3, \; b = 4, \; \gamma = 40°$

$$c^2 = a^2 + b^2 - 2ab \cos\gamma$$

$$c^2 = 3^2 + 4^2 - 2 \cdot 3 \cdot 4 \cos(40°) \approx 6.6149$$

$$c \approx 2.57$$

$$a^2 = b^2 + c^2 - 2bc \cos\alpha \;\; \rightarrow \;\; \cos\alpha = \frac{b^2 + c^2 - a^2}{2bc} = \frac{4^2 + 2.57^2 - 3^2}{2(4)(2.57)} \approx 0.6617 \rightarrow \alpha \approx 48.6°$$

$$b^2 = a^2 + c^2 - 2ac \cos\beta \;\; \rightarrow \;\; \cos\beta = \frac{a^2 + c^2 - b^2}{2ac} = \frac{3^2 + 2.57^2 - 4^2}{2(3)(2.57)} \approx -0.0256 \rightarrow \beta \approx 91.4°$$

11. $b = 1, \; c = 3, \; \alpha = 80°$

$$a^2 = b^2 + c^2 - 2bc \cos\alpha$$

$$a^2 = 1^2 + 3^2 - 2 \cdot 1 \cdot 3 \cos(80°) \approx 8.9581$$

$$a \approx 2.99$$

$$c^2 = a^2 + b^2 - 2ab \cos\gamma \;\; \rightarrow \;\; \cos\gamma = \frac{a^2 + b^2 - c^2}{2ab} = \frac{2.99^2 + 1^2 - 3^2}{2(2.99)(1)} \approx 0.1572 \rightarrow \gamma \approx 80.8°$$

$$b^2 = a^2 + c^2 - 2ac \cos\beta \;\; \rightarrow \;\; \cos\beta = \frac{a^2 + c^2 - b^2}{2ac} = \frac{2.99^2 + 3^2 - 1^2}{2(2.99)(3)} \approx 0.9443 \rightarrow \beta = 19.2°$$

13. $a = 3, \; c = 2, \; \beta = 110°$

$$b^2 = a^2 + c^2 - 2ac \cos\beta$$

$$b^2 = 3^2 + 2^2 - 2 \cdot 3 \cdot 2 \cos(110°) \approx 17.1042$$

$$b \approx 4.14$$

$$a^2 = b^2 + c^2 - 2bc \cos\alpha \;\; \rightarrow \;\; \cos\alpha = \frac{b^2 + c^2 - a^2}{2bc} = \frac{4.14^2 + 2^2 - 3^2}{2(4.14)(2)} \approx 0.7331 \rightarrow \alpha \approx 43.0°$$

$$c^2 = a^2 + b^2 - 2ab \cos\gamma \;\; \rightarrow \;\; \cos\gamma = \frac{a^2 + b^2 - c^2}{2ab} = \frac{3^2 + 4.14^2 - 2^2}{2(3)(4.14)} \approx 0.8913 \rightarrow \gamma \approx 27.0°$$

15. $a = 2, \; b = 2, \; \gamma = 50°$

$$c^2 = a^2 + b^2 - 2ab \cos\gamma$$

$$c^2 = 2^2 + 2^2 - 2 \cdot 2 \cdot 2 \cos(50°) \approx 2.8577$$

$$c \approx 1.69$$

$$a^2 = b^2 + c^2 - 2bc\cos\alpha \quad \rightarrow \quad \cos\alpha = \frac{b^2+c^2-a^2}{2bc} = \frac{2^2+1.69^2-2^2}{2(2)(1.69)} \approx 0.4225 \rightarrow \alpha \approx 65.0°$$

$$b^2 = a^2 + c^2 - 2ac\cos\beta \quad \rightarrow \quad \cos\beta = \frac{a^2+c^2-b^2}{2ac} = \frac{2^2+1.69^2-2^2}{2(2)(1.69)} \approx 0.4225 \rightarrow \beta \approx 65.0°$$

17. $a = 12, \ b = 13, \ c = 5$

$$a^2 = b^2 + c^2 - 2bc\cos\alpha \quad \rightarrow \quad \cos\alpha = \frac{b^2+c^2-a^2}{2bc} = \frac{13^2+5^2-12^2}{2(13)(5)} \approx 0.3846 \rightarrow \alpha \approx 67.4°$$

$$b^2 = a^2 + c^2 - 2ac\cos\beta \quad \rightarrow \quad \cos\beta = \frac{a^2+c^2-b^2}{2ac} = \frac{12^2+5^2-13^2}{2(12)(5)} = 0 \rightarrow \beta = 90°$$

$$c^2 = a^2 + b^2 - 2ab\cos\gamma \quad \rightarrow \quad \cos\gamma = \frac{a^2+b^2-c^2}{2ab} = \frac{12^2+13^2-5^2}{2(12)(13)} \approx 0.9231 \rightarrow \gamma \approx 22.6°$$

19. $a = 2, \ b = 2, \ c = 2$

$$a^2 = b^2 + c^2 - 2bc\cos\alpha \quad \rightarrow \quad \cos\alpha = \frac{b^2+c^2-a^2}{2bc} = \frac{2^2+2^2-2^2}{2(2)(2)} = 0.5 \rightarrow \alpha = 60°$$

$$b^2 = a^2 + c^2 - 2ac\cos\beta \quad \rightarrow \quad \cos\beta = \frac{a^2+c^2-b^2}{2ac} = \frac{2^2+2^2-2^2}{2(2)(2)} = 0.5 \rightarrow \beta = 60°$$

$$c^2 = a^2 + b^2 - 2ab\cos\gamma \quad \rightarrow \quad \cos\gamma = \frac{a^2+b^2-c^2}{2ab} = \frac{2^2+2^2-2^2}{2(2)(2)} = 0.5 \rightarrow \gamma = 60°$$

21. $a = 5, \ b = 8, \ c = 9$

$$a^2 = b^2 + c^2 - 2bc\cos\alpha \quad \rightarrow \quad \cos\alpha = \frac{b^2+c^2-a^2}{2bc} = \frac{8^2+9^2-5^2}{2(8)(9)} \approx 0.8333 \rightarrow \alpha \approx 33.6°$$

$$b^2 = a^2 + c^2 - 2ac\cos\beta \quad \rightarrow \quad \cos\beta = \frac{a^2+c^2-b^2}{2ac} = \frac{5^2+9^2-8^2}{2\cdot5\cdot9} \approx 0.4667 \rightarrow \beta \approx 62.2°$$

$$c^2 = a^2 + b^2 - 2ab\cos\gamma \quad \rightarrow \quad \cos\gamma = \frac{a^2+b^2-c^2}{2ab} = \frac{5^2+8^2-9^2}{2\cdot5\cdot8} = 0.1000 \rightarrow \gamma \approx 84.3°$$

23. $a = 10, \ b = 8, \ c = 5$

$$a^2 = b^2 + c^2 - 2bc\cos\alpha \quad \rightarrow \quad \cos\alpha = \frac{b^2+c^2-a^2}{2bc} = \frac{8^2+5^2-10^2}{2(8)(5)} \approx -0.1375 \rightarrow \alpha \approx 97.9°$$

$$b^2 = a^2 + c^2 - 2ac\cos\beta \quad \rightarrow \quad \cos\beta = \frac{a^2+c^2-b^2}{2ac} = \frac{10^2+5^2-8^2}{2(10)(5)} \approx 0.6100 \rightarrow \beta \approx 52.4°$$

$$c^2 = a^2 + b^2 - 2ab\cos\gamma \quad \rightarrow \quad \cos\gamma = \frac{a^2+b^2-c^2}{2ab} = \frac{10^2+8^2-5^2}{2(10)(8)} \approx 0.8688 \rightarrow \gamma \approx 29.7°$$

25.   Find the third side of the triangle using the Law of Cosines:

$a = 50, \ b = 70, \ \gamma = 70°$

$$c^2 = a^2 + b^2 - 2ab\cos\gamma = 50^2 + 70^2 - 2\cdot50\cdot70\cos(70°) \approx 5005.86 \rightarrow c \approx 70.75$$

The houses are approximately 70.75 feet apart.

27. (a) After 15 minutes, the plane would have flown $220(0.25) = 55$ miles.
Find the third side of the triangle:

$a = 55, \quad b = 330, \quad \gamma = 10°$

$c^2 = a^2 + b^2 - 2ab\cos\gamma = 55^2 + 330^2 - 2 \cdot 55 \cdot 330\cos(10°) \approx 76176.48 \to c \approx 276$

Find the measure of the angle opposite the 330 side:

$\cos\beta = \dfrac{a^2 + c^2 - b^2}{2ac} = \dfrac{55^2 + 276^2 - 330^2}{2(55)(276)} \approx -0.9782 \to \beta \approx 168°$

The pilot should turn through an angle of $180° - 168° = 12°$.

(b) If the total trip is to be done in 90 minutes, and 15 minutes were used already, then there are 75 minutes or 1.25 hours to complete the trip. The plane must travel 276 miles in 1.25 hours.

$r = \dfrac{276}{1.25} = 220.8$ miles / hour

The pilot must maintain a speed of 220.8 mi/hr to complete the whole trip in 90 minutes.

29. (a) Find x in the figure:

$x^2 = 60.5^2 + 90^2 - 2(60.5)90\cos(45°) \approx 4059.86$

$x \approx 63.7$ feet

It is about 63.7 feet from the pitching rubber

to first base.

(b) Use the Pythagorean Theorem to find y in the figure:

$90^2 + 90^2 = (60.5 + y)^2$

$8100 + 8100 = (60.5 + y)^2$

$16200 = (60.5 + y)^2$

$60.5 + y \approx 127.3$

$y \approx 66.8$ feet

It is about 66.8 feet from the pitching rubber to second base.

(c) Find β in the figure by using the Law of Cosines:

$\cos\beta = \dfrac{60.5^2 + 63.7^2 - 90^2}{2(60.5)(63.7)} \approx -0.0496 \to \beta \approx 92.8°$

The pitcher needs to turn through an angle of 92.8° to face first base.

31. (a) Find x by using the Law of Cosines:

$x^2 = 500^2 + 100^2 - 2(500)100\cos(80°) \approx 242,635$

$x \approx 492.6$ feet

The guy wire needs to be about 492.6 feet long.

(b) Use the Pythagorean Theorem to find the value of y:

$y^2 = 100^2 + 250^2 = 72500$

$y = 269.3$ feet

The guy wire needs to be about 269.3 feet long.

33. Find x by using the Law of Cosines:

$$x^2 = 400^2 + 90^2 - 2(400)90\cos(45°) \approx 117,188.3$$

$$x \approx 342.3 \text{ feet}$$

It is approximately 342.3 feet from dead center to third base.

35. Use the Law of Cosines:

$$L^2 = x^2 + r^2 - 2xr\cos\theta$$

$$x^2 - 2xr\cos\theta + r^2 - L^2 = 0$$

$$x = \frac{2r\cos\theta + \sqrt{(2r\cos\theta)^2 - 4(1)(r^2 - L^2)}}{2(1)}$$

$$x = \frac{2r\cos\theta + \sqrt{4r^2\cos^2\theta - 4(r^2 - L^2)}}{2}$$

$$x = r\cos\theta + \sqrt{r^2\cos^2\theta + L^2 - r^2}$$

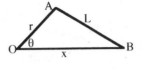

37. $$\cos\left(\frac{\gamma}{2}\right) = \sqrt{\frac{1+\cos\gamma}{2}} = \sqrt{\frac{\left(1 + \dfrac{a^2 + b^2 - c^2}{2ab}\right)}{2}} = \sqrt{\frac{2ab + a^2 + b^2 - c^2}{4ab}} = \sqrt{\frac{(a+b)^2 - c^2}{4ab}}$$

$$= \sqrt{\frac{(a+b+c)(a+b-c)}{4ab}} = \sqrt{\frac{2s(2s-c-c)}{4ab}} = \sqrt{\frac{4s(s-c)}{4ab}} = \sqrt{\frac{s(s-c)}{ab}}$$

39. $$\frac{\cos\alpha}{a} + \frac{\cos\beta}{b} + \frac{\cos\gamma}{c} = \frac{b^2 + c^2 - a^2}{2bca} + \frac{a^2 + c^2 - b^2}{2acb} + \frac{a^2 + b^2 - c^2}{2abc}$$

$$= \frac{b^2 + c^2 - a^2 + a^2 + c^2 - b^2 + a^2 + b^2 - c^2}{2abc} = \frac{a^2 + b^2 + c^2}{2abc}$$

Applications of Trigonometric Functions

7.4 The Area of a Triangle

1. $a = 2$, $c = 4$, $\beta = 45°$

    $A = \dfrac{1}{2}ac\sin\beta = \dfrac{1}{2}(2)(4)\sin(45°) \approx 2.83$

3. $a = 2$, $b = 3$, $\gamma = 95°$

    $A = \dfrac{1}{2}ab\sin\gamma = \dfrac{1}{2}(2)(3)\sin(95°) \approx 2.99$

5. $a = 6$, $b = 5$, $c = 8$

    $s = \dfrac{1}{2}(a + b + c) = \dfrac{1}{2}(6 + 5 + 8) = \dfrac{19}{2}$

    $A = \sqrt{s(s-a)(s-b)(s-c)} = \sqrt{\dfrac{19}{2}\left(\dfrac{7}{2}\right)\left(\dfrac{9}{2}\right)\left(\dfrac{3}{2}\right)} = \sqrt{\dfrac{3591}{16}} \approx 14.98$

7. $a = 9$, $b = 6$, $c = 4$

    $s = \dfrac{1}{2}(a + b + c) = \dfrac{1}{2}(9 + 6 + 4) = \dfrac{19}{2}$

    $A = \sqrt{s(s-a)(s-b)(s-c)} = \sqrt{\dfrac{19}{2}\left(\dfrac{1}{2}\right)\left(\dfrac{7}{2}\right)\left(\dfrac{11}{2}\right)} = \sqrt{\dfrac{1463}{16}} \approx 9.56$

9. $a = 3$, $b = 4$, $\gamma = 40°$

    $A = \dfrac{1}{2}ab\sin\gamma = \dfrac{1}{2}(3)(4)\sin(40°) \approx 3.86$

11. $b = 1$, $c = 3$, $\alpha = 80°$

    $A = \dfrac{1}{2}bc\sin\alpha = \dfrac{1}{2}(1)(3)\sin(80°) \approx 1.48$

13. $a = 3$, $c = 2$, $\beta = 110°$

    $A = \dfrac{1}{2}ac\sin\beta = \dfrac{1}{2}(3)(2)\sin(110°) \approx 2.82$

15. $a = 2$, $b = 2$, $\gamma = 50°$

    $A = \dfrac{1}{2}ab\sin\gamma = \dfrac{1}{2}(2)(2)\sin(50°) \approx 1.53$

17. $a = 12, \ b = 13, \ c = 5$

$s = \dfrac{1}{2}(a + b + c) = \dfrac{1}{2}(12 + 13 + 5) = 15$

$A = \sqrt{s(s - a)(s - b)(s - c)} = \sqrt{15(3)(2)(10)} = \sqrt{900} = 30$

19. $a = 2, \ b = 2, \ c = 2$

$s = \dfrac{1}{2}(a + b + c) = \dfrac{1}{2}(2 + 2 + 2) = 3$

$A = \sqrt{s(s - a)(s - b)(s - c)} = \sqrt{3(1)(1)(1)} = \sqrt{3} \approx 1.73$

21. $a = 5, \ b = 8, \ c = 9$

$s = \dfrac{1}{2}(a + b + c) = \dfrac{1}{2}(5 + 8 + 9) = 11$

$A = \sqrt{s(s - a)(s - b)(s - c)} = \sqrt{11(6)(3)(2)} = \sqrt{396} \approx 19.90$

23. $a = 10, \ b = 8, \ c = 5$

$s = \dfrac{1}{2}(a + b + c) = \dfrac{1}{2}(10 + 8 + 5) = \dfrac{23}{2}$

$A = \sqrt{s(s - a)(s - b)(s - c)} = \sqrt{\dfrac{23}{2}\left(\dfrac{3}{2}\right)\left(\dfrac{7}{2}\right)\left(\dfrac{13}{2}\right)} = \sqrt{\dfrac{6279}{16}} \approx 19.81$

25. Area of a sector $= \dfrac{1}{2}r^2\theta$ where θ is in radians.

$\theta = 70° \cdot \dfrac{\pi}{180} = \dfrac{7\pi}{18}$

Area of the sector $= \dfrac{1}{2} \cdot 8^2 \cdot \dfrac{7\pi}{18} = \dfrac{112\pi}{9} \approx 39.10$ square feet

Area of the triangle $= \dfrac{1}{2} \cdot 8 \cdot 8 \sin(70°) = 32\sin(70°) \approx 30.07$ square feet

Area of the segment $= 39.10 - 30.07 = 9.03$ square feet

27. Find the area of the lot using Heron's Formula:

$a = 100, \ b = 50, \ c = 75$

$s = \dfrac{1}{2}(a + b + c) = \dfrac{1}{2}(100 + 50 + 75) = \dfrac{225}{2}$

$A = \sqrt{s(s - a)(s - b)(s - c)} = \sqrt{\dfrac{225}{2}\left(\dfrac{25}{2}\right)\left(\dfrac{125}{2}\right)\left(\dfrac{75}{2}\right)} = \sqrt{\dfrac{52,734,375}{16}} \approx 1815.46$

The cost is $3 times the area: Cost $= \$3(1815.46) = \$5446.38$

29. The area of the shaded region = the area of the semicircle – the area of the triangle.

Area of the semicircle $= \dfrac{1}{2}\pi r^2 = \dfrac{1}{2}\pi(4)^2 = 8\pi$ square centimeters

The triangle is a right triangle. Find the other leg:

$6^2 + b^2 = 8^2 \ \rightarrow \ b^2 = 64 - 36 = 28 \ \rightarrow \ b = \sqrt{28} = 2\sqrt{7}$

Area of the triangle $= \dfrac{1}{2} \cdot 6 \cdot 2\sqrt{7} = 6\sqrt{7}$ square centimeters

Area of the shaded region $= 8\pi - 6\sqrt{7} \approx 9.26$ square centimeters

31. Use the Law of Sines in the area of the triangle formula:

$$A = \frac{1}{2}ab\sin\gamma = \frac{1}{2}a\sin\gamma\left(\frac{a\sin\beta}{\sin\alpha}\right) = \frac{a^2\sin\beta\sin\gamma}{2\sin\alpha}$$

33. $\alpha = 40°$, $\beta = 20°$, $a = 2$ $\gamma = 180° - \alpha - \beta = 180° - 40° - 20° = 120°$

$$A = \frac{a^2\sin\beta\sin\gamma}{2\sin\alpha} = \frac{2^2\sin(20°)\sin(120°)}{2\sin(40°)} \approx \frac{4(0.3420)(0.8660)}{2(0.6428)} \approx 0.92$$

35. $\beta = 70°$, $\gamma = 10°$, $b = 5$ $\alpha = 180° - \beta - \gamma = 180° - 70° - 10° = 100°$

$$A = \frac{b^2\sin\alpha\sin\gamma}{2\sin\beta} = \frac{5^2\sin(100°)\sin(10°)}{2\sin(70°)} \approx \frac{25(0.9848)(0.1736)}{2(0.9397)} \approx 2.27$$

37. $\alpha = 110°$, $\gamma = 30°$, $c = 3$ $\beta = 180° - \alpha - \gamma = 180° - 110° - 30° = 40°$

$$A = \frac{c^2\sin\alpha\sin\beta}{2\sin\gamma} = \frac{3^2\sin(110°)\sin(40°)}{2\sin(30°)} \approx \frac{9(0.9397)(0.6428)}{2(0.5000)} \approx 5.44$$

39. The area is the sum of the area of a triangle and a sector.

Area of the triangle $= \frac{1}{2}r \cdot r\sin(\pi - \theta) = \frac{1}{2}r^2\sin(\pi - \theta)$

Area of the sector $= \frac{1}{2}r^2\theta$

$$A = \frac{1}{2}r^2\sin(\pi - \theta) + \frac{1}{2}r^2\theta = \frac{1}{2}r^2(\sin(\pi - \theta) + \theta)$$

$$= \frac{1}{2}r^2(\sin\pi\cos\theta - \cos\pi\sin\theta + \theta) = \frac{1}{2}r^2(0 + \sin\theta + \theta) = \frac{1}{2}r^2(\theta + \sin\theta)$$

41. The grazing area must be considered in sections. A_1 represents $\frac{3}{4}$ of a circle:

$$A_1 = \frac{3}{4}\pi(100)^2 = 7500\pi \approx 23{,}562 \text{ square feet}$$

Angles are needed to find A_2 and A_3: (see the figure)

     In $\triangle ABC$, $\angle CBA = 45°$, $AB = 10$, $AC = 90$

     Find $\angle BCA$:

$$\frac{\sin\angle CBA}{90} = \frac{\sin\angle BCA}{10} \rightarrow \frac{\sin(45°)}{90} = \frac{\sin\angle BCA}{10}$$

$$\sin\angle BCA = \frac{10\sin(45°)}{90} \approx 0.0786$$

$$\angle BCA \approx 4.5°$$

$m\angle BAC = 180° - 45° - 4.5° = 130.5°$

$m\angle DAC = 130.5° - 90° = 40.5°$

Area of $A_3 = \dfrac{1}{2}(10)(90)\sin(40.5°) \approx 292$ square feet

Area of sector $A_2 = \dfrac{1}{2}(90)^2\left(49.5° \cdot \dfrac{\pi}{180}\right) \approx 3499$ square feet

Since the cow can go in either direction around the barn, A_2 and A_3 must be doubled. Total grazing area is: $23,562 + 2(3499) + 2(292) = 31,144$ square feet

43. $h_1 = \dfrac{2K}{a}, \quad h_2 = \dfrac{2K}{b}, \quad h_3 = \dfrac{2K}{c}$ where K is the area of the triangle.

$\dfrac{1}{h_1} + \dfrac{1}{h_2} + \dfrac{1}{h_3} = \dfrac{a}{2K} + \dfrac{b}{2K} + \dfrac{c}{2K} = \dfrac{a+b+c}{2K} = \dfrac{2s}{2K} = \dfrac{s}{K}$

45. $h = \dfrac{a\sin\beta \sin\gamma}{\sin\alpha}$ where h is the altitude to side a.

In $\triangle OAB$, c is opposite angle AOB. The two adjacent angles are $\dfrac{\alpha}{2}$ and $\dfrac{\beta}{2}$.

Then $r = \dfrac{c \cdot \sin\left(\dfrac{\alpha}{2}\right)\sin\left(\dfrac{\beta}{2}\right)}{\sin(\angle AOB)}; \quad \angle AOB = \pi - \left(\dfrac{\alpha}{2} + \dfrac{\beta}{2}\right)$

$\sin(\angle AOB) = \sin\left(\pi - \left(\dfrac{\alpha}{2} + \dfrac{\beta}{2}\right)\right) = \sin\left(\dfrac{\alpha}{2} + \dfrac{\beta}{2}\right) = \sin\left(\dfrac{\alpha+\beta}{2}\right) = \cos\left(\dfrac{\pi}{2} - \left(\dfrac{\alpha+\beta}{2}\right)\right)$

$= \cos\left(\dfrac{\pi - (\alpha+\beta)}{2}\right) = \cos\left(\dfrac{\gamma}{2}\right)$

Thus, $r = \dfrac{c \cdot \sin\left(\dfrac{\alpha}{2}\right)\sin\left(\dfrac{\beta}{2}\right)}{\cos\left(\dfrac{\gamma}{2}\right)}$

47. Use the result of Problem 46:

$\cot\left(\dfrac{\alpha}{2}\right) + \cot\left(\dfrac{\beta}{2}\right) + \cot\left(\dfrac{\gamma}{2}\right) = \dfrac{s-a}{r} + \dfrac{s-b}{r} + \dfrac{s-c}{r} = \dfrac{s-a+s-b+s-c}{r}$

$= \dfrac{3s - (a+b+c)}{r} = \dfrac{3s - 2s}{r} = \dfrac{s}{r}$

49. (a) Area $\triangle OAC = \dfrac{1}{2}|OC| \cdot |AC| = \dfrac{1}{2} \cdot \dfrac{|OC|}{1} \cdot \dfrac{|AC|}{1} = \dfrac{1}{2}\cos\alpha \sin\alpha = \dfrac{1}{2}\sin\alpha \cos\alpha$

(b) Area $\triangle OCB = \dfrac{1}{2}|OC| \cdot |BC| = \dfrac{1}{2} \cdot |OB|^2 \cdot \dfrac{|OC|}{|OB|} \cdot \dfrac{|BC|}{|OB|} = \dfrac{1}{2}|OB|^2 \cos\beta \sin\beta$

$= \dfrac{1}{2}|OB|^2 \sin\beta\cos\beta$

(c)   Area $\Delta OAB = \frac{1}{2}|BD|\cdot|OA| = \frac{1}{2}|BD|\cdot 1 = \frac{1}{2}\cdot|OB|\cdot\frac{|BD|}{|OB|} = \frac{1}{2}|OB|\sin(\alpha+\beta)$

(d) $\dfrac{\cos\alpha}{\cos\beta} = \dfrac{\dfrac{|OC|}{|OA|}}{\dfrac{|OC|}{|OB|}} = \dfrac{|OC|}{1}\cdot\dfrac{|OB|}{|OC|} = |OB|$

(e)   Area ΔOAB = Area ΔOAC + Area ΔOCB

$\frac{1}{2}|OB|\sin(\alpha+\beta) = \frac{1}{2}\sin\alpha\cos\alpha + \frac{1}{2}|OB|^2\sin\beta\cos\beta$

$\frac{\cos\alpha}{\cos\beta}\sin(\alpha+\beta) = \sin\alpha\cos\alpha + \frac{\cos^2\alpha}{\cos^2\beta}\sin\beta\cos\beta$

$\sin(\alpha+\beta) = \frac{\cos\beta}{\cos\alpha}\sin\alpha\cos\alpha + \frac{\cos\alpha}{\cos\beta}\sin\beta\cos\beta$

$\sin(\alpha+\beta) = \sin\alpha\cos\beta + \cos\alpha\sin\beta$

Applications of Trigonometric Functions

7.5   Simple Harmonic Motion; Damped Motion

1. $d = -5\cos(\pi t)$ 3. $d = -6\cos(2t)$ 5. $d = -5\sin(\pi t)$ 7. $d = -6\sin(2t)$

9. $d = 5\sin(3t)$
   (a)   Simple harmonic
   (b)   5 meters
   (c) $\dfrac{2\pi}{3}$ seconds
   (d) $\dfrac{3}{2\pi}$ oscillation/second

11. $d = 6\cos(\pi t)$
   (a)   Simple harmonic
   (b)   6 meters
   (c)   2 seconds
   (d) $\dfrac{1}{2}$ oscillation/second

13. $d = -3\sin\left(\dfrac{1}{2}t\right)$
   (a)   Simple harmonic
   (b)   3 meters
   (c) 4π seconds
   (d) $\dfrac{1}{4\pi}$ oscillation/second

15. $d = 6 + 2\cos(2\pi t)$
   (a)   Simple harmonic
   (b)   2 meters
   (c)   1 second
   (d)   1 oscillation/second

17. $d(t) = e^{-t/\pi}\cos(2t), \qquad 0 \le t \le 2\pi$

19. $d(t) = e^{-t/2\pi}\cos(t), \qquad 0 \le t \le 2\pi$

21. $f(x) = x + \cos x$

23. $f(x) = x - \sin x$

25. $f(x) = \sin x + \cos x$

27. $f(x) = \sin x + \sin(2x)$

29. (a) Graph:

(b) The graph of V touches the graph of $y = e^{-1.9\,t}$ when $t = 0, 2$.

The graph of V touches the graph of $y = -e^{-1.9\,t}$ when $t = 1, 3$.

(c) To solve the inequality $-0.4 < e^{-t/3} \cdot \cos(\pi t) < 0.4$ on the interval $0 \le t \le 3$, we consider the graphs of $y = -0.4$; $y = e^{-t/3} \cdot \cos(\pi t)$; and $y = 0.4$.

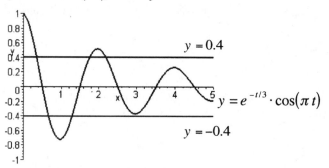

On the interval $0 \le t \le 3$, we can use the INTERSECT feature on a calculator to determine that $y = e^{-t/3} \cdot \cos(\pi t)$ intersects $y = 0.4$ when $t \approx 0.35$, $t \approx 1.75$ and $t \approx 2.19$, $y = e^{-t/3} \cdot \cos(\pi t)$ intersects $y = -0.4$ when $t \approx 0.67$ and $t \approx 1.28$ and the graph shows that $-0.4 < e^{-t/3} \cdot \cos(\pi t) < 0.4$ when $t = 3$. Therefore, the voltage V is between -0.4 and 0.4 on the intervals $0.35 < t < 0.67$, $1.28 < t < 1.75$ and $2.19 < t \le 3$.

31. $y = \sin(2\pi(852)t) + \sin(2\pi(1209)t)$

33. Graph $f(x) = \dfrac{\sin x}{x}$:

As x approaches 0, $\dfrac{\sin x}{x}$ approaches 1.

35. Graphing:

$y = \dfrac{1}{x}\sin x$ $y = \dfrac{1}{x^2}\sin x$ $y = \dfrac{1}{x^3}\sin x$

As x gets larger, the graph of $y = \dfrac{1}{x^n}\sin x$ gets closer to $y = 0$.

Chapter 7

Applications of Trigonometric Functions

7.R Chapter Review

1. $c = 10, \ \beta = 20°$

$$\sin\beta = \frac{b}{c} \ \rightarrow \ \sin(20°) = \frac{b}{10} \ \rightarrow \ b = 10\sin(20°) \approx 10(0.3420) \approx 3.42$$

$$\cos\beta = \frac{a}{c} \ \rightarrow \ \cos(20°) = \frac{a}{10} \ \rightarrow \ a = 10\cos(20°) \approx 10(0.9397) \approx 9.40$$

$$\alpha = 90°-\beta = 90°-20° = 70°$$

3. $b = 2, \ c = 5$

$$c^2 = a^2 + b^2 \ \rightarrow \ a^2 = c^2 - b^2 = 5^2 - 2^2 = 25 - 4 = 21 \ \rightarrow \ a = \sqrt{21} \approx 4.58$$

$$\sin\beta = \frac{b}{c} = \frac{2}{5} = 0.4000 \ \rightarrow \ \beta \approx 23.6°$$

$$\alpha = 90° - \beta = 90° - 23.6° = 66.4°$$

5. $\alpha = 50°, \ \beta = 30°, \ a = 1$

$$\gamma = 180° - \alpha - \beta = 180° - 50° - 30° = 100°$$

$$\frac{\sin\alpha}{a} = \frac{\sin\beta}{b} \ \rightarrow \ \frac{\sin(50°)}{1} = \frac{\sin(30°)}{b} \ \rightarrow \ b = \frac{1\sin(30°)}{\sin(50°)} \approx 0.65$$

$$\frac{\sin\gamma}{c} = \frac{\sin\alpha}{a} \ \rightarrow \ \frac{\sin(100°)}{c} = \frac{\sin(50°)}{1} \ \rightarrow \ c = \frac{1\sin(100°)}{\sin(50°)} \approx 1.29$$

7. $a = 5, \ c = 2, \ \alpha = 100°$

$$\frac{\sin\gamma}{c} = \frac{\sin\alpha}{a} \ \rightarrow \ \frac{\sin\gamma}{2} = \frac{\sin(100°)}{5} \ \rightarrow \ \sin\gamma = \frac{2\sin(100°)}{5} \approx 0.3939$$

$$\rightarrow \ \gamma = 23.2° \ \text{or} \ \gamma = 156.8°$$

The second value is discarded because $\alpha + \gamma > 180°$.

$$\beta = 180° - \alpha - \gamma = 180° - 100° - 23.2° = 56.8°$$

$$\frac{\sin\beta}{b} = \frac{\sin\alpha}{a} \ \rightarrow \ \frac{\sin(56.8°)}{b} = \frac{\sin(100°)}{5} \ \rightarrow \ b = \frac{5\sin(56.8°)}{\sin(100°)} \approx 4.25$$

9. $a = 3, \ c = 1, \ \gamma = 110°$

$$\frac{\sin\gamma}{c} = \frac{\sin\alpha}{a} \ \rightarrow \ \frac{\sin(110°)}{1} = \frac{\sin\alpha}{3} \ \rightarrow \ \sin\alpha = \frac{3\sin(110°)}{1} \approx 2.8191$$

There is no angle α for which $\sin\alpha > 1$. Therefore, there is no triangle with the given measurements.

11. $a = 3, \ c = 1, \ \beta = 100°$

$$b^2 = a^2 + c^2 - 2ac\cos\beta = 3^2 + 1^2 - 2\cdot 3\cdot 1\cos(100°) \approx 11.0419 \rightarrow b \approx 3.32$$

$$a^2 = b^2 + c^2 - 2bc\cos\alpha \quad\rightarrow\quad \cos\alpha = \frac{b^2+c^2-a^2}{2bc} = \frac{3.32^2+1^2-3^2}{2(3.32)(1)} \approx 0.4552 \rightarrow \alpha \approx 62.8°$$

$$c^2 = a^2 + b^2 - 2ab\cos\gamma \rightarrow \cos\gamma = \frac{a^2+b^2-c^2}{2ab} = \frac{3^2+3.32^2-1^2}{2(3)(3.32)} \approx 0.9549 \rightarrow \gamma \approx 17.2°$$

13. $a = 2, \ b = 3, \ c = 1$

$$a^2 = b^2 + c^2 - 2bc\cos\alpha \quad\rightarrow\quad \cos\alpha = \frac{b^2+c^2-a^2}{2bc} = \frac{3^2+1^2-2^2}{2(3)(1)} = 1.000 \rightarrow \alpha \approx 0°$$

No triangle exists with an angle of 0°.

15. $a = 1, \ b = 3, \ \gamma = 40°$

$$c^2 = a^2 + b^2 - 2ab\cos\gamma$$

$$c^2 = 1^2 + 3^2 - 2\cdot 1\cdot 3\cos(40°) \approx 5.4037$$

$$c \approx 2.32$$

$$a^2 = b^2 + c^2 - 2bc\cos\alpha \quad\rightarrow\quad \cos\alpha = \frac{b^2+c^2-a^2}{2bc} = \frac{3^2+2.32^2-1^2}{2(3)(2.32)} \approx 0.9614 \rightarrow \alpha \approx 16.1°$$

$$b^2 = a^2 + c^2 - 2ac\cos\beta \quad\rightarrow\quad \cos\beta = \frac{a^2+c^2-b^2}{2ac} = \frac{1^2+2.32^2-3^2}{2(1)(2.32)} \approx -0.5641 \rightarrow \beta \approx 123.9°$$

17. $a = 5, \ b = 3, \ \alpha = 80°$

$$\frac{\sin\beta}{b} = \frac{\sin\alpha}{a} \quad\rightarrow\quad \frac{\sin\beta}{3} = \frac{\sin(80°)}{5} \quad\rightarrow\quad \sin\beta = \frac{3\sin(80°)}{5} \approx 0.5909$$

$$\rightarrow \quad \beta = 36.2° \ \text{ or } \ \beta = 143.8°$$

The second value is discarded because $\alpha + \beta > 180°$.

$$\gamma = 180° - \alpha - \beta = 180° - 80° - 36.2° = 63.8°$$

$$\frac{\sin\gamma}{c} = \frac{\sin\alpha}{a}$$

$$\frac{\sin(63.8°)}{c} = \frac{\sin(80°)}{5}$$

$$c = \frac{5\sin(63.8°)}{\sin(80°)} \approx 4.55$$

19. $a = 1, \ b = \dfrac{1}{2}, \ c = \dfrac{4}{3}$

$$a^2 = b^2 + c^2 - 2bc\cos\alpha \rightarrow \cos\alpha = \frac{b^2+c^2-a^2}{2bc} = \frac{\left(\left(\frac{1}{2}\right)^2 + \left(\frac{4}{3}\right)^2 - 1^2\right)}{2\left(\frac{1}{2}\right)\left(\frac{4}{3}\right)} \approx 0.7708 \rightarrow \alpha \approx 39.6°$$

$$b^2 = a^2 + c^2 - 2ac\cos\beta \rightarrow \cos\beta = \frac{a^2+c^2-b^2}{2ac} = \frac{\left(1^2 + \left(\frac{4}{3}\right)^2 - \left(\frac{1}{2}\right)^2\right)}{2(1)\left(\frac{4}{3}\right)} \approx 0.9479 \rightarrow \beta \approx 18.6°$$

$$\gamma = 180° - \alpha - \beta \approx 180° - 39.6° - 18.6° \approx 121.9°$$

21. $a = 3, \ b = 4, \ \alpha = 10°$

$$\frac{\sin\beta}{b} = \frac{\sin\alpha}{a} \ \rightarrow \ \frac{\sin\beta}{4} = \frac{\sin(10°)}{3} \ \rightarrow \ \sin\beta = \frac{4\sin(10°)}{3} \approx 0.2315$$

$$\rightarrow \ \beta_1 \approx 13.4° \ \text{or} \ \beta_2 \approx 166.6°$$

For both values, $\alpha + \beta < 180°$. Therefore, there are two triangles.

$$\gamma_1 = 180° - \alpha - \beta_1 = 180° - 10° - 13.4° \approx 156.6°$$

$$\frac{\sin\alpha}{a} = \frac{\sin\gamma_1}{c_1} \ \rightarrow \ \frac{\sin(10°)}{3} = \frac{\sin(156.6°)}{c_1} \ \rightarrow \ c_1 = \frac{3\sin(156.6°)}{\sin(10°)} \approx 6.86$$

$$\gamma_2 = 180° - \alpha - \beta_2 = 180° - 10° - 166.6° \approx 3.4°$$

$$\frac{\sin\alpha}{a} = \frac{\sin\gamma_2}{c_2} \ \rightarrow \ \frac{\sin(10°)}{3} = \frac{\sin(3.4°)}{c_2} \ \rightarrow \ c_2 = \frac{3\sin(3.4°)}{\sin(10°)} \approx 1.02$$

Two triangles: $\beta_1 \approx 13.4°, \ \gamma_1 \approx 156.6°, \ c_1 \approx 6.86$
or $\beta_2 \approx 166.6°, \ \gamma_2 \approx 3.4°, \ c_2 \approx 1.02$

23. $b = 4, \ c = 5, \ \alpha = 70°$

$$a^2 = b^2 + c^2 - 2bc\cos\alpha$$

$$a^2 = 4^2 + 5^2 - 2 \cdot 4 \cdot 5\cos(70°) \approx 27.3192$$

$$a \approx 5.23$$

$$c^2 = a^2 + b^2 - 2ab\cos\gamma \ \rightarrow \ \cos\gamma = \frac{a^2+b^2-c^2}{2ab} = \frac{5.23^2 + 4^2 - 5^2}{2(5.23)(4)} \approx 0.4386 \rightarrow \gamma \approx 64.0°$$

$$\beta = 180° - \alpha - \gamma = 180° - 70° - 64° \approx 46.0°$$

25. $a = 2, \ b = 3, \ \gamma = 40°$

$$A = \frac{1}{2}ab\sin\gamma = \frac{1}{2}(2)(3)\sin(40°) \approx 1.93$$

27. $b = 4, \ c = 10, \ \alpha = 70°$

$$A = \frac{1}{2}bc\sin\alpha = \frac{1}{2}(4)(10)\sin(70°) \approx 18.79$$

29. $a = 4, \ b = 3, \ c = 5$

$$s = \frac{1}{2}(a+b+c) = \frac{1}{2}(4+3+5) = 6$$

$$A = \sqrt{s(s-a)(s-b)(s-c)} = \sqrt{6(2)(3)(1)} = \sqrt{36} = 6$$

31. $a = 4$, $b = 2$, $c = 5$

$$s = \frac{1}{2}(a + b + c) = \frac{1}{2}(4 + 2 + 5) = \frac{11}{2}$$

$$A = \sqrt{s(s - a)(s - b)(s - c)} = \sqrt{\frac{11}{2}\left(\frac{3}{2}\right)\left(\frac{7}{2}\right)\left(\frac{1}{2}\right)} = \sqrt{\frac{231}{16}} \approx 3.80$$

33. $\alpha = 50°$, $\beta = 30°$, $a = 1$ $\gamma = 180° - \alpha - \beta = 180° - 50° - 30° = 100°$

$$A = \frac{a^2 \sin\beta \sin\gamma}{2\sin\alpha} = \frac{1^2 \sin(30°)\sin(100°)}{2\sin(50°)} \approx \frac{1(0.5000)(0.9848)}{2(0.7660)} \approx 0.32$$

35. Use right triangle methods:

$$\tan(65°) = \frac{500}{b} \quad \rightarrow \quad b = \frac{500}{\tan(65°)} \approx 233.15$$

$$\tan(25°) = \frac{500}{a + b} \quad \rightarrow \quad a + b = \frac{500}{\tan(25°)} \approx 1072.25$$

$a = 1072.25 - 233.15 = 839.1$ feet

The lake is approximately 839 feet long.

37. $\tan(25°) = \dfrac{b}{50} \quad \rightarrow \quad b = 50\tan(25°) \approx 50(0.4663) \approx 23.32$ feet

39. 1454 ft ≈ 0.2754 miles

$$\tan(5°) \approx \frac{0.2754}{a + 1}$$

$$a + 1 = \frac{0.2754}{\tan(5°)} \approx 3.15 \quad \rightarrow \quad a \approx 2.15 \text{ miles}$$

The boat is about 2.15 miles offshore.

41. $\angle ABC = 180° - 20° = 160°$

Find the angle at city C:

$$\frac{\sin C}{100} = \frac{\sin(160°)}{300}$$

$$\sin C = \frac{100\sin(160°)}{300} \approx 0.1140 \quad \rightarrow \quad C \approx 6.55°$$

Find the angle at city A :

$$A = 180° - 160° - 6.55° = 13.45°$$

$$\frac{\sin(13.45°)}{y} = \frac{\sin(160°)}{300}$$

$$y = \frac{300\sin(13.45°)}{\sin(160°)} \approx 204.07 \text{ miles}$$

The distance from city B to city C is approximately 204 miles.

43.   Draw a line perpendicular to the shore:

   (a) $\angle ACB = 12° + 30° = 42°$

        $\angle ABC = 90° - 30° = 60°$

        $\angle CAB = 90° - 12° = 78°$

$$\frac{\sin(60°)}{b} = \frac{\sin(42°)}{2} \quad \rightarrow \quad b = \frac{2\sin(60°)}{\sin(42°)} \approx 2.59 \text{ miles}$$

   (b) $\dfrac{\sin(78°)}{a} = \dfrac{\sin(42°)}{2} \quad \rightarrow \quad a = \dfrac{2\sin(78°)}{\sin(42°)} \approx 2.92 \text{ miles}$

   (c) $\cos 12° = \dfrac{x}{2.59} \quad \rightarrow \quad x = 2.59\cos 12° \approx 2.53 \text{ miles}$

45.   (a)   After 4 hours, the yacht would have sailed $18(4) = 72$ miles.
        Find the third side of the triangle determines the distance from the island:
        $a = 72$, $b = 200$, $\gamma = 15°$

        $c^2 = a^2 + b^2 - 2ab\cos\gamma$

        $c^2 = 72^2 + 200^2 - 2 \cdot 72 \cdot 200\cos(15°) \approx 17365.34$

        $c \approx 131.8$ miles
        The yacht is about 131.8 miles from the island.

   (b)   Find the measure of the angle opposite the 200 side:

$$\cos\beta = \frac{a^2 + c^2 - b^2}{2ac} \rightarrow \cos\beta = \frac{72^2 + 131.8^2 - 200^2}{2(72)(131.8)} \approx -0.9192 \rightarrow \beta \approx 156.8°$$

        The yacht should turn through an angle of $180° - 156.8° = 23.1°$ to correct its course.

   (c)   The original trip would have taken: $t = \dfrac{200}{18} \approx 11.1$ hours.

        The actual trip takes: $t = 4 + \dfrac{131.8}{18} \approx 4 + 7.3 \approx 11.3$ hours.

        The trip takes about 0.21 hour or 12 minutes longer.

47.   Find the lengths of the two unknown sides of the middle triangle:

        $x^2 = 100^2 + 125^2 - 2(100)(125)\cos(50°) \approx 9555.31 \rightarrow x \approx 97.75$ feet

        $y^2 = 70^2 + 50^2 - 2(70)(50)\cos(100°) \approx 8615.54 \rightarrow y \approx 92.82$ feet

   Find the areas of the three triangles:

        $A_1 = \dfrac{1}{2}(100)(125)\sin(50°) \approx 4787.78 \text{ ft}^2$

        $A_2 = \dfrac{1}{2}(50)(70)\sin(100°) \approx 1723.41 \text{ ft}^2$

        $s = \dfrac{1}{2}(50 + 97.75 + 92.82) = 120.285$

        $A_3 = \sqrt{120.285(70.285)(22.535)(27.465)} \approx 2287.47$

   The approximate area of the lake is $4787.78 + 1723.41 + 2287.47 = 8798.67$ sq.ft.

49. Area of the segment = area of the sector - area of the triangle.

$$\text{Area of sector} = \frac{1}{2}r^2\theta = \frac{1}{2}\cdot 6^2\left(50\cdot\frac{\pi}{180}\right) \approx 15.708 \text{ in}^2$$

$$\text{Area of triangle} = \frac{1}{2}ab\sin\theta = \frac{1}{2}\cdot 6\cdot 6\sin(50°) \approx 13.789 \text{ in}^2$$

Area of segment $= 15.708 - 13.789 = 1.92 \text{ in}^2$

51. Extend the tangent line until it meets a line extended through the centers of the pulleys. Label these extensions x and y. The distance between the points of tangency is z. Two similar triangles are formed. Therefore:

$$\frac{24+y}{y} = \frac{6.5}{2.5}$$

where $24 + y$ is the hypotenuse of the larger triangle and y is the hypotenuse of the smaller triangle.

Solve for y:

$6.5y = 2.5(24 + y)$

$6.5y = 60 + 2.5y$

$4y = 60 \rightarrow y = 15$

Use the Pythagorean Theorem to find x:

$x^2 + 2.5^2 = 15^2$

$x^2 = 225 - 6.25 = 218.75$

$x \approx 14.79$

Use the Pythagorean Theorem to find z:

$$(z+14.79)^2 + 6.5^2 = (24+15)^2 \rightarrow (z+14.79)^2 = 1521 - 42.25 = 1478.75$$

$$z + 14.79 = 38.45 \rightarrow z = 23.66$$

Find α:

$$\cos\alpha = \frac{2.5}{15} \approx 0.1667 \quad \rightarrow \quad \alpha \approx 1.4033 \text{ radians}$$

$\beta = \pi - 1.4033 \approx 1.7383$ radians

The arc length on the larger pulley is: $6.5(1.7383) = 11.30$ inches.
The arc length on the smaller pulley is $2.5(1.4033) = 3.51$ inches.
The distance between the points of tangency is 23.66 inches.
The length of the belt is: $2(11.30 + 3.51 + 23.66) = 76.94$ inches.

53. $d = 6\sin(2t)$
    (a) Simple harmonic
    (b) 6 feet
    (c) π seconds
    (d) $\frac{1}{\pi}$ oscillation/second

55. $d = -2\cos(\pi t)$
    (a) Simple harmonic
    (b) 2 feet
    (c) 2 seconds
    (d) $\frac{1}{2}$ oscillation/second

57. $y = e^{-x/2\pi} \sin(2x)$, $0 \le x \le 2\pi$

59. $y = x \cos(x)$, $0 \le x \le 2\pi$

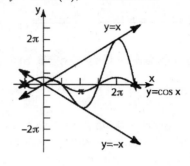

61. $y = 2\sin(x) + \cos(2x)$, $0 \le x \le 2\pi$

Chapter 8

Polar Coordinates; Vectors

8.1 Polar Coordinates

1. A   3. C   5. B   7. A

9. $(3, 90°)$ 11. $(-2, 0)$

13. $\left(6, \dfrac{\pi}{6}\right)$ 15. $(-2, 135°)$

17. $\left(-1, -\dfrac{\pi}{3}\right)$ 19. $(-2, -\pi)$

21. $\left(5, \frac{2\pi}{3}\right)$

(a) $r > 0, \; -2\pi \le \theta < 0$ $\left(5, -\frac{4\pi}{3}\right)$

(b) $r < 0, \; 0 \le \theta < 2\pi$ $\left(-5, \frac{5\pi}{3}\right)$

(c) $r > 0, \; 2\pi \le \theta < 4\pi$ $\left(5, \frac{8\pi}{3}\right)$

23. $(-2, 3\pi)$

(a) $r > 0, \; -2\pi \le \theta < 0$ $(2, -2\pi)$

(b) $r < 0, \; 0 \le \theta < 2\pi$ $(-2, \pi)$

(c) $r > 0, \; 2\pi \le \theta < 4\pi$ $(2, 2\pi)$

25. $\left(1, \frac{\pi}{2}\right)$

(a) $r > 0, \; -2\pi \le \theta < 0$ $\left(1, -\frac{3\pi}{2}\right)$

(b) $r < 0, \; 0 \le \theta < 2\pi$ $\left(-1, \frac{3\pi}{2}\right)$

(c) $r > 0, \; 2\pi \le \theta < 4\pi$ $\left(1, \frac{5\pi}{2}\right)$

27. $\left(-3, -\frac{\pi}{4}\right)$

(a) $r > 0, \; -2\pi \le \theta < 0$ $\left(3, -\frac{5\pi}{4}\right)$

(b) $r < 0, \; 0 \le \theta < 2\pi$ $\left(-3, \frac{7\pi}{4}\right)$

(c) $r > 0, \; 2\pi \le \theta < 4\pi$ $\left(3, \frac{11\pi}{4}\right)$

29. $x = r\cos\theta = 3\cos\left(\frac{\pi}{2}\right) = 3 \cdot 0 = 0$

$y = r\sin\theta = 3\sin\left(\frac{\pi}{2}\right) = 3 \cdot 1 = 3$

The rectangular coordinates of the

point $\left(3, \frac{\pi}{2}\right)$ are $(0, 3)$.

31. $x = r\cos\theta = -2\cos(0) = -2 \cdot 1 = -2$

$y = r\sin\theta = -2\sin(0) = -2 \cdot 0 = 0$

The rectangular coordinates of the

point $(-2, 0)$ are $(-2, 0)$.

33. $x = r\cos\theta = 6\cos(150°)$

$$= 6 \cdot \left(-\frac{\sqrt{3}}{2}\right) = -3\sqrt{3}$$

$$y = r\sin\theta = 6\sin(150°) = 6 \cdot \frac{1}{2} = 3$$

The rectangular coordinates of the point $(6, 150°)$ are $(-3\sqrt{3}, 3)$.

35. $x = r\cos\theta = -2\cos\left(\frac{3\pi}{4}\right) = -2 \cdot \left(-\frac{\sqrt{2}}{2}\right) = \sqrt{2}$

$$y = r\sin\theta = -2\sin\left(\frac{3\pi}{4}\right) = -2 \cdot \frac{\sqrt{2}}{2} = -\sqrt{2}$$

The rectangular coordinates of the point $\left(-2, \frac{3\pi}{4}\right)$ are $\left(\sqrt{2}, -\sqrt{2}\right)$.

37. $x = r\cos\theta = -1\cos\left(-\frac{\pi}{3}\right) = -1 \cdot \frac{1}{2} = -\frac{1}{2}$

$$y = r\sin\theta = -1\sin\left(-\frac{\pi}{3}\right) = -1 \cdot \left(-\frac{\sqrt{3}}{2}\right) = \frac{\sqrt{3}}{2}$$

The rectangular coordinates of the point $\left(-1, -\frac{\pi}{3}\right)$ are $\left(-\frac{1}{2}, \frac{\sqrt{3}}{2}\right)$.

39. $x = r\cos\theta = -2\cos(-180°) = -2 \cdot -1 = 2$
$y = r\sin\theta = -2\sin(-180°) = -2 \cdot 0 = 0$

The rectangular coordinates of the point $(-2, -180°)$ are $(2, 0)$.

41. $x = r\cos\theta = 7.5\cos(110°) \approx 7.5(-0.3420) \approx -2.57$
$y = r\sin\theta = 7.5\sin(110°) \approx 7.5(0.9397) \approx 7.05$

The rectangular coordinates of the point $(7.5, 110°)$ are $(-2.57, 7.05)$.

43. $x = r\cos\theta = 6.3\cos(3.8) \approx 6.3(-0.7910) \approx -4.98$
$y = r\sin\theta = 6.3\sin(3.8) \approx 6.3(-0.6119) \approx -3.85$

The rectangular coordinates of the point $(6.3, 3.8)$ are $(-4.98, -3.85)$.

45. $r = \sqrt{x^2 + y^2} = \sqrt{3^2 + 0^2} = \sqrt{9} = 3$ $\theta = \tan^{-1}\left(\frac{y}{x}\right) = \tan^{-1}\left(\frac{0}{3}\right) = \tan^{-1}(0) = 0$

Polar coordinates of the point $(3, 0)$ are $(3, 0)$.

47. $r = \sqrt{x^2 + y^2} = \sqrt{(-1)^2 + 0^2} = \sqrt{1} = 1$ $\theta = \tan^{-1}\left(\frac{y}{x}\right) = \tan^{-1}\left(\frac{0}{-1}\right) = \tan^{-1}(0) = 0$

The point lies on the negative x-axis thus $\theta = \pi$.
Polar coordinates of the point $(-1, 0)$ are $(1, \pi)$.

49. The point $(1, -1)$ lies in quadrant IV.

$r = \sqrt{x^2 + y^2} = \sqrt{1^2 + (-1)^2} = \sqrt{2}$ $\theta = \tan^{-1}\left(\frac{y}{x}\right) = \tan^{-1}\left(\frac{-1}{1}\right) = \tan^{-1}(-1) = -\frac{\pi}{4}$

Polar coordinates of the point $(1, -1)$ are $\left(\sqrt{2}, -\frac{\pi}{4}\right)$.

51. The point $\left(\sqrt{3}, 1\right)$ lies in quadrant I.

$r = \sqrt{x^2 + y^2} = \sqrt{\left(\sqrt{3}\right)^2 + 1^2} = \sqrt{4} = 2$ $\theta = \tan^{-1}\left(\frac{y}{x}\right) = \tan^{-1}\left(\frac{1}{\sqrt{3}}\right) = \frac{\pi}{6}$

Polar coordinates of the point $\left(\sqrt{3}, 1\right)$ are $\left(2, \frac{\pi}{6}\right)$.

53. The point $(1.3, -2.1)$ lies in quadrant IV.

$r = \sqrt{x^2 + y^2} = \sqrt{1.3^2 + (-2.1)^2} = \sqrt{6.1} \approx 2.47$

$\theta = \tan^{-1}\left(\frac{y}{x}\right) = \tan^{-1}\left(\frac{-2.1}{1.3}\right) \approx \tan^{-1}(-1.6154) \approx -1.02$

Polar coordinates of the point $(1.3, -2.1)$ are $(2.47, -1.02)$.

55. The point $(8.3, 4.2)$ lies in quadrant I.

$r = \sqrt{x^2 + y^2} = \sqrt{8.3^2 + 4.2^2} = \sqrt{86.53} \approx 9.30$

$\theta = \tan^{-1}\left(\frac{y}{x}\right) = \tan^{-1}\left(\frac{4.2}{8.3}\right) \approx \tan^{-1}(0.5060) \approx 0.47$

Polar coordinates of the point $(8.3, 4.2)$ are $(9.30, 0.47)$.

57. $2x^2 + 2y^2 = 3$
$2(x^2 + y^2) = 3$

$2r^2 = 3 \rightarrow r^2 = \frac{3}{2}$

59. $$x^2 = 4y$$
$$(r\cos\theta)^2 = 4r\sin\theta$$
$$r^2\cos^2\theta - 4r\sin\theta = 0$$

61. $2xy = 1$
$2(r\cos\theta)(r\sin\theta) = 1$

$2r^2\sin\theta\cos\theta = 1 \rightarrow r^2\sin2\theta = 1$

62. $4x^2 y = 1$
$4(r\cos\theta)^2 r\sin\theta = 1$

$4r^2\cos^2\theta\, r\sin\theta = 1 \rightarrow 4r^3\cos^2\theta\sin\theta = 1$

63. $x = 4$
$r\cos\theta = 4$

65. $r = \cos\theta$
$r^2 = r\cos\theta$

$x^2 + y^2 = x \rightarrow x^2 - x + y^2 = 0$

67. $r^2 = \cos\theta$
$r^3 = r\cos\theta$

$\left(x^2 + y^2\right)^{3/2} = x \rightarrow \left(x^2 + y^2\right)^{3/2} - x = 0$

69. $r = 2$
$\sqrt{x^2 + y^2} = 2 \rightarrow x^2 + y^2 = 4$

71. $r = \dfrac{4}{1 - \cos\theta}$
$r(1 - \cos\theta) = 4 \rightarrow r - r\cos\theta = 4$

$\sqrt{x^2 + y^2} - x = 4 \rightarrow \sqrt{x^2 + y^2} = x + 4$

$x^2 + y^2 = x^2 + 8x + 16$

$y^2 = 8(x + 2)$

73.   Rewrite the polar coordinates in rectangular form:

$P_1 = (r_1, \theta_1) \quad \rightarrow \quad P_1 = (r_1 \cos \theta_1, r_1 \sin \theta_1)$

$P_2 = (r_2, \theta_2) \quad \rightarrow \quad P_2 = (r_2 \cos \theta_2, r_2 \sin \theta_2)$

$d = \sqrt{(r_2 \cos \theta_2 - r_1 \cos \theta_1)^2 + (r_2 \sin \theta_2 - r_1 \sin \theta_1)^2}$

$= \sqrt{r_2^2 \cos^2 \theta_2 - 2r_1 r_2 \cos \theta_2 \cos \theta_1 + r_1^2 \cos^2 \theta_1 + r_2^2 \sin^2 \theta_2 - 2r_1 r_2 \sin \theta_2 \sin \theta_1 + r_1^2 \sin^2 \theta_1}$

$= \sqrt{r_2^2 (\cos^2 \theta_2 + \sin^2 \theta_2) + r_1^2 (\cos^2 \theta_1 + \sin^2 \theta_1) - 2r_1 r_2 (\cos \theta_2 \cos \theta_1 + \sin \theta_2 \sin \theta_1)}$

$= \sqrt{r_2^2 + r_1^2 - 2r_1 r_2 \cos(\theta_2 - \theta_1)}$

Chapter 8

Polar Coordinates; Vectors

8.2   Polar Equations and Graphs

1. $r = 4$

The equation is of the form $r = a, \ a > 0$.
It is a circle, center at the pole and radius 4.
Transform to rectangular form:

$$r = 4$$
$$r^2 = 16$$
$$x^2 + y^2 = 16$$

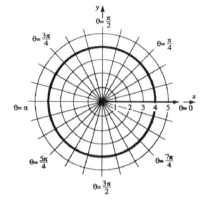

3. $\theta = \dfrac{\pi}{3}$

The equation is of the form $\theta = \alpha$. It is a
line, passing through the pole at an angle of
$\dfrac{\pi}{3}$.

Transform to rectangular form:

$$\theta = \frac{\pi}{3}$$

$$\tan\theta = \tan\left(\frac{\pi}{3}\right)$$

$$\frac{y}{x} = \sqrt{3} \ \rightarrow \ y = \sqrt{3}x$$

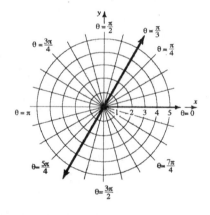

5. $r\sin\theta = 4$

The equation is of the form $r\sin\theta = b$. It is
a horizontal line, 4 units above the pole.
Transform to rectangular form:

$$r\sin\theta = 4$$
$$y = 4$$

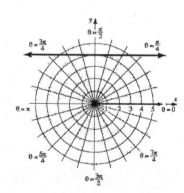

7. $r \cos \theta = -2$

The equation is of the form $r \cos \theta = a$. It is a vertical line, 2 units to the left of the pole. Transform to rectangular form:

$$r \cos \theta = -2$$
$$x = -2$$

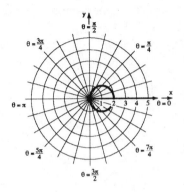

9. $r = 2 \cos \theta$

The equation is of the form $r = 2a \cos \theta$, $a > 0$. It is a circle, passing through the pole, and center on the polar axis.

Transform to rectangular form:

$$r = 2 \cos \theta$$
$$r^2 = 2r \cos \theta$$
$$x^2 + y^2 = 2x$$
$$x^2 - 2x + y^2 = 0$$
$$(x-1)^2 + y^2 = 1$$

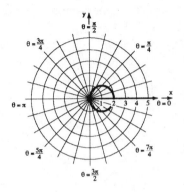

11. $r = -4 \sin \theta$

The equation is of the form $r = 2a \sin \theta$, $a > 0$. It is a circle, passing through the pole, and center on the line $\theta = \dfrac{\pi}{2}$.

Transform to rectangular form:

$$r = -4 \sin \theta$$
$$r^2 = -4r \sin \theta$$
$$x^2 + y^2 = -4y$$
$$x^2 + y^2 + 4y = 0$$
$$x^2 + (y+2)^2 = 4$$

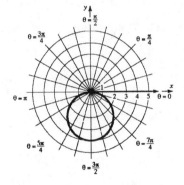

13. $r\sec\theta = 4$

Transform to rectangular form:
$$r\sec\theta = 4$$
$$r\cdot\frac{1}{\cos\theta} = 4$$
$$r = 4\cos\theta$$
$$r^2 = 4r\cos\theta$$
$$x^2 + y^2 = 4x$$
$$x^2 - 4x + y^2 = 0$$
$$(x-2)^2 + y^2 = 4$$

The equation is a circle, passing through the pole, center on the polar axis and radius 2.

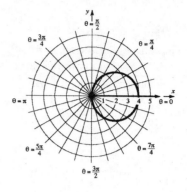

15. $r\csc\theta = -2$

Transform to rectangular form:
$$r\csc\theta = -2$$
$$r\cdot\frac{1}{\sin\theta} = -2$$
$$r = -2\sin\theta$$
$$r^2 = -2r\sin\theta$$
$$x^2 + y^2 = -2y$$
$$x^2 + y^2 + 2y = 0$$
$$x^2 + (y+1)^2 = 1$$

The equation is a circle, passing through the pole, center on the line $\theta = \dfrac{\pi}{2}$ and radius 1.

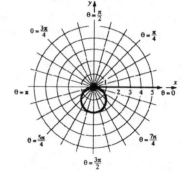

17. E   19. F   21. H   23. D

25. $r = 2 + 2\cos\theta$ The graph will be a cardioid. Check for symmetry:

Polar axis: Replace θ by $-\theta$. The result is $r = 2 + 2\cos(-\theta) = 2 + 2\cos\theta$.
  The graph is symmetric with respect to the polar axis.

The line $\theta = \dfrac{\pi}{2}$: Replace θ by $\pi - \theta$.

$$r = 2 + 2\cos(\pi - \theta) = 2 + 2(\cos(\pi)\cos\theta + \sin(\pi)\sin\theta)$$
$$= 2 + 2(-\cos\theta + 0) = 2 - 2\cos\theta$$

  The test fails.

The pole: Replace r by $-r$. $-r = 2 + 2\cos\theta$. The test fails.

Due to symmetry to the polar axis, assign values to θ from 0 to π.

θ	0	$\dfrac{\pi}{6}$	$\dfrac{\pi}{3}$	$\dfrac{\pi}{2}$	$\dfrac{2\pi}{3}$	$\dfrac{5\pi}{6}$	π
$r = 2 + 2\cos\theta$	4	$2+\sqrt{3} \approx 3.7$	3	2	1	$2-\sqrt{3} \approx 0.3$	0

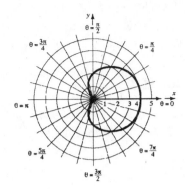

27. $r = 3 - 3\sin\theta$ The graph will be a cardioid. Check for symmetry:
Polar axis: Replace θ by $-\theta$. The result is $r = 3 - 3\sin(-\theta) = 3 + 3\sin\theta$.
The test fails.

The line $\theta = \dfrac{\pi}{2}$: Replace θ by $\pi - \theta$.

$$r = 3 - 3\sin(\pi - \theta) = 3 - 3(\sin(\pi)\cos\theta - \cos(\pi)\sin\theta)$$

$$= 3 - 3(0 + \sin\theta) = 3 - 3\sin\theta$$

The graph is symmetric with respect to the line $\theta = \dfrac{\pi}{2}$.

The pole: Replace r by $-r$. $-r = 3 - 3\sin\theta$. The test fails.

Due to symmetry to the line $\theta = \dfrac{\pi}{2}$, assign values to θ from $-\dfrac{\pi}{2}$ to $\dfrac{\pi}{2}$.

θ	$-\dfrac{\pi}{2}$	$-\dfrac{\pi}{3}$	$-\dfrac{\pi}{6}$	0	$\dfrac{\pi}{6}$	$\dfrac{\pi}{3}$	$\dfrac{\pi}{2}$
$r = 3 - 3\sin\theta$	6	$3 + \dfrac{3\sqrt{3}}{2} \approx 5.6$	$\dfrac{9}{2}$	3	$\dfrac{3}{2}$	$3 - \dfrac{3\sqrt{3}}{2} \approx 0.4$	0

29. $r = 2 + \sin\theta$ The graph will be a limacon without an inner loop.
Check for symmetry:
Polar axis: Replace θ by $-\theta$. The result is $r = 2 + \sin(-\theta) = 2 - \sin\theta$.
The test fails.

The line $\theta = \dfrac{\pi}{2}$: Replace θ by $\pi - \theta$.

$$r = 2 + \sin(\pi - \theta) = 2 + (\sin(\pi)\cos\theta - \cos(\pi)\sin\theta)$$

$$= 2 + (0 + \sin\theta) = 2 + \sin\theta$$

The graph is symmetric with respect to the line $\theta = \frac{\pi}{2}$.

The pole: Replace r by $-r$. $-r = 2 + \sin\theta$. The test fails.

Due to symmetry to the line $\theta = \frac{\pi}{2}$, assign values to θ from $-\frac{\pi}{2}$ to $\frac{\pi}{2}$.

θ	$-\frac{\pi}{2}$	$-\frac{\pi}{3}$	$-\frac{\pi}{6}$	0	$\frac{\pi}{6}$	$\frac{\pi}{3}$	$\frac{\pi}{2}$
$r = 2 + \sin\theta$	1	$2 - \frac{\sqrt{3}}{2} \approx 1.1$	$\frac{3}{2}$	2	$\frac{5}{2}$	$2 + \frac{\sqrt{3}}{2} \approx 2.9$	3

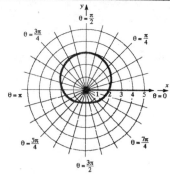

31. $r = 4 - 2\cos\theta$ The graph will be a limacon without an inner loop.
Check for symmetry:
Polar axis: Replace θ by $-\theta$. The result is $r = 4 - 2\cos(-\theta) = 4 - 2\cos\theta$.

The graph is symmetric with respect to the polar axis.

The line $\theta = \frac{\pi}{2}$: Replace θ by $\pi - \theta$.

$$r = 4 - 2\cos(\pi - \theta) = 4 - 2(\cos(\pi)\cos\theta + \sin(\pi)\sin\theta)$$

$$= 4 - 2(-\cos\theta + 0) = 4 + 2\cos\theta$$

The test fails.

The pole: Replace r by $-r$. $-r = 4 - 2\cos\theta$. The test fails.

Due to symmetry to the polar axis, assign values to θ from 0 to π.

θ	0	$\frac{\pi}{6}$	$\frac{\pi}{3}$	$\frac{\pi}{2}$	$\frac{2\pi}{3}$	$\frac{5\pi}{6}$	π
$r = 4 - 2\cos\theta$	2	$4 - \sqrt{3} \approx 2.3$	3	4	5	$4 + \sqrt{3} \approx 5.7$	6

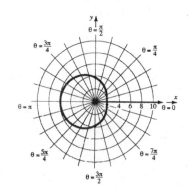

33. $r = 1 + 2\sin\theta$ The graph will be a limacon with an inner loop.
Check for symmetry:
Polar axis: Replace θ by $-\theta$. The result is $r = 1 + 2\sin(-\theta) = 1 - 2\sin\theta$.
              The test fails.

The line $\theta = \dfrac{\pi}{2}$: Replace θ by $\pi - \theta$.

$$r = 1 + 2\sin(\pi - \theta) = 1 + 2(\sin(\pi)\cos\theta - \cos(\pi)\sin\theta)$$

$$= 1 + 2(0 + \sin\theta) = 1 + 2\sin\theta$$

The graph is symmetric with respect to the line $\theta = \dfrac{\pi}{2}$.

The pole: Replace r by $-r$. $-r = 1 + 2\sin\theta$. The test fails.

Due to symmetry to the line $\theta = \dfrac{\pi}{2}$, assign values to θ from $-\dfrac{\pi}{2}$ to $\dfrac{\pi}{2}$.

θ	$-\dfrac{\pi}{2}$	$-\dfrac{\pi}{3}$	$-\dfrac{\pi}{6}$	0	$\dfrac{\pi}{6}$	$\dfrac{\pi}{3}$	$\dfrac{\pi}{2}$
$r = 1 + 2\sin\theta$	-1	$1 - \sqrt{3} \approx -0.7$	0	1	2	$1 + \sqrt{3} \approx 2.7$	3

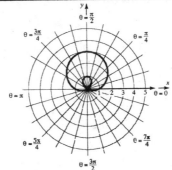

35. $r = 2 - 3\cos\theta$ The graph will be a limacon with an inner loop.
Check for symmetry:
Polar axis: Replace θ by $-\theta$. The result is $r = 2 - 3\cos(-\theta) = 2 - 3\cos\theta$.
              The graph is symmetric with respect to the polar axis.

The line $\theta = \dfrac{\pi}{2}$: Replace θ by $\pi - \theta$.

$$r = 2 - 3\cos(\pi - \theta) = 2 - 3(\cos(\pi)\cos\theta + \sin(\pi)\sin\theta)$$

$$= 2 - 3(-\cos\theta + 0) = 2 + 3\cos\theta$$

The test fails.

The pole: Replace r by $-r$. $-r = 2 - 3\cos\theta$. The test fails.

Due to symmetry to the polar axis, assign values to θ from 0 to π.

θ	0	$\dfrac{\pi}{6}$	$\dfrac{\pi}{3}$	$\dfrac{\pi}{2}$	$\dfrac{2\pi}{3}$	$\dfrac{5\pi}{6}$	π
$r = 2 - 3\cos\theta$	-1	$2 - \dfrac{3\sqrt{3}}{2} \approx -0.6$	$\dfrac{1}{2}$	2	$\dfrac{7}{2}$	$2 + \dfrac{3\sqrt{3}}{2} \approx 4.6$	5

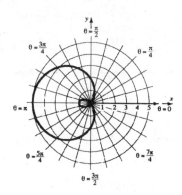

37. $r = 3\cos(2\theta)$ The graph will be a rose with four petals. Check for symmetry:
Polar axis: Replace θ by $-\theta$. $r = 3\cos(2(-\theta)) = 3\cos(-2\theta) = 3\cos(2\theta)$.
        The graph is symmetric with respect to the polar axis.

The line $\theta = \dfrac{\pi}{2}$: Replace θ by $\pi - \theta$.

$$r = 3\cos(2(\pi - \theta)) = 3\cos(2\pi - 2\theta)$$

$$= 3(\cos(2\pi)\cos(2\theta) + \sin(2\pi)\sin(2\theta)) = 3(\cos 2\theta + 0) = 3\cos(2\theta)$$

The graph is symmetric with respect to the line $\theta = \dfrac{\pi}{2}$.

The pole: Since the graph is symmetric to both the polar axis and the line
            $\theta = \dfrac{\pi}{2}$, it is also symmetric to the pole.

Due to symmetry, assign values to θ from 0 to $\dfrac{\pi}{2}$.

θ	0	$\dfrac{\pi}{6}$	$\dfrac{\pi}{4}$	$\dfrac{\pi}{3}$	$\dfrac{\pi}{2}$
$r = 3\cos 2\theta$	3	$\dfrac{3}{2}$	0	$-\dfrac{3}{2}$	-3

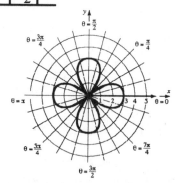

39. $r = 4\sin(5\theta)$ The graph will be a rose with five petals. Check for symmetry:
Polar axis: Replace θ by $-\theta$. $r = 4\sin(5(-\theta)) = 4\sin(-5\theta) = -4\sin(5\theta)$.
            The test fails.

The line $\theta = \dfrac{\pi}{2}$: Replace θ by $\pi - \theta$.

$$r = 4\sin(5(\pi - \theta)) = 4\sin(5\pi - 5\theta)$$

$$= 4(\sin(5\pi)\cos(5\theta) - \cos(5\pi)\sin(5\theta)) = 4(0 + \sin(5\theta)) = 4\sin(5\theta)$$

The graph is symmetric with respect to the line $\theta = \frac{\pi}{2}$.

The pole: Replace r by $-r$. $-r = 4\sin(5\theta)$. The test fails.

Due to symmetry to the line $\theta = \frac{\pi}{2}$, assign values to θ from $-\frac{\pi}{2}$ to $\frac{\pi}{2}$.

θ	$-\frac{\pi}{2}$	$-\frac{\pi}{3}$	$-\frac{\pi}{4}$	$-\frac{\pi}{6}$	0	$\frac{\pi}{6}$	$\frac{\pi}{4}$	$\frac{\pi}{3}$	$\frac{\pi}{2}$
$r = 4\sin(5\theta)$	-4	$2\sqrt{3} \approx 3.5$	$-2\sqrt{2} \approx -2.8$	-2	0	2	$-2\sqrt{2} \approx -2.8$	$-2\sqrt{3} \approx -3.5$	4

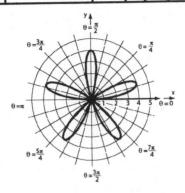

41. $r^2 = 9\cos(2\theta)$ The graph will be a lemniscate. Check for symmetry:
Polar axis: Replace θ by $-\theta$. $r^2 = 9\cos(2(-\theta)) = 9\cos(-2\theta) = 9\cos(2\theta)$.
          The graph is symmetric with respect to the polar axis.

The line $\theta = \frac{\pi}{2}$: Replace θ by $\pi - \theta$.

$$r^2 = 9\cos(2(\pi - \theta)) = 9\cos(2\pi - 2\theta)$$
$$= 9(\cos(2\pi)\cos 2\theta + \sin(2\pi)\sin 2\theta) = 9(\cos 2\theta + 0) = 9\cos(2\theta)$$

The graph is symmetric with respect to the line $\theta = \frac{\pi}{2}$.

The pole: Since the graph is symmetric to both the polar axis and the line
          $\theta = \frac{\pi}{2}$, it is also symmetric to the pole.

Due to symmetry, assign values to θ from 0 to $\frac{\pi}{2}$.

θ	0	$\frac{\pi}{6}$	$\frac{\pi}{4}$	$\frac{\pi}{3}$	$\frac{\pi}{2}$
$r = \pm\sqrt{9\cos(2\theta)}$	± 3	$\pm\frac{3\sqrt{2}}{2} \approx \pm 2.1$	0	not defined	not defined

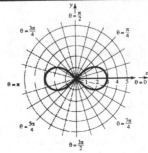

43. $r = 2^{\theta}$ The graph will be a spiral. Check for symmetry:
Polar axis: Replace θ by $-\theta$. $r = 2^{-\theta}$. The test fails.

The line $\theta = \dfrac{\pi}{2}$: Replace θ by $\pi - \theta$. $r = 2^{\pi-\theta}$. The test fails.

The pole: Replace r by $-r$. $-r = 2^{\theta}$. The test fails.

θ	$-\pi$	$-\dfrac{\pi}{2}$	$-\dfrac{\pi}{4}$	0	$\dfrac{\pi}{4}$	$\dfrac{\pi}{2}$	π	$\dfrac{3\pi}{2}$	2π
$r = 2^{\theta}$	0.1	0.3	0.6	1	1.7	3.0	8.8	26.2	77.9

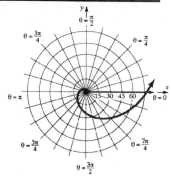

45. $r = 1 - \cos\theta$ The graph will be a cardioid. Check for symmetry:
Polar axis: Replace θ by $-\theta$. The result is $r = 1 - \cos(-\theta) = 1 - \cos\theta$.
          The graph is symmetric with respect to the polar axis.

The line $\theta = \dfrac{\pi}{2}$: Replace θ by $\pi - \theta$.

$$r = 1 - \cos(\pi - \theta) = 1 - (\cos(\pi)\cos\theta + \sin(\pi)\sin\theta)$$

$$= 1 - (-\cos\theta + 0) = 1 + \cos\theta. \quad \text{The test fails}$$

The pole: Replace r by $-r$. $-r = 1 - \cos\theta$. The test fails.
Due to symmetry to the polar axis, assign values to θ from 0 to π.

θ	0	$\dfrac{\pi}{6}$	$\dfrac{\pi}{3}$	$\dfrac{\pi}{2}$	$\dfrac{2\pi}{3}$	$\dfrac{5\pi}{6}$	π
$r = 1 - \cos\theta$	0	$1 - \dfrac{\sqrt{3}}{2} \approx 0.1$	$\dfrac{1}{2}$	1	$\dfrac{3}{2}$	$1 + \dfrac{\sqrt{3}}{2} \approx 1.9$	2

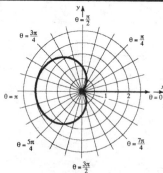

47. $r = 1 - 3\cos\theta$ The graph will be a limacon with an inner loop.
Check for symmetry:
Polar axis: Replace θ by $-\theta$. The result is $r = 1 - 3\cos(-\theta) = 1 - 3\cos\theta$.
          The graph is symmetric with respect to the polar axis.

The line $\theta = \frac{\pi}{2}$: Replace θ by $\pi - \theta$.

$$r = 1 - 3\cos(\pi - \theta) = 1 - 3(\cos(\pi)\cos\theta + \sin(\pi)\sin\theta)$$

$$= 1 - 3(-\cos\theta + 0) = 1 + 3\cos\theta$$

The test fails.

The pole:　　　Replace r by $-r$.　$-r = 1 - 3\cos\theta$.　The test fails.

Due to symmetry to the polar axis, assign values to θ from 0 to π.

θ	0	$\frac{\pi}{6}$	$\frac{\pi}{3}$	$\frac{\pi}{2}$	$\frac{2\pi}{3}$	$\frac{5\pi}{6}$	π
$r = 1 - 3\cos\theta$	-2	$1 - \frac{3\sqrt{3}}{2} \approx -1.6$	$-\frac{1}{2}$	1	$\frac{5}{2}$	$1 + \frac{3\sqrt{3}}{2} \approx 3.6$	4

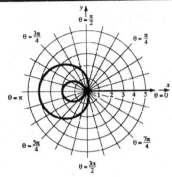

49.　$r = \dfrac{2}{1 - \cos\theta}$　　Check for symmetry:

Polar axis: Replace θ by $-\theta$. The result is $r = \dfrac{2}{1 - \cos(-\theta)} = \dfrac{2}{1 - \cos\theta}$.

　　　　　　The graph is symmetric with respect to the polar axis.

The line $\theta = \frac{\pi}{2}$: Replace θ by $\pi - \theta$.

$$r = \frac{2}{1 - \cos(\pi - \theta)} = \frac{2}{1 - (\cos(\pi)\cos\theta + \sin(\pi)\sin\theta)}$$

$$= \frac{2}{1 - (-\cos\theta + 0)} = \frac{2}{1 + \cos\theta}$$

The test fails.

The pole:　　　Replace r by $-r$.　$-r = \dfrac{2}{1 - \cos\theta}$. The test fails.

Due to symmetry to the polar axis, assign values to θ from 0 to π.

θ	0	$\frac{\pi}{6}$	$\frac{\pi}{3}$	$\frac{\pi}{2}$	$\frac{2\pi}{3}$	$\frac{5\pi}{6}$	π
$r = \dfrac{2}{1 - \cos\theta}$	undefined	$\dfrac{2}{\left(1 - \dfrac{\sqrt{3}}{2}\right)} \approx 14.9$	4	2	$\dfrac{4}{3}$	$\dfrac{2}{\left(1 + \dfrac{\sqrt{3}}{2}\right)} \approx 1.1$	1

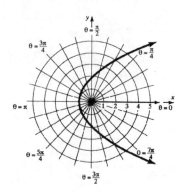

51. $r = \dfrac{1}{3 - 2\cos\theta}$ Check for symmetry:

Polar axis: Replace θ by $-\theta$. The result is $r = \dfrac{1}{3 - 2\cos(-\theta)} = \dfrac{1}{3 - 2\cos\theta}$.

The graph is symmetric with respect to the polar axis.

The line $\theta = \dfrac{\pi}{2}$: Replace θ by $\pi - \theta$.

$$r = \frac{1}{3 - 2\cos(\pi - \theta)} = \frac{1}{3 - 2(\cos(\pi)\cos\theta + \sin(\pi)\sin\theta)}$$

$$= \frac{1}{3 - 2(-\cos\theta + 0)} = \frac{1}{3 + 2\cos\theta}$$

The test fails.

The pole: Replace r by $-r$. $-r = \dfrac{1}{3 - 2\cos\theta}$. The test fails.

Due to symmetry to the polar axis, assign values to θ from 0 to π.

θ	0	$\dfrac{\pi}{6}$	$\dfrac{\pi}{3}$	$\dfrac{\pi}{2}$	$\dfrac{2\pi}{3}$	$\dfrac{5\pi}{6}$	π
$r = \dfrac{1}{3 - 2\cos\theta}$	1	$\dfrac{1}{3 - \sqrt{3}} \approx 0.8$	$\dfrac{1}{2}$	$\dfrac{1}{3}$	$\dfrac{1}{4}$	$\dfrac{1}{3 + \sqrt{3}} \approx 0.2$	$\dfrac{1}{5}$

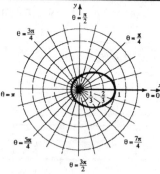

53. $r = \theta, \ \theta \geq 0$ Check for symmetry:

Polar axis: Replace θ by $-\theta$. $r = -\theta$. The test fails.

The line $\theta = \dfrac{\pi}{2}$: Replace θ by $\pi - \theta$. $r = \pi - \theta$. The test fails.

The pole: Replace r by $-r$. $-r = \theta$. The test fails.

θ	0	$\dfrac{\pi}{6}$	$\dfrac{\pi}{3}$	$\dfrac{\pi}{2}$	π	$\dfrac{3\pi}{2}$	2π
$r = \theta$	0	$\dfrac{\pi}{6} \approx 0.5$	$\dfrac{\pi}{3} \approx 1.0$	$\dfrac{\pi}{2} \approx 1.6$	$\pi \approx 3.1$	$\dfrac{3\pi}{2} \approx 4.7$	$2\pi \approx 6.3$

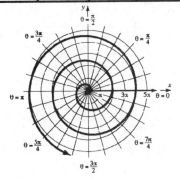

55. $r = \csc\theta - 2 = \dfrac{1}{\sin\theta} - 2, \quad 0 < \theta < \pi$ Check for symmetry:

Polar axis: Replace θ by $-\theta$. $r = \csc(-\theta) - 2 = -\csc\theta - 2$.

　　　　　　　　The test fails.

The line $\theta = \dfrac{\pi}{2}$: Replace θ by $\pi - \theta$.

$$r = \csc(\pi - \theta) - 2 = \frac{1}{\sin(\pi - \theta)} - 2$$

$$= \frac{1}{\sin(\pi)\cos\theta - \cos(\pi)\sin\theta} - 2 = \frac{1}{\sin\theta} - 2 = \csc\theta - 2$$

The graph is symmetric with respect to the line $\theta = \dfrac{\pi}{2}$.

The pole:　　　Replace r by $-r$. $-r = \csc\theta - 2$. The test fails.

Due to symmetry, assign values to θ from 0 to $\dfrac{\pi}{2}$.

θ	0	$\dfrac{\pi}{6}$	$\dfrac{\pi}{4}$	$\dfrac{\pi}{3}$	$\dfrac{\pi}{2}$
$r = \csc\theta - 2$	not defined	0	$\sqrt{2} - 2 \approx -0.6$	$\dfrac{2\sqrt{3}}{3} - 2 \approx -0.8$	-1

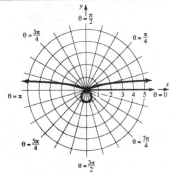

57. $r = \tan\theta$, $-\dfrac{\pi}{2} < \theta < \dfrac{\pi}{2}$ Check for symmetry:

Polar axis: Replace θ by $-\theta$. $r = \tan(-\theta) = -\tan\theta$. The test fails.

The line $\theta = \dfrac{\pi}{2}$: Replace θ by $\pi - \theta$.

$$r = \tan(\pi - \theta) = \frac{\tan(\pi) - \tan\theta}{1 + \tan(\pi)\tan\theta} = \frac{-\tan\theta}{1} = -\tan\theta$$

The test fails.

The pole: Replace r by $-r$. $-r = \tan\theta$. The test fails.

θ	$-\dfrac{\pi}{3}$	$-\dfrac{\pi}{4}$	$-\dfrac{\pi}{6}$	0	$\dfrac{\pi}{6}$	$\dfrac{\pi}{4}$	$\dfrac{\pi}{3}$
$r = \tan\theta$	$-\sqrt{3} \approx -1.7$	-1	$-\dfrac{\sqrt{3}}{3} \approx -0.6$	0	$\dfrac{\sqrt{3}}{3} \approx 0.6$	1	$\sqrt{3} \approx 1.7$

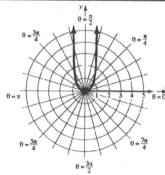

59.   Convert the equation to rectangular form:
$$r\sin\theta = a \rightarrow y = a$$
The graph of $r\sin\theta = a$ is a horizontal line a units above the pole if $a > 0$, and $|a|$ units below the pole, if $a < 0$.

61.   Convert the equation to rectangular form:
$$r = 2a\sin\theta, a > 0$$
$$r^2 = 2ar\sin\theta$$
$$x^2 + y^2 = 2ay \rightarrow x^2 + y^2 - 2ay = 0 \rightarrow x^2 + (y - a)^2 = a^2$$
Circle: radius a, center at rectangular coordinates $(0, a)$.

63.   Convert the equation to rectangular form:
$$r = 2a\cos\theta, a > 0$$
$$r^2 = 2ar\cos\theta$$
$$x^2 + y^2 = 2ax \rightarrow x^2 - 2ax + y^2 = 0 \rightarrow (x - a)^2 + y^2 = a^2$$
Circle: radius a, center at rectangular coordinates $(a, 0)$.

65.   (a) $r^2 = \cos\theta$: $r^2 = \cos(\pi - \theta)$ $\rightarrow$ $r^2 = -\cos\theta$ Test fails.
$\qquad\qquad\qquad (-r)^2 = \cos(-\theta)$ $\rightarrow$ $r^2 = \cos\theta$ New test works.

(b) $r^2 = \sin\theta$: $r^2 = \sin(\pi - \theta)$ $\rightarrow$ $r^2 = \sin\theta$ Test works.
$\qquad\qquad\qquad (-r)^2 = \sin(-\theta)$ $\rightarrow$ $r^2 = -\sin\theta$ New test fails

Chapter 8

Polar Coordinates; Vectors

8.3 The Complex Plane; De Moivre's Theorem

1. $r = \sqrt{x^2 + y^2} = \sqrt{1^2 + 1^2} = \sqrt{2}$

    $\tan\theta = \dfrac{y}{x} = 1 \;\rightarrow\; \theta = 45°$

    The polar form of $z = 1 + i$ is
    $z = r(\cos\theta + i\sin\theta)$
    $\quad = \sqrt{2}\left(\cos(45°) + i\sin(45°)\right)$

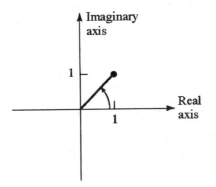

3. $r = \sqrt{x^2 + y^2} = \sqrt{\left(\sqrt{3}\right)^2 + (-1)^2} = \sqrt{4} = 2$

    $\tan\theta = \dfrac{y}{x} = \dfrac{-1}{\sqrt{3}} = -\dfrac{\sqrt{3}}{3} \;\rightarrow\; \theta = 330°$

    The polar form of $z = \sqrt{3} - i$ is
    $z = r(\cos\theta + i\sin\theta)$
    $\quad = 2\left(\cos(330°) + i\sin(330°)\right)$

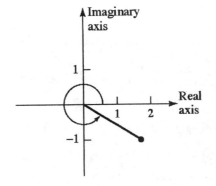

5. $r = \sqrt{x^2 + y^2} = \sqrt{0^2 + (-3)^2} = \sqrt{9} = 3$

    $\tan\theta = \dfrac{y}{x} = \dfrac{-3}{0} = \text{undefined} \rightarrow \theta = 270°$

    The polar form of $z = -3i$ is
    $z = r(\cos\theta + i\sin\theta)$
    $\quad = 3\left(\cos(270°) + i\sin(270°)\right)$

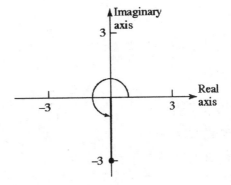

7. $r = \sqrt{x^2 + y^2} = \sqrt{4^2 + (-4)^2} = \sqrt{32} = 4\sqrt{2}$

$\tan\theta = \dfrac{y}{x} = \dfrac{-4}{4} = -1 \;\rightarrow\; \theta = 315°$

The polar form of $z = 4 - 4i$ is

$z = r(\cos\theta + i\sin\theta)$

$\quad = 4\sqrt{2}\left(\cos(315°) + i\sin(315°)\right)$

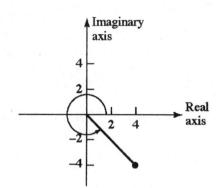

9. $r = \sqrt{x^2 + y^2} = \sqrt{3^2 + (-4)^2} = \sqrt{25} = 5$

$\tan\theta = \dfrac{y}{x} = \dfrac{-4}{3} \;\rightarrow\; \theta \approx 306.9°$

The polar form of $z = 3 - 4i$ is

$z = r(\cos\theta + i\sin\theta)$

$\quad = 5\left(\cos(306.9°) + i\sin(306.9°)\right)$

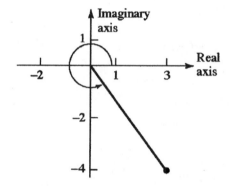

11. $r = \sqrt{x^2 + y^2} = \sqrt{(-2)^2 + 3^2} = \sqrt{13}$

$\tan\theta = \dfrac{y}{x} = \dfrac{3}{-2} = -\dfrac{3}{2} \;\rightarrow\; \theta \approx 123.7°$

The polar form of $z = -2 + 3i$ is

$z = r(\cos\theta + i\sin\theta)$

$\quad = \sqrt{13}\left(\cos(123.7°) + i\sin(123.7°)\right)$

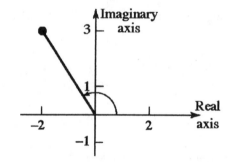

13. $2\left(\cos(120°) + i\sin(120°)\right) = 2\left(-\dfrac{1}{2} + \dfrac{\sqrt{3}}{2}i\right) = -1 + \sqrt{3}\,i$

15. $4\left(\cos\left(\dfrac{7\pi}{4}\right) + i\sin\left(\dfrac{7\pi}{4}\right)\right) = 4\left(\dfrac{\sqrt{2}}{2} - \dfrac{\sqrt{2}}{2}i\right) = 2\sqrt{2} - 2\sqrt{2}\,i$

17. $3\left(\cos\left(\dfrac{3\pi}{2}\right) + i\sin\left(\dfrac{3\pi}{2}\right)\right) = 3(0 - 1i) = -3i$

19. $0.2\left(\cos(100°) + i\sin(100°)\right) \approx 0.2(-0.1736 + 0.9848i) = -0.0347 + 0.1970\,i$

21. $2\left(\cos\left(\dfrac{\pi}{18}\right)+i\sin\left(\dfrac{\pi}{18}\right)\right)\approx 2(0.9848+0.1736i)=1.9696+0.3472i$

23. $z\cdot w = 2(\cos(40°)+i\sin(40°))\cdot 4(\cos(20°)+i\sin(20°))$
$= 2\cdot 4(\cos(40°+20°)+i\sin(40°+20°)) = 8(\cos(60°)+i\sin(60°))$

$\dfrac{z}{w} = \dfrac{2(\cos(40°)+i\sin(40°))}{4(\cos(20°)+i\sin(20°))} = \dfrac{2}{4}(\cos(40°-20°)+i\sin(40°-20°))$

$= \dfrac{1}{2}(\cos(20°)+i\sin(20°))$

25. $z\cdot w = 3(\cos(130°)+i\sin(130°))\cdot 4(\cos(270°)+i\sin(270°))$
$= 3\cdot 4(\cos(130°+270°)+i\sin(130°+270°)) = 12(\cos(400°)+i\sin(400°))$
$= 12(\cos(400°-360°)+i\sin(400°-360°)) = 12(\cos(40°)+i\sin(40°))$

$\dfrac{z}{w} = \dfrac{3(\cos(130°)+i\sin(130°))}{4(\cos(270°)+i\sin(270°))} = \dfrac{3}{4}(\cos(130°-270°)+i\sin(130°-270°))$

$= \dfrac{3}{4}(\cos(-140°)+i\sin(-140°)) = \dfrac{3}{4}(\cos(220°)+i\sin(220°))$

27. $z\cdot w = 2\left(\cos\left(\dfrac{\pi}{8}\right)+i\sin\left(\dfrac{\pi}{8}\right)\right)\cdot 2\left(\cos\left(\dfrac{\pi}{10}\right)+i\sin\left(\dfrac{\pi}{10}\right)\right) = 2\cdot 2\left(\cos\left(\dfrac{\pi}{8}+\dfrac{\pi}{10}\right)+i\sin\left(\dfrac{\pi}{8}+\dfrac{\pi}{10}\right)\right)$

$= 4\left(\cos\left(\dfrac{9\pi}{40}\right)+i\sin\left(\dfrac{9\pi}{40}\right)\right)$

$\dfrac{z}{w} = \dfrac{2\left(\cos\left(\dfrac{\pi}{8}\right)+i\sin\left(\dfrac{\pi}{8}\right)\right)}{2\left(\cos\left(\dfrac{\pi}{10}\right)+i\sin\left(\dfrac{\pi}{10}\right)\right)} = \dfrac{2}{2}\left(\cos\left(\dfrac{\pi}{8}-\dfrac{\pi}{10}\right)+i\sin\left(\dfrac{\pi}{8}-\dfrac{\pi}{10}\right)\right) = \cos\left(\dfrac{\pi}{40}\right)+i\sin\left(\dfrac{\pi}{40}\right)$

29. $z = 2+2i$ $r = \sqrt{2^2+2^2} = \sqrt{8} = 2\sqrt{2}$ $\tan\theta = \dfrac{2}{2}=1$ $\theta = 45°$
$z = 2\sqrt{2}(\cos(45°)+i\sin(45°))$

$w = \sqrt{3}-i$ $r = \sqrt{\left(\sqrt{3}\right)^2+(-1)^2} = \sqrt{4} = 2$ $\tan\theta = \dfrac{-1}{\sqrt{3}} = -\dfrac{\sqrt{3}}{3}$ $\theta = 330°$
$w = 2(\cos(330°)+i\sin(330°))$

$$z \cdot w = 2\sqrt{2}\left(\cos(45°) + i\sin(45°)\right) \cdot 2\left(\cos(330°) + i\sin(330°)\right)$$

$$= 2\sqrt{2} \cdot 2\left(\cos(45° + 330°) + i\sin(45° + 330°)\right) = 4\sqrt{2}\left(\cos(375°) + i\sin(375°)\right)$$

$$= 4\sqrt{2}\left(\cos(375° - 360°) + i\sin(375° - 360°)\right) = 4\sqrt{2}\left(\cos(15°) + i\sin(15°)\right)$$

$$\frac{z}{w} = \frac{2\sqrt{2}\left(\cos(45°) + i\sin(45°)\right)}{2\left(\cos(330°) + i\sin(330°)\right)} = \frac{2\sqrt{2}}{2}\left(\cos(45° - 330°) + i\sin(45° - 330°)\right)$$

$$= \sqrt{2}\left(\cos(-285°) + i\sin(-285°)\right) = \sqrt{2}\left(\cos(75°) + i\sin(75°)\right)$$

31. $\left[4\left(\cos(40°) + i\sin(40°)\right)\right]^3 = 4^3\left(\cos(3 \cdot 40°) + i\sin(3 \cdot 40°)\right) = 64\left(\cos(120°) + i\sin(120°)\right)$

$$= 64\left(-\frac{1}{2} + \frac{\sqrt{3}}{2}i\right) = -32 + 32\sqrt{3}\,i$$

33. $\left[2\left(\cos\left(\frac{\pi}{10}\right) + i\sin\left(\frac{\pi}{10}\right)\right)\right]^5 = 2^5\left(\cos\left(5 \cdot \frac{\pi}{10}\right) + i\sin\left(5 \cdot \frac{\pi}{10}\right)\right) = 32\left(\cos\left(\frac{\pi}{2}\right) + i\sin\left(\frac{\pi}{2}\right)\right)$

$$= 32(0 + 1\,i) = 0 + 32\,i$$

35. $\left[\sqrt{3}\left(\cos(10°) + i\sin(10°)\right)\right]^6 = \sqrt{3}^6\left(\cos(6 \cdot 10°) + i\sin(6 \cdot 10°)\right) = 27\left(\cos(60°) + i\sin(60°)\right)$

$$= 27\left(\frac{1}{2} + \frac{\sqrt{3}}{2}i\right) = \frac{27}{2} + \frac{27\sqrt{3}}{2}i$$

37. $\left[\sqrt{5}\left(\cos\left(\frac{3\pi}{16}\right) + i\sin\left(\frac{3\pi}{16}\right)\right)\right]^4 = \left(\sqrt{5}\right)^4\left(\cos\left(4 \cdot \frac{3\pi}{16}\right) + i\sin\left(4 \cdot \frac{3\pi}{16}\right)\right)$

$$= 25\left(\cos\left(\frac{3\pi}{4}\right) + i\sin\left(\frac{3\pi}{4}\right)\right) = 25\left(-\frac{\sqrt{2}}{2} + \frac{\sqrt{2}}{2}i\right) = -\frac{25\sqrt{2}}{2} + \frac{25\sqrt{2}}{2}i$$

39. $1 - i \qquad r = \sqrt{1^2 + (-1)^2} = \sqrt{2} \qquad \tan\theta = \frac{-1}{1} = -1 \qquad \theta = \frac{7\pi}{4}$

$$1 - i = \sqrt{2}\left(\cos\left(\frac{7\pi}{4}\right) + i\sin\left(\frac{7\pi}{4}\right)\right)$$

$$(1 - i)^5 = \left[\sqrt{2}\left(\cos\left(\frac{7\pi}{4}\right) + i\sin\left(\frac{7\pi}{4}\right)\right)\right]^5 = \left(\sqrt{2}\right)^5\left(\cos\left(5 \cdot \frac{7\pi}{4}\right) + i\sin\left(5 \cdot \frac{7\pi}{4}\right)\right)$$

$$= 4\sqrt{2}\left(\cos\left(\frac{35\pi}{4}\right) + i\sin\left(\frac{35\pi}{4}\right)\right) = 4\sqrt{2}\left(-\frac{\sqrt{2}}{2} + \frac{\sqrt{2}}{2}i\right) = -4 + 4\,i$$

41. $\sqrt{2}-i$ $r=\sqrt{\left(\sqrt{2}\right)^2+(-1)^2}=\sqrt{3}$ $\tan\theta=\dfrac{-1}{\sqrt{2}}=-\dfrac{\sqrt{2}}{2}$ $\theta\approx324.7°$

$\sqrt{2}-i\approx\sqrt{3}\left(\cos(324.7°)+i\sin(324.7°)\right)$

$\left(\sqrt{2}-i\right)^6\approx\left[\sqrt{3}\left(\cos(324.7°)+i\sin(324.7°)\right)\right]^6=\left(\sqrt{3}\right)^6\left(\cos(6\cdot324.7°)+i\sin(6\cdot324.7°)\right)$

$\qquad\qquad=27\left(\cos(1948.2°)+i\sin(1948.2°)\right)\approx27\left(-0.8499+0.5270\,i\right)$

$\qquad\qquad=-22.95+14.23\,i$

43. $1+i$ $r=\sqrt{1^2+1^2}=\sqrt{2}$ $\tan\theta=\dfrac{1}{1}=1$ $\theta=45°$

$1+i=\sqrt{2}\left(\cos45°+i\sin45°\right)$

The three complex cube roots of $1+i=\sqrt{2}\left(\cos(45°)+i\sin(45°)\right)$ are:

$z_k=\sqrt[3]{\sqrt{2}}\left[\cos\left(\dfrac{45°}{3}+\dfrac{360°\,k}{3}\right)+i\sin\left(\dfrac{45°}{3}+\dfrac{360°\,k}{3}\right)\right]$

$\qquad=\sqrt[6]{2}\left[\cos(15°+120°k)+i\sin(15°+120°k)\right]$

$z_0=\sqrt[6]{2}\left[\cos(15°+120°\cdot0)+i\sin(15°+120°\cdot0)\right]=\sqrt[6]{2}\left(\cos(15°)+i\sin(15°)\right)$

$z_1=\sqrt[6]{2}\left[\cos(15°+120°\cdot1)+i\sin(15°+120°\cdot1)\right]=\sqrt[6]{2}\left(\cos(135°)+i\sin(135°)\right)$

$z_2=\sqrt[6]{2}\left[\cos(15°+120°\cdot2)+i\sin(15°+120°\cdot2)\right]=\sqrt[6]{2}\left(\cos(255°)+i\sin(255°)\right)$

45. $4-4\sqrt{3}i$ $r=\sqrt{4^2+\left(-4\sqrt{3}\right)^2}=\sqrt{64}=8$ $\tan\theta=\dfrac{-4\sqrt{3}}{4}=-\sqrt{3}$ $\theta=300°$

$4-4\sqrt{3}\,i=8\left(\cos(300°)+i\sin(300°)\right)$

The four complex fourth roots of $4-4\sqrt{3}\,i=8\left(\cos(300°)+i\sin(300°)\right)$ are:

$z_k=\sqrt[4]{8}\left[\cos\left(\dfrac{300°}{4}+\dfrac{360°\,k}{4}\right)+i\sin\left(\dfrac{300°}{4}+\dfrac{360°\,k}{4}\right)\right]$

$\qquad=\sqrt[4]{8}\left[\cos(75°+90°k)+i\sin(75°+90°k)\right]$

$z_0=\sqrt[4]{8}\left[\cos(75°+90°\cdot0)+i\sin(75°+90°\cdot0)\right]=\sqrt[4]{8}\left(\cos(75°)+i\sin(75°)\right)$

$z_1=\sqrt[4]{8}\left[\cos(75°+90°\cdot1)+i\sin(75°+90°\cdot1)\right]=\sqrt[4]{8}\left(\cos(165°)+i\sin(165°)\right)$

$z_2=\sqrt[4]{8}\left[\cos(75°+90°\cdot2)+i\sin(75°+90°\cdot2)\right]=\sqrt[4]{8}\left(\cos(255°)+i\sin(255°)\right)$

$z_3=\sqrt[4]{8}\left[\cos(75°+90°\cdot3)+i\sin(75°+90°\cdot3)\right]=\sqrt[4]{8}\left(\cos(345°)+i\sin(345°)\right)$

47. $-16i$ $r=\sqrt{0^2+(-16)^2}=\sqrt{256}=16$ $\tan\theta=\dfrac{-16}{0}=$ undefined $\theta=270°$

$-16i=16\left(\cos(270°)+i\sin(270°)\right)$

The four complex fourth roots of $-16i=16\left(\cos270°+i\sin270°\right)$ are:

$z_k=\sqrt[4]{16}\left[\cos\left(\dfrac{270°}{4}+\dfrac{360°\,k}{4}\right)+i\sin\left(\dfrac{270°}{4}+\dfrac{360°\,k}{4}\right)\right]$

$\qquad=2\left[\cos(67.5°+90°k)+i\sin(67.5°+90°k)\right]$

$$z_0 = 2\left[\cos(67.5° + 90° \cdot 0) + i\sin(67.5° + 90° \cdot 0)\right] = 2\left(\cos(67.5°) + i\sin(67.5°)\right)$$

$$z_1 = 2\left[\cos(67.5° + 90° \cdot 1) + i\sin(67.5° + 90° \cdot 1)\right] = 2\left(\cos(157.5°) + i\sin(157.5°)\right)$$

$$z_2 = 2\left[\cos(67.5° + 90° \cdot 2) + i\sin(67.5° + 90° \cdot 2)\right] = 2\left(\cos(247.5°) + i\sin(247.5°)\right)$$

$$z_3 = 2\left[\cos(67.5° + 90° \cdot 3) + i\sin(67.5° + 90° \cdot 3)\right] = 2\left(\cos(337.5°) + i\sin(337.5°)\right)$$

49. i $r = \sqrt{0^2 + 1^2} = \sqrt{1} = 1$ $\tan\theta = \dfrac{1}{0} =$ undefined $\theta = 90°$

$i = 1\left(\cos(90°) + i\sin(90°)\right)$

The five complex fifth roots of $i = 1\left(\cos(90°) + i\sin(90°)\right)$ are:

$$z_k = \sqrt[5]{1}\left[\cos\left(\frac{90°}{5} + \frac{360°k}{5}\right) + i\sin\left(\frac{90°}{5} + \frac{360°k}{5}\right)\right] = 1\left[\cos(18° + 72°k) + i\sin(18° + 72°k)\right]$$

$$z_0 = 1\left[\cos(18° + 72° \cdot 0) + i\sin(18° + 72° \cdot 0)\right] = \cos(18°) + i\sin(18°)$$

$$z_1 = 1\left[\cos(18° + 72° \cdot 1) + i\sin(18° + 72° \cdot 1)\right] = \cos(90°) + i\sin(90°)$$

$$z_2 = 1\left[\cos(18° + 72° \cdot 2) + i\sin(18° + 72° \cdot 2)\right] = \cos(162°) + i\sin(162°)$$

$$z_3 = 1\left[\cos(18° + 72° \cdot 3) + i\sin(18° + 72° \cdot 3)\right] = \cos(234°) + i\sin(234°)$$

$$z_4 = 1\left[\cos(18° + 72° \cdot 4) + i\sin(18° + 72° \cdot 4)\right] = \cos(306°) + i\sin(306°)$$

51. $1 = 1 + 0i$ $r = \sqrt{1^2 + 0^2} = \sqrt{1} = 1$ $\tan\theta = \dfrac{0}{1} = 0$ $\theta = 0°$

$1 + 0i = 1\left(\cos(0°) + i\sin(0°)\right)$

The four complex fourth roots of $1 + 0i = 1\left(\cos(0°) + i\sin(0°)\right)$ are:

$$z_k = \sqrt[4]{1}\left[\cos\left(\frac{0°}{4} + \frac{360°k}{4}\right) + i\sin\left(\frac{0°}{4} + \frac{360°k}{4}\right)\right] = 1\left[\cos(90°k) + i\sin(90°k)\right]$$

$$z_0 = \cos(90° \cdot 0) + i\sin(90° \cdot 0) = \cos(0°) + i\sin(0°) = 1 + 0i = 1$$

$$z_1 = \cos(90° \cdot 1) + i\sin(90° \cdot 1) = \cos(90°) + i\sin(90°) = 0 + 1i = i$$

$$z_2 = \cos(90° \cdot 2) + i\sin(90° \cdot 2) = \cos(180°) + i\sin(180°) = -1 + 0i = -1$$

$$z_3 = \cos(90° \cdot 3) + i\sin(90° \cdot 3) = \cos(270°) + i\sin(270°) = 0 - 1i = -i$$

The complex roots are: $1, i, -1, -i$.

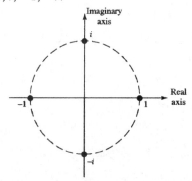

53. Let $w = r(\cos\theta + i\sin\theta)$ be a complex number. If $w \neq 0$, there are n distinct nth roots of w, given by the formula:

$$z_k = \sqrt[n]{r}\left(\cos\left(\frac{\theta}{n} + \frac{2k\pi}{n}\right) + i\sin\left(\frac{\theta}{n} + \frac{2k\pi}{n}\right)\right), \text{ where } k = 0, 1, 2, \ldots, n-1$$

$|z_k| = \sqrt[n]{r}$ for all k

55. Examining the formula for the distinct complex nth roots of the complex number $w = r(\cos\theta + i\sin\theta)$,

$$z_k = \sqrt[n]{r}\left(\cos\left(\frac{\theta}{n} + \frac{2k\pi}{n}\right) + i\sin\left(\frac{\theta}{n} + \frac{2k\pi}{n}\right)\right), \text{ where } k = 0, 1, 2, \ldots, n-1$$

we see that the z_k are spaced apart by an angle of $\frac{2\pi}{n}$.

Polar Coordinates; Vectors

8.4   Vectors

1.   **v + w**

3.   **3v**

5.   **v − w**

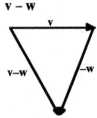

7.   **3v + u − 2w**

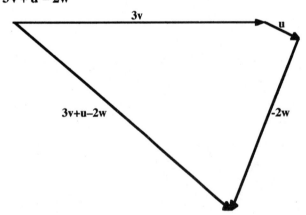

9.   True

11.   False $\mathbf{C} = -\mathbf{F} + \mathbf{E} - \mathbf{D}$

13.   False $\mathbf{D} - \mathbf{E} = \mathbf{H} + \mathbf{G}$

15.   True

17.   If $\|\mathbf{v}\| = 4$, then $\|3\mathbf{v}\| = |3| \|\mathbf{v}\| = 3(4) = 12$.

19. $P = (0, 0), Q = (3, 4)$ $\mathbf{v} = (3 - 0)\mathbf{i} + (4 - 0)\mathbf{j} = 3\mathbf{i} + 4\mathbf{j}$

21. $P = (3, 2), Q = (5, 6)$ $\mathbf{v} = (5 - 3)\mathbf{i} + (6 - 2)\mathbf{j} = 2\mathbf{i} + 4\mathbf{j}$

23. $P = (-2, -1), Q = (6, -2)$ $\mathbf{v} = (6 - (-2))\mathbf{i} + (-2 - (-1))\mathbf{j} = 8\mathbf{i} - \mathbf{j}$

25. $P = (1, 0), Q = (0, 1)$ $\mathbf{v} = (0 - 1)\mathbf{i} + (1 - 0)\mathbf{j} = -\mathbf{i} + \mathbf{j}$

27. For $\mathbf{v} = 3\mathbf{i} - 4\mathbf{j}$, $\|\mathbf{v}\| = \sqrt{3^2 + (-4)^2} = \sqrt{25} = 5$

29. For $\mathbf{v} = \mathbf{i} - \mathbf{j}$, $\|\mathbf{v}\| = \sqrt{1^2 + (-1)^2} = \sqrt{2}$

31. For $\mathbf{v} = -2\mathbf{i} + 3\mathbf{j}$, $\|\mathbf{v}\| = \sqrt{(-2)^2 + 3^2} = \sqrt{13}$

33. $\mathbf{v} = 3\mathbf{i} - 5\mathbf{j}$, $\mathbf{w} = -2\mathbf{i} + 3\mathbf{j}$
$$2\mathbf{v} + 3\mathbf{w} = 2(3\mathbf{i} - 5\mathbf{j}) + 3(-2\mathbf{i} + 3\mathbf{j}) = 6\mathbf{i} - 10\mathbf{j} - 6\mathbf{i} + 9\mathbf{j} = -\mathbf{j}$$

35. $\mathbf{v} = 3\mathbf{i} - 5\mathbf{j}$, $\mathbf{w} = -2\mathbf{i} + 3\mathbf{j}$
$$\|\mathbf{v} - \mathbf{w}\| = \|(3\mathbf{i} - 5\mathbf{j}) - (-2\mathbf{i} + 3\mathbf{j})\| = \|5\mathbf{i} - 8\mathbf{j}\| = \sqrt{5^2 + (-8)^2} = \sqrt{89}$$

37. $\mathbf{v} = 3\mathbf{i} - 5\mathbf{j}$, $\mathbf{w} = -2\mathbf{i} + 3\mathbf{j}$
$$\|\mathbf{v}\| - \|\mathbf{w}\| = \|3\mathbf{i} - 5\mathbf{j}\| - \|-2\mathbf{i} + 3\mathbf{j}\| = \sqrt{3^2 + (-5)^2} - \sqrt{(-2)^2 + 3^2} = \sqrt{34} - \sqrt{13}$$

39. $\mathbf{u} = \dfrac{\mathbf{v}}{\|\mathbf{v}\|} = \dfrac{5\mathbf{i}}{\|5\mathbf{i}\|} = \dfrac{5\mathbf{i}}{\sqrt{25 + 0}} = \dfrac{5\mathbf{i}}{5} = \mathbf{i}$

41. $\mathbf{u} = \dfrac{\mathbf{v}}{\|\mathbf{v}\|} = \dfrac{3\mathbf{i} - 4\mathbf{j}}{\|3\mathbf{i} - 4\mathbf{j}\|} = \dfrac{3\mathbf{i} - 4\mathbf{j}}{\sqrt{3^2 + (-4)^2}} = \dfrac{3\mathbf{i} - 4\mathbf{j}}{\sqrt{25}} = \dfrac{3\mathbf{i} - 4\mathbf{j}}{5} = \dfrac{3}{5}\mathbf{i} - \dfrac{4}{5}\mathbf{j}$

43. $\mathbf{u} = \dfrac{\mathbf{v}}{\|\mathbf{v}\|} = \dfrac{\mathbf{i} - \mathbf{j}}{\|\mathbf{i} - \mathbf{j}\|} = \dfrac{\mathbf{i} - \mathbf{j}}{\sqrt{1^2 + (-1)^2}} = \dfrac{\mathbf{i} - \mathbf{j}}{\sqrt{2}} = \dfrac{1}{\sqrt{2}}\mathbf{i} - \dfrac{1}{\sqrt{2}}\mathbf{j} = \dfrac{\sqrt{2}}{2}\mathbf{i} - \dfrac{\sqrt{2}}{2}\mathbf{j}$

45. Let $\mathbf{v} = a\mathbf{i} + b\mathbf{j}$. We want $\|\mathbf{v}\| = 4$ and $a = 2b$.
$$\|\mathbf{v}\| = \sqrt{a^2 + b^2} = \sqrt{(2b)^2 + b^2} = \sqrt{5b^2}$$
$$\sqrt{5b^2} = 4 \;\rightarrow\; 5b^2 = 16 \;\rightarrow\; b^2 = \frac{16}{5} \;\rightarrow\; b = \pm\sqrt{\frac{16}{5}} = \pm\frac{4}{\sqrt{5}} = \pm\frac{4\sqrt{5}}{5}$$
$$a = 2b = \pm\frac{8\sqrt{5}}{5}$$
$$\mathbf{v} = \frac{8\sqrt{5}}{5}\mathbf{i} + \frac{4\sqrt{5}}{5}\mathbf{j} \text{ or } \mathbf{v} = -\frac{8\sqrt{5}}{5}\mathbf{i} - \frac{4\sqrt{5}}{5}\mathbf{j}$$

47. $\mathbf{v} = 2\mathbf{i} - \mathbf{j}, \ \mathbf{w} = x\mathbf{i} + 3\mathbf{j} \quad \|\mathbf{v} + \mathbf{w}\| = 5$

$$\|\mathbf{v} + \mathbf{w}\| = \|2\mathbf{i} - \mathbf{j} + x\mathbf{i} + 3\mathbf{j}\| = \|(2 + x)\mathbf{i} + 2\mathbf{j}\| = \sqrt{(2 + x)^2 + 2^2}$$
$$= \sqrt{x^2 + 4x + 4 + 4} = \sqrt{x^2 + 4x + 8}$$

Solve for x:

$$\sqrt{x^2 + 4x + 8} = 5 \rightarrow x^2 + 4x + 8 = 25 \rightarrow x^2 + 4x - 17 = 0$$

$$x = \frac{-4 \pm \sqrt{16 - 4(1)(-17)}}{2(1)} = \frac{-4 \pm \sqrt{84}}{2} = \frac{-4 \pm 2\sqrt{21}}{2} = -2 \pm \sqrt{21}$$

$$x = -2 + \sqrt{21} \approx 2.58 \ \text{ or } \ x = -2 - \sqrt{21} \approx -6.58$$

49. $\|\mathbf{v}\| = 5, \quad \alpha = 60°$

$$\mathbf{v} = \|\mathbf{v}\|(\cos \alpha \mathbf{i} + \sin \alpha \mathbf{j}) = 5(\cos(60°)\mathbf{i} + \sin(60°)\mathbf{j}) = 5\left(\frac{1}{2}\mathbf{i} + \frac{\sqrt{3}}{2}\mathbf{j}\right) = \frac{5}{2}\mathbf{i} + \frac{5\sqrt{3}}{2}\mathbf{j}$$

51. $\|\mathbf{v}\| = 14, \quad \alpha = 120°$

$$\mathbf{v} = \|\mathbf{v}\|(\cos \alpha \mathbf{i} + \sin \alpha \mathbf{j}) = 14(\cos(120°)\mathbf{i} + \sin(120°)\mathbf{j}) = 14\left(-\frac{1}{2}\mathbf{i} + \frac{\sqrt{3}}{2}\mathbf{j}\right) = -7\mathbf{i} + 7\sqrt{3}\mathbf{j}$$

53. $\|\mathbf{v}\| = 25, \quad \alpha = 330°$

$$\mathbf{v} = \|\mathbf{v}\|(\cos \alpha \mathbf{i} + \sin \alpha \mathbf{j}) = 25(\cos(330°)\mathbf{i} + \sin(330°)\mathbf{j}) = 25\left(\frac{\sqrt{3}}{2}\mathbf{i} - \frac{1}{2}\mathbf{j}\right) = \frac{25\sqrt{3}}{2}\mathbf{i} - \frac{25}{2}\mathbf{j}$$

55. $\mathbf{F} = 40(\cos(30°)\mathbf{i} + \sin(30°)\mathbf{j}) = 40\left(\frac{\sqrt{3}}{2}\mathbf{i} + \frac{1}{2}\mathbf{j}\right) = 20\sqrt{3}\mathbf{i} + 20\mathbf{j}$

57. $\mathbf{F}_1 = 40(\cos(30°)\mathbf{i} + \sin(30°)\mathbf{j}) = 40\left(\frac{\sqrt{3}}{2}\mathbf{i} + \frac{1}{2}\mathbf{j}\right) = 20\sqrt{3}\mathbf{i} + 20\mathbf{j}$

$$\mathbf{F}_2 = 60(\cos(-45°)\mathbf{i} + \sin(-45°)\mathbf{j}) = 60\left(\frac{\sqrt{2}}{2}\mathbf{i} - \frac{\sqrt{2}}{2}\mathbf{j}\right) = 30\sqrt{2}\mathbf{i} - 30\sqrt{2}\mathbf{j}$$

$$\mathbf{F}_1 + \mathbf{F}_2 = 20\sqrt{3}\mathbf{i} + 20\mathbf{j} + 30\sqrt{2}\mathbf{i} - 30\sqrt{2}\mathbf{j} = \left(20\sqrt{3} + 30\sqrt{2}\right)\mathbf{i} + \left(20 - 30\sqrt{2}\right)\mathbf{j}$$

59. Let $\mathbf{F}_1$ be the tension on the left cable and $\mathbf{F}_2$ be the tension on the right cable.
Let $\mathbf{F}_3$ represent the force of the weight of the box.

$\mathbf{F}_1 = \|\mathbf{F}_1\|(\cos(155°)\mathbf{i} + \sin(155°)\mathbf{j}) \approx \|\mathbf{F}_1\|(-0.9063\mathbf{i} + 0.4226\mathbf{j})$

$\mathbf{F}_2 = \|\mathbf{F}_2\|(\cos(40°)\mathbf{i} + \sin(40°)\mathbf{j}) \approx \|\mathbf{F}_2\|(0.7660\mathbf{i} + 0.6428\mathbf{j})$

$\mathbf{F}_3 = -1000\mathbf{j}$

For equilibrium, the sum of the force vectors must be zero.

$\mathbf{F}_1 + \mathbf{F}_2 + \mathbf{F}_3 = -0.9063\|\mathbf{F}_1\|\mathbf{i} + 0.4226\|\mathbf{F}_1\|\mathbf{j} + 0.7660\|\mathbf{F}_2\|\mathbf{i} + 0.6428\|\mathbf{F}_2\|\mathbf{j} - 1000\mathbf{j}$

$\qquad = \left(-0.9063\|\mathbf{F}_1\| + 0.7660\|\mathbf{F}_2\|\right)\mathbf{i} + \left(0.4226\|\mathbf{F}_1\| + 0.6428\|\mathbf{F}_2\| - 1000\right)\mathbf{j} = 0$

Set the **i** and **j** components equal to zero and solve:

$$\begin{cases} -0.9063\|\mathbf{F}_1\| + 0.7660\|\mathbf{F}_2\| = 0 & \to \quad \|\mathbf{F}_2\| = \dfrac{0.9063}{0.7660}\|\mathbf{F}_1\| = 1.1832\|\mathbf{F}_1\| \\ 0.4226\|\mathbf{F}_1\| + 0.6428\|\mathbf{F}_2\| - 1000 = 0 \end{cases}$$

$$0.4226\|\mathbf{F}_1\| + 0.6428(1.1832\|\mathbf{F}_1\|) - 1000 = 0 \to 1.1832\|\mathbf{F}_1\| = 1000$$

$$\|\mathbf{F}_1\| = 845.2 \text{ pounds}$$

$$\|\mathbf{F}_2\| = 1.1832(845.2) = 1000 \text{ pounds}$$

The tension in the left cable is about 845.2 pounds and the tension in the right cable is about 1000 pounds.

61. Let $\mathbf{F}_1$ be the tension on the left end of the rope and $\mathbf{F}_2$ be the tension on the right end of the rope. Let $\mathbf{F}_3$ represent the force of the weight of the tightrope walker.

$$\mathbf{F}_1 = \|\mathbf{F}_1\|(\cos(175.8°)\mathbf{i} + \sin(175.8°)\mathbf{j}) \approx \|\mathbf{F}_1\|(-0.9973\mathbf{i} + 0.0732\mathbf{j})$$

$$\mathbf{F}_2 = \|\mathbf{F}_2\|(\cos(3.7°)\mathbf{i} + \sin(3.7°)\mathbf{j}) \approx \|\mathbf{F}_2\|(0.9979\mathbf{i} + 0.0645\mathbf{j})$$

$$\mathbf{F}_3 = -150\mathbf{j}$$

For equilibrium, the sum of the force vectors must be zero.

$$\mathbf{F}_1 + \mathbf{F}_2 + \mathbf{F}_3 = -0.9973\|\mathbf{F}_1\|\mathbf{i} + 0.0732\|\mathbf{F}_1\|\mathbf{j} + 0.9979\|\mathbf{F}_2\|\mathbf{i} + 0.0645\|\mathbf{F}_2\|\mathbf{j} - 150\mathbf{j}$$

$$= (-0.9973\|\mathbf{F}_1\| + 0.9979\|\mathbf{F}_2\|)\mathbf{i} + (0.0732\|\mathbf{F}_1\| + 0.0645\|\mathbf{F}_2\| - 150)\mathbf{j} = 0$$

Set the **i** and **j** components equal to zero and solve:

$$\begin{cases} -0.9973\|\mathbf{F}_1\| + 0.9979\|\mathbf{F}_2\| = 0 & \to \quad \|\mathbf{F}_2\| = \dfrac{0.9973}{0.9979}\|\mathbf{F}_1\| = 0.9994\|\mathbf{F}_1\| \\ 0.0732\|\mathbf{F}_1\| + 0.0645\|\mathbf{F}_2\| - 150 = 0 \end{cases}$$

$$0.0732\|\mathbf{F}_1\| + 0.0645(0.9994\|\mathbf{F}_1\|) - 150 = 0 \to 0.1377\|\mathbf{F}_1\| = 150$$

$$\|\mathbf{F}_1\| = 1089.3 \text{ pounds}; \quad \|\mathbf{F}_2\| = 0.9994(1089.3) = 1088.6 \text{ pounds}$$

The tension in the left end of the rope is about 1089.3 pounds and the tension in the right end of the rope is about 1088.6 pounds.

63. The given forces are:

$$\mathbf{F}_1 = -3\mathbf{i}; \quad \mathbf{F}_2 = -\mathbf{i} + 4\mathbf{j}; \quad \mathbf{F}_3 = 4\mathbf{i} - 2\mathbf{j}; \quad \mathbf{F}_4 = -4\mathbf{j}$$

A vector $\mathbf{x} = a\mathbf{i} + b\mathbf{j}$ needs to be added for equilibrium. Find vector $\mathbf{x} = a\mathbf{i} + b\mathbf{j}$:

$$\mathbf{F}_1 + \mathbf{F}_2 + \mathbf{F}_3 + \mathbf{F}_4 + \mathbf{x} = 0$$

$$-3\mathbf{i} + (-\mathbf{i} + 4\mathbf{j}) + (4\mathbf{i} - 2\mathbf{j}) + (-4\mathbf{j}) + (a\mathbf{i} + b\mathbf{j}) = 0$$

$$0\mathbf{i} - 2\mathbf{j} + (a\mathbf{i} + b\mathbf{j}) = 0$$

$$a\mathbf{i} + (-2 + b)\mathbf{j} = 0$$

$$a = 0$$

$$-2 + b = 0 \quad \to \quad b = 2$$

Therefore, $\mathbf{x} = 2\mathbf{j}$.

Polar Coordinates; Vectors

8.5   The Dot Product

1. $\mathbf{v} = \mathbf{i} - \mathbf{j}, \quad \mathbf{w} = \mathbf{i} + \mathbf{j}$
    (a) $\mathbf{v} \cdot \mathbf{w} = 1(1) + (-1)(1) = 1 - 1 = 0$
    (b) $\cos\theta = \dfrac{\mathbf{v} \cdot \mathbf{w}}{\|\mathbf{v}\|\|\mathbf{w}\|} = \dfrac{0}{\sqrt{1^2 + (-1)^2}\sqrt{1^2 + 1^2}} = \dfrac{0}{\sqrt{2}\sqrt{2}} = \dfrac{0}{2} = 0 \quad \rightarrow \quad \theta = 90°$
    (c)   The vectors are orthogonal.

3. $\mathbf{v} = 2\mathbf{i} + \mathbf{j}, \quad \mathbf{w} = \mathbf{i} + 2\mathbf{j}$
    (a) $\mathbf{v} \cdot \mathbf{w} = 2(1) + 1(2) = 2 + 2 = 4$
    (b) $\cos\theta = \dfrac{\mathbf{v} \cdot \mathbf{w}}{\|\mathbf{v}\|\|\mathbf{w}\|} = \dfrac{4}{\sqrt{2^2 + 1^2}\sqrt{1^2 + 2^2}} = \dfrac{4}{\sqrt{5}\sqrt{5}} = \dfrac{4}{5} = 0.8 \quad \rightarrow \quad \theta \approx 36.87°$
    (c)   The vectors are neither parallel nor orthogonal.

5. $\mathbf{v} = \sqrt{3}\,\mathbf{i} - \mathbf{j}, \quad \mathbf{w} = \mathbf{i} + \mathbf{j}$
    (a) $\mathbf{v} \cdot \mathbf{w} = \sqrt{3}(1) + (-1)(1) = \sqrt{3} - 1$
    (b) $\cos\theta = \dfrac{\mathbf{v} \cdot \mathbf{w}}{\|\mathbf{v}\|\|\mathbf{w}\|} = \dfrac{\sqrt{3} - 1}{\sqrt{\left(\sqrt{3}\right)^2 + (-1)^2}\sqrt{1^2 + 1^2}} = \dfrac{\sqrt{3} - 1}{\sqrt{4}\sqrt{2}} = \dfrac{\sqrt{3} - 1}{2\sqrt{2}} = \dfrac{\sqrt{6} - \sqrt{2}}{4}$
          $\theta \approx 75°$
    (c)   The vectors are neither parallel nor orthogonal.

7. $\mathbf{v} = 3\mathbf{i} + 4\mathbf{j}, \quad \mathbf{w} = 4\mathbf{i} + 3\mathbf{j}$
    (a) $\mathbf{v} \cdot \mathbf{w} = 3(4) + 4(3) = 12 + 12 = 24$
    (b) $\cos\theta = \dfrac{\mathbf{v} \cdot \mathbf{w}}{\|\mathbf{v}\|\|\mathbf{w}\|} = \dfrac{24}{\sqrt{3^2 + 4^2}\sqrt{4^2 + 3^2}} = \dfrac{24}{\sqrt{25}\sqrt{25}} = \dfrac{24}{25} = 0.96 \quad \rightarrow \quad \theta \approx 16.26°$
    (c)   The vectors are neither parallel nor orthogonal.

9. $\mathbf{v} = 4\mathbf{i}, \quad \mathbf{w} = \mathbf{j}$
    (a) $\mathbf{v} \cdot \mathbf{w} = 4(0) + 0(1) = 0 + 0 = 0$
    (b) $\cos\theta = \dfrac{\mathbf{v} \cdot \mathbf{w}}{\|\mathbf{v}\|\|\mathbf{w}\|} = \dfrac{0}{\sqrt{4^2 + 0^2}\sqrt{0^2 + 1^2}} = \dfrac{0}{4 \cdot 1} = \dfrac{0}{4} = 0 \quad \rightarrow \quad \theta = 90°$
    (c)   The vectors are orthogonal.

11. $\mathbf{v} = \mathbf{i} - a\mathbf{j},$ $\mathbf{w} = 2\mathbf{i} + 3\mathbf{j}$

Two vectors are orthogonal if the dot product is zero. Solve for a:
$$\mathbf{v} \bullet \mathbf{w} = 1(2) + (-a)(3) = 2 - 3a$$

$$2 - 3a = 0 \rightarrow 3a = 2 \rightarrow a = \frac{2}{3}$$

13. $\mathbf{v} = 2\mathbf{i} - 3\mathbf{j},$ $\mathbf{w} = \mathbf{i} - \mathbf{j}$

$$\mathbf{v}_1 = \text{proj}_\mathbf{w}\mathbf{v} = \frac{\mathbf{v} \bullet \mathbf{w}}{\left(\|\mathbf{w}\|\right)^2}\mathbf{w} = \frac{2(1) + (-3)(-1)}{\left(\sqrt{1^2 + (-1)^2}\right)^2}(\mathbf{i} - \mathbf{j}) = \frac{5}{2}(\mathbf{i} - \mathbf{j}) = \frac{5}{2}\mathbf{i} - \frac{5}{2}\mathbf{j}$$

$$\mathbf{v}_2 = \mathbf{v} - \mathbf{v}_1 = (2\mathbf{i} - 3\mathbf{j}) - \left(\frac{5}{2}\mathbf{i} - \frac{5}{2}\mathbf{j}\right) = -\frac{1}{2}\mathbf{i} - \frac{1}{2}\mathbf{j}$$

15. $\mathbf{v} = \mathbf{i} - \mathbf{j},$ $\mathbf{w} = \mathbf{i} + 2\mathbf{j}$

$$\mathbf{v}_1 = \text{proj}_\mathbf{w}\mathbf{v} = \frac{\mathbf{v} \bullet \mathbf{w}}{\left(\|\mathbf{w}\|\right)^2}\mathbf{w} = \frac{1(1) + (-1)(2)}{\left(\sqrt{1^2 + 2^2}\right)^2}(\mathbf{i} + 2\mathbf{j}) = -\frac{1}{5}(\mathbf{i} + 2\mathbf{j}) = -\frac{1}{5}\mathbf{i} - \frac{2}{5}\mathbf{j}$$

$$\mathbf{v}_2 = \mathbf{v} - \mathbf{v}_1 = (\mathbf{i} - \mathbf{j}) - \left(-\frac{1}{5}\mathbf{i} - \frac{2}{5}\mathbf{j}\right) = \frac{6}{5}\mathbf{i} - \frac{3}{5}\mathbf{j}$$

17. $\mathbf{v} = 3\mathbf{i} + \mathbf{j},$ $\mathbf{w} = -2\mathbf{i} - \mathbf{j}$

$$\mathbf{v}_1 = \text{proj}_\mathbf{w}\mathbf{v} = \frac{\mathbf{v} \bullet \mathbf{w}}{\left(\|\mathbf{w}\|\right)^2}\mathbf{w} = \frac{3(-2) + 1(-1)}{\left(\sqrt{(-2)^2 + (-1)^2}\right)^2}(-2\mathbf{i} - \mathbf{j}) = -\frac{7}{5}(-2\mathbf{i} - \mathbf{j}) = \frac{14}{5}\mathbf{i} + \frac{7}{5}\mathbf{j}$$

$$\mathbf{v}_2 = \mathbf{v} - \mathbf{v}_1 = (3\mathbf{i} + \mathbf{j}) - \left(\frac{14}{5}\mathbf{i} + \frac{7}{5}\mathbf{j}\right) = \frac{1}{5}\mathbf{i} - \frac{2}{5}\mathbf{j}$$

19.   Let $\mathbf{v}_a$ = the velocity of the plane in still air.

   $\mathbf{v}_w$ = the velocity of the wind.

   $\mathbf{v}_g$ = the velocity of the plane relative to the ground.

   $\mathbf{v}_g = \mathbf{v}_a + \mathbf{v}_w$

$$\mathbf{v}_a = 550\left(\cos(225°)\mathbf{i} + \sin(225°)\mathbf{j}\right) = 550\left(-\frac{\sqrt{2}}{2}\mathbf{i} - \frac{\sqrt{2}}{2}\mathbf{j}\right) = -275\sqrt{2}\,\mathbf{i} - 275\sqrt{2}\,\mathbf{j}$$

   $\mathbf{v}_w = 80\mathbf{i}$

$$\mathbf{v}_g = \mathbf{v}_a + \mathbf{v}_w = -275\sqrt{2}\,\mathbf{i} - 275\sqrt{2}\,\mathbf{j} + 80\mathbf{i} = \left(80 - 275\sqrt{2}\right)\mathbf{i} - 275\sqrt{2}\,\mathbf{j}$$

The speed of the plane relative to the ground is:
$$\left|\mathbf{v}_g\right| = \sqrt{\left(80 - 275\sqrt{2}\right)^2 + \left(-275\sqrt{2}\right)^2} = \sqrt{6400 - 44000\sqrt{2} + 151250 + 151250}$$

$$= \sqrt{246674.6} \approx 496.7 \text{ miles per hour}$$

To find the direction, find the angle between $\mathbf{v}_g$ and a convenient vector such as due south, $-\mathbf{j}$.

$$\cos\theta = \frac{\mathbf{v}_g \bullet -\mathbf{j}}{\|\mathbf{v}_g\| \|-\mathbf{j}\|} = \frac{\left(80 - 275\sqrt{2}\right)\cdot 0 + \left(-275\sqrt{2}\right)(-1)}{496.7\sqrt{0^2 + (-1)^2}} = \frac{275\sqrt{2}}{496.7} \approx 0.7829$$

$$\theta \approx 38.5°$$

The plane is traveling with a ground speed of about 496.7 miles per hour in a direction of 38.5° west of south.

21. Let the positive x-axis point downstream, so that the velocity of the current is $\mathbf{v}_c = 3\mathbf{i}$.

Let $\mathbf{v}_w$ = the velocity of the boat in the water.

Let $\mathbf{v}_g$ = the velocity of the boat relative to the land.

Then $\mathbf{v}_g = \mathbf{v}_w + \mathbf{v}_c$

The speed of the boat is $\|\mathbf{v}_w\| = 20$; we need to find the direction.

Let $\mathbf{v}_w = a\mathbf{i} + b\mathbf{j}$ so $\|\mathbf{v}_w\| = \sqrt{a^2 + b^2} = 20 \rightarrow a^2 + b^2 = 400$.

Let $\mathbf{v}_g = k\mathbf{j}$.

Since $\mathbf{v}_g = \mathbf{v}_w + \mathbf{v}_c$, $k\mathbf{j} = a\mathbf{i} + b\mathbf{j} + 3\mathbf{i} \rightarrow k\mathbf{j} = (a+3)\mathbf{i} + b\mathbf{j}$

$a + 3 = 0$ and $k = b \rightarrow a = -3$

$a^2 + b^2 = 400 \rightarrow 9 + b^2 = 400 \rightarrow b^2 = 391 \rightarrow k = b \approx 19.77$

$\mathbf{v}_w = -3\mathbf{i} + 19.77\mathbf{j}$ and $\mathbf{v}_g = 19.77\mathbf{j}$

Find the angle between $\mathbf{v}_w$ and $\mathbf{j}$:

$$\cos\theta = \frac{\mathbf{v}_w \bullet \mathbf{j}}{\|\mathbf{v}_w\| \|\mathbf{j}\|} = \frac{-3\cdot 0 + 19.77(1)}{20\sqrt{0^2 + 1^2}} = \frac{19.77}{20} \approx 0.9885$$

$$\theta \approx 8.7°$$

The heading of the boat needs to be 8.7° upstream.

The velocity of the boat directly across the river is 19.77 kilometers per hour. The time to cross the river is: $t = \frac{0.5}{19.77} \approx 0.025$ hours or $t \approx 1.5$ minutes.

23. Split the force into the components going down the hill and perpendicular to the hill.

$\mathbf{F}_d = \mathbf{F}\sin(8°) = 5300\sin(8°)$

$= 5300(0.1392) \approx 738$ pounds

$\mathbf{F}_p = \mathbf{F}\cos(8°) = 5300\cos(8°)$

$= 5300(0.9903) \approx 5249$ pounds

The force required to keep the car from rolling down the hill is about 738 pounds.

The force perpendicular to the hill is approximately 5249 pounds.

25. Let $\mathbf{v}_a$ = the velocity of the plane in still air.

$\mathbf{v}_w$ = the velocity of the wind.

$\mathbf{v}_g$ = the velocity of the plane relative to the ground.

$\mathbf{v}_g = \mathbf{v}_a + \mathbf{v}_w$

$$\mathbf{v_a} = 500\big(\cos(45°)\mathbf{i} + \sin(45°)\mathbf{j}\big) = 500\left(\frac{\sqrt{2}}{2}\mathbf{i} + \frac{\sqrt{2}}{2}\mathbf{j}\right) = 250\sqrt{2}\,\mathbf{i} + 250\sqrt{2}\,\mathbf{j}$$

$$\mathbf{v_w} = 60\big(\cos(120°)\mathbf{i} + \sin(120°)\mathbf{j}\big) = 60\left(-\frac{1}{2}\mathbf{i} + \frac{\sqrt{3}}{2}\mathbf{j}\right) = -30\mathbf{i} + 30\sqrt{3}\,\mathbf{j}$$

$$\mathbf{v_g} = \mathbf{v_a} + \mathbf{v_w} = 250\sqrt{2}\,\mathbf{i} + 250\sqrt{2}\,\mathbf{j} - 30\mathbf{i} + 30\sqrt{3}\,\mathbf{j}$$
$$= \big(-30 + 250\sqrt{2}\big)\mathbf{i} + \big(250\sqrt{2} + 30\sqrt{3}\big)\mathbf{j}$$

The speed of the plane relative to the ground is:

$$\left\|\mathbf{v_g}\right\| = \sqrt{\big(-30 + 250\sqrt{2}\big)^2 + \big(250\sqrt{2} + 30\sqrt{3}\big)^2}$$
$$= \sqrt{269129.1} \approx 518.8 \text{ kilometers per hour}$$

To find the direction, find the angle between $\mathbf{v_g}$ and a convenient vector such as due north, $\mathbf{j}$.

$$\cos\theta = \frac{\mathbf{v_g} \bullet \mathbf{j}}{\left\|\mathbf{v_g}\right\|\left\|\mathbf{j}\right\|} = \frac{\big(-30 + 250\sqrt{2}\big)\cdot 0 + \big(250\sqrt{2} + 30\sqrt{3}\big)(1)}{518.8\sqrt{0^2 + 1^2}} = \frac{250\sqrt{2} + 30\sqrt{3}}{518.8}$$
$$= \frac{405.5}{518.8} \approx 0.7816 \quad \rightarrow \quad \theta \approx 38.6°$$

The plane is traveling with a ground speed of about 518.8 kilometers per hour in a direction of 38.6° east of north.

27. Let the positive x-axis point downstream, so that the velocity of the current is $\mathbf{v_c} = 3\mathbf{i}$.

Let $\mathbf{v_w}$ = the velocity of the boat in the water.

Let $\mathbf{v_g}$ = the velocity of the boat relative to the land.

Then $\mathbf{v_g} = \mathbf{v_w} + \mathbf{v_c}$

The speed of the boat is $\left\|\mathbf{v_w}\right\| = 20$; its direction is directly across the river, so

Let $\mathbf{v_w} = 20\mathbf{j}$.
$$\mathbf{v_g} = \mathbf{v_w} + \mathbf{v_c} = 20\mathbf{j} + 3\mathbf{i} = 3\mathbf{i} + 20\mathbf{j}$$
Let $\left\|\mathbf{v_g}\right\| = \sqrt{3^2 + 20^2} = \sqrt{409} \approx 20.2$ miles per hour

Find the angle between $\mathbf{v_g}$ and $\mathbf{j}$:

$$\cos\theta = \frac{\mathbf{v_g} \bullet \mathbf{j}}{\left\|\mathbf{v_g}\right\|\left\|\mathbf{j}\right\|} = \frac{3\cdot 0 + 20(1)}{20.2\sqrt{0^2 + 1^2}} = \frac{20}{20.2} \approx 0.9901$$
$$\theta \approx 8.1°$$

The heading of the boat will be 8.1° downstream.

29. $\mathbf{F} = 3\big(\cos(60°)\mathbf{i} + \sin(60°)\mathbf{j}\big) = 3\left(\frac{1}{2}\mathbf{i} + \frac{\sqrt{3}}{2}\mathbf{j}\right) = \frac{3}{2}\mathbf{i} + \frac{3\sqrt{3}}{2}\mathbf{j}$

$W = \mathbf{F} \bullet AB = \left(\frac{3}{2}\mathbf{i} + \frac{3\sqrt{3}}{2}\mathbf{j}\right) \bullet 2\mathbf{i} = \frac{3}{2}(2) + \frac{3\sqrt{3}}{2}\cdot 0 = 3 \text{ foot-pounds}$

31. $\mathbf{F} = 20\left(\cos\left(30°\right)\mathbf{i} + \sin\left(30°\right)\mathbf{j}\right) = 20\left(\dfrac{\sqrt{3}}{2}\mathbf{i} + \dfrac{1}{2}\mathbf{j}\right) = 10\sqrt{3}\mathbf{i} + 10\mathbf{j}$

$W = \mathbf{F} \bullet AB = \left(10\sqrt{3}\mathbf{i} + 10\mathbf{j}\right) \bullet 100\mathbf{i} = 10\sqrt{3}(100) + 10 \cdot 0 = 1732 \text{ foot - pounds}$

33.   Let $\mathbf{u} = a_1\mathbf{i} + b_1\mathbf{j}, \quad \mathbf{v} = a_2\mathbf{i} + b_2\mathbf{j}, \quad \mathbf{w} = a_3\mathbf{i} + b_3\mathbf{j}$

$\mathbf{u} \bullet (\mathbf{v} + \mathbf{w}) = \left(a_1\mathbf{i} + b_1\mathbf{j}\right) \bullet \left(a_2\mathbf{i} + b_2\mathbf{j} + a_3\mathbf{i} + b_3\mathbf{j}\right) = \left(a_1\mathbf{i} + b_1\mathbf{j}\right) \bullet \left(a_2\mathbf{i} + a_3\mathbf{i} + b_2\mathbf{j} + b_3\mathbf{j}\right)$

$\quad = \left(a_1\mathbf{i} + b_1\mathbf{j}\right) \bullet \left((a_2 + a_3)\mathbf{i} + (b_2 + b_3)\mathbf{j}\right) = a_1(a_2 + a_3) + b_1(b_2 + b_3)$

$\quad = a_1a_2 + a_1a_3 + b_1b_2 + b_1b_3 = a_1a_2 + b_1b_2 + a_1a_3 + b_1b_3$

$\quad = \left(a_1\mathbf{i} + b_1\mathbf{j}\right) \bullet \left(a_2\mathbf{i} + b_2\mathbf{j}\right) + \left(a_1\mathbf{i} + b_1\mathbf{j}\right) \bullet \left(a_3\mathbf{i} + b_3\mathbf{j}\right) = \mathbf{u} \bullet \mathbf{v} + \mathbf{u} \bullet \mathbf{w}$

35.   Let $\mathbf{v} = a\mathbf{i} + b\mathbf{j}$.

Since $\mathbf{v}$ is a unit vector, $\|\mathbf{v}\| = \sqrt{a^2 + b^2} = 1$ or $a^2 + b^2 = 1$

If α is the angle between $\mathbf{v}$ and $\mathbf{i}$, then $\cos\alpha = \dfrac{\mathbf{v} \bullet \mathbf{i}}{\|\mathbf{v}\|\|\mathbf{i}\|}$ or $\cos\alpha = \dfrac{(a\mathbf{i} + b\mathbf{j}) \bullet \mathbf{i}}{1 \cdot 1} = a$.

$a^2 + b^2 = 1$

$\cos^2\alpha + b^2 = 1 \rightarrow b^2 = 1 - \cos^2\alpha \rightarrow b^2 = \sin^2\alpha \rightarrow b = \sin\alpha$

Thus, $\mathbf{v} = \cos\alpha \, \mathbf{i} + \sin\alpha \, \mathbf{j}$

37.   Let $\mathbf{v} = a\mathbf{i} + b\mathbf{j}$.

$\text{proj}_\mathbf{i}\,\mathbf{v} = \dfrac{\mathbf{v} \bullet \mathbf{i}}{\left(\|\mathbf{i}\|\right)^2}\mathbf{i} = \dfrac{(a\mathbf{i} + b\mathbf{j}) \bullet \mathbf{i}}{\left(\sqrt{1^2 + 0^2}\right)^2}\mathbf{i} = \dfrac{a(1) + b(0)}{1^2}\mathbf{i} = a\mathbf{i}$

$\mathbf{v} \bullet \mathbf{i} = a, \quad \mathbf{v} \bullet \mathbf{j} = b,$

$\mathbf{v} = (\mathbf{v} \bullet \mathbf{i})\mathbf{i} + (\mathbf{v} \bullet \mathbf{j})\mathbf{j}$

39. $(\mathbf{v} - \alpha\mathbf{w}) \bullet \mathbf{w} = \mathbf{v} \bullet \mathbf{w} - \alpha\mathbf{w} \bullet \mathbf{w} = \mathbf{v} \bullet \mathbf{w} - \alpha\left(\|\mathbf{w}\|\right)^2 = \mathbf{v} \bullet \mathbf{w} - \dfrac{\mathbf{v} \bullet \mathbf{w}}{\left(\|\mathbf{w}\|\right)^2}\left(\|\mathbf{w}\|\right)^2 = 0$

Therefore the vectors are orthogonal.

41.   If $\mathbf{F}$ is orthogonal to $\mathbf{AB}$, then $W = \mathbf{F} \bullet \mathbf{AB} = (\mathbf{i} + \mathbf{j})(\mathbf{i} - \mathbf{j}) = 1 \cdot 1 + 1(-1) = 0$

Polar Coordinates; Vectors

8.6 Vectors in Space

1. $y = 0$ is the set of all points in the xz - plane.

3. $z = 2$ is the set of all points of the form $(x, y, 2)$; the plane two units above the xy - plane.

5. $x = -4$ is the set of all points of the form $(-4, y, z)$; the plane four units to the left of yz - plane.

7. $x = 1$ and $y = 2$ is the set of all points of the form $(1, 2, z)$; a line parallel to the z - axis.

9. $d = \sqrt{(4-0)^2 + (1-0)^2 + (2-0)^2} = \sqrt{16+1+4} = \sqrt{21}$

11. $d = \sqrt{(0-(-1))^2 + (-2-2)^2 + (1-(-3))^2} = \sqrt{1+16+16} = \sqrt{33}$

13. $d = \sqrt{(3-4)^2 + (2-(-2))^2 + (1-(-2))^2} = \sqrt{1+16+9} = \sqrt{26}$

15. The bottom of the box is formed by the vertices $(0, 0, 0)$, $(2, 0, 0)$, $(0, 1, 0)$, and $(2, 1, 0)$. The top of the box is formed by the vertices $(0, 0, 3)$, $(2, 0, 3)$, $(0, 1, 3)$, and $(2, 1, 3)$.

17. The bottom of the box is formed by the vertices $(1, 2, 3)$, $(3, 2, 3)$, $(3, 4, 3)$, and $(1, 4, 3)$. The top of the box is formed by the vertices $(3, 4, 5)$, $(1, 2, 5)$, $(3, 2, 5)$, and $(1, 4, 5)$.

19. The bottom of the box is formed by the vertices $(-1, 0, 2)$, $(4, 0, 2)$, $(-1, 2, 2)$, and $(4, 2, 2)$. The top of the box is formed by the vertices $(4, 2, 5)$, $(-1, 0, 5)$, $(4, 0, 5)$, and $(-1, 2, 5)$.

21. $\mathbf{v} = (3-0)\mathbf{i} + (4-0)\mathbf{j} + (-1-0)\mathbf{k} = 3\mathbf{i} + 4\mathbf{j} - 1\mathbf{k}$

23. $\mathbf{v} = (5-3)\mathbf{i} + (6-2)\mathbf{j} + (0-(-1))\mathbf{k} = 2\mathbf{i} + 4\mathbf{j} + 1\mathbf{k}$

25. $\mathbf{v} = (6-(-2))\mathbf{i} + (-2-(-1))\mathbf{j} + (4-4)\mathbf{k} = 8\mathbf{i} - 1\mathbf{j}$

27. $\|\mathbf{v}\| = \sqrt{3^2 + (-6)^2 + (-2)^2} = \sqrt{9+36+4} = \sqrt{49} = 7$

29. $\|\mathbf{v}\| = \sqrt{1^2 + (-1)^2 + 1^2} = \sqrt{1+1+1} = \sqrt{3}$

31. $\|\mathbf{v}\| = \sqrt{(-2)^2 + 3^2 + (-3)^2} = \sqrt{4+9+9} = \sqrt{22}$

33. $2\mathbf{v} + 3\mathbf{w} = 2(3\mathbf{i} - 5\mathbf{j} + 2\mathbf{k}) + 3(-2\mathbf{i} + 3\mathbf{j} - 2\mathbf{k}) = 6\mathbf{i} - 10\mathbf{j} + 4\mathbf{k} - 6\mathbf{i} + 9\mathbf{j} - 6\mathbf{k}$
$\qquad = 0\mathbf{i} - 1\mathbf{j} - 2\mathbf{k}$

35. $|\mathbf{v} - \mathbf{w}| = |(3\mathbf{i} - 5\mathbf{j} + 2\mathbf{k}) - (-2\mathbf{i} + 3\mathbf{j} - 2\mathbf{k})| = |3\mathbf{i} - 5\mathbf{j} + 2\mathbf{k} + 2\mathbf{i} - 3\mathbf{j} + 2\mathbf{k}|$
$\qquad = |5\mathbf{i} - 8\mathbf{j} + 4\mathbf{k}| = \sqrt{5^2 + (-8)^2 + 4^2} = \sqrt{25 + 64 + 16} = \sqrt{105}$

37. $|\mathbf{v}| - |\mathbf{w}| = |3\mathbf{i} - 5\mathbf{j} + 2\mathbf{k}| - |-2\mathbf{i} + 3\mathbf{j} - 2\mathbf{k}|$
$\qquad = \sqrt{3^2 + (-5)^2 + 2^2} - \sqrt{(-2)^2 + 3^2 + (-2)^2} = \sqrt{38} - \sqrt{17}$

39. $\mathbf{u} = \dfrac{\mathbf{v}}{|\mathbf{v}|} = \dfrac{5\mathbf{i}}{\sqrt{5^2 + 0^2 + 0^2}} = \dfrac{5\mathbf{i}}{5} = \mathbf{i}$

41. $\mathbf{u} = \dfrac{\mathbf{v}}{|\mathbf{v}|} = \dfrac{3\mathbf{i} - 6\mathbf{j} - 2\mathbf{k}}{\sqrt{3^2 + (-6)^2 + (-2)^2}} = \dfrac{3\mathbf{i} - 6\mathbf{j} - 2\mathbf{k}}{7} = \dfrac{3}{7}\mathbf{i} - \dfrac{6}{7}\mathbf{j} - \dfrac{2}{7}\mathbf{k}$

43. $\mathbf{u} = \dfrac{\mathbf{v}}{|\mathbf{v}|} = \dfrac{\mathbf{i} + \mathbf{j} + \mathbf{k}}{\sqrt{1^2 + 1^2 + 1^2}} = \dfrac{\mathbf{i} + \mathbf{j} + \mathbf{k}}{\sqrt{3}} = \dfrac{1}{\sqrt{3}}\mathbf{i} + \dfrac{1}{\sqrt{3}}\mathbf{j} + \dfrac{1}{\sqrt{3}}\mathbf{k} = \dfrac{\sqrt{3}}{3}\mathbf{i} + \dfrac{\sqrt{3}}{3}\mathbf{j} + \dfrac{\sqrt{3}}{3}\mathbf{k}$

45. $\mathbf{v} \cdot \mathbf{w} = (\mathbf{i} - \mathbf{j}) \cdot (\mathbf{i} + \mathbf{j} + \mathbf{k}) = 1 \cdot 1 + (-1)(1) + 0 \cdot 1 = 1 - 1 + 0 = 0$
$\qquad \cos\theta = \dfrac{\mathbf{v} \cdot \mathbf{w}}{|\mathbf{v}||\mathbf{w}|} = \dfrac{0}{\sqrt{1^2 + (-1)^2 + 0^2}\sqrt{1^2 + 1^2 + 1^2}} = \dfrac{0}{\sqrt{2}\sqrt{3}} = \dfrac{0}{\sqrt{6}} = 0$
$\qquad \theta = \dfrac{\pi}{2} = 90°$

47. $\mathbf{v} \cdot \mathbf{w} = (2\mathbf{i} + \mathbf{j} - 3\mathbf{k}) \cdot (\mathbf{i} + 2\mathbf{j} + 2\mathbf{k}) = 2 \cdot 1 + 1(2) + (-3)(2) = 2 + 2 - 6 = -2$
$\qquad \cos\theta = \dfrac{\mathbf{v} \cdot \mathbf{w}}{|\mathbf{v}||\mathbf{w}|} = \dfrac{-2}{\sqrt{2^2 + 1^2 + (-3)^2}\sqrt{1^2 + 2^2 + 2^2}} = \dfrac{-2}{\sqrt{14}\sqrt{9}} = \dfrac{-2}{3\sqrt{14}} \approx -0.1782$
$\qquad \theta = 1.75 = 100.3°$

49. $\mathbf{v} \cdot \mathbf{w} = (3\mathbf{i} - \mathbf{j} + 2\mathbf{k}) \cdot (\mathbf{i} + \mathbf{j} - \mathbf{k}) = 3 \cdot 1 + (-1)(1) + 2(-1) = 3 - 1 - 2 = 0$
$\qquad \cos\theta = \dfrac{\mathbf{v} \cdot \mathbf{w}}{|\mathbf{v}||\mathbf{w}|} = \dfrac{0}{\sqrt{3^2 + (-1)^2 + 2^2}\sqrt{1^2 + 1^2 + (-1)^2}} = \dfrac{0}{\sqrt{14}\sqrt{3}} = 0$
$\qquad \theta = \dfrac{\pi}{2} = 90°$

51. $\mathbf{v} \cdot \mathbf{w} = (3\mathbf{i} + 4\mathbf{j} + \mathbf{k}) \cdot (6\mathbf{i} + 8\mathbf{j} + 2\mathbf{k}) = 3 \cdot 6 + 4 \cdot 8 + 1 \cdot 2 = 18 + 32 + 2 = 52$
$\qquad \cos\theta = \dfrac{\mathbf{v} \cdot \mathbf{w}}{|\mathbf{v}||\mathbf{w}|} = \dfrac{52}{\sqrt{3^2 + 4^2 + 1^2}\sqrt{6^2 + 8^2 + 2^2}} = \dfrac{52}{\sqrt{26}\sqrt{104}} = \dfrac{52}{52} = 1$
$\qquad \theta = 0 = 0°$

53. $\cos\alpha = \dfrac{a}{\|\mathbf{v}\|} = \dfrac{3}{\sqrt{3^2+(-6)^2+(-2)^2}} = \dfrac{3}{\sqrt{49}} = \dfrac{3}{7} \;\rightarrow\; \alpha \approx 64.6°$

$\cos\beta = \dfrac{b}{\|\mathbf{v}\|} = \dfrac{-6}{\sqrt{3^2+(-6)^2+(-2)^2}} = \dfrac{-6}{\sqrt{49}} = -\dfrac{6}{7} \;\rightarrow\; \beta \approx 149.0°$

$\cos\gamma = \dfrac{c}{\|\mathbf{v}\|} = \dfrac{-2}{\sqrt{3^2+(-6)^2+(-2)^2}} = \dfrac{-2}{\sqrt{49}} = -\dfrac{2}{7} \;\rightarrow\; \gamma \approx 106.6°$

$\mathbf{v} = 7\big(\cos(64.6°)\mathbf{i} + \cos(149.0°)\mathbf{j} + \cos(106.6°)\mathbf{k}\big)$

55. $\cos\alpha = \dfrac{a}{\|\mathbf{v}\|} = \dfrac{1}{\sqrt{1^2+1^2+1^2}} = \dfrac{1}{\sqrt{3}} = \dfrac{\sqrt{3}}{3} \;\rightarrow\; \alpha \approx 54.7°$

$\cos\beta = \dfrac{b}{\|\mathbf{v}\|} = \dfrac{1}{\sqrt{1^2+1^2+1^2}} = \dfrac{1}{\sqrt{3}} = \dfrac{\sqrt{3}}{3} \;\rightarrow\; \beta \approx 54.7°$

$\cos\gamma = \dfrac{c}{\|\mathbf{v}\|} = \dfrac{1}{\sqrt{1^2+1^2+1^2}} = \dfrac{1}{\sqrt{3}} = \dfrac{\sqrt{3}}{3} \;\rightarrow\; \gamma \approx 54.7°$

$\mathbf{v} = \sqrt{3}\big(\cos(54.7°)\mathbf{i} + \cos(54.7°)\mathbf{j} + \cos(54.7°)\mathbf{k}\big)$

57. $\cos\alpha = \dfrac{a}{\|\mathbf{v}\|} = \dfrac{1}{\sqrt{1^2+1^2+0^2}} = \dfrac{1}{\sqrt{2}} = \dfrac{\sqrt{2}}{2} \;\rightarrow\; \alpha = 45°$

$\cos\beta = \dfrac{b}{\|\mathbf{v}\|} = \dfrac{1}{\sqrt{1^2+1^2+0^2}} = \dfrac{1}{\sqrt{2}} = \dfrac{\sqrt{2}}{2} \;\rightarrow\; \beta = 45°$

$\cos\gamma = \dfrac{c}{\|\mathbf{v}\|} = \dfrac{0}{\sqrt{1^2+1^2+0^2}} = \dfrac{0}{\sqrt{2}} = 0 \;\rightarrow\; \gamma = 90°$

$\mathbf{v} = \sqrt{2}\big(\cos(45°)\mathbf{i} + \cos(45°)\mathbf{j} + \cos(90°)\mathbf{k}\big)$

59. $\cos\alpha = \dfrac{a}{\|\mathbf{v}\|} = \dfrac{3}{\sqrt{3^2+(-5)^2+2^2}} = \dfrac{3}{\sqrt{38}} \;\rightarrow\; \alpha \approx 60.9°$

$\cos\beta = \dfrac{b}{\|\mathbf{v}\|} = \dfrac{-5}{\sqrt{3^2+(-5)^2+2^2}} = -\dfrac{5}{\sqrt{38}} \;\rightarrow\; \beta \approx 144.2°$

$\cos\gamma = \dfrac{c}{\|\mathbf{v}\|} = \dfrac{2}{\sqrt{3^2+(-5)^2+2^2}} = \dfrac{2}{\sqrt{38}} \;\rightarrow\; \gamma \approx 71.1°$

$\mathbf{v} = \sqrt{38}\big(\cos(60.9°)\mathbf{i} + \cos(144.2°)\mathbf{j} + \cos(71.1°)\mathbf{k}\big)$

61. $d(P_0, P) = \sqrt{(x-x_0)^2 + (y-y_0)^2 + (z-z_0)^2} = r$

$(x-x_0)^2 + (y-y_0)^2 + (z-z_0)^2 = r^2$

63. $(x-1)^2 + (y-2)^2 + (z-2)^2 = 4$

65. $x^2 + y^2 + z^2 + 2x - 2y = 2$

          $x^2 + 2x + y^2 - 2y + z^2 = 2$

      $x^2 + 2x + 1 + y^2 - 2y + 1 + z^2 = 2 + 1 + 1$

      $(x + 1)^2 + (y - 1)^2 + (z - 0)^2 = 4$

      Center: $(-1, 1, 0)$; Radius: 2

67. $x^2 + y^2 + z^2 - 4x + 4y + 2z = 0$

          $x^2 - 4x + y^2 + 4y + z^2 + 2z = 0$

  $x^2 - 4x + 4 + y^2 + 4y + 4 + z^2 + 2z + 1 = 4 + 4 + 1$

      $(x - 2)^2 + (y + 2)^2 + (z + 1)^2 = 9$

      Center: $(2, -2, -1)$; Radius: 3

69. $2x^2 + 2y^2 + 2z^2 - 8x + 4z = -1$

          $x^2 - 4x + y^2 + z^2 + 2z = \dfrac{-1}{2}$

  $x^2 - 4x + 4 + y^2 + z^2 + 2z + 1 = \dfrac{-1}{2} + 4 + 1$

      $(x - 2)^2 + (y - 0)^2 + (z + 1)^2 = \dfrac{9}{2}$

      Center: $(2, 0, -1)$; Radius: $\dfrac{3\sqrt{2}}{2}$

71.   Write the force as a vector:

$$\cos\alpha = \frac{2}{\sqrt{2^2 + 1^2 + 2^2}} = \frac{2}{\sqrt{9}} = \frac{2}{3}; \quad \cos\beta = \frac{1}{3}; \quad \cos\gamma = \frac{2}{3}$$

$$\mathbf{F} = 3\left(\frac{2}{3}\mathbf{i} + \frac{1}{3}\mathbf{j} + \frac{2}{3}\mathbf{k}\right)$$

$$W = 3\left(\frac{2}{3}\mathbf{i} + \frac{1}{3}\mathbf{j} + \frac{2}{3}\mathbf{k}\right) \bullet 2\mathbf{j} = 3\left(\frac{1}{3}\cdot 2\right) = 2 \text{ joules}$$

73. $W = \mathbf{F} \bullet \mathbf{AB} = (2\mathbf{i} - \mathbf{j} - \mathbf{k}) \bullet (3\mathbf{i} + 2\mathbf{j} - 5\mathbf{k}) = 2 \cdot 3 + (-1)(2) + (-1)(-5) = 9$

Chapter 8

Polar Coordinates; Vectors

8.7   The Cross Product

1. $\begin{vmatrix} 3 & 4 \\ 1 & 2 \end{vmatrix} = 3 \cdot 2 - 1 \cdot 4 = 6 - 4 = 2$

3. $\begin{vmatrix} 6 & 5 \\ -2 & -1 \end{vmatrix} = 6(-1) - (-2)(5) = -6 + 10 = 4$

5. $\begin{vmatrix} A & B & C \\ 2 & 1 & 4 \\ 1 & 3 & 1 \end{vmatrix} = \begin{vmatrix} 1 & 4 \\ 3 & 1 \end{vmatrix} A - \begin{vmatrix} 2 & 4 \\ 1 & 1 \end{vmatrix} B + \begin{vmatrix} 2 & 1 \\ 1 & 3 \end{vmatrix} C = (1 - 12)A - (2 - 4)B + (6 - 1)C$

$= -11A + 2B + 5C$

7. $\begin{vmatrix} A & B & C \\ -1 & 3 & 5 \\ 5 & 0 & -2 \end{vmatrix} = \begin{vmatrix} 3 & 5 \\ 0 & -2 \end{vmatrix} A - \begin{vmatrix} -1 & 5 \\ 5 & -2 \end{vmatrix} B + \begin{vmatrix} -1 & 3 \\ 5 & 0 \end{vmatrix} C = (-6 - 0)A - (2 - 25)B + (0 - 15)C$

$= -6A + 23B - 15C$

9. (a) $\mathbf{v} \times \mathbf{w} = \begin{vmatrix} \mathbf{i} & \mathbf{j} & \mathbf{k} \\ 2 & -3 & 1 \\ 3 & -2 & -1 \end{vmatrix} = \begin{vmatrix} -3 & 1 \\ -2 & -1 \end{vmatrix} \mathbf{i} - \begin{vmatrix} 2 & 1 \\ 3 & -1 \end{vmatrix} \mathbf{j} + \begin{vmatrix} 2 & -3 \\ 3 & -2 \end{vmatrix} \mathbf{k} = 5\mathbf{i} + 5\mathbf{j} + 5\mathbf{k}$

(b) $\mathbf{w} \times \mathbf{v} = \begin{vmatrix} \mathbf{i} & \mathbf{j} & \mathbf{k} \\ 3 & -2 & -1 \\ 2 & -3 & 1 \end{vmatrix} = \begin{vmatrix} -2 & -1 \\ -3 & 1 \end{vmatrix} \mathbf{i} - \begin{vmatrix} 3 & -1 \\ 2 & 1 \end{vmatrix} \mathbf{j} + \begin{vmatrix} 3 & -2 \\ 2 & -3 \end{vmatrix} \mathbf{k} = -5\mathbf{i} - 5\mathbf{j} - 5\mathbf{k}$

(c) $\mathbf{w} \times \mathbf{w} = \begin{vmatrix} \mathbf{i} & \mathbf{j} & \mathbf{k} \\ 3 & -2 & -1 \\ 3 & -2 & -1 \end{vmatrix} = \begin{vmatrix} -2 & -1 \\ -2 & -1 \end{vmatrix} \mathbf{i} - \begin{vmatrix} 3 & -1 \\ 3 & -1 \end{vmatrix} \mathbf{j} + \begin{vmatrix} 3 & -2 \\ 3 & -2 \end{vmatrix} \mathbf{k} = 0\mathbf{i} + 0\mathbf{j} + 0\mathbf{k} = \mathbf{0}$

(d) $\mathbf{v} \times \mathbf{v} = \begin{vmatrix} \mathbf{i} & \mathbf{j} & \mathbf{k} \\ 2 & -3 & 1 \\ 2 & -3 & 1 \end{vmatrix} = \begin{vmatrix} -3 & 1 \\ -3 & 1 \end{vmatrix} \mathbf{i} - \begin{vmatrix} 2 & 1 \\ 2 & 1 \end{vmatrix} \mathbf{j} + \begin{vmatrix} 2 & -3 \\ 2 & -3 \end{vmatrix} \mathbf{k} = 0\mathbf{i} + 0\mathbf{j} + 0\mathbf{k} = \mathbf{0}$

11. (a) $\mathbf{v} \times \mathbf{w} = \begin{vmatrix} \mathbf{i} & \mathbf{j} & \mathbf{k} \\ 1 & 1 & 0 \\ 2 & 1 & 1 \end{vmatrix} = \begin{vmatrix} 1 & 0 \\ 1 & 1 \end{vmatrix} \mathbf{i} - \begin{vmatrix} 1 & 0 \\ 2 & 1 \end{vmatrix} \mathbf{j} + \begin{vmatrix} 1 & 1 \\ 2 & 1 \end{vmatrix} \mathbf{k} = 1\mathbf{i} - 1\mathbf{j} - 1\mathbf{k}$

(b) $\mathbf{w} \times \mathbf{v} = \begin{vmatrix} \mathbf{i} & \mathbf{j} & \mathbf{k} \\ 2 & 1 & 1 \\ 1 & 1 & 0 \end{vmatrix} = \begin{vmatrix} 1 & 1 \\ 1 & 0 \end{vmatrix} \mathbf{i} - \begin{vmatrix} 2 & 1 \\ 1 & 0 \end{vmatrix} \mathbf{j} + \begin{vmatrix} 2 & 1 \\ 1 & 1 \end{vmatrix} \mathbf{k} = -1\mathbf{i} + 1\mathbf{j} + 1\mathbf{k}$

(c) $\mathbf{w} \times \mathbf{w} = \begin{vmatrix} \mathbf{i} & \mathbf{j} & \mathbf{k} \\ 2 & 1 & 1 \\ 2 & 1 & 1 \end{vmatrix} = \begin{vmatrix} 1 & 1 \\ 1 & 1 \end{vmatrix} \mathbf{i} - \begin{vmatrix} 2 & 1 \\ 2 & 1 \end{vmatrix} \mathbf{j} + \begin{vmatrix} 2 & 1 \\ 2 & 1 \end{vmatrix} \mathbf{k} = 0\mathbf{i} + 0\mathbf{j} + 0\mathbf{k} = \mathbf{0}$

(d) $\mathbf{v} \times \mathbf{v} = \begin{vmatrix} \mathbf{i} & \mathbf{j} & \mathbf{k} \\ 1 & 1 & 0 \\ 1 & 1 & 0 \end{vmatrix} = \begin{vmatrix} 1 & 0 \\ 1 & 0 \end{vmatrix}\mathbf{i} - \begin{vmatrix} 1 & 0 \\ 1 & 0 \end{vmatrix}\mathbf{j} + \begin{vmatrix} 1 & 1 \\ 1 & 1 \end{vmatrix}\mathbf{k} = 0\mathbf{i} + 0\mathbf{j} + 0\mathbf{k} = \mathbf{0}$

13. (a) $\mathbf{v} \times \mathbf{w} = \begin{vmatrix} \mathbf{i} & \mathbf{j} & \mathbf{k} \\ 2 & -1 & 2 \\ 0 & 1 & -1 \end{vmatrix} = \begin{vmatrix} -1 & 2 \\ 1 & -1 \end{vmatrix}\mathbf{i} - \begin{vmatrix} 2 & 2 \\ 0 & -1 \end{vmatrix}\mathbf{j} + \begin{vmatrix} 2 & -1 \\ 0 & 1 \end{vmatrix}\mathbf{k} = -1\mathbf{i} + 2\mathbf{j} + 2\mathbf{k}$

(b) $\mathbf{w} \times \mathbf{v} = \begin{vmatrix} \mathbf{i} & \mathbf{j} & \mathbf{k} \\ 0 & 1 & -1 \\ 2 & -1 & 2 \end{vmatrix} = \begin{vmatrix} 1 & -1 \\ -1 & 2 \end{vmatrix}\mathbf{i} - \begin{vmatrix} 0 & -1 \\ 2 & 2 \end{vmatrix}\mathbf{j} + \begin{vmatrix} 0 & 1 \\ 2 & -1 \end{vmatrix}\mathbf{k} = 1\mathbf{i} - 2\mathbf{j} - 2\mathbf{k}$

(c) $\mathbf{w} \times \mathbf{w} = \begin{vmatrix} \mathbf{i} & \mathbf{j} & \mathbf{k} \\ 0 & 1 & -1 \\ 0 & 1 & -1 \end{vmatrix} = \begin{vmatrix} 1 & -1 \\ 1 & -1 \end{vmatrix}\mathbf{i} - \begin{vmatrix} 0 & -1 \\ 0 & -1 \end{vmatrix}\mathbf{j} + \begin{vmatrix} 0 & 1 \\ 0 & 1 \end{vmatrix}\mathbf{k} = 0\mathbf{i} + 0\mathbf{j} + 0\mathbf{k} = \mathbf{0}$

(d) $\mathbf{v} \times \mathbf{v} = \begin{vmatrix} \mathbf{i} & \mathbf{j} & \mathbf{k} \\ 2 & -1 & 2 \\ 2 & -1 & 2 \end{vmatrix} = \begin{vmatrix} -1 & 2 \\ -1 & 2 \end{vmatrix}\mathbf{i} - \begin{vmatrix} 2 & 2 \\ 2 & 2 \end{vmatrix}\mathbf{j} + \begin{vmatrix} 2 & -1 \\ 2 & -1 \end{vmatrix}\mathbf{k} = 0\mathbf{i} + 0\mathbf{j} + 0\mathbf{k} = \mathbf{0}$

15. (a) $\mathbf{v} \times \mathbf{w} = \begin{vmatrix} \mathbf{i} & \mathbf{j} & \mathbf{k} \\ 1 & -1 & -1 \\ 4 & 0 & -3 \end{vmatrix} = \begin{vmatrix} -1 & -1 \\ 0 & -3 \end{vmatrix}\mathbf{i} - \begin{vmatrix} 1 & -1 \\ 4 & -3 \end{vmatrix}\mathbf{j} + \begin{vmatrix} 1 & -1 \\ 4 & 0 \end{vmatrix}\mathbf{k} = 3\mathbf{i} - 1\mathbf{j} + 4\mathbf{k}$

(b) $\mathbf{w} \times \mathbf{v} = \begin{vmatrix} \mathbf{i} & \mathbf{j} & \mathbf{k} \\ 4 & 0 & -3 \\ 1 & -1 & -1 \end{vmatrix} = \begin{vmatrix} 0 & -3 \\ -1 & -1 \end{vmatrix}\mathbf{i} - \begin{vmatrix} 4 & -3 \\ 1 & -1 \end{vmatrix}\mathbf{j} + \begin{vmatrix} 4 & 0 \\ 1 & -1 \end{vmatrix}\mathbf{k} = -3\mathbf{i} + 1\mathbf{j} - 4\mathbf{k}$

(c) $\mathbf{w} \times \mathbf{w} = \begin{vmatrix} \mathbf{i} & \mathbf{j} & \mathbf{k} \\ 4 & 0 & -3 \\ 4 & 0 & -3 \end{vmatrix} = \begin{vmatrix} 0 & -3 \\ 0 & -3 \end{vmatrix}\mathbf{i} - \begin{vmatrix} 4 & -3 \\ 4 & -3 \end{vmatrix}\mathbf{j} + \begin{vmatrix} 4 & 0 \\ 4 & 0 \end{vmatrix}\mathbf{k} = 0\mathbf{i} + 0\mathbf{j} + 0\mathbf{k} = \mathbf{0}$

(d) $\mathbf{v} \times \mathbf{v} = \begin{vmatrix} \mathbf{i} & \mathbf{j} & \mathbf{k} \\ 1 & -1 & -1 \\ 1 & -1 & -1 \end{vmatrix} = \begin{vmatrix} -1 & -1 \\ -1 & -1 \end{vmatrix}\mathbf{i} - \begin{vmatrix} 1 & -1 \\ 1 & -1 \end{vmatrix}\mathbf{j} + \begin{vmatrix} 1 & -1 \\ 1 & -1 \end{vmatrix}\mathbf{k} = 0\mathbf{i} + 0\mathbf{j} + 0\mathbf{k} = \mathbf{0}$

17. $\mathbf{u} \times \mathbf{v} = \begin{vmatrix} \mathbf{i} & \mathbf{j} & \mathbf{k} \\ 2 & -3 & 1 \\ -3 & 3 & 2 \end{vmatrix} = \begin{vmatrix} -3 & 1 \\ 3 & 2 \end{vmatrix}\mathbf{i} - \begin{vmatrix} 2 & 1 \\ -3 & 2 \end{vmatrix}\mathbf{j} + \begin{vmatrix} 2 & -3 \\ -3 & 3 \end{vmatrix}\mathbf{k} = -9\mathbf{i} - 7\mathbf{j} - 3\mathbf{k}$

19. $\mathbf{v} \times \mathbf{u} = \begin{vmatrix} \mathbf{i} & \mathbf{j} & \mathbf{k} \\ -3 & 3 & 2 \\ 2 & -3 & 1 \end{vmatrix} = \begin{vmatrix} 3 & 2 \\ -3 & 1 \end{vmatrix}\mathbf{i} - \begin{vmatrix} -3 & 2 \\ 2 & 1 \end{vmatrix}\mathbf{j} + \begin{vmatrix} -3 & 3 \\ 2 & -3 \end{vmatrix}\mathbf{k} = 9\mathbf{i} + 7\mathbf{j} + 3\mathbf{k}$

21. $\mathbf{v} \times \mathbf{v} = \begin{vmatrix} \mathbf{i} & \mathbf{j} & \mathbf{k} \\ -3 & 3 & 2 \\ -3 & 3 & 2 \end{vmatrix} = \begin{vmatrix} 3 & 2 \\ 3 & 2 \end{vmatrix}\mathbf{i} - \begin{vmatrix} -3 & 2 \\ -3 & 2 \end{vmatrix}\mathbf{j} + \begin{vmatrix} -3 & 3 \\ -3 & 3 \end{vmatrix}\mathbf{k} = 0\mathbf{i} + 0\mathbf{j} + 0\mathbf{k} = \mathbf{0}$

23. $(3\mathbf{u}) \times \mathbf{v} = \begin{vmatrix} \mathbf{i} & \mathbf{j} & \mathbf{k} \\ 6 & -9 & 3 \\ -3 & 3 & 2 \end{vmatrix} = \begin{vmatrix} -9 & 3 \\ 3 & 2 \end{vmatrix}\mathbf{i} - \begin{vmatrix} 6 & 3 \\ -3 & 2 \end{vmatrix}\mathbf{j} + \begin{vmatrix} 6 & -9 \\ -3 & 3 \end{vmatrix}\mathbf{k} = -27\mathbf{i} - 21\mathbf{j} - 9\mathbf{k}$

25. $\mathbf{u} \times (2\mathbf{v}) = \begin{vmatrix} \mathbf{i} & \mathbf{j} & \mathbf{k} \\ 2 & -3 & 1 \\ -6 & 6 & 4 \end{vmatrix} = \begin{vmatrix} -3 & 1 \\ 6 & 4 \end{vmatrix}\mathbf{i} - \begin{vmatrix} 2 & 1 \\ -6 & 4 \end{vmatrix}\mathbf{j} + \begin{vmatrix} 2 & -3 \\ -6 & 6 \end{vmatrix}\mathbf{k} = -18\mathbf{i} - 14\mathbf{j} - 6\mathbf{k}$

27. $\mathbf{u} \bullet (\mathbf{u} \times \mathbf{v}) = \mathbf{u} \bullet \begin{vmatrix} \mathbf{i} & \mathbf{j} & \mathbf{k} \\ 2 & -3 & 1 \\ -3 & 3 & 2 \end{vmatrix} = \mathbf{u} \cdot \left(\begin{vmatrix} -3 & 1 \\ 3 & 2 \end{vmatrix}\mathbf{i} - \begin{vmatrix} 2 & 1 \\ -3 & 2 \end{vmatrix}\mathbf{j} + \begin{vmatrix} 2 & -3 \\ -3 & 3 \end{vmatrix}\mathbf{k} \right)$

    $= (2\mathbf{i} - 3\mathbf{j} + \mathbf{k}) \bullet (-9\mathbf{i} - 7\mathbf{j} - 3\mathbf{k}) = 2(-9) + (-3)(-7) + 1(-3) = -18 + 21 - 3 = 0$

29. $\mathbf{u} \bullet (\mathbf{v} \times \mathbf{w}) = \mathbf{u} \bullet \begin{vmatrix} \mathbf{i} & \mathbf{j} & \mathbf{k} \\ -3 & 3 & 2 \\ 1 & 1 & 3 \end{vmatrix} = \mathbf{u} \bullet \left(\begin{vmatrix} 3 & 2 \\ 1 & 3 \end{vmatrix}\mathbf{i} - \begin{vmatrix} -3 & 2 \\ 1 & 3 \end{vmatrix}\mathbf{j} + \begin{vmatrix} -3 & 3 \\ 1 & 1 \end{vmatrix}\mathbf{k} \right)$

    $= (2\mathbf{i} - 3\mathbf{j} + \mathbf{k}) \bullet (7\mathbf{i} + 11\mathbf{j} - 6\mathbf{k}) = 2 \cdot 7 + (-3)(11) + 1(-6) = 14 - 33 - 6 = -25$

31. $\mathbf{v} \bullet (\mathbf{u} \times \mathbf{w}) = \mathbf{v} \bullet \begin{vmatrix} \mathbf{i} & \mathbf{j} & \mathbf{k} \\ 2 & -3 & 1 \\ 1 & 1 & 3 \end{vmatrix} = \mathbf{v} \bullet \left(\begin{vmatrix} -3 & 1 \\ 1 & 3 \end{vmatrix}\mathbf{i} - \begin{vmatrix} 2 & 1 \\ 1 & 3 \end{vmatrix}\mathbf{j} + \begin{vmatrix} 2 & -3 \\ 1 & 1 \end{vmatrix}\mathbf{k} \right)$

    $= (-3\mathbf{i} + 3\mathbf{j} + 2\mathbf{k}) \bullet (-10\mathbf{i} - 5\mathbf{j} + 5\mathbf{k}) = -3(-10) + 3(-5) + 2 \cdot 5 = 30 - 15 + 10 = 25$

33. $\mathbf{u} \times (\mathbf{v} \times \mathbf{v}) = \mathbf{u} \times \begin{vmatrix} \mathbf{i} & \mathbf{j} & \mathbf{k} \\ -3 & 3 & 2 \\ -3 & 3 & 2 \end{vmatrix} = \mathbf{u} \times \left(\begin{vmatrix} 3 & 2 \\ 3 & 2 \end{vmatrix}\mathbf{i} - \begin{vmatrix} -3 & 2 \\ -3 & 2 \end{vmatrix}\mathbf{j} + \begin{vmatrix} -3 & 3 \\ -3 & 3 \end{vmatrix}\mathbf{k} \right)$

    $= (2\mathbf{i} - 3\mathbf{j} + \mathbf{k}) \times (0\mathbf{i} + 0\mathbf{j} + 0\mathbf{k}) = \begin{vmatrix} \mathbf{i} & \mathbf{j} & \mathbf{k} \\ 2 & -3 & 1 \\ 0 & 0 & 0 \end{vmatrix}$

    $= \begin{vmatrix} -3 & 1 \\ 0 & 0 \end{vmatrix}\mathbf{i} - \begin{vmatrix} 2 & 1 \\ 0 & 0 \end{vmatrix}\mathbf{j} + \begin{vmatrix} 2 & -3 \\ 0 & 0 \end{vmatrix}\mathbf{k} = 0\mathbf{i} + 0\mathbf{j} + 0\mathbf{k} = \mathbf{0}$

35. $\mathbf{u} \times \mathbf{v} = \begin{vmatrix} \mathbf{i} & \mathbf{j} & \mathbf{k} \\ 2 & -3 & 1 \\ -3 & 3 & 2 \end{vmatrix} = \begin{vmatrix} -3 & 1 \\ 3 & 2 \end{vmatrix}\mathbf{i} - \begin{vmatrix} 2 & 1 \\ -3 & 2 \end{vmatrix}\mathbf{j} + \begin{vmatrix} 2 & -3 \\ -3 & 3 \end{vmatrix}\mathbf{k} = -9\mathbf{i} - 7\mathbf{j} - 3\mathbf{k}$ is orthogonal to both

    $\mathbf{u}$ and $\mathbf{v}$.

37. A vector that is orthogonal to both $\mathbf{u}$ and $\mathbf{i} + \mathbf{j}$ is $\mathbf{u} \times (\mathbf{i} + \mathbf{j})$.

    $\mathbf{u} \times (\mathbf{i} + \mathbf{j}) = \begin{vmatrix} \mathbf{i} & \mathbf{j} & \mathbf{k} \\ 2 & -3 & 1 \\ 1 & 1 & 0 \end{vmatrix} = \begin{vmatrix} -3 & 1 \\ 1 & 0 \end{vmatrix}\mathbf{i} - \begin{vmatrix} 2 & 1 \\ 1 & 0 \end{vmatrix}\mathbf{j} + \begin{vmatrix} 2 & -3 \\ 1 & 1 \end{vmatrix}\mathbf{k} = -1\mathbf{i} + 1\mathbf{j} + 5\mathbf{k}$

39. $\mathbf{u} = P_1P_2 = 1\mathbf{i} + 2\mathbf{j} + 3\mathbf{k}$ $\mathbf{v} = P_1P_3 = -2\mathbf{i} + 3\mathbf{j} + 0\mathbf{k}$

    $\mathbf{u} \times \mathbf{v} = \begin{vmatrix} \mathbf{i} & \mathbf{j} & \mathbf{k} \\ 1 & 2 & 3 \\ -2 & 3 & 0 \end{vmatrix} = \begin{vmatrix} 2 & 3 \\ 3 & 0 \end{vmatrix}\mathbf{i} - \begin{vmatrix} 1 & 3 \\ -2 & 0 \end{vmatrix}\mathbf{j} + \begin{vmatrix} 1 & 2 \\ -2 & 3 \end{vmatrix}\mathbf{k} = -9\mathbf{i} - 6\mathbf{j} + 7\mathbf{k}$

    Area $= \|\mathbf{u} \times \mathbf{v}\| = \sqrt{(-9)^2 + (-6)^2 + 7^2} = \sqrt{166} \approx 12.9$

41. $\mathbf{u} = P_1P_2 = -3\mathbf{i} + 1\mathbf{j} + 4\mathbf{k}$ $\mathbf{v} = P_1P_3 = -1\mathbf{i} - 4\mathbf{j} + 3\mathbf{k}$

$$\mathbf{u} \times \mathbf{v} = \begin{vmatrix} \mathbf{i} & \mathbf{j} & \mathbf{k} \\ -3 & 1 & 4 \\ -1 & -4 & 3 \end{vmatrix} = \begin{vmatrix} 1 & 4 \\ -4 & 3 \end{vmatrix}\mathbf{i} - \begin{vmatrix} -3 & 4 \\ -1 & 3 \end{vmatrix}\mathbf{j} + \begin{vmatrix} -3 & 1 \\ -1 & -4 \end{vmatrix}\mathbf{k} = 19\mathbf{i} + 5\mathbf{j} + 13\mathbf{k}$$

Area = $\|\mathbf{u} \times \mathbf{v}\| = \sqrt{19^2 + 5^2 + 13^2} = \sqrt{555} \approx 23.6$

43. $\mathbf{u} = P_1P_2 = 0\mathbf{i} + 1\mathbf{j} + 1\mathbf{k}$ $\mathbf{v} = P_1P_3 = -3\mathbf{i} + 2\mathbf{j} - 2\mathbf{k}$

$$\mathbf{u} \times \mathbf{v} = \begin{vmatrix} \mathbf{i} & \mathbf{j} & \mathbf{k} \\ 0 & 1 & 1 \\ -3 & 2 & -2 \end{vmatrix} = \begin{vmatrix} 1 & 1 \\ 2 & -2 \end{vmatrix}\mathbf{i} - \begin{vmatrix} 0 & 1 \\ -3 & -2 \end{vmatrix}\mathbf{j} + \begin{vmatrix} 0 & 1 \\ -3 & 2 \end{vmatrix}\mathbf{k} = -4\mathbf{i} - 3\mathbf{j} + 3\mathbf{k}$$

Area = $\|\mathbf{u} \times \mathbf{v}\| = \sqrt{(-4)^2 + (-3)^2 + 3^2} = \sqrt{34} \approx 5.8$

45. $\mathbf{u} = P_1P_2 = 3\mathbf{i} + 0\mathbf{j} - 2\mathbf{k}$ $\mathbf{v} = P_1P_3 = 5\mathbf{i} - 7\mathbf{j} + 3\mathbf{k}$

$$\mathbf{u} \times \mathbf{v} = \begin{vmatrix} \mathbf{i} & \mathbf{j} & \mathbf{k} \\ 3 & 0 & -2 \\ 5 & -7 & 3 \end{vmatrix} = \begin{vmatrix} 0 & -2 \\ -7 & 3 \end{vmatrix}\mathbf{i} - \begin{vmatrix} 3 & -2 \\ 5 & 3 \end{vmatrix}\mathbf{j} + \begin{vmatrix} 3 & 0 \\ 5 & -7 \end{vmatrix}\mathbf{k} = -14\mathbf{i} - 19\mathbf{j} - 21\mathbf{k}$$

Area = $\|\mathbf{u} \times \mathbf{v}\| = \sqrt{(-14)^2 + (-19)^2 + (-21)^2} = \sqrt{998} \approx 31.6$

47. $\mathbf{v} \times \mathbf{w} = \begin{vmatrix} \mathbf{i} & \mathbf{j} & \mathbf{k} \\ 1 & 3 & -2 \\ -2 & 1 & 3 \end{vmatrix} = \begin{vmatrix} 3 & -2 \\ 1 & 3 \end{vmatrix}\mathbf{i} - \begin{vmatrix} 1 & -2 \\ -2 & 3 \end{vmatrix}\mathbf{j} + \begin{vmatrix} 1 & 3 \\ -2 & 1 \end{vmatrix}\mathbf{k} = 11\mathbf{i} + 1\mathbf{j} + 7\mathbf{k}$

$\|\mathbf{v} \times \mathbf{w}\| = \sqrt{11^2 + 1^2 + 7^2} = \sqrt{171}$

$\mathbf{u} = \dfrac{\mathbf{v} \times \mathbf{w}}{\|\mathbf{v} \times \mathbf{w}\|} = \dfrac{11\mathbf{i} + 1\mathbf{j} + 7\mathbf{k}}{\sqrt{171}} = \dfrac{11}{\sqrt{171}}\mathbf{i} + \dfrac{1}{\sqrt{171}}\mathbf{j} + \dfrac{7}{\sqrt{171}}\mathbf{k}$

49. Prove: $\mathbf{u} \times \mathbf{v} = -(\mathbf{v} \times \mathbf{u})$

Let $\mathbf{u} = a_1\mathbf{i} + b_1\mathbf{j} + c_1\mathbf{k}$ and $\mathbf{v} = a_2\mathbf{i} + b_2\mathbf{j} + c_2\mathbf{k}$

$$\mathbf{u} \times \mathbf{v} = \begin{vmatrix} \mathbf{i} & \mathbf{j} & \mathbf{k} \\ a_1 & b_1 & c_1 \\ a_2 & b_2 & c_2 \end{vmatrix} = \begin{vmatrix} b_1 & c_1 \\ b_2 & c_2 \end{vmatrix}\mathbf{i} - \begin{vmatrix} a_1 & c_1 \\ a_2 & c_2 \end{vmatrix}\mathbf{j} + \begin{vmatrix} a_1 & b_1 \\ a_2 & b_2 \end{vmatrix}\mathbf{k}$$

$= (b_1c_2 - b_2c_1)\mathbf{i} - (a_1c_2 - a_2c_1)\mathbf{j} + (a_1b_2 - a_2b_1)\mathbf{k}$

$= -(b_2c_1 - b_1c_2)\mathbf{i} + (a_2c_1 - a_1c_2)\mathbf{j} - (a_2b_1 - a_1b_2)\mathbf{k}$

$= -((b_2c_1 - b_1c_2)\mathbf{i} - (a_2c_1 - a_1c_2)\mathbf{j} + (a_2b_1 - a_1b_2)\mathbf{k})$

$= -\left(\begin{vmatrix} b_2 & c_2 \\ b_1 & c_1 \end{vmatrix}\mathbf{i} - \begin{vmatrix} a_2 & c_2 \\ a_1 & c_1 \end{vmatrix}\mathbf{j} + \begin{vmatrix} a_2 & b_2 \\ a_1 & b_1 \end{vmatrix}\mathbf{k} \right)$

$= -\begin{vmatrix} \mathbf{i} & \mathbf{j} & \mathbf{k} \\ a_2 & b_2 & c_2 \\ a_1 & b_1 & c_1 \end{vmatrix}$

$= -(\mathbf{v} \times \mathbf{u})$

51. Prove: $\|\mathbf{u} \times \mathbf{v}\|^2 = \|\mathbf{u}\|^2 \|\mathbf{v}\|^2 - (\mathbf{u} \cdot \mathbf{v})^2$

Let $\mathbf{u} = a_1\mathbf{i} + b_1\mathbf{j} + c_1\mathbf{k}$ and $\mathbf{v} = a_2\mathbf{i} + b_2\mathbf{j} + c_2\mathbf{k}$

$$\mathbf{u} \times \mathbf{v} = \begin{vmatrix} \mathbf{i} & \mathbf{j} & \mathbf{k} \\ a_1 & b_1 & c_1 \\ a_2 & b_2 & c_2 \end{vmatrix} = \begin{vmatrix} b_1 & c_1 \\ b_2 & c_2 \end{vmatrix}\mathbf{i} - \begin{vmatrix} a_1 & c_1 \\ a_2 & c_2 \end{vmatrix}\mathbf{j} + \begin{vmatrix} a_1 & b_1 \\ a_2 & b_2 \end{vmatrix}\mathbf{k}$$

$$= (b_1c_2 - b_2c_1)\mathbf{i} - (a_1c_2 - a_2c_1)\mathbf{j} + (a_1b_2 - a_2b_1)\mathbf{k}$$

$$\|\mathbf{u} \times \mathbf{v}\|^2 = (b_1c_2 - b_2c_1)^2 + (a_1c_2 - a_2c_1)^2 + (a_1b_2 - a_2b_1)^2$$

$$= b_1^2c_2^2 - 2b_1b_2c_1c_2 + b_2^2c_1^2 + a_1^2c_2^2 - 2a_1a_2c_1c_2 + a_2^2c_1^2$$
$$+ a_1^2b_2^2 - 2a_1a_2b_1b_2 + a_2^2b_1^2$$

$$= a_1^2b_2^2 + a_1^2c_2^2 + a_2^2b_1^2 + a_2^2c_1^2 + b_1^2c_2^2 + b_2^2c_1^2 - 2a_1a_2b_1b_2$$
$$- 2a_1a_2c_1c_2 - 2b_1b_2c_1c_2$$

$$\|\mathbf{u}\|^2 = a_1^2 + b_1^2 + c_1^2$$

$$\|\mathbf{v}\|^2 = a_2^2 + b_2^2 + c_2^2$$

$$(\mathbf{u} \cdot \mathbf{v})^2 = (a_1a_2 + b_1b_2 + c_1c_2)^2$$

$$\|\mathbf{u}\|^2\|\mathbf{v}\|^2 - (\mathbf{u} \cdot \mathbf{v})^2 = \left(a_1^2 + b_1^2 + c_1^2\right)\left(a_2^2 + b_2^2 + c_2^2\right) - (a_1a_2 + b_1b_2 + c_1c_2)^2$$

$$= a_1^2a_2^2 + a_1^2b_2^2 + a_1^2c_2^2 + b_1^2a_2^2 + b_1^2b_2^2 + b_1^2c_2^2 + c_1^2a_2^2 + c_1^2b_2^2 + c_1^2c_2^2$$

$$- \left(\begin{array}{l} a_1^2a_2^2 + a_1a_2b_1b_2 + a_1a_2c_1c_2 + a_1a_2b_1b_2 + b_1^2b_2^2 + b_1b_2c_1c_2 \\ + a_1a_2c_1c_2 + b_1b_2c_1c_2 + c_1^2c_2^2 \end{array}\right)$$

$$= a_1^2b_2^2 + a_1^2c_2^2 + a_2^2b_1^2 + a_2^2c_1^2 + b_1^2c_2^2 + b_2^2c_1^2 - 2a_1a_2b_1b_2$$
$$- 2a_1a_2c_1c_2 - 2b_1b_2c_1c_2$$

53. If $\mathbf{u}$ and $\mathbf{v}$ are orthogonal, then $\mathbf{u} \cdot \mathbf{v} = 0$. From problem 51, then:

$$\|\mathbf{u} \times \mathbf{v}\|^2 = \|\mathbf{u}\|^2\|\mathbf{v}\|^2 - (\mathbf{u} \cdot \mathbf{v})^2 = \|\mathbf{u}\|^2\|\mathbf{v}\|^2 - (0)^2 = \|\mathbf{u}\|^2\|\mathbf{v}\|^2$$

$$\|\mathbf{u} \times \mathbf{v}\| = \|\mathbf{u}\|\|\mathbf{v}\|$$

Polar Coordinates; Vectors

8.R   Chapter Review

1. $\left(3, \frac{\pi}{6}\right)$

3. $\left(-2, \frac{4\pi}{3}\right)$

$$x = 3\cos\left(\frac{\pi}{6}\right) = \frac{3\sqrt{3}}{2}$$

$$y = 3\sin\left(\frac{\pi}{6}\right) = \frac{3}{2}$$

$$\left(\frac{3\sqrt{3}}{2}, \frac{3}{2}\right)$$

$$x = -2\cos\left(\frac{4\pi}{3}\right) = 1$$

$$y = -2\sin\left(\frac{4\pi}{3}\right) = \sqrt{3}$$

$$\left(1, \sqrt{3}\right)$$

5. $\left(-3, -\frac{\pi}{2}\right)$

$$x = -3\cos\left(-\frac{\pi}{2}\right) = 0$$

$$y = -3\sin\left(-\frac{\pi}{2}\right) = 3$$

$$(0, 3)$$

7.   The point $(-3, 3)$ lies in quadrant II.

$$r = \sqrt{x^2 + y^2} = \sqrt{(-3)^2 + 3^2} = 3\sqrt{2} \qquad \theta = \tan^{-1}\left(\frac{y}{x}\right) = \tan^{-1}\left(\frac{3}{-3}\right) = \tan^{-1}(-1) = -\frac{\pi}{4}$$

Polar coordinates of the point $(-3, 3)$ are $\left(-3\sqrt{2}, -\frac{\pi}{4}\right)$ or $\left(3\sqrt{2}, \frac{3\pi}{4}\right)$.

9.   The point $(0, -2)$ lies on the negative y-axis.

$$r = \sqrt{x^2 + y^2} = \sqrt{0^2 + (-2)^2} = 2 \qquad \theta = \tan^{-1}\left(\frac{y}{x}\right) = \tan^{-1}\left(\frac{-2}{0}\right) = -\frac{\pi}{2}$$

Polar coordinates of the point $(0, -2)$ are $\left(2, -\frac{\pi}{2}\right)$ or $\left(-2, \frac{\pi}{2}\right)$.

11.   The point $(3, 4)$ lies in quadrant I.

$$r = \sqrt{x^2 + y^2} = \sqrt{3^2 + 4^2} = 5 \qquad \theta = \tan^{-1}\left(\frac{y}{x}\right) = \tan^{-1}\left(\frac{4}{3}\right) \approx 0.93$$

Polar coordinates of the point $(3, 4)$ are $(5, 0.93)$ or $(-5, 4.07)$.

12. The point $(-5, 12)$ lies in quadrant II.

$r = \sqrt{x^2 + y^2} = \sqrt{(-5)^2 + 12^2} = 13 \qquad \theta = \tan^{-1}\left(\frac{y}{x}\right) = \tan^{-1}\left(\frac{12}{-5}\right) \approx -1.18$

Polar coordinates of the point $(-5, 12)$ are $(13, 1.96)$ or $(-13, 5.10)$.

13.
$$3x^2 + 3y^2 = 6y$$
$$x^2 + y^2 = 2y$$
$$r^2 = 2r\sin\theta$$
$$r^2 - 2r\sin\theta = 0$$

15.
$$2x^2 - y^2 = \frac{y}{x}$$
$$2x^2 + 2y^2 - 3y^2 = \frac{y}{x}$$
$$2(x^2 + y^2) - 3y^2 = \frac{y}{x}$$
$$2r^2 - 3(r\sin\theta)^2 = \tan\theta$$
$$2r^2 - 3r^2\sin^2\theta - \tan\theta = 0$$
$$r^2(2 - 3\sin\theta) - \tan\theta = 0$$

17.
$$x(x^2 + y^2) = 4$$
$$r\cos\theta(r^2) = 4$$
$$r^3\cos\theta = 4$$

19.
$$r = 2\sin\theta$$
$$r^2 = 2r\sin\theta$$
$$x^2 + y^2 = 2y$$
$$x^2 + y^2 - 2y = 0$$

21.
$$r = 5$$
$$r^2 = 25$$
$$x^2 + y^2 = 25$$

23.
$$r\cos\theta + 3r\sin\theta = 6$$
$$x + 3y = 6$$

25. $r = 4\cos\theta$ The graph will be a circle. Check for symmetry:
Polar axis: Replace θ by $-\theta$. The result is $r = 4\cos(-\theta) = 4\cos\theta$.
       The graph is symmetric with respect to the polar axis.

The line $\theta = \frac{\pi}{2}$: Replace θ by $\pi - \theta$.

$$r = 4\cos(\pi - \theta) = 4(\cos(\pi)\cos\theta + \sin(\pi)\sin\theta)$$
$$= 4(-\cos\theta + 0) = -4\cos\theta$$

The test fails.

The pole: Replace r by $-r$. $-r = 4\cos\theta$. The test fails.
Due to symmetry to the polar axis, assign values to θ from 0 to π.

θ	0	$\frac{\pi}{6}$	$\frac{\pi}{3}$	$\frac{\pi}{2}$	$\frac{2\pi}{3}$	$\frac{5\pi}{6}$	π
$r = 4\cos\theta$	4	$2\sqrt{3} \approx 3.5$	2	0	-2	$-2\sqrt{3} \approx -3.5$	-4

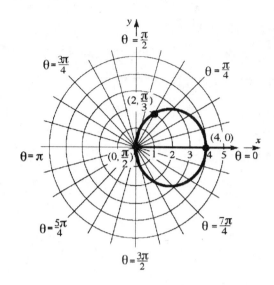

27. $r = 3 - 3\sin\theta$ The graph will be a cardioid. Check for symmetry:
     Polar axis: Replace θ by $-\theta$. The result is $r = 3 - 3\sin(-\theta) = 3 + 3\sin\theta$.
                 The test fails.

The line $\theta = \dfrac{\pi}{2}$: Replace θ by $\pi - \theta$.

$$r = 3 - 3\sin(\pi - \theta) = 3 - 3(\sin(\pi)\cos\theta - \cos(\pi)\sin\theta)$$

$$= 3 - 3(0 + \sin\theta) = 3 - 3\sin\theta$$

The graph is symmetric with respect to the line $\theta = \dfrac{\pi}{2}$.

The pole: Replace r by $-r$. $-r = 3 - 3\sin\theta$. The test fails.

Due to symmetry to the line $\theta = \dfrac{\pi}{2}$, assign values to θ from $-\dfrac{\pi}{2}$ to $\dfrac{\pi}{2}$.

θ	$-\dfrac{\pi}{2}$	$-\dfrac{\pi}{3}$	$-\dfrac{\pi}{6}$	0	$\dfrac{\pi}{6}$	$\dfrac{\pi}{3}$	$\dfrac{\pi}{2}$
$r = 3 - 3\sin\theta$	6	$3 + \dfrac{3\sqrt{3}}{2} \approx 5.6$	$\dfrac{9}{2}$	3	$\dfrac{3}{2}$	$3 - \dfrac{3\sqrt{3}}{2} \approx 0.4$	0

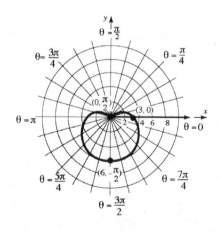

29. $r = 4 - \cos\theta$ The graph will be a limacon without an inner loop.
Check for symmetry:
Polar axis: Replace θ by $-\theta$. The result is $r = 4 - \cos(-\theta) = 4 - \cos\theta$.
         The graph is symmetric with respect to the polar axis.
The line $\theta = \dfrac{\pi}{2}$: Replace θ by $\pi - \theta$.

$$r = 4 - \cos(\pi - \theta) = 4 - (\cos(\pi)\cos\theta + \sin(\pi)\sin\theta)$$
$$= 4 - (-\cos\theta + 0) = 4 + \cos\theta$$

         The test fails.
The pole: Replace r by $-r$. $-r = 4 - \cos\theta$. The test fails.
Due to symmetry to the polar axis, assign values to θ from 0 to π.

θ	0	$\dfrac{\pi}{6}$	$\dfrac{\pi}{3}$	$\dfrac{\pi}{2}$	$\dfrac{2\pi}{3}$	$\dfrac{5\pi}{6}$	π
$r = 4 - \cos\theta$	3	$4 - \dfrac{\sqrt{3}}{2} \approx 3.1$	$\dfrac{7}{2}$	4	$\dfrac{9}{2}$	$4 + \dfrac{\sqrt{3}}{2} \approx 4.9$	5

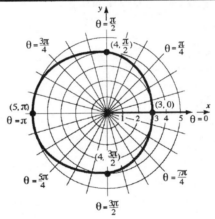

31. $r = \sqrt{x^2 + y^2} = \sqrt{(-1)^2 + (-1)^2} = \sqrt{2}$
$\tan\theta = \dfrac{y}{x} = \dfrac{-1}{-1} = 1 \ \rightarrow \ \theta = 225°$
The polar form of $z = -1 - i$ is $z = r(\cos\theta + i\sin\theta) = \sqrt{2}(\cos(225°) + i\sin(225°))$.

33. $r = \sqrt{x^2 + y^2} = \sqrt{4^2 + (-3)^2} = \sqrt{25} = 5$
$\tan\theta = \dfrac{y}{x} = -\dfrac{3}{4} \ \rightarrow \ \theta \approx 323.1°$
The polar form of $z = 4 - 3i$ is $z = r(\cos\theta + i\sin\theta) = 5(\cos(323.1°) + i\sin(323.1°))$.

35. $2(\cos(150°) + i\sin(150°)) = 2\left(-\dfrac{\sqrt{3}}{2} + \dfrac{1}{2}i\right) = -\sqrt{3} + i$

37. $3\left(\cos\left(\dfrac{2\pi}{3}\right) + i\sin\left(\dfrac{2\pi}{3}\right)\right) = 3\left(-\dfrac{1}{2} + \dfrac{\sqrt{3}}{2}i\right) = -\dfrac{3}{2} + \dfrac{3\sqrt{3}}{2}i$

39. $0.1(\cos(350^\circ) + i\sin(350^\circ)) \approx 0.1(0.9848 - 0.1736i) = 0.0985 - 0.0174i$

41. $z \cdot w = (\cos(80^\circ) + i\sin(80^\circ)) \cdot (\cos(50^\circ) + i\sin(50^\circ))$

$= 1 \cdot 1(\cos(80^\circ + 50^\circ) + i\sin(80^\circ + 50^\circ)) = \cos(130^\circ) + i\sin(130^\circ)$

$\dfrac{z}{w} = \dfrac{(\cos(80^\circ) + i\sin(80^\circ))}{(\cos(50^\circ) + i\sin(50^\circ))} = \dfrac{1}{1}(\cos(80^\circ - 50^\circ) + i\sin(80^\circ - 50^\circ))$

$= \cos(30^\circ) + i\sin(30^\circ)$

43. $z \cdot w = 3\left(\cos\left(\dfrac{9\pi}{5}\right) + i\sin\left(\dfrac{9\pi}{5}\right)\right) \cdot 2\left(\cos\dfrac{\pi}{5} + i\sin\dfrac{\pi}{5}\right) = 3 \cdot 2\left(\cos\left(\dfrac{9\pi}{5} + \dfrac{\pi}{5}\right) + i\sin\left(\dfrac{9\pi}{5} + \dfrac{\pi}{5}\right)\right)$

$= 6 \cdot (\cos(2\pi) + i\sin(2\pi)) = 6(\cos(0) + i\sin(0))$

$\dfrac{z}{w} = \dfrac{3 \cdot \left(\cos\left(\dfrac{9\pi}{5}\right) + i\sin\left(\dfrac{9\pi}{5}\right)\right)}{2 \cdot \left(\cos\left(\dfrac{\pi}{5}\right) + i\sin\left(\dfrac{\pi}{5}\right)\right)} = \dfrac{3}{2}\left(\cos\left(\dfrac{9\pi}{5} - \dfrac{\pi}{5}\right) + i\sin\left(\dfrac{9\pi}{5} - \dfrac{\pi}{5}\right)\right) = \dfrac{3}{2}\left(\cos\left(\dfrac{8\pi}{5}\right) + i\sin\left(\dfrac{8\pi}{5}\right)\right)$

45. $z \cdot w = 5(\cos(10^\circ) + i\sin(10^\circ)) \cdot (\cos(355^\circ) + i\sin(355^\circ))$

$= 5 \cdot 1(\cos(10^\circ + 355^\circ) + i\sin(10^\circ + 355^\circ)) = 5 \cdot (\cos(365^\circ) + i\sin(365^\circ))$

$= 5 \cdot (\cos(5^\circ) + i\sin(5^\circ))$

$\dfrac{z}{w} = \dfrac{5(\cos(10^\circ) + i\sin(10^\circ))}{(\cos(355^\circ) + i\sin(355^\circ))} = \dfrac{5}{1}(\cos(10^\circ - 355^\circ) + i\sin(10^\circ - 355^\circ))$

$= 5 \cdot (\cos(-345^\circ) + i\sin(-345^\circ)) = 5 \cdot (\cos(15^\circ) + i\sin(15^\circ))$

47. $[3(\cos(20^\circ) + i\sin(20^\circ))]^3 = 3^3(\cos(3 \cdot 20^\circ) + i\sin(3 \cdot 20^\circ)) = 27(\cos(60^\circ) + i\sin(60^\circ))$

$= 27\left(\dfrac{1}{2} + \dfrac{\sqrt{3}}{2}i\right) = \dfrac{27}{2} + \dfrac{27\sqrt{3}}{2}i$

49. $\left[\sqrt{2} \cdot \left(\cos\left(\dfrac{5\pi}{8}\right) + i\sin\left(\dfrac{5\pi}{8}\right)\right)\right]^4 = (\sqrt{2})^4\left(\cos\left(4 \cdot \dfrac{5\pi}{8}\right) + i\sin\left(4 \cdot \dfrac{5\pi}{8}\right)\right)$

$= 4 \cdot \left(\cos\left(\dfrac{5\pi}{2}\right) + i\sin\left(\dfrac{5\pi}{2}\right)\right) = 4(0 + 1i) = 4i$

51. $1-\sqrt{3}\,i$ $r=\sqrt{1^2+\left(-\sqrt{3}\right)^2}=2$ $\tan\theta=\dfrac{-\sqrt{3}}{1}=-\sqrt{3}$ $\theta=300°$

$1-\sqrt{3}\,i=2(\cos(300°)+i\sin(300°))$

$\left(1-\sqrt{3}\,i\right)^6=\left[2(\cos(300°)+i\sin(300°))\right]^6=2^6\cdot\left(\cos(6\cdot300°)+i\sin(6\cdot300°)\right)$

$=64\cdot\left(\cos(1800°)+i\sin(1800°)\right)=64\cdot\left(\cos(0°)+i\sin(0°)\right)$

$=64+0\,i=64$

53. $3+4i$ $r=\sqrt{3^2+4^2}=5$ $\tan\theta=\dfrac{4}{3}$ $\theta\approx53.1°$

$3+4i=5\cdot\left(\cos(53.1°)+i\sin(53.1°)\right)$

$(3+4i)^4=\left[5^4\cdot\left(\cos(4\cdot53.1°)+i\sin(4\cdot53.1°)\right)\right]^4=625\cdot\left(\cos(212.4°)+i\sin(212.4°)\right)$

$\approx625\cdot\left(-0.8443+i(-0.5358)\right)=-527.7-334.9i$

55. $27+0i$ $r=\sqrt{27^2+0^2}=27$ $\tan\theta=\dfrac{0}{27}=0$ $\theta=0°$

$27+0i=27(\cos(0°)+i\sin(0°))$

The three complex cube roots of $27=27(\cos0°+i\sin0°)$ are:

$z_k=\sqrt[3]{27}\cdot\left[\cos\left(\dfrac{0°}{3}+\dfrac{360°\cdot k}{3}\right)+i\sin\left(\dfrac{0°}{3}+\dfrac{360°\cdot k}{3}\right)\right]$

$=3\left[\cos(120°\,k)+i\sin(120°\,k)\right]$

$z_0=3\left[\cos(120°\cdot0)+i\sin(120°\cdot0)\right]=3\cdot\left(\cos(0°)+i\sin(0°)\right)=3$

$z_1=3\left[\cos(120°\cdot1)+i\sin(120°\cdot1)\right]=3\cdot\left(\cos(120°)+i\sin(120°)\right)=-\dfrac{3}{2}+\dfrac{3\sqrt{3}}{2}i$

$z_2=3\left[\cos(120°\cdot2)+i\sin(120°\cdot2)\right]=3\cdot\left(\cos(240°)+i\sin(240°)\right)=-\dfrac{3}{2}-\dfrac{3\sqrt{3}}{2}i$

57. $P=(1,-2), Q=(3,-6)$ $\mathbf{v}=(3-1)\mathbf{i}+(-6-(-2))\mathbf{j}=2\mathbf{i}-4\mathbf{j}$

$\|\mathbf{v}\|=\sqrt{2^2+(-4)^2}=\sqrt{20}=2\sqrt{5}$

59. $P=(0,-2), Q=(-1,1)$ $\mathbf{v}=(-1-0)\mathbf{i}+(1-(-2))\mathbf{j}=-1\mathbf{i}+3\mathbf{j}$

$\|\mathbf{v}\|=\sqrt{(-1)^2+3^2}=\sqrt{10}$

61. $\mathbf{v}=(3-6)\mathbf{i}+(0-2)\mathbf{j}+(2-1)\mathbf{k}=-3\mathbf{i}-2\mathbf{j}+\mathbf{k}$

$\|\mathbf{v}\|=\sqrt{(-3)^2+(-2)^2+1^2}=\sqrt{14}$

63. $\mathbf{v}=(2-(-1))\mathbf{i}+(0-0)\mathbf{j}+(0-1)\mathbf{k}=3\mathbf{i}-\mathbf{k}$

$\|\mathbf{v}\|=\sqrt{3^2+0^2+(-1)^2}=\sqrt{10}$

65. $\mathbf{v}=-2\mathbf{i}+\mathbf{j},\ \mathbf{w}=4\mathbf{i}-3\mathbf{j}$

$4\mathbf{v}-3\mathbf{w}=4(-2\mathbf{i}+\mathbf{j})-3(4\mathbf{i}-3\mathbf{j})=-8\mathbf{i}+4\mathbf{j}-12\mathbf{i}+9\mathbf{j}=-20\mathbf{i}+13\mathbf{j}$

67. $\mathbf{v} = -2\mathbf{i} + \mathbf{j}$

$$\|\mathbf{v}\| = \|-2\mathbf{i} + \mathbf{j}\| = \sqrt{(-2)^2 + 1^2} = \sqrt{5}$$

69. $\mathbf{v} = -2\mathbf{i} + \mathbf{j}, \ \mathbf{w} = 4\mathbf{i} - 3\mathbf{j}$

$$\|\mathbf{v}\| + \|\mathbf{w}\| = \|-2\mathbf{i} + \mathbf{j}\| + \|4\mathbf{i} - 3\mathbf{j}\| = \sqrt{(-2)^2 + 1^2} + \sqrt{4^2 + (-3)^2} = \sqrt{5} + 5$$

71. $\mathbf{u} = \dfrac{\mathbf{v}}{\|\mathbf{v}\|} = \dfrac{-2\mathbf{i} + \mathbf{j}}{\|-2\mathbf{i} + \mathbf{j}\|} = \dfrac{-2\mathbf{i} + \mathbf{j}}{\sqrt{(-2)^2 + 1^2}} = \dfrac{-2\mathbf{i} + \mathbf{j}}{\sqrt{5}} = -\dfrac{2\sqrt{5}}{5}\mathbf{i} + \dfrac{\sqrt{5}}{5}\mathbf{j}$

73. $4\mathbf{v} - 3\mathbf{w} = 4(3\mathbf{i} + \mathbf{j} - 2\mathbf{k}) - 3(-3\mathbf{i} + 2\mathbf{j} - \mathbf{k}) = 12\mathbf{i} + 4\mathbf{j} - 8\mathbf{k} + 9\mathbf{i} - 6\mathbf{j} + 3\mathbf{k}$

$$= 21\mathbf{i} - 2\mathbf{j} - 5\mathbf{k}$$

75. $\|\mathbf{v} - \mathbf{w}\| = \|(3\mathbf{i} + \mathbf{j} - 2\mathbf{k}) - (-3\mathbf{i} + 2\mathbf{j} - \mathbf{k})\| = \|3\mathbf{i} + \mathbf{j} - 2\mathbf{k} + 3\mathbf{i} - 2\mathbf{j} + \mathbf{k}\|$

$$= \|6\mathbf{i} - \mathbf{j} - \mathbf{k}\| = \sqrt{6^2 + (-1)^2 + (-1)^2} = \sqrt{36 + 1 + 1} = \sqrt{38}$$

77. $\|\mathbf{v}\| - \|\mathbf{w}\| = \|3\mathbf{i} + \mathbf{j} - 2\mathbf{k}\| - \|-3\mathbf{i} + 2\mathbf{j} - \mathbf{k}\|$

$$= \sqrt{3^2 + 1^2 + (-2)^2} - \sqrt{(-3)^2 + 2^2 + (-1)^2} = \sqrt{14} - \sqrt{14} = 0$$

79. $\mathbf{v} \times \mathbf{w} = \begin{vmatrix} \mathbf{i} & \mathbf{j} & \mathbf{k} \\ 3 & 1 & -2 \\ -3 & 2 & -1 \end{vmatrix} = \begin{vmatrix} 1 & -2 \\ 2 & -1 \end{vmatrix}\mathbf{i} - \begin{vmatrix} 3 & -2 \\ -3 & -1 \end{vmatrix}\mathbf{j} + \begin{vmatrix} 3 & 1 \\ -3 & 2 \end{vmatrix}\mathbf{k} = 3\mathbf{i} + 9\mathbf{j} + 9\mathbf{k}$

81. Same direction:

$$\frac{\mathbf{v}}{\|\mathbf{v}\|} = \frac{3\mathbf{i} + \mathbf{j} - 2\mathbf{k}}{\sqrt{3^2 + 1^2 + (-2)^2}} = \frac{3\mathbf{i} + \mathbf{j} - 2\mathbf{k}}{\sqrt{14}} = \frac{3\sqrt{14}}{14}\mathbf{i} + \frac{\sqrt{14}}{14}\mathbf{j} - \frac{\sqrt{14}}{7}\mathbf{k}$$

Opposite direction:

$$\frac{-\mathbf{v}}{\|\mathbf{v}\|} = -\frac{3\sqrt{14}}{14}\mathbf{i} - \frac{\sqrt{14}}{14}\mathbf{j} + \frac{\sqrt{14}}{7}\mathbf{k}$$

83. $\mathbf{v} = -2\mathbf{i} + \mathbf{j}, \ \mathbf{w} = 4\mathbf{i} - 3\mathbf{j}$

$\mathbf{v} \cdot \mathbf{w} = -2(4) + 1(-3) = -8 - 3 = -11$

$$\cos\theta = \frac{\mathbf{v} \cdot \mathbf{w}}{\|\mathbf{v}\|\|\mathbf{w}\|} = \frac{-11}{\sqrt{(-2)^2 + 1^2}\sqrt{4^2 + (-3)^2}} = \frac{-11}{\sqrt{5} \cdot 5} = -\frac{11}{5\sqrt{5}} \approx -0.9839$$

$\theta \approx 169.7°$

85. $\mathbf{v} = \mathbf{i} - 3\mathbf{j}, \ \mathbf{w} = -\mathbf{i} + \mathbf{j}$

$\mathbf{v} \cdot \mathbf{w} = 1(-1) + (-3)(1) = -1 - 3 = -4$

$$\cos\theta = \frac{\mathbf{v} \cdot \mathbf{w}}{\|\mathbf{v}\|\|\mathbf{w}\|} = \frac{-4}{\sqrt{1^2 + (-3)^2}\sqrt{(-1)^2 + 1^2}} = \frac{-4}{\sqrt{10}\sqrt{2}} = -\frac{2}{\sqrt{5}} \approx -0.8944$$

$\theta \approx 153.4°$

87. $\mathbf{v} \bullet \mathbf{w} = (\mathbf{i}+\mathbf{j}+\mathbf{k}) \bullet (\mathbf{i}-\mathbf{j}+\mathbf{k}) = 1\cdot1 + 1(-1) + 1\cdot1 = 1 - 1 + 1 = 1$

$\cos\theta = \dfrac{\mathbf{v}\bullet\mathbf{w}}{\|\mathbf{v}\|\|\mathbf{w}\|} = \dfrac{1}{\sqrt{1^2+1^2+1^2}\sqrt{1^2+(-1)^2+1^2}} = \dfrac{1}{\sqrt{3}\sqrt{3}} = \dfrac{1}{3}$

$\theta \approx 70.5°$

89. $\mathbf{v} \bullet \mathbf{w} = (4\mathbf{i}-\mathbf{j}+2\mathbf{k}) \bullet (\mathbf{i}-2\mathbf{j}-3\mathbf{k}) = 4\cdot1 + (-1)(-2) + 2(-3) = 4 + 2 - 6 = 0$

$\cos\theta = \dfrac{\mathbf{v}\bullet\mathbf{w}}{\|\mathbf{v}\|\|\mathbf{w}\|} = \dfrac{0}{\sqrt{4^2+(-1)^2+2^2}\sqrt{1^2+(-2)^2+(-3)^2}} = 0$

$\theta = 90°$

91. $\text{proj}_{\mathbf{w}}\mathbf{v} = \dfrac{\mathbf{v}\bullet\mathbf{w}}{(\|\mathbf{w}\|)^2}\mathbf{w} = \dfrac{(2\mathbf{i}+3\mathbf{j})\bullet(3\mathbf{i}+\mathbf{j})}{\left(\sqrt{3^2+1^2}\right)^2}(3\mathbf{i}+\mathbf{j}) = \dfrac{2\cdot3+3\cdot1}{10}(3\mathbf{i}+\mathbf{j}) = \dfrac{27}{10}\mathbf{i} + \dfrac{9}{10}\mathbf{j}$

93. $\cos\alpha = \dfrac{a}{\|\mathbf{v}\|} = \dfrac{3}{\sqrt{3^2+(-4)^2+2^2}} = \dfrac{3}{\sqrt{29}} \rightarrow \alpha \approx 56.1°$

$\cos\beta = \dfrac{b}{\|\mathbf{v}\|} = \dfrac{-4}{\sqrt{3^2+(-4)^2+2^2}} = -\dfrac{4}{\sqrt{29}} \rightarrow \beta \approx 138.0°$

$\cos\gamma = \dfrac{c}{\|\mathbf{v}\|} = \dfrac{2}{\sqrt{3^2+(-4)^2+2^2}} = \dfrac{2}{\sqrt{29}} \rightarrow \gamma \approx 68.2°$

95. $\mathbf{u} = P_1P_2 = 1\mathbf{i}+2\mathbf{j}+3\mathbf{k} \qquad \mathbf{v} = P_1P_3 = 5\mathbf{i}+4\mathbf{j}+1\mathbf{k}$

$\mathbf{u}\times\mathbf{v} = \begin{vmatrix} \mathbf{i} & \mathbf{j} & \mathbf{k} \\ 1 & 2 & 3 \\ 5 & 4 & 1 \end{vmatrix} = \begin{vmatrix} 2 & 3 \\ 4 & 1 \end{vmatrix}\mathbf{i} - \begin{vmatrix} 1 & 3 \\ 5 & 1 \end{vmatrix}\mathbf{j} + \begin{vmatrix} 1 & 2 \\ 5 & 4 \end{vmatrix}\mathbf{k} = -10\mathbf{i}+14\mathbf{j}-6\mathbf{k}$

Area $= \|\mathbf{u}\times\mathbf{v}\| = \sqrt{(-10)^2+14^2+(-6)^2} = \sqrt{332} \approx 18.2$

97. Let the positive x-axis point downstream, so that the velocity of the current is $\mathbf{v}_c = 2\mathbf{i}$.

Let $\mathbf{v}_w$ = the velocity of the swimmer in the water.

Let $\mathbf{v}_g$ = the velocity of the swimmer relative to the land.

Then $\mathbf{v}_g = \mathbf{v}_w + \mathbf{v}_c$

The speed of the swimmer is $\|\mathbf{v}_w\| = 5$; its direction is directly across the river, so

Let $\mathbf{v}_w = 5\mathbf{j}$.

$\mathbf{v}_g = \mathbf{v}_w + \mathbf{v}_c = 5\mathbf{j} + 2\mathbf{i} = 2\mathbf{i}+5\mathbf{j}$

Let $\|\mathbf{v}_g\| = \sqrt{2^2+5^2} = \sqrt{29} \approx 5.4$ miles per hour

Since the river is 1 mile wide, it takes the swimmer 0.2 hours to cross the river. The swimmer will end up $(0.2)(2) = 0.4$ miles downstream.

99. Let $\mathbf{F}_1$ be the tension on the left cable and $\mathbf{F}_2$ be the tension on the right cable.
Let $\mathbf{F}_3$ represent the force of the weight of the box.

$\mathbf{F}_1 = \|\mathbf{F}_1\|(\cos(140°)\mathbf{i} + \sin(140°)\mathbf{j}) \approx \|\mathbf{F}_1\|(-0.7660\mathbf{i} + 0.6428\mathbf{j})$

$\mathbf{F}_2 = \|\mathbf{F}_2\|(\cos(30°)\mathbf{i} + \sin(30°)\mathbf{j}) \approx \|\mathbf{F}_2\|(0.8660\mathbf{i} + 0.5000\mathbf{j})$

$\mathbf{F}_3 = -2000\mathbf{j}$

For equilibrium, the sum of the force vectors must be zero.

$\mathbf{F}_1 + \mathbf{F}_2 + \mathbf{F}_3 = -0.7660\|\mathbf{F}_1\|\mathbf{i} + 0.6428\|\mathbf{F}_1\|\mathbf{j} + 0.8660\|\mathbf{F}_2\|\mathbf{i} + 0.5000\|\mathbf{F}_2\|\mathbf{j} - 2000\mathbf{j}$

$= \left(-0.7660\|\mathbf{F}_1\| + 0.8660\|\mathbf{F}_2\|\right)\mathbf{i} + \left(0.6428\|\mathbf{F}_1\| + 0.5000\|\mathbf{F}_2\| - 2000\right)\mathbf{j}$

$= 0$

Set the $\mathbf{i}$ and $\mathbf{j}$ components equal to zero and solve:

$$\begin{cases} -0.7660\|\mathbf{F}_1\| + 0.8660\|\mathbf{F}_2\| = 0 \quad \rightarrow \quad \|\mathbf{F}_2\| = \dfrac{0.7660}{0.8660}\|\mathbf{F}_1\| = 0.8845\|\mathbf{F}_1\| \\ 0.6428\|\mathbf{F}_1\| + 0.5000\|\mathbf{F}_2\| - 2000 = 0 \end{cases}$$

$0.6428\|\mathbf{F}_1\| + 0.5000(0.8845\|\mathbf{F}_1\|) - 2000 = 0$

$1.0851\|\mathbf{F}_1\| = 2000$

$\|\mathbf{F}_1\| \approx 1843$ pounds

$\|\mathbf{F}_2\| = 0.8845(1843) \approx 1630$ pounds

The tension in the left cable is about 1843 pounds and the tension in the right cable is about 1630 pounds.

Chapter 9

Analytic Geometry

9.2   The Parabola

1. B   3. E   5. H   7. C

9. The focus is (4, 0) and the vertex is (0, 0). Both lie
    on the horizontal line $y = 0$. $a = 4$ and since (4, 0)
    is to the right of (0, 0), the parabola opens to the
    right. The equation of the parabola is:
    $$y^2 = 4ax$$
    $$y^2 = 4 \cdot 4 \cdot x$$
    $$y^2 = 16x$$
    Letting $x = 4$, we find $y^2 = 64$ or $y = \pm 8$.
    The points (4, 8) and (4, –8) define the latus rectum.

11. The focus is (0, –3) and the vertex is (0, 0). Both lie
    on the vertical line $x = 0$. $a = 3$ and since (0, –3) is
    below (0, 0), the parabola opens down. The
    equation of the parabola is:
    $$x^2 = -4ay$$
    $$x^2 = -4 \cdot 3 \cdot y$$
    $$x^2 = -12y$$
    Letting $y = -3$, we find $x^2 = 36$ or $x = \pm 6$.
    The points (6, 3) and (6, –3) define the latus rectum.

13. The focus is (–2, 0) and the directrix is $x = 2$. The
    vertex is (0, 0). $a = 2$ and since (–2, 0) is to the left
    of (0, 0), the parabola opens to the left. The
    equation of the parabola is:
    $$y^2 = -4ax$$
    $$y^2 = -4 \cdot 2 \cdot x$$
    $$y^2 = -8x$$
    Letting $x = -2$, we find $y^2 = 16$ or $y = \pm 4$. The
    points (–2, 4) and (–2, –4) define the latus rectum.

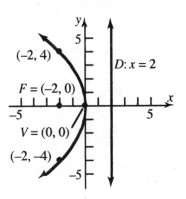

15. The directrix is $y = -\dfrac{1}{2}$ and the vertex is $(0, 0)$. The focus is $\left(0, \dfrac{1}{2}\right)$. $a = \dfrac{1}{2}$ and since $\left(0, \dfrac{1}{2}\right)$ is above $(0, 0)$, the parabola opens up. The equation of the parabola is:

$$x^2 = 4ay$$

$$x^2 = 4 \cdot \dfrac{1}{2} \cdot y \rightarrow x^2 = 2y$$

Letting $y = \dfrac{1}{2}$, we find $x^2 = 1$ or $x = \pm 1$.

The points $\left(1, \dfrac{1}{2}\right)$ and $\left(-1, \dfrac{1}{2}\right)$ define the latus rectum.

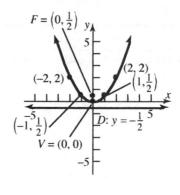

17. The focus is $(2, -5)$ and the vertex is $(2, -3)$. Both lie on the vertical line $x = 2$. $a = 2$ and since $(2, -5)$ is below $(2, -3)$, the parabola opens down. The equation of the parabola is:

$$(x - h)^2 = -4a(y - k)$$
$$(x - 2)^2 = -4 \cdot 2 \cdot (y - (-3))$$
$$(x - 2)^2 = -8(y + 3)$$

Letting $y = -5$, we find $(x - 2)^2 = 16$ or $x - 2 = \pm 4$. So, $x = 6$ or $x = -2$. The points $(6, -5)$ and $(-2, -5)$ define the latus rectum.

19. Vertex: $(0,0)$. Since the axis of symmetry is vertical, the parabola opens up or down. Since $(2, 3)$ is above $(0, 0)$, the parabola opens up. The equation has the form $x^2 = 4ay$. Substitute the coordinates of $(2, 3)$ into the equation to find a:

$$2^2 = 4a \cdot 3$$

$$4 = 12a \rightarrow a = \dfrac{1}{3}$$

The equation of the parabola is: $x^2 = \dfrac{4}{3}y$.

The focus is $\left(0, \dfrac{1}{3}\right)$. Letting $y = \dfrac{1}{3}$,

we find $x^2 = \dfrac{4}{9}$ or $x = \pm \dfrac{2}{3}$. The points $\left(\dfrac{2}{3}, \dfrac{1}{3}\right)$

and $\left(-\dfrac{2}{3}, \dfrac{1}{3}\right)$ define the latus rectum.

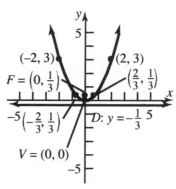

21. The directrix is $y = 2$ and the focus is $(-3, 4)$.
This is a vertical case, so the vertex is $(-3, 3)$.
$a = 1$ and since $(-3, 4)$ is above $y = 2$, the
parabola opens up. The equation of the parabola
is:
$$(x - h)^2 = 4a(y - k)$$
$$(x - (-3))^2 = 4 \cdot 1 \cdot (y - 3)$$
$$(x + 3)^2 = 4(y - 3)$$
Letting $y = 4$, we find $(x + 3)^2 = 4$ or $x + 3 = \pm 2$.
So, $x = -1$ or $x = -5$. The points $(-1, 4)$ and
$(-5, 4)$ define the latus rectum.

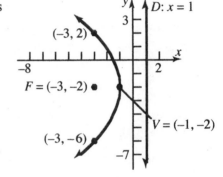

23. The directrix is $x = 1$ and the focus is $(-3, -2)$. This
is a horizontal case, so the vertex is $(-1, -2)$. $a = 2$
and since $(-3, -2)$ is to the left of $x = 1$, the
parabola opens to the left. The equation of the
parabola is:
$$(y - k)^2 = -4a(x - h)$$
$$(y - (-2))^2 = -4 \cdot 2 \cdot (x - (-1))$$
$$(y + 2)^2 = -8(x + 1)$$
Letting $x = -3$, we find $(y + 2)^2 = 16$ or $y + 2 = \pm 4$.
So, $y = 2$ or $y = -6$. The points $(-3, 2)$ and
$(-3, -6)$ define the latus rectum.

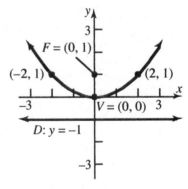

25. The equation $x^2 = 4y$ is in the form $x^2 = 4ay$
where $4a = 4$ or $a = 1$. Thus, we have:

      Vertex: $(0, 0)$
      Focus: $(0, 1)$
      Directrix: $y = -1$

To graph, enter: $y_1 = x^2 / 4$

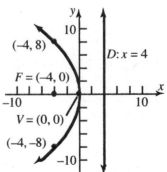

27. The equation $y^2 = -16x$ is in the form $y^2 = -4ax$
where $-4a = -16$ or $a = 4$. Thus, we have:

      Vertex: $(0, 0)$
      Focus: $(-4, 0)$
      Directrix: $x = 4$

To graph, enter: $y_1 = \sqrt{-16x}$; $y_2 = -\sqrt{-16x}$

29. The equation $(y - 2)^2 = 8(x + 1)$ is in the form
$(y - k)^2 = 4a(x - h)$ where $4a = 8$ or $a = 2$,
$h = -1$, and $k = 2$. Thus, we have:

   Vertex: $(-1, 2)$
   Focus: $(1, 2)$
   Directrix: $x = -3$

To graph, enter:
$$y_1 = 2 + \sqrt{8(x + 1)}; \quad y_2 = 2 - \sqrt{8(x + 1)}$$

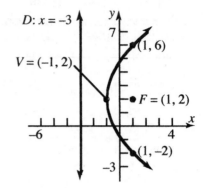

31. The equation $(x - 3)^2 = -(y + 1)$ is in the form
$(x - h)^2 = -4a(y - k)$ where $-4a = -1$ or $a = \dfrac{1}{4}$,
$h = 3$, and $k = -1$. Thus, we have:

   Vertex: $(3, -1)$
   Focus: $\left(3, -\dfrac{5}{4}\right)$
   Directrix: $y = -\dfrac{3}{4}$

To graph, enter: $y_1 = -1 - (x - 3)^2$

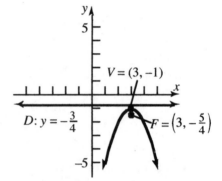

33. The equation $(y + 3)^2 = 8(x - 2)$ is in the form
$(y - k)^2 = 4a(x - h)$ where $4a = 8$ or $a = 2$,
$h = 2$, and $k = -3$. Thus, we have:

   Vertex: $(2, -3)$
   Focus: $(4, -3)$
   Directrix: $x = 0$

To graph, enter:
$$y_1 = -3 + \sqrt{8(x - 2)}; \quad y_2 = -3 - \sqrt{8(x - 2)}$$

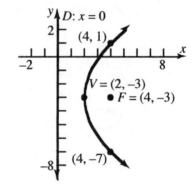

35. Complete the square to put in standard form:
$$y^2 - 4y + 4x + 4 = 0$$
$$y^2 - 4y + 4 = -4x$$
$$(y - 2)^2 = -4x$$

The equation is in the form $(y - k)^2 = -4a(x - h)$
where $-4a = -4$ or $a = 1$, $h = 0$, and $k = 2$.
Thus, we have:

   Vertex: $(0, 2)$
   Focus: $(-1, 2)$
   Directrix: $x = 1$

To graph, enter: $y_1 = 2 + \sqrt{-4x}; \quad y_2 = 2 - \sqrt{-4x}$

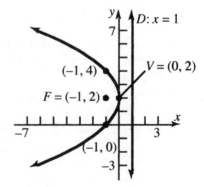

37. Complete the square to put in standard form:
$$x^2 + 8x = 4y - 8$$
$$x^2 + 8x + 16 = 4y - 8 + 16$$
$$(x + 4)^2 = 4(y + 2)$$
The equation is in the form $(x - h)^2 = 4a(y - k)$
where $4a = 4$ or $a = 1$, $h = -4$, and $k = -2$.
Thus, we have:
   Vertex: $(-4, -2)$
   Focus: $(-4, -1)$
   Directrix: $y = -3$
To graph, enter: $y_1 = -2 + (x + 4)^2 / 4$

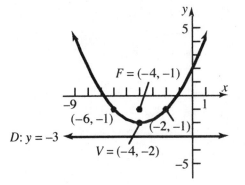

39. Complete the square to put in standard form:
$$y^2 + 2y - x = 0$$
$$y^2 + 2y + 1 = x + 1$$
$$(y + 1)^2 = x + 1$$
The equation is in the form

$(y - k)^2 = 4a(x - h)$ where $4a = 1$ or $a = \dfrac{1}{4}$,

$h = -1$, and $k = -1$.
Thus, we have:
   Vertex: $(-1, -1)$

   Focus: $\left(-\dfrac{3}{4}, -1\right)$

   Directrix: $x = -\dfrac{5}{4}$
To graph, enter:
$y_1 = -1 + \sqrt{x + 1}$; $y_2 = -1 - \sqrt{x + 1}$

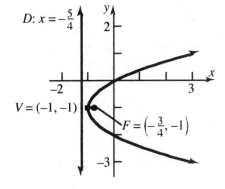

41. Complete the square to put in standard form:
$$x^2 - 4x = y + 4$$
$$x^2 - 4x + 4 = y + 4 + 4$$
$$(x - 2)^2 = y + 8$$
The equation is in the form $(x - h)^2 = 4a(y - k)$

where $4a = 1$ or $a = \dfrac{1}{4}$,

$h = 2$, and $k = -8$. Thus, we have:
   Vertex: $(2, -8)$

   Focus: $\left(2, -\dfrac{31}{4}\right)$

   Directrix: $y = -\dfrac{33}{4}$
To graph, enter: $y_1 = -8 + (x - 2)^2$

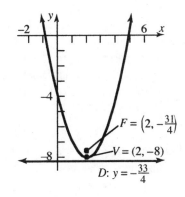

43. $(y-1)^2 = c(x-0)$

$(y-1)^2 = cx$

$(2-1)^2 = c(1) \rightarrow 1 = c$

$(y-1)^2 = x$

45. $(y-1)^2 = c(x-2)$

$(0-1)^2 = c(1-2)$

$1 = -c \rightarrow c = -1$

$(y-1)^2 = -(x-2)$

47. $(x-0)^2 = c(y-1)$

$(2-0)^2 = c(2-1) \rightarrow 4 = c$

$x^2 = 4(y-1)$

49. $(y-0)^2 = c(x-(-2))$

$y^2 = c(x+2)$

$1^2 = c(0+2) \rightarrow 1 = 2c \rightarrow c = \dfrac{1}{2}$

$y^2 = \dfrac{1}{2}(x+2)$

51. Set up the problem so that the vertex of the parabola is at (0, 0) and it opens up. Then the equation of the parabola has the form: $x^2 = 4ay$. Since the parabola is 10 feet across and 4 feet deep, the points (5, 4) and (–5, 4) are on the parabola.
Substitute and solve for a:

$$5^2 = 4a(4) \rightarrow 25 = 16a \rightarrow a = \dfrac{25}{16}$$

a is the distance from the vertex to the focus. Thus, the receiver (located at the focus) is $\dfrac{25}{16} = 1.5625$ feet, or 18.75 inches from the base of the dish, along the axis of the parabola.

53. Set up the problem so that the vertex of the parabola is at (0, 0) and it opens up. Then the equation of the parabola has the form: $x^2 = 4ay$. Since the parabola is 4 inches across and 1 inch deep, the points (2, 1) and (–2, 1) are on the parabola.
Substitute and solve for a: $2^2 = 4a(1)$ $\rightarrow$ $4 = 4a$ $\rightarrow$ $a = 1$, a is the distance from the vertex to the focus. Thus, the bulb (located at the focus) should be 1 inch, from the vertex.

55. Set up the problem so that the vertex of the parabola is at (0, 0) and it opens up. Then the equation of the parabola has the form: $x^2 = cy$.
The point (300, 80) is a point on the parabola.
Solve for c and find the equation:

$$300^2 = c(80) \rightarrow c = 1125$$

$$x^2 = 1125y$$

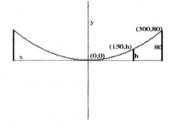

Since the height of the cable, 150 feet from the center, is to be found, the point (150, h) is a point on the parabola. Solve for h:

$$150^2 = 1125h \rightarrow 22500 = 1125h \rightarrow h = 20$$

The height of the cable, 150 feet from the center, is 20 feet.

57. Set up the problem so that the vertex of the parabola is at (0, 0) and it opens up. Then the equation of the parabola has the form: $x^2 = 4ay$. a is the distance from the vertex to the focus (where the source is located), so $a = 2$. Since the opening is 5 feet across, there is a point (2.5, y) on the parabola. Solve for y:

$$x^2 = 8y \rightarrow 2.5^2 = 8y \rightarrow 6.25 = 8y \rightarrow y = 0.78125 \text{ feet}$$

The depth of the searchlight should be 0.78125 feet.

59. Set up the problem so that the vertex of the parabola is at (0, 0) and it opens up. Then the equation of the parabola has the form: $x^2 = 4ay$. Since the parabola is 20 feet across and 6 feet deep, the points (10, 6) and (−10, 6) are on the parabola.
Substitute and solve for a:

$$10^2 = 4a(6) \rightarrow 100 = 24a \rightarrow a \approx 4.17 \text{ feet}$$

The heat source will be concentrated 4.17 feet from the base, along the axis of symmetry.

61. Set up the problem so that the vertex of the parabola is at (0, 0) and it opens down. Then the equation of the parabola has the form: $x^2 = cy$.
The point (60, −25) is a point on the parabola.
Solve for c and find the equation:

$$60^2 = c(-25) \rightarrow c = -144$$
$$x^2 = -144y$$

To find the height of the bridge, 10 feet from the center, the point (10, y) is a point on the parabola. Solve for y:

$$10^2 = -144y \rightarrow 100 = -144y \rightarrow y = -0.69$$

The height of the bridge, 10 feet from the center, is $25 - 0.69 = 24.31$ feet.
To find the height of the bridge, 30 feet from the center, the point (30, y) is a point on the parabola. Solve for y:

$$30^2 = -144y \rightarrow 900 = -144y \rightarrow y = -6.25$$

The height of the bridge, 30 feet from the center, is $25 - 6.25 = 18.75$ feet.
To find the height of the bridge, 50 feet from the center, the point (50, y) is a point on the parabola. Solve for y:

$$50^2 = -144y$$
$$2500 = -144y$$
$$y = -17.36$$

The height of the bridge, 50 feet from the center, is $25 - 17.36 = 7.64$ feet.

63. $Ax^2 + Ey = 0 \quad A \neq 0, \ E \neq 0$
$$Ax^2 = -Ey$$
$$x^2 = \frac{-E}{A}y$$

This is the equation of a parabola with vertex at (0, 0) and axis of symmetry being the y-axis. The focus is $\left(0, \frac{-E}{4A}\right)$. The directrix is $y = \frac{E}{4A}$.

65. $Ax^2 + Dx + Ey + F = 0$　　　$A \neq 0$
   (a)　If $E \neq 0$, then:

   $$Ax^2 + Dx = -Ey - F$$

   $$A\left(x^2 + \frac{D}{A}x + \frac{D^2}{4A^2}\right) = -Ey - F + \frac{D^2}{4A}$$

   $$\left(x + \frac{D}{2A}\right)^2 = \frac{1}{A}\left(-Ey - F + \frac{D^2}{4A}\right)$$

   $$\left(x + \frac{D}{2A}\right)^2 = \frac{-E}{A}\left(y + \frac{F}{E} - \frac{D^2}{4AE}\right)$$

   $$\left(x + \frac{D}{2A}\right)^2 = \frac{-E}{A}\left(y - \frac{D^2 - 4AF}{4AE}\right)$$

   This is the equation of a parabola whose vertex is $\left(\dfrac{-D}{2A}, \dfrac{D^2 - 4AF}{4AE}\right)$.

   (b)　If $E = 0$, then

   $$Ax^2 + Dx + F = 0$$

   $$x = \frac{-D \pm \sqrt{D^2 - 4AF}}{2A}$$

   If $D^2 - 4AF = 0$, then $x = \dfrac{-D}{2A}$ is a vertical line.

   (c)　If $E = 0$, then

   $$Ax^2 + Dx + F = 0$$

   $$x = \frac{-D \pm \sqrt{D^2 - 4AF}}{2A}$$

   If $D^2 - 4AF > 0$,

   then $x = \dfrac{-D + \sqrt{D^2 - 4AF}}{2A}$ or $x = \dfrac{-D - \sqrt{D^2 - 4AF}}{2A}$ are two vertical lines.

   (d)　If $E = 0$, then

   $$Ax^2 + Dx + F = 0$$

   $$x = \frac{-D \pm \sqrt{D^2 - 4AF}}{2A}$$

   If $D^2 - 4AF < 0$, there is no real solution. The graph contains no points.

Analytic Geometry

9.3 The Ellipse

1. C   3. B

5. $\dfrac{x^2}{25} + \dfrac{y^2}{4} = 1$

The center of the ellipse is at the origin.
$a = 5$, $b = 2$. The vertices are $(5, 0)$ and
$(-5, 0)$. Find the value of c:
$$c^2 = a^2 - b^2 = 25 - 4 = 21$$
$$c = \sqrt{21}$$
The foci are $\left(\sqrt{21}, 0\right)$ and $\left(-\sqrt{21}, 0\right)$.
To graph, enter:
$$y_1 = 2\sqrt{(1 - x^2/25)};$$
$$y_1 = -2\sqrt{(1 - x^2/25)}$$

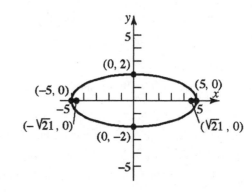

7. $\dfrac{x^2}{9} + \dfrac{y^2}{25} = 1$

The center of the ellipse is at the origin.
$a = 5$, $b = 3$. The vertices are $(0, 5)$ and $(0, -5)$.
Find the value of c:
$$c^2 = a^2 - b^2 = 25 - 9 = 16$$
$$c = 4$$
The foci are $(0, 4)$ and $(0, -4)$.
To graph, enter:
$$y_1 = 5\sqrt{(1 - x^2/9)};$$
$$y_1 = -5\sqrt{(1 - x^2/9)}$$

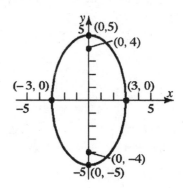

9. $4x^2 + y^2 = 16$

Divide by 16 to put in standard form:

$$\frac{4x^2}{16} + \frac{y^2}{16} = \frac{16}{16} \rightarrow \frac{x^2}{4} + \frac{y^2}{16} = 1$$

The center of the ellipse is at the origin.
$a = 4$, $b = 2$. The vertices are $(0, 4)$ and $(0, -4)$.
Find the value of c:
$$c^2 = a^2 - b^2 = 16 - 4 = 12$$
$$c = \sqrt{12} = 2\sqrt{3}$$
The foci are $\left(0, 2\sqrt{3}\right)$ and $\left(0, -2\sqrt{3}\right)$.

To graph, enter:
$$y_1 = \sqrt{(16 - 4x^2)}; \ y_1 = -\sqrt{(16 - 4x^2)}$$

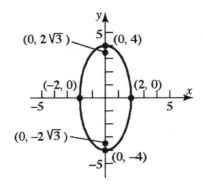

11. $4y^2 + x^2 = 8$

Divide by 8 to put in standard form:

$$\frac{4y^2}{8} + \frac{x^2}{8} = \frac{8}{8} \rightarrow \frac{x^2}{8} + \frac{y^2}{2} = 1$$

The center of the ellipse is at the origin.
$a = \sqrt{8} = 2\sqrt{2}$, $b = \sqrt{2}$. The vertices are
$\left(2\sqrt{2}, 0\right)$ and $\left(-2\sqrt{2}, 0\right)$. Find the value of c:
$$c^2 = a^2 - b^2 = 8 - 2 = 6$$
$$c = \sqrt{6}$$
The foci are $\left(\sqrt{6}, 0\right)$ and $\left(-\sqrt{6}, 0\right)$.

To graph, enter:
$$y_1 = \sqrt{(2 - x^2/4)}; \ y_1 = -\sqrt{(2 - x^2/4)}$$

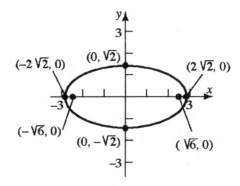

13. $x^2 + y^2 = 16$

This is the equation of a circle whose center is at
$(0, 0)$ and radius $= 4$.

To graph, enter: $y_1 = \sqrt{(16 - x^2)}; \ y_1 = -\sqrt{(16 - x^2)}$

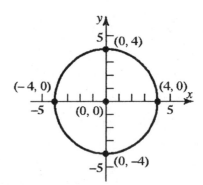

15. Center: $(0, 0)$; Focus: $(3, 0)$; Vertex: $(5, 0)$;
Major axis is the x-axis; $a = 5$; $c = 3$. Find b:
$$b^2 = a^2 - c^2 = 25 - 9 = 16$$
$$b = 4$$
Write the equation: $\dfrac{x^2}{25} + \dfrac{y^2}{16} = 1$

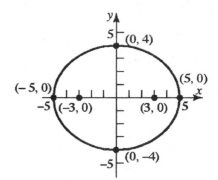

17. Center: $(0, 0)$; Focus: $(0, -4)$; Vertex: $(0, 5)$;
Major axis is the y-axis; $a = 5$; $c = 4$. Find b:
$$b^2 = a^2 - c^2 = 25 - 16 = 9$$
$$b = 3$$
Write the equation: $\dfrac{x^2}{9} + \dfrac{y^2}{25} = 1$

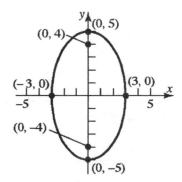

19. Foci: $(\pm 2, 0)$; Length of major axis is 6.
Center: $(0, 0)$; Major axis is the x-axis;
$a = 3$; $c = 2$. Find b:
$$b^2 = a^2 - c^2 = 9 - 4 = 5$$
$$b = \sqrt{5}$$
Write the equation: $\dfrac{x^2}{9} + \dfrac{y^2}{5} = 1$

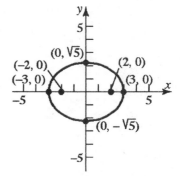

21. Foci: $(0, \pm 3)$; x-intercepts are ± 2. Center: $(0, 0)$;
Major axis is the y-axis; $c = 3$; $b = 2$. Find a:
$$a^2 = b^2 + c^2 = 4 + 9 = 13$$
$$a = \sqrt{13}$$
Write the equation: $\dfrac{x^2}{4} + \dfrac{y^2}{13} = 1$

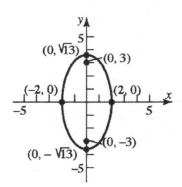

23.   Center: $(0, 0)$; Vertex: $(0, 4)$; $b = 1$; Major axis is
the y-axis; $a = 4$; $b = 1$.

Write the equation: $\dfrac{x^2}{1} + \dfrac{y^2}{16} = 1$

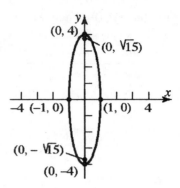

25. $\dfrac{(x+1)^2}{4} + \dfrac{(y-1)^2}{1} = 1$ 27. $\dfrac{(x-1)^2}{1} + \dfrac{y^2}{4} = 1$

29.   The equation $\dfrac{(x-3)^2}{4} + \dfrac{(y+1)^2}{9} = 1$ is in the form $\dfrac{(x-h)^2}{b^2} + \dfrac{(y-k)^2}{a^2} = 1$
(major axis parallel to the y-axis) where $a = 3$, $b = 2$, $h = 3$, and $k = -1$.
Solving for c:

$$c^2 = a^2 - b^2 = 9 - 4 = 5 \rightarrow c = \sqrt{5}$$

Thus, we have:

Center: $(3, -1)$

Foci: $\left(3, -1+\sqrt{5}\right), \left(3, -1-\sqrt{5}\right)$

Vertices: $(3, 2), (3, -4)$

To graph, enter: $y_1 = -1 + 3\sqrt{1 - (x-3)^2 / 4}$;

$y_2 = -1 - 3\sqrt{1 - (x-3)^2 / 4}$

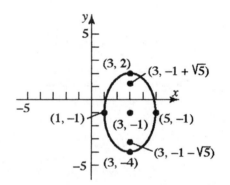

31.   Divide by 16 to put the equation in standard form:

$(x+5)^2 + 4(y-4)^2 = 16$

$$\dfrac{(x+5)^2}{16} + \dfrac{4(y-4)^2}{16} = \dfrac{16}{16}$$

$$\dfrac{(x+5)^2}{16} + \dfrac{(y-4)^2}{4} = 1$$

The equation is in the form $\dfrac{(x-h)^2}{a^2} + \dfrac{(y-k)^2}{b^2} = 1$ (major axis parallel to the x-axis) where
$a = 4$, $b = 2$, $h = -5$, and $k = 4$.
Solving for c:

$$c^2 = a^2 - b^2 = 16 - 4 = 12$$

$$c = \sqrt{12} = 2\sqrt{3}$$

Thus, we have:

Center: $(-5, 4)$

Foci: $\left(-5 - 2\sqrt{3}, 4\right)$, $\left(-5 + 2\sqrt{3}, 4\right)$

Vertices: $(-9, 4)$, $(-1, 4)$

To graph, enter:

$y_1 = 4 + 2\sqrt{1 - (x+5)^2 / 16}$;

$y_2 = 4 - 2\sqrt{1 - (x+5)^2 / 16}$

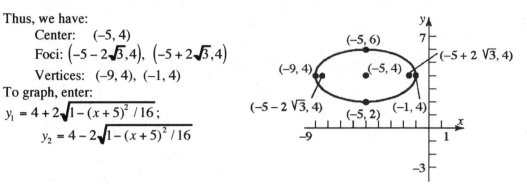

33. Complete the square to put the equation in standard form:

$$x^2 + 4x + 4y^2 - 8y + 4 = 0$$

$$(x^2 + 4x + 4) + 4(y^2 - 2y + 1) = -4 + 4 + 4 \rightarrow (x+2)^2 + 4(y-1)^2 = 4$$

$$\frac{(x+2)^2}{4} + \frac{4(y-1)^2}{4} = \frac{4}{4} \rightarrow \frac{(x+2)^2}{4} + \frac{(y-1)^2}{1} = 1$$

The equation is in the form $\dfrac{(x-h)^2}{a^2} + \dfrac{(y-k)^2}{b^2} = 1$ (major axis parallel to the x-axis) where

$a = 2$, $b = 1$, $h = -2$, and $k = 1$.

Solving for c:

$$c^2 = a^2 - b^2 = 4 - 1 = 3 \rightarrow c = \sqrt{3}$$

Thus, we have:

Center: $(-2, 1)$

Foci: $\left(-2 - \sqrt{3}, 1\right)$, $\left(-2 + \sqrt{3}, 1\right)$

Vertices: $(-4, 1)$, $(0, 1)$

To graph, enter: $y_1 = 1 + \sqrt{1 - (x+2)^2 / 4}$;

$y_2 = 1 - \sqrt{1 - (x+2)^2 / 4}$

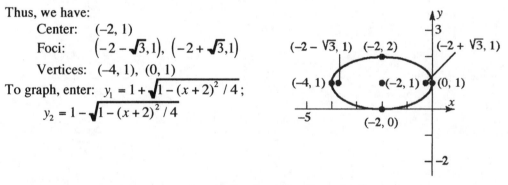

35. Complete the square to put the equation in standard form:

$$2x^2 + 3y^2 - 8x + 6y + 5 = 0 \rightarrow 2(x^2 - 4x) + 3(y^2 + 2y) = -5$$

$$2(x^2 - 4x + 4) + 3(y^2 + 2y + 1) = -5 + 8 + 3 \rightarrow 2(x-2)^2 + 3(y+1)^2 = 6$$

$$\frac{2(x-2)^2}{6} + \frac{3(y+1)^2}{6} = \frac{6}{6} \rightarrow \frac{(x-2)^2}{3} + \frac{(y+1)^2}{2} = 1$$

The equation is in the form $\dfrac{(x-h)^2}{a^2} + \dfrac{(y-k)^2}{b^2} = 1$ (major axis parallel to the x-axis) where

$a = \sqrt{3}$, $b = \sqrt{2}$, $h = 2$, and $k = -1$.

Solving for c:

$$c^2 = a^2 - b^2 = 3 - 2 = 1 \rightarrow c = 1$$

Thus, we have:
  Center: $(2, -1)$
  Foci: $(1, -1),\ (3, -1)$
  Vertices: $\left(2 - \sqrt{3}, -1\right),\ \left(2 + \sqrt{3}, -1\right)$

To graph, enter: $y_1 = -1 + \sqrt{2 - 2(x-2)^2 / 3}$;

$y_2 = -1 - \sqrt{2 - 2(x-2)^2 / 3}$

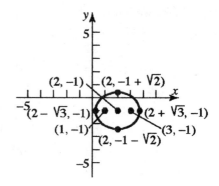

37. Complete the square to put the equation in standard form:

$$9x^2 + 4y^2 - 18x + 16y - 11 = 0 \rightarrow 9(x^2 - 2x) + 4(y^2 + 4y) = 11$$

$$9(x^2 - 2x + 1) + 4(y^2 + 4y + 4) = 11 + 9 + 16 \rightarrow 9(x-1)^2 + 4(y+2)^2 = 36$$

$$\frac{9(x-1)^2}{36} + \frac{4(y+2)^2}{36} = \frac{36}{36} \rightarrow \frac{(x-1)^2}{4} + \frac{(y+2)^2}{9} = 1$$

The equation is in the form $\dfrac{(x-h)^2}{b^2} + \dfrac{(y-k)^2}{a^2} = 1$ (major axis parallel to the y-axis) where

$a = 3,\ b = 2,\ h = 1,$ and $k = -2$.
Solving for c:

$$c^2 = a^2 - b^2 = 9 - 4 = 5 \rightarrow c = \sqrt{5}$$

Thus, we have:
  Center: $(1, -2)$
  Foci: $\left(1, -2 + \sqrt{5}\right),\ \left(1, -2 - \sqrt{5}\right)$
  Vertices: $(1, 1),\ (1, -5)$

To graph, enter: $y_1 = -2 + 3\sqrt{1 - (x-1)^2 / 4}$;

$y_2 = -2 - 3\sqrt{1 - (x-1)^2 / 4}$

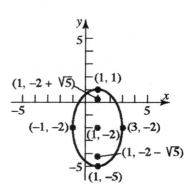

39. Complete the square to put the equation in standard form:

$$4x^2 + y^2 + 4y = 0 \rightarrow 4x^2 + y^2 + 4y + 4 = 4$$

$$4x^2 + (y+2)^2 = 4 \rightarrow \frac{4x^2}{4} + \frac{(y+2)^2}{4} = \frac{4}{4} \rightarrow \frac{x^2}{1} + \frac{(y+2)^2}{4} = 1$$

The equation is in the form $\dfrac{(x-h)^2}{b^2} + \dfrac{(y-k)^2}{a^2} = 1$ (major axis parallel to the y-axis) where

$a = 2,\ b = 1,\ h = 0,$ and $k = -2$.
 Solving for c:

$$c^2 = a^2 - b^2 = 4 - 1 = 3 \rightarrow c = \sqrt{3}$$

Thus, we have:

    Center: (0,–2)

    Foci: $\left(0, -2 + \sqrt{3}\right)$, $\left(0, -2 - \sqrt{3}\right)$

    Vertices: (0, 0), (0, –4)

To graph, enter: $y_1 = -2 + 2\sqrt{1 - x^2}$;

        $y_2 = -2 - 2\sqrt{1 - x^2}$

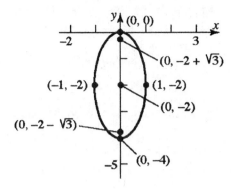

41.   Center: (2, –2); Vertex: (7, –2);
Focus: (4, –2); Major axis parallel
to the x-axis; $a = 5$; $c = 2$.

    Find b:

       $b^2 = a^2 - c^2 = 25 - 4 = 21$

         $b = \sqrt{21}$

    Write the equation:

    $\dfrac{(x - 2)^2}{25} + \dfrac{(y + 2)^2}{21} = 1$

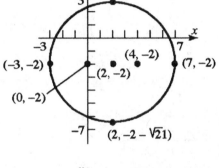

43.   Vertices: (4, 3), (4, 9); Focus: (4, 8);
Center: (4, 6); Major axis parallel to the
y-axis; $a = 3$; $c = 2$.

    Find b:

       $b^2 = a^2 - c^2 = 9 - 4 = 5$

         $b = \sqrt{5}$

    Write the equation: $\dfrac{(x - 4)^2}{5} + \dfrac{(y - 6)^2}{9} = 1$

45.   Foci: (5, 1), (–1, 1); length of the major axis = 8;
Center: (2, 1); Major axis parallel to the x-axis;
$a = 4$; $c = 3$.   Find b:

       $b^2 = a^2 - c^2 = 16 - 9 = 7$

         $b = \sqrt{7}$

    Write the equation: $\dfrac{(x - 2)^2}{16} + \dfrac{(y - 1)^2}{7} = 1$

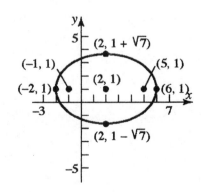

47. Center: $(1, 2)$; Focus: $(4, 2)$; contains the point $(1, 3)$; Major axis parallel to the x-axis; $c = 3$.

The equation has the form:
$$\frac{(x-1)^2}{a^2} + \frac{(y-2)^2}{b^2} = 1$$

Since the point $(1, 3)$ is on the curve:
$$\frac{0}{a^2} + \frac{1}{b^2} = 1$$

$$\frac{1}{b^2} = 1 \;\rightarrow\; b^2 = 1 \;\rightarrow\; b = 1$$

Find a:
$$a^2 = b^2 + c^2 = 1 + 9 = 10 \;\rightarrow\; a = \sqrt{10}$$

Write the equation: $\dfrac{(x-1)^2}{10} + \dfrac{(y-2)^2}{1} = 1$

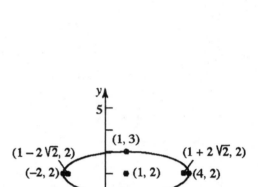

49. Center: $(1, 2)$; Vertex: $(4, 2)$; contains the point $(1, 3)$; Major axis parallel to the x-axis; $a = 3$.

The equation has the form:
$$\frac{(x-1)^2}{a^2} + \frac{(y-2)^2}{b^2} = 1$$

Since the point $(1, 3)$ is on the curve:
$$\frac{0}{9} + \frac{1}{b^2} = 1$$

$$\frac{1}{b^2} = 1 \;\rightarrow\; b^2 = 1 \;\rightarrow\; b = 1$$

Write the equation: $\dfrac{(x-1)^2}{9} + \dfrac{(y-2)^2}{1} = 1$

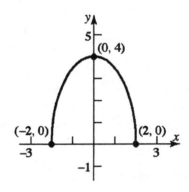

51. Rewrite the equation:
$$y = \sqrt{16 - 4x^2}$$
$$y^2 = 16 - 4x^2, \quad y \geq 0$$
$$4x^2 + y^2 = 16, \quad y \geq 0$$
$$\frac{x^2}{4} + \frac{y^2}{16} = 1, \quad y \geq 0$$

53. Rewrite the equation:

$$y = -\sqrt{64 - 16x^2}$$
$$y^2 = 64 - 16x^2, \quad y \le 0$$
$$16x^2 + y^2 = 64, \quad\quad y \le 0$$
$$\frac{x^2}{4} + \frac{y^2}{64} = 1, \quad\quad y \le 0$$

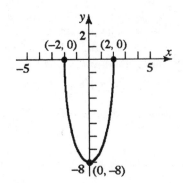

55. The center of the ellipse is $(0, 0)$. The length of the major axis is 20, so $a = 10$. The length of half the minor axis is 6, so $b = 6$. The ellipse is situated with its major axis on the x-axis. The equation is: $\dfrac{x^2}{100} + \dfrac{y^2}{36} = 1$.

57. Assume that the half ellipse formed by the gallery is centered at $(0, 0)$. Since the hall is 100 feet long, $2a = 100$ or $a = 50$. The distance from the center to the foci is 25 feet, so $c = 25$. Find the height of the gallery which is b:

$$b^2 = a^2 - c^2 = 2500 - 625 = 1875 \;\rightarrow\; b = \sqrt{1875} \approx 43.3$$

The ceiling will be 43.3 feet high in the center.

59. Place the semielliptical arch so that the x-axis coincides with the water and the y-axis passes through the center of the arch. Since the bridge has a span of 120 feet, the length of the major axis is 120, or $2a = 120$ or $a = 60$. The maximum height of the bridge is 25 feet, so $b = 25$. The equation is: $\dfrac{x^2}{3600} + \dfrac{y^2}{625} = 1$.

The height 10 feet from the center:

$$\frac{10^2}{3600} + \frac{y^2}{625} = 1 \rightarrow \frac{y^2}{625} = 1 - \frac{100}{3600} \rightarrow y^2 = 625 \cdot \frac{3500}{3600} \rightarrow y \approx 24.65 \text{ feet}$$

The height 30 feet from the center:

$$\frac{30^2}{3600} + \frac{y^2}{625} = 1 \rightarrow \frac{y^2}{625} = 1 - \frac{900}{3600} \rightarrow y^2 = 625 \cdot \frac{2700}{3600} \rightarrow y \approx 21.65 \text{ feet}$$

The height 50 feet from the center:

$$\frac{50^2}{3600} + \frac{y^2}{625} = 1 \rightarrow \frac{y^2}{625} = 1 - \frac{2500}{3600} \rightarrow y^2 = 625 \cdot \frac{1100}{3600} \rightarrow y \approx 13.82 \text{ feet}$$

61. Place the semielliptical arch so that the x-axis coincides with the major axis and the y-axis passes through the center of the arch. Since the ellipse is 40 feet wide, the length of the major axis is 40, or $2a = 40$ or $a = 20$. The height is 15 feet at the center, so $b = 15$. The equation is: $\dfrac{x^2}{400} + \dfrac{y^2}{225} = 1$.

The height 10 feet either side of the center:

$$\frac{10^2}{400} + \frac{y^2}{225} = 1 \rightarrow \frac{y^2}{225} = 1 - \frac{100}{400} \rightarrow y^2 = 225 \cdot \frac{3}{4} \rightarrow y \approx 12.99 \text{ feet}$$

The height 20 feet either side of the center:

$$\frac{20^2}{400} + \frac{y^2}{225} = 1 \rightarrow \frac{y^2}{225} = 1 - \frac{400}{400} \rightarrow y^2 = 225 \cdot 0 \rightarrow y \approx 0 \text{ feet}$$

63. Since the mean distance is 93 million miles, $a = 93$ million. The length of the major axis is 186 million. The perihelion is 186 million $- 94.5$ million $= 91.5$ million miles. The distance from the center of the ellipse to the sun (focus) is 93 million $- 91.5$ million $= 1.5$ million miles; therefore, $c = 1.5$ million. Find b:

$$b^2 = a^2 - c^2 = \left(93 \times 10^6\right)^2 - \left(1.5 \times 10^6\right)^2 = 8.64675 \times 10^{15} \rightarrow b = 92.99 \times 10^6$$

The equation of the orbit is: $\dfrac{x^2}{\left(93 \times 10^6\right)^2} + \dfrac{y^2}{\left(92.99 \times 10^6\right)^2} = 1$.

65. The mean distance is 507 million $- 23.2$ million $= 483.8$ million miles.
The perihelion is 483.8 million $- 23.2$ million $= 460.6$ million miles.
Since $a = 483.8 \times 10^6$ and $c = 23.2 \times 10^6$, we can find b:

$$b^2 = a^2 - c^2 = \left(483.8 \times 10^6\right)^2 - \left(23.2 \times 10^6\right)^2 = 2.335242 \times 10^{17} \rightarrow b = 483.2 \times 10^6$$

The equation of the orbit of Jupiter is: $\dfrac{x^2}{\left(483.8 \times 10^6\right)^2} + \dfrac{y^2}{\left(483.2 \times 10^6\right)^2} = 1$.

67. If the x-axis is placed along the 100 foot length and the y-axis is placed along the 50 foot

length, the equation for the ellipse is: $\dfrac{x^2}{50^2} + \dfrac{y^2}{25^2} = 1$.

Find y when x = 40:

$$\dfrac{40^2}{50^2} + \dfrac{y^2}{25^2} = 1 \rightarrow \dfrac{y^2}{625} = 1 - \dfrac{1600}{2500} \rightarrow y^2 = 625 \cdot \dfrac{9}{25} \rightarrow y \approx 15 \text{ feet}$$

The width 10 feet from the side is 30 feet.

69. (a) Put the equation in standard ellipse form:
$$Ax^2 + Cy^2 + F = 0 \qquad A \neq 0, C \neq 0, F \neq 0$$
$$Ax^2 + Cy^2 = -F$$
$$\dfrac{Ax^2}{-F} + \dfrac{Cy^2}{-F} = 1$$
$$\dfrac{x^2}{(-F/A)} + \dfrac{y^2}{(-F/C)} = 1 \qquad \text{where } -F/A \text{ and } -F/C \text{ are positive}$$

This is the equation of an ellipse with center at (0, 0).

(b) If $A = C$, the equation becomes:
$$Ax^2 + Ay^2 = -F \rightarrow x^2 + y^2 = \dfrac{-F}{A}$$

This is the equation of a circle with center at (0, 0) and radius of $\sqrt{\dfrac{-F}{A}}$.

71. Answers will vary.

Chapter 9

Analytic Geometry

9.4   The Hyperbola

1.   B   3.   A

5.   Center: $(0,0)$; Focus: $(3,0)$; Vertex: $(1,0)$; Transverse axis is the x-axis; $a=1$; $c=3$. Find b:
$$b^2 = c^2 - a^2 = 9 - 1 = 8$$
$$b = \sqrt{8} = 2\sqrt{2}$$
Write the equation: $\dfrac{x^2}{1} - \dfrac{y^2}{8} = 1$
To graph, enter:
$$y_1 = \sqrt{8(x^2 - 1)};\ y_2 = -\sqrt{8(x^2 - 1)}$$

7.   Center: $(0,0)$; Focus: $(0,-6)$; Vertex: $(0,4)$ Transverse axis is the y-axis; $a=4$; $c=6$. Find b:
$$b^2 = c^2 - a^2 = 36 - 16 = 20$$
$$b = \sqrt{20} = 2\sqrt{5}$$
Write the equation: $\dfrac{y^2}{16} - \dfrac{x^2}{20} = 1$
To graph, enter:
$$y_1 = 4\sqrt{1 + x^2 / 20};\ y_2 = -4\sqrt{1 + x^2 / 20}$$

9.   Foci: $(-5,0)$, $(5,0)$; Vertex: $(3,0)$ Center: $(0,0)$; Transverse axis is the x-axis; $a=3$; $c=5$. Find b:
$$b^2 = c^2 - a^2 = 25 - 9 = 16 \rightarrow b = 4$$
Write the equation: $\dfrac{x^2}{9} - \dfrac{y^2}{16} = 1$
To graph, enter:
$$y_1 = 4\sqrt{x^2 / 9 - 1};\ y_2 = -4\sqrt{x^2 / 9 - 1}$$

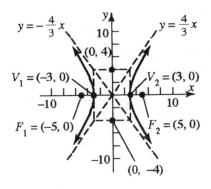

11. Vertices: $(0, -6)$, $(0, 6)$; Asymptote: $y = 2x$;
Center: $(0, 0)$; Transverse axis is the y-axis;
$a = 6$. Find b using the slope of the
asymptote:

$$\frac{a}{b} = \frac{6}{b} = 2 \;\rightarrow\; 2b = 6 \;\rightarrow\; b = 3$$

Write the equation: $\dfrac{y^2}{36} - \dfrac{x^2}{9} = 1$

To graph, enter:

$$y_1 = 6\sqrt{1 + x^2/9}\,;\, y_2 = -6\sqrt{1 + x^2/9}$$

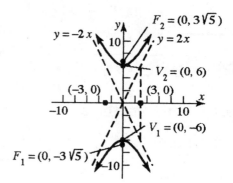

13. Foci: $(-4, 0)$, $(4, 0)$; Asymptote: $y = -x$;
Center: $(0, 0)$; Transverse axis is the
x-axis; $c = 4$. Using the slope of the
asymptote:

$$-\frac{b}{a} = -1 \;\rightarrow\; -b = -a \;\rightarrow\; b = a$$

Find b:

$$b^2 = c^2 - a^2 \rightarrow a^2 + b^2 = c^2 \qquad (c = 4)$$

$$b^2 + b^2 = 16 \rightarrow 2b^2 = 16 \rightarrow b^2 = 8$$

$$b = \sqrt{8} = 2\sqrt{2}$$

$$a = \sqrt{8} = 2\sqrt{2} \qquad (a = b)$$

Write the equation: $\dfrac{x^2}{8} - \dfrac{y^2}{8} = 1$

To graph, enter:

$$y_1 = \sqrt{x^2 - 8}\,;\, y_2 = -\sqrt{x^2 - 8}$$

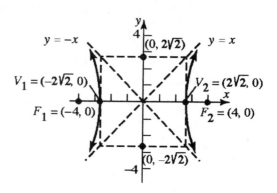

15. $\dfrac{x^2}{25} - \dfrac{y^2}{9} = 1$

The center of the hyperbola is at $(0, 0)$.
$a = 5$, $b = 3$. The vertices are $(5, 0)$ and
$(-5, 0)$. Find the value of c:

$$c^2 = a^2 + b^2 = 25 + 9 = 34 \rightarrow c = \sqrt{34}$$

The foci are $\left(\sqrt{34}, 0\right)$ and $\left(-\sqrt{34}, 0\right)$.

The transverse axis is $y = 0$.

The asymptotes are

$$y = \frac{3}{5}x \text{ and } y = -\frac{3}{5}x\,.$$

To graph, enter:

$$y_1 = 3\sqrt{(x^2/25 - 1)};\; y_2 = -3\sqrt{(x^2/25 - 1)}$$

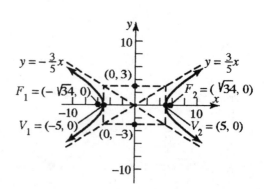

17. $4x^2 - y^2 = 16$

Divide both sides by 16 to put in standard form: $\dfrac{4x^2}{16} - \dfrac{y^2}{16} = \dfrac{16}{16} \rightarrow \dfrac{x^2}{4} - \dfrac{y^2}{16} = 1$

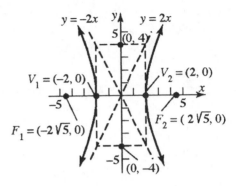

The center of the hyperbola is at (0, 0).
$a = 2$, $b = 4$. The vertices are (2, 0)
and (–2, 0). Find the value of c:
$$c^2 = a^2 + b^2 = 4 + 16 = 20$$
$$c = \sqrt{20} = 2\sqrt{5}$$
The foci are $\left(2\sqrt{5}, 0\right)$ and $\left(-2\sqrt{5}, 0\right)$.
The transverse axis is $y = 0$.
The asymptotes are $y = 2x$ and $y = -2x$.
To graph, enter:
$$y_1 = 4\sqrt{(x^2/4 - 1)}; \quad y_2 = -4\sqrt{(x^2/4 - 1)}$$

19. $y^2 - 9x^2 = 9$

Divide both sides by 9 to put in standard form: $\dfrac{y^2}{9} - \dfrac{9x^2}{9} = \dfrac{9}{9} \rightarrow \dfrac{y^2}{9} - \dfrac{x^2}{1} = 1$

The center of the hyperbola is at (0, 0).
$a = 3$, $b = 1$. The vertices are (0, 3) and
(0, –3). Find the value of c:
$$c^2 = a^2 + b^2 = 9 + 1 = 10$$
$$c = \sqrt{10}$$
The foci are $\left(0, \sqrt{10}\right)$ and $\left(0, -\sqrt{10}\right)$.
The transverse axis is $x = 0$.
The asymptotes are $y = 3x$ and $y = -3x$.
To graph, enter:
$$y_1 = \sqrt{(9x^2 + 9)}; \quad y_2 = -\sqrt{(9x^2 + 9)}$$

21. $y^2 - x^2 = 25$

Divide both sides by 25 to put in standard form: $\dfrac{y^2}{25} - \dfrac{x^2}{25} = 1$

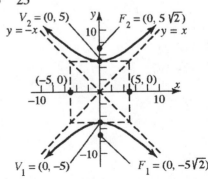

The center of the hyperbola is at (0, 0).
$a = 5$, $b = 5$. The vertices are (0, 5) and (0, –5).
Find the value of c:
$$c^2 = a^2 + b^2 = 25 + 25 = 50$$
$$c = \sqrt{50} = 5\sqrt{2}$$
The foci are $\left(0, 5\sqrt{2}\right)$ and $\left(0, -5\sqrt{2}\right)$.
The transverse axis is $x = 0$.
The asymptotes are $y = x$ and $y = -x$.
To graph, enter:
$$y_1 = \sqrt{(x^2 + 25)}; \quad y_2 = -\sqrt{(x^2 + 25)}$$

23. $x^2 - y^2 = 1$

25. $\dfrac{y^2}{36} - \dfrac{x^2}{9} = 1$

27. Center: $(4, -1)$; Focus: $(7, -1)$;
Vertex: $(6, -1)$; Transverse axis is
parallel to the x-axis; $a = 2$; $c = 3$.
Find b:

$$b^2 = c^2 - a^2 = 9 - 4 = 5 \rightarrow b = \sqrt{5}$$

Write the equation:

$$\dfrac{(x-4)^2}{4} - \dfrac{(y+1)^2}{5} = 1$$

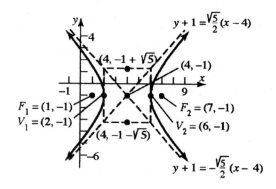

29. Center: $(-3, -4)$;
Focus: $(-3, -8)$;
Vertex: $(-3, -2)$;
Transverse axis is parallel to the
y-axis; $a = 2$; $c = 4$.
Find b:

$$b^2 = c^2 - a^2 = 16 - 4 = 12$$

$$b = \sqrt{12} = 2\sqrt{3}$$

Write the equation:

$$\dfrac{(y+4)^2}{4} - \dfrac{(x+3)^2}{12} = 1$$

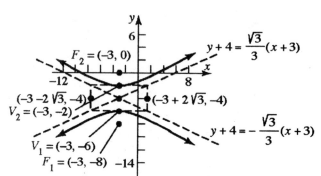

31. Foci: $(3, 7)$, $(7, 7)$; Vertex: $(6, 7)$;
Center: $(5, 7)$; Transverse axis is parallel to
the x-axis; $a = 1$; $c = 2$.
Find b:

$$b^2 = c^2 - a^2 = 4 - 1 = 3$$

$$b = \sqrt{3}$$

Write the equation: $\dfrac{(x-5)^2}{1} - \dfrac{(y-7)^2}{3} = 1$

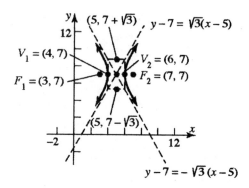

33. Vertices: $(-1, -1)$, $(3, -1)$;
Center: $(1, -1)$; Transverse
axis is parallel to the x-axis;
$a = 2$.

Asymptote: $\dfrac{x-1}{2} = \dfrac{y+1}{3}$

Using the slope of the
asymptote:

$$\dfrac{b}{a} = \dfrac{b}{2} = \dfrac{3}{2} \;\rightarrow\; b = 3$$

Write the equation:

$$\dfrac{(x-1)^2}{4} - \dfrac{(y+1)^2}{9} = 1$$

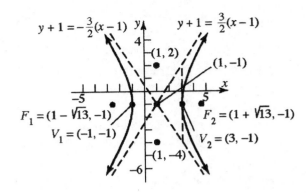

35. $\dfrac{(x-2)^2}{4} - \dfrac{(y+3)^2}{9} = 1$

The center of the hyperbola is at $(2, -3)$.
$a = 2$, $b = 3$. The vertices are $(0, -3)$ and
$(4, -3)$. Find the value of c:

$$c^2 = a^2 + b^2 = 4 + 9 = 13 \rightarrow c = \sqrt{13}$$

Foci: $\left(2 - \sqrt{13}, -3\right)$ and $\left(2 + \sqrt{13}, -3\right)$.

Transverse axis: $y = -3$.

Asymptotes:

$$y + 3 = \dfrac{3}{2}(x - 2); \;\; y + 3 = -\dfrac{3}{2}(x - 2).$$

To graph, enter:

$$y_1 = -3 + 3\sqrt{((x-2)^2 / 4 - 1)};$$
$$y_2 = -3 - 3\sqrt{((x-2)^2 / 4 - 1)}$$

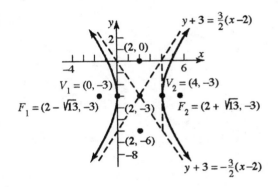

37. $(y - 2)^2 - 4(x + 2)^2 = 4$

Divide both sides by 4 to put in standard form: $\dfrac{(y-2)^2}{4} - \dfrac{(x+2)^2}{1} = 1$

The center of the hyperbola is at $(-2, 2)$.
$a = 2$, $b = 1$. The vertices are $(-2, 4)$ and
$(-2, 0)$. Find the value of c:

$$c^2 = a^2 + b^2 = 4 + 1 = 5 \rightarrow c = \sqrt{5}$$

Foci: $\left(-2, 2 - \sqrt{5}\right)$ and $\left(-2, 2 + \sqrt{5}\right)$.

Transverse axis: $x = -2$.

Asymptotes:

$$y - 2 = 2(x + 2); \, y - 2 = -2(x + 2).$$

To graph, enter: $y_1 = 2 + 2\sqrt{((x+2)^2 + 1)};$

$$y_2 = 2 - 2\sqrt{((x+2)^2 + 1)}$$

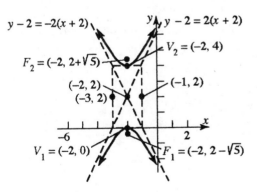

39. $(x+1)^2 - (y+2)^2 = 4$

Divide both sides by 4 to put in standard form: $\dfrac{(x+1)^2}{4} - \dfrac{(y+2)^2}{4} = 1$

The center of the hyperbola is at $(-1, -2)$. $a = 2,\ b = 2$.

The vertices are $(-3, -2)$ and $(1, -2)$.

Find the value of c:

$$c^2 = a^2 + b^2 = 4 + 4 = 8 \rightarrow c = \sqrt{8} = 2\sqrt{2}$$

Foci: $\left(-1-2\sqrt{2}, -2\right)$ and $\left(-1+2\sqrt{2}, -2\right)$.

Transverse axis: $y = -2$.

Asymptotes: $y + 2 = x + 1,\ y + 2 = -(x+1)$.

To graph, enter: $y_1 = -2 + 2\sqrt{((x+1)^2/4 - 1)}$; $y_2 = -2 - 2\sqrt{((x+1)^2/4 - 1)}$

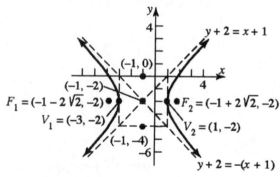

41.   Complete the square to put in standard form:

$$x^2 - y^2 - 2x - 2y - 1 = 0$$
$$(x^2 - 2x + 1) - (y^2 + 2y + 1) = 1 + 1 - 1$$
$$(x-1)^2 - (y+1)^2 = 1$$

The center of the hyperbola is at
$(1, -1)$. $a = 1,\ b = 1$.

The vertices are $(0, -1)$ and $(2, -1)$.

Find the value of c:

$$c^2 = a^2 + b^2 = 1 + 1 = 2$$

$$c = \sqrt{2}$$

Foci: $\left(1 - \sqrt{2}, -1\right)$ and $\left(1 + \sqrt{2}, -1\right)$.

Transverse axis: $y = -1$.

Asymptotes:
$y + 1 = x - 1,\ y + 1 = -(x - 1)$.

To graph, enter:

$$y_1 = -1 + \sqrt{((x-1)^2 - 1)};$$
$$y_2 = -1 - \sqrt{((x-1)^2 - 1)}$$

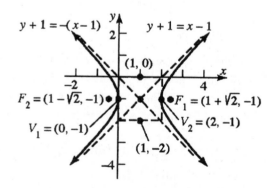

43. Complete the square to put in standard form:
$$y^2 - 4x^2 - 4y - 8x - 4 = 0$$
$$(y^2 - 4y + 4) - 4(x^2 + 2x + 1) = 4 + 4 - 4$$
$$(y - 2)^2 - 4(x + 1)^2 = 4$$
$$\frac{(y - 2)^2}{4} - \frac{(x + 1)^2}{1} = 1$$

The center of the hyperbola is at
$(-1, 2)$. $a = 2$, $b = 1$.
The vertices are $(-1, 4)$ and $(-1, 0)$.
Find the value of c:
$$c^2 = a^2 + b^2 = 4 + 1 = 5$$
$$c = \sqrt{5}$$
Foci: $\left(-1, 2 - \sqrt{5}\right)$ and $\left(-1, 2 + \sqrt{5}\right)$.
Transverse axis: $x = -1$.
Asymptotes:
$y - 2 = 2(x + 1)$, $y - 2 = -2(x + 1)$.
To graph, enter:
$$y_1 = 2 + 2\sqrt{((x + 1)^2 + 1)};$$
$$y_2 = 2 - 2\sqrt{((x + 1)^2 + 1)}$$

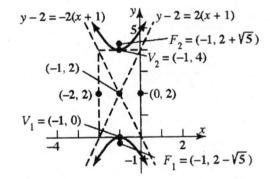

45. Complete the square to put in standard form:
$$4x^2 - y^2 - 24x - 4y + 16 = 0 \rightarrow 4(x^2 - 6x + 9) - (y^2 + 4y + 4) = -16 + 36 - 4$$
$$4(x - 3)^2 - (y + 2)^2 = 16 \rightarrow \frac{(x - 3)^2}{4} - \frac{(y + 2)^2}{16} = 1$$

The center of the hyperbola is at
$(3, -2)$. $a = 2$, $b = 4$.
The vertices are $(1, -2)$ and $(5, -2)$.
Find the value of c:
$$c^2 = a^2 + b^2 = 4 + 16 = 20$$
$$c = \sqrt{20} = 2\sqrt{5}$$
Foci:
$\left(3 - 2\sqrt{5}, -2\right)$ and $\left(3 + 2\sqrt{5}, -2\right)$.
Transverse axis: $y = -2$.
Asymptotes:
$y + 2 = 2(x - 3)$, $y + 2 = -2(x - 3)$.
To graph, enter:
$$y_1 = -2 + 4\sqrt{((x - 3)^2 / 4 - 1)};$$
$$y_2 = -2 - 4\sqrt{((x - 3)^2 / 4 - 1)}$$

47. Complete the square to put in standard form:
$$y^2 - 4x^2 - 16x - 2y - 19 = 0 \rightarrow (y^2 - 2y + 1) - 4(x^2 + 4x + 4) = 19 + 1 - 16$$
$$(y - 1)^2 - 4(x + 2)^2 = 4 \rightarrow \frac{(y - 1)^2}{4} - \frac{(x + 2)^2}{1} = 1$$

The center of the hyperbola is at $(-2, 1)$. $a = 2$, $b = 1$. The vertices are $(-2, 3)$ and $(-2, -1)$. Find the value of c:
$$c^2 = a^2 + b^2 = 4 + 1 = 5$$
$$c = \sqrt{5}$$
Foci: $\left(-2, 1 - \sqrt{5}\right)$ and $\left(-2, 1 + \sqrt{5}\right)$.
Transverse axis: $x = -2$.
Asymptotes:
$$y - 1 = 2(x + 2); y - 1 = -2(x + 2).$$
To graph, enter:
$$y_1 = 1 + 2\sqrt{((x + 2)^2 + 1)};$$
$$y_2 = 1 - 2\sqrt{((x + 2)^2 + 1)}$$

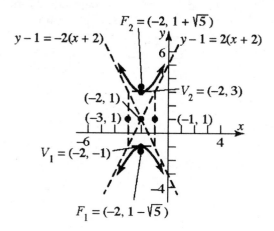

49. Rewrite the equation:
$$y = \sqrt{16 + 4x^2}$$
$$y^2 = 16 + 4x^2, \quad y \geq 0$$
$$y^2 - 4x^2 = 16, \quad y \geq 0$$
$$\frac{y^2}{16} - \frac{x^2}{4} = 1, \quad y \geq 0$$

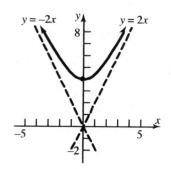

51. Rewrite the equation:
$$y = -\sqrt{-25 + x^2}$$
$$y^2 = -25 + x^2, \quad y \leq 0$$
$$x^2 - y^2 = 25, \quad y \leq 0$$
$$\frac{x^2}{25} - \frac{y^2}{25} = 1, \quad y \leq 0$$

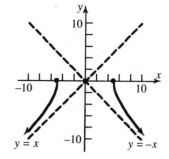

53. (a) Set up a coordinate system so that the two stations lie on the x-axis and the origin is midway between them. The ship lies on a hyperbola whose foci are the locations of the two stations. Since the time difference is 0.00038 seconds and the speed of the signal is 186,000 miles per second, the difference in the distances of the ships from each station is:

distance = (186,000)(0.00038) $\approx$ 70.68 miles

458

The difference of the distances from the ship to each station, 70.68, equals $2a$, so $a = 35.34$ and the vertex of the corresponding hyperbola is at $(35.34, 0)$. Since the focus is at $(100, 0)$, following this hyperbola, the ship would reach shore 64.66 miles from the master station.

(b)   The ship should follow a hyperbola with a vertex at $(80, 0)$. For this hyperbola, $a = 80$, so the constant difference of the distances from the ship to each station is 160. The time difference the ship should look for is:

$$\text{time} = \frac{160}{186{,}000} \approx 0.00086 \text{ seconds}$$

(c)   Find the equation of the hyperbola with vertex at $(80, 0)$ and a focus at $(100, 0)$. The form of the equation of the hyperbola is:

$$\frac{x^2}{a^2} - \frac{y^2}{b^2} = 1 \quad \text{where } a = 80.$$

Since $c = 100$ and $b^2 = c^2 - a^2 \rightarrow b^2 = 100^2 - 80^2 = 3600$.

The equation of the hyperbola is: $\dfrac{x^2}{6400} - \dfrac{y^2}{3600} = 1$.

Since the ship is 50 miles off shore, we have $y = 50$. Solve the equation for x:

$$\frac{x^2}{6400} - \frac{50^2}{3600} = 1 \rightarrow \frac{x^2}{6400} = 1 + \frac{2500}{3600} = \frac{61}{36} \rightarrow x^2 = 6400 \cdot \frac{61}{36}$$

$$x \approx 104 \text{ miles}$$

The ship's location is $(104, 50)$.

55.   (a)   Set up a rectangular coordinate system so that the two devices lie on the x-axis and the origin is midway between them. The devices serve as foci to the hyperbola so $c = \dfrac{2000}{2} = 1000$. Since the explosion occurs 200 feet from point B, the vertex of the hyperbola is $(800, 0)$; therefore, $a = 800$. Finding b:

$$b^2 = c^2 - a^2 \rightarrow b^2 = 1000^2 - 800^2 = 360000 \rightarrow b = 600$$

The equation of the hyperbola is:

$$\frac{x^2}{800^2} - \frac{y^2}{600^2} = 1$$

If $x = 1000$, find y:

$$\frac{1000^2}{800^2} - \frac{y^2}{600^2} = 1$$

$$\frac{y^2}{600^2} = \frac{1000^2}{800^2} - 1 = \frac{600^2}{800^2}$$

$$y^2 = 600^2 \cdot \frac{600^2}{800^2}$$

$$y = 450 \text{ feet}$$

The second detonation should take place 450 feet north of point B.

57.   If the eccentricity is close to 1, then $c \approx a$ and $b \approx 0$. When b is close to 0, the hyperbola is very narrow, because the slopes of the asymptotes are close to 0.

If the eccentricity is very large, then c is much larger than a and b is very large.

The result is a hyperbola that is very wide.

59. $\dfrac{x^2}{4} - y^2 = 1$ $(a = 2, \; b = 1)$

is a hyperbola with horizontal transverse axis, centered at $(0, 0)$ and has asymptotes: $y = \pm \dfrac{1}{2}x$

$y^2 - \dfrac{x^2}{4} = 1$ $(a = 1, \; b = 2)$

is a hyperbola with vertical transverse axis, centered at $(0, 0)$ and has asymptotes: $y = \pm \dfrac{1}{2}x$

Since the two hyperbolas have the same asymptotes, they are conjugate.

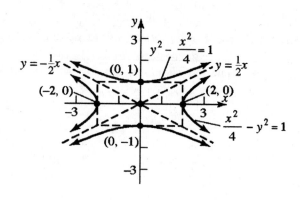

61. Put the equation in standard hyperbola form:

$$Ax^2 + Cy^2 + F = 0 \qquad A \neq 0, \; C \neq 0, \; F \neq 0$$

$$Ax^2 + Cy^2 = -F \rightarrow \dfrac{Ax^2}{-F} + \dfrac{Cy^2}{-F} = 1$$

$$\dfrac{x^2}{-F/A} + \dfrac{y^2}{-F/C} = 1$$

Since $-F/A$ and $-F/C$ have opposite signs, this is a hyperbola with center at $(0, 0)$.

Chapter 9

Analytic Geometry

9.5 Rotation of Axes; General Form of a Conic

1. $x^2 + 4x + y + 3 = 0$
    $A = 1$ and $C = 0$; $AC = (1)(0) = 0$. Since $AC = 0$, the equation defines a parabola.

3. $6x^2 + 3y^2 - 12x + 6y = 0$
    $A = 6$ and $C = 3$; $AC = (6)(3) = 18$. Since $AC > 0$ and $A \neq C$, the equation defines an ellipse.

5. $3x^2 - 2y^2 + 6x + 4 = 0$
    $A = 3$ and $C = -2$; $AC = (3)(-2) = -6$. Since $AC < 0$, the equation defines a hyperbola.

7. $2y^2 - x^2 - y + x = 0$
    $A = -1$ and $C = 2$; $AC = (-1)(2) = -2$. Since $AC < 0$, the equation defines a hyperbola.

9. $x^2 + y^2 - 8x + 4y = 0$
    $A = 1$ and $C = 1$; $AC = (1)(1) = 1$. Since $AC > 0$ and $A = C$, the equation defines a circle.

11. $x^2 + 4xy + y^2 - 3 = 0$
    $A = 1, B = 4,$ and $C = 1$; $\cot(2\theta) = \dfrac{A - C}{B} = \dfrac{1 - 1}{4} = \dfrac{0}{4} = 0 \;\rightarrow\; 2\theta = \dfrac{\pi}{2} \;\rightarrow\; \theta = \dfrac{\pi}{4}$

    $x = x'\cos\left(\dfrac{\pi}{4}\right) - y'\sin\left(\dfrac{\pi}{4}\right) = \dfrac{\sqrt{2}}{2}x' - \dfrac{\sqrt{2}}{2}y' = \dfrac{\sqrt{2}}{2}(x' - y')$

    $y = x'\sin\left(\dfrac{\pi}{4}\right) + y'\cos\left(\dfrac{\pi}{4}\right) = \dfrac{\sqrt{2}}{2}x' + \dfrac{\sqrt{2}}{2}y' = \dfrac{\sqrt{2}}{2}(x' + y')$

13. $5x^2 + 6xy + 5y^2 - 8 = 0$
    $A = 5, B = 6,$ and $C = 5$; $\cot(2\theta) = \dfrac{A - C}{B} = \dfrac{5 - 5}{6} = \dfrac{0}{6} = 0 \;\rightarrow\; 2\theta = \dfrac{\pi}{2} \;\rightarrow\; \theta = \dfrac{\pi}{4}$

    $x = x'\cos\left(\dfrac{\pi}{4}\right) - y'\sin\left(\dfrac{\pi}{4}\right) = \dfrac{\sqrt{2}}{2}x' - \dfrac{\sqrt{2}}{2}y' = \dfrac{\sqrt{2}}{2}(x' - y')$

    $y = x'\sin\left(\dfrac{\pi}{4}\right) + y'\cos\left(\dfrac{\pi}{4}\right) = \dfrac{\sqrt{2}}{2}x' + \dfrac{\sqrt{2}}{2}y' = \dfrac{\sqrt{2}}{2}(x' + y')$

15. $13x^2 - 6\sqrt{3}xy + 7y^2 - 16 = 0$

$A = 13, B = -6\sqrt{3}, \text{ and } C = 7; \quad \cot(2\theta) = \dfrac{A-C}{B} = \dfrac{13-7}{-6\sqrt{3}} = \dfrac{6}{-6\sqrt{3}} = -\dfrac{\sqrt{3}}{3}$

$2\theta = \dfrac{2\pi}{3} \quad \rightarrow \quad \theta = \dfrac{\pi}{3}$

$x = x'\cos\left(\dfrac{\pi}{3}\right) - y'\sin\left(\dfrac{\pi}{3}\right) = \dfrac{1}{2}x' - \dfrac{\sqrt{3}}{2}y' = \dfrac{1}{2}\left(x' - \sqrt{3}y'\right)$

$y = x'\sin\left(\dfrac{\pi}{3}\right) + y'\cos\left(\dfrac{\pi}{3}\right) = \dfrac{\sqrt{3}}{2}x' + \dfrac{1}{2}y' = \dfrac{1}{2}\left(\sqrt{3}x' + y'\right)$

17. $4x^2 - 4xy + y^2 - 8\sqrt{5}x - 16\sqrt{5}y = 0$

$A = 4, B = -4, \text{ and } C = 1; \quad \cot(2\theta) = \dfrac{A-C}{B} = \dfrac{4-1}{-4} = -\dfrac{3}{4}; \quad \cos 2\theta = -\dfrac{3}{5}$

$\sin\theta = \sqrt{\dfrac{1 - \left(-\dfrac{3}{5}\right)}{2}} = \sqrt{\dfrac{4}{5}} = \dfrac{2}{\sqrt{5}} = \dfrac{2\sqrt{5}}{5};$

$\cos\theta = \sqrt{\dfrac{1 + \left(-\dfrac{3}{5}\right)}{2}} = \sqrt{\dfrac{1}{5}} = \dfrac{1}{\sqrt{5}} = \dfrac{\sqrt{5}}{5}$

$x = x'\cos\theta - y'\sin\theta = \dfrac{\sqrt{5}}{5}x' - \dfrac{2\sqrt{5}}{5}y' = \dfrac{\sqrt{5}}{5}\left(x' - 2y'\right)$

$y = x'\sin\theta + y'\cos\theta = \dfrac{2\sqrt{5}}{5}x' + \dfrac{\sqrt{5}}{5}y' = \dfrac{\sqrt{5}}{5}\left(2x' + y'\right)$

19. $25x^2 - 36xy + 40y^2 - 12\sqrt{13}x - 8\sqrt{13}y = 0$

$A = 25, B = -36, \text{ and } C = 40; \quad \cot(2\theta) = \dfrac{A-C}{B} = \dfrac{25-40}{-36} = \dfrac{5}{12}; \quad \cos 2\theta = \dfrac{5}{13}$

$\sin\theta = \sqrt{\dfrac{\left(1 - \dfrac{5}{13}\right)}{2}} = \sqrt{\dfrac{4}{13}} = \dfrac{2}{\sqrt{13}} = \dfrac{2\sqrt{13}}{13};$

$\cos\theta = \sqrt{\dfrac{\left(1 + \dfrac{5}{13}\right)}{2}} = \sqrt{\dfrac{9}{13}} = \dfrac{3}{\sqrt{13}} = \dfrac{3\sqrt{13}}{13}$

$x = x'\cos\theta - y'\sin\theta = \dfrac{3\sqrt{13}}{13}x' - \dfrac{2\sqrt{13}}{13}y' = \dfrac{\sqrt{13}}{13}\left(3x' - 2y'\right)$

$y = x'\sin\theta + y'\cos\theta = \dfrac{2\sqrt{13}}{13}x' + \dfrac{3\sqrt{13}}{13}y' = \dfrac{\sqrt{13}}{13}\left(2x' + 3y'\right)$

21. $x^2 + 4xy + y^2 - 3 = 0$; $\theta = 45°$ (see Problem 11)

$$\left(\frac{\sqrt{2}}{2}(x'-y')\right)^2 + 4\left(\frac{\sqrt{2}}{2}(x'-y')\right)\left(\frac{\sqrt{2}}{2}(x'+y')\right) + \left(\frac{\sqrt{2}}{2}(x'+y')\right)^2 - 3 = 0$$

$$\frac{1}{2}\left(x'^2 - 2x'y' + y'^2\right) + 2\left(x'^2 - y'^2\right) + \frac{1}{2}\left(x'^2 + 2x'y' + y'^2\right) - 3 = 0$$

$$\frac{1}{2}x'^2 - x'y' + \frac{1}{2}y'^2 + 2x'^2 - 2y'^2 + \frac{1}{2}x'^2 + x'y' + \frac{1}{2}y'^2 = 3$$

$$3x'^2 - y'^2 = 3$$

$$\frac{x'^2}{1} - \frac{y'^2}{3} = 1$$

Hyperbola; center at the origin,
transverse axis is the x'-axis, vertices at (± 1, 0).

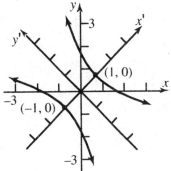

23. $5x^2 + 6xy + 5y^2 - 8 = 0$; $\theta = 45°$ (see Problem 13)

$$5\left(\frac{\sqrt{2}}{2}(x'-y')\right)^2 + 6\left(\frac{\sqrt{2}}{2}(x'-y')\right)\left(\frac{\sqrt{2}}{2}(x'+y')\right) + 5\left(\frac{\sqrt{2}}{2}(x'+y')\right)^2 - 8 = 0$$

$$\frac{5}{2}\left(x'^2 - 2x'y' + y'^2\right) + 3\left(x'^2 - y'^2\right) + \frac{5}{2}\left(x'^2 + 2x'y' + y'^2\right) - 8 = 0$$

$$\frac{5}{2}x'^2 - 5x'y' + \frac{5}{2}y'^2 + 3x'^2 - 3y'^2 + \frac{5}{2}x'^2 + 5x'y' + \frac{5}{2}y'^2 = 8$$

$$8x'^2 + 2y'^2 = 8$$

$$\frac{x'^2}{1} + \frac{y'^2}{4} = 1$$

Ellipse; center at the origin,
major axis is the y'-axis, vertices at (0, ± 2).

25. $13x^2 - 6\sqrt{3}\,xy + 7y^2 - 16 = 0$; $\theta = 60°$ (see Problem 15)

$$13\left(\frac{1}{2}(x' - \sqrt{3}y')\right)^2 - 6\sqrt{3}\left(\frac{1}{2}(x' - \sqrt{3}y')\right)\left(\frac{1}{2}(\sqrt{3}\,x' + y')\right) + 7\left(\frac{1}{2}(\sqrt{3}\,x' + y')\right)^2 - 16 = 0$$

$$\frac{13}{4}\left(x'^2 - 2\sqrt{3}x'y' + 3y'^2\right) - \frac{3\sqrt{3}}{2}\left(\sqrt{3}x'^2 - 2x'y' - \sqrt{3}y'^2\right) + \frac{7}{4}\left(3x'^2 + 2\sqrt{3}x'y' + y'^2\right) = 16$$

$$\frac{13}{4}x'^2 - \frac{13\sqrt{3}}{2}x'y' + \frac{39}{4}y'^2 - \frac{9}{2}x'^2 + 3\sqrt{3}x'y' + \frac{9}{2}y'^2 + \frac{21}{4}x'^2 + \frac{7\sqrt{3}}{2}x'y' + \frac{7}{4}y'^2 = 16$$

$$4x'^2 + 16y'^2 = 16$$

$$\frac{x'^2}{4} + \frac{y'^2}{1} = 1$$

Ellipse; center at the origin,
major axis is the x'-axis, vertices at (±2, 0).

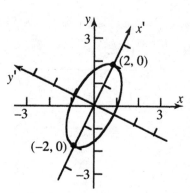

27. $4x^2 - 4xy + y^2 - 8\sqrt{5}\,x - 16\sqrt{5}\,y = 0$; $\theta = 63.4°$ (see Problem 17)

$$4\left(\frac{\sqrt{5}}{5}(x' - 2y')\right)^2 - 4\left(\frac{\sqrt{5}}{5}(x' - 2y')\right)\left(\frac{\sqrt{5}}{5}(2x' + y')\right) + \left(\frac{\sqrt{5}}{5}(2x' + y')\right)^2$$

$$- 8\sqrt{5}\left(\frac{\sqrt{5}}{5}(x' - 2y')\right) - 16\sqrt{5}\left(\frac{\sqrt{5}}{5}(2x' + y')\right) = 0$$

$$\frac{4}{5}\left(x'^2 - 4x'y' + 4y'^2\right) - \frac{4}{5}\left(2x'^2 - 3x'y' - 2y'^2\right) + \frac{1}{5}\left(4x'^2 + 4x'y' + y'^2\right)$$

$$- 8x' + 16y' - 32x' - 16y' = 0$$

$$\frac{4}{5}x'^2 - \frac{16}{5}x'y' + \frac{16}{5}y'^2 - \frac{8}{5}x'^2 + \frac{12}{5}x'y' + \frac{8}{5}y'^2 + \frac{4}{5}x'^2 + \frac{4}{5}x'y' + \frac{1}{5}y'^2 - 40x' = 0$$

$$5y'^2 - 40x' = 0$$

$$y'^2 = 8x'$$

Parabola; vertex at the origin, focus at (2, 0).

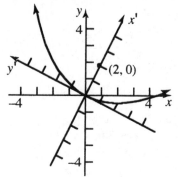

29. $25x^2 - 36xy + 40y^2 - 12\sqrt{13}\,x - 8\sqrt{13}\,y = 0$; $\theta \approx 33.7°$ (see Problem 19)

$$25\left(\frac{\sqrt{13}}{13}(3x' - 2y')\right)^2 - 36\left(\frac{\sqrt{13}}{13}(3x' - 2y')\right)\left(\frac{\sqrt{13}}{13}(2x' + 3y')\right) + 40\left(\frac{\sqrt{13}}{13}(2x' + 3y')\right)^2$$

$$- 12\sqrt{13}\left(\frac{\sqrt{13}}{13}(3x' - 2y')\right) - 8\sqrt{13}\left(\frac{\sqrt{13}}{13}(2x' + 3y')\right) = 0$$

$$\frac{25}{13}\left(9x'^2 - 12x'y' + 4y'^2\right) - \frac{36}{13}\left(6x'^2 + 5x'y' - 6y^2\right) + \frac{40}{13}\left(4x'^2 + 12x'y' + 9y^2\right)$$

$$- 36x' + 24y' - 16x' - 24y' = 0$$

$$\frac{225}{13}x'^2 - \frac{300}{13}x'y' + \frac{100}{13}y'^2 - \frac{216}{13}x'^2 - \frac{180}{13}x'y' + \frac{216}{13}y'^2$$

$$+ \frac{160}{13}x'^2 + \frac{480}{13}x'y' + \frac{360}{13}y'^2 - 52x' = 0$$

$$13x'^2 + 52y'^2 - 52x' = 0$$

$$x'^2 - 4x' + 4y'^2 = 0$$

$$(x'-2)^2 + 4y'^2 = 4$$

$$\frac{(x'-2)^2}{4} + \frac{y'^2}{1} = 1$$

Ellipse; center at $(2, 0)$,
major axis is the x'-axis, vertices at $(4, 0)$ and $(0, 0)$.

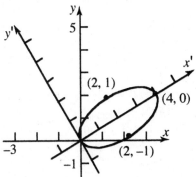

31. $16x^2 + 24xy + 9y^2 - 130x + 90y = 0$

$A = 16, B = 24,$ and $C = 9$; $\cot(2\theta) = \dfrac{A-C}{B} = \dfrac{16-9}{24} = \dfrac{7}{24} \;\rightarrow\; \cos(2\theta) = \dfrac{7}{25}$

$\sin\theta = \sqrt{\dfrac{\left(1-\dfrac{7}{25}\right)}{2}} = \sqrt{\dfrac{9}{25}} = \dfrac{3}{5};\quad \cos\theta = \sqrt{\dfrac{\left(1+\dfrac{7}{25}\right)}{2}} = \sqrt{\dfrac{16}{25}} = \dfrac{4}{5} \;\rightarrow\; \theta \approx 36.9°$

$x = x'\cos\theta - y'\sin\theta = \dfrac{4}{5}x' - \dfrac{3}{5}y' = \dfrac{1}{5}(4x' - 3y')$

$y = x'\sin\theta + y'\cos\theta = \dfrac{3}{5}x' + \dfrac{4}{5}y' = \dfrac{1}{5}(3x' + 4y')$

$16\left(\dfrac{1}{5}(4x'-3y')\right)^2 + 24\left(\dfrac{1}{5}(4x'-3y')\right)\left(\dfrac{1}{5}(3x'+4y')\right) + 9\left(\dfrac{1}{5}(3x'+4y')\right)^2$

$$-130\left(\dfrac{1}{5}(4x'-3y')\right) + 90\left(\dfrac{1}{5}(3x'+4y')\right) = 0$$

$\dfrac{16}{25}\left(16x'^2 - 24x'y' + 9y'^2\right) + \dfrac{24}{25}\left(12x'^2 + 7x'y' - 12y'^2\right) + \dfrac{9}{25}\left(9x'^2 + 24x'y' + 16y'^2\right)$

$$-104x' + 78y' + 54x' + 72y' = 0$$

$\dfrac{256}{25}x'^2 - \dfrac{384}{25}x'y' + \dfrac{144}{25}y'^2 + \dfrac{288}{25}x'^2 + \dfrac{168}{25}x'y' - \dfrac{288}{25}y'^2$

$$+ \dfrac{81}{25}x'^2 + \dfrac{216}{25}x'y' + \dfrac{144}{25}y'^2 - 50x' + 150y' = 0$$

$25x'^2 - 50x' + 150y' = 0$

$x'^2 - 2x' = -6y'$

$(x'-1)^2 = -6y' + 1$

$(x'-1)^2 = -6\left(y' - \dfrac{1}{6}\right)$

Parabola; vertex at $\left(1, \dfrac{1}{6}\right)$, focus at $\left(1, -\dfrac{4}{3}\right)$.

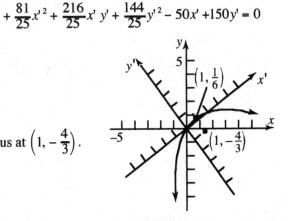

33. $A = 1$. $B = 3$, $C = -2$ $B^2 - 4AC = 3^2 - 4(1)(-2) = 17 > 0$;   hyperbola

35. $A = 1$. $B = -7$, $C = 3$ $B^2 - 4AC = (-7)^2 - 4(1)(3) = 37 > 0$;   hyperbola

37. $A = 9$. $B = 12$, $C = 4$ $B^2 - 4AC = 12^2 - 4(9)(4) = 0$;   parabola

39. $A = 10$. $B = -12$, $C = 4$ $B^2 - 4AC = (-12)^2 - 4(10)(4) = -16 < 0$;   ellipse

41. $A = 3$. $B = -2$, $C = 1$ $B^2 - 4AC = (-2)^2 - 4(3)(1) = -8 < 0$;   ellipse

43.   See equation 6 on page 770.

$$A' = A\cos^2\theta + B\sin\theta\cos\theta + C\sin^2\theta; \qquad D' = D\cos\theta + E\sin\theta$$
$$B' = B(\cos^2\theta - \sin^2\theta) + 2(C - A)(\sin\theta\cos\theta); \quad E' = -D\sin\theta + E\cos\theta$$
$$C' = A\sin^2\theta - B\sin\theta\cos\theta + C\cos^2\theta; \qquad F' = F$$

45. $B'^2 - 4A'C'$

$$= \left[B(\cos^2\theta - \sin^2\theta) + 2(C - A)\sin\theta\cos\theta \right]^2$$
$$\qquad - 4\left(A\cos^2\theta + B\sin\theta\cos\theta + C\sin^2\theta \right)\left(A\sin^2\theta - B\sin\theta\cos\theta + C\cos^2\theta \right)$$
$$= B^2\left(\cos^4\theta - 2\cos^2\theta\sin^2\theta + \sin^4\theta \right) + 4B(C - A)\sin\theta\cos\theta(\cos^2\theta - \sin^2\theta)$$
$$\qquad + 4(C - A)^2\sin^2\theta\cos^2\theta - 4\big[A^2\sin^2\theta\cos^2\theta - AB\sin\theta\cos^3\theta + AC\cos^4\theta$$
$$\qquad + AB\sin^3\theta\cos\theta - B^2\sin^2\theta\cos^2\theta + BC\sin\theta\cos^3\theta + AC\sin^4\theta$$
$$\qquad - BC\sin^3\theta\cos\theta + C^2\sin^2\theta\cos^2\theta \big]$$
$$= B^2\left(\cos^4\theta - 2\cos^2\theta\sin^2\theta + \sin^4\theta + 4\sin^2\theta\cos^2\theta \right)$$
$$\qquad + BC\left(4\sin\theta\cos\theta(\cos^2\theta - \sin^2\theta) - 4\sin\theta\cos^3\theta + 4\sin^3\theta\cos\theta \right)$$
$$\qquad - AB\left(4\sin\theta\cos\theta(\cos^2\theta - \sin^2\theta) - 4\sin\theta\cos^3\theta + 4\sin^3\theta\cos\theta \right)$$
$$\qquad + 4C^2\left(\sin^2\theta\cos^2\theta - \sin^2\theta\cos^2\theta \right) - 4AC\left(2\sin^2\theta\cos^2\theta + \cos^4\theta + \sin^4\theta \right)$$
$$\qquad + 4A^2\left(\sin^2\theta\cos^2\theta - \sin^2\theta\cos^2\theta \right)$$
$$= B^2\left(\cos^4\theta + 2\sin^2\theta\cos^2\theta + \sin^4\theta \right) - 4AC\left(\cos^4\theta + 2\sin^2\theta\cos^2\theta + \sin^4\theta \right)$$
$$= B^2\left(\cos^2\theta + \sin^2\theta \right)^2 - 4AC\left(\cos^2\theta + \sin^2\theta \right)^2 = B^2 - 4AC$$

47. $d^2 = (y_2 - y_1)^2 + (x_2 - x_1)^2$

$$= \left(x_2'\sin\theta + y_2'\cos\theta - x_1'\sin\theta - y_1'\cos\theta \right)^2 + \left(x_2'\cos\theta - y_2'\sin\theta - x_1'\cos\theta + y_1'\sin\theta \right)^2$$
$$= \left((x_2'-x_1')\sin\theta + (y_2'-y_1')\cos\theta \right)^2 + \left((x_2'-x_1')\cos\theta - (y_2'-y_1')\sin\theta \right)^2$$
$$= (x_2'-x_1')^2\sin^2\theta + 2(x_2'-x_1')(y_2'-y_1')\sin\theta\cos\theta + (y_2'-y_1')^2\cos^2\theta$$
$$\qquad + (x_2'-x_1')^2\cos^2\theta - 2(x_2'-x_1')(y_2'-y_1')\sin\theta\cos\theta + (y_2'-y_1')^2\sin^2\theta$$
$$= (x_2'-x_1')^2\sin^2\theta + (x_2'-x_1')^2\cos^2\theta + (y_2'-y_1')^2\cos^2\theta + (y_2'-y_1')^2\sin^2\theta = (x_2'-x_1')^2 + (y_2'-y_1')^2$$

Chapter 9

Analytic Geometry

9.6   Polar Equations of Conics

1. $e = 1$; $p = 1$; parabola; directrix is perpendicular to the polar axis and 1 unit to the right of the pole.

$$r = \frac{4}{2 - 3\sin\theta} = \frac{4}{2\left(1 - \frac{3}{2}\sin\theta\right)} = \frac{2}{\left(1 - \frac{3}{2}\sin\theta\right)};$$

3.
$$ep = 2, \ e = \frac{3}{2}; \ p = \frac{4}{3}$$

Hyperbola; directrix is parallel to the polar axis and $\frac{4}{3}$ units below the pole.

$$r = \frac{3}{4 - 2\cos\theta} = \frac{3}{4\left(1 - \frac{1}{2}\cos\theta\right)} = \frac{\left(\frac{3}{4}\right)}{\left(1 - \frac{1}{2}\cos\theta\right)};$$

5.
$$ep = \frac{3}{4}, \ e = \frac{1}{2}; \ p = \frac{3}{2}$$

Ellipse; directrix is perpendicular to the polar axis and $\frac{3}{2}$ units to the left of the pole.

7. $r = \frac{1}{1 + \cos\theta}$

$ep = 1, \ e = 1, \ p = 1$
Parabola; directrix is perpendicular to the polar axis 1 unit to the right of the pole;
vertex is $\left(\frac{1}{2}, 0\right)$.

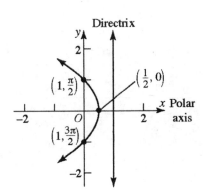

9. $r = \dfrac{8}{4 + 3\sin\theta}$

$$r = \dfrac{8}{4\left(1 + \dfrac{3}{4}\sin\theta\right)} = \dfrac{2}{\left(1 + \dfrac{3}{4}\sin\theta\right)}$$

$ep = 2,\ \ e = \dfrac{3}{4},\ \ p = \dfrac{8}{3}$

Ellipse; directrix is parallel to the polar axis $\dfrac{8}{3}$ units above the pole; vertices are

$\left(\dfrac{8}{7}, \dfrac{\pi}{2}\right)$ and $\left(8, \dfrac{3\pi}{2}\right)$.

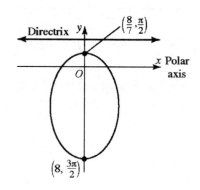

11. $r = \dfrac{9}{3 - 6\cos\theta}$

$$r = \dfrac{9}{3(1 - 2\cos\theta)} = \dfrac{3}{1 - 2\cos\theta}$$

$ep = 3,\ \ e = 2,\ \ p = \dfrac{3}{2}$

Hyperbola; directrix is perpendicular to the polar axis $\dfrac{3}{2}$ units to the left of the pole; vertices are $(-3, 0)$ and $(1, \pi)$.

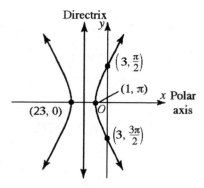

13. $r = \dfrac{8}{2 - \sin\theta}$

$$r = \dfrac{8}{2\left(1 - \dfrac{1}{2}\sin\theta\right)} = \dfrac{4}{1 - \dfrac{1}{2}\sin\theta}$$

$ep = 4,\ \ e = \dfrac{1}{2},\ \ p = 8$

Ellipse; directrix is parallel to the polar axis 8 units below the pole; vertices are

$\left(8, \dfrac{\pi}{2}\right)$ and $\left(\dfrac{8}{3}, \dfrac{3\pi}{2}\right)$.

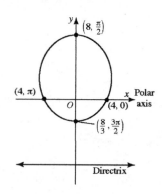

15. $r(3 - 2\sin\theta) = 6 \quad \rightarrow \quad r = \dfrac{6}{3 - 2\sin\theta}$

$r = \dfrac{6}{3\left(1 - \dfrac{2}{3}\sin\theta\right)} = \dfrac{2}{1 - \dfrac{2}{3}\sin\theta}$

$ep = 2, \;\; e = \dfrac{2}{3}, \;\; p = 3$

Ellipse; directrix is parallel to the polar axis 3 units below the pole; vertices are $\left(6, \dfrac{\pi}{2}\right)$ and $\left(\dfrac{6}{5}, \dfrac{3\pi}{2}\right)$.

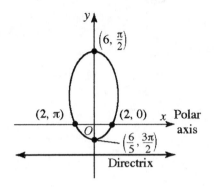

17. $r = \dfrac{6\sec\theta}{2\sec\theta - 1} = \dfrac{6}{2 - \cos\theta}$

$r = \dfrac{6}{2\left(1 - \dfrac{1}{2}\cos\theta\right)} = \dfrac{3}{1 - \dfrac{1}{2}\cos\theta}$

$ep = 3, \;\; e = \dfrac{1}{2}, \;\; p = 6$

Ellipse; directrix is perpendicular to the polar axis 6 units to the left of the pole; vertices are $(6, 0)$ and $(2, \pi)$.

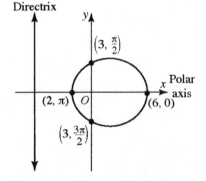

19. $r = \dfrac{1}{1 + \cos\theta}$

$r + r\cos\theta = 1$

$r = 1 - r\cos\theta$

$r^2 = (1 - r\cos\theta)^2$

$x^2 + y^2 = (1 - x)^2$

$x^2 + y^2 = 1 - 2x + x^2$

$y^2 + 2x - 1 = 0$

21. $r = \dfrac{8}{4 + 3\sin\theta}$

$4r + 3r\sin\theta = 8$

$4r = 8 - 3r\sin\theta$

$16r^2 = (8 - 3r\sin\theta)^2$

$16(x^2 + y^2) = (8 - 3y)^2$

$16x^2 + 16y^2 = 64 - 48y + 9y^2$

$16x^2 + 7y^2 + 48y - 64 = 0$

23. $r = \dfrac{9}{3 - 6\cos\theta}$

$3r - 6r\cos\theta = 9$

$3r = 9 + 6r\cos\theta$

$r = 3 + 2r\cos\theta$

$r^2 = (3 + 2r\cos\theta)^2$

$x^2 + y^2 = (3 + 2x)^2$

$x^2 + y^2 = 9 + 12x + 4x^2$

$3x^2 - y^2 + 12x + 9 = 0$

25. $r = \dfrac{8}{2 - \sin\theta}$

$2r - r\sin\theta = 8$

$2r = 8 + r\sin\theta$

$4r^2 = (8 + r\sin\theta)^2$

$4(x^2 + y^2) = (8 + y)^2$

$4x^2 + 4y^2 = 64 + 16y + y^2$

$4x^2 + 3y^2 - 16y - 64 = 0$

27. $r(3 - 2\sin\theta) = 6$
$$3r - 2r\sin\theta = 6$$
$$3r = 6 + 2r\sin\theta$$
$$9r^2 = (6 + 2r\sin\theta)^2$$
$$9(x^2 + y^2) = (6 + 2y)^2$$
$$9x^2 + 9y^2 = 36 + 24y + 4y^2$$
$$9x^2 + 5y^2 - 24y - 36 = 0$$

29. $r = \dfrac{6\sec\theta}{2\sec\theta - 1}$
$$r = \dfrac{6}{2 - \cos\theta}$$
$$2r - r\cos\theta = 6$$
$$2r = 6 + r\cos\theta$$
$$4r^2 = (6 + r\cos\theta)^2$$
$$4(x^2 + y^2) = (6 + x)^2$$
$$4x^2 + 4y^2 = 36 + 12x + x^2$$
$$3x^2 + 4y^2 - 12x - 36 = 0$$

31. $r = \dfrac{ep}{1 + e\sin\theta}$
$e = 1; \quad p = 1$
$$r = \dfrac{1}{1 + \sin\theta}$$

33. $r = \dfrac{ep}{1 - e\cos\theta}$
$e = \dfrac{4}{5}; \quad p = 3$
$$r = \dfrac{\left(\dfrac{12}{5}\right)}{\left(1 - \dfrac{4}{5}\cos\theta\right)} = \dfrac{12}{5 - 4\cos\theta}$$

35. $r = \dfrac{ep}{1 - e\sin\theta}$
$e = 6; \quad p = 2$
$$r = \dfrac{12}{1 - 6\sin\theta}$$

37. $d(F, P) = e \cdot d(D, P)$ $d(D, P) = p - r\cos\theta$
$$r = e(p - r\cos\theta)$$
$$r = ep - er\cos\theta$$
$$r + er\cos\theta = ep$$
$$r(1 + e\cos\theta) = ep$$
$$r = \dfrac{ep}{1 + e\cos\theta}$$

39. $d(F, P) = e \cdot d(D, P)$ $d(D, P) = p + r\sin\theta$
$$r = e(p + r\sin\theta)$$
$$r = ep + er\sin\theta$$
$$r - er\sin\theta = ep$$
$$r(1 - e\sin\theta) = ep$$
$$r = \dfrac{ep}{1 - e\sin\theta}$$

Chapter 9

Analytic Geometry

9.7 Plane Curves and Parametric Equations

1. $x = 3t + 2$, $y = t + 1$, $0 \le t \le 4$

$x = 3(y - 1) + 2$
$x = 3y - 3 + 2$
$x = 3y - 1$
$x - 3y + 1 = 0$

3. $x = t + 2$, $y = \sqrt{t}$, $t \ge 0$
$y = \sqrt{x - 2}$

5. $x = t^2 + 4$, $y = t^2 - 4$, $-\infty < t < \infty$
$y = (x - 4) - 4$
$y = x - 8$

For $-\infty < t < 0$ the movement is to the left. For $0 < t < \infty$ the movement is to the right.

7. $x = 3t^2$, $y = t + 1$, $-\infty < t < \infty$
$x = 3(y - 1)^2$

9. $x = 2e^t$, $y = 1 + e^t$, $t \ge 0$
$y = 1 + \dfrac{x}{2}$
$2y = 2 + x$

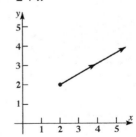

11. $x = \sqrt{t}$, $y = t^{3/2}$, $t \ge 0$

$y = (x^2)^{3/2}$

$y = x^3$

13. $x = 2\cos t$, $y = 3\sin t$, $0 \le t \le 2\pi$

$\dfrac{x}{2} = \cos t$ $\dfrac{y}{3} = \sin t$

$\left(\dfrac{x}{2}\right)^2 + \left(\dfrac{y}{3}\right)^2 = \cos^2 t + \sin^2 t = 1$

$\dfrac{x^2}{4} + \dfrac{y^2}{9} = 1$

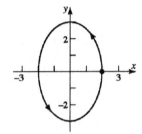

15. $x = 2\cos t$, $y = 3\sin t$, $-\pi \le t \le 0$

$\dfrac{x}{2} = \cos t$ $\dfrac{y}{3} = \sin t$

$\left(\dfrac{x}{2}\right)^2 + \left(\dfrac{y}{3}\right)^2 = \cos^2 t + \sin^2 t = 1$

$\dfrac{x^2}{4} + \dfrac{y^2}{9} = 1$

17. $x = \sec t$, $y = \tan t$, $0 \le t \le \dfrac{\pi}{4}$

$\sec^2 t = 1 + \tan^2 t$

$x^2 = 1 + y^2$

$x^2 - y^2 = 1$

$(\sqrt{2}, 1)$

19. $x = \sin^2 t$, $y = \cos^2 t$, $0 \le t \le 2\pi$

$\sin^2 t + \cos^2 t = 1$

$x + y = 1$

back and
forth twice

21. (a) Use equation (2):
$$x = 50\cos(90°) = 0$$

$$y = -\frac{1}{2}(32)\,t^2 + (50\sin(90°))\,t + 6 = -16t^2 + 50t + 6$$

(b) The ball is in the air until $y = 0$. Solve:
$$-16t^2 + 50t + 6 = 0$$

$$t = \frac{-50 \pm \sqrt{50^2 - 4(-16)(6)}}{2(-16)} = \frac{-50 \pm \sqrt{2884}}{-32} \approx -0.12 \text{ or } 3.24$$

The ball is in the air for about 3.24 seconds. (The negative solution is extraneous.)

(c) The maximum height occurs at the vertex of the quadratic function.
$$t = \frac{-b}{2a} = \frac{-50}{2(-16)} = 1.5625 \text{ seconds}$$

Evaluate the function to find the maximum height:
$$-16(1.5625)^2 + 50(1.5625) + 6 = 45.0625$$

The maximum height is 45.0625 feet.

(d) We use $x = 3$ so that the line is not on top of the y-axis.

23. (a) Train: Use equation (2) with $g = 2$, $v_0 = 0$, $h = 0$
$$x_1 = \frac{1}{2}(2)t^2 + 0 \cdot t + 0 = t^2$$

$$y_1 = 1$$

Bill:
$$x_2 = 5(t - 5)$$

$$y_2 = 3$$

(b) Bill will catch the train if $x_1 = x_2$.

$$t^2 = 5(t - 5) \rightarrow t^2 = 5t - 25 \rightarrow t^2 - 5t + 25 = 0$$

Since $b^2 - 4ac = (-5)^2 - 4(1)(25) = 25 - 100 = -75 < 0$, the equation has no real solution. Thus, Bill will not catch the train.

(c)

473

25. (a) Use equation (2):

$$x = (145\cos(20°))\,t$$

$$y = -\frac{1}{2}(32)\,t^2 + (145\sin(20°))\,t + 5$$

(b) The ball is in the air until $y = 0$. Solve:

$$-16t^2 + (145\sin 20°)t + 5 = 0$$

$$t = \frac{-145\sin(20°) \pm \sqrt{(145\sin(20°))^2 - 4(-16)(5)}}{2(-16)}$$

$$\approx \frac{-49.59 \pm \sqrt{2779.46}}{-32} \approx -0.10 \text{ or } 3.20$$

The ball is in the air for about 3.20 seconds. (The negative solution is extraneous.)

(c) The maximum height occurs at the vertex of the quadratic function.

$$t = \frac{-b}{2a} = \frac{-145\sin(20°)}{2(-16)} \approx 1.55 \text{ seconds}$$

Evaluate the function to find the maximum height:

$$-16(1.55)^2 + 145\sin(20°)(1.55) + 5 = 43.43$$

The maximum height is about 43.43 feet.

(d) Find the horizontal displacement:

$$x = (145\cos(20°))(3.20) \approx 436 \text{ feet}$$

(e)

27. (a) Use equation (2):

$$x = (40\cos(45°))\,t$$

$$y = -\frac{1}{2}(9.8)t^2 + (40\sin(45°))t + 300$$

(b) The ball is in the air until $y = 0$. Solve:

$$-4.9t^2 + (40\sin(45°))\,t + 300 = 0$$

$$t = \frac{-20\sqrt{2} \pm \sqrt{(20\sqrt{2})^2 - 4(-4.9)(300)}}{2(-4.9)} = \frac{-20\sqrt{2} \pm \sqrt{6680}}{-9.8} \approx -5.45 \text{ or } 11.23$$

The ball is in the air for about 11.23 seconds. (The negative solution is extraneous.)

(c) The maximum height occurs at the vertex of the quadratic function.

$$t = \frac{-b}{2a} = \frac{-20\sqrt{2}}{2(-4.9)} \approx 2.89 \text{ seconds}$$

Evaluate the function to find the maximum height:

$$-4.9(2.89)^2 + 20\sqrt{2}(2.89) + 300 = 340.8 \text{ meters}$$

(d)   Find the horizontal displacement:
$$x = (40\cos(45°))(11.23) \approx 317.6 \text{ meters}$$

(e)

29.   (a)   At $t = 0$, the Paseo is 5 miles from the
intersection (at (0, 0)) traveling east (along the
x-axis) at 40 mph. Thus, $x = 40t - 5$
describes the position of the Paseo as a
function of time. The Bonneville, at $t = 0$, is
4 miles from the intersection traveling north
(along the y-axis) at 30 mph. Thus,
$y = 30t - 4$ describes the position of the
Bonneville as a function of time.

Let d represent the distance between the cars. Use the Pythagorean Theorem to find
the distance:
$$d = \sqrt{(40t - 5)^2 + (30t - 4)^2}$$

(b)   From part (a):

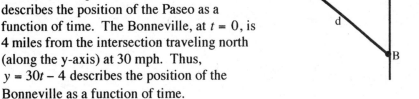

Paseo:	$x = 40t - 5$	Bonneville:	$x = 0$
	$y = 0$		$y = 30t - 4$

(c)   Note this is a function graph not a parametric graph.

(d)   The minimum distance between the cars is 0.2 miles and occurs at 0.128 seconds.

(e)

31. $x = t$, $y = 4t - 1$ $x = t + 1$, $y = 4t + 3$

33. $x = t$, $y = t^2 + 1$ $x = t - 1$, $y = t^2 - 2t + 2$

35. $x = t, \ y = t^3$ $x = \sqrt[3]{t}, \ y = t$

37. $x = t^{3/2}, \ y = t$ $x = t, \ y = t^{2/3}$

39. $x = t + 2, \ y = t; \ \ 0 \le t \le 5$

41. $x = 3\cos t, \ y = 2\sin t; \ \ 0 \le t \le 2\pi$

43. $x = 2\cos(\omega t), \ y = -3\sin(\omega t)$

$\dfrac{2\pi}{\omega} = 2 \ \rightarrow \ \omega = \pi$

$x = 2\cos(\pi t), \ y = -3\sin(\pi t), \ \ 0 \le t \le 2$

45. $x = -2\sin(\omega)t, \ y = 3\cos(\omega t)$

$\dfrac{2\pi}{\omega} = 1 \ \rightarrow \ \omega = 2\pi$

$x = -2\sin(2\pi t), \ y = 3\cos(2\pi t), \ \ 0 \le t \le 1$

47. C_1

C_2

C_3

C_4

49. $x = (x_2 - x_1)t + x_1, \quad y = (y_2 - y_1)t + y_1, \ \ -\infty < t < \infty$

$\dfrac{x - x_1}{x_2 - x_1} = t$

$y = (y_2 - y_1)\left(\dfrac{x - x_1}{x_2 - x_1}\right) + y_1 \rightarrow y - y_1 = \left(\dfrac{y_2 - y_1}{x_2 - x_1}\right)(x - x_1)$

This is the two point form for the equation of a line.
Its orientation is from (x_1, y_1) to (x_2, y_2).

51. $x = t\sin t$, $y = t\cos t$

53. $x = 4\sin t - 2\sin(2t)$
     $y = 4\cos t - 2\cos(2t)$

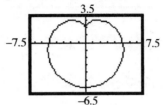

55. (a) $x(t) = \cos^3 t$, $y(t) = \sin^3 t$, $0 \le t \le 2\pi$

$$\cos t = x^{2/3}, \quad \sin t = y^{2/3}$$

(b) $\cos t = x^{2/3}$, $\sin t = y^{2/3}$

$$\cos^2 t + \sin^2 t = \left(x^{1/3}\right)^2 + \left(y^{1/3}\right)^2$$

$$x^{2/3} + y^{2/3} = 1$$

Chapter 9

Analytic Geometry

9.R   Chapter Review

1. $y^2 = -16x$
   This is a parabola.
   $a = 4$
   Vertex: $(0, 0)$
   Focus: $(-4, 0)$
   Directrix: $x = 4$

3. $\dfrac{x^2}{25} - y^2 = 1$
   This is a hyperbola.
   $a = 5, \ b = 1$.
   Find the value of c:
   $c^2 = a^2 + b^2 = 25 + 1 = 26$
   $c = \sqrt{26}$
   Center: $(0, 0)$
   Vertices: $(5, 0), (-5, 0)$
   Foci: $\left(\sqrt{26}, 0\right), \left(-\sqrt{26}, 0\right)$
   Asymptotes: $y = \dfrac{1}{5}x; \ \ y = -\dfrac{1}{5}x$

5. $\dfrac{y^2}{25} + \dfrac{x^2}{16} = 1$
   This is an ellipse.
   $a = 5, \ b = 4$.

   Find the value of c:
   $c^2 = a^2 - b^2 = 25 - 16 = 9$
   $c = 3$
   Center: $(0, 0)$
   Vertices: $(0, 5), (0, -5)$
   Foci: $(0, 3), (0, -3)$

7. $x^2 + 4y = 4$
   This is a parabola.
   Write in standard form:
   $x^2 = -4y + 4$
   $x^2 = -4(y - 1)$

   $a = 1$
   Vertex: $(0, 1)$
   Focus: $(0, 0)$
   Directrix: $y = 2$

9. $4x^2 - y^2 = 8$
   This is a hyperbola.
   Write in standard form:
   $\dfrac{x^2}{2} - \dfrac{y^2}{8} = 1$
   $a = \sqrt{2}, \ b = \sqrt{8} = 2\sqrt{2}$.

   Find the value of c:
   $c^2 = a^2 + b^2 = 2 + 8 = 10$
   $c = \sqrt{10}$
   Center: $(0, 0)$
   Vertices: $\left(-\sqrt{2}, 0\right), \left(\sqrt{2}, 0\right)$
   Foci: $\left(-\sqrt{10}, 0\right), \left(\sqrt{10}, 0\right)$
   Asymptotes: $y = 2x; \ \ y = -2x$

11. $x^2 - 4x = 2y$
This is a parabola.
Write in standard form:
$$x^2 - 4x + 4 = 2y + 4$$
$$(x-2)^2 = 2(y+2)$$

$a = \dfrac{1}{2}$

Vertex: $(2, -2)$

Focus: $\left(2, -\dfrac{3}{2}\right)$

Directrix: $y = -\dfrac{5}{2}$

13. $y^2 - 4y - 4x^2 + 8x = 4$
This is a hyperbola.
Write in standard form:
$$(y^2 - 4y + 4) - 4(x^2 - 2x + 1) = 4 + 4 - 4$$
$$(y-2)^2 - 4(x-1)^2 = 4$$
$$\frac{(y-2)^2}{4} - \frac{(x-1)^2}{1} = 1$$
$a = 2, \; b = 1.$
Find the value of c:

$c^2 = a^2 + b^2 = 4 + 1 = 5$

$c = \sqrt{5}$

Center: $(1, 2)$
Vertices: $(1, 0), (1, 4)$
Foci: $\left(1, 2 - \sqrt{5}\right), \left(1, 2 + \sqrt{5}\right)$
Asymptotes:
$y - 2 = 2(x - 1); \;\; y - 2 = -2(x - 1)$

15. $4x^2 + 9y^2 - 16x - 18y = 11$
This is an ellipse.
Write in standard form:
$$4x^2 + 9y^2 - 16x - 18y = 11$$
$$4(x^2 - 4x + 4) + 9(y^2 - 2y + 1) = 11 + 16 + 9$$
$$4(x-2)^2 + 9(y-1)^2 = 36$$
$$\frac{(x-2)^2}{9} + \frac{(y-1)^2}{4} = 1$$
$a = 3, \; b = 2.$

Find the value of c:
$$c^2 = a^2 - b^2 = 9 - 4 = 5$$
$$c = \sqrt{5}$$
Center: $(2, 1)$
Vertices: $(-1, 1), (5, 1)$
Foci: $\left(2 - \sqrt{5}, 1\right), \left(2 + \sqrt{5}, 1\right)$

17. $4x^2 - 16x + 16y + 32 = 0$
This is a parabola.
Write in standard form:
$$4(x^2 - 4x + 4) = -16y - 32 + 16$$
$$4(x-2)^2 = -16(y+1)$$
$$(x-2)^2 = -4(y+1)$$

$a = -1$
Vertex: $(2, -1)$
Focus: $(2, -2)$
Directrix: $y = 0$

19. $9x^2 + 4y^2 - 18x + 8y = 23$
This is an ellipse.
Write in standard form:
$$9(x^2 - 2x + 1) + 4(y^2 + 2y + 1) = 23 + 9 + 4$$
$$9(x-1)^2 + 4(y+1)^2 = 36$$
$$\frac{(x-1)^2}{4} + \frac{(y+1)^2}{9} = 1$$
$a = 3, \; b = 2.$

Find the value of c:
$$c^2 = a^2 - b^2 = 9 - 4 = 5$$
$$c = \sqrt{5}$$
Center: $(1, -1)$
Vertices: $(1, -4), (1, 2)$
Foci: $\left(1, -1 - \sqrt{5}\right), \left(1, -1 + \sqrt{5}\right)$

21. Parabola: The focus is $(-2, 0)$ and the directrix is
     $x = 2$. The vertex is $(0, 0)$. $a = 2$ and since $(-2, 0)$
     is to the left of $(0, 0)$, the parabola opens to the left.
     The equation of the parabola is:

     $$y^2 = -4ax$$
     $$y^2 = -4 \cdot 2 \cdot x$$
     $$y^2 = -8x$$

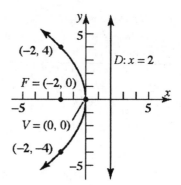

23. Hyperbola: Center: $(0, 0)$;
     Focus: $(0, 4)$;   Vertex: $(0, -2)$;
     Transverse axis is the y-axis;
     $a = 2;\ c = 4$.
     Find b:
     $$b^2 = c^2 - a^2 = 16 - 4 = 12$$
     $$b = \sqrt{12} = 2\sqrt{3}$$
     Write the equation: $\dfrac{y^2}{4} - \dfrac{x^2}{12} = 1$

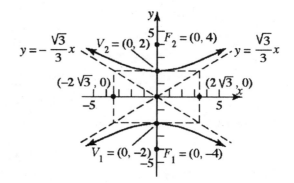

25. Ellipse: Foci: $(-3, 0)$, $(3, 0)$; Vertex: $(4, 0)$;
     Center: $(0, 0)$; Major axis is the x-axis;
     $a = 4;\ c = 3$. Find b:
     $$b^2 = a^2 - c^2 = 16 - 9 = 7$$
     $$b = \sqrt{7}$$
     Write the equation: $\dfrac{x^2}{16} + \dfrac{y^2}{7} = 1$

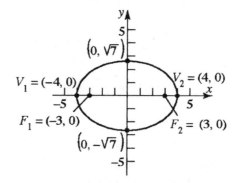

27. Parabola: The focus is $(2, -4)$ and the vertex is
     $(2, -3)$. Both lie on the vertical line $x = 2$. $a = 1$
     and since $(2, -4)$ is below $(2, -3)$, the parabola
     opens down. The equation of the parabola is:
     $$(x - h)^2 = -4a(y - k)$$
     $$(x - 2)^2 = -4 \cdot 1 \cdot (y - (-3))$$
     $$(x - 2)^2 = -4(y + 3)$$

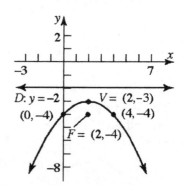

29. Hyperbola: Center: $(-2, -3)$; Focus: $(-4, -3)$;
    Vertex: $(-3, -3)$; Transverse axis is parallel to the
    x-axis; $a = 1$; $c = 2$. Find b:
    $$b^2 = c^2 - a^2 = 4 - 1 = 3$$
    $$b = \sqrt{3}$$
    Write the equation: $\dfrac{(x+2)^2}{1} - \dfrac{(y+3)^2}{3} = 1$

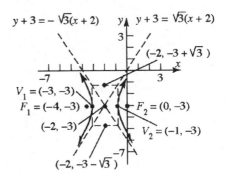

31. Ellipse: Foci: $(-4, 2)$, $(-4, 8)$; Vertex: $(-4, 10)$;
    Center: $(-4, 5)$; Major axis is parallel to the y-axis;
    $a = 5$; $c = 3$. Find b:
    $$b^2 = a^2 - c^2 = 25 - 9 = 16$$
    $$b = 4$$
    Write the equation: $\dfrac{(x+4)^2}{16} + \dfrac{(y-5)^2}{25} = 1$

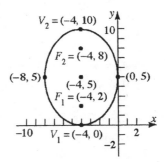

33. Hyperbola: Center: $(-1, 2)$;
    $a = 3$; $c = 4$; Transverse axis parallel to
    the x-axis;
    Find b:
    $$b^2 = c^2 - a^2 = 16 - 9 = 7$$
    $$b = \sqrt{7}$$
    Write the equation:
    $\dfrac{(x+1)^2}{9} - \dfrac{(y-2)^2}{7} = 1$

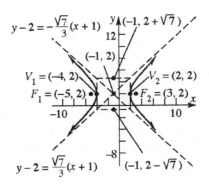

35. Hyperbola:
    Vertices: $(0, 1)$, $(6, 1)$;
    Asymptote: $3y + 2x - 9 = 0$;
    Center: $(3, 1)$;

    Transverse axis is parallel to the x- axis; The slope of the asymptote is $-\dfrac{2}{3}$.

    Find b:
    $$\frac{-b}{a} = \frac{-b}{3} = \frac{-2}{3}$$
    $$-3b = -6$$
    $$b = 2$$

Write the equation: $\dfrac{(x-3)^2}{9} - \dfrac{(y-1)^2}{4} = 1$

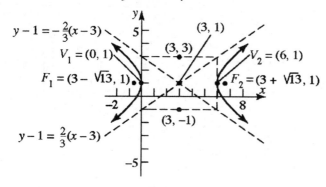

37. $y^2 + 4x + 3y - 8 = 0$
    $A = 0$ and $C = 1$; $AC = (0)(1) = 0$. Since $AC = 0$, the equation defines a parabola.

39. $x^2 + 2y^2 + 4x - 8y + 2 = 0$
    $A = 1$ and $C = 2$; $AC = (1)(2) = 2$. Since $AC > 0$ and $A \neq C$, the equation defines an ellipse.

41. $9x^2 - 12xy + 4y^2 + 8x + 12y = 0$
    $A = 9$, $B = -12$, $C = 4$ $B^2 - 4AC = (-12)^2 - 4(9)(4) = 0$; parabola

43. $4x^2 + 10xy + 4y^2 - 9 = 0$
    $A = 4$, $B = 10$, $C = 4$ $B^2 - 4AC = 10^2 - 4(4)(4) = 36 > 0$; hyperbola

45. $x^2 - 2xy + 3y^2 + 2x + 4y - 1 = 0$
    $A = 1$, $B = -2$, $C = 3$ $B^2 - 4AC = (-2)^2 - 4(1)(3) = -8 < 0$; ellipse

47. $2x^2 + 5xy + 2y^2 - \dfrac{9}{2} = 0$

$A = 2, B = 5,$ and $C = 2$; $\cot(2\theta) = \dfrac{A - C}{B} = \dfrac{2-2}{5} = 0 \;\rightarrow\; 2\theta = \dfrac{\pi}{2} \;\rightarrow\; \theta = \dfrac{\pi}{4}$

$x = x'\cos\theta - y'\sin\theta = \dfrac{\sqrt{2}}{2}x' - \dfrac{\sqrt{2}}{2}y' = \dfrac{\sqrt{2}}{2}(x' - y')$

$y = x'\sin\theta + y'\cos\theta = \dfrac{\sqrt{2}}{2}x' + \dfrac{\sqrt{2}}{2}y' = \dfrac{\sqrt{2}}{2}(x' + y')$

$2\left(\dfrac{\sqrt{2}}{2}(x' - y')\right)^2 + 5\left(\dfrac{\sqrt{2}}{2}(x' - y')\right)\left(\dfrac{\sqrt{2}}{2}(x' + y')\right) + 2\left(\dfrac{\sqrt{2}}{2}(x' + y')\right)^2 - \dfrac{9}{2} = 0$

$\left(x'^2 - 2x'y' + y'^2\right) + \dfrac{5}{2}\left(x'^{2} - y'^2\right) + \left(x'^2 + 2x'y' + y'^2\right) - \dfrac{9}{2} = 0$

$\dfrac{9}{2}x'^2 - \dfrac{1}{2}y'^2 = \dfrac{9}{2} \rightarrow 9x'^2 - y'^2 = 9 \rightarrow \dfrac{x'^2}{1} - \dfrac{y'^2}{9} = 1$

Hyperbola; center at (0, 0), transverse axis is the x'-axis, vertices at (±1, 0).

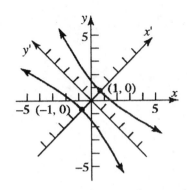

49. $6x^2 + 4xy + 9y^2 - 20 = 0$

$A = 6, B = 4,$ and $C = 9;$ $\cot(2\theta) = \dfrac{A - C}{B} = \dfrac{6 - 9}{4} = -\dfrac{3}{4}$ $\rightarrow$ $\cos(2\theta) = -\dfrac{3}{5}$

$\sin\theta = \sqrt{\dfrac{\left(1 - \left(-\dfrac{3}{5}\right)\right)}{2}} = \sqrt{\dfrac{4}{5}} = \dfrac{2\sqrt{5}}{5};$ $\cos\theta = \sqrt{\dfrac{\left(1 + \left(-\dfrac{3}{5}\right)\right)}{2}} = \sqrt{\dfrac{1}{5}} = \dfrac{\sqrt{5}}{5}$ $\rightarrow$ $\theta \approx 63.4°$

$x = x'\cos\theta - y'\sin\theta = \dfrac{\sqrt{5}}{5}x' - \dfrac{2\sqrt{5}}{5}y' = \dfrac{\sqrt{5}}{5}(x' - 2y')$

$y = x'\sin\theta + y'\cos\theta = \dfrac{2\sqrt{5}}{5}x' + \dfrac{\sqrt{5}}{5}y' = \dfrac{\sqrt{5}}{5}(2x' + y')$

$6\left(\dfrac{\sqrt{5}}{5}(x' - 2y')\right)^2 + 4\left(\dfrac{\sqrt{5}}{5}(x' - 2y')\right)\left(\dfrac{\sqrt{5}}{5}(2x' + y')\right) + 9\left(\dfrac{\sqrt{5}}{5}(2x' + y')\right)^2 - 20 = 0$

$\dfrac{6}{5}\left(x'^2 - 4x'y' + 4y'^2\right) + \dfrac{4}{5}\left(2x'^2 - 3x'y' - 2y'^2\right) + \dfrac{9}{5}\left(4x'^2 + 4x'y' + y'^2\right) - 20 = 0$

$\dfrac{6}{5}x'^2 - \dfrac{24}{5}x'y' + \dfrac{24}{5}y'^2 + \dfrac{8}{5}x'^2 - \dfrac{12}{5}x'y' - \dfrac{8}{5}y'^2 + \dfrac{36}{5}x'^2 + \dfrac{36}{5}x'y' + \dfrac{9}{5}y'^2 = 20$

$10x'^2 + 5y'^2 = 20 \rightarrow \dfrac{x'^2}{2} + \dfrac{y'^2}{4} = 1$

Ellipse; center at the origin, major axis is the y'-axis, vertices at (0, ±2).

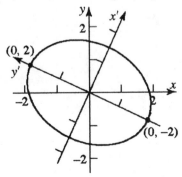

51. $4x^2 - 12xy + 9y^2 + 12x + 8y = 0$

$A = 4, B = -12,$ and $C = 9$; $\cot(2\theta) = \dfrac{A-C}{B} = \dfrac{4-9}{-12} = \dfrac{5}{12}$ $\rightarrow$ $\cos(2\theta) = \dfrac{5}{13}$

$\sin\theta = \sqrt{\dfrac{\left(1 - \dfrac{5}{13}\right)}{2}} = \sqrt{\dfrac{4}{13}} = \dfrac{2\sqrt{13}}{13}$; $\cos\theta = \sqrt{\dfrac{\left(1 + \dfrac{5}{13}\right)}{2}} = \sqrt{\dfrac{9}{13}} = \dfrac{3\sqrt{13}}{13}$ $\rightarrow$ $\theta \approx 33.7°$

$x = x'\cos\theta - y'\sin\theta = \dfrac{3\sqrt{13}}{13}x' - \dfrac{2\sqrt{13}}{13}y' = \dfrac{\sqrt{13}}{13}(3x' - 2y')$

$y = x'\sin\theta + y'\cos\theta = \dfrac{2\sqrt{13}}{13}x' + \dfrac{3\sqrt{13}}{13}y' = \dfrac{\sqrt{13}}{13}(2x' + 3y')$

$4\left(\dfrac{\sqrt{13}}{13}(3x' - 2y')\right)^2 - 12\left(\dfrac{\sqrt{13}}{13}(3x' - 2y')\right)\left(\dfrac{\sqrt{13}}{13}(2x' + 3y')\right) + 9\left(\dfrac{\sqrt{13}}{13}(2x' + 3y')\right)^2$

$+ 12\left(\dfrac{\sqrt{13}}{13}(3x' - 2y')\right) + 8\left(\dfrac{\sqrt{13}}{13}(2x' + 3y')\right) = 0$

$\dfrac{4}{13}\left(9x'^2 - 12x'y' + 4y'^2\right) - \dfrac{12}{13}\left(6x'^2 + 5x'y' - 6y'^2\right) + \dfrac{9}{13}\left(4x'^2 + 12x'y' + 9y'^2\right)$

$+ \dfrac{36\sqrt{13}}{13}x' - \dfrac{24\sqrt{13}}{13}y' + \dfrac{16\sqrt{13}}{13}x' + \dfrac{24\sqrt{13}}{13}y' = 0$

$\dfrac{36}{13}x'^2 - \dfrac{48}{13}x'y' + \dfrac{16}{13}y'^2 - \dfrac{72}{13}x'^2 - \dfrac{60}{13}x'y' + \dfrac{72}{13}y'^2$

$+ \dfrac{36}{13}x'^2 + \dfrac{108}{13}x'y' + \dfrac{81}{13}y'^2 + 4\sqrt{13}x' = 0$

$13y'^2 + 4\sqrt{13}x' = 0$

$y'^2 = -\dfrac{4\sqrt{13}}{13}x'$

Parabola; vertex at the origin,

focus at $\left(-\dfrac{\sqrt{13}}{13}, 0\right)$.

53. $r = \dfrac{4}{1 - \cos\theta}$

$ep = 4,\ e = 1,\ p = 4$

Parabola; directrix is perpendicular to the
polar axis 4 units to the left of the pole;
vertex is $(2, \pi)$.

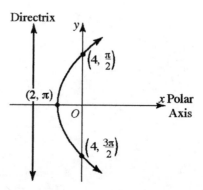

55. $r = \dfrac{6}{2 - \sin\theta} = \dfrac{3}{\left(1 - \frac{1}{2}\sin\theta\right)}$

$ep = 3, \quad e = \dfrac{1}{2}, \quad p = 6$

Ellipse; directrix is parallel to the polar axis 6 units below the pole; vertices are

$\left(6, \dfrac{\pi}{2}\right)$ and $\left(2, \dfrac{3\pi}{2}\right)$.

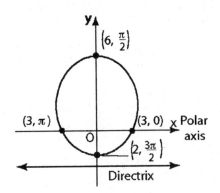

57. $r = \dfrac{8}{4 + 8\cos\theta} = \dfrac{2}{1 + 2\cos\theta}$

$ep = 2, \quad e = 2, \quad p = 1$

Hyperbola; directrix is perpendicular to the polar axis 1 unit to the right of the pole;

vertices are $\left(\dfrac{2}{3}, 0\right)$ and $(-2, \pi)$.

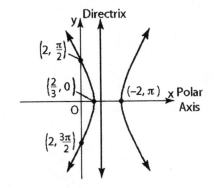

59. $r = \dfrac{4}{1 - \cos\theta}$

$r - r\cos\theta = 4$

$\qquad r = 4 + r\cos\theta$

$\qquad r^2 = (4 + r\cos\theta)^2$

$\qquad x^2 + y^2 = (4 + x)^2$

$\qquad x^2 + y^2 = 16 + 8x + x^2$

$\quad y^2 - 8x - 16 = 0$

61. $r = \dfrac{8}{4 + 8\cos\theta}$

$4r + 8r\cos\theta = 8$

$\qquad 4r = 8 - 8r\cos\theta$

$\qquad\quad r = 2 - 2r\cos\theta$

$\qquad\quad r^2 = (2 - 2r\cos\theta)^2$

$\qquad x^2 + y^2 = (2 - 2x)^2$

$\qquad x^2 + y^2 = 4 - 8x + 4x^2$

$\quad 3x^2 - y^2 - 8x + 4 = 0$

63. $x = 4t - 2, \quad y = 1 - t, \quad -\infty < t < \infty$

$x = 4(1 - y) - 2$

$x = 4 - 4y - 2$

$x + 4y = 2$

65. $x = 3\sin t,\ y = 4\cos t + 2,\ 0 \le t \le 2\pi$

$$\frac{x}{3} = \sin t, \quad \frac{y-2}{4} = \cos t$$

$$\sin^2 t + \cos^2 t = 1$$

$$\left(\frac{x}{3}\right)^2 + \left(\frac{y-2}{4}\right)^2 = 1$$

$$\frac{x^2}{9} + \frac{(y-2)^2}{16} = 1$$

67. $x = \sec^2 t,\ y = \tan^2 t,\ 0 \le t \le \dfrac{\pi}{4}$

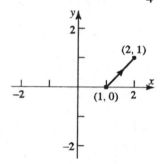

$$\tan^2 t + 1 = \sec^2 t \rightarrow y + 1 = x$$

69.   Write the equation in standard form:

$$4x^2 + 9y^2 = 36 \ \rightarrow\ \frac{x^2}{9} + \frac{y^2}{4} = 1$$

The center of the ellipse is (0, 0). The major axis is the x-axis.
$a = 3;\ b = 2;\ c^2 = a^2 - b^2 = 9 - 4 = 5 \ \rightarrow\ c = \sqrt{5}$.
For the ellipse:
     Vertices: (−3, 0), (3, 0)
     Foci: $\left(-\sqrt{5}, 0\right), \left(\sqrt{5}, 0\right)$
For the hyperbola:
     Foci: (−3, 0), (3, 0)
     Vertices: $\left(-\sqrt{5}, 0\right), \left(\sqrt{5}, 0\right)$
     Center: (0, 0)
     $a = \sqrt{5};\ c = 3;\ b^2 = c^2 - a^2 = 9 - 5 = 4 \ \rightarrow\ b = 2$

The equation of the hyperbola is: $\dfrac{x^2}{5} - \dfrac{y^2}{4} = 1$

71.   Let (x, y) be any point in the collection of points.

The distance from (x, y) to (3, 0) $= \sqrt{(x-3)^2 + y^2}$.

The distance from (x, y) to the line $x = \dfrac{16}{3}$ is $\left| x - \dfrac{16}{3} \right|$.

Relating the distances, we have:

$$\sqrt{(x-3)^2+y^2}=\frac{3}{4}\left|x-\frac{16}{3}\right| \rightarrow (x-3)^2+y^2=\frac{9}{16}\left(x-\frac{16}{3}\right)^2$$

$$x^2-6x+9+y^2=\frac{9}{16}\left(x^2-\frac{32}{3}x+\frac{256}{9}\right)$$

$$16x^2-96x+144+16y^2=9x^2-96x+256$$

$$7x^2+16y^2=112 \rightarrow \frac{7x^2}{112}+\frac{16y^2}{112}=1 \rightarrow \frac{x^2}{16}+\frac{y^2}{7}=1$$

The set of points is an ellipse.

73. Locate the parabola so that the vertex is at $(0, 0)$ and opens up. It then has the equation: $x^2 = 4ay$. Since the light source is located at the focus and is 1 foot from the base, $a = 1$. The diameter is 2, so the point $(1, y)$ is located on the parabola. Solve for y:

$$1^2 = 4(1)y \rightarrow 1 = 4y \rightarrow y = 0.25 \text{ feet}$$

The mirror should be 0.25 feet deep or 3 inches deep.

75. Place the semielliptical arch so that the x-axis coincides with the water and the y-axis passes through the center of the arch. Since the bridge has a span of 60 feet, the length of the major axis is 60, or $2a = 60$ or $a = 30$. The maximum height of the bridge is 20 feet, so $b = 20$. The equation is: $\dfrac{x^2}{900}+\dfrac{y^2}{400}=1$.

The height 5 feet from the center:

$$\frac{5^2}{900}+\frac{y^2}{400}=1 \rightarrow \frac{y^2}{400}=1-\frac{25}{900} \rightarrow y^2=400\cdot\frac{875}{900} \rightarrow y\approx 19.72 \text{ feet}$$

The height 10 feet from the center:

$$\frac{10^2}{900}+\frac{y^2}{400}=1 \rightarrow \frac{y^2}{400}=1-\frac{100}{900} \rightarrow y^2=400\cdot\frac{800}{900} \rightarrow y\approx 18.86 \text{ feet}$$

The height 20 feet from the center:

$$\frac{20^2}{900}+\frac{y^2}{400}=1 \rightarrow \frac{y^2}{400}=1-\frac{400}{900} \rightarrow y^2=400\cdot\frac{500}{900} \rightarrow y\approx 14.91 \text{ feet}$$

77. (a) Set up a coordinate system so that the two stations lie on the x-axis and the origin is midway between them. The ship lies on a hyperbola whose foci are the locations of the two stations. Since the time difference is 0.00032 seconds and the speed of the signal is 186,000 miles per second, the difference in the distances of the ships from each station is:

distance $= (186,000)(0.00032) = 59.52$ miles

The difference of the distances from the ship to each station, 59.52, equals $2a$, so $a = 29.76$ and the vertex of the corresponding hyperbola is at $(29.76, 0)$. Since the focus is at $(75, 0)$, following this hyperbola, the ship would reach shore 45.24 miles from the master station.

(b) The ship should follow a hyperbola with a vertex at $(60, 0)$. For this hyperbola, $a = 60$, so the constant difference of the distances from the ship to each station is 120. The time difference the ship should look for is:

$$\text{time} = \frac{120}{186{,}000} = 0.000645 \text{ seconds}$$

(c) Find the equation of the hyperbola with vertex at $(60, 0)$ and a focus at $(75, 0)$. The form of the equation of the hyperbola is:

$$\frac{x^2}{a^2} - \frac{y^2}{b^2} = 1 \quad \text{where } a = 60.$$

Since $c = 75$ and $b^2 = c^2 - a^2 \;\rightarrow\; b^2 = 75^2 - 60^2 = 2025$.

The equation of the hyperbola is: $\dfrac{x^2}{3600} - \dfrac{y^2}{2025} = 1$.

Since the ship is 20 miles off shore, we have $y = 20$.

Solve the equation for x:

$$\frac{x^2}{3600} - \frac{20^2}{2025} = 1 \;\rightarrow\; \frac{x^2}{3600} = 1 + \frac{400}{2025} = \frac{97}{81} \;\rightarrow\; x^2 = 3600 \cdot \frac{97}{81}$$
$$x \approx 66 \text{ miles}$$

The ship's location is $(66, 20)$.

79. (a) Use equation (2):

$$x = (100\cos(35°))t; \qquad y = -\frac{1}{2}(32)t^2 + (100\sin(35°))t + 6$$

(b) The ball is in the air until $y = 0$. Solve:

$$-16t^2 + (100\sin(35°))t + 6 = 0$$

$$t = \frac{-100\sin(35°) \pm \sqrt{(100\sin(35°))^2 - 4(-16)(6)}}{2(-16)} \approx \frac{-57.36 \pm \sqrt{3673.9}}{-32} \approx -0.10 \text{ or } 3.69$$

The ball is in the air for about 3.69 seconds. (The negative solution is extraneous.)

(c) The maximum height occurs at the vertex of the quadratic function.

$$t = \frac{-b}{2a} = \frac{-100\sin(35°)}{2(-16)} \approx 1.79 \text{ seconds}$$

Evaluate the function to find the maximum height:

$$-16(1.79)^2 + 100\sin(35°)(1.79) + 6 \approx 57.4 \text{ feet}$$

(d) Find the horizontal displacement:

$$x = (100\cos(35°))(3.69) \approx 302 \text{ feet}$$

(e)

488

Chapter 10

Systems of Equations and Inequalities

10.1 Systems of Linear Equations: Two Equations Containing Two Variables

1. Substituting the values of the variables:
$$\begin{cases} 2x - y = 5 \\ 5x + 2y = 8 \end{cases} \begin{array}{l} \rightarrow \quad 2(2) - (-1) = 4 + 1 = 5 \\ \rightarrow \quad 5(2) + 2(-1) = 10 - 2 = 8 \end{array}$$
Each equation is satisfied, so $x = 2$, $y = -1$ is a solution to the system of equations.

3. Substituting the values of the variables:
$$\begin{cases} 3x - 4y = 4 \quad \rightarrow \quad 3(2) - 4\left(\dfrac{1}{2}\right) = 6 - 2 = 4 \\[4mm] \dfrac{1}{2}x - 3y = -\dfrac{1}{2} \quad \rightarrow \quad \dfrac{1}{2}(2) - 3\left(\dfrac{1}{2}\right) = 1 - \dfrac{3}{2} = -\dfrac{1}{2} \end{cases}$$
Each equation is satisfied, so $x = 2, y = \dfrac{1}{2}$ is a solution to the system of equations.

5. Substituting the values of the variables:
$$\begin{cases} x^2 - y^2 = 3 \quad \rightarrow \quad 2^2 - 1^2 = 4 - 1 = 3 \\ xy = 2 \quad \rightarrow \quad (2)(1) = 2 \end{cases}$$
Each equation is satisfied, so $x = 2, y = 1$ is a solution to the system of equations.

7. Substituting the values of the variables:
$$\begin{cases} \dfrac{x}{1+x} + 3y = 6 \quad \rightarrow \quad \dfrac{0}{1+0} + 3(2) = 0 + 6 = 6 \\[4mm] x + 9y^2 = 36 \quad \rightarrow \quad 0 + 9(2)^2 = 0 + 9 \cdot 4 = 36 \end{cases}$$
Each equation is satisfied, so $x = 0, y = 2$ is a solution to the system.

9. Solve the first equation for y, substitute into the second equation and solve:
$$\begin{cases} x + y = 8 \quad \rightarrow \quad y = 8 - x \\ x - y = 4 \end{cases}$$
$$x - (8 - x) = 4 \rightarrow x - 8 + x = 4$$
$$2x = 12 \rightarrow x = 6$$
Since $x = 6$, $y = 8 - 6 = 2$.
The solution of the system is $x = 6$, $y = 2$.

11.   Multiply each side of the first equation by 3 and add the equations:

$$\begin{cases} 5x - y = 13 \\ 2x + 3y = 12 \end{cases} \xrightarrow{3} \begin{aligned} 15x - 3y &= 39 \\ 2x + 3y &= 12 \end{aligned}$$

$$\begin{aligned} 17x &= 51 \\ x &= 3 \end{aligned}$$

Substitute and solve for y:

$$5(3) - y = 13 \rightarrow 15 - y = 13 \rightarrow -y = -2 \rightarrow y = 2$$

The solution of the system is $x = 3, \ y = 2$.

13.   Solve the first equation for x and substitute into the second equation:

$$\begin{cases} 3x = 24 \\ x + 2y = 0 \end{cases} \rightarrow \quad x = 8$$

$$8 + 2y = 0 \rightarrow 2y = -8 \rightarrow y = -4$$

The solution of the system is $x = 8, \ y = -4$.

15.   Multiply each side of the first equation by 2 and each side of the second equation by 3 to eliminate y:

$$\begin{cases} 3x - 6y = 2 \\ 5x + 4y = 1 \end{cases} \begin{aligned} \xrightarrow{2} & \ 6x - 12y = 4 \\ \xrightarrow{3} & \ 15x + 12y = 3 \end{aligned}$$

$$\begin{aligned} 21x &= 7 \\ x &= \frac{1}{3} \end{aligned}$$

Substitute and solve for y:

$$3\left(\frac{1}{3}\right) - 6y = 2 \rightarrow 1 - 6y = 2 \rightarrow -6y = 1 \rightarrow y = -\frac{1}{6}$$

The solution of the system is $x = \frac{1}{3}, \ y = -\frac{1}{6}$.

17.   Solve the first equation for y, substitute into the second equation and solve:

$$\begin{cases} 2x + y = 1 \\ 4x + 2y = 3 \end{cases} \rightarrow \quad y = 1 - 2x$$

$$4x + 2(1 - 2x) = 3 \rightarrow 4x + 2 - 4x = 3 \rightarrow 0x = 1$$

This has no solution, so the system is inconsistent.

19.   Solve the first equation for y, substitute into the second equation and solve:

$$\begin{cases} 2x - y = 0 \\ 3x + 2y = 7 \end{cases} \rightarrow \quad 2x = y$$

$$3x + 2(2x) = 7 \rightarrow 3x + 4x = 7 \rightarrow 7x = 7 \rightarrow x = 1$$

Since $x = 1, \ y = 2(1) = 2$ The solution of the system is $x = 1, \ y = 2$.

21. Solve the first equation for x, substitute into the second equation and solve:
$$\begin{cases} x+2y=4 & \rightarrow \quad x=4-2y \\ 2x+4y=8 \end{cases}$$

$$2(4-2y)+4y=8 \rightarrow 8-4y+4y=8 \rightarrow 0y=0$$
These equations are dependent. Any real number is a solution for y.
The solution of the system is $x=4-2y$, where y is any real number.

23. Multiply each side of the first equation by –5, and add the equations to eliminate x:
$$\begin{cases} 2x-3y=-1 & \xrightarrow{-5} & -10x+15y=5 \\ 10x+\ y=11 & \longrightarrow & \underline{\ 10x+\ y=11} \end{cases}$$
$$16y=16$$
$$y=1$$

Substitute and solve for x:

$$2x-3(1)=-1 \rightarrow 2x-3=-1 \rightarrow 2x=2 \rightarrow x=1$$
The solution of the system is $x=1,\ y=1$.

25. Solve the second equation for x, substitute into the first equation and solve:
$$\begin{cases} 2x+3y=6 \\ x-\ y=\dfrac{1}{2} & \rightarrow \quad x=y+\dfrac{1}{2} \end{cases}$$
$$2\left(y+\dfrac{1}{2}\right)+3y=6 \rightarrow 2y+1+3y=6 \rightarrow 5y=5 \rightarrow y=1$$

Since $y=1,\ x=1+\dfrac{1}{2}=\dfrac{3}{2}$ The solution of the system is $x=\dfrac{3}{2},\ y=1$.

27. Multiply each side of the first equation by –6 and each side of the second equation by 12 to eliminate x:
$$\begin{cases} \dfrac{1}{2}x+\dfrac{1}{3}y=3 & \xrightarrow{-6} & -3x-2y=-18 \\ \dfrac{1}{4}x-\dfrac{2}{3}y=-1 & \xrightarrow{12} & \underline{\ 3x-8y=-12} \end{cases}$$
$$-10y=-30$$
$$y=3$$

Substitute and solve for x:

$$\dfrac{1}{2}x+\dfrac{1}{3}(3)=3$$
$$\dfrac{1}{2}x+1=3$$
$$\dfrac{1}{2}x=2$$
$$x=4$$
The solution of the system is $x=4,\ y=3$.

29. Add the equations to eliminate y and solve for x:

$$\begin{cases} 3x - 5y = 3 \\ 15x + 5y = 21 \end{cases}$$

$$\overline{18x \quad\quad = 24}$$

$$x \quad = \frac{4}{3}$$

Substitute and solve for y:

$$3\left(\frac{4}{3}\right) - 5y = 3$$

$$4 - 5y = 3$$

$$-5y = -1$$

$$y = \frac{1}{5}$$

The solution of the system is $x = \frac{4}{3},\ y = \frac{1}{5}$.

31. Rewrite letting $a = \frac{1}{x},\ b = \frac{1}{y}$:

$$\begin{cases} \dfrac{1}{x} + \dfrac{1}{y} = 8 \\ \dfrac{3}{x} - \dfrac{5}{y} = 0 \end{cases} \begin{array}{l} \longrightarrow \quad a + b = 8 \\ \\ \longrightarrow \quad 3a - 5b = 0 \end{array}$$

Solve the first equation for a, substitute into the second equation and solve:

$$\begin{cases} a + b = 8 \quad \rightarrow \quad a = 8 - b \\ 3a - 5b = 0 \end{cases}$$

$$3(8 - b) - 5b = 0 \rightarrow 24 - 3b - 5b = 0 \rightarrow -8b = -24 \rightarrow b = 3$$

Since $b = 3,\ a = 8 - 3 = 5$

Thus, $x = \frac{1}{a} = \frac{1}{5},\ y = \frac{1}{b} = \frac{1}{3}$

The solution of the system is $x = \frac{1}{5},\ y = \frac{1}{3}$.

33. Graph the two equations as y_1 and y_2, and use INTERSECT to solve:

$$\begin{cases} y_1 = \sqrt{2}x - 20\sqrt{7} \\ y_2 = -0.1x + 20 \end{cases}$$

The solution of the system is $x = 48.15,\ y = 15.18$.

35. Solve for y in each equation, graph the two equations as y_1 and y_2, and use INTERSECT to solve:

$$\begin{cases} \sqrt{2}x + \sqrt{3}y + \sqrt{6} = 0 \\ \sqrt{3}x - \sqrt{2}y + 60 = 0 \end{cases}$$

$$\begin{cases} y_1 = \dfrac{-\sqrt{2}x - \sqrt{6}}{\sqrt{3}} \\ \\ y_2 = \dfrac{\sqrt{3}x + 60}{\sqrt{2}} \end{cases}$$

The solution of the system is $x = -21.48$, $y = 16.12$.

37. Solve for y in each equation, graph the two equations as y_1 and y_2, and use INTERSECT to solve:

$$\begin{cases} \sqrt{3}x + \sqrt{2}y = \sqrt{0.3} \\ 100x - 95y = 20 \end{cases}$$

$$\begin{cases} y_1 = \dfrac{-\sqrt{3}x + \sqrt{0.3}}{\sqrt{2}} \\ \\ y_2 = \dfrac{100x - 20}{95} \end{cases}$$

The solution of the system is $x = 0.26$, $y = 0.07$.

39. Solve the system by substitution:

$$Q_s = Q_d$$

$$-200 + 50p = 1000 - 25p \rightarrow 75p = 1200 \rightarrow p = 16$$

Therefore, $Q_s = -200 + 50(16) = -200 + 800 = 600$

The equilibrium price is $16 and the equilibrium quantity is 600 T-shirts.

41. Let l be the length of the rectangle and w be the width of the rectangle. Then:

$$2l + 2w = 90$$

$$l = 2w$$

Solve by substitution:

$$2(2w) + 2w = 90$$

$$4w + 2w = 90 \rightarrow 6w = 90 \rightarrow w = 15 \text{ feet}$$

$$l = 2(15) = 30 \text{ feet}$$

The dimensions of the floor are 15 feet by 30 feet.

43. Let x = the cost of one cheeseburger and y = the cost of one shake. Then:

$$4x + 2y = 790$$

$$2y = x + 15$$

Solve by substitution:

$$4x + x + 15 = 790 \rightarrow 5x = 775 \rightarrow x = 155$$

$$2y = 155 + 15 \rightarrow 2y = 170 \rightarrow y = 85$$

A cheeseburger cost $1.55 and a shake costs $0.85.

45.   Let x = the number of pounds of cashews.
Then $x + 30$ is the number of pounds in the mixture.
The value of the cashews is $5x$.
The value of the peanuts is $1.50(30) = 45$.
The value of the mixture is $3(x + 30)$.
Setting up a value equation:
$$5x + 45 = 3(x + 30)$$
$$5x + 45 = 3x + 90$$
$$2x = 45$$
$$x = 22.5$$
22.5 pounds of cashews should be used in the mixture.

47.   Let x = the plane's air speed and y = the wind speed.

	Rate	Time	Distance
With Wind	$x + y$	3	600
Against	$x - y$	4	600

$(x + y)(3) = 600 \quad \rightarrow \quad x + y = 200$
$(x - y)(4) = 600 \quad \rightarrow \quad x - y = 150$
Solving by elimination:
$$2x = 350$$
$$x = 175$$
$$y = 200 - x = 200 - 175 = 25$$
The airspeed of the plane is 175 mph, and the wind speed is 25 mph.

49.   Let x = the number of one design.
Let y = the number of the second design.
Then $x + y$ = the total number of sets of dishes.
$25x + 45y$ = the cost of the dishes.
Setting up the equations and solving by substitution:
$$\begin{cases} x + y = 200 \quad \rightarrow \quad y = 200 - x \\ 25x + 45y = 7400 \end{cases}$$
$$25x + 45(200 - x) = 7400$$
$$25x + 9000 - 45x = 7400$$
$$-20x = -1600$$
$$x = 80$$
$$y = 200 - 80 = 120$$
80 sets of the \$25 dishes and 120 sets of the \$45 dishes should be ordered.

51.   Let x = the cost per package of bacon.
Let y = the cost of a carton of eggs.
Set up a system of equations for the problem:
$$\begin{cases} 3x + 2y = 7.45 \\ 2x + 3y = 6.45 \end{cases}$$
Multiply each side of the first equation by 3 and each side of the second equation by -2 and solve by elimination:

$$\begin{cases} 3x + 2y = 7.45 \\ 2x + 3y = 6.45 \end{cases} \xrightarrow{\;\;3\;\;} \begin{array}{r} 9x + 6y = 22.35 \\ -4x - 6y = -12.90 \\ \hline 5x = 9.45 \\ x = 1.89 \end{array}$$

Substitute and solve for y:
$$3(1.89) + 2y = 7.45$$

$$5.67 + 2y = 7.45 \rightarrow 2y = 1.78 \rightarrow y = 0.89$$

A package of bacon costs $1.89 and a carton of eggs cost $0.89.
The refund for 2 packages of bacon and 2 cartons of eggs will be $5.56.

53.   Let x = the # of mg of liquid 1.
Let y = the # of mg of liquid 2.
Setting up the equations and solving by substitution:
$$\begin{cases} 0.2x + 0.4y = 40 \quad \text{vitamin C} \\ 0.3x + 0.2y = 30 \ \text{vitamin D} \end{cases}$$
multiplying each equation by 10 yields
$$\begin{cases} 2x + 4y = 400 \\ 3x + 2y = 300 \end{cases} \begin{array}{l} \xrightarrow{} \quad 2x + 4y = 400 \\ \xrightarrow{\;2\;} \quad 6x + 4y = 600 \end{array}$$
subtracting the bottom equation from the top equation yields
$$2x + 4y - (6x + 4y) = -200$$

$$2x - 6x = -200 \rightarrow -4x = -200 \rightarrow x = 50$$

$$\rightarrow 2(50) + 4y = 400 \rightarrow 100 + 4y = 400$$

$$\rightarrow 4y = 300 \rightarrow y = \frac{300}{4} = 75$$

So 50 mg of liquid 1 should be mixed with 75 mg of liquid 2.

55.   Solve the system by substitution:
$$R = C$$

$$8x = 4.5x + 17500 \rightarrow 3.5x = 17500 \rightarrow x = 5000$$
5000 units must be produced and sold for the firm to break-even.

57. $\begin{cases} y = m_1 x + b_1 \\ y = m_2 x + b_2 \end{cases}$ subtracting the 2 equations yields

$$0 = m_1 x + b_1 - m_2 x + b_2$$

$$-b_2 = x(m_1 - m_2)$$

$$\frac{-b_2}{m_1 - m_2} = x, \ \text{ provided } m_1 \neq m_2$$

$$\text{so } y = m_1\left(\frac{-b_2}{m_1 - m_2}\right) + b_1 = \left(\frac{-b_2 m_1}{m_1 - m_2}\right) + b_1 = \frac{-b_2 m_1 + b_1(m_1 - m_2)}{m_1 - m_2}$$

Therefore the solution set is $\left\{ x = \dfrac{-b_2}{m_1 - m_2}, y = \dfrac{-b_2 m_1 + b_1(m_1 - m_2)}{m_1 - m_2} \right\}$, provided $m_1 \neq m_2$.

59. $\begin{cases} y = m_1 x + b_1 \\ y = m_2 x + b_2 \end{cases}$ where $m_1 = m_2 = m$ and $b_1 = b_2$

$\begin{cases} y = mx + b_1 \\ y = mx + b_1 \end{cases}$ subtracting the 2 equations yields

$0 = 0$

Since this statement is always true, there are infinitely many solutions to the system. That is, the system is dependent.

61. Answers will vary.

Chapter 10

Systems of Equations and Inequalities

10.2 Systems of Linear Equations: Three Equations Containing Three Variables

1. Substituting the values of the variables:

$$\begin{cases} 3x+3y+2z = 4 & \longrightarrow \quad 3(1)+3(-1)+2(2)= 3-3+4=4 \\ x-y-z =0 & \longrightarrow \quad 1-(-1)-2 = 1+1-2 =0 \\ 2y-3z = -8 & \longrightarrow \quad 2(-1)-3(2) = -2-6 = -8 \end{cases}$$

Each equation is satisfied, so $x =1, y =-1, z =2$ is a solution to the system of equations.

3. Multiply each side of the first equation by –2 and add to the second equation to eliminate x:

$$\begin{cases} x-\ y = 6 & \xrightarrow{-2} \quad -2x+2y \qquad = -12 \\ 2x-3z = 16 & \longrightarrow \qquad 2x \qquad -3z = \ 16 \\ 2y+\ z = 4 & \qquad\qquad\qquad 2y-3z = \ \ 4 \end{cases}$$

Multiply each side of the result by –1 and add to the original third equation to eliminate y:

$$\begin{array}{ll} 2y-3z = 4 & \xrightarrow{-1} \quad -2y+3z = -4 \\ 2y+\ z = 4 & \longrightarrow \qquad 2y+\ z = \ \ 4 \\ & \qquad\qquad\qquad\quad 4z = \ \ 0 \\ & \qquad\qquad\qquad\quad\ z = 0 \end{array}$$

Substituting and solving for the other variables:

$$\begin{array}{ll} 2y+0 = 4 & \qquad 2x-3(0) = 16 \\ 2y = 4 & \qquad\quad 2x = 16 \\ y = 2 & \qquad\quad\ x = 8 \end{array}$$

The solution is $x = 8,\ y = 2,\ z = 0$.

5. Multiply each side of the first equation by –2 and add to the second equation to eliminate x; and multiply each side of the first equation by 3 and add to the third equation to eliminate x:

$$\begin{cases} x-2y+3z = \quad 7 & \xrightarrow{-2} \quad -2x+4y-6z = -14 \\ 2x+\ y+\ z = \quad 4 & \longrightarrow \qquad 2x+\ y+\ z = \ \ 4 \\ -3x+2y-2z = -10 & \qquad\qquad\qquad\quad 5y-5z = -10 \xrightarrow{1/5} \quad y-z = -2 \\ & \xrightarrow{3} \qquad\quad 3x-6y+9z = \ \ 21 \\ & \longrightarrow \quad -3x+2y-2z = -10 \\ & \qquad\qquad\qquad\quad -4y+7z = \ \ 11 \end{cases}$$

Multiply each side of the first result by 4 and add to the second result to eliminate y:

$$y - z = -2 \quad \xrightarrow{\ 4\ } \quad 4y - 4z = -8$$

$$-4y + 7z = 11 \quad \longrightarrow \quad \underline{-4y + 7z = \ \ 11}$$

$$3z = \ \ 3 \to z = 1$$

Substituting and solving for the other variables:

$$y - 1 = -2$$
$$y = -1 \qquad\qquad x - 2(-1) + 3(1) = 7$$
$$x + 2 + 3 = 7 \to x = 2$$

The solution is $x = 2$, $y = -1$, $z = 1$.

7.   Add the first and second equations to eliminate z:

$$\begin{cases} x - y - z = 1 \quad \longrightarrow \quad x - y - z = 1 \\ 2x + 3y + z = 2 \quad \longrightarrow \quad \underline{2x + 3y + z = 2} \\ 3x + 2y \quad = 0 \qquad\qquad 3x + 2y \quad = 3 \end{cases}$$

Multiply each side of the result by –1 and add to the original third equation to eliminate y:

$$3x + 2y = 3 \quad \xrightarrow{\ -1\ } \quad -3x - 2y = -3$$

$$3x + 2y = 0 \quad \longrightarrow \quad \underline{3x + 2y = \ \ 0}$$

$$0 = -3$$

This result has no solution, so the system is inconsistent.

9.   Add the first and second equations to eliminate x; and multiply the first equation by –3 and add to the third equation to eliminate x:

$$\begin{cases} x - y - z = \ \ 1 \quad \longrightarrow \quad x - y - z = \ \ 1 \\ -x + 2y - 3z = -4 \quad \longrightarrow \quad \underline{-x + 2y - 3z = -4} \\ 3x - 2y - 7z = \ \ 0 \qquad\qquad y - 4z = -3 \end{cases}$$

$$\xrightarrow{\ -3\ } \quad -3x + 3y + 3z = -3$$

$$\longrightarrow \quad \underline{3x - 2y - 7z = \ \ 0}$$

$$y - 4z = -3$$

Multiply each side of the first result by –1 and add to the second result to eliminate y:

$$y - 4z = -3 \quad \xrightarrow{\ -1\ } \quad -y + 4z = \ \ 3$$

$$y - 4z = -3 \quad \longrightarrow \quad \underline{y - 4z = -3}$$

$$0 = \ \ 0$$

The system is dependent. If z is any real number, then $y = 4z - 3$.
Solving for x in terms of z in the first equation:

$$x - (4z - 3) - z = 1$$
$$x - 4z + 3 - z = 1$$
$$x - 5z + 3 = 1$$
$$x = 5z - 2$$

The solution is $x = 5z - 2$, $y = 4z - 3$, z is any real number.

11. Multiply the first equation by –2 and add to the second equation to eliminate x; and add the first and third equations to eliminate x:

$$\begin{cases} 2x - 2y + 3z = 6 \\ 4x - 3y + 2z = 0 \\ -2x + 3y - 7z = 1 \end{cases}$$

$$\xrightarrow{-2} \quad \begin{aligned} -4x + 4y - 6z &= -12 \\ 4x - 3y + 2z &= 0 \\ \hline y - 4z &= -12 \end{aligned}$$

$$\xrightarrow{} \quad \begin{aligned} 2x - 2y + 3z &= 6 \\ -2x + 3y - 7z &= 1 \\ \hline y - 4z &= 7 \end{aligned}$$

Multiply each side of the first result by –1 and add to the second result to eliminate y:

$$\begin{aligned} y - 4z &= -12 \\ y - 4z &= 7 \end{aligned} \quad \xrightarrow{-1} \quad \begin{aligned} -y + 4z &= 12 \\ y - 4z &= 7 \\ \hline 0 &= 19 \end{aligned}$$

This result has no solution, so the system is inconsistent.

13. Add the first and second equations to eliminate z; and multiply the second equation by 2 and add to the third equation to eliminate z:

$$\begin{cases} x + y - z = 6 \\ 3x - 2y + z = -5 \\ x + 3y - 2z = 14 \end{cases}$$

$$\xrightarrow{} \quad \begin{aligned} x + y - z &= 6 \\ 3x - 2y + z &= -5 \\ \hline 4x - y &= 1 \end{aligned}$$

$$\xrightarrow{2} \quad \begin{aligned} 6x - 4y + 2z &= -10 \\ x + 3y - 2z &= 14 \\ \hline 7x - y &= 4 \end{aligned}$$

Multiply each side of the first result by –1 and add to the second result to eliminate y:

$$\begin{aligned} 4x - y &= 1 \\ 7x - y &= 4 \end{aligned} \quad \xrightarrow{-1} \quad \begin{aligned} -4x + y &= -1 \\ 7x - y &= 4 \\ \hline 3x &= 3 \\ x &= 1 \end{aligned}$$

Substituting and solving for the other variables:

$$\begin{aligned} 4(1) - y &= 1 & 3(1) - 2(3) + z &= -5 \\ -y &= -3 & 3 - 6 + z &= -5 \\ y &= 3 & z &= -2 \end{aligned}$$

The solution is $x = 1$, $y = 3$, $z = -2$.

15. Add the first and second equations to eliminate z; and multiply the second equation by 3 and add to the third equation to eliminate z:

$$\begin{cases} x + 2y - z = -3 \\ 2x - 4y + z = -7 \\ -2x + 2y - 3z = 4 \end{cases}$$

$$\xrightarrow{} \quad \begin{aligned} x + 2y - z &= -3 \\ 2x - 4y + z &= -7 \\ \hline 3x - 2y &= -10 \end{aligned}$$

$$\xrightarrow{3} \quad 6x - 12y + 3z = -21$$
$$\xrightarrow{} \quad -2x + 2y - 3z = 4$$
$$4x - 10y = -17$$

Multiply each side of the first result by –5 and add to the second result to eliminate y:

$$3x - 2y = -10 \xrightarrow{-5} -15x + 10y = 50$$
$$4x - 10y = -17 \xrightarrow{} 4x - 10y = -17$$
$$-11x = 33$$
$$x = -3$$

Substituting and solving for the other variables:

$$3(-3) - 2y = -10 \qquad\qquad -3 + 2\left(\frac{1}{2}\right) - z = -3$$
$$-9 - 2y = -10 \qquad\qquad -3 + 1 - z = -3$$
$$-2y = -1 \qquad\qquad -z = -1$$
$$y = \frac{1}{2} \qquad\qquad z = 1$$

The solution is $x = -3,\ y = \frac{1}{2},\ z = 1$.

17. $y = ax^2 + bx + c$

At (–1, 4) the equation becomes:
$$4 = a(-1)^2 + b(-1) + c$$
$$4 = a - b + c$$
$$a - b + c = 4$$

At (2, 3) the equation becomes:
$$3 = a(2)^2 + b(2) + c$$
$$3 = 4a + 2b + c$$
$$4a + 2b + c = 3$$

At (0, 1) the equation becomes:
$$1 = a(0)^2 + b(0) + c$$
$$c = 1$$

The system of equations is:
$$\begin{cases} a - b + c = 4 \\ 4a + 2b + c = 3 \\ \phantom{4a + 2b + {}} c = 1 \end{cases}$$

Substitute $c = 1$ into the first and second equations and simplify:
$$\begin{cases} a - b + 1 = 4 \quad \rightarrow \quad a - b = 3 \quad \rightarrow \quad a = b + 3 \\ 4a + 2b + 1 = 3 \quad \rightarrow \quad 4a + 2b = 2 \end{cases}$$

Solve the first equation for a, substitute into the second equation and solve:
$$4(b + 3) + 2b = 2$$
$$4b + 12 + 2b = 2$$

$$6b = -10 \rightarrow b = -\frac{5}{3} \rightarrow a = -\frac{5}{3} + 3 = \frac{4}{3}$$

The solution is $a = \frac{4}{3},\ b = -\frac{5}{3},\ c = 1$. So the equation is $y = \frac{4}{3}x^2 - \frac{5}{3}x + 1$.

19.   Substitute the expression for I_2 into the second and third equations and simplify:

$$\begin{cases} I_2 = I_1 + I_3 \\ 5 - 3I_1 - 5I_2 = 0 \\ 10 - 5I_2 - 7I_3 = 0 \end{cases}$$

$$5 - 3I_1 - 5(I_1 + I_3) = 0 \;\rightarrow\; -8I_1 - 5I_3 = -5$$

$$10 - 5(I_1 + I_3) - 7I_3 = 0 \;\rightarrow\; -5I_1 - 12I_3 = -10$$

Multiply both sides of the second equation by 5 and multiply both sides of the third equation by –8 to eliminate I_1:

$$-8I_1 - 5I_3 = -5 \;\xrightarrow{\;5\;}\; -40I_1 - 25I_3 = -25$$

$$-5I_1 - 12I_3 = -10 \;\xrightarrow{\;-8\;}\; \underline{\;\;40I_1 + 96I_3 = \;\;\;80\;}$$

$$71I_3 = \;\;55$$

$$I_3 = \frac{55}{71}$$

Substituting and solving for the other variables:

$$-8I_1 - 5\left(\frac{55}{71}\right) = -5 \qquad\qquad I_2 = \frac{10}{71} + \frac{55}{71}$$

$$-8I_1 - \frac{275}{71} = -5 \qquad\qquad I_2 = \frac{65}{71}$$

$$-8I_1 = -\frac{80}{71}$$

$$I_1 = \frac{10}{71}$$

The solution is $I_1 = \dfrac{10}{71}$, $I_2 = \dfrac{65}{71}$, $I_3 = \dfrac{55}{71}$.

21.   Let x = the number of orchestra seats.
Let y = the number of main seats.
Let z = the number of balcony seats.
Since the total number of seats is 500, $x + y + z = 500$.
Since the total revenue is \$17,100 if all seats are sold, $50x + 35y + 25z = 17,100$.
If only half of the orchestra seats are sold, the revenue is \$14,600. So,

$$50\left(\tfrac{1}{2}x\right) + 35y + 25z = 14,600.$$

Multiply each side of the first equation by –25 and add to the second equation to eliminate z; and multiply each side of the third equation by –1 and add to the second equation to eliminate z:

$$\begin{cases} x + y + z = 500 \\ 50x + 35y + 25z = 17100 \\ 25x + 35y + 25z = 14600 \end{cases}$$

$$\xrightarrow{\;-25\;}\; -25x - 25y - 25z = -12500$$

$$\longrightarrow\; \underline{\;\;50x + 35y + 25z = \;\;17100\;}$$

$$25x + 10y \;\;\;\;\;\;\;= \;\;\;4600$$

$$\longrightarrow\; 50x + 35y + 25z = \;\;17100$$

$$\xrightarrow{\;-1\;}\; \underline{\;-25x - 35y - 25z = -14600\;}$$

$$25x \;\;\;\;\;\;\;\;\;\;\;\; = \;\;\;2500$$

$$x = \;\;100$$

Substituting and solving for the other variables:

$$25(100) + 10y = 4600 \qquad\qquad 100 + 210 + z = 500$$
$$2500 + 10y = 4600 \qquad\qquad 310 + z = 500$$
$$10y = 2100 \qquad\qquad z = 190$$
$$y = 210$$

There are 100 orchestra seats, 210 main seats, and 190 balcony seats.

23. Let x = the number of servings of chicken.
    Let y = the number of servings of corn.
    Let z = the number of servings of 2% milk.
    Protein equation: $30x + 3y + 9z = 66$
    Carbohydrate equation: $35x + 16y + 13z = 94.5$
    Calcium equation: $200x + 10y + 300z = 910$

    Multiply each side of the first equation by –16 and multiply each side of the second equation by 3 and add them to eliminate y; and multiply each side of the second equation by –5 and multiply each side of the third equation by 8 and add to eliminate y:

$$
\begin{cases}
30x + 3y + 9z = 66 \\
35x + 16y + 13z = 94.5 \\
200x + 10y + 300z = 910
\end{cases}
$$

$$\xrightarrow{-16}\quad -480x - 48y - 144z = -1056$$
$$\xrightarrow{3}\quad \underline{\;\;105x + 48y + 39z = \;\;283.5\;}$$
$$-375x \qquad\quad -105z = -772.5$$

$$\xrightarrow{-5}\quad -175x - 80y - 65z = -472.5$$
$$\xrightarrow{8}\quad \underline{\;1600x + 80y + 2400z = 7280\;}$$
$$1425x \qquad\quad +2335z = 6807.5$$

Multiply each side of the first result by 19 and multiply each side of the second result by 5 to eliminate x:

$$-375x - 105z = -772.5 \xrightarrow{19} -7125x - 1995z = -14677.5$$
$$1425x + 2335z = 6807.5 \xrightarrow{5} \underline{\;7125x + 11675z = \;\;34037.5\;}$$
$$9680z = 19360$$
$$z = 2$$

Substituting and solving for the other variables:

$$-375x - 105(2) = -772.5 \qquad\qquad 30(1.5) + 3y + 9(2) = 66$$
$$-375x - 210 = -772.5 \qquad\qquad 45 + 3y + 18 = 66$$
$$-375x = -562.5 \qquad\qquad 3y = 3$$
$$x = 1.5 \qquad\qquad y = 1$$

The dietitian should serve 1.5 servings of chicken, 1 serving of corn, and 2 servings of 2% milk.

25. Let x = the price of 1 hamburger.
    Let y = the price of 1 order of fries.
    Let z = the price of 1 drink.
    We can construct the system

$$
\begin{cases}
8x + 6y + 6z = 26.10 \\
10x + 6y + 8z = 31.60
\end{cases}
$$

A system involving only 2 equations that contain 3 or more unknowns cannot be solved uniquely. In other words, we can create as many solutions as we want by choosing a specific value for one of the variables and then solving the resulting 2 x 2 system.

For example, suppose we know that

$1.75 < hamburger price < $2.25

$0.75 < fries price < $1.00

$0.60 < fries price < $0.90

Pick a specific value for x, y or z	2x2 system	Solution
$x = \$2.00$	$\begin{cases} 8(2)+6y+6z = 26.10 \\ 10(2)+6y+8z = 31.60 \end{cases}$	$x = \$2.00$
	$\Downarrow$	$y = \$0.93$
	$\begin{cases} 16+6y+6z = 26.10 \\ 20+6y+8z = 31.60 \end{cases}$	$z = \$0.75$
	$\Downarrow$	
	$\begin{cases} 6y+6z = 10.10 \\ 6y+8z = 11.60 \end{cases}$	
$y = \$0.90$	$\begin{cases} 8x+6(0.9)+6z = 26.10 \\ 10x+6(0.9)+8z = 31.60 \end{cases}$	$x = \$2.00$
	$\Downarrow$	$y = \$0.90$
	$\begin{cases} 8x+5.4+6z = 26.10 \\ 10x+5.4+8z = 31.60 \end{cases}$	$z = \$0.65$
	$\Downarrow$	
	$\begin{cases} 8x+6z = 20.7 \\ 10x+8z = 26.2 \end{cases}$	
$z = \$0.80$	$\begin{cases} 8x+6y+6(.8) = 26.10 \\ 10x+6y+8(.8) = 31.60 \end{cases}$	$x = \$1.95$
	$\Downarrow$	$y = \$0.95$
	$\begin{cases} 8x+6y+4.8 = 26.10 \\ 10x+6y+6.4 = 31.60 \end{cases}$	$z = \$0.80$
	$\Downarrow$	
	$\begin{cases} 8x+6y = 21.3 \\ 10x+6y = 25.2 \end{cases}$	

27. Let x = Beth's time working alone.
     Let y = Bill's time working alone.
     Let z = Edie's time working alone.

We can use the following tables to organize our work:

	Beth	Bill	Edie	Together
Hours to do job	x	y	z	10
Part of job done in 1 hour	$\dfrac{1}{x}$	$\dfrac{1}{y}$	$\dfrac{1}{z}$	$\dfrac{1}{10}$

Equation $\dfrac{1}{x}+\dfrac{1}{y}+\dfrac{1}{z}=\dfrac{1}{10}$

	Bill	Edie	Together
Hours to do job	y	z	15
Part of job done in 1 hour	$\dfrac{1}{y}$	$\dfrac{1}{z}$	$\dfrac{1}{15}$

Equation $\dfrac{1}{y}+\dfrac{1}{z}=\dfrac{1}{15}$

	Beth	Bill	Edie	All three	Beth and Bill
Hours to do job	x	y	z	4	8
Part of job done in 1 hour	$\dfrac{1}{x}$	$\dfrac{1}{y}$	$\dfrac{1}{z}$	$\dfrac{1}{4}$	$\dfrac{1}{8}$

Equation

$$4\left(\frac{1}{x}+\frac{1}{y}+\frac{1}{z}\right)+8\left(\frac{1}{x}+\frac{1}{y}\right)=1 \rightarrow \frac{12}{x}+\frac{12}{y}+\frac{4}{z}=1$$

We can construct the system

$$\begin{cases} \dfrac{1}{x}+\dfrac{1}{y}+\dfrac{1}{z}=\dfrac{1}{10} \\ \dfrac{1}{y}+\dfrac{1}{z}=\dfrac{1}{15} \\ \dfrac{12}{x}+\dfrac{12}{y}+\dfrac{4}{z}=1 \end{cases} \longrightarrow$$

subtracting the second equation from the first equation yields

$$\frac{1}{x}+\frac{1}{y}+\frac{1}{z}=\frac{1}{10}$$
$$\underline{\frac{1}{y}+\frac{1}{z}=\frac{1}{15}}$$
$$\frac{1}{x}=\frac{1}{10}-\frac{1}{15}$$

$\rightarrow \dfrac{1}{x}=\dfrac{1}{30} \rightarrow x=30$

Plugging $x = 30$ into the original system yields

$$\begin{cases} \dfrac{1}{30}+\dfrac{1}{y}+\dfrac{1}{z}=\dfrac{1}{10} \longrightarrow \dfrac{1}{y}+\dfrac{1}{z}=\dfrac{1}{10}-\dfrac{1}{30} \longrightarrow \dfrac{1}{y}+\dfrac{1}{z}=\dfrac{1}{15} \\ \dfrac{1}{y}+\dfrac{1}{z}=\dfrac{1}{15} \\ \dfrac{12}{30}+\dfrac{12}{y}+\dfrac{4}{z}=1 \longrightarrow \dfrac{12}{y}+\dfrac{4}{z}=1-\dfrac{12}{30} \longrightarrow \dfrac{12}{y}+\dfrac{4}{z}=\dfrac{3}{5} \end{cases}$$

Now consider the system

$$\begin{cases} \dfrac{1}{y}+\dfrac{1}{z}=\dfrac{1}{15} \xrightarrow{-12} \dfrac{-12}{y}+\dfrac{-12}{z}=\dfrac{-12}{15} \\ \dfrac{12}{y}+\dfrac{4}{z}=\dfrac{3}{5} \longrightarrow \dfrac{12}{y}+\dfrac{4}{z}=\dfrac{3}{5} \end{cases}$$

adding these 2 equations yields

$$\dfrac{-12}{y}+\dfrac{-12}{z}=\dfrac{-12}{15}$$

$$\dfrac{12}{y}+\dfrac{4}{z}=\dfrac{3}{5}$$

$$\dfrac{-12}{z}+\dfrac{4}{z}=\dfrac{-12}{15}+\dfrac{3}{5} \quad \rightarrow \dfrac{-8}{z}=\dfrac{-3}{15} \rightarrow \dfrac{8}{z}=\dfrac{1}{5} \rightarrow z=40$$

plugging $z = 40$ into the equation

$$\dfrac{12}{y}+\dfrac{4}{z}=\dfrac{3}{5} \rightarrow \dfrac{12}{y}+\dfrac{4}{40}=\dfrac{3}{5}$$

$$\dfrac{12}{y}+\dfrac{1}{10}=\dfrac{3}{5} \rightarrow \dfrac{12}{y}=\dfrac{3}{5}-\dfrac{1}{10} \rightarrow \dfrac{12}{y}=\dfrac{1}{2} \rightarrow y=24$$

So, working alone, it would take Beth 30 hours, Bill 24 hours and Edie 40 hours to finish the job.

Chapter 10

Systems of Equations and Inequalities

10.3 Systems of Linear Equations: Matrices

1. Writing the augmented matrix for the system of equations:
$$\begin{cases} x - 5y = 5 \\ 4x + 3y = 6 \end{cases} \rightarrow \begin{bmatrix} 1 & -5 & | & 5 \\ 4 & 3 & | & 6 \end{bmatrix}$$

3. Writing the augmented matrix for the system of equations:
$$\begin{cases} 2x + 3y - 6 = 0 \\ 4x - 6y + 2 = 0 \end{cases} \rightarrow \begin{cases} 2x + 3y = 6 \\ 4x - 6y = -2 \end{cases} \rightarrow \begin{bmatrix} 2 & 3 & | & 6 \\ 4 & -6 & | & -2 \end{bmatrix}$$

5. Writing the augmented matrix for the system of equations:
$$\begin{cases} 0.01x - 0.03y = 0.06 \\ 0.13x + 0.10y = 0.20 \end{cases} \rightarrow \begin{bmatrix} 0.01 & -0.03 & | & 0.06 \\ 0.13 & 0.10 & | & 0.20 \end{bmatrix}$$

7. Writing the augmented matrix for the system of equations:
$$\begin{cases} x - y + z = 10 \\ 3x + 3y = 5 \\ x + y + 2z = 2 \end{cases} \rightarrow \begin{bmatrix} 1 & -1 & 1 & | & 10 \\ 3 & 3 & 0 & | & 5 \\ 1 & 1 & 2 & | & 2 \end{bmatrix}$$

9. Writing the augmented matrix for the system of equations:
$$\begin{cases} x + y - z = 2 \\ 3x - 2y = 2 \\ 5x + 3y - z = 1 \end{cases} \rightarrow \begin{bmatrix} 1 & 1 & -1 & | & 2 \\ 3 & -2 & 0 & | & 2 \\ 5 & 3 & -1 & | & 1 \end{bmatrix}$$

11. Writing the augmented matrix for the system of equations:
$$\begin{cases} x - y - z = 10 \\ 2x + y + 2z = -1 \\ -3x + 4y = 5 \\ 4x - 5y + z = 0 \end{cases} \rightarrow \begin{bmatrix} 1 & -1 & -1 & | & 10 \\ 2 & 1 & 2 & | & -1 \\ -3 & 4 & 0 & | & 5 \\ 4 & -5 & 1 & | & 0 \end{bmatrix}$$

13. $\begin{bmatrix} 1 & -3 & | & -2 \\ 2 & -5 & | & 5 \end{bmatrix} \rightarrow \begin{bmatrix} 1 & -3 & | & -2 \\ 0 & 1 & | & 9 \end{bmatrix}$

$$R_2 = -2r_1 + r_2$$

15. $\begin{bmatrix} 1 & -3 & 4 & | & 3 \\ 2 & -5 & 6 & | & 6 \\ -3 & 3 & 4 & | & 6 \end{bmatrix} \rightarrow \begin{bmatrix} 1 & -3 & 4 & | & 3 \\ 0 & 1 & -2 & | & 0 \\ -3 & 3 & 4 & | & 6 \end{bmatrix} \rightarrow \begin{bmatrix} 1 & -3 & 4 & | & 3 \\ 0 & 1 & -2 & | & 0 \\ 0 & -6 & 16 & | & 15 \end{bmatrix}$

(a) $R_2 = -2r_1 + r_2$ (b) $R_3 = 3r_1 + r_3$

17. $\begin{bmatrix} 1 & -3 & 2 & | & -6 \\ 2 & -5 & 3 & | & -4 \\ -3 & -6 & 4 & | & 6 \end{bmatrix} \rightarrow \begin{bmatrix} 1 & -3 & 2 & | & -6 \\ 0 & 1 & -1 & | & 8 \\ -3 & -6 & 4 & | & 6 \end{bmatrix} \rightarrow \begin{bmatrix} 1 & -3 & 2 & | & -6 \\ 0 & 1 & -1 & | & 8 \\ 0 & -15 & 10 & | & -12 \end{bmatrix}$

(a) $R_2 = -2r_1 + r_2$ (b) $R_3 = 3r_1 + r_3$

19. $\begin{bmatrix} 1 & -3 & 1 & | & -2 \\ 2 & -5 & 6 & | & -2 \\ -3 & 1 & 4 & | & 6 \end{bmatrix} \rightarrow \begin{bmatrix} 1 & -3 & 1 & | & -2 \\ 0 & 1 & 4 & | & 2 \\ -3 & 1 & 4 & | & 6 \end{bmatrix} \rightarrow \begin{bmatrix} 1 & -3 & 1 & | & -2 \\ 0 & 1 & 4 & | & 2 \\ 0 & -8 & 7 & | & 0 \end{bmatrix}$

(a) $R_2 = -2r_1 + r_2$ (b) $R_3 = 3r_1 + r_3$

21. $\begin{cases} x = 5 \\ y = -1 \end{cases}$ consistent $x = 5, y = -1$

23. $\begin{cases} x = 1 \\ y = 2 \\ 0 = 3 \end{cases}$ inconsistent

25. $\begin{cases} x + 2z = -1 \\ y - 4z = -2 \\ \quad 0 = 0 \end{cases}$ consistent $x = -1 - 2z, y = -2 + 4z, z$ is any real number

27.
$\begin{cases} x_1 = 1 \\ x_2 + x_4 = 2 \\ x_3 + 2x_4 = 3 \end{cases}$ consistent $x_1 = 1, x_2 = 2 - x_4, x_3 = 3 - 2x_4, x_4$ is any real number

29.
$\begin{cases} x_1 + 4x_4 = 2 \\ x_2 + + x_3 + 3x_4 = 3 \\ \qquad 0 = 0 \end{cases}$ consistent $x_1 = 2 - 4x_4, x_2 = 3 - x_3 - 3x_4,$
x_3, x_4 are any real numbers

31.

$$\begin{cases} x_1 + x_4 = -2 \\ x_2 + 2x_4 = 2 \\ x_3 - x_4 = 0 \\ \quad\quad 0 = 0 \end{cases}$$ consistent $x_1 = -2 - x_4, x_2 = 2 - 2x_4, x_3 = x_4,$ x_4 is any real number

33. $\begin{cases} x + y = 8 \\ x - y = 4 \end{cases}$ can be written as $\begin{bmatrix} 1 & 1 & | & 8 \\ 1 & -1 & | & 4 \end{bmatrix}$

$\rightarrow \begin{bmatrix} 1 & 1 & | & 8 \\ 0 & -2 & | & -4 \end{bmatrix} \rightarrow \begin{bmatrix} 1 & 1 & | & 8 \\ 0 & 1 & | & 2 \end{bmatrix} \rightarrow \begin{bmatrix} 1 & 0 & | & 6 \\ 0 & 1 & | & 2 \end{bmatrix}$

$\quad R_2 = -r_1 + r_2 \quad R_2 = -\frac{1}{2}r_2 \quad R_1 = -r_2 + r_1$

The solution is $x = 6$, $y = 2$.

35. $\begin{cases} 2x - 4y = -2 \\ 3x + 2y = 3 \end{cases}$ can be written as $\begin{bmatrix} 2 & -4 & | & -2 \\ 3 & 2 & | & 3 \end{bmatrix}$

$\rightarrow \begin{bmatrix} 1 & -2 & | & -1 \\ 3 & 2 & | & 3 \end{bmatrix} \rightarrow \begin{bmatrix} 1 & -2 & | & -1 \\ 0 & 8 & | & 6 \end{bmatrix} \rightarrow \begin{bmatrix} 1 & -2 & | & -1 \\ 0 & 1 & | & \frac{3}{4} \end{bmatrix} \rightarrow \begin{bmatrix} 1 & 0 & | & \frac{1}{2} \\ 0 & 1 & | & \frac{3}{4} \end{bmatrix}$

$\quad R_1 = \frac{1}{2}r_1 \quad\quad R_2 = -3r_1 + r_2 \quad R_2 = \frac{1}{8}r_2 \quad\quad R_1 = 2r_2 + r_1$

The solution is $x = \dfrac{1}{2}, y = \dfrac{3}{4}$.

37. $\begin{cases} x + 2y = 4 \\ 2x + 4y = 8 \end{cases}$ can be written as $\begin{bmatrix} 1 & 2 & | & 4 \\ 2 & 4 & | & 8 \end{bmatrix}$

$\rightarrow \begin{bmatrix} 1 & 2 & | & 4 \\ 0 & 0 & | & 0 \end{bmatrix}$

$\quad R_2 = -2r_1 + r_2$

This is a dependent system and the solution is $x = -2y + 4$, y is any real number.

39. $\begin{cases} 2x + 3y = 6 \\ x - y = \dfrac{1}{2} \end{cases}$ can be written as: $\begin{bmatrix} 2 & 3 & | & 6 \\ 1 & -1 & | & \frac{1}{2} \end{bmatrix}$

$\rightarrow \begin{bmatrix} 1 & \frac{3}{2} & | & 3 \\ 1 & -1 & | & \frac{1}{2} \end{bmatrix} \rightarrow \begin{bmatrix} 1 & \frac{3}{2} & | & 3 \\ 0 & -\frac{5}{2} & | & -\frac{5}{2} \end{bmatrix} \rightarrow \begin{bmatrix} 1 & \frac{3}{2} & | & 3 \\ 0 & 1 & | & 1 \end{bmatrix} \rightarrow \begin{bmatrix} 1 & 0 & | & \frac{3}{2} \\ 0 & 1 & | & 1 \end{bmatrix}$

$\quad R_1 = \frac{1}{2}r_1 \quad\quad R_2 = -r_1 + r_2 \quad R_2 = -\frac{2}{5}r_2 \quad R_1 = -\frac{3}{2}r_2 + r_1$

The solution is $x = \dfrac{3}{2}, y = 1$.

41. $\begin{cases} 3x - 5y = 3 \\ 15x + 5y = 21 \end{cases}$ can be written as $\begin{bmatrix} 3 & -5 & | & 3 \\ 15 & 5 & | & 21 \end{bmatrix}$

$\rightarrow \begin{bmatrix} 1 & -\frac{5}{3} & | & 1 \\ 15 & 5 & | & 21 \end{bmatrix} \rightarrow \begin{bmatrix} 1 & -\frac{5}{3} & | & 1 \\ 0 & 30 & | & 6 \end{bmatrix} \rightarrow \begin{bmatrix} 1 & -\frac{5}{3} & | & 1 \\ 0 & 1 & | & \frac{1}{5} \end{bmatrix} \rightarrow \begin{bmatrix} 1 & 0 & | & \frac{4}{3} \\ 0 & 1 & | & \frac{1}{5} \end{bmatrix}$

$\quad R_1 = \frac{1}{3}r_1 \qquad R_2 = -15r_1 + r_2 \quad R_2 = \frac{1}{30}r_2 \quad R_1 = \frac{5}{3}r_2 + r_1$

The solution is $x = \dfrac{4}{3}, y = \dfrac{1}{5}$.

43. $\begin{cases} x - y = 6 \\ 2x - 3z = 16 \\ 2y + z = 4 \end{cases}$ can be written as $\begin{bmatrix} 1 & -1 & 0 & | & 6 \\ 2 & 0 & -3 & | & 16 \\ 0 & 2 & 1 & | & 4 \end{bmatrix}$

$\rightarrow \begin{bmatrix} 1 & -1 & 0 & | & 6 \\ 0 & 2 & -3 & | & 4 \\ 0 & 2 & 1 & | & 4 \end{bmatrix} \rightarrow \begin{bmatrix} 1 & -1 & 0 & | & 6 \\ 0 & 1 & -\frac{3}{2} & | & 2 \\ 0 & 2 & 1 & | & 4 \end{bmatrix} \rightarrow \begin{bmatrix} 1 & 0 & -\frac{3}{2} & | & 8 \\ 0 & 1 & -\frac{3}{2} & | & 2 \\ 0 & 0 & 4 & | & 0 \end{bmatrix} \rightarrow \begin{bmatrix} 1 & 0 & -\frac{3}{2} & | & 8 \\ 0 & 1 & -\frac{3}{2} & | & 2 \\ 0 & 0 & 1 & | & 0 \end{bmatrix}$

$\quad R_2 = -2r_1 + r_2 \quad R_2 = \frac{1}{2}r_2 \qquad\quad R_1 = r_2 + r_1 \qquad\qquad R_3 = \frac{1}{4}r_3$

$\qquad\qquad\qquad\qquad\qquad\qquad\quad R_3 = -2r_2 + r_3$

$\rightarrow \begin{bmatrix} 1 & 0 & 0 & | & 8 \\ 0 & 1 & 0 & | & 2 \\ 0 & 0 & 1 & | & 0 \end{bmatrix}$

$\quad R_1 = \frac{3}{2}r_3 + r_1$

$\quad R_2 = \frac{3}{2}r_3 + r_2$

The solution is $x = 8, y = 2, z = 0$.

45. $\begin{cases} x - 2y + 3z = 7 \\ 2x + y + z = 4 \\ -3x + 2y - 2z = -10 \end{cases}$ can be written as $\begin{bmatrix} 1 & -2 & 3 & | & 7 \\ 2 & 1 & 1 & | & 4 \\ -3 & 2 & -2 & | & -10 \end{bmatrix}$

$\rightarrow \begin{bmatrix} 1 & -2 & 3 & | & 7 \\ 0 & 5 & -5 & | & -10 \\ 0 & -4 & 7 & | & 11 \end{bmatrix} \rightarrow \begin{bmatrix} 1 & -2 & 3 & | & 7 \\ 0 & 1 & -1 & | & -2 \\ 0 & -4 & 7 & | & 11 \end{bmatrix} \rightarrow \begin{bmatrix} 1 & 0 & 1 & | & 3 \\ 0 & 1 & -1 & | & -2 \\ 0 & 0 & 3 & | & 3 \end{bmatrix}$

$\quad R_2 = -2r_1 + r_2 \qquad\quad R_2 = \frac{1}{5}r_2 \qquad\qquad R_1 = 2r_2 + r_1$

$\quad R_3 = 3r_1 + r_3 \qquad\qquad\qquad\qquad\qquad R_3 = 4r_2 + r_3$

$\rightarrow \begin{bmatrix} 1 & 0 & 1 & | & 3 \\ 0 & 1 & -1 & | & -2 \\ 0 & 0 & 1 & | & 1 \end{bmatrix} \rightarrow \begin{bmatrix} 1 & 0 & 0 & | & 2 \\ 0 & 1 & 0 & | & -1 \\ 0 & 0 & 1 & | & 1 \end{bmatrix}$

$\quad R_3 = \frac{1}{3}r_3 \qquad\qquad R_1 = -r_3 + r_1$

$\qquad\qquad\qquad\qquad R_2 = r_3 + r_2$

The solution is $x = 2, y = -1, z = 1$.

47. $\begin{cases} 2x - 2y - 2z = 2 \\ 2x + 3y + z = 2 \\ 3x + 2y = 0 \end{cases}$ can be written as $\begin{bmatrix} 2 & -2 & -2 & | & 2 \\ 2 & 3 & 1 & | & 2 \\ 3 & 2 & 0 & | & 0 \end{bmatrix}$

$$\rightarrow \begin{bmatrix} 1 & -1 & -1 & | & 1 \\ 2 & 3 & 1 & | & 2 \\ 3 & 2 & 0 & | & 0 \end{bmatrix} \rightarrow \begin{bmatrix} 1 & -1 & -1 & | & 1 \\ 0 & 5 & 3 & | & 0 \\ 0 & 5 & 3 & | & -3 \end{bmatrix} \rightarrow \begin{bmatrix} 1 & -1 & -1 & | & 1 \\ 0 & 5 & 3 & | & 0 \\ 0 & 0 & 0 & | & -3 \end{bmatrix}$$

$R_1 = \frac{1}{2} r_1 \qquad\qquad R_2 = -2r_1 + r_2 \qquad R_3 = -r_2 + r_3$
$\qquad\qquad\qquad\qquad R_3 = -3r_1 + r_3$

There is no solution. The system is inconsistent.

49. $\begin{cases} -x + y + z = -1 \\ -x + 2y - 3z = -4 \\ 3x - 2y - 7z = 0 \end{cases}$ can be written as $\begin{bmatrix} -1 & 1 & 1 & | & -1 \\ -1 & 2 & -3 & | & -4 \\ 3 & -2 & -7 & | & 0 \end{bmatrix}$

$$\rightarrow \begin{bmatrix} 1 & -1 & -1 & | & 1 \\ -1 & 2 & -3 & | & -4 \\ 3 & -2 & -7 & | & 0 \end{bmatrix} \rightarrow \begin{bmatrix} 1 & -1 & -1 & | & 1 \\ 0 & 1 & -4 & | & -3 \\ 0 & 1 & -4 & | & -3 \end{bmatrix} \rightarrow \begin{bmatrix} 1 & 0 & -5 & | & -2 \\ 0 & 1 & -4 & | & -3 \\ 0 & 0 & 0 & | & 0 \end{bmatrix} \rightarrow \begin{matrix} x - 5z = -2 \\ y - 4z = -3 \end{matrix}$$

$R_1 = -r_1 \qquad\qquad R_2 = r_1 + r_2 \qquad R_1 = r_2 + r_1$
$\qquad\qquad\qquad\qquad R_3 = -3r_1 + r_3 \qquad R_3 = -r_2 + r_3$

The solution is $x = 5z - 2, y = 4z - 3, z$ is any real number.

51. $\begin{cases} 2x - 2y + 3z = 6 \\ 4x - 3y + 2z = 0 \\ -2x + 3y - 7z = 1 \end{cases}$ can be written as $\begin{bmatrix} 2 & -2 & 3 & | & 6 \\ 4 & -3 & 2 & | & 0 \\ -2 & 3 & -7 & | & 1 \end{bmatrix}$

$$\rightarrow \begin{bmatrix} 1 & -1 & \frac{3}{2} & | & 3 \\ 4 & -3 & 2 & | & 0 \\ -2 & 3 & -7 & | & 1 \end{bmatrix} \rightarrow \begin{bmatrix} 1 & -1 & \frac{3}{2} & | & 3 \\ 0 & 1 & -4 & | & -12 \\ 0 & 1 & -4 & | & 7 \end{bmatrix} \rightarrow \begin{bmatrix} 1 & 0 & -\frac{5}{2} & | & -9 \\ 0 & 1 & -4 & | & -12 \\ 0 & 0 & 0 & | & 19 \end{bmatrix}$$

$R_1 = \frac{1}{2} r_1 \qquad\qquad R_2 = -4r_1 + r_2 \qquad R_1 = r_2 + r_1$
$\qquad\qquad\qquad\qquad R_3 = 2r_1 + r_3 \qquad R_3 = -r_2 + r_3$

There is no solution. The system is inconsistent.

53. $\begin{cases} x + y - z = 6 \\ 3x - 2y + z = -5 \\ x + 3y - 2z = 14 \end{cases}$ can be written as: $\begin{bmatrix} 1 & 1 & -1 & | & 6 \\ 3 & -2 & 1 & | & -5 \\ 1 & 3 & -2 & | & 14 \end{bmatrix}$

$$\rightarrow \begin{bmatrix} 1 & 1 & -1 & | & 6 \\ 0 & -5 & 4 & | & -23 \\ 0 & 2 & -1 & | & 8 \end{bmatrix} \rightarrow \begin{bmatrix} 1 & 1 & -1 & | & 6 \\ 0 & 1 & -\frac{4}{5} & | & \frac{23}{5} \\ 0 & 2 & -1 & | & 8 \end{bmatrix} \rightarrow \begin{bmatrix} 1 & 0 & -\frac{1}{5} & | & \frac{7}{5} \\ 0 & 1 & -\frac{4}{5} & | & \frac{23}{5} \\ 0 & 0 & \frac{3}{5} & | & -\frac{6}{5} \end{bmatrix}$$

$R_2 = -3r_1 + r_2 \qquad R_2 = -\frac{1}{5} r_2 \qquad R_1 = -r_2 + r_1$
$R_3 = -r_1 + r_3 \qquad\qquad\qquad\qquad R_3 = -2r_2 + r_3$

$$\rightarrow \begin{bmatrix} 1 & 0 & -\frac{1}{5} & \Big| & \frac{7}{5} \\ 0 & 1 & -\frac{4}{5} & \Big| & \frac{23}{5} \\ 0 & 0 & 1 & \Big| & -2 \end{bmatrix} \rightarrow \begin{bmatrix} 1 & 0 & 0 & \Big| & 1 \\ 0 & 1 & 0 & \Big| & 3 \\ 0 & 0 & 1 & \Big| & -2 \end{bmatrix}$$

$$R_3 = \frac{5}{3} r_3 \qquad\qquad R_1 = \frac{1}{5} r_3 + r_1$$
$$R_2 = \frac{4}{5} r_3 + r_2$$

The solution is $x = 1, y = 3, z = -2$.

55. $\begin{cases} x + 2y - z = -3 \\ 2x - 4y + z = -7 \\ -2x + 2y - 3z = 4 \end{cases}$ can be written as $\begin{bmatrix} 1 & 2 & -1 & \Big| & -3 \\ 2 & -4 & 1 & \Big| & -7 \\ -2 & 2 & -3 & \Big| & 4 \end{bmatrix}$

$$\rightarrow \begin{bmatrix} 1 & 2 & -1 & \Big| & -3 \\ 0 & -8 & 3 & \Big| & -1 \\ 0 & 6 & -5 & \Big| & -2 \end{bmatrix} \rightarrow \begin{bmatrix} 1 & 2 & -1 & \Big| & -3 \\ 0 & 1 & -\frac{3}{8} & \Big| & \frac{1}{8} \\ 0 & 6 & -5 & \Big| & -2 \end{bmatrix} \rightarrow \begin{bmatrix} 1 & 0 & -\frac{1}{4} & \Big| & -\frac{13}{4} \\ 0 & 1 & -\frac{3}{8} & \Big| & \frac{1}{8} \\ 0 & 0 & -\frac{11}{4} & \Big| & -\frac{11}{4} \end{bmatrix}$$

$$R_2 = -2r_1 + r_2 \qquad R_2 = -\frac{1}{8} r_2 \qquad R_1 = -2r_2 + r_1$$
$$R_3 = 2r_1 + r_3 \qquad\qquad\qquad\qquad R_3 = -6r_2 + r_3$$

$$\rightarrow \begin{bmatrix} 1 & 0 & -\frac{1}{4} & \Big| & -\frac{13}{4} \\ 0 & 1 & -\frac{3}{8} & \Big| & \frac{1}{8} \\ 0 & 0 & 1 & \Big| & 1 \end{bmatrix} \rightarrow \begin{bmatrix} 1 & 0 & 0 & \Big| & -3 \\ 0 & 1 & 0 & \Big| & \frac{1}{2} \\ 0 & 0 & 1 & \Big| & 1 \end{bmatrix}$$

$$R_3 = -\frac{4}{11} r_3 \qquad\qquad R_1 = \frac{1}{4} r_3 + r_1$$
$$R_2 = \frac{3}{8} r_3 + r_2$$

The solution is $x = -3, y = \dfrac{1}{2}, z = 1$.

57. $\begin{cases} 3x + y - z = \dfrac{2}{3} \\ 2x - y + z = 1 \\ 4x + 2y = \dfrac{8}{3} \end{cases}$ can be written as: $\begin{bmatrix} 3 & 1 & -1 & \Big| & \frac{2}{3} \\ 2 & -1 & 1 & \Big| & 1 \\ 4 & 2 & 0 & \Big| & \frac{8}{3} \end{bmatrix}$

$$\rightarrow \begin{bmatrix} 1 & \frac{1}{3} & -\frac{1}{3} & \Big| & \frac{2}{9} \\ 2 & -1 & 1 & \Big| & 1 \\ 4 & 2 & 0 & \Big| & \frac{8}{3} \end{bmatrix} \rightarrow \begin{bmatrix} 1 & \frac{1}{3} & -\frac{1}{3} & \Big| & \frac{2}{9} \\ 0 & -\frac{5}{3} & \frac{5}{3} & \Big| & \frac{5}{9} \\ 0 & \frac{2}{3} & \frac{4}{3} & \Big| & \frac{16}{9} \end{bmatrix} \rightarrow \begin{bmatrix} 1 & \frac{1}{3} & -\frac{1}{3} & \Big| & \frac{2}{9} \\ 0 & 1 & -1 & \Big| & -\frac{1}{3} \\ 0 & \frac{2}{3} & \frac{4}{3} & \Big| & \frac{16}{9} \end{bmatrix} \rightarrow \begin{bmatrix} 1 & 0 & 0 & \Big| & \frac{1}{3} \\ 0 & 1 & -1 & \Big| & -\frac{1}{3} \\ 0 & 0 & 2 & \Big| & 2 \end{bmatrix}$$

$$R_1 = \frac{1}{3} r_1 \qquad\quad R_2 = -2r_1 + r_2 \qquad R_2 = -\frac{3}{5} r_2 \qquad R_1 = -\frac{1}{3} r_2 + r_1$$
$$\qquad\qquad\qquad R_3 = -4r_1 + r_3 \qquad\qquad\qquad\qquad R_3 = -\frac{2}{3} r_2 + r_3$$

$$\rightarrow \begin{bmatrix} 1 & 0 & 0 & \Big| & \frac{1}{3} \\ 0 & 1 & -1 & \Big| & -\frac{1}{3} \\ 0 & 0 & 1 & \Big| & 1 \end{bmatrix} \rightarrow \begin{bmatrix} 1 & 0 & 0 & \Big| & \frac{1}{3} \\ 0 & 1 & 0 & \Big| & \frac{2}{3} \\ 0 & 0 & 1 & \Big| & 1 \end{bmatrix}$$

$$R_3 = \frac{1}{2} r_3 \qquad\qquad R_2 = r_3 + r_2$$

The solution is $x = \dfrac{1}{3}, y = \dfrac{2}{3}, z = 1$.

59.
$$\begin{cases} x + y + z + w = 4 \\ 2x - y + z \quad\;\;\; = 0 \\ 3x + 2y + z - w = 6 \\ x - 2y - 2z + 2w = -1 \end{cases}$$
can be written as
$$\begin{bmatrix} 1 & 1 & 1 & 1 & 4 \\ 2 & -1 & 1 & 0 & 0 \\ 3 & 2 & 1 & -1 & 6 \\ 1 & -2 & -2 & 2 & -1 \end{bmatrix}$$

$$\rightarrow \begin{bmatrix} 1 & 1 & 1 & 1 & 4 \\ 0 & -3 & -1 & -2 & -8 \\ 0 & -1 & -2 & -4 & -6 \\ 0 & -3 & -3 & 1 & -5 \end{bmatrix} \rightarrow \begin{bmatrix} 1 & 1 & 1 & 1 & 4 \\ 0 & -1 & -2 & -4 & -6 \\ 0 & -3 & -1 & -2 & -8 \\ 0 & -3 & -3 & 1 & -5 \end{bmatrix} \rightarrow \begin{bmatrix} 1 & 1 & 1 & 1 & 4 \\ 0 & 1 & 2 & 4 & 6 \\ 0 & -3 & -1 & -2 & -8 \\ 0 & -3 & -3 & 1 & -5 \end{bmatrix}$$

$R_2 = -2r_1 + r_2$ Interchange r_2 and r_3 $R_2 = -r_2$
$R_3 = -3r_1 + r_3$
$R_4 = -r_1 + r_4$

$$\rightarrow \begin{bmatrix} 1 & 0 & -1 & -3 & -2 \\ 0 & 1 & 2 & 4 & 6 \\ 0 & 0 & 5 & 10 & 10 \\ 0 & 0 & 3 & 13 & 13 \end{bmatrix} \rightarrow \begin{bmatrix} 1 & 0 & -1 & -3 & -2 \\ 0 & 1 & 2 & 4 & 6 \\ 0 & 0 & 1 & 2 & 2 \\ 0 & 0 & 3 & 13 & 13 \end{bmatrix} \rightarrow \begin{bmatrix} 1 & 0 & 0 & -1 & 0 \\ 0 & 1 & 0 & 0 & 2 \\ 0 & 0 & 1 & 2 & 2 \\ 0 & 0 & 0 & 7 & 7 \end{bmatrix}$$

$R_1 = -r_2 + r_1$ $R_3 = \frac{1}{5}r_3$ $R_1 = r_3 + r_1$
$R_3 = 3r_2 + r_3$ $R_2 = -2r_3 + r_2$
$R_4 = 3r_2 + r_4$ $R_4 = -3r_3 + r_4$

$$\rightarrow \begin{bmatrix} 1 & 0 & 0 & -1 & 0 \\ 0 & 1 & 0 & 0 & 2 \\ 0 & 0 & 1 & 2 & 2 \\ 0 & 0 & 0 & 1 & 1 \end{bmatrix} \rightarrow \begin{bmatrix} 1 & 0 & 0 & 0 & 1 \\ 0 & 1 & 0 & 0 & 2 \\ 0 & 0 & 1 & 0 & 0 \\ 0 & 0 & 0 & 1 & 1 \end{bmatrix}$$

$R_4 = \frac{1}{7}r_4$ $R_1 = r_4 + r_1$
$R_3 = -2r_4 + r_3$

The solution is $x = 1, y = 2, z = 0, w = 1$.

61.
$$\begin{cases} x + 2y + z = 1 \\ 2x - y + 2z = 2 \\ 3x + y + 3z = 3 \end{cases}$$
can be written as
$$\begin{bmatrix} 1 & 2 & 1 & 1 \\ 2 & -1 & 2 & 2 \\ 3 & 1 & 3 & 3 \end{bmatrix}$$

$$\rightarrow \begin{bmatrix} 1 & 2 & 1 & 1 \\ 0 & -5 & 0 & 0 \\ 0 & -5 & 0 & 0 \end{bmatrix} \rightarrow \begin{bmatrix} 1 & 2 & 1 & 1 \\ 0 & -5 & 0 & 0 \\ 0 & 0 & 0 & 0 \end{bmatrix} \rightarrow \begin{cases} x + 2y + z = 1 \\ -5y \quad\;\; = 0 \end{cases}$$

$R_2 = -2r_1 + r_2$ $R_3 = -r_2 + r_3$
$R_3 = -3r_1 + r_3$

Substitute and solve:
$$y = 0$$
$$x + 2(0) + z = 1$$
$$x + z = 1$$
$$x = 1 - z$$
The solution is $y = 0,\ x = 1 - z,\ z$ is any real number.

63. $\begin{cases} x - y + z = 5 \\ 3x + 2y - 2z = 0 \end{cases}$ can be written as: $\begin{bmatrix} 1 & -1 & 1 & | & 5 \\ 3 & 2 & -2 & | & 0 \end{bmatrix}$

$\rightarrow \begin{bmatrix} 1 & -1 & 1 & | & 5 \\ 0 & 5 & -5 & | & -15 \end{bmatrix} \rightarrow \begin{bmatrix} 1 & -1 & 1 & | & 5 \\ 0 & 1 & -1 & | & -3 \end{bmatrix} \rightarrow \begin{bmatrix} 1 & 0 & 0 & | & 2 \\ 0 & 1 & -1 & | & -3 \end{bmatrix}$

$\quad R_2 = -3r_1 + r_2 \qquad\quad R_2 = \frac{1}{5}r_2 \qquad\qquad R_1 = r_2 + r_1$

The matrix in the third step represents the system $\begin{cases} x = 2 \\ y - z = -3 \end{cases}$

Therefore the solution is $x = 2;\ y = -3 + z;\ z$ is any real number

or

$$x = 2;\ z = y + 3;\ y \text{ is any real number}$$

65.

$\begin{cases} 2x + 3y - z = 3 \\ x - y - z = 0 \\ -x + y + z = 0 \\ x + y + 3z = 5 \end{cases}$ can be written as: $\begin{bmatrix} 2 & 3 & -1 & | & 3 \\ 1 & -1 & -1 & | & 0 \\ -1 & 1 & 1 & | & 0 \\ 1 & 1 & 3 & | & 5 \end{bmatrix}$

$\rightarrow \begin{bmatrix} 1 & -1 & -1 & | & 0 \\ 2 & 3 & -1 & | & 3 \\ -1 & 1 & 1 & | & 0 \\ 1 & 1 & 3 & | & 5 \end{bmatrix} \longrightarrow \begin{bmatrix} 1 & -1 & -1 & | & 0 \\ 0 & 5 & 1 & | & 3 \\ 0 & 0 & 0 & | & 0 \\ 0 & 2 & 4 & | & 5 \end{bmatrix} \rightarrow \begin{bmatrix} 1 & -1 & -1 & | & 0 \\ 0 & 5 & 1 & | & 3 \\ 0 & 2 & 4 & | & 5 \\ 0 & 0 & 0 & | & 0 \end{bmatrix}$

interchange r_1 and r_2 $\qquad R_2 = -2r_1 + r_2 \qquad$ interchange r_3 and r_4
$\qquad\qquad\qquad\qquad\qquad R_3 = r_1 + r_3$
$\qquad\qquad\qquad\qquad\qquad R_4 = -r_1 + r_4$

$\rightarrow \begin{bmatrix} 1 & -1 & -1 & | & 0 \\ 0 & 1 & -7 & | & -7 \\ 0 & 2 & 4 & | & 5 \\ 0 & 0 & 0 & | & 0 \end{bmatrix} \longrightarrow \begin{bmatrix} 1 & 0 & -8 & | & -7 \\ 0 & 1 & -7 & | & -7 \\ 0 & 1 & 18 & | & 19 \\ 0 & 0 & 0 & | & 0 \end{bmatrix} \rightarrow \begin{bmatrix} 1 & 0 & -8 & | & -7 \\ 0 & 1 & -7 & | & -7 \\ 0 & 0 & 1 & | & \frac{19}{18} \\ 0 & 0 & 0 & | & 0 \end{bmatrix}$

$\quad R_2 = -2r_3 + r_2 \qquad\qquad R_1 = r_2 + r_1 \qquad R_3 = \frac{1}{18}r_3$
$\qquad\qquad\qquad\qquad\qquad R_3 = -2r_2 + r_3$

513

The matrix in the last step represents the system
$$\begin{cases} x - 8z = -7 \\ y - 7z = -7 \\ \quad z = \dfrac{19}{18} \end{cases}$$

Therefore the solution is

$$z = \frac{19}{18}$$

$$x = -7 + 8z = -7 + 8\left(\frac{19}{18}\right) = \frac{13}{9}; \quad y = -7 + 7z = -7 + 7\left(\frac{19}{18}\right) = \frac{7}{18}$$

67. $\begin{cases} 4x + y + z - w = 4 \\ x - y + 2z + 3w = 3 \end{cases}$ can be written as: $\begin{bmatrix} 4 & 1 & 1 & -1 & | & 4 \\ 1 & -1 & 2 & 3 & | & 3 \end{bmatrix}$

$$\rightarrow \begin{bmatrix} 1 & -1 & 2 & 3 & | & 3 \\ 4 & 1 & 1 & -1 & | & 4 \end{bmatrix} \longrightarrow \begin{bmatrix} 1 & -1 & 2 & 3 & | & 3 \\ 0 & 5 & -7 & -13 & | & -8 \end{bmatrix}$$

interchange r_1 and r_2 $\qquad$ $R_2 = -4r_1 + r_2$

The matrix in the last step represents the system $\qquad \begin{cases} x - y + 2z + 3w = 3 \\ 5y - 7z - 13w = -8 \end{cases}$

The second equation yields

$$5y - 7z - 13w = -8 \rightarrow 5y = -8 + 7z + 13w \rightarrow y = -\frac{8}{5} + \frac{7}{5}z + \frac{13}{5}w$$

The first equation yields

$$x - y + 2z + 3w = 3 \rightarrow x = 3 + y - 2z - 3w$$

substituting for y

$$x = 3 + \left(-\frac{8}{5} + \frac{7}{5}z + \frac{13}{5}w\right) - 2z - 3w$$

$$x = -\frac{3}{5}z - \frac{2}{5}w + \frac{7}{5}$$

Therefore the solution is

$$x = -\frac{3}{5}z - \frac{2}{5}w + \frac{7}{5}$$

$$y = -\frac{8}{5} + \frac{7}{5}z + \frac{13}{5}w$$

z and w are any real numbers

69. Each of the points must satisfy the equation $y = ax^2 + bx + c$.

$\quad$ (1,2): $\qquad 2 = a + b + c$

$\quad$ (−2,−7): $\quad -7 = 4a - 2b + c$

$\quad$ (2,−3): $\qquad -3 = 4a + 2b + c$

Set up a matrix and solve:

$$\begin{bmatrix} 1 & 1 & 1 & | & 2 \\ 4 & -2 & 1 & | & -7 \\ 4 & 2 & 1 & | & -3 \end{bmatrix} \rightarrow \begin{bmatrix} 1 & 1 & 1 & | & 2 \\ 0 & -6 & -3 & | & -15 \\ 0 & -2 & -3 & | & -11 \end{bmatrix} \rightarrow \begin{bmatrix} 1 & 1 & 1 & | & 2 \\ 0 & 1 & \frac{1}{2} & | & \frac{5}{2} \\ 0 & -2 & -3 & | & -11 \end{bmatrix} \rightarrow \begin{bmatrix} 1 & 0 & \frac{1}{2} & | & -\frac{1}{2} \\ 0 & 1 & \frac{1}{2} & | & \frac{5}{2} \\ 0 & 0 & -2 & | & -6 \end{bmatrix}$$

$$\begin{array}{ll} R_2 = -4r_1 + r_2 & R_2 = -\frac{1}{6}r_2 \qquad R_1 = -r_2 + r_1 \\ R_3 = -4r_1 + r_3 & \qquad\qquad R_3 = 2r_2 + r_3 \end{array}$$

$$\rightarrow \begin{bmatrix} 1 & 0 & \frac{1}{2} & | & -\frac{1}{2} \\ 0 & 1 & \frac{1}{2} & | & \frac{5}{2} \\ 0 & 0 & 1 & | & 3 \end{bmatrix} \rightarrow \begin{bmatrix} 1 & 0 & 0 & | & -2 \\ 0 & 1 & 0 & | & 1 \\ 0 & 0 & 1 & | & 3 \end{bmatrix}$$

$$\begin{array}{ll} R_3 = -\frac{1}{2}r_3 & R_1 = -\frac{1}{2}r_3 + r_1 \\ & R_2 = -\frac{1}{2}r_3 + r_2 \end{array}$$

The solution is $a = -2, b = 1, c = 3$; so the equation is $y = -2x^2 + x + 3$.

71. Each of the points must satisfy the equation $f(x) = ax^3 + bx^2 + cx + d$.

$$\begin{array}{ll} f(-3) = -112\text{:} & -27a + 9b - 3c + d = -112 \\ f(-1) = -2\text{:} & -a + b - c + d = -2 \\ f(1) = 4\text{:} & a + b + c + d = 4 \\ f(2) = 13\text{:} & 8a + 4b + 2c + d = 13 \end{array}$$

Set up a matrix and solve:

$$\begin{bmatrix} -27 & 9 & -3 & 1 & | & -112 \\ -1 & 1 & -1 & 1 & | & -2 \\ 1 & 1 & 1 & 1 & | & 4 \\ 8 & 4 & 2 & 1 & | & 13 \end{bmatrix} \rightarrow \begin{bmatrix} 1 & 1 & 1 & 1 & | & 4 \\ -1 & 1 & -1 & 1 & | & -2 \\ -27 & 9 & -3 & 1 & | & -112 \\ 8 & 4 & 2 & 1 & | & 13 \end{bmatrix} \rightarrow \begin{bmatrix} 1 & 1 & 1 & 1 & | & 4 \\ 0 & 2 & 0 & 2 & | & 2 \\ 0 & 36 & 24 & 28 & | & -4 \\ 0 & -4 & -6 & -7 & | & -19 \end{bmatrix}$$

$$\begin{array}{ll} \text{Interchange } r_3 \text{ and } r_1 & R_2 = r_1 + r_2 \\ & R_3 = 27r_1 + r_3 \\ & R_4 = -8r_1 + r_4 \end{array}$$

$$\rightarrow \begin{bmatrix} 1 & 1 & 1 & 1 & | & 4 \\ 0 & 1 & 0 & 1 & | & 1 \\ 0 & 36 & 24 & 28 & | & -4 \\ 0 & -4 & -6 & -7 & | & -19 \end{bmatrix} \rightarrow \begin{bmatrix} 1 & 0 & 1 & 0 & | & 3 \\ 0 & 1 & 0 & 1 & | & 1 \\ 0 & 0 & 24 & -8 & | & -40 \\ 0 & 0 & -6 & -3 & | & -15 \end{bmatrix} \rightarrow \begin{bmatrix} 1 & 0 & 1 & 0 & | & 3 \\ 0 & 1 & 0 & 1 & | & 1 \\ 0 & 0 & 1 & -\frac{1}{3} & | & -\frac{5}{3} \\ 0 & 0 & -6 & -3 & | & -15 \end{bmatrix}$$

$$\begin{array}{lll} R_2 = \frac{1}{2}r_2 & R_1 = -r_2 + r_1 & R_3 = \frac{1}{24}r_3 \\ & R_3 = -36r_2 + r_3 & \\ & R_4 = 4r_2 + r_4 & \end{array}$$

$$\rightarrow \begin{bmatrix} 1 & 0 & 0 & \frac{1}{3} & \frac{14}{3} \\ 0 & 1 & 0 & 1 & 1 \\ 0 & 0 & 1 & -\frac{1}{3} & -\frac{5}{3} \\ 0 & 0 & 0 & -5 & -25 \end{bmatrix} \rightarrow \begin{bmatrix} 1 & 0 & 0 & \frac{1}{3} & \frac{14}{3} \\ 0 & 1 & 0 & 1 & 1 \\ 0 & 0 & 1 & -\frac{1}{3} & -\frac{5}{3} \\ 0 & 0 & 0 & 1 & 5 \end{bmatrix} \rightarrow \begin{bmatrix} 1 & 0 & 0 & 0 & 3 \\ 0 & 1 & 0 & 0 & -4 \\ 0 & 0 & 1 & 0 & 0 \\ 0 & 0 & 0 & 1 & 5 \end{bmatrix}$$

$$R_1 = -r_3 + r_1 \qquad R_4 = -\frac{1}{5}r_4 \qquad R_1 = -\frac{1}{3}r_4 + r_1$$
$$R_4 = 6r_3 + r_4 \qquad\qquad\qquad\qquad R_2 = -r_4 + r_2$$
$$\qquad\qquad\qquad\qquad\qquad\qquad\qquad R_3 = \frac{1}{3}r_4 + r_3$$

The solution is $a = 3, b = -4, c = 0, d = 5$; so the equation is $f(x) = 3x^3 - 4x^2 + 5$.

73. Let x = the number of servings of salmon steak.
Let y = the number of servings of baked eggs.
Let z = the number of servings of acorn squash.
Protein equation: $30x + 15y + 3z = 78$
Carbohydrate equation: $20x + 2y + 25z = 59$
Vitamin A equation: $2x + 20y + 32z = 75$
Set up a matrix and solve:

$$\begin{bmatrix} 30 & 15 & 3 & 78 \\ 20 & 2 & 25 & 59 \\ 2 & 20 & 32 & 75 \end{bmatrix} \rightarrow \begin{bmatrix} 2 & 20 & 32 & 75 \\ 20 & 2 & 25 & 59 \\ 30 & 15 & 3 & 78 \end{bmatrix} \rightarrow \begin{bmatrix} 1 & 10 & 16 & 37.5 \\ 20 & 2 & 25 & 59 \\ 30 & 15 & 3 & 78 \end{bmatrix}$$

Interchange r_3 and r_1 $R_1 = \frac{1}{2}r_1$

$$\rightarrow \begin{bmatrix} 1 & 10 & 16 & 37.5 \\ 0 & -198 & -295 & -691 \\ 0 & -285 & -477 & -1047 \end{bmatrix} \rightarrow \begin{bmatrix} 1 & 10 & 16 & 37.5 \\ 0 & -198 & -295 & -691 \\ 0 & 0 & -\frac{3457}{66} & -\frac{3457}{66} \end{bmatrix}$$

$$R_2 = -20r_1 + r_2 \qquad\qquad R_3 = -\frac{95}{66}r_2 + r_3$$
$$R_3 = -30r_1 + r_3$$

$$\rightarrow \begin{bmatrix} 1 & 10 & 16 & 37.5 \\ 0 & -198 & -295 & -691 \\ 0 & 0 & 1 & 1 \end{bmatrix}$$

$$R_3 = -\frac{66}{3457}r_3$$

Substitute $z = 1$ and solve:
$$-198y - 295(1) = -691 \qquad\qquad x + 10(2) + 16(1) = 37.5$$
$$-198y = -396 \qquad\qquad\qquad\qquad x + 36 = 37.5$$
$$y = 2 \qquad\qquad\qquad\qquad\qquad x = 1.5$$

The dietitian should serve 1.5 servings of salmon steak, 2 servings of baked eggs, and 1 serving of acorn squash.

75. Let x = the amount invested in Treasury bills.
Let y = the amount invested in Treasury bonds.
Let z = the amount invested in corporate bonds.
Total investment equation: $x + y + z = 10000$
Annual income equation: $0.06x + 0.07y + 0.08z = 680$

Condition on investment equation : $z = \dfrac{1}{2}x$

Set up a matrix and solve:

$$\begin{bmatrix} 1 & 1 & 1 & | & 10000 \\ 0.06 & 0.07 & 0.08 & | & 680 \\ 1 & 0 & -2 & | & 0 \end{bmatrix} \rightarrow \begin{bmatrix} 1 & 1 & 1 & | & 10000 \\ 0 & 0.01 & 0.02 & | & 80 \\ 0 & -1 & -3 & | & -10000 \end{bmatrix} \rightarrow \begin{bmatrix} 1 & 1 & 1 & | & 10000 \\ 0 & 1 & 2 & | & 8000 \\ 0 & -1 & -3 & | & -10000 \end{bmatrix}$$

$$R_2 = -0.06r_1 + r_2 \qquad\qquad R_2 = 100r_2$$
$$R_3 = -r_1 + r_3$$

$$\rightarrow \begin{bmatrix} 1 & 0 & -1 & | & 2000 \\ 0 & 1 & 2 & | & 8000 \\ 0 & 0 & -1 & | & -2000 \end{bmatrix} \rightarrow \begin{bmatrix} 1 & 0 & -1 & | & 2000 \\ 0 & 1 & 2 & | & 8000 \\ 0 & 0 & 1 & | & 2000 \end{bmatrix} \rightarrow \begin{bmatrix} 1 & 0 & 0 & | & 4000 \\ 0 & 1 & 0 & | & 4000 \\ 0 & 0 & 1 & | & 2000 \end{bmatrix}$$

$$R_1 = -r_2 + r_1 \qquad R_3 = -r_3 \qquad R_1 = r_3 + r_1$$
$$R_3 = r_2 + r_3 \qquad\qquad\qquad R_2 = -2r_3 + r_2$$

Carletta should invest \$4000 in Treasury bills, \$4000 in Treasury bonds, and \$2000 in corporate bonds.

77. Let x = the number of Deltas produced.
Let y = the number of Betas produced.
Let z = the number of Sigmas produced.
Painting equation: $10x + 16y + 8z = 240$
Drying equation: $3x + 5y + 2z = 69$
Polishing equation: $2x + 3y + z = 41$
Set up a matrix and solve:

$$\begin{bmatrix} 10 & 16 & 8 & | & 240 \\ 3 & 5 & 2 & | & 69 \\ 2 & 3 & 1 & | & 41 \end{bmatrix} \rightarrow \begin{bmatrix} 1 & 1 & 2 & | & 33 \\ 3 & 5 & 2 & | & 69 \\ 2 & 3 & 1 & | & 41 \end{bmatrix} \rightarrow \begin{bmatrix} 1 & 1 & 2 & | & 33 \\ 0 & 2 & -4 & | & -30 \\ 0 & 1 & -3 & | & -25 \end{bmatrix} \rightarrow \begin{bmatrix} 1 & 1 & 2 & | & 33 \\ 0 & 1 & -2 & | & -15 \\ 0 & 1 & -3 & | & -25 \end{bmatrix}$$

$$R_1 = -3r_2 + r_1 \qquad R_2 = -3r_1 + r_2 \qquad R_2 = \tfrac{1}{2}r_2$$
$$R_3 = -2r_1 + r_3$$

$$\rightarrow \begin{bmatrix} 1 & 0 & 4 & | & 48 \\ 0 & 1 & -2 & | & -15 \\ 0 & 0 & -1 & | & -10 \end{bmatrix} \rightarrow \begin{bmatrix} 1 & 0 & 4 & | & 48 \\ 0 & 1 & -2 & | & -15 \\ 0 & 0 & 1 & | & 10 \end{bmatrix} \rightarrow \begin{bmatrix} 1 & 0 & 0 & | & 8 \\ 0 & 1 & 0 & | & 5 \\ 0 & 0 & 1 & | & 10 \end{bmatrix}$$

$$R_1 = -r_2 + r_1 \qquad R_3 = -r_3 \qquad R_1 = -4r_3 + r_1$$
$$R_3 = -r_2 + r_3 \qquad\qquad\qquad R_2 = 2r_3 + r_2$$

The company should produce 8 Deltas, 5 Betas, and 10 Sigmas.

79. Rewrite the system as set up and solve the matrix:

$$\begin{cases} -4 + 8 - 2I_2 = 0 \\ 8 = 5I_4 + I_1 \\ 4 = 3I_3 + I_1 \\ I_3 + I_4 = I_1 \end{cases} \rightarrow \begin{cases} 2I_2 = 4 \\ I_1 + 5I_4 = 8 \\ I_1 + 3I_3 = 4 \\ I_1 - I_3 - I_4 = 0 \end{cases}$$

$$\begin{bmatrix} 0 & 2 & 0 & 0 & | & 4 \\ 1 & 0 & 0 & 5 & | & 8 \\ 1 & 0 & 3 & 0 & | & 4 \\ 1 & 0 & -1 & -1 & | & 0 \end{bmatrix} \rightarrow \begin{bmatrix} 1 & 0 & 0 & 5 & | & 8 \\ 0 & 2 & 0 & 0 & | & 4 \\ 1 & 0 & 3 & 0 & | & 4 \\ 1 & 0 & -1 & -1 & | & 0 \end{bmatrix} \rightarrow \begin{bmatrix} 1 & 0 & 0 & 5 & | & 8 \\ 0 & 1 & 0 & 0 & | & 2 \\ 0 & 0 & 3 & -5 & | & -4 \\ 0 & 0 & -1 & -6 & | & -8 \end{bmatrix}$$

$$\text{Interchange } r_2 \text{ and } r_1 \qquad R_2 = \tfrac{1}{2} r_2$$
$$R_3 = -r_1 + r_3$$
$$R_4 = -r_1 + r_4$$

$$\rightarrow \begin{bmatrix} 1 & 0 & 0 & 5 & | & 8 \\ 0 & 1 & 0 & 0 & | & 2 \\ 0 & 0 & -1 & -6 & | & -8 \\ 0 & 0 & 3 & -5 & | & -4 \end{bmatrix} \rightarrow \begin{bmatrix} 1 & 0 & 0 & 5 & | & 8 \\ 0 & 1 & 0 & 0 & | & 2 \\ 0 & 0 & 1 & 6 & | & 8 \\ 0 & 0 & 0 & -23 & | & -28 \end{bmatrix} \rightarrow \begin{bmatrix} 1 & 0 & 0 & 5 & | & 8 \\ 0 & 1 & 0 & 0 & | & 2 \\ 0 & 0 & 1 & 6 & | & 8 \\ 0 & 0 & 0 & 1 & | & \frac{28}{23} \end{bmatrix}$$

$$\text{Interchange } r_3 \text{ and } r_4 \qquad R_3 = -r_3 \qquad\qquad R_4 = -\tfrac{1}{23} r_4$$
$$R_4 = -3 r_3 + r_4$$

$$\rightarrow \begin{bmatrix} 1 & 0 & 0 & 0 & | & \frac{44}{23} \\ 0 & 1 & 0 & 0 & | & 2 \\ 0 & 0 & 1 & 0 & | & \frac{16}{23} \\ 0 & 0 & 0 & 1 & | & \frac{28}{23} \end{bmatrix}$$

$$R_1 = -5 r_4 + r_1$$
$$R_3 = -6 r_4 + r_3$$

The solution is $I_1 = \dfrac{44}{23}$, $I_2 = 2$, $I_3 = \dfrac{16}{23}$, $I_4 = \dfrac{28}{23}$.

81.   Let x = the amount invested in Treasury bills.
      Let y = the amount invested in Treasury bonds.
      Let z = the amount invested in corporate bonds.

(a)   Total investment equation: $x + y + z = 20000$
       Annual income equation: $0.07x + 0.09y + 0.11z = 2000$

Set up a matrix and solve:

$$\begin{bmatrix} 1 & 1 & 1 & | & 20000 \\ .07 & .09 & .11 & | & 2000 \end{bmatrix} \rightarrow \begin{bmatrix} 1 & 1 & 1 & | & 20000 \\ 7 & 9 & 11 & | & 200000 \end{bmatrix} \rightarrow \begin{bmatrix} 1 & 1 & 1 & | & 20000 \\ 0 & 2 & 4 & | & 60000 \end{bmatrix}$$

$$R_2 = 100 r_2 \qquad\qquad R_2 = r_2 - 7 r_1$$

$$\rightarrow \begin{bmatrix} 1 & 1 & 1 & | & 20000 \\ 0 & 1 & 2 & | & 30000 \end{bmatrix} \rightarrow \begin{bmatrix} 1 & 0 & -1 & | & -10000 \\ 0 & 1 & 2 & | & 30000 \end{bmatrix}$$

$$R_2 = \tfrac{1}{2} r_2 \qquad\qquad R_1 = r_1 - r_2$$

The matrix in the last step represents the system $\begin{cases} x - z = -10000 \\ y + 2z = 30000 \end{cases}$

Therefore the solution is

$$x = -10000 + z, \quad y = 30000 - 2z; \quad z \text{ is any real number}$$

Possible investment strategies:

Amount invested at		
7%	9%	11%
0	10000	10000
1000	8000	11000
2000	6000	12000
3000	4000	13000
4000	2000	14000
5000	0	15000

(b)   Total investment equation: $x + y + z = 25000$

Annual income equation: $0.07x + 0.09y + 0.11z = 2000$

Set up a matrix and solve:

$$\begin{bmatrix} 1 & 1 & 1 & | & 25000 \\ .07 & .09 & .11 & | & 2000 \end{bmatrix} \rightarrow \begin{bmatrix} 1 & 1 & 1 & | & 25000 \\ 7 & 9 & 11 & | & 200000 \end{bmatrix} \rightarrow \begin{bmatrix} 1 & 1 & 1 & | & 25000 \\ 0 & 2 & 4 & | & 25000 \end{bmatrix}$$

$$R_2 = 100r_2 \qquad\qquad R_2 = r_2 - 7r_1$$

$$\rightarrow \begin{bmatrix} 1 & 1 & 1 & | & 25000 \\ 0 & 1 & 2 & | & 12500 \end{bmatrix} \rightarrow \begin{bmatrix} 1 & 0 & -1 & | & 12500 \\ 0 & 1 & 2 & | & 12500 \end{bmatrix}$$

$$R_2 = \tfrac{1}{2}r_2 \qquad\qquad R_1 = r_1 - r_2$$

The matrix in the last step represents the system $\begin{cases} x - z = 12500 \\ y + 2z = 12500 \end{cases}$

Therefore the solution is

$$x = 12500 + z, \quad y = 12500 - 2z, \quad z \text{ is any real number}$$

Possible investment strategies:

Amount invested at		
7%	9%	11%
12500	12500	0
14500	8500	2000
16500	4500	4000
18750	0	6250

(c)   Total investment equation: $x + y + z = 30000$

Annual income equation: $0.07x + 0.09y + 0.11z = 2000$

Set up a matrix and solve:

$$\begin{bmatrix} 1 & 1 & 1 & | & 30000 \\ .07 & .09 & .11 & | & 2000 \end{bmatrix} \rightarrow \begin{bmatrix} 1 & 1 & 1 & | & 30000 \\ 7 & 9 & 11 & | & 200000 \end{bmatrix} \rightarrow \begin{bmatrix} 1 & 1 & 1 & | & 30000 \\ 0 & 2 & 4 & | & -10000 \end{bmatrix}$$

$$R_2 = 100r_2 \qquad\qquad R_1 = r_2 - 7r_1$$

$$\rightarrow \begin{bmatrix} 1 & 1 & 1 & | & 30000 \\ 0 & 1 & 2 & | & -5000 \end{bmatrix} \rightarrow \begin{bmatrix} 1 & 0 & -1 & | & 35000 \\ 0 & 1 & 2 & | & -5000 \end{bmatrix}$$

$$R_2 = \tfrac{1}{2}r_2 \qquad\qquad R_1 = r_1 - r_2$$

The matrix in the last step represents the system $\begin{cases} x - z = 35000 \\ y + 2z = -5000 \end{cases}$

Therefore the solution is

$$x = 35000 + z, \ y = -5000 - 2z; \ z \text{ is any real number}$$

One possible investment strategy

	Amount invested at	
7%	9%	11%
30000	0	0

This will yield ($30000)(.07) = $2100, which is more than the required income.

83. Let x = the amount of liquid 1.
Let y = the amount of liquid 2.
Let z = the amount of liquid 3.
$$.20x + .40y + .30z = 40 \ \text{Vitamin C}$$
$$.30x + .20y + .50z = 30 \ \text{Vitamin D}$$
multiplying each equation by 10 yields
$$2x + 4y + 3z = 400$$
$$3x + 2y + 5z = 300$$

Set up a matrix and solve: $\begin{bmatrix} 2 & 4 & 3 & | & 400 \\ 3 & 2 & 5 & | & 300 \end{bmatrix} \rightarrow \begin{bmatrix} 1 & 2 & \frac{3}{2} & | & 200 \\ 3 & 2 & 5 & | & 300 \end{bmatrix} \rightarrow \begin{bmatrix} 1 & 2 & \frac{3}{2} & | & 200 \\ 0 & -4 & \frac{1}{2} & | & -300 \end{bmatrix}$

$$R_1 = \tfrac{1}{2}r_1 \qquad R_2 = r_2 - 3r_1$$

$$\rightarrow \begin{bmatrix} 1 & 2 & \frac{3}{2} & | & 200 \\ 0 & 1 & -\frac{1}{8} & | & 75 \end{bmatrix} \rightarrow \begin{bmatrix} 1 & 0 & \frac{7}{4} & | & 50 \\ 0 & 1 & -\frac{1}{8} & | & 75 \end{bmatrix}$$

$$R_2 = -\tfrac{1}{4}r_2 \qquad R_1 = r_1 - 2r_2$$

The matrix in the last step represents the system $\begin{cases} x + \frac{7}{4}z = 50 \\ y - \frac{1}{8}z = 75 \end{cases}$

Therefore the solution is

$$x = 50 - \tfrac{7}{4}z, \ y = 75 + \tfrac{1}{8}z, \ z \text{ is any real number}$$

Possible combinations:

Liquid 1	Liquid 2	Liquid 3
50mg	75mg	0mg
36mg	76mg	8mg
22mg	77mg	16mg
8mg	78mg	24mg

85 – 87.   Answers will vary.

Systems of Equations and Inequalities

10.4 Systems of Linear Equations: Determinants

1. Evaluating the determinant:
$$\begin{vmatrix} 3 & 1 \\ 4 & 2 \end{vmatrix} = 3(2) - 4(1) = 6 - 4 = 2$$

3. Evaluating the determinant:
$$\begin{vmatrix} 6 & 4 \\ -1 & 3 \end{vmatrix} = 6(3) - (-1)(4) = 18 + 4 = 22$$

5. Evaluating the determinant:
$$\begin{vmatrix} -3 & -1 \\ 4 & 2 \end{vmatrix} = -3(2) - 4(-1) = -6 + 4 = -2$$

7. Evaluating the determinant:
$$\begin{vmatrix} 3 & 4 & 2 \\ 1 & -1 & 5 \\ 1 & 2 & -2 \end{vmatrix} = 3\begin{vmatrix} -1 & 5 \\ 2 & -2 \end{vmatrix} - 4\begin{vmatrix} 1 & 5 \\ 1 & -2 \end{vmatrix} + 2\begin{vmatrix} 1 & -1 \\ 1 & 2 \end{vmatrix}$$
$$= 3[(-1)(-2) - 2(5)] - 4[1(-2) - 1(5)] + 2[1(2) - 1(-1)]$$
$$= 3(2 - 10) - 4(-2 - 5) + 2(2 + 1)$$
$$= 3(-8) - 4(-7) + 2(3)$$
$$= -24 + 28 + 6$$
$$= 10$$

9. Evaluating the determinant:
$$\begin{vmatrix} 4 & -1 & 2 \\ 6 & -1 & 0 \\ 1 & -3 & 4 \end{vmatrix} = 4\begin{vmatrix} -1 & 0 \\ -3 & 4 \end{vmatrix} - (-1)\begin{vmatrix} 6 & 0 \\ 1 & 4 \end{vmatrix} + 2\begin{vmatrix} 6 & -1 \\ 1 & -3 \end{vmatrix}$$
$$= 4[-1(4) - 0(-3)] + 1[6(4) - 1(0)] + 2[6(-3) - 1(-1)]$$
$$= 4(-4) + 1(24) + 2(-17)$$
$$= -16 + 24 - 34$$
$$= -26$$

11. Set up and evaluate the determinants to use Cramer's Rule:

$$\begin{cases} x + y = 8 \\ x - y = 4 \end{cases}$$

$$D = \begin{vmatrix} 1 & 1 \\ 1 & -1 \end{vmatrix} = 1(-1) - 1(1) = -1 - 1 = -2$$

$$D_x = \begin{vmatrix} 8 & 1 \\ 4 & -1 \end{vmatrix} = 8(-1) - 4(1) = -8 - 4 = -12$$

$$D_y = \begin{vmatrix} 1 & 8 \\ 1 & 4 \end{vmatrix} = 1(4) - 1(8) = 4 - 8 = -4$$

Find the solutions by Cramer's Rule:

$$x = \frac{D_x}{D} = \frac{-12}{-2} = 6 \qquad y = \frac{D_y}{D} = \frac{-4}{-2} = 2$$

13. Set up and evaluate the determinants to use Cramer's Rule:

$$\begin{cases} 5x - y = 13 \\ 2x + 3y = 12 \end{cases}$$

$$D = \begin{vmatrix} 5 & -1 \\ 2 & 3 \end{vmatrix} = 5(3) - 2(-1) = 15 + 2 = 17$$

$$D_x = \begin{vmatrix} 13 & -1 \\ 12 & 3 \end{vmatrix} = 13(3) - 12(-1) = 39 + 12 = 51$$

$$D_y = \begin{vmatrix} 5 & 13 \\ 2 & 12 \end{vmatrix} = 5(12) - 2(13) = 60 - 26 = 34$$

Find the solutions by Cramer's Rule:

$$x = \frac{D_x}{D} = \frac{51}{17} = 3 \qquad y = \frac{D_y}{D} = \frac{34}{17} = 2$$

15. Set up and evaluate the determinants to use Cramer's Rule:

$$\begin{cases} 3x \qquad = 24 \\ x + 2y = 0 \end{cases}$$

$$D = \begin{vmatrix} 3 & 0 \\ 1 & 2 \end{vmatrix} = 6 - 0 = 6; \quad D_x = \begin{vmatrix} 24 & 0 \\ 0 & 2 \end{vmatrix} = 48 - 0 = 48; \quad D_y = \begin{vmatrix} 3 & 24 \\ 1 & 0 \end{vmatrix} = 0 - 24 = -24$$

Find the solutions by Cramer's Rule:

$$x = \frac{D_x}{D} = \frac{48}{6} = 8 \qquad y = \frac{D_y}{D} = \frac{-24}{6} = -4$$

17. Set up and evaluate the determinants to use Cramer's Rule:

$$\begin{cases} 3x - 6y = 24 \\ 5x + 4y = 12 \end{cases}$$

$$D = \begin{vmatrix} 3 & -6 \\ 5 & 4 \end{vmatrix} = 12 - (-30) = 42; \quad D_x = \begin{vmatrix} 24 & -6 \\ 12 & 4 \end{vmatrix} = 96 - (-72) = 168; \quad D_y = \begin{vmatrix} 3 & 24 \\ 5 & 12 \end{vmatrix} = 36 - 120 = -84$$

Find the solutions by Cramer's Rule:

$$x = \frac{D_x}{D} = \frac{168}{42} = 4 \qquad y = \frac{D_y}{D} = \frac{-84}{42} = -2$$

19. Set up and evaluate the determinants to use Cramer's Rule:

$$\begin{cases} 3x - 2y = 4 \\ 6x - 4y = 0 \end{cases}$$

$$D = \begin{vmatrix} 3 & -2 \\ 6 & -4 \end{vmatrix} = -12 - (-12) = 0$$

Since $D = 0$, Cramer's Rule does not apply.

21. Set up and evaluate the determinants to use Cramer's Rule:

$$\begin{cases} 2x - 4y = -2 \\ 3x + 2y = 3 \end{cases}$$

$$D = \begin{vmatrix} 2 & -4 \\ 3 & 2 \end{vmatrix} = 4 - (-12) = 16$$

$$D_x = \begin{vmatrix} -2 & -4 \\ 3 & 2 \end{vmatrix} = -4 - (-12) = 8$$

$$D_y = \begin{vmatrix} 2 & -2 \\ 3 & 3 \end{vmatrix} = 6 - (-6) = 12$$

Find the solutions by Cramer's Rule:

$$x = \frac{D_x}{D} = \frac{8}{16} = \frac{1}{2} \qquad y = \frac{D_y}{D} = \frac{12}{16} = \frac{3}{4}$$

23. Set up and evaluate the determinants to use Cramer's Rule:

$$\begin{cases} 2x - 3y = -1 \\ 10x + 10y = 5 \end{cases}$$

$$D = \begin{vmatrix} 2 & -3 \\ 10 & 10 \end{vmatrix} = 20 - (-30) = 50$$

$$D_x = \begin{vmatrix} -1 & -3 \\ 5 & 10 \end{vmatrix} = -10 - (-15) = 5$$

$$D_y = \begin{vmatrix} 2 & -1 \\ 10 & 5 \end{vmatrix} = 10 - (-10) = 20$$

Find the solutions by Cramer's Rule:

$$x = \frac{D_x}{D} = \frac{5}{50} = \frac{1}{10} \qquad y = \frac{D_y}{D} = \frac{20}{50} = \frac{2}{5}$$

25. Set up and evaluate the determinants to use Cramer's Rule:

$$\begin{cases} 2x + 3y = 6 \\ x - y = \dfrac{1}{2} \end{cases}$$

$$D = \begin{vmatrix} 2 & 3 \\ 1 & -1 \end{vmatrix} = -2 - 3 = -5$$

$$D_x = \begin{vmatrix} 6 & 3 \\ \frac{1}{2} & -1 \end{vmatrix} = -6 - \frac{3}{2} = -\frac{15}{2}$$

$$D_y = \begin{vmatrix} 2 & 6 \\ 1 & \frac{1}{2} \end{vmatrix} = 1 - 6 = -5$$

Find the solutions by Cramer's Rule:

$$x = \frac{D_x}{D} = \frac{\left(-\frac{15}{2}\right)}{-5} = \frac{3}{2} \qquad y = \frac{D_y}{D} = \frac{-5}{-5} = 1$$

27. Set up and evaluate the determinants to use Cramer's Rule:

$$\begin{cases} 3x - 5y = 3 \\ 15x + 5y = 21 \end{cases}$$

$$D = \begin{vmatrix} 3 & -5 \\ 15 & 5 \end{vmatrix} = 15 - (-75) = 90$$

$$D_x = \begin{vmatrix} 3 & -5 \\ 21 & 5 \end{vmatrix} = 15 - (-105) = 120$$

$$D_y = \begin{vmatrix} 3 & 3 \\ 15 & 21 \end{vmatrix} = 63 - 45 = 18$$

Find the solutions by Cramer's Rule:

$$x = \frac{D_x}{D} = \frac{120}{90} = \frac{4}{3} \qquad y = \frac{D_y}{D} = \frac{18}{90} = \frac{1}{5}$$

29. Set up and evaluate the determinants to use Cramer's Rule:

$$\begin{cases} x + y - z = 6 \\ 3x - 2y + z = -5 \\ x + 3y - 2z = 14 \end{cases}$$

$$D = \begin{vmatrix} 1 & 1 & -1 \\ 3 & -2 & 1 \\ 1 & 3 & -2 \end{vmatrix} = 1\begin{vmatrix} -2 & 1 \\ 3 & -2 \end{vmatrix} - 1\begin{vmatrix} 3 & 1 \\ 1 & -2 \end{vmatrix} + (-1)\begin{vmatrix} 3 & -2 \\ 1 & 3 \end{vmatrix}$$

$$= 1(4 - 3) - 1(-6 - 1) - 1(9 + 2) = 1 + 7 - 11 = -3$$

$$D_x = \begin{vmatrix} 6 & 1 & -1 \\ -5 & -2 & 1 \\ 14 & 3 & -2 \end{vmatrix} = 6\begin{vmatrix} -2 & 1 \\ 3 & -2 \end{vmatrix} - 1\begin{vmatrix} -5 & 1 \\ 14 & -2 \end{vmatrix} + (-1)\begin{vmatrix} -5 & -2 \\ 14 & 3 \end{vmatrix}$$

$$= 6(4 - 3) - 1(10 - 14) - 1(-15 + 28) = 6 + 4 - 13 = -3$$

$$D_y = \begin{vmatrix} 1 & 6 & -1 \\ 3 & -5 & 1 \\ 1 & 14 & -2 \end{vmatrix} = 1\begin{vmatrix} -5 & 1 \\ 14 & -2 \end{vmatrix} - 6\begin{vmatrix} 3 & 1 \\ 1 & -2 \end{vmatrix} + (-1)\begin{vmatrix} 3 & -5 \\ 1 & 14 \end{vmatrix}$$

$$= 1(10 - 14) - 6(-6 - 1) - 1(42 + 5) = -4 + 42 - 47 = -9$$

$$D_z = \begin{vmatrix} 1 & 1 & 6 \\ 3 & -2 & -5 \\ 1 & 3 & 14 \end{vmatrix} = 1\begin{vmatrix} -2 & -5 \\ 3 & 14 \end{vmatrix} - 1\begin{vmatrix} 3 & -5 \\ 1 & 14 \end{vmatrix} + 6\begin{vmatrix} 3 & -2 \\ 1 & 3 \end{vmatrix}$$

$$= 1(-28 + 15) - 1(42 + 5) + 6(9 + 2) = -13 - 47 + 66 = 6$$

Find the solutions by Cramer's Rule:

$$x = \frac{D_x}{D} = \frac{-3}{-3} = 1 \qquad y = \frac{D_y}{D} = \frac{-9}{-3} = 3 \qquad z = \frac{D_z}{D} = \frac{6}{-3} = -2$$

31. Set up and evaluate the determinants to use Cramer's Rule:

$$\begin{cases} x + 2y - z = -3 \\ 2x - 4y + z = -7 \\ -2x + 2y - 3z = 4 \end{cases}$$

$$D = \begin{vmatrix} 1 & 2 & -1 \\ 2 & -4 & 1 \\ -2 & 2 & -3 \end{vmatrix} = 1 \begin{vmatrix} -4 & 1 \\ 2 & -3 \end{vmatrix} - 2 \begin{vmatrix} 2 & 1 \\ -2 & -3 \end{vmatrix} + (-1) \begin{vmatrix} 2 & -4 \\ -2 & 2 \end{vmatrix}$$

$$= 1(12 - 2) - 2(-6 + 2) - 1(4 - 8) = 10 + 8 + 4 = 22$$

$$D_x = \begin{vmatrix} -3 & 2 & -1 \\ -7 & -4 & 1 \\ 4 & 2 & -3 \end{vmatrix} = -3 \begin{vmatrix} -4 & 1 \\ 2 & -3 \end{vmatrix} - 2 \begin{vmatrix} -7 & 1 \\ 4 & -3 \end{vmatrix} + (-1) \begin{vmatrix} -7 & -4 \\ 4 & 2 \end{vmatrix}$$

$$= -3(12 - 2) - 2(21 - 4) - 1(-14 + 16) = -30 - 34 - 2 = -66$$

$$D_y = \begin{vmatrix} 1 & -3 & -1 \\ 2 & -7 & 1 \\ -2 & 4 & -3 \end{vmatrix} = 1 \begin{vmatrix} -7 & 1 \\ 4 & -3 \end{vmatrix} - (-3) \begin{vmatrix} 2 & 1 \\ -2 & -3 \end{vmatrix} + (-1) \begin{vmatrix} 2 & -7 \\ -2 & 4 \end{vmatrix}$$

$$= 1(21 - 4) + 3(-6 + 2) - 1(8 - 14) = 17 - 12 + 6 = 11$$

$$D_z = \begin{vmatrix} 1 & 2 & -3 \\ 2 & -4 & -7 \\ -2 & 2 & 4 \end{vmatrix} = 1 \begin{vmatrix} -4 & -7 \\ 2 & 4 \end{vmatrix} - 2 \begin{vmatrix} 2 & -7 \\ -2 & 4 \end{vmatrix} + (-3) \begin{vmatrix} 2 & -4 \\ -2 & 2 \end{vmatrix}$$

$$= 1(-16 + 14) - 2(8 - 14) - 3(4 - 8) = -2 + 12 + 12 = 22$$

Find the solutions by Cramer's Rule:

$$x = \frac{D_x}{D} = \frac{-66}{22} = -3 \qquad y = \frac{D_y}{D} = \frac{11}{22} = \frac{1}{2} \qquad z = \frac{D_z}{D} = \frac{22}{22} = 1$$

33. Set up and evaluate the determinants to use Cramer's Rule:

$$\begin{cases} x - 2y + 3z = 1 \\ 3x + y - 2z = 0 \\ 2x - 4y + 6z = 2 \end{cases}$$

$$D = \begin{vmatrix} 1 & -2 & 3 \\ 3 & 1 & -2 \\ 2 & -4 & 6 \end{vmatrix} = 1 \begin{vmatrix} 1 & -2 \\ -4 & 6 \end{vmatrix} - (-2) \begin{vmatrix} 3 & -2 \\ 2 & 6 \end{vmatrix} + 3 \begin{vmatrix} 3 & 1 \\ 2 & -4 \end{vmatrix}$$

$$= 1(6 - 8) + 2(18 + 4) + 3(-12 - 2) = -2 + 44 - 42 = 0$$

Since $D = 0$, Cramer's Rule does not apply.

35. Set up and evaluate the determinants to use Cramer's Rule:

$$\begin{cases} x + 2y - z = 0 \\ 2x - 4y + z = 0 \\ -2x + 2y - 3z = 0 \end{cases}$$

$$D = \begin{vmatrix} 1 & 2 & -1 \\ 2 & -4 & 1 \\ -2 & 2 & -3 \end{vmatrix} = 1\begin{vmatrix} -4 & 1 \\ 2 & -3 \end{vmatrix} - 2\begin{vmatrix} 2 & 1 \\ -2 & -3 \end{vmatrix} + (-1)\begin{vmatrix} 2 & -4 \\ -2 & 2 \end{vmatrix}$$

$$= 1(12 - 2) - 2(-6 + 2) - 1(4 - 8) = 10 + 8 + 4 = 22$$

$$D_x = \begin{vmatrix} 0 & 2 & -1 \\ 0 & -4 & 1 \\ 0 & 2 & -3 \end{vmatrix} = 0 \quad \text{(By Theorem 12)}$$

$$D_y = \begin{vmatrix} 1 & 0 & -1 \\ 2 & 0 & 1 \\ -2 & 0 & -3 \end{vmatrix} = 0 \quad \text{(By Theorem 12)}$$

$$D_z = \begin{vmatrix} 1 & 2 & 0 \\ 2 & -4 & 0 \\ -2 & 2 & 0 \end{vmatrix} = 0 \quad \text{(By Theorem 12)}$$

Find the solutions by Cramer's Rule:

$$x = \frac{D_x}{D} = \frac{0}{22} = 0 \qquad y = \frac{D_y}{D} = \frac{0}{22} = 0 \qquad z = \frac{D_z}{D} = \frac{0}{22} = 0$$

37. Set up and evaluate the determinants to use Cramer's Rule:

$$\begin{cases} x - 2y + 3z = 0 \\ 3x + y - 2z = 0 \\ 2x - 4y + 6z = 0 \end{cases}$$

$$D = \begin{vmatrix} 1 & -2 & 3 \\ 3 & 1 & -2 \\ 2 & -4 & 6 \end{vmatrix} = 1\begin{vmatrix} 1 & -2 \\ -4 & 6 \end{vmatrix} - (-2)\begin{vmatrix} 3 & -2 \\ 2 & 6 \end{vmatrix} + 3\begin{vmatrix} 3 & 1 \\ 2 & -4 \end{vmatrix}$$

$$= 1(6 - 8) + 2(18 + 4) + 3(-12 - 2) = -2 + 44 - 42 = 0$$

Since $D = 0$, Cramer's Rule does not apply.

39. Rewrite the system letting $u = \dfrac{1}{x}$ and $v = \dfrac{1}{y}$:

$$\begin{cases} \dfrac{1}{x} + \dfrac{1}{y} = 8 \\ \dfrac{3}{x} - \dfrac{5}{y} = 0 \end{cases} \quad \rightarrow \quad \begin{cases} u + v = 8 \\ 3u - 5v = 0 \end{cases}$$

Set up and evaluate the determinants to use Cramer's Rule:

$$D = \begin{vmatrix} 1 & 1 \\ 3 & -5 \end{vmatrix} = -5 - 3 = -8$$

$$D_u = \begin{vmatrix} 8 & 1 \\ 0 & -5 \end{vmatrix} = -40 - 0 = -40$$

$$D_v = \begin{vmatrix} 1 & 8 \\ 3 & 0 \end{vmatrix} = 0 - 24 = -24$$

Find the solutions by Cramer's Rule:

$$u = \frac{D_u}{D} = \frac{-40}{-8} = 5 \qquad v = \frac{D_v}{D} = \frac{-24}{-8} = 3$$

The solutions are $x = \dfrac{1}{5}, \; y = \dfrac{1}{3}$

41. Solve for x:

$$\begin{vmatrix} x & x \\ 4 & 3 \end{vmatrix} = 3x - 4x = -x$$

$$-x = 5 \rightarrow x = -5$$

43. Solve for x:

$$\begin{vmatrix} x & 1 & 1 \\ 4 & 3 & 2 \\ -1 & 2 & 5 \end{vmatrix} = x\begin{vmatrix} 3 & 2 \\ 2 & 5 \end{vmatrix} - 1\begin{vmatrix} 4 & 2 \\ -1 & 5 \end{vmatrix} + 1\begin{vmatrix} 4 & 3 \\ -1 & 2 \end{vmatrix}$$

$$= x(15 - 4) - (20 + 2) + (8 + 3) = 11x - 22 + 11 = 11x - 11$$

So, $11x - 11 = 2 \rightarrow 11x = 13 \rightarrow x = \dfrac{13}{11}$

45. Solve for x:

$$\begin{vmatrix} x & 2 & 3 \\ 1 & x & 0 \\ 6 & 1 & -2 \end{vmatrix} = x\begin{vmatrix} x & 0 \\ 1 & -2 \end{vmatrix} - 2\begin{vmatrix} 1 & 0 \\ 6 & -2 \end{vmatrix} + 3\begin{vmatrix} 1 & x \\ 6 & 1 \end{vmatrix}$$

$$= x(-2x - 0) - 2(-2 - 0) + 3(1 - 6x)$$

$$= -2x^2 + 4 + 3 - 18x = -2x^2 - 18x + 7$$

So, $-2x^2 - 18x + 7 = 7$

$$-2x^2 - 18x = 0$$

$$-2x(x + 9) = 0$$

$$x = 0 \text{ or } x = -9$$

47. Let $\begin{vmatrix} x & y & z \\ u & v & w \\ 1 & 2 & 3 \end{vmatrix} = 4$

Then $\begin{vmatrix} 1 & 2 & 3 \\ u & v & w \\ x & y & z \end{vmatrix} = -4$ by Theorem 11

The value of the determinant changes sign when two rows are interchanged.

Problems 49 – 53 use the Laws for Determinants in reverse order.

49. Let $\begin{vmatrix} x & y & z \\ u & v & w \\ 1 & 2 & 3 \end{vmatrix} = 4$

$$\begin{vmatrix} x & y & z \\ -3 & -6 & -9 \\ u & v & w \end{vmatrix} = -3\begin{vmatrix} x & y & z \\ 1 & 2 & 3 \\ u & v & w \end{vmatrix} = -3(-1)\begin{vmatrix} x & y & z \\ u & v & w \\ 1 & 2 & 3 \end{vmatrix} = 3(4) = 12$$

Theorem 14   Theorem 11

527

51.   Let $\begin{vmatrix} x & y & z \\ u & v & w \\ 1 & 2 & 3 \end{vmatrix} = 4$

$$\begin{vmatrix} 1 & 2 & 3 \\ x-3 & y-6 & z-9 \\ 2u & 2v & 2w \end{vmatrix} = 2\begin{vmatrix} 1 & 2 & 3 \\ x-3 & y-6 & z-9 \\ u & v & w \end{vmatrix} = 2(-1)\begin{vmatrix} x-3 & y-6 & z-9 \\ 1 & 2 & 3 \\ u & v & w \end{vmatrix}$$

$$\underbrace{\qquad\qquad\qquad}_{\text{Theorem 14}} \qquad \underbrace{\qquad\qquad\qquad}_{\text{Theorem 11}}$$

$$= 2(-1)(-1)\begin{vmatrix} x-3 & y-6 & z-9 \\ u & v & w \\ 1 & 2 & 3 \end{vmatrix} = 2(-1)(-1)\begin{vmatrix} x & y & z \\ u & v & w \\ 1 & 2 & 3 \end{vmatrix} = 2(-1)(-1)(4) = 8$$

$$\underbrace{\qquad\qquad}_{\text{Theorem 11}} \qquad\qquad \underbrace{\qquad\qquad}_{\text{Theorem 15 } (R_1 = -3r_3 + r_1)}$$

53.   Let $\begin{vmatrix} x & y & z \\ u & v & w \\ 1 & 2 & 3 \end{vmatrix} = 4$

$$\begin{vmatrix} 1 & 2 & 3 \\ 2x & 2y & 2z \\ u-1 & v-2 & w-3 \end{vmatrix} = 2\begin{vmatrix} 1 & 2 & 3 \\ x & y & z \\ u-1 & v-2 & w-3 \end{vmatrix} = 2(-1)\begin{vmatrix} x & y & z \\ 1 & 2 & 3 \\ u-1 & v-2 & w-3 \end{vmatrix}$$

$$\underbrace{\qquad\qquad\qquad}_{\text{Theorem 14}} \qquad \underbrace{\qquad\qquad\qquad}_{\text{Theorem 11}}$$

$$= 2(-1)(-1)\begin{vmatrix} x & y & z \\ u-1 & v-2 & w-3 \\ 1 & 2 & 3 \end{vmatrix} = 2(-1)(-1)\begin{vmatrix} x & y & z \\ u & v & w \\ 1 & 2 & 3 \end{vmatrix} = 2(-1)(-1)(4) = 8$$

$$\underbrace{\qquad\qquad}_{\text{Theorem 11}} \qquad\qquad \underbrace{\qquad\qquad}_{\text{Theorem 15 } (R_2 = -r_3 + r_2)}$$

55.   Expanding the determinant:

$$\begin{vmatrix} x & y & 1 \\ x_1 & y_1 & 1 \\ x_2 & y_2 & 1 \end{vmatrix} = x\begin{vmatrix} y_1 & 1 \\ y_2 & 1 \end{vmatrix} - y\begin{vmatrix} x_1 & 1 \\ x_2 & 1 \end{vmatrix} + 1\begin{vmatrix} x_1 & y_1 \\ x_2 & y_2 \end{vmatrix}$$

$$= x(y_1 - y_2) - y(x_1 - x_2) + (x_1 y_2 - x_2 y_1) = 0$$

$$x(y_1 - y_2) + y(x_2 - x_1) = x_2 y_1 - x_1 y_2$$

$$y(x_2 - x_1) = x_2 y_1 - x_1 y_2 + x(y_2 - y_1)$$

$$y(x_2 - x_1) - y_1(x_2 - x_1) = x_2 y_1 - x_1 y_2 + x(y_2 - y_1) - y_1(x_2 - x_1)$$

$$(x_2 - x_1)(y - y_1) = x(y_2 - y_1) + x_2 y_1 - x_1 y_2 - y_1 x_2 + y_1 x_1$$

$$(x_2 - x_1)(y - y_1) = (y_2 - y_1)x - (y_2 - y_1)x_1$$

$$(x_2 - x_1)(y - y_1) = (y_2 - y_1)(x - x_1)$$

$$(y - y_1) = \frac{(y_2 - y_1)}{(x_2 - x_1)}(x - x_1)$$

57. Expanding the determinant:

$$\begin{vmatrix} x^2 & x & 1 \\ y^2 & y & 1 \\ z^2 & z & 1 \end{vmatrix} = x^2 \begin{vmatrix} y & 1 \\ z & 1 \end{vmatrix} - x \begin{vmatrix} y^2 & 1 \\ z^2 & 1 \end{vmatrix} + 1 \begin{vmatrix} y^2 & y \\ z^2 & z \end{vmatrix}$$

$$= x^2(y-z) - x(y^2 - z^2) + 1(y^2 z - z^2 y)$$
$$= x^2(y-z) - x(y-z)(y+z) + yz(y-z)$$
$$= (y-z)\left[x^2 - xy - xz + yz\right]$$
$$= (y-z)\left[x(x-y) - z(x-y)\right]$$
$$= (y-z)(x-y)(x-z)$$

59. Evaluating the determinant to show the relationship:

$$\begin{vmatrix} a_{13} & a_{12} & a_{11} \\ a_{23} & a_{22} & a_{21} \\ a_{33} & a_{32} & a_{31} \end{vmatrix} = a_{13} \begin{vmatrix} a_{22} & a_{21} \\ a_{32} & a_{31} \end{vmatrix} - a_{12} \begin{vmatrix} a_{23} & a_{21} \\ a_{33} & a_{31} \end{vmatrix} + a_{11} \begin{vmatrix} a_{23} & a_{22} \\ a_{33} & a_{32} \end{vmatrix}$$

$$= a_{13}(a_{22}a_{31} - a_{21}a_{32}) - a_{12}(a_{23}a_{31} - a_{21}a_{33}) + a_{11}(a_{23}a_{32} - a_{22}a_{33})$$
$$= a_{13}a_{22}a_{31} - a_{13}a_{21}a_{32} - a_{12}a_{23}a_{31} + a_{12}a_{21}a_{33} + a_{11}a_{23}a_{32} - a_{11}a_{22}a_{33}$$
$$= -a_{11}a_{22}a_{33} + a_{11}a_{23}a_{32} + a_{12}a_{21}a_{33} - a_{12}a_{23}a_{31} - a_{13}a_{21}a_{32} + a_{13}a_{22}a_{31}$$
$$= -a_{11}(a_{22}a_{33} - a_{23}a_{32}) + a_{12}(a_{21}a_{33} - a_{23}a_{31}) - a_{13}(a_{21}a_{32} - a_{22}a_{31})$$
$$= -a_{11} \begin{vmatrix} a_{22} & a_{23} \\ a_{32} & a_{33} \end{vmatrix} + a_{12} \begin{vmatrix} a_{21} & a_{23} \\ a_{31} & a_{33} \end{vmatrix} - a_{13} \begin{vmatrix} a_{21} & a_{22} \\ a_{31} & a_{32} \end{vmatrix}$$
$$= -\left[a_{11} \begin{vmatrix} a_{22} & a_{23} \\ a_{32} & a_{33} \end{vmatrix} - a_{12} \begin{vmatrix} a_{21} & a_{23} \\ a_{31} & a_{33} \end{vmatrix} + a_{13} \begin{vmatrix} a_{21} & a_{22} \\ a_{31} & a_{32} \end{vmatrix} \right]$$
$$= -\begin{vmatrix} a_{11} & a_{12} & a_{13} \\ a_{21} & a_{22} & a_{23} \\ a_{31} & a_{32} & a_{33} \end{vmatrix}$$

61. Set up a 3 by 3 determinant in which the first column and third column are the same and evaluate:

$$\begin{vmatrix} a & b & a \\ c & d & c \\ e & f & e \end{vmatrix} = -b \begin{vmatrix} c & c \\ e & e \end{vmatrix} + d \begin{vmatrix} a & a \\ e & e \end{vmatrix} - f \begin{vmatrix} a & a \\ c & c \end{vmatrix}$$

$$= -b(ce - ce) + d(ae - ae) - f(ac - ac) = -b(0) + d(0) - f(0) = 0$$

Chapter 10

Systems of Equations and Inequalities

10.5 Matrix Algebra

1. $A + B = \begin{bmatrix} 0 & 3 & -5 \\ 1 & 2 & 6 \end{bmatrix} + \begin{bmatrix} 4 & 1 & 0 \\ -2 & 3 & -2 \end{bmatrix} = \begin{bmatrix} 0+4 & 3+1 & -5+0 \\ 1+(-2) & 2+3 & 6+(-2) \end{bmatrix} = \begin{bmatrix} 4 & 4 & -5 \\ -1 & 5 & 4 \end{bmatrix}$

3. $4A = 4\begin{bmatrix} 0 & 3 & -5 \\ 1 & 2 & 6 \end{bmatrix} = \begin{bmatrix} 4\cdot 0 & 4\cdot 3 & 4(-5) \\ 4\cdot 1 & 4\cdot 2 & 4\cdot 6 \end{bmatrix} = \begin{bmatrix} 0 & 12 & -20 \\ 4 & 8 & 24 \end{bmatrix}$

5. $3A - 2B = 3\begin{bmatrix} 0 & 3 & -5 \\ 1 & 2 & 6 \end{bmatrix} - 2\begin{bmatrix} 4 & 1 & 0 \\ -2 & 3 & -2 \end{bmatrix}$

$= \begin{bmatrix} 0 & 9 & -15 \\ 3 & 6 & 18 \end{bmatrix} - \begin{bmatrix} 8 & 2 & 0 \\ -4 & 6 & -4 \end{bmatrix} = \begin{bmatrix} -8 & 7 & -15 \\ 7 & 0 & 22 \end{bmatrix}$

7. $AC = \begin{bmatrix} 0 & 3 & -5 \\ 1 & 2 & 6 \end{bmatrix} \cdot \begin{bmatrix} 4 & 1 \\ 6 & 2 \\ -2 & 3 \end{bmatrix}$

$= \begin{bmatrix} 0(4)+3(6)+(-5)(-2) & 0(1)+3(2)+(-5)(3) \\ 1(4)+2(6)+6(-2) & 1(1)+2(2)+6(3) \end{bmatrix} = \begin{bmatrix} 28 & -9 \\ 4 & 23 \end{bmatrix}$

9. $CA = \begin{bmatrix} 4 & 1 \\ 6 & 2 \\ -2 & 3 \end{bmatrix} \cdot \begin{bmatrix} 0 & 3 & -5 \\ 1 & 2 & 6 \end{bmatrix}$

$= \begin{bmatrix} 4(0)+1(1) & 4(3)+1(2) & 4(-5)+1(6) \\ 6(0)+2(1) & 6(3)+2(2) & 6(-5)+2(6) \\ -2(0)+3(1) & -2(3)+3(2) & -2(-5)+3(6) \end{bmatrix} = \begin{bmatrix} 1 & 14 & -14 \\ 2 & 22 & -18 \\ 3 & 0 & 28 \end{bmatrix}$

11. $C(A+B) = \begin{bmatrix} 4 & 1 \\ 6 & 2 \\ -2 & 3 \end{bmatrix} \left(\begin{bmatrix} 0 & 3 & -5 \\ 1 & 2 & 6 \end{bmatrix} + \begin{bmatrix} 4 & 1 & 0 \\ -2 & 3 & -2 \end{bmatrix} \right)$

$= \begin{bmatrix} 4 & 1 \\ 6 & 2 \\ -2 & 3 \end{bmatrix} \cdot \begin{bmatrix} 4 & 4 & -5 \\ -1 & 5 & 4 \end{bmatrix} = \begin{bmatrix} 15 & 21 & -16 \\ 22 & 34 & -22 \\ -11 & 7 & 22 \end{bmatrix}$

13. $AC - 3I_2 = \begin{bmatrix} 0 & 3 & -5 \\ 1 & 2 & 6 \end{bmatrix} \cdot \begin{bmatrix} 4 & 1 \\ 6 & 2 \\ -2 & 3 \end{bmatrix} - 3\begin{bmatrix} 1 & 0 \\ 0 & 1 \end{bmatrix} = \begin{bmatrix} 28 & -9 \\ 4 & 23 \end{bmatrix} - \begin{bmatrix} 3 & 0 \\ 0 & 3 \end{bmatrix} = \begin{bmatrix} 25 & -9 \\ 4 & 20 \end{bmatrix}$

15. $CA - CB = \begin{bmatrix} 4 & 1 \\ 6 & 2 \\ -2 & 3 \end{bmatrix} \cdot \begin{bmatrix} 0 & 3 & -5 \\ 1 & 2 & 6 \end{bmatrix} - \begin{bmatrix} 4 & 1 \\ 6 & 2 \\ -2 & 3 \end{bmatrix} \cdot \begin{bmatrix} 4 & 1 & 0 \\ -2 & 3 & -2 \end{bmatrix}$

$= \begin{bmatrix} 1 & 14 & -14 \\ 2 & 22 & -18 \\ 3 & 0 & 28 \end{bmatrix} - \begin{bmatrix} 14 & 7 & -2 \\ 20 & 12 & -4 \\ -14 & 7 & -6 \end{bmatrix} = \begin{bmatrix} -13 & 7 & -12 \\ -18 & 10 & -14 \\ 17 & -7 & 34 \end{bmatrix}$

17. $\begin{bmatrix} 2 & -2 \\ 1 & 0 \end{bmatrix} \begin{bmatrix} 2 & 1 & 4 & 6 \\ 3 & -1 & 3 & 2 \end{bmatrix}$

$= \begin{bmatrix} 2(2)+(-2)(3) & 2(1)+(-2)(-1) & 2(4)+(-2)(3) & 2(6)+(-2)(2) \\ 1(2)+0(3) & 1(1)+0(-1) & 1(4)+0(3) & 1(6)+0(2) \end{bmatrix}$

$= \begin{bmatrix} -2 & 4 & 2 & 8 \\ 2 & 1 & 4 & 6 \end{bmatrix}$

19. $\begin{bmatrix} 1 & 0 & 1 \\ 2 & 4 & 1 \\ 3 & 6 & 1 \end{bmatrix} \begin{bmatrix} 1 & 3 \\ 6 & 2 \\ 8 & -1 \end{bmatrix} = \begin{bmatrix} 1(1)+0(6)+1(8) & 1(3)+0(2)+1(-1) \\ 2(1)+4(6)+1(8) & 2(3)+4(2)+1(-1) \\ 3(1)+6(6)+1(8) & 3(3)+6(2)+1(-1) \end{bmatrix} = \begin{bmatrix} 9 & 2 \\ 34 & 13 \\ 47 & 20 \end{bmatrix}$

21. Augment the matrix with the identity and use row operations to find the inverse:

$A = \begin{bmatrix} 2 & 1 \\ 1 & 1 \end{bmatrix} \rightarrow \begin{bmatrix} 2 & 1 & | & 1 & 0 \\ 1 & 1 & | & 0 & 1 \end{bmatrix}$

$\rightarrow \begin{bmatrix} 1 & 1 & | & 0 & 1 \\ 2 & 1 & | & 1 & 0 \end{bmatrix} \rightarrow \begin{bmatrix} 1 & 1 & | & 0 & 1 \\ 0 & -1 & | & 1 & -2 \end{bmatrix} \rightarrow \begin{bmatrix} 1 & 1 & | & 0 & 1 \\ 0 & 1 & | & -1 & 2 \end{bmatrix} \rightarrow \begin{bmatrix} 1 & 0 & | & 1 & -1 \\ 0 & 1 & | & -1 & 2 \end{bmatrix}$

Interchange $R_2 = -2r_1 + r_2$ $R_2 = -r_2$ $R_1 = -r_2 + r_1$
r_1 and r_2

$A^{-1} = \begin{bmatrix} 1 & -1 \\ -1 & 2 \end{bmatrix}$

23. Augment the matrix with the identity and use row operations to find the inverse:

$A = \begin{bmatrix} 6 & 5 \\ 2 & 2 \end{bmatrix} \rightarrow \begin{bmatrix} 6 & 5 & | & 1 & 0 \\ 2 & 2 & | & 0 & 1 \end{bmatrix}$

$\rightarrow \begin{bmatrix} 2 & 2 & | & 0 & 1 \\ 6 & 5 & | & 1 & 0 \end{bmatrix} \rightarrow \begin{bmatrix} 2 & 2 & | & 0 & 1 \\ 0 & -1 & | & 1 & -3 \end{bmatrix} \rightarrow \begin{bmatrix} 1 & 1 & | & 0 & \frac{1}{2} \\ 0 & 1 & | & -1 & 3 \end{bmatrix} \rightarrow \begin{bmatrix} 1 & 0 & | & 1 & -\frac{5}{2} \\ 0 & 1 & | & -1 & 3 \end{bmatrix}$

Interchange $R_2 = -3r_1 + r_2$ $R_1 = \frac{1}{2}r_1$ $R_1 = -r_2 + r_1$
r_1 and r_2 $R_2 = -r_2$

$A^{-1} = \begin{bmatrix} 1 & -\frac{5}{2} \\ -1 & 3 \end{bmatrix}$

25. Augment the matrix with the identity and use row operations to find the inverse:

$A = \begin{bmatrix} 2 & 1 \\ a & a \end{bmatrix} \rightarrow \begin{bmatrix} 2 & 1 & | & 1 & 0 \\ a & a & | & 0 & 1 \end{bmatrix}$ where $a \neq 0$.

$$\rightarrow \begin{bmatrix} 1 & \frac{1}{2} & \frac{1}{2} & 0 \\ a & a & 0 & 1 \end{bmatrix} \rightarrow \begin{bmatrix} 1 & \frac{1}{2} & \frac{1}{2} & 0 \\ 0 & \frac{1}{2}a & -\frac{1}{2}a & 1 \end{bmatrix} \rightarrow \begin{bmatrix} 1 & \frac{1}{2} & \frac{1}{2} & 0 \\ 0 & 1 & -1 & \frac{2}{a} \end{bmatrix} \rightarrow \begin{bmatrix} 1 & 0 & 1 & -\frac{1}{a} \\ 0 & 1 & -1 & \frac{2}{a} \end{bmatrix}$$

$R_1 = \frac{1}{2}r_1 \qquad R_2 = -ar_1 + r_2 \qquad R_2 = \left(\frac{2}{a}\right)r_2 \qquad R_1 = -\frac{1}{2}r_2 + r_1$

$$A^{-1} = \begin{bmatrix} 1 & -\frac{1}{a} \\ -1 & \frac{2}{a} \end{bmatrix}$$

27. Augment the matrix with the identity and use row operations to find the inverse:

$$A = \begin{bmatrix} 1 & -1 & 1 \\ 0 & -2 & 1 \\ -2 & -3 & 0 \end{bmatrix} \rightarrow \begin{bmatrix} 1 & -1 & 1 & 1 & 0 & 0 \\ 0 & -2 & 1 & 0 & 1 & 0 \\ -2 & -3 & 0 & 0 & 0 & 1 \end{bmatrix}$$

$$\rightarrow \begin{bmatrix} 1 & -1 & 1 & 1 & 0 & 0 \\ 0 & -2 & 1 & 0 & 1 & 0 \\ 0 & -5 & 2 & 2 & 0 & 1 \end{bmatrix} \rightarrow \begin{bmatrix} 1 & -1 & 1 & 1 & 0 & 0 \\ 0 & 1 & -\frac{1}{2} & 0 & -\frac{1}{2} & 0 \\ 0 & -5 & 2 & 2 & 0 & 1 \end{bmatrix} \rightarrow \begin{bmatrix} 1 & 0 & \frac{1}{2} & 1 & -\frac{1}{2} & 0 \\ 0 & 1 & -\frac{1}{2} & 0 & -\frac{1}{2} & 0 \\ 0 & 0 & -\frac{1}{2} & 2 & -\frac{5}{2} & 1 \end{bmatrix}$$

$R_3 = 2r_1 + r_3 \qquad R_2 = -\frac{1}{2}r_2 \qquad R_1 = r_2 + r_1$
$\qquad\qquad\qquad\qquad\qquad\qquad\qquad\qquad R_3 = 5r_2 + r_3$

$$\rightarrow \begin{bmatrix} 1 & 0 & \frac{1}{2} & 1 & -\frac{1}{2} & 0 \\ 0 & 1 & -\frac{1}{2} & 0 & -\frac{1}{2} & 0 \\ 0 & 0 & 1 & -4 & 5 & -2 \end{bmatrix} \rightarrow \begin{bmatrix} 1 & 0 & 0 & 3 & -3 & 1 \\ 0 & 1 & 0 & -2 & 2 & -1 \\ 0 & 0 & 1 & -4 & 5 & -2 \end{bmatrix}$$

$R_3 = -2r_3 \qquad\qquad R_1 = -\frac{1}{2}r_3 + r_1$
$\qquad\qquad\qquad\qquad R_2 = \frac{1}{2}r_3 + r_2$

$$A^{-1} = \begin{bmatrix} 3 & -3 & 1 \\ -2 & 2 & -1 \\ -4 & 5 & -2 \end{bmatrix}$$

29. Augment the matrix with the identity and use row operations to find the inverse:

$$A = \begin{bmatrix} 1 & 1 & 1 \\ 3 & 2 & -1 \\ 3 & 1 & 2 \end{bmatrix} \rightarrow \begin{bmatrix} 1 & 1 & 1 & 1 & 0 & 0 \\ 3 & 2 & -1 & 0 & 1 & 0 \\ 3 & 1 & 2 & 0 & 0 & 1 \end{bmatrix}$$

$$\rightarrow \begin{bmatrix} 1 & 1 & 1 & 1 & 0 & 0 \\ 0 & -1 & -4 & -3 & 1 & 0 \\ 0 & -2 & -1 & -3 & 0 & 1 \end{bmatrix} \rightarrow \begin{bmatrix} 1 & 1 & 1 & 1 & 0 & 0 \\ 0 & 1 & 4 & 3 & -1 & 0 \\ 0 & -2 & -1 & -3 & 0 & 1 \end{bmatrix} \rightarrow \begin{bmatrix} 1 & 0 & -3 & -2 & 1 & 0 \\ 0 & 1 & 4 & 3 & -1 & 0 \\ 0 & 0 & 7 & 3 & -2 & 1 \end{bmatrix}$$

$R_2 = -3r_1 + r_2 \qquad R_2 = -r_2 \qquad R_1 = -r_2 + r_1$
$R_3 = -3r_1 + r_3 \qquad\qquad\qquad\qquad R_3 = 2r_2 + r_3$

$$\rightarrow \begin{bmatrix} 1 & 0 & -3 & -2 & 1 & 0 \\ 0 & 1 & 4 & 3 & -1 & 0 \\ 0 & 0 & 1 & \frac{3}{7} & -\frac{2}{7} & \frac{1}{7} \end{bmatrix} \rightarrow \begin{bmatrix} 1 & 0 & 0 & -\frac{5}{7} & \frac{1}{7} & \frac{3}{7} \\ 0 & 1 & 0 & \frac{9}{7} & \frac{1}{7} & -\frac{4}{7} \\ 0 & 0 & 1 & \frac{3}{7} & -\frac{2}{7} & \frac{1}{7} \end{bmatrix}$$

$R_3 = \frac{1}{7}r_3 \qquad\qquad R_1 = 3r_3 + r_1$
$\qquad\qquad\qquad R_2 = -4r_3 + r_2$

$$A^{-1} = \begin{bmatrix} -\frac{5}{7} & \frac{1}{7} & \frac{3}{7} \\ \frac{9}{7} & \frac{1}{7} & -\frac{4}{7} \\ \frac{3}{7} & -\frac{2}{7} & \frac{1}{7} \end{bmatrix}$$

31. Rewrite the system of equations in matrix form:

$$\begin{cases} 2x + y = 8 \\ x + y = 5 \end{cases} \qquad A = \begin{bmatrix} 2 & 1 \\ 1 & 1 \end{bmatrix}, \quad X = \begin{bmatrix} x \\ y \end{bmatrix}, \quad B = \begin{bmatrix} 8 \\ 5 \end{bmatrix}$$

Find the inverse of A and solve $X = A^{-1}B$:

From Problem 21, $A^{-1} = \begin{bmatrix} 1 & -1 \\ -1 & 2 \end{bmatrix}$ and $X = A^{-1}B = \begin{bmatrix} 1 & -1 \\ -1 & 2 \end{bmatrix}\begin{bmatrix} 8 \\ 5 \end{bmatrix} = \begin{bmatrix} 3 \\ 2 \end{bmatrix}$.

The solution is $x = 3, y = 2$.

33. Rewrite the system of equations in matrix form:

$$\begin{cases} 2x + y = 0 \\ x + y = 5 \end{cases} \qquad A = \begin{bmatrix} 2 & 1 \\ 1 & 1 \end{bmatrix}, \quad X = \begin{bmatrix} x \\ y \end{bmatrix}, \quad B = \begin{bmatrix} 0 \\ 5 \end{bmatrix}$$

Find the inverse of A and solve $X = A^{-1}B$:

From Problem 21, $A^{-1} = \begin{bmatrix} 1 & -1 \\ -1 & 2 \end{bmatrix}$ and $X = A^{-1}B = \begin{bmatrix} 1 & -1 \\ -1 & 2 \end{bmatrix}\begin{bmatrix} 0 \\ 5 \end{bmatrix} = \begin{bmatrix} -5 \\ 10 \end{bmatrix}$.

The solution is $x = -5, y = 10$.

35. Rewrite the system of equations in matrix form:

$$\begin{cases} 6x + 5y = 7 \\ 2x + 2y = 2 \end{cases} \qquad A = \begin{bmatrix} 6 & 5 \\ 2 & 2 \end{bmatrix}, \quad X = \begin{bmatrix} x \\ y \end{bmatrix}, \quad B = \begin{bmatrix} 7 \\ 2 \end{bmatrix}$$

Find the inverse of A and solve $X = A^{-1}B$:

From Problem 23, $A^{-1} = \begin{bmatrix} 1 & -\frac{5}{2} \\ -1 & 3 \end{bmatrix}$ and $X = A^{-1}B = \begin{bmatrix} 1 & -\frac{5}{2} \\ -1 & 3 \end{bmatrix}\begin{bmatrix} 7 \\ 2 \end{bmatrix} = \begin{bmatrix} 2 \\ -1 \end{bmatrix}$.

The solution is $x = 2, y = -1$.

37. Rewrite the system of equations in matrix form:

$$\begin{cases} 6x + 5y = 13 \\ 2x + 2y = 5 \end{cases} \qquad A = \begin{bmatrix} 6 & 5 \\ 2 & 2 \end{bmatrix}, \quad X = \begin{bmatrix} x \\ y \end{bmatrix}, \quad B = \begin{bmatrix} 13 \\ 5 \end{bmatrix}$$

Find the inverse of A and solve $X = A^{-1}B$:

From Problem 23, $A^{-1} = \begin{bmatrix} 1 & -\frac{5}{2} \\ -1 & 3 \end{bmatrix}$ and $X = A^{-1}B = \begin{bmatrix} 1 & -\frac{5}{2} \\ -1 & 3 \end{bmatrix}\begin{bmatrix} 13 \\ 5 \end{bmatrix} = \begin{bmatrix} \frac{1}{2} \\ 2 \end{bmatrix}$.

The solution is $x = \dfrac{1}{2}, y = 2$.

39. Rewrite the system of equations in matrix form:

$$\begin{cases} 2x + y = -3 \\ ax + ay = -a \end{cases} \quad a \neq 0 \qquad A = \begin{bmatrix} 2 & 1 \\ a & a \end{bmatrix}, \quad X = \begin{bmatrix} x \\ y \end{bmatrix}, \quad B = \begin{bmatrix} -3 \\ -a \end{bmatrix}$$

Find the inverse of A and solve $X = A^{-1}B$:

From Problem 25, $A^{-1} = \begin{bmatrix} 1 & -\frac{1}{a} \\ -1 & \frac{2}{a} \end{bmatrix}$ and $X = A^{-1}B = \begin{bmatrix} 1 & -\frac{1}{a} \\ -1 & \frac{2}{a} \end{bmatrix}\begin{bmatrix} -3 \\ -a \end{bmatrix} = \begin{bmatrix} -2 \\ 1 \end{bmatrix}$.

The solution is $x = -2, y = 1$.

41.   Rewrite the system of equations in matrix form:

$$\begin{cases} 2x + y = \dfrac{7}{a} \quad a \neq 0 \\ ax + ay = 5 \end{cases} \qquad A = \begin{bmatrix} 2 & 1 \\ a & a \end{bmatrix}, \quad X = \begin{bmatrix} x \\ y \end{bmatrix}, \quad B = \begin{bmatrix} \frac{7}{a} \\ 5 \end{bmatrix}$$

Find the inverse of A and solve $X = A^{-1}B$:

From Problem 25, $A^{-1} = \begin{bmatrix} 1 & -\frac{1}{a} \\ -1 & \frac{2}{a} \end{bmatrix}$ and $X = A^{-1}B = \begin{bmatrix} 1 & -\frac{1}{a} \\ -1 & \frac{2}{a} \end{bmatrix}\begin{bmatrix} \frac{7}{a} \\ 5 \end{bmatrix} = \begin{bmatrix} \frac{2}{a} \\ \frac{3}{a} \end{bmatrix}$.

The solution is $x = \dfrac{2}{a}, y = \dfrac{3}{a}$.

43.   Rewrite the system of equations in matrix form:

$$\begin{cases} x - y + z = 0 \\ -2y + z = -1 \\ -2x - 3y = -5 \end{cases} \qquad A = \begin{bmatrix} 1 & -1 & 1 \\ 0 & -2 & 1 \\ -2 & -3 & 0 \end{bmatrix}, \quad X = \begin{bmatrix} x \\ y \\ z \end{bmatrix}, \quad B = \begin{bmatrix} 0 \\ -1 \\ -5 \end{bmatrix}$$

Find the inverse of A and solve $X = A^{-1}B$:

From Problem 27,

$$A^{-1} = \begin{bmatrix} 3 & -3 & 1 \\ -2 & 2 & -1 \\ -4 & 5 & -2 \end{bmatrix} \text{ and } X = A^{-1}B = \begin{bmatrix} 3 & -3 & 1 \\ -2 & 2 & -1 \\ -4 & 5 & -2 \end{bmatrix}\begin{bmatrix} 0 \\ -1 \\ -5 \end{bmatrix} = \begin{bmatrix} -2 \\ 3 \\ 5 \end{bmatrix}$$

The solution is $x = -2, y = 3, z = 5$.

45.   Rewrite the system of equations in matrix form:

$$\begin{cases} x - y + z = 2 \\ -2y + z = 2 \\ -2x - 3y = \dfrac{1}{2} \end{cases} \qquad A = \begin{bmatrix} 1 & -1 & 1 \\ 0 & -2 & 1 \\ -2 & -3 & 0 \end{bmatrix}, \quad X = \begin{bmatrix} x \\ y \\ z \end{bmatrix}, \quad B = \begin{bmatrix} 2 \\ 2 \\ \frac{1}{2} \end{bmatrix}$$

Find the inverse of A and solve $X = A^{-1}B$:
From Problem 27,

$$A^{-1} = \begin{bmatrix} 3 & -3 & 1 \\ -2 & 2 & -1 \\ -4 & 5 & -2 \end{bmatrix} \text{ and } X = A^{-1}B = \begin{bmatrix} 3 & -3 & 1 \\ -2 & 2 & -1 \\ -4 & 5 & -2 \end{bmatrix}\begin{bmatrix} 2 \\ 2 \\ \frac{1}{2} \end{bmatrix} = \begin{bmatrix} \frac{1}{2} \\ -\frac{1}{2} \\ 1 \end{bmatrix}$$

The solution is $x = \dfrac{1}{2}, y = -\dfrac{1}{2}, z = 1$.

47.   Rewrite the system of equations in matrix form:

$$\begin{cases} x + y + z = 9 \\ 3x + 2y - z = 8 \\ 3x + y + 2z = 1 \end{cases} \qquad A = \begin{bmatrix} 1 & 1 & 1 \\ 3 & 2 & -1 \\ 3 & 1 & 2 \end{bmatrix}, \quad X = \begin{bmatrix} x \\ y \\ z \end{bmatrix}, \quad B = \begin{bmatrix} 9 \\ 8 \\ 1 \end{bmatrix}$$

Find the inverse of A and solve $X = A^{-1}B$:

From Problem 29,

$$A^{-1} = \begin{bmatrix} -\frac{5}{7} & \frac{1}{7} & \frac{3}{7} \\ \frac{9}{7} & \frac{1}{7} & -\frac{4}{7} \\ \frac{3}{7} & -\frac{2}{7} & \frac{1}{7} \end{bmatrix} \text{ and } X = A^{-1}B = \begin{bmatrix} -\frac{5}{7} & \frac{1}{7} & \frac{3}{7} \\ \frac{9}{7} & \frac{1}{7} & -\frac{4}{7} \\ \frac{3}{7} & -\frac{2}{7} & \frac{1}{7} \end{bmatrix}\begin{bmatrix} 9 \\ 8 \\ 1 \end{bmatrix} = \begin{bmatrix} -\frac{34}{7} \\ \frac{85}{7} \\ \frac{12}{7} \end{bmatrix}.$$

The solution is $x = -\dfrac{34}{7}, y = \dfrac{85}{7}, z = \dfrac{12}{7}$.

49. Rewrite the system of equations in matrix form:

$$\begin{cases} x + y + z = 2 \\ 3x + 2y - z = \dfrac{7}{3} \\ 3x + y + 2z = \dfrac{10}{3} \end{cases} \qquad A = \begin{bmatrix} 1 & 1 & 1 \\ 3 & 2 & -1 \\ 3 & 1 & 2 \end{bmatrix}, \quad X = \begin{bmatrix} x \\ y \\ z \end{bmatrix}, \quad B = \begin{bmatrix} 2 \\ \frac{7}{3} \\ \frac{10}{3} \end{bmatrix}$$

Find the inverse of A and solve $X = A^{-1}B$:

From Problem 29,

$$A^{-1} = \begin{bmatrix} -\frac{5}{7} & \frac{1}{7} & \frac{3}{7} \\ \frac{9}{7} & \frac{1}{7} & -\frac{4}{7} \\ \frac{3}{7} & -\frac{2}{7} & \frac{1}{7} \end{bmatrix} \text{ and } X = A^{-1}B = \begin{bmatrix} -\frac{5}{7} & \frac{1}{7} & \frac{3}{7} \\ \frac{9}{7} & \frac{1}{7} & -\frac{4}{7} \\ \frac{3}{7} & -\frac{2}{7} & \frac{1}{7} \end{bmatrix}\begin{bmatrix} 2 \\ \frac{7}{3} \\ \frac{10}{3} \end{bmatrix} = \begin{bmatrix} \frac{1}{3} \\ 1 \\ \frac{2}{3} \end{bmatrix}.$$

The solution is $x = \dfrac{1}{3}, y = 1, z = \dfrac{2}{3}$.

51. Augment the matrix with the identity and use row operations to find the inverse:

$$A = \begin{bmatrix} 4 & 2 \\ 2 & 1 \end{bmatrix} \rightarrow \begin{bmatrix} 4 & 2 & | & 1 & 0 \\ 2 & 1 & | & 0 & 1 \end{bmatrix}$$

$$\rightarrow \begin{bmatrix} 4 & 2 & | & 1 & 0 \\ 0 & 0 & | & -\frac{1}{2} & 1 \end{bmatrix} \rightarrow \begin{bmatrix} 1 & \frac{1}{2} & | & \frac{1}{4} & 0 \\ 0 & 0 & | & -\frac{1}{2} & 1 \end{bmatrix}$$

$$R_2 = -\tfrac{1}{2}r_1 + r_2 \qquad R_1 = \tfrac{1}{4}r_1$$

There is no way to obtain the identity matrix on the left; thus, there is no inverse.

53. Augment the matrix with the identity and use row operations to find the inverse:

$$A = \begin{bmatrix} 15 & 3 \\ 10 & 2 \end{bmatrix} \rightarrow \begin{bmatrix} 15 & 3 & | & 1 & 0 \\ 10 & 2 & | & 0 & 1 \end{bmatrix}$$

$$\rightarrow \begin{bmatrix} 15 & 3 & | & 1 & 0 \\ 0 & 0 & | & -\frac{2}{3} & 1 \end{bmatrix} \rightarrow \begin{bmatrix} 1 & \frac{1}{5} & | & \frac{1}{15} & 0 \\ 0 & 0 & | & -\frac{2}{3} & 1 \end{bmatrix}$$

$$R_2 = -\tfrac{2}{3}r_1 + r_2 \qquad R_1 = \tfrac{1}{15}r_1$$

There is no way to obtain the identity matrix on the left; thus, there is no inverse.

55. Augment the matrix with the identity and use row operations to find the inverse:

$$A = \begin{bmatrix} -3 & 1 & -1 \\ 1 & -4 & -7 \\ 1 & 2 & 5 \end{bmatrix} \rightarrow \begin{bmatrix} -3 & 1 & -1 & | & 1 & 0 & 0 \\ 1 & -4 & -7 & | & 0 & 1 & 0 \\ 1 & 2 & 5 & | & 0 & 0 & 1 \end{bmatrix}$$

$$\rightarrow \begin{bmatrix} 1 & 2 & 5 & | & 0 & 0 & 1 \\ 1 & -4 & -7 & | & 0 & 1 & 0 \\ -3 & 1 & -1 & | & 1 & 0 & 0 \end{bmatrix} \rightarrow \begin{bmatrix} 1 & 2 & 5 & | & 0 & 0 & 1 \\ 0 & -6 & -12 & | & 0 & 1 & -1 \\ 0 & 7 & 14 & | & 1 & 0 & 3 \end{bmatrix} \rightarrow \begin{bmatrix} 1 & 2 & 5 & | & 0 & 0 & 1 \\ 0 & 1 & 2 & | & 0 & -\frac{1}{6} & \frac{1}{6} \\ 0 & 7 & 14 & | & 1 & 0 & 3 \end{bmatrix}$$

Interchange r_1 and r_3 $R_2 = -r_1 + r_2$ $R_2 = -\frac{1}{6}r_2$
$$R_3 = 3r_1 + r_3$$

$$\rightarrow \begin{bmatrix} 1 & 0 & 1 & | & 0 & \frac{1}{3} & \frac{2}{3} \\ 0 & 1 & 2 & | & 0 & -\frac{1}{6} & \frac{1}{6} \\ 0 & 0 & 0 & | & 1 & \frac{7}{6} & \frac{11}{6} \end{bmatrix}$$

$$R_1 = -2r_2 + r_1$$
$$R_3 = -7r_2 + r_3$$

There is no way to obtain the identity matrix on the left; thus, there is no inverse.

57.
$$\begin{bmatrix} 0.01 & 0.05 & -0.01 \\ 0.01 & -0.02 & 0.01 \\ -0.02 & 0.01 & 0.03 \end{bmatrix}$$

59.
$$\begin{bmatrix} 0.02 & -0.04 & -0.01 & 0.01 \\ -0.02 & 0.05 & 0.03 & -0.03 \\ 0.02 & 0.01 & -0.04 & 0.00 \\ -0.02 & 0.06 & 0.07 & 0.06 \end{bmatrix}$$

61. $x = 4.57,\ y = -6.44,\ z = -24.07$ 63. $x = -1.19,\ y = 2.46,\ z = 8.27$

65.   (a)   The rows of the 2 by 3 matrix represent stainless steel and aluminum. The columns represent 10-gallon, 5-gallon, and 1-gallon.

           The 2 by 3 matrix is: The 3 by 2 matrix is:
$$\begin{bmatrix} 500 & 350 & 400 \\ 700 & 500 & 850 \end{bmatrix} \qquad \begin{bmatrix} 500 & 700 \\ 350 & 500 \\ 400 & 850 \end{bmatrix}$$

       (b)   The 3 by 1 matrix representing the amount of material is:
$$\begin{bmatrix} 15 \\ 8 \\ 3 \end{bmatrix}$$

       (c)   The days usage of materials is:
$$\begin{bmatrix} 500 & 350 & 400 \\ 700 & 500 & 850 \end{bmatrix} \cdot \begin{bmatrix} 15 \\ 8 \\ 3 \end{bmatrix} = \begin{bmatrix} 11,500 \\ 17,050 \end{bmatrix}$$

           11,500 pounds of stainless steel and 17,050 pounds of aluminum are used each day.

       (d)   The 1 by 2 matrix representing cost is:
$$\begin{bmatrix} 0.10 & 0.05 \end{bmatrix}$$

       (e)   The total cost of the days production was:
$$\begin{bmatrix} 0.10 & 0.05 \end{bmatrix} \cdot \begin{bmatrix} 11,500 \\ 17,050 \end{bmatrix} = \begin{bmatrix} 2002.50 \end{bmatrix}$$

           The total cost of the days production was $2,002.50.

67. We need to consider 2 different cases.
Case 1: $a \neq 0$, $D = ad - bc \neq 0$.

$$A = \begin{bmatrix} a & b & | & 1 & 0 \\ c & d & | & 0 & 1 \end{bmatrix} \longrightarrow \begin{bmatrix} 1 & \dfrac{b}{a} & | & \dfrac{1}{a} & 0 \\ c & d & | & 0 & 1 \end{bmatrix} \longrightarrow \begin{bmatrix} 1 & \dfrac{b}{a} & | & \dfrac{1}{a} & 0 \\ c & d - \dfrac{cb}{a} & | & -\dfrac{c}{a} & 1 \end{bmatrix} = \begin{bmatrix} 1 & \dfrac{b}{b} & | & \dfrac{1}{a} & 0 \\ c & \dfrac{ad - cb}{a} & | & -\dfrac{c}{a} & 1 \end{bmatrix}$$

$$R_1 = \tfrac{1}{a} \cdot r_1 \qquad\qquad R_2 = -c \cdot r_1 + r_2$$

$$\longrightarrow \begin{bmatrix} 1 & \dfrac{b}{a} & | & \dfrac{1}{a} & 0 \\ 0 & 1 & | & -\dfrac{c}{ad - bc} & \dfrac{a}{ad - bc} \end{bmatrix} \longrightarrow \begin{bmatrix} 1 & 0 & | & \dfrac{1}{a} + \dfrac{bc}{a(ad - bc)} & \dfrac{-b}{ad - bc} \\ 0 & 1 & | & -\dfrac{c}{ad - bc} & \dfrac{a}{ad - bc} \end{bmatrix}$$

$$R_2 = \left(\dfrac{a}{ad - bc}\right) \cdot r_2 \qquad\qquad R_1 = \left(-\dfrac{b}{a}\right) r_2 + r_1$$

Note that

$$\frac{1}{a} + \frac{bc}{a(ad - bc)} = \frac{1 \cdot (ad - bc) + bc}{a(ad - bc)}$$

$$= \frac{ad - bc + bc}{a(ad - bc)} = \frac{ad}{a(ad - bc)}$$

$$= \frac{d}{ad - bc}$$

So, we have $\begin{bmatrix} 1 & 0 & | & \dfrac{1}{a} + \dfrac{bc}{a(ad - bc)} & \dfrac{-b}{ad - bc} \\ 0 & 1 & | & -\dfrac{c}{ad - bc} & \dfrac{a}{ad - bc} \end{bmatrix} = \begin{bmatrix} 1 & 0 & | & \dfrac{d}{ad - bc} & \dfrac{-b}{ad - bc} \\ 0 & 1 & | & -\dfrac{c}{ad - bc} & \dfrac{a}{ad - bc} \end{bmatrix}$.

Therefore

$$A^{-1} = \begin{bmatrix} \dfrac{d}{ad - bc} & \dfrac{-b}{ad - bc} \\ -\dfrac{c}{ad - bc} & \dfrac{a}{ad - bc} \end{bmatrix}$$

$$= \left(\frac{1}{ad - bc}\right) \begin{bmatrix} d & -b \\ -c & a \end{bmatrix},$$

$$= \left(\frac{1}{D}\right) \begin{bmatrix} d & -b \\ -c & a \end{bmatrix}$$

where $D = ad - bc$.

Case 2: $a = 0,\ D = ad - bc \neq 0$.

First note that $D = ad - bc = 0 \cdot d - bc = -bc \neq 0 \Rightarrow b \neq 0$ and $c \neq 0$.

$$A = \begin{bmatrix} 0 & b & 1 & 0 \\ c & d & 0 & 1 \end{bmatrix} \longrightarrow \begin{bmatrix} c & d & 0 & 1 \\ 0 & b & 1 & 0 \end{bmatrix} \longrightarrow \begin{bmatrix} 1 & \dfrac{d}{c} & 0 & \dfrac{1}{c} \\ 0 & b & 1 & 0 \end{bmatrix} \longrightarrow \begin{bmatrix} 1 & \dfrac{d}{c} & 0 & \dfrac{1}{c} \\ 0 & 1 & \dfrac{1}{b} & 0 \end{bmatrix}$$

$$\text{interchange } r_1 \text{ and } r_2 \qquad R_1 = \frac{1}{c} \cdot r_1 \qquad R_2 = \frac{1}{b} \cdot r_2$$

$$\longrightarrow \begin{bmatrix} 1 & 0 & \dfrac{-d}{bc} & \dfrac{1}{c} \\ 0 & 1 & \dfrac{1}{b} & 0 \end{bmatrix}$$

$$R_1 = \left(\frac{-d}{c} \right) \cdot r_2 + r_1$$

Therefore $\quad A^{-1} = \begin{bmatrix} \dfrac{-d}{bc} & \dfrac{1}{c} \\ \dfrac{1}{b} & 0 \end{bmatrix} = \left(\dfrac{1}{-bc} \right) \begin{bmatrix} d & -b \\ -c & 0 \end{bmatrix} = \left(\dfrac{1}{D} \right) \begin{bmatrix} d & -b \\ -c & 0 \end{bmatrix},$

where $D = ad - bc = 0 \cdot d - bc = -bc$.

Chapter 10

Systems of Equations and Inequalities

10.6 Partial Fraction Decomposition

1. The rational expression $\dfrac{x}{x^2-1}$ is proper, since the degree of the numerator is less than the degree of the denominator.

3. The rational expression $\dfrac{x^2+5}{x^2-4}$ is improper, so perform the division:

$$\begin{array}{r} 1 \\ x^2-4\overline{)x^2+5} \\ \underline{x^2-4} \\ 9 \end{array}$$

The proper rational expression is:
$$\frac{x^2+5}{x^2-4}=1+\frac{9}{x^2-4}$$

5. The rational expression $\dfrac{5x^3+2x-1}{x^2-4}$ is improper, so perform the division:

$$\begin{array}{r} 5x \\ x^2-4\overline{)5x^3+0x^2+2x-1} \\ \underline{5x^3\qquad-20x} \\ 22x-1 \end{array}$$

The proper rational expression is:
$$\frac{5x^3+2x-1}{x^2-4}=5x+\frac{22x-1}{x^2-4}$$

7. The rational expression $\dfrac{x(x-1)}{(x+4)(x-3)}=\dfrac{x^2-x}{x^2+x-12}$ is improper, so perform the division:

$$\begin{array}{r} 1 \\ x^2+x-12\overline{)x^2-x+0} \\ \underline{x^2+x-12} \\ -2x+12 \end{array}$$

The proper rational expression is: $\qquad \dfrac{x(x-1)}{(x+4)(x-3)}=1+\dfrac{-2x+12}{x^2+x-12}$

9. Find the partial fraction decomposition:

$$\frac{4}{x(x-1)}=\frac{A}{x}+\frac{B}{x-1}$$

$$4=A(x-1)+Bx \quad \text{(Multiply both sides by } x(x-1).)$$

Let $x=1$: then $4=A(0)+B \rightarrow B=4$

Let $x=0$: then $4=A(-1)+B(0) \rightarrow A=-4$

$$\frac{4}{x(x-1)}=\frac{-4}{x}+\frac{4}{x-1}$$

11.   Find the partial fraction decomposition:
$$\frac{1}{x(x^2+1)} = \frac{A}{x} + \frac{Bx+C}{x^2+1}$$
$$1 = A(x^2+1) + (Bx+C)x \quad \text{(Multiply both sides by } x(x^2+1).)$$
Let $x=0$: then $1 = A(1) + (B(0)+C)(0) \rightarrow A = 1$

Let $x=1$: then $1 = A(1+1) + (B(1)+C)(1) \rightarrow 1 = 2A + B + C$

$\qquad\qquad$ so $1 = 2(1) + B + C \rightarrow -1 = B + C$

Let $x=-1$: then $1 = A(1+1) + (B(-1)+C)(-1) \rightarrow 1 = 2A + B - C$

$\qquad\qquad$ so $1 = 2(1) + B - C \rightarrow -1 = B - C$

Solve the system of equations:
$$\begin{array}{rl} B + C &= -1 \\ B - C &= -1 \\ \hline 2B &= -2 \qquad -1 + C = -1 \\ B &= -1 \qquad\quad C = 0 \end{array}$$
$$\frac{1}{x(x^2+1)} = \frac{1}{x} + \frac{-x}{x^2+1}$$

13.   Find the partial fraction decomposition:
$$\frac{x}{(x-1)(x-2)} = \frac{A}{x-1} + \frac{B}{x-2}$$
$$x = A(x-2) + B(x-1) \quad \text{(Multiply both sides by } (x-1)(x-2).)$$
Let $x = 1$: then $1 = A(1-2) + B(1-1) \rightarrow 1 = -A \rightarrow A = -1$
Let $x = 2$: then $2 = A(2-2) + B(2-1) \rightarrow 2 = B \rightarrow B = 2$
$$\frac{x}{(x-1)(x-2)} = \frac{-1}{x-1} + \frac{2}{x-2}$$

15.   Find the partial fraction decomposition:
$$\frac{x^2}{(x-1)^2(x+1)} = \frac{A}{x-1} + \frac{B}{(x-1)^2} + \frac{C}{x+1}$$
$$\text{(Multiply both sides by } (x-1)^2(x+1).)$$
$$x^2 = A(x-1)(x+1) + B(x+1) + C(x-1)^2$$
Let $x=1$: then $1^2 = A(1-1)(1+1) + B(1+1) + C(1-1)^2$
$$\rightarrow 1 = 2B \rightarrow B = \frac{1}{2}$$
Let $x=-1$: then $(-1)^2 = A(-1-1)(-1+1) + B(-1+1) + C(-1-1)^2$
$$\rightarrow 1 = 4C \rightarrow C = \frac{1}{4}$$
Let $x=0$: then $0^2 = A(0-1)(0+1) + B(0+1) + C(0-1)^2$
$$\rightarrow 0 = -A + B + C$$
$$A = \frac{1}{2} + \frac{1}{4} = \frac{3}{4}$$

$$\frac{x^2}{(x-1)^2(x+1)} = \frac{\left(\frac{3}{4}\right)}{x-1} + \frac{\left(\frac{1}{2}\right)}{(x-1)^2} + \frac{\left(\frac{1}{4}\right)}{x+1}$$

17. Find the partial fraction decomposition:

$$\frac{1}{x^3-8} = \frac{1}{(x-2)(x^2+2x+4)} = \frac{A}{x-2} + \frac{Bx+C}{x^2+2x+4}$$

(Multiply both sides by $(x-2)(x^2+2x+4)$.)

$$1 = A(x^2+2x+4) + (Bx+C)(x-2)$$

Let $x=2$: then $1 = A(2^2+2(2)+4) + (B(2)+C)(2-2)$

$$\rightarrow 1 = 12A \rightarrow A = \frac{1}{12}$$

Let $x=0$: then $1 = A(0^2+2(0)+4) + (B(0)+C)(0-2)$

$$\rightarrow 1 = 4A - 2C \rightarrow 1 = 4\left(\frac{1}{12}\right) - 2C$$

$$\rightarrow -2C = \frac{2}{3} \rightarrow C = -\frac{1}{3}$$

Let $x=1$: then $1 = A(1^2+2(1)+4) + (B(1)+C)(1-2)$

$$\rightarrow 1 = 7A - B - C \rightarrow 1 = 7\left(\frac{1}{12}\right) - B + \frac{1}{3} \rightarrow B = -\frac{1}{12}$$

$$\frac{1}{x^3-8} = \frac{\left(\frac{1}{12}\right)}{x-2} + \frac{-\left(\frac{1}{12}\right)x - \frac{1}{3}}{x^2+2x+4}$$

19. Find the partial fraction decomposition:

$$\frac{x^2}{(x-1)^2(x+1)^2} = \frac{A}{x-1} + \frac{B}{(x-1)^2} + \frac{C}{x+1} + \frac{D}{(x+1)^2}$$

(Multiply both sides by $(x-1)^2(x+1)^2$.)

$$x^2 = A(x-1)(x+1)^2 + B(x+1)^2 + C(x-1)^2(x+1) + D(x-1)^2$$

Let $x=1$: then $1^2 = A(1-1)(1+1)^2 + B(1+1)^2 + C(1-1)^2(1+1) + D(1-1)^2$

$$\rightarrow 1 = 4B \rightarrow B = \frac{1}{4}$$

Let $x=-1$: then

$$(-1)^2 = A(-1-1)(-1+1)^2 + B(-1+1)^2 + C(-1-1)^2(-1+1) + D(-1-1)^2$$

$$\rightarrow 1 = 4D \rightarrow D = \frac{1}{4}$$

Let $x=0$: then

$$0^2 = A(0-1)(0+1)^2 + B(0+1)^2 + C(0-1)^2(0+1) + D(0-1)^2$$

$$\rightarrow 0 = -A + B + C + D \rightarrow A - C = \frac{1}{4} + \frac{1}{4} = \frac{1}{2}$$

Let $x = 2:$ then

$$2^2 = A(2-1)(2+1)^2 + B(2+1)^2 + C(2-1)^2(2+1) + D(2-1)^2$$

$$\rightarrow 4 = 9A + 9B + 3C + D \rightarrow 9A + 3C = 4 - \frac{9}{4} - \frac{1}{4} = \frac{3}{2}$$

$$\rightarrow 3A + C = \frac{1}{2}$$

Solve the system of equations:

$$A - C = \frac{1}{2}$$

$$3A + C = \frac{1}{2}$$

$$\overline{ \;\; 4A \;\;\;\; = 1}$$

$$A \;\; = \frac{1}{4} \;\; \rightarrow \;\; \frac{3}{4} + C = \frac{1}{2} \;\; \rightarrow \;\; C = -\frac{1}{4}$$

$$\frac{x^2}{(x-1)^2(x+1)^2} = \frac{\left(\frac{1}{4}\right)}{x-1} + \frac{\left(\frac{1}{4}\right)}{(x-1)^2} + \frac{\left(-\frac{1}{4}\right)}{x+1} + \frac{\left(\frac{1}{4}\right)}{(x+1)^2}$$

21. Find the partial fraction decomposition:

$$\frac{x-3}{(x+2)(x+1)^2} = \frac{A}{x+2} + \frac{B}{x+1} + \frac{C}{(x+1)^2}$$

(Multiply both sides by $(x+2)(x+1)^2$.)

$$x - 3 = A(x+1)^2 + B(x+2)(x+1) + C(x+2)$$

Let $x = -2:$ then $-2 - 3 = A(-2+1)^2 + B(-2+2)(-2+1) + C(-2+2)$

$$\rightarrow -5 = A \;\; \rightarrow \;\; A = -5$$

Let $x = -1:$ then $-1 - 3 = A(-1+1)^2 + B(-1+2)(-1+1) + C(-1+2)$

$$\rightarrow -4 = C \;\; \rightarrow \;\; C = -4$$

Let $x = 0:$ then

$$0 - 3 = A(0+1)^2 + B(0+2)(0+1) + C(0+2) \;\; \rightarrow \;\; -3 = A + 2B + 2C$$

$$\rightarrow -3 = -5 + 2B + 2(-4) \;\; \rightarrow \;\; 2B = 10 \;\; \rightarrow \;\; B = 5$$

$$\frac{x-3}{(x+2)(x+1)^2} = \frac{-5}{x+2} + \frac{5}{x+1} + \frac{-4}{(x+1)^2}$$

23. Find the partial fraction decomposition:

$$\frac{x+4}{x^2(x^2+4)} = \frac{A}{x} + \frac{B}{x^2} + \frac{Cx+D}{x^2+4} \text{ (Multiply both sides by } x^2(x^2+4).)$$

$$x + 4 = Ax(x^2+4) + B(x^2+4) + (Cx+D)x^2$$

Let $x = 0:$ then $0 + 4 = A(0)(0^2+4) + B(0^2+4) + (C0+D)(0)^2$

$$\rightarrow 4 = 4B \;\; \rightarrow \;\; B = 1$$

Let $x = 1:$ then $1 + 4 = A(1)(1^2+4) + B(1^2+4) + (C(1)+D)(1)^2$

$$\rightarrow 5 = 5A + 5B + C + D \;\; \rightarrow \;\; 5 = 5A + 5 + C + D$$

$$\rightarrow 5A + C + D = 0$$

Let $x = -1$: then
$$-1 + 4 = A(-1)((-1)^2 + 4) + B((-1)^2 + 4) + (C(-1) + D)(-1)^2$$
$$\rightarrow \ 3 = -5A + 5B - C + D \ \rightarrow \ 3 = -5A + 5 - C + D$$
$$\rightarrow \ -5A - C + D = -2$$

Let $x = 2$: then $2 + 4 = A(2)(2^2 + 4) + B(2^2 + 4) + (C(2) + D)(2)^2$
$$\rightarrow \ 6 = 16A + 8B + 8C + 4D \ \rightarrow \ 6 = 16A + 8 + 8C + 4D$$
$$\rightarrow \ 16A + 8C + 4D = -2$$

Solve the system of equations:
$$5A + C + D = \quad 0$$
$$\underline{-5A - C + D = -2}$$
$$2D = -2 \qquad\qquad 5A + C - 1 = 0$$
$$D = -1 \qquad\qquad C = 1 - 5A$$
$$16A + 8(1 - 5A) + 4(-1) = -2$$
$$16A + 8 - 40A - 4 = -2 \qquad\qquad C = 1 - 5\left(\frac{1}{4}\right)$$
$$-24A = -6$$
$$A = \frac{1}{4} \qquad\qquad C = 1 - \frac{5}{4} = -\frac{1}{4}$$

$$\frac{x+4}{x^2(x^2+4)} = \frac{\left(\dfrac{1}{4}\right)}{x} + \frac{1}{x^2} + \frac{\left(-\dfrac{1}{4}x - 1\right)}{x^2+4}$$

25. Find the partial fraction decomposition:
$$\frac{x^2 + 2x + 3}{(x+1)(x^2+2x+4)} = \frac{A}{x+1} + \frac{Bx+C}{x^2+2x+4}$$
(Multiply both sides by $(x+1)(x^2+2x+4)$.)
$$x^2 + 2x + 3 = A(x^2 + 2x + 4) + (Bx + C)(x+1)$$

Let $x = -1$: then
$$(-1)^2 + 2(-1) + 3 = A((-1)^2 + 2(-1) + 4) + (B(-1) + C)(-1 + 1)$$
$$\rightarrow \ 2 = 3A \ \rightarrow \ A = \frac{2}{3}$$

Let $x = 0$: then $0^2 + 2(0) + 3 = A(0^2 + 2(0) + 4) + (B(0) + C)(0 + 1)$
$$\rightarrow \ 3 = 4A + C \ \rightarrow \ 3 = 4\left(\frac{2}{3}\right) + C \ \rightarrow \ C = \frac{1}{3}$$

Let $x = 1$: then
$$1^2 + 2(1) + 3 = A(1^2 + 2(1) + 4) + (B(1) + C)(1 + 1)$$
$$\rightarrow \ 6 = 7A + 2B + 2C \ \rightarrow \ 6 = 7\left(\frac{2}{3}\right) + 2B + 2\left(\frac{1}{3}\right)$$
$$\rightarrow \ 2B = 6 - \frac{14}{3} - \frac{2}{3} \ \rightarrow \ 2B = \frac{2}{3} \ \rightarrow \ B = \frac{1}{3}$$

$$\frac{x^2 + 2x + 3}{(x+1)(x^2+2x+4)} = \frac{\left(\dfrac{2}{3}\right)}{x+1} + \frac{\left(\dfrac{1}{3}x + \dfrac{1}{3}\right)}{x^2+2x+4}$$

27. Find the partial fraction decomposition:

$$\frac{x}{(3x-2)(2x+1)} = \frac{A}{3x-2} + \frac{B}{2x+1} \quad \text{(Multiply both sides by } (3x-2)(2x+1).)$$

$$x = A(2x+1) + B(3x-2)$$

Let $x = -\dfrac{1}{2}$: then $-\dfrac{1}{2} = A\left(2\left(-\dfrac{1}{2}\right)+1\right) + B\left(3\left(-\dfrac{1}{2}\right)-2\right)$

$$\rightarrow \quad -\frac{1}{2} = -\frac{7}{2}B \quad \rightarrow \quad B = \frac{1}{7}$$

Let $x = \dfrac{2}{3}$: then $\dfrac{2}{3} = A\left(2\left(\dfrac{2}{3}\right)+1\right) + B\left(3\left(\dfrac{2}{3}\right)-2\right) \quad \rightarrow \quad \dfrac{2}{3} = \dfrac{7}{3}A \quad \rightarrow \quad A = \dfrac{2}{7}$

$$\frac{x}{(3x-2)(2x+1)} = \frac{\left(\dfrac{2}{7}\right)}{3x-2} + \frac{\left(\dfrac{1}{7}\right)}{2x+1}$$

29. Find the partial fraction decomposition:

$$\frac{x}{x^2+2x-3} = \frac{x}{(x+3)(x-1)} = \frac{A}{x+3} + \frac{B}{x-1}$$

$$\text{(Multiply both sides by } (x+3)(x-1).)$$

$$x = A(x-1) + B(x+3)$$

Let $x = 1$: then $1 = A(1-1) + B(1+3) \quad \rightarrow \quad 1 = 4B \quad \rightarrow \quad B = \dfrac{1}{4}$

Let $x = -3$: then $-3 = A(-3-1) + B(-3+3) \quad \rightarrow \quad -3 = -4A \quad \rightarrow \quad A = \dfrac{3}{4}$

$$\frac{x}{x^2+2x-3} = \frac{\left(\dfrac{3}{4}\right)}{x+3} + \frac{\left(\dfrac{1}{4}\right)}{x-1}$$

31. Find the partial fraction decomposition:

$$\frac{x^2+2x+3}{(x^2+4)^2} = \frac{Ax+B}{x^2+4} + \frac{Cx+D}{(x^2+4)^2}$$

$$\text{(Multiply both sides by } (x^2+4)^2.)$$

$$x^2 + 2x + 3 = (Ax+B)(x^2+4) + Cx + D$$

$$x^2 + 2x + 3 = Ax^3 + Bx^2 + 4Ax + 4B + Cx + D$$

$$x^2 + 2x + 3 = Ax^3 + Bx^2 + (4A+C)x + 4B + D$$

$$A = 0$$

$$B = 1$$

$$4A + C = 2 \quad \rightarrow \quad 4(0) + C = 2 \quad \rightarrow \quad C = 2$$

$$4B + D = 3 \quad \rightarrow \quad 4(1) + D = 3 \quad \rightarrow \quad D = -1$$

$$\frac{x^2 + 2x + 3}{(x^2 + 4)^2} = \frac{1}{x^2 + 4} + \frac{2x - 1}{(x^2 + 4)^2}$$

33. Find the partial fraction decomposition:

$$\frac{7x + 3}{x^3 - 2x^2 - 3x} = \frac{7x + 3}{x(x - 3)(x + 1)} = \frac{A}{x} + \frac{B}{x - 3} + \frac{C}{x + 1}$$

(Multiply both sides by $x(x - 3)(x + 1)$.)

$$7x + 3 = A(x - 3)(x + 1) + Bx(x + 1) + Cx(x - 3)$$

Let $x = 0$:   then $7(0) + 3 = A(0 - 3)(0 + 1) + B(0)(0 + 1) + C(0)(0 - 3)$

$\rightarrow \ 3 = -3A \ \rightarrow \ A = -1$

Let $x = 3$:   then $7(3) + 3 = A(3 - 3)(3 + 1) + B(3)(3 + 1) + C(3)(3 - 3)$

$\rightarrow \ 24 = 12B \ \rightarrow \ B = 2$

Let $x = -1$:   then

$$7(-1) + 3 = A(-1 - 3)(-1 + 1) + B(-1)(-1 + 1) + C(-1)(-1 - 3)$$

$\rightarrow \ -4 = 4C \ \rightarrow \ C = -1$

$$\frac{7x + 3}{x^3 - 2x^2 - 3x} = \frac{-1}{x} + \frac{2}{x - 3} + \frac{-1}{x + 1}$$

35. Perform synthetic division to find a factor:

$$2\overline{)1 \ -4 \ \ \ 5 \ -2}$$
$$\ \ \ \ \ \ 2 \ -4 \ \ \ 2$$
$$\overline{\ \ 1 \ -2 \ \ \ 1 \ \ \ 0}$$

$$x^3 - 4x^2 + 5x - 2 = (x - 2)(x^2 - 2x + 1) = (x - 2)(x - 1)^2$$

Find the partial fraction decomposition:

$$\frac{x^2}{x^3 - 4x^2 + 5x - 2} = \frac{x^2}{(x - 2)(x - 1)^2} = \frac{A}{x - 2} + \frac{B}{x - 1} + \frac{C}{(x - 1)^2}$$

(Multiply both sides by $(x - 2)(x - 1)^2$.)

$$x^2 = A(x - 1)^2 + B(x - 2)(x - 1) + C(x - 2)$$

Let $x = 2$:   then $2^2 = A(2 - 1)^2 + B(2 - 2)(2 - 1) + C(2 - 2)$

$\rightarrow \ 4 = A \ \rightarrow \ A = 4$

Let $x = 1$:   then $1^2 = A(1 - 1)^2 + B(1 - 2)(1 - 1) + C(1 - 2)$

$\rightarrow \ 1 = -C \ \rightarrow \ C = -1$

Let $x = 0$:   then $0^2 = A(0 - 1)^2 + B(0 - 2)(0 - 1) + C(0 - 2)$

$\rightarrow \ 0 = A + 2B - 2C \ \rightarrow \ 0 = 4 + 2B - 2(-1)$

$\rightarrow \ 2B = -6 \ \rightarrow \ B = -3$

$$\frac{x^2}{x^3 - 4x^2 + 5x - 2} = \frac{4}{x - 2} + \frac{-3}{x - 1} + \frac{-1}{(x - 1)^2}$$

37. Find the partial fraction decomposition:

$$\frac{x^3}{(x^2 + 16)^3} = \frac{Ax + B}{x^2 + 16} + \frac{Cx + D}{(x^2 + 16)^2} + \frac{Ex + F}{(x^2 + 16)^3}$$

(Multiply both sides by $(x^2 + 16)^3$.)

$$x^3 = (Ax + B)(x^2 + 16)^2 + (Cx + D)(x^2 + 16) + Ex + F$$

$$x^3 = (Ax + B)(x^4 + 32x^2 + 256) + Cx^3 + Dx^2 + 16Cx + 16D + Ex + F$$

$$x^3 = Ax^5 + Bx^4 + 32Ax^3 + 32Bx^2 + 256Ax + 256B + Cx^3 + Dx^2$$
$$+ 16Cx + 16D + Ex + F$$

$$x^3 = Ax^5 + Bx^4 + (32A + C)x^3 + (32B + D)x^2 + (256A + 16C + E)x$$
$$+ (256B + 16D + F)$$

$$A = 0$$
$$B = 0$$
$$32A + C = 1 \rightarrow 32(0) + C = 1 \rightarrow C = 1$$
$$32B + D = 0 \rightarrow 32(0) + D = 0 \rightarrow D = 0$$
$$256A + 16C + E = 0 \rightarrow 256(0) + 16(1) + E = 0 \rightarrow E = -16$$
$$256B + 16D + F = 0 \rightarrow 256(0) + 16(0) + F = 0 \rightarrow F = 0$$

$$\frac{x^3}{(x^2 + 16)^3} = \frac{x}{(x^2 + 16)^2} + \frac{-16x}{(x^2 + 16)^3}$$

39. Find the partial fraction decomposition:

$$\frac{4}{2x^2 - 5x - 3} = \frac{4}{(x - 3)(2x + 1)} = \frac{A}{x - 3} + \frac{B}{2x + 1}$$

(Multiply both sides by $(x - 3)(2x + 1)$.)

$$4 = A(2x + 1) + B(x - 3)$$

Let $x = -\frac{1}{2}$: then $4 = A\left[2\left(-\frac{1}{2}\right) + 1\right] + B\left(-\frac{1}{2} - 3\right)$

$$\rightarrow 4 = -\frac{7}{2}B \rightarrow B = -\frac{8}{7}$$

Let $x = 3$: then $4 = A(2(3) + 1) + B(3 - 3) \rightarrow 4 = 7A \rightarrow A = \frac{4}{7}$

$$\frac{4}{2x^2 - 5x - 3} = \frac{4}{(x - 3)(2x + 1)} = \frac{\left(\frac{4}{7}\right)}{x - 3} + \frac{\left(-\frac{8}{7}\right)}{2x + 1}$$

41. Find the partial fraction decomposition:

$$\frac{2x + 3}{x^4 - 9x^2} = \frac{2x + 3}{x^2(x - 3)(x + 3)} = \frac{A}{x} + \frac{B}{x^2} + \frac{C}{x - 3} + \frac{D}{x + 3}$$

(Multiply both sides by $x^2(x - 3)(x + 3)$.)

$$2x + 3 = Ax(x - 3)(x + 3) + B(x - 3)(x + 3) + Cx^2(x + 3) + Dx^2(x - 3)$$

Let $x = 0$: then

$$2 \cdot 0 + 3 = A \cdot 0(0 - 3)(0 + 3) + B(0 - 3)(0 + 3) + C \cdot 0^2(0 + 3) + D \cdot 0^2(0 - 3)$$

$$\rightarrow 3 = -9B \rightarrow B = -\frac{1}{3}$$

Let $x = 3$: then

$$2 \cdot 3 + 3 = A \cdot 3(3 - 3)(3 + 3) + B(3 - 3)(3 + 3) + C \cdot 3^2(3 + 3) + D \cdot 3^2(3 - 3)$$

$$\rightarrow 9 = 54C \rightarrow C = \frac{1}{6}$$

Let $x = -3$: then

$$2(-3) + 3 = A(-3)(-3-3)(-3+3) + B(-3-3)(-3+3) + C(-3)^2(-3+3)$$
$$+ D(-3)^2(-3-3)$$

$$\rightarrow -3 = -54D \rightarrow D = \frac{1}{18}$$

Let $x = 1$: then

$$2 \cdot 1 + 3 = A \cdot 1(1-3)(1+3) + B(1-3)(1+3) + C \cdot 1^2(1+3) + D \cdot 1^2(1-3)$$

$$\rightarrow 5 = -8A - 8B + 4C - 2D$$

$$\rightarrow 5 = -8A - 8\left(-\frac{1}{3}\right) + 4\left(\frac{1}{6}\right) - 2\left(\frac{1}{18}\right)$$

$$\rightarrow 5 = -8A + \frac{8}{3} + \frac{2}{3} - \frac{1}{9} \rightarrow -8A = \frac{16}{9} \rightarrow A = -\frac{2}{9}$$

$$\frac{2x+3}{x^4 - 9x^2} = \frac{2x+3}{x^2(x-3)(x+3)} = \frac{\left(-\frac{2}{9}\right)}{x} + \frac{\left(-\frac{1}{3}\right)}{x^2} + \frac{\left(\frac{1}{6}\right)}{x-3} + \frac{\left(\frac{1}{18}\right)}{x+3}$$

Chapter 10

Systems of Equations and Inequalities

10.7 Systems of Nonlinear Equations

1. $\begin{cases} y = x^2 + 1 \\ y = x + 1 \end{cases}$

Graph: $y_1 = x^2 + 1$; $y_2 = x + 1$

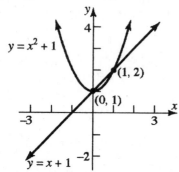

(0, 1) and (1, 2) are the intersection points.

Solve by substitution:
$$x^2 + 1 = x + 1$$
$$x^2 - x = 0$$
$$x(x - 1) = 0$$
$$x = 0 \text{ or } x = 1$$
$$y = 1 \qquad y = 2$$
Solutions: (0, 1) and (1, 2)

3. $\begin{cases} y = \sqrt{36 - x^2} \\ y = 8 - x \end{cases}$

Graph: $y_1 = \sqrt{36 - x^2}$; $y_2 = 8 - x$

(2.59, 5.41) and (5.41, 2.59) are the intersection points.

Solve by substitution:
$$\sqrt{36 - x^2} = 8 - x$$
$$36 - x^2 = 64 - 16x + x^2$$
$$2x^2 - 16x + 28 = 0$$
$$x^2 - 8x + 14 = 0$$
$$x = \frac{8 \pm \sqrt{64 - 56}}{2}$$
$$x = \frac{8 \pm 2\sqrt{2}}{2}$$
$$x = 4 \pm \sqrt{2}$$
If $x = 4 + \sqrt{2}$, $y = 8 - \left(4 + \sqrt{2}\right) = 4 - \sqrt{2}$
If $x = 4 - \sqrt{2}$, $y = 8 - \left(4 - \sqrt{2}\right) = 4 + \sqrt{2}$

Solutions:
$$\left(4 + \sqrt{2}, 4 - \sqrt{2}\right) \text{ and } \left(4 - \sqrt{2}, 4 + \sqrt{2}\right)$$

5. $\begin{cases} y = \sqrt{x} \\ y = 2 - x \end{cases}$

Graph: $y_1 = \sqrt{x}$; $y_2 = 2 - x$

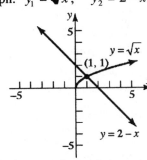

$(1, 1)$ is the intersection point.

Solve by substitution:
$$\sqrt{x} = 2 - x$$
$$x = 4 - 4x + x^2$$
$$x^2 - 5x + 4 = 0$$
$$(x - 4)(x - 1) = 0$$
$$x = 4 \quad \text{or } x = 1$$
$$y = -2 \text{ or } y = 1$$

Eliminate $(4, -2)$; it does not check.

Solution: $(1, 1)$

7. $\begin{cases} x = 2y \\ x = y^2 - 2y \end{cases}$

Solve each equation for y in order to enter it into the graphing utility:
$$y^2 - 2y + 1 = x + 1 \rightarrow (y - 1)^2 = x + 1 \rightarrow y - 1 = \pm\sqrt{x + 1} \rightarrow y = 1 \pm \sqrt{x + 1}$$

Graph: $y_1 = \dfrac{x}{2}$; $y_2 = 1 + \sqrt{x + 1}$; $y_3 = 1 - \sqrt{x + 1}$

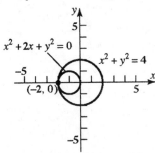

$(0, 0)$ and $(8, 4)$ are the intersection points.

Solve by substitution:
$$2y = y^2 - 2y$$
$$y^2 - 4y = 0$$
$$y(y - 4) = 0$$
$$y = 0 \quad \text{or } y = 4$$
$$x = 0 \quad \text{or } x = 8$$

Solutions: $(0, 0)$ and $(8, 4)$

9. $\begin{cases} x^2 + y^2 = 4 \\ x^2 + 2x + y^2 = 0 \end{cases}$

Graph: $y_1 = \sqrt{4 - x^2}$; $y_2 = -\sqrt{4 - x^2}$; $y_3 = \sqrt{-x^2 - 2x}$; $y_4 = -\sqrt{-x^2 - 2x}$

Substitute 4 for $x^2 + y^2$ in the second equation:
$$2x + 4 = 0$$
$$2x = -4$$
$$x = -2$$
$$y = \sqrt{4 - (-2)^2} = 0$$

Solution: $(-2, 0)$

$(-2, 0)$ is the intersection point.   Note: This intersection point is impossible to find on your graphing utility unless you have just the right window and make an excellent guess.

11. $\begin{cases} y = 3x - 5 \\ x^2 + y^2 = 5 \end{cases}$

Graph: $y_1 = 3x - 5$; $y_2 = \sqrt{5 - x^2}$;

$y_3 = -\sqrt{5 - x^2}$

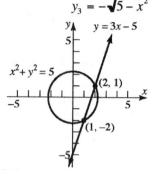

$(1, -2)$ and $(2, 1)$ are the intersection points.

Solve by substitution:

$$x^2 + (3x - 5)^2 = 5$$
$$x^2 + 9x^2 - 30x + 25 = 5$$
$$10x^2 - 30x + 20 = 0$$
$$x^2 - 3x + 2 = 0$$
$$(x - 1)(x - 2) = 0$$
$$x = 1 \qquad \text{or } x = 2$$
$$y = 3(1) - 5 \qquad y = 3(2) - 5$$
$$y = -2 \qquad y = 1$$

Solutions: $(1, -2)$ and $(2, 1)$

13. $\begin{cases} x^2 + y^2 = 4 \\ y^2 - x = 4 \end{cases}$

Graph: $y_1 = \sqrt{4 - x^2}$; $y_2 = -\sqrt{4 - x^2}$; $y_3 = \sqrt{x + 4}$; $y_4 = -\sqrt{x + 4}$

$(-1, 1.73)$, $(-1, -1.73)$, $(0, 2)$, and $(0, -2)$ are the intersection points.

Substitute $x + 4$ for y^2 in the first equation:

$$x^2 + x + 4 = 4$$
$$x^2 + x = 0$$
$$x(x + 1) = 0$$
$$x = 0 \qquad \text{or } x = -1$$
$$y^2 = 4 \qquad y^2 = 3$$
$$y = \pm 2 \qquad y^2 = \pm\sqrt{3}$$

Solutions:

$$(0, -2), (0, 2), \left(-1, \sqrt{3}\right), \left(-1, -\sqrt{3}\right)$$

15. $\begin{cases} xy = 4 \\ x^2 + y^2 = 8 \end{cases}$

Graph: $y_1 = \dfrac{4}{x}$; $y_2 = \sqrt{8 - x^2}$;

$y_3 = -\sqrt{8 - x^2}$

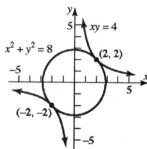

$(-2, -2)$ and $(2, 2)$ are the intersection points.

Solve by substitution:

$$x^2 + \left(\frac{4}{x}\right)^2 = 8$$
$$x^2 + \frac{16}{x^2} = 8$$
$$x^4 + 16 = 8x^2$$
$$x^4 - 8x^2 + 16 = 0$$
$$\left(x^2 - 4\right)^2 = 0$$
$$x^2 - 4 = 0$$
$$x^2 = 4$$
$$x = 2 \text{ or } x = -2$$
$$y = 2 \qquad y = -2$$

Solutions: $(-2, -2)$ and $(2, 2)$

17. $\begin{cases} x^2 + y^2 = 4 \\ \quad\quad y = x^2 - 9 \end{cases}$

Graph: $y_1 = x^2 - 9$; $y_2 = \sqrt{4 - x^2}$;

$y_3 = -\sqrt{4 - x^2}$

No solution; Inconsistent.

Solve by substitution:

$$x^2 + (x^2 - 9)^2 = 4$$
$$x^2 + x^4 - 18x^2 + 81 = 4$$
$$x^4 - 17x^2 + 77 = 0$$
$$x^2 = \frac{17 \pm \sqrt{289 - 4(77)}}{2}$$
$$x^2 = \frac{17 \pm \sqrt{-19}}{2}$$

There are no real solutions to this expression; Inconsistent.

19. $\begin{cases} y = x^2 - 4 \\ y = 6x - 13 \end{cases}$

Graph: $y_1 = x^2 - 4$; $y_2 = 6x - 13$

(3,5) is the intersection point.

Solve by substitution:

$$x^2 - 4 = 6x - 13$$
$$x^2 - 6x + 9 = 0$$
$$(x - 3)^2 = 0$$
$$x - 3 = 0$$
$$x = 3$$
$$y = 6(3) - 13 = 5$$

Solutions: (3,5)

21. Solve the second equation for y, substitute into the first equation and solve:

$$\begin{cases} 2x^2 + y^2 = 18 \\ \quad\quad xy = 4 \;\rightarrow\; y = \dfrac{4}{x} \end{cases}$$

$$2x^2 + \left(\frac{4}{x}\right)^2 = 18$$

$$2x^2 + \frac{16}{x^2} = 18$$

$$2x^4 + 16 = 18x^2$$

$$2x^4 - 18x^2 + 16 = 0$$

$$x^4 - 9x^2 + 8 = 0 \rightarrow (x^2 - 8)(x^2 - 1) = 0$$

$$x^2 = 8 \;\rightarrow\; x = \pm\sqrt{8} = \pm 2\sqrt{2}$$

$$\text{or}\;\; x^2 = 1 \rightarrow x = \pm 1$$

$$\text{If } x = 2\sqrt{2}: \qquad y = \frac{4}{2\sqrt{2}} = \sqrt{2}$$

$$\text{If } x = -2\sqrt{2}: \qquad y = \frac{4}{-2\sqrt{2}} = -\sqrt{2}$$

$$\text{If } x = 1: \qquad y = \frac{4}{1} = 4$$

$$\text{If } x = -1: \qquad y = \frac{4}{-1} = -4$$

Solutions: $\left(2\sqrt{2}, \sqrt{2}\right), \left(-2\sqrt{2}, -\sqrt{2}\right), (1, 4), (-1, -4)$

23. Substitute the first equation into the second equation and solve:

$$\begin{cases} y = 2x + 1 \\ 2x^2 + y^2 = 1 \end{cases}$$

$$2x^2 + (2x+1)^2 = 1 \rightarrow 2x^2 + 4x^2 + 4x + 1 = 1 \rightarrow 6x^2 + 4x = 0 \rightarrow 2x(3x+2) = 0$$

$$2x = 0 \rightarrow x = 0$$

$$\text{or} \quad 3x + 2 = 0 \rightarrow x = -\frac{2}{3}$$

If $x = 0$: $\qquad y = 2(0) + 1 = 1$

If $x = -\frac{2}{3}$: $\qquad y = 2\left(-\frac{2}{3}\right) + 1 = -\frac{4}{3} + 1 = -\frac{1}{3}$

Solutions: $(0, 1), \left(-\frac{2}{3}, -\frac{1}{3}\right)$

25. Solve the first equation for y, substitute into the second equation and solve:

$$\begin{cases} x + y + 1 = 0 \quad \rightarrow \quad y = -x - 1 \\ x^2 + y^2 + 6y - x = -5 \end{cases}$$

$$x^2 + (-x-1)^2 + 6(-x-1) - x = -5$$

$$x^2 + x^2 + 2x + 1 - 6x - 6 - x = -5$$

$$2x^2 - 5x = 0$$

$$x(2x - 5) = 0 \rightarrow x = 0 \quad \text{or} \quad x = \frac{5}{2}$$

If $x = 0$: $\qquad y = -(0) - 1 = -1$

If $x = \frac{5}{2}$: $\qquad y = -\frac{5}{2} - 1 = -\frac{7}{2}$

Solutions: $(0, -1), \left(\frac{5}{2}, -\frac{7}{2}\right)$

27. Solve the second equation for y, substitute into the first equation and solve:

$$\begin{cases} 4x^2 - 3xy + 9y^2 = 15 \\ \\ 2x + 3y = 5 \quad \rightarrow \quad y = -\frac{2}{3}x + \frac{5}{3} \end{cases}$$

$$4x^2 - 3x\left(-\frac{2}{3}x + \frac{5}{3}\right) + 9\left(-\frac{2}{3}x + \frac{5}{3}\right)^2 = 15 \rightarrow 4x^2 + 2x^2 - 5x + 4x^2 - 20x + 25 = 15$$

$$10x^2 - 25x + 10 = 0 \rightarrow 2x^2 - 5x + 2 = 0$$

$$(2x - 1)(x - 2) = 0 \rightarrow x = \frac{1}{2} \text{ or } x = 2$$

If $x = \frac{1}{2}$: $y = -\frac{2}{3}\left(\frac{1}{2}\right) + \frac{5}{3} = \frac{4}{3}$

If $x = 2$: $y = -\frac{2}{3}(2) + \frac{5}{3} = \frac{1}{3}$

Solutions: $\left(\frac{1}{2}, \frac{4}{3}\right), \left(2, \frac{1}{3}\right)$

29. Multiply each side of the second equation by 4 and add the equations to eliminate y:

$$\begin{cases} x^2 - 4y^2 = -7 \longrightarrow \quad x^2 - 4y^2 = -7 \\ 3x^2 + y^2 = 31 \xrightarrow{\ 4\ } 12x^2 + 4y^2 = 124 \end{cases}$$

$$13x^2 \qquad = 117 \rightarrow x^2 = 9 \rightarrow x = \pm 3$$

If $x = 3$: $3(3)^2 + y^2 = 31 \rightarrow y^2 = 4 \rightarrow y = \pm 2$

If $x = -3$: $3(-3)^2 + y^2 = 31 \rightarrow y^2 = 4 \rightarrow y = \pm 2$

Solutions: $(3, 2), (3, -2), (-3, 2), (-3, -2)$

31. Multiply each side of the first equation by 5 and each side of the second equation by 3 to eliminate y:

$$\begin{cases} 7x^2 - 3y^2 = -5 \xrightarrow{\ 5\ } 35x^2 - 15y^2 = -25 \\ 3x^2 + 5y^2 = 12 \xrightarrow{\ 3\ } \quad 9x^2 + 15y^2 = \quad 36 \end{cases}$$

$$44x^2 \qquad = 11 \rightarrow x^2 = \frac{1}{4} \rightarrow x = \pm\frac{1}{2}$$

If $x = \frac{1}{2}$: $3\left(\frac{1}{2}\right)^2 + 5y^2 = 12 \rightarrow 5y^2 = \frac{45}{4} \rightarrow y^2 = \frac{9}{4} \rightarrow y = \pm\frac{3}{2}$

If $x = -\frac{1}{2}$: $3\left(-\frac{1}{2}\right)^2 + 5y^2 = 12 \rightarrow 5y^2 = \frac{45}{4} \rightarrow y^2 = \frac{9}{4} \rightarrow y = \pm\frac{3}{2}$

Solutions: $\left(\frac{1}{2}, \frac{3}{2}\right), \left(\frac{1}{2}, -\frac{3}{2}\right), \left(-\frac{1}{2}, \frac{3}{2}\right), \left(-\frac{1}{2}, -\frac{3}{2}\right)$

33. Multiply each side of the second equation by 2 and add to eliminate xy:

$$\begin{cases} x^2 + 2xy = 10 \longrightarrow \quad x^2 + 2xy = 10 \\ 3x^2 - xy = 2 \xrightarrow{\ 2\ } 6x^2 - 2xy = 4 \end{cases}$$

$$7x^2 \qquad = 14 \rightarrow x^2 = 2 \rightarrow x = \pm\sqrt{2}$$

If $x = \sqrt{2}$: $3\left(\sqrt{2}\right)^2 - \sqrt{2} \cdot y = 2 \rightarrow -\sqrt{2} \cdot y = -4 \rightarrow y = \frac{4}{\sqrt{2}} \rightarrow y = 2\sqrt{2}$

If $x = -\sqrt{2}$: $3\left(-\sqrt{2}\right)^2 - \left(-\sqrt{2}\right)y = 2 \rightarrow \sqrt{2} \cdot y = -4 \rightarrow y = \frac{-4}{\sqrt{2}} \rightarrow y = -2\sqrt{2}$

Solutions: $\left(\sqrt{2}, 2\sqrt{2}\right), \left(-\sqrt{2}, -2\sqrt{2}\right)$

35. Multiply each side of the first equation by 2 and add the equations to eliminate y:

$$\begin{cases} 2x^2 + y^2 = 2 & \xrightarrow{\;2\;} 4x^2 + 2y^2 = 4 \\ x^2 - 2y^2 = -8 & \longrightarrow \quad x^2 - 2y^2 = -8 \end{cases}$$

$$5x^2 \qquad = -4 \rightarrow x^2 = -\frac{4}{5}$$

No solution. The system is inconsistent.

37. Multiply each side of the second equation by 2 and add the equations to eliminate y:

$$\begin{cases} x^2 + 2y^2 = 16 & \longrightarrow \quad x^2 + 2y^2 = 16 \\ 4x^2 - y^2 = 24 & \xrightarrow{\;2\;} 8x^2 - 2y^2 = 48 \end{cases}$$

$$9x^2 \qquad = 64 \rightarrow x^2 = \frac{64}{9} \quad \rightarrow \quad x = \pm\frac{8}{3}$$

If $x = \frac{8}{3}$: $\left(\frac{8}{3}\right)^2 + 2y^2 = 16 \rightarrow 2y^2 = \frac{80}{9} \rightarrow y^2 = \frac{40}{9} \rightarrow y = \pm\frac{2\sqrt{10}}{3}$

If $x = -\frac{8}{3}$: $\left(-\frac{8}{3}\right)^2 + 2y^2 = 16 \rightarrow 2y^2 = \frac{80}{9} \rightarrow y^2 = \frac{40}{9} \rightarrow y = \pm\frac{2\sqrt{10}}{3}$

Solutions: $\left(\frac{8}{3}, \frac{2\sqrt{10}}{3}\right), \left(\frac{8}{3}, \frac{-2\sqrt{10}}{3}\right), \left(-\frac{8}{3}, \frac{2\sqrt{10}}{3}\right), \left(-\frac{8}{3}, \frac{-2\sqrt{10}}{3}\right)$

39. Multiply each side of the second equation by 2 and add the equations to eliminate y:

$$\begin{cases} \dfrac{5}{x^2} - \dfrac{2}{y^2} = -3 & \longrightarrow \quad \dfrac{5}{x^2} - \dfrac{2}{y^2} = -3 \\ \dfrac{3}{x^2} + \dfrac{1}{y^2} = 7 & \xrightarrow{\;2\;} \dfrac{6}{x^2} + \dfrac{2}{y^2} = 14 \end{cases}$$

$$\frac{11}{x^2} \qquad = 11 \rightarrow 11 = 11x^2 \rightarrow x^2 = 1 \rightarrow x = \pm 1$$

If $x = 1$: $\dfrac{3}{(1)^2} + \dfrac{1}{y^2} = 7 \rightarrow \dfrac{1}{y^2} = 4$

$$y^2 = \frac{1}{4} \rightarrow y = \pm\frac{1}{2}$$

If $x = -1$: $\dfrac{3}{(-1)^2} + \dfrac{1}{y^2} = 7 \rightarrow \dfrac{1}{y^2} = 4$

$$y^2 = \frac{1}{4} \rightarrow y = \pm\frac{1}{2}$$

Solutions: $\left(1, \frac{1}{2}\right), \left(1, -\frac{1}{2}\right), \left(-1, \frac{1}{2}\right), \left(-1, -\frac{1}{2}\right)$

41. Multiply each side of the first equation by –2 and add the equations to eliminate x:

$$\begin{cases} \dfrac{1}{x^4} + \dfrac{6}{y^4} = 6 \\ \dfrac{2}{x^4} - \dfrac{2}{y^4} = 19 \end{cases} \xrightarrow{-2} \begin{array}{l} \dfrac{-2}{x^4} - \dfrac{12}{y^4} = -12 \\ \dfrac{2}{x^4} - \dfrac{2}{y^4} = 19 \end{array}$$

$$\dfrac{-14}{y^4} = 7 \to -14 = 7y^4 \to y^4 = -2$$

There are no real solutions. The system is inconsistent.

43. Factor the first equation, solve for x, substitute into the second equation and solve:

$$\begin{cases} x^2 - 3xy + 2y^2 = 0 \to (x - 2y)(x - y) = 0 \to x = 2y \text{ or } x = y \\ x^2 + xy = 6 \end{cases}$$

Substitute $x = 2y$ and solve: Substitute $x = y$ and solve:

$$x^2 + xy = 6 \qquad\qquad x^2 + xy = 6$$
$$(2y)^2 + (2y)y = 6 \qquad\qquad y^2 + y \cdot y = 6$$
$$4y^2 + 2y^2 = 6 \to 6y^2 = 6 \qquad y^2 + y^2 = 6 \to 2y^2 = 6$$
$$y^2 = 1 \to y = \pm 1 \qquad\qquad y^2 = 3 \to y = \pm\sqrt{3}$$

If $y = 1$: $x = 2 \cdot 1 = 2$ If $y = \sqrt{3}$: $x = \sqrt{3}$

If $y = -1$: $x = 2(-1) = -2$ If $y = -\sqrt{3}$: $x = -\sqrt{3}$

Solutions: $(2, 1), (-2, -1), \left(\sqrt{3}, \sqrt{3}\right), \left(-\sqrt{3}, -\sqrt{3}\right)$

45. Multiply each side of the second equation by –y and add the equations to eliminate y:

$$\begin{cases} y^2 + y + x^2 - x - 2 = 0 \\ y + 1 + \dfrac{x - 2}{y} = 0 \end{cases} \longrightarrow \begin{array}{l} y^2 + y + x^2 - x - 2 = 0 \\ -y^2 - y \qquad - x + 2 = 0 \end{array}$$

$$x^2 - 2x \qquad = 0 \to x(x - 2) = 0$$
$$x = 0 \text{ or } x = 2$$

If $x = 0$: $y^2 + y + 0^2 - 0 - 2 = 0 \to y^2 + y - 2 = 0 \to (y + 2)(y - 1) = 0$
$$\to y = -2 \text{ or } y = 1$$

If $x = 2$: $y^2 + y + 2^2 - 2 - 2 = 0 \to y^2 + y = 0 \to y(y + 1) = 0$
$$\to y = 0 \text{ or } y = -1$$

Solutions: $(0, -2), (0, 1), (2, 0), (2, -1)$

47. Rewrite each equation in exponential form:

$$\begin{cases} \log_x y = 3 \to y = x^3 \\ \log_x(4y) = 5 \to 4y = x^5 \end{cases}$$

Substitute the first equation into the second and solve:

$$4x^3 = x^5$$
$$x^5 - 4x^3 = 0 \to x^3(x^2 - 4) = 0 \to x^3 = 0 \text{ or } x^2 = 4 \to x = 0 \text{ or } x = \pm 2$$

The base of a logarithm must be positive, thus $x \neq 0$ and $x \neq -2$.

If $x = 2$: $y = 2^3 = 8$
Solution: $(2, 8)$

49. Rewrite each equation in exponential form:

$$\begin{cases} \ln x = 4\ln y \;\rightarrow\; x = e^{4\ln y} = e^{\ln y^4} = y^4 \\ \log_3 x = 2 + 2\log_3 y \;\rightarrow\; x = 3^{2+2\log_3 y} = 3^2 \cdot 3^{2\log_3 y} = 3^2 \cdot 3^{\log_3 y^2} = 9y^2 \end{cases}$$

So we have the system

$$\begin{cases} x = y^4 \\ x = 9y^2 \end{cases}$$

Therefore we have
$$9y^2 = y^4$$
$$9y^2 - y^4 = 0$$
$$y^2(9 - y^2) = 0$$
$$y^2(3 + y)(3 - y) = 0$$
$$y = 0 \text{ or } y = -3 \text{ or } y = 3$$

Since $\ln y$ is undefined when $y \le 0$, the only solution is $y = 3$.
     If $y = 3$: $x = y^4 \rightarrow x = 3^4 = 81$
Solution: $(81, 3)$

51. Solve the first equation for x, substitute into the second equation and solve:
$$\begin{cases} x + 2y = 0 \;\rightarrow\; x = -2y \\ (x-1)^2 + (y-1)^2 = 5 \end{cases}$$
$$(-2y-1)^2 + (y-1)^2 = 5$$
$$4y^2 + 4y + 1 + y^2 - 2y + 1 = 5 \rightarrow 5y^2 + 2y - 3 = 0$$
$$(5y-3)(y+1) = 0$$
$$y = \frac{3}{5} = 0.6 \text{ or } y = -1$$
$$x = -\frac{6}{5} = -1.2 \text{ or } x = 2$$

The points of intersection are $(-1.2, 0.6), (2, -1)$.

53. Complete the square on the second equation, substitute into the first equation and solve:
$$\begin{cases} (x-1)^2 + (y+2)^2 = 4 \\ y^2 + 4y - x + 1 = 0 \;\rightarrow\; y^2 + 4y + 4 = x - 1 + 4 \rightarrow (y+2)^2 = x + 3 \end{cases}$$
$$(x-1)^2 + x + 3 = 4$$
$$x^2 - 2x + 1 + x + 3 = 4$$
$$x^2 - x = 0$$
$$x(x-1) = 0$$
$$x = 0 \text{ or } x = 1$$

If $x = 0$: $(y + 2)^2 = 0 + 3$

$y + 2 = \pm\sqrt{3} \rightarrow y = -2 \pm \sqrt{3}$

If $x = 1$: $(y + 2)^2 = 1 + 3$

$y + 2 = \pm 2 \rightarrow y = -2 \pm 2$

The points of intersection are:

$\left(0, -2 - \sqrt{3}\right), \left(0, -2 + \sqrt{3}\right), (1, -4), (1, 0).$

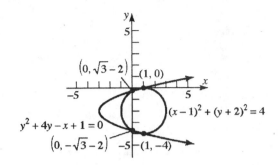

55. Solve the first equation for x, substitute into the second equation and solve:

$$\begin{cases} y = \dfrac{4}{x - 3} \\ x^2 - 6x + y^2 + 1 = 0 \end{cases} \rightarrow x - 3 = \dfrac{4}{y} \rightarrow x = \dfrac{4}{y} + 3$$

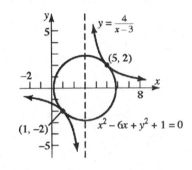

$$\left(\dfrac{4}{y} + 3\right)^2 - 6\left(\dfrac{4}{y} + 3\right) + y^2 + 1 = 0$$

$$\dfrac{16}{y^2} + \dfrac{24}{y} + 9 - \dfrac{24}{y} - 18 + y^2 + 1 = 0$$

$$\dfrac{16}{y^2} + y^2 - 8 = 0$$

$$16 + y^4 - 8y^2 = 0$$

$$y^4 - 8y^2 + 16 = 0$$

$$(y^2 - 4)^2 = 0$$

$$y^2 - 4 = 0$$

$$y^2 = 4$$

$$y = \pm 2$$

If $y = 2$: $x = \dfrac{4}{2} + 3 = 5$

If $y = -2$: $x = \dfrac{4}{-2} + 3 = 1$

The points of intersection are: $(1, -2)$, $(5, 2)$.

57. Graph: $y_1 = x \wedge (2/3)$; $y_2 = e \wedge (-x)$
Use INTERSECT to solve:

Solution: $(0.48, 0.62)$

59. Graph: $y_1 = \sqrt[3]{(2 - x^2)}$; $y_2 = 4/x^3$
Use INTERSECT to solve:

Solution: $(-1.65, -0.89)$

61. Graph:

$$y_1 = \sqrt[4]{(12 - x^4)}; \quad y_2 = -\sqrt[4]{(12 - x^4)};$$

$$y_3 = \sqrt{2/x}; \quad y_4 = -\sqrt{2/x}$$

Use INTERSECT to solve:

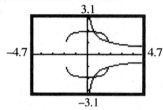

Solutions: (0.58, 1.86), (1.81, 1.05),
(1.81, −1.05), (0.58, −1.86)

63. Graph: $y_1 = 2/x; \quad y_2 = \ln x$

Use INTERSECT to solve:

Solution: (2.35, 0.85)

65. Let x and y be the two numbers. The system of equations is:

$$\begin{cases} x - y = 2 \\ x^2 + y^2 = 10 \end{cases}$$

Solve the first equation for x, substitute into the second equation and solve:

$$(y + 2)^2 + y^2 = 10 \rightarrow y^2 + 4y + 4 + y^2 = 10$$

$$2y^2 + 4y - 6 = 0 \rightarrow y^2 + 2y - 3 = 0 \rightarrow (y + 3)(y - 1) = 0 \rightarrow y = -3 \text{ or } y = 1$$

If $y = -3$: $x = -3 + 2 = -1$

If $y = 1$: $x = 1 + 2 = 3$

The two numbers are 1 and 3 or −1 and −3.

67. Let x and y be the two numbers. The system of equations is:

$$\begin{cases} xy = 4 \\ x^2 + y^2 = 8 \end{cases}$$

Solve the first equation for x, substitute into the second equation and solve:

$$\left(\frac{4}{y}\right)^2 + y^2 = 8 \rightarrow \frac{16}{y^2} + y^2 = 8 \rightarrow 16 + y^4 = 8y^2$$

$$y^4 - 8y^2 + 16 = 0 \rightarrow (y^2 - 4)^2 = 0 \rightarrow y^2 - 4 = 0 \ \rightarrow \ y^2 = 4 \ \rightarrow \ y = \pm 2$$

If $y = 2$: $x = \dfrac{4}{2} = 2$

If $y = -2$: $x = \dfrac{4}{-2} = -2$

The two numbers are 2 and 2 or −2 and −2.

69. Let x and y be the two numbers. The system of equations is:

$$\begin{cases} x - y = xy \\ \dfrac{1}{x} + \dfrac{1}{y} = 5 \end{cases}$$

Solve the first equation for x, substitute into the second equation and solve:

$$x - xy = y \rightarrow x(1-y) = y \rightarrow x = \frac{y}{1-y}$$

$$\frac{1}{\left(\dfrac{y}{1-y}\right)} + \frac{1}{y} = 5$$

$$\frac{1-y}{y} + \frac{1}{y} = 5$$

$$\frac{2-y}{y} = 5$$

$$2 - y = 5y \rightarrow 6y = 2 \rightarrow y = \frac{1}{3}$$

If $y = \dfrac{1}{3}$: $x = \dfrac{\left(\dfrac{1}{3}\right)}{\left(1-\dfrac{1}{3}\right)} = \dfrac{\left(\dfrac{1}{3}\right)}{\left(\dfrac{2}{3}\right)} = \dfrac{1}{2}$ $\therefore$ The two numbers are $\dfrac{1}{2}$ and $\dfrac{1}{3}$.

71. $\begin{cases} \dfrac{a}{b} = \dfrac{2}{3} \\ a + b = 10 \end{cases}$

Solve the second equation for a, substitute into the first equation and solve:

$$\frac{10-b}{b} = \frac{2}{3} \rightarrow 3(10-b) = 2b$$

$$30 - 3b = 2b$$

$$30 = 5b$$

$$b = 6 \rightarrow a = 4$$

$a + b = 10;\ \ b - a = 2$ $\therefore$ The ratio of $a+b$ to $b-a$ is $\dfrac{10}{2} = 5.$

73. Let x = the width of the rectangle.
Let y = the length of the rectangle.

$$\begin{cases} 2x + 2y = 16 \\ xy = 15 \end{cases}$$

Solve the first equation for y, substitute into the second equation and solve:

$$2x + 2y = 16$$

$$2y = 16 - 2x$$

$$y = 8 - x$$

$$x(8-x) = 15 \rightarrow 8x - x^2 = 15 \rightarrow x^2 - 8x + 15 = 0 \rightarrow (x-5)(x-3) = 0$$

$$x = 5 \ \text{ or } \ x = 3$$

$$y = 3 \qquad y = 5$$

The dimensions of the rectangle are 3 inches by 5 inches.

75.   Let x = the radius of the first circle.
      Let y = the radius of the second circle.
$$\begin{cases} 2\pi x + 2\pi y = 12\pi \\ \pi x^2 + \pi y^2 = 20\pi \end{cases}$$
      Solve the first equation for y, substitute into the second equation and solve:

$$2\pi x + 2\pi y = 12\pi$$
$$x + y = 6$$
$$y = 6 - x$$

$$\pi x^2 + \pi y^2 = 20\pi$$
$$x^2 + y^2 = 20$$
$$x^2 + (6 - x)^2 = 20$$
$$x^2 + 36 - 12x + x^2 = 20$$
$$2x^2 - 12x + 16 = 0$$
$$x^2 - 6x + 8 = 0$$
$$(x - 4)(x - 2) = 0$$
$$x = 4 \ \text{ or } \ x = 2$$
$$y = 2 \qquad y = 4$$

The radii of the circles are 2 centimeters and 4 centimeters.

77.   The tortoise takes $9 + 3 = 12$ minutes or 0.2 hour longer to complete the race than the hare.
      Let r = the rate of the hare.
      Let t = the time for the hare to complete the race.
      Then $t + 0.2$ = the time for the tortoise and $r - 0.5$ = the rate for the tortoise.
      Since the length of the race is 21 meters, the distance equations are:
$$\begin{cases} rt = 21 \\ (r - 0.5)(t + 0.2) = 21 \end{cases}$$
      Solve the first equation for r, substitute into the second equation and solve:
$$\left(\frac{21}{t} - 0.5\right)(t + 0.2) = 21 \rightarrow 21 + \frac{4.2}{t} - 0.5t - 0.1 = 21$$

$$10t \cdot \left(21 + \frac{4.2}{t} - 0.5t - 0.1\right) = 10t \cdot (21)$$

$$210t + 42 - 5t^2 - t = 210t \rightarrow 5t^2 + t - 42 = 0 \rightarrow (5t - 14)(t + 3) = 0$$

$$t = \frac{14}{5} = 2.8 \ \text{ or } \ t = -3$$

$t = -3$ makes no sense, since time cannot be negative.
Solve for r:
$$r = \frac{21}{2.8} = 7.5$$
The average speed of the hare is 7.5 meters per hour, and the average speed for the tortoise
is 7 meters per hour.

79.   Let x = the width of the cardboard.
      Let y = the length of the cardboard.
      The width of the box will be $x - 4$, the length of the box will be $y - 4$, and the height is 2.
      The volume is $V = (x - 4)(y - 4)(2)$.

Solve the system of equations:
$$\begin{cases} xy = 216 \\ 2(x-4)(y-4) = 224 \end{cases}$$
Solve the first equation for y, substitute into the second equation and solve:
$$(2x-8)\left(\frac{216}{x}-4\right) = 224 \rightarrow 432 - 8x - \frac{1728}{x} + 32 = 224$$
$$432x - 8x^2 - 1728 + 32x = 224x \rightarrow -8x^2 + 240x - 1728 = 0$$
$$x^2 - 30x + 216 = 0 \rightarrow (x-12)(x-18) = 0$$
$$x = 12 \text{ or } x = 18$$
$$y = 18 \qquad y = 12$$
The cardboard should be 12 centimeters by 18 centimeters.

81. Find equations relating area and perimeter:
$$\begin{cases} x^2 + y^2 = 4500 \\ 3x + 3y + (x - y) = 300 \end{cases}$$
Solve the second equation for y, substitute into the first equation and solve:

$$4x + 2y = 300 \qquad\qquad x^2 + (150 - 2x)^2 = 4500$$
$$2y = 300 - 4x \qquad x^2 + 22500 - 600x + 4x^2 = 4500$$
$$y = 150 - 2x \qquad\qquad 5x^2 - 600x + 18000 = 0$$
$$x^2 - 120x + 3600 = 0$$
$$(x - 60)^2 = 0$$
$$x - 60 = 0$$
$$x = 60$$
$$y = 150 - 2(60) = 30$$

The sides of the squares are 30 feet and 60 feet.

83. Solve the system for l and w:
$$\begin{cases} 2l + 2w = P \\ lw = A \end{cases}$$
Solve the first equation for l, substitute into the second equation and solve:
$$2l = P - 2w \rightarrow l = \frac{P}{2} - w$$
$$\left(\frac{P}{2} - w\right)w = A$$
$$\frac{P}{2}w - w^2 = A \rightarrow w^2 - \frac{P}{2}w + A = 0$$
$$w = \frac{\left(\frac{P}{2} \pm \sqrt{\frac{P^2}{4} - 4A}\right)}{2} = \frac{\left(\frac{P}{2} \pm \sqrt{\frac{P^2 - 16A}{4}}\right)}{2} = \frac{\left(\frac{P}{2} \pm \frac{\sqrt{P^2 - 16A}}{2}\right)}{2}$$
$$w = \frac{P \pm \sqrt{P^2 - 16A}}{4}$$

$$\text{If } w = \frac{P + \sqrt{P^2 - 16A}}{4} \text{ then } l = \frac{P}{2} - \frac{P + \sqrt{P^2 - 16A}}{4} = \frac{P - \sqrt{P^2 - 16A}}{4}$$

$$\text{If } w = \frac{P - \sqrt{P^2 - 16A}}{4} \text{ then } l = \frac{P}{2} - \frac{P - \sqrt{P^2 - 16A}}{4} = \frac{P + \sqrt{P^2 - 16A}}{4}$$

If it is required that length be greater than width, then the solution is:

$$w = \frac{P - \sqrt{P^2 - 16A}}{4} \text{ and } l = \frac{P + \sqrt{P^2 - 16A}}{4}$$

85. Solve the equation:
$$m^2 - 4(2m - 4) = 0 \rightarrow m^2 - 8m + 16 = 0 \rightarrow (m - 4)^2 = 0 \rightarrow m - 4 = 0 \rightarrow m = 4$$
Use the point-slope equation with slope 4 and the point (2, 4) to obtain the equation of the tangent line:

$$y - 4 = 4(x - 2) \rightarrow y - 4 = 4x - 8 \rightarrow y = 4x - 4$$

87. Solve the system:
$$\begin{cases} y = x^2 + 2 \\ y = mx + b \end{cases}$$
Solve the system by substitution:
$$x^2 + 2 = mx + b \rightarrow x^2 - mx + 2 - b = 0$$
Note that the tangent line passes through (1, 3). Find the relation between m and b:
$$3 = m(1) + b \rightarrow b = 3 - m$$
Substitute into the quadratic to eliminate b:
$$x^2 - mx + 2 - (3 - m) = 0 \rightarrow x^2 - mx + (m - 1) = 0$$
Find when the discriminant is 0:
$$(-m)^2 - 4(1)(m - 1) = 0$$
$$m^2 - 4m + 4 = 0$$
$$(m - 2)^2 = 0$$
$$m - 2 = 0 \rightarrow m = 2 \quad \rightarrow \quad b = 3 - 2 = 1$$
The equation of the tangent line is $y = 2x + 1$.

89. Solve the system:
$$\begin{cases} 2x^2 + 3y^2 = 14 \\ \quad\quad y = mx + b \end{cases}$$
Solve the system by substitution:
$$2x^2 + 3(mx + b)^2 = 14$$
$$2x^2 + 3m^2x^2 + 6mbx + 3b^2 = 14$$
$$(3m^2 + 2)x^2 + 6mbx + 3b^2 - 14 = 0$$
Note that the tangent line passes through (1, 2). Find the relation between m and b:
$$2 = m(1) + b \rightarrow b = 2 - m$$
Substitute into the quadratic to eliminate b:
$$(3m^2 + 2)x^2 + 6m(2 - m)x + 3(2 - m)^2 - 14 = 0$$
$$(3m^2 + 2)x^2 + (12m - 6m^2)x + 12 - 12m + 3m^2 - 14 = 0$$
$$(3m^2 + 2)x^2 + (12m - 6m^2)x + (3m^2 - 12m - 2) = 0$$

Find when the discriminant is 0:

$$(12m - 6m^2)^2 - 4(3m^2 + 2)(3m^2 - 12m - 2) = 0$$
$$144m^2 - 144m^3 + 36m^4 - 4(9m^4 - 36m^3 - 24m - 4) = 0$$
$$144m^2 - 144m^3 + 36m^4 - 36m^4 + 144m^3 + 96m + 16 = 0$$
$$144m^2 + 96m + 16 = 0$$
$$9m^2 + 6m + 1 = 0$$
$$(3m + 1)^2 = 0$$
$$3m + 1 = 0$$
$$m = -\frac{1}{3} \qquad b = 2 - \left(-\frac{1}{3}\right) = \frac{7}{3}$$

The equation of the tangent line is $y = -\frac{1}{3}x + \frac{7}{3}$.

91. Solve the system:

$$\begin{cases} x^2 - y^2 = 3 \\ \qquad y = mx + b \end{cases}$$

Solve the system by substitution:

$$x^2 - (mx + b)^2 = 3 \rightarrow x^2 - m^2x^2 - 2mbx - b^2 = 3 \rightarrow (1 - m^2)x^2 - 2mbx - b^2 - 3 = 0$$

Note that the tangent line passes through (2, 1). Find the relation between m and b:

$$1 = m(2) + b \rightarrow b = 1 - 2m$$

Substitute into the quadratic to eliminate b:

$$(1 - m^2)x^2 - 2m(1 - 2m)x - (1 - 2m)^2 - 3 = 0$$
$$(1 - m^2)x^2 + (-2m + 4m^2)x - 1 + 4m - 4m^2 - 3 = 0$$
$$(1 - m^2)x^2 + (-2m + 4m^2)x + (-4m^2 + 4m - 4) = 0$$

Find when the discriminant is 0:

$$(-2m + 4m^2)^2 - 4(1 - m^2)(-4m^2 + 4m - 4) = 0$$
$$4m^2 - 16m^3 + 16m^4 - 4(4m^4 - 4m^3 + 4m - 4) = 0$$
$$4m^2 - 16m^3 + 16m^4 - 16m^4 + 16m^3 - 16m + 16 = 0$$
$$4m^2 - 16m + 16 = 0 \rightarrow m^2 - 4m + 4 = 0$$
$$(m - 2)^2 = 0 \rightarrow m - 2 = 0 \rightarrow m = 2 \quad \rightarrow \quad b = 1 - 2(2) = -3$$

The equation of the tangent line is $y = 2x - 3$.

93. Solve for r_1 and r_2:

$$\begin{cases} r_1 + r_2 = -\dfrac{b}{a} \\ \qquad r_1 r_2 = \dfrac{c}{a} \end{cases}$$

Substitute and solve:

$$r_1 = -r_2 - \frac{b}{a} \rightarrow \left(-r_2 - \frac{b}{a}\right)r_2 = \frac{c}{a}$$

$$-r_2^2 - \frac{b}{a}r_2 - \frac{c}{a} = 0$$

$$ar_2^2 + br_2 + c = 0$$

$$r_2 = \frac{-b \pm \sqrt{b^2 - 4ac}}{2a}$$

$$r_1 = -r_2 - \frac{b}{a} = -\left(\frac{-b \pm \sqrt{b^2 - 4ac}}{2a}\right) - \frac{2b}{2a} = \frac{-b \mp \sqrt{b^2 - 4ac}}{2a}$$

The solutions are: $\dfrac{-b + \sqrt{b^2 - 4ac}}{2a}$ and $\dfrac{-b - \sqrt{b^2 - 4ac}}{2a}$.

95. Since the area of the square piece of sheet metal is 100 square feet, the sheet's dimensions are 10 feet by 10 feet. Let x = the length of the cut.

The dimensions of the box are length $= 10 - 2x$; width $= 10 - 2x$; height $= x$
Note that each of these expressions must be positive. So we must have
$\qquad x > 0$ and $10 - 2x > 0 \rightarrow x < 5$, that is, $0 < x < 5$.
So the volume of the box is given by
$$V = (length) \cdot (width) \cdot (height) = (10 - 2x)(10 - 2x)(x) = (10 - 2x)^2 (x)$$

(a) In order to get a volume equal to 9 cubic feet, we solve $(10 - 2x)^2 (x) = 9$.
$$(10 - 2x)^2 (x) = 9 \rightarrow (100 - 40x + 4x^2) x = 9 \rightarrow 100x - 40x^2 + 4x^3 = 9$$
So we need to solve the equation $4x^3 - 40x^2 + 100x - 9 = 0$.
Graphing the function $y_1 = 4x^3 - 40x^2 + 100x - 9$ on a calculator yields the graph

The graph indicates that there three real zeros on the interval [0,6].
Using the ZERO feature of a graphing calculator, we find that the three roots shown occur at $x \approx 0.09$, $x \approx 4.27$ and $x \approx 5.63$.
But we've already noted that we must have $0 < x < 5$, so the only practical values for the cut are $x \approx 0.09$ feet and $x \approx 4.27$ feet.
(b) If the sheet metal has dimensions k feet by k feet, then the volume equation becomes
$$V = (k - 2x)(k - 2x)(x) = (k - 2x)^2 (x) = 9$$
Solving for k we get the quadratic equation
$$xk^2 - 4x^2 k + 4x^3 - 9 = 0$$

Using the quadratic formula we get:

$$k = \frac{-\left(-4x^2\right) \pm \sqrt{\left(-4x^2\right)^2 - 4(x)\left(4x^3 - 9\right)}}{2x} = \frac{4x^2 \pm \sqrt{16x^4 - 16x^4 + 36x}}{2x}$$

$$= \frac{4x^2 \pm \sqrt{36x}}{2x} = \frac{4x^2 \pm 6\sqrt{x}}{2x}$$

Therefore, we get a real solution for k provided $x > 0$ and $4x^2 \pm 6\sqrt{x} \geq 0$.

$$4x^2 \pm 6\sqrt{x} \geq 0 \rightarrow 4x^2 \geq 6\sqrt{x} \rightarrow 16x^4 \geq 36x$$

$$16x^4 - 36x \geq 0 \rightarrow 4x\left(4x^3 - 9\right) \geq 0$$

This last inequality holds provided $x \geq \sqrt[3]{\dfrac{9}{4}}$.

Systems of Equations and Inequalities

10.8 Systems of Inequalities

1. $x \geq 0$

Graph the line $x = 0$. Use a solid line since the inequality uses $\geq$.
Choose a test point not on the line, such as (2, 0). Since $2 \geq 0$ is true, shade the side of the line containing (2, 0).

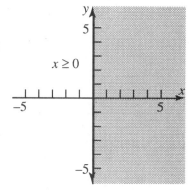

3. $x \geq 4$

Graph the line $x = 4$. Use a solid line since the inequality uses $\geq$.
Choose a test point not on the line, such as (5, 0).
Since $5 \geq 0$ is true, shade the side of the line containing (5, 0).

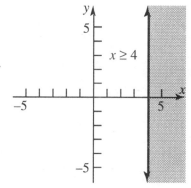

5. $2x + y \geq 6$

Graph the line $2x + y = 6$. Use a solid line since the inequality uses $\geq$.
Choose a test point not on the line, such as (0, 0).
Since $2(0) + 0 \geq 6$ is false, shade the opposite side of the Line from (0, 0).

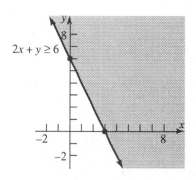

7. $x^2 + y^2 > 1$

Graph the circle $x^2 + y^2 > 1$. Use a dashed line since the inequality uses >.

Choose a test point not on the circle, such as (0, 0).

Since $0^2 + 0^2 > 1$ is false, shade the opposite side of the circle from (0, 0).

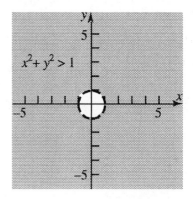

9. $y \leq x^2 - 1$

Graph the parabola $y = x^2 - 1$. Use a solid line since the inequality uses $\leq$.

Choose a test point not on the parabola, such as (0, 0).

Since $0 \leq 0^2 - 1$ is false, shade the opposite side of the parabola from (0, 0).

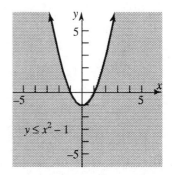

11. $xy \geq 4$

Graph the hyperbola $xy = 4$. Use a solid line since the inequality uses $\geq$.

Choose a test point not on the hyperbola, such as (0, 0). Since $0 \cdot 0 \geq 4$ is false, shade the opposite side of the hyperbola from (0, 0).

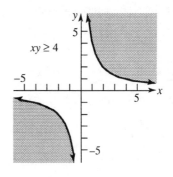

13. $\begin{cases} x + y \leq 2 \\ 2x + y \geq 4 \end{cases}$

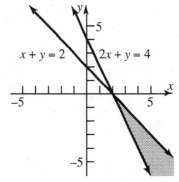

(a) Graph the line $x + y = 2$. Use a solid line since the inequality uses $\leq$.
Choose a test point not on the line, such as $(0, 0)$. Since $0 + 0 \leq 2$ is true, shade the side of the line containing $(0, 0)$.

(b) Graph the line $2x + y = 4$. Use a solid line since the inequality uses $\geq$.
Choose a test point not on the line, such as $(0, 0)$. Since $2(0) + 0 \geq 4$ is false, shade the opposite side of the line from $(0, 0)$.

(c) The overlapping region is the solution.

15. $\begin{cases} 2x - y \leq 4 \\ 3x + 2y \geq -6 \end{cases}$

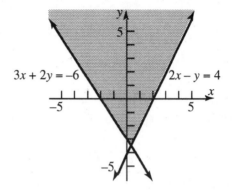

(a) Graph the line $2x - y = 4$. Use a solid line since the inequality uses $\leq$.
Choose a test point not on the line, such as $(0, 0)$. Since $2(0) - 0 \leq 4$ is true, shade the side of the line containing $(0, 0)$.

(b) Graph the line $3x + 2y = -6$. Use a solid line since the inequality uses $\geq$.
Choose a test point not on the line, such as $(0, 0)$. Since $3(0) + 2(0) \geq -6$ is true, shade the side of the line containing $(0, 0)$.

(c) The overlapping region is the solution.

17. $\begin{cases} 2x - 3y \leq 0 \\ 3x + 2y \leq 6 \end{cases}$

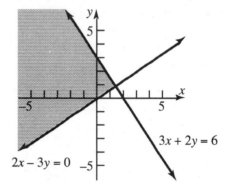

(a) Graph the line $2x - 3y = 0$. Use a solid line since the inequality uses $\leq$.
Choose a test point not on the line, such as $(0, 3)$. Since $2(0) - 3(3) \leq 0$ is true, shade the side of the line containing $(0, 3)$.

(b) Graph the line $3x + 2y = 6$. Use a solid line since the inequality uses $\leq$.
Choose a test point not on the line, such as $(0, 0)$. Since $3(0) + 2(0) \leq 6$ is true, shade the side of the line containing $(0, 0)$.

(c) The overlapping region is the solution.

19. $\begin{cases} x^2 + y^2 \le 9 \\ \quad x + y \ge 3 \end{cases}$

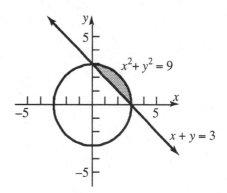

(a)   Graph the circle $x^2 + y^2 = 9$. Use a solid line since the inequality uses $\ge$. Choose a test point not on the circle, such as (0, 0). Since $0^2 + 0^2 \le 9$ is true, shade the same side of the circle as (0, 0).

(b)   Graph the line $x + y = 3$. Use a solid line since the inequality uses $\ge$. Choose a test point not on the line, such as (0, 0). Since $0 + 0 \ge 3$ is false, shade the opposite side of the line from (0, 0).

(c)   The overlapping region is the solution.

21. $\begin{cases} y \ge x^2 - 4 \\ y \le x - 2 \end{cases}$

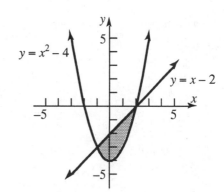

(a)   Graph the parabola $y = x^2 - 4$.
Use a solid line since the inequality uses $\ge$. Choose a test point not on the parabola, such as (0, 0). Since $0 \ge 0^2 - 4$ is true, shade the same side of the parabola as (0, 0).

(b)   Graph the line $y = x - 2$. Use a solid line since the inequality uses $\le$. Choose a test point not on the line, such as (0, 0). Since $0 \le 0 - 2$ is false, shade the opposite side of the line from (0, 0).

(c)   The overlapping region is the solution.

23. $\begin{cases} xy \geq 4 \\ y \geq x^2 + 1 \end{cases}$

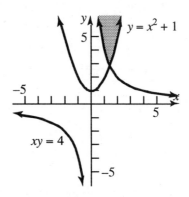

(a)   Graph the hyperbola $xy = 4$.
      Use a solid line since the inequality uses $\geq$. Choose a test point not on the
      parabola, such as (0, 0). Since $0 \cdot 0 \geq 4$ is false, shade the opposite side of the hyperbola
      from (0, 0).

(b)   Graph the parabola $y = x^2 + 1$. Use a solid line since the inequality uses $\geq$. Choose a test
      point not on the parabola, such as (0, 0). Since $0 \geq 0^2 + 1$ is false, shade the opposite
      side of the parabola from (0, 0).

(c)   The overlapping region is the solution.

25. $\begin{cases} x - 2y \leq 6 \\ 2x - 4y \geq 0 \end{cases}$

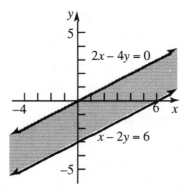

(a)   Graph the line $x - 2y = 6$. Use a solid line since the inequality uses $\leq$.
      Choose a test point not on the line, such as (0, 0). Since $0 - 2(0) \leq 6$ is true, shade
      the side of the line containing (0, 0).

(b)   Graph the line $2x - 4y = 0$. Use a solid line since the inequality uses $\geq$.
      Choose a test point not on the line, such as (0, 2). Since $2(0) - 4(2) \geq 0$ is false,
      shade the opposite side of the line from (0, 2).

(c)   The overlapping region is the solution.

27. $\begin{cases} 2x + y \geq -2 \\ 2x + y \geq 2 \end{cases}$

(a) Graph the line $2x + y = -2$. Use a solid line since the inequality uses $\geq$.
Choose a test point not on the line, such as $(0, 0)$. Since $2(0) + 0 \geq -2$ is true, shade the side of the line containing $(0, 0)$.

(b) Graph the line $2x + y = 2$. Use a solid line since the inequality uses $\geq$.
Choose a test point not on the line, such as $(0, 0)$. Since $2(0) + 0 \geq 2$ is false, shade the opposite side of the line from $(0, 0)$.

(c) The overlapping region is the solution.

29. $\begin{cases} 2x + 3y \geq 6 \\ 2x + 3y \leq 0 \end{cases}$

(a) Graph the line $2x + 3y = 6$. Use a solid line since the inequality uses $\geq$.
Choose a test point not on the line, such as $(0, 0)$. Since $2(0) + 3(0) \geq 6$ is false, shade the opposite side of the line from $(0, 0)$.

(b) Graph the line $2x + 3y = 0$. Use a solid line since the inequality uses $\leq$.
Choose a test point not on the line, such as $(0, 2)$. Since $2(0) + 3(2) \leq 0$ is false, shade the opposite side of the line from $(0, 2)$.

(c) Since the regions do not overlap, the solution is an empty set.

31. Graph the system of linear inequalities:

$\begin{cases} x \geq 0 \\ y \geq 0 \\ 2x + y \leq 6 \\ x + 2y \leq 6 \end{cases}$

571

(a)   Graph $x \geq 0$; $y \geq 0$. Shaded region is the first quadrant.
(b)   Graph the line $2x + y = 6$. Use a solid line since the inequality uses $\leq$.
       Choose a test point not on the line, such as $(0, 0)$. Since $2(0)+0 \leq 6$ is true, shade the side of the line containing $(0, 0)$.
(c)   Graph the line $x + 2y = 6$. Use a solid line since the inequality uses $\leq$.
       Choose a test point not on the line, such as $(0, 0)$. Since $0 + 2(0) \leq 6$ is true, shade the side of the line containing $(0, 0)$.
(d)   The overlapping region is the solution.
(e)   The graph is bounded.
(f)   Find the vertices:
       The x-axis and y-axis intersect at $(0, 0)$.
       The intersection of $x + 2y = 6$ and the y-axis is $(0, 3)$.
       The intersection of $2x + y = 6$ and the x-axis is $(3, 0)$.
       To find the intersection of $x + 2y = 6$ and $2x + y = 6$, solve the system:
       $$\begin{cases} x + 2y = 6 \quad \rightarrow \quad x = 6 - 2y \\ 2x + y = 6 \end{cases}$$
       Substitute and solve:
       $$2(6 - 2y) + y = 6 \rightarrow 12 - 4y + y = 6 \rightarrow -3y = -6 \rightarrow y = 2$$
       $$x = 6 - 2(2) = 6 - 4 = 2$$
       The point of intersection is $(2, 2)$.
       The four corner points are $(0, 0)$, $(0, 3)$, $(3, 0)$, and $(2, 2$

33.   Graph the system of linear inequalities:

$$\begin{cases} x \geq 0 \\ y \geq 0 \\ x + y \geq 2 \\ 2x + y \geq 4 \end{cases}$$

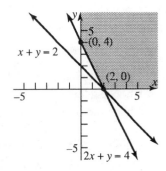

(a)   Graph $x \geq 0$; $y \geq 0$. Shaded region is the first quadrant.
(b)   Graph the line $x + y = 2$. Use a solid line since the inequality uses $\geq$.
       Choose a test point not on the line, such as $(0, 0)$. Since $0 + 0 \geq 2$ is false, shade the opposite side of the line from $(0, 0)$.
(c)   Graph the line $2x + y = 4$. Use a solid line since the inequality uses $\geq$.
       Choose a test point not on the line, such as $(0, 0)$. Since $2(0)+0 \geq 4$ is false, shade the opposite side of the line from $(0, 0)$.
(d)   The overlapping region is the solution.
(e)   The graph is unbounded.
(f)   Find the vertices:
       The intersection of $x + y = 2$ and the x-axis is $(2, 0)$.
       The intersection of $2x + y = 4$ and the y-axis is $(0, 4)$.
       The two corner points are $(2, 0)$, and $(0, 4)$.

35.   Graph the system of linear inequalities:

$$\begin{cases} x \geq 0 \\ y \geq 0 \\ x+y \geq 2 \\ 2x+3y \leq 12 \\ 3x+y \leq 12 \end{cases}$$

(a)   Graph $x \geq 0$; $y \geq 0$. Shaded region is the first quadrant.
(b)   Graph the line $x+y = 2$. Use a solid line since the inequality uses $\geq$.
      Choose a test point not on the line, such as $(0, 0)$. Since $0+0 \geq 2$ is false, shade the
      opposite side of the line from $(0, 0)$.
(c)   Graph the line $2x+3y = 12$. Use a solid line since the inequality uses $\leq$.
      Choose a test point not on the line, such as $(0, 0)$. Since $2(0)+3(0) \leq 12$ is true, shade
      the side of the line containing $(0, 0)$.
(d)   Graph the line $3x+y = 12$. Use a solid line since the inequality uses $\leq$.
      Choose a test point not on the line, such as $(0, 0)$. Since $3(0)+0 \leq 12$ is true, shade
      the side of the line containing $(0, 0)$.
(e)   The overlapping region is the solution.
(f)   The graph is bounded.
(g)   Find the vertices:
      The intersection of $x+y = 2$ and the y-axis is $(0, 2)$.
      The intersection of $x+y = 2$ and the x-axis is $(2, 0)$.
      The intersection of $2x+3y = 12$ and the y-axis is $(0, 4)$.
      The intersection of $3x+y = 12$ and the x-axis is $(4, 0)$.
      To find the intersection of $2x+3y = 12$ and $3x+y = 12$, solve the system:
      $$\begin{cases} 2x+3y = 12 \\ 3x+\ y = 12 \end{cases} \rightarrow \quad y = 12-3x$$
      Substitute and solve:
      $$2x+3(12-3x) = 12 \rightarrow 2x+36-9x = 12$$
      $$-7x = -24 \rightarrow x = \frac{24}{7}$$
      $$y = 12-3\left(\frac{24}{7}\right) = 12-\frac{72}{2} = \frac{12}{7}$$
      The point of intersection is $\left(\frac{24}{7}, \frac{12}{7}\right)$.

      The five corner points are $(0, 2)$, $(0, 4)$, $(2, 0)$, $(4, 0)$, and $\left(\frac{24}{7}, \frac{12}{7}\right)$.

37. Graph the system of linear inequalities:

$$\begin{cases} x \geq 0 \\ y \geq 0 \\ x+y \geq 2 \\ x+y \leq 8 \\ 2x+y \leq 10 \end{cases}$$

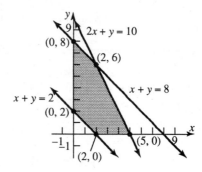

(a) Graph $x \geq 0$; $y \geq 0$. Shaded region is the first quadrant.
(b) Graph the line $x+y = 2$. Use a solid line since the inequality uses $\geq$.
     Choose a test point not on the line, such as (0, 0). Since $0+0 \geq 2$ is false, shade the opposite side of the line from (0, 0).
(c) Graph the line $x+y = 8$. Use a solid line since the inequality uses $\leq$.
     Choose a test point not on the line, such as (0, 0). Since $0+0 \leq 8$ is true, shade the side of the line containing (0, 0).
(d) Graph the line $2x+y = 10$. Use a solid line since the inequality uses $\leq$.
     Choose a test point not on the line, such as (0, 0). Since $2(0)+0 \leq 10$ is true, shade the side of the line containing (0, 0).
(e) The overlapping region is the solution.
(f) The graph is bounded.
(g) Find the vertices:
     The intersection of $x+y = 2$ and the y-axis is (0, 2).
     The intersection of $x+y = 2$ and the x-axis is (2, 0).
     The intersection of $x+y = 8$ and the y-axis is (0, 8).
     The intersection of $2x+y = 10$ and the x-axis is (5, 0).
     To find the intersection of $x+y = 8$ and $2x+y = 10$, solve the system:
     $$\begin{cases} x+y = 8 & \rightarrow \quad y = 8-x \\ 2x+y = 10 \end{cases}$$
     Substitute and solve:
     $$2x+8-x = 10 \rightarrow x = 2$$
     $$y = 8-2 = 6$$
     The point of intersection is (2, 6).
     The five corner points are (0, 2), (0, 8), (2, 0), (5, 0), and (2, 6).

39. Graph the system of linear inequalities:

$$\begin{cases} x \geq 0 \\ y \geq 0 \\ x+2y \geq 1 \\ x+2y \leq 10 \end{cases}$$

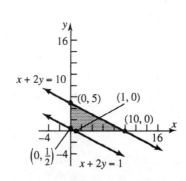

574

(a)   Graph $x \geq 0$; $y \geq 0$. Shaded region is the first quadrant.
(b)   Graph the line $x + 2y = 1$. Use a solid line since the inequality uses $\geq$.
      Choose a test point not on the line, such as $(0, 0)$. Since $0 + 2(0) \geq 1$ is false, shade
      the opposite side of the line from $(0, 0)$.
(c)   Graph the line $x + 2y = 10$. Use a solid line since the inequality uses $\leq$.
      Choose a test point not on the line, such as $(0, 0)$. Since $0 + 2(0) \leq 10$ is true,
      shade the side of the line containing $(0, 0)$.
(d)   The overlapping region is the solution.
(e)   The graph is bounded.
(f)   Find the vertices:
      The intersection of $x + 2y = 1$ and the y-axis is $(0, 0.5)$.
      The intersection of $x + 2y = 1$ and the x-axis is $(1, 0)$.
      The intersection of $x + 2y = 10$ and the y-axis is $(0, 5)$.
      The intersection of $x + 2y = 10$ and the x-axis is $(10, 0)$.
      The four corner points are $(0, 0.5)$, $(0, 5)$, $(1, 0)$, and $(10, 0)$.

41.   The system of linear inequalities is:

$$\begin{cases} x \geq 0 \\ y \geq 0 \\ x \leq 4 \\ x + y \leq 6 \end{cases}$$

43.   The system of linear inequalities is:

$$\begin{cases} x \geq 0 \\ y \geq 15 \\ x \leq 20 \\ x + y \leq 50 \\ x - y \leq 0 \end{cases}$$

45.   (a)   Let x = the amount invested in Treasury bills.
            Let y = the amount invested in corporate bonds.
      The constraints are:

      $x \geq 0, y \geq 0$ A non-negative amount must be invested.
      $x + y \leq 50000$ Total investment cannot exceed \$50,000.
      $y \leq 10000$ Amount invested in corporate bonds must not exceed
                                  \$10,000.
      $x \geq 35000$ Amount invested in Treasury bills must be at least \$35,000.
      $x > y$ Amount invested in Treasury bills must be greater than the amount
                                  invested in corporate bonds.

      (b)   Graph the system.

$$\begin{cases} x \geq 0 \\ y \geq 0 \\ x + y \leq 50000 \\ y \leq 10000 \\ x \geq 35000 \\ x > y \end{cases}$$

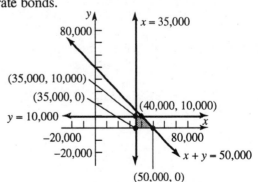

The corner points are $(35000, 0)$, $(35000, 10000)$, $(40000, 10000)$, $(50000, 0)$.

47.   (a)   Let x = the # of packages of the economy blend.
            Let y = the # of packages of the superior blend.
      The constraints are:

$x \geq 0, y \geq 0$	A non-negative # of packages must be produced.
$4x + 8y \leq 75 \cdot 16$	Total amount of grade A coffee cannot exceed 75 pounds. (Note: 75 pounds = (75)(16) ounces.)
$12x + 8y \leq 120 \cdot 16$	Total amount of grade B coffee cannot exceed 120 pounds. (Note: 120 pounds = (120)(16) ounces.)

      We can simplify the equations
$$4x + 8y \leq 75 \cdot 16 \rightarrow x + 2y \leq 75 \cdot 4 \rightarrow x + 2y \leq 300$$
$$12x + 8y \leq 120 \cdot 16 \rightarrow 3x + 2y \leq 120 \cdot 4 \rightarrow 3x + 2y \leq 480$$

      (b)   Graph the system.
$$\begin{cases} x \geq 0 \\ y \geq 0 \\ x + 2y \leq 300 \\ 3x + 2y \leq 480 \end{cases}$$

      The corner points are (0, 0), (0, 150), (90, 105), (160, 0).

49.   (a)   Let x = the # of microwaves.
            Let y = the # of printers.
      The constraints are:

$x \geq 0, y \geq 0$	A non-negative # of items must be shipped.
$30x + 20y \leq 1600$	Total cargo weight cannot exceed 1600 pounds.
$2x + 3y \leq 150$	Total cargo volume cannot exceed 150 cubic feet.

      (b)   Graph the system.
$$\begin{cases} x \geq 0 \\ y \geq 0 \\ 30x + 20y \leq 1600 \\ 2x + 3y \leq 150 \end{cases}$$

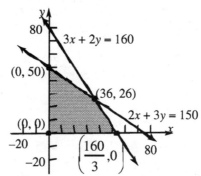

      The corner points are (0, 0), (0, 50), (36, 26), (53.3, 0).

Chapter 10

Systems of Equations and Inequalities

10.9 Linear Programming

1. $z = x + y$

Vertex	Value of $z = x + y$
(0, 3)	$z = 0 + 3 = 3$
(0, 6)	$z = 0 + 6 = 6$
(5, 6)	$z = 5 + 6 = 11$
(5, 2)	$z = 5 + 2 = 7$
(4, 0)	$z = 4 + 0 = 4$

The maximum value is 11 at (5, 6), and the minimum value is 3 at (0, 3).

3. $z = x + 10y$

Vertex	Value of $z = x + 10y$
(0, 3)	$z = 0 + 10(3) = 30$
(0, 6)	$z = 0 + 10(6) = 60$
(5, 6)	$z = 5 + 10(6) = 65$
(5, 2)	$z = 5 + 10(2) = 25$
(4, 0)	$z = 4 + 10(0) = 4$

The maximum value is 65 at (5, 6), and the minimum value is 4 at (4, 0).

5. $z = 5x + 7y$

Vertex	Value of $z = 5x + 7y$
(0, 3)	$z = 5(0) + 7(3) = 21$
(0, 6)	$z = 5(0) + 7(6) = 42$
(5, 6)	$z = 5(5) + 7(6) = 67$
(5, 2)	$z = 5(5) + 7(2) = 39$
(4, 0)	$z = 5(4) + 7(0) = 20$

The maximum value is 67 at (5, 6), and the minimum value is 20 at (4, 0).

7. Maximize $z = 2x + y$
    Subject to $x \geq 0,\ y \geq 0,\ x + y \leq 6,\ x + y \geq 1$
    Graph the constraints.

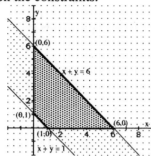

The corner points are $(0, 1)$, $(1, 0)$, $(0, 6)$, $(6, 0)$.
Evaluate the objective function:

Vertex	Value of $z = 2x + y$
$(0, 1)$	$z = 2(0) + 1 = 1$
$(0, 6)$	$z = 2(0) + 6 = 6$
$(1, 0)$	$z = 2(1) + 0 = 2$
$(6, 0)$	$z = 2(6) + 0 = 12$

The maximum value is 12 at $(6, 0)$.

9. Minimize $z = 2x + 5y$
    Subject to $x \geq 0,\ y \geq 0,\ x + y \geq 2,\ x \leq 5,\ y \leq 3$
    Graph the constraints.

The corner points are $(0, 2)$, $(2, 0)$, $(0, 3)$, $(5, 0)$, $(5, 3)$.
Evaluate the objective function:

Vertex	Value of $z = 2x + 5y$
$(0, 2)$	$z = 2(0) + 5(2) = 10$
$(0, 3)$	$z = 2(0) + 5(3) = 15$
$(2, 0)$	$z = 2(2) + 5(0) = 4$
$(5, 0)$	$z = 2(5) + 5(0) = 10$
$(5, 3)$	$z = 2(5) + 5(3) = 25$

The minimum value is 4 at $(2, 0)$.

11.   Maximize $z = 3x + 5y$

Subject to $x \geq 0, \ y \geq 0, \ x + y \geq 2, \ 2x + 3y \leq 12, \ 3x + 2y \leq 12$

Graph the constraints.

To find the intersection of $2x + 3y = 12$ and $3x + 2y = 12$, solve the system:

$$\begin{cases} 2x + 3y = 12 \\ 3x + 2y = 12 \quad \rightarrow \quad y = 6 - \dfrac{3}{2}x \end{cases}$$

Substitute and solve:

$$2x + 3\left(6 - \frac{3}{2}x\right) = 12 \rightarrow 2x + 18 - \frac{9}{2}x = 12 \rightarrow -\frac{5}{2}x = -6$$

$$x = \frac{12}{5} \qquad y = 6 - \frac{3}{2}\left(\frac{12}{5}\right) = 6 - \frac{18}{5} = \frac{12}{5}$$

The point of intersection is $(2.4, 2.4)$.

The corner points are $(0, 2), (2, 0), (0, 4), (4, 0), (2.4, 2.4)$.

Evaluate the objective function:

Vertex	Value of $z = 3x + 5y$
(0, 2)	z = 3(0) + 5(2) = 10
(0, 4)	z = 3(0) + 5(4) = 20
(2, 0)	z = 3(2) + 5(0) = 6
(4, 0)	z = 3(4) + 5(0) = 12
(2.4, 2.4)	z = 3(2.4) + 5(2.4) = 19.2

The maximum value is 20 at $(0, 4)$.

13.   Minimize $z = 5x + 4y$

Subject to $x \geq 0, \ y \geq 0, \ x + y \geq 2, \ 2x + 3y \leq 12, \ 3x + y \leq 12$

Graph the constraints.

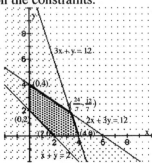

To find the intersection of $2x + 3y = 12$ and $3x + y = 12$, solve the system:

$$\begin{cases} 2x + 3y = 12 \\ 3x + \ y = 12 \end{cases} \rightarrow \ y = 12 - 3x$$

Substitute and solve:

$$2x + 3(12 - 3x) = 12 \rightarrow 2x + 36 - 9x = 12 \rightarrow -7x = -24 \rightarrow x = \frac{24}{7}$$

$$y = 12 - 3\left(\frac{24}{7}\right) = 12 - \frac{72}{7} = \frac{12}{7}$$

The point of intersection is $\left(\frac{24}{7}, \frac{12}{7}\right)$.

The corner points are $(0, 2)$, $(2, 0)$, $(0, 4)$, $(4, 0)$, $\left(\frac{24}{7}, \frac{12}{7}\right)$.

Evaluate the objective function:

Vertex	Value of $z = 5x + 4y$
$(0, 2)$	$z = 5(0) + 4(2) = 8$
$(0, 4)$	$z = 5(0) + 4(4) = 16$
$(2, 0)$	$z = 5(2) + 4(0) = 10$
$(4, 0)$	$z = 5(4) + 4(0) = 20$
$\left(\frac{24}{7}, \frac{12}{7}\right)$	$z = 5\left(\frac{24}{7}\right) + 4\left(\frac{12}{7}\right) = \frac{120}{7} + \frac{48}{7} = \frac{168}{7} = 24$

The minimum value is 8 at $(0, 2)$.

15. Maximize $z = 5x + 2y$

Subject to $x \geq 0$, $y \geq 0$, $x + y \leq 10$, $2x + y \geq 10$, $x + 2y \geq 10$

Graph the constraints.

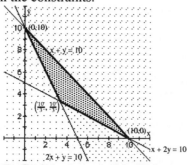

To find the intersection of $2x + y = 10$ and $x + 2y = 10$, solve the system:

$$\begin{cases} 2x + \ y = 10 \\ \ x + 2y = 10 \end{cases} \rightarrow \ y = 10 - 2x$$

Substitute and solve:

$$x + 2(10 - 2x) = 10 \rightarrow x + 20 - 4x = 10 \rightarrow -3x = -10 \rightarrow x = \frac{10}{3}$$

$$y = 10 - 2\left(\frac{10}{3}\right) = 10 - \frac{20}{3} = \frac{10}{3}$$

The point of intersection is $\left(\dfrac{10}{3}, \dfrac{10}{3}\right)$.

The corner points are $(0, 10)$, $(10, 0)$, $\left(\dfrac{10}{3}, \dfrac{10}{3}\right)$.

Evaluate the objective function:

Vertex	Value of $z = 5x + 2y$
$(0, 10)$	$z = 5(0) + 2(10) = 20$
$(10, 0)$	$z = 5(10) + 2(0) = 50$
$\left(\dfrac{10}{3}, \dfrac{10}{3}\right)$	$z = 5\left(\dfrac{10}{3}\right) + 2\left(\dfrac{10}{3}\right) = \dfrac{50}{3} + \dfrac{20}{3} = \dfrac{70}{3} = 23\tfrac{1}{3}$

The maximum value is 50 at $(10, 0)$.

17. Let $x =$ the number of downhill skis produced.
Let $y =$ the number of cross-country skis produced.
The total profit is: $P = 70x + 50y$. Profit is to be maximized; thus, this is the objective function.
The constraints are:

$x \geq 0$, $y \geq 0$ A positive number of skis must be produced.

$2x + y \leq 40$ Only 40 hours of manufacturing time is available.

$x + y \leq 32$ Only 32 hours of finishing time is available.

Graph the constraints.

To find the intersection of $x + y = 32$ and $2x + y = 40$, solve the system:

$$\begin{cases} x + y = 32 & \to \quad y = 32 - x \\ 2x + y = 40 \end{cases}$$

Substitute and solve:

$$2x + 32 - x = 40 \to x = 8 \to y = 32 - 8 = 24$$

The point of intersection is $(8, 24)$.

The corner points are $(0, 0)$, $(0, 32)$, $(20, 0)$, $(8, 24)$.

Evaluate the objective function:

Vertex	Value of $P = 70x + 50y$
$(0, 0)$	$P = 70(0) + 50(0) = 0$
$(0, 32)$	$P = 70(0) + 50(32) = 1600$
$(20, 0)$	$P = 70(20) + 50(0) = 1400$
$(8, 24)$	$P = 70(8) + 50(24) = 1760$

The maximum profit is $1760, when 8 downhill skis and 24 cross-country skis are produced.

With the increase of the manufacturing time to 48 hours, we do the following:

The constraints are:

$x \geq 0, \quad y \geq 0$ A positive number of skis must be produced.

$2x + y \leq 48$ Only 48 hours of manufacturing time is available.

$x + y \leq 32$ Only 32 hours of finishing time is available.

Graph the constraints.

To find the intersection of $x + y = 32$ and $2x + y = 48$, solve the system:

$$\begin{cases} x + y = 32 \\ 2x + y = 48 \end{cases} \rightarrow \quad y = 32 - x$$

Substitute and solve:

$$2x + 32 - x = 48$$

$$x = 16$$

$$y = 32 - 16 = 16$$

The point of intersection is (16, 16).

The corner points are (0, 0), (0, 32), (24, 0), (16, 16).

Evaluate the objective function:

Vertex	Value of $P = 70x + 50y$
(0, 0)	$P = 70(0) + 50(0) = 0$
(0, 32)	$P = 70(0) + 50(32) = 1600$
(24, 0)	$P = 70(24) + 50(0) = 1680$
(16, 16)	$P = 70(16) + 50(16) = 1920$

The maximum profit is $1920, when 16 downhill skis and 16 cross-country skis are produced.

19. Let $x =$ the number of acres of corn planted.

Let $y =$ the number of acres of soybeans planted.

The total profit is: $P = 250x + 200y$.

Profit is to be maximized; thus, this is the objective function.

The constraints are:

$x \geq 0, \quad y \geq 0$ A non-negative number of acres must be planted.

$x + y \leq 100$ Acres available to plant.

$60x + 40y \leq 1800$ Money available for cultivation costs.

$60x + 60y \leq 2400$ Money available for labor costs.

Graph the constraints.

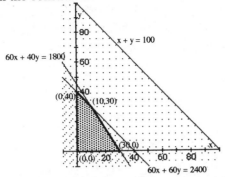

To find the intersection of $60x + 40y = 1800$ and $60x + 60y = 2400$, solve the system:
$$\begin{cases} 60x + 40y = 1800 \\ 60x + 60y = 2400 \end{cases} \rightarrow 60x = 1800 - 40y$$

Substitute and solve:
$$1800 - 40y + 60y = 2400$$
$$20y = 600$$
$$y = 30$$
$$60x = 1800 - 40(30)$$
$$60x = 600$$
$$x = 10$$

The point of intersection is (10, 30).
The corner points are (0, 0), (0, 40), (30, 0), (10, 30).
Evaluate the objective function:

Vertex	Value of $P = 250x + 200y$
(0, 0)	$P = 250(0) + 200(0) = 0$
(0, 40)	$P = 250(0) + 200(40) = 8000$
(30, 0)	$P = 250(30) + 200(0) = 7500$
(10, 30)	$P = 250(10) + 200(30) = 8500$

The maximum profit is $8500, when 10 acres of corn and 30 acres of soybeans are planted.

21. Let x = the number of hours that machine 1 is operated.
Let y = the number of hours that machine 2 is operated.
The total cost is: $C = 50x + 30y$. Cost is to be minimized; thus, this is the objective function.
The constraints are:
$$x \geq 0, \quad y \geq 0 \quad \text{A positive number of hours must be used.}$$
$$x \leq 10 \qquad \text{10 hours available on machine 1.}$$
$$y \leq 10 \qquad \text{10 hours available on machine 2.}$$
$$60x + 40y \geq 240 \qquad \text{At least 240 8-inch plyers must be produced.}$$
$$70x + 20y \geq 140 \qquad \text{At least 140 6-inch plyers must be produced.}$$

Graph the constraints.

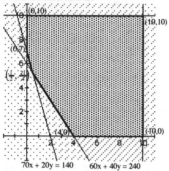

$$70x + 20y = 140 \qquad 60x + 40y = 240$$

To find the intersection of $60x + 40y = 240$ and $70x + 20y = 140$, solve the system:

$$\begin{cases} 60x + 40y = 240 \\ 70x + 20y = 140 \end{cases} \rightarrow \quad 20y = 140 - 70x$$

Substitute and solve:

$$60x + 2(140 - 70x) = 240$$
$$60x + 280 - 140x = 240$$
$$-80x = -40$$
$$x = 0.5$$
$$20y = 140 - 70(0.5)$$
$$20y = 105$$
$$y = 5.25$$

The point of intersection is $(0.5, 5.25)$.

The corner points are $(0, 7)$, $(0, 10)$, $(4, 0)$, $(10, 0)$, $(10, 10)$, $(0.5, 5.25)$.

Evaluate the objective function:

Vertex	Value of $C = 50x + 30y$
$(0, 7)$	$C = 50(0) + 30(7) = 210$
$(0, 10)$	$C = 50(0) + 30(10) = 300$
$(4, 0)$	$C = 50(4) + 30(0) = 200$
$(10, 0)$	$C = 50(10) + 30(0) = 500$
$(10, 10)$	$C = 50(10) + 30(10) = 800$
$(0.5, 5.25)$	$C = 50(0.5) + 30(5.25) = 182.50$

The minimum cost is $182.50, when machine 1 is used for 0.5 hours and machine 2 is used for 5.25 hours.

23. Let $x =$ the number of pounds of ground beef.
     Let $y =$ the number of pounds of ground pork.
     The total cost is: $C = 0.75x + 0.45y$. Cost is to be minimized; thus, this is the objective
                                                        function.
     The constraints are:

     $x \geq 0, \; y \geq 0$ \qquad A positive number of pounds must be used.
     $x \leq 200$ \qquad Only 200 pounds of ground beef are available.
     $y \geq 50$ \qquad At least 50 pounds of ground pork must be used.
     $0.75x + 0.60y \geq 0.70(x + y) \rightarrow 0.05x \geq 0.10y$ \quad Leanness condition to be met.

Graph the constraints.

The corner points are (100, 50), (200, 50), (200, 100).

Evaluate the objective function:

Vertex	Value of $C = 0.75x + 0.45y$
(100, 50)	$C = 0.75(100) + 0.45(50) = 97.50$
(200, 50)	$C = 0.75(200) + 0.45(50) = 172.50$
(200,100)	$C = 0.75(200) + 0.45(100) = 195.00$

The minimum cost is $97.50, when 100 pounds of ground beef and 50 pounds of ground pork are used.

25.   Let x = the number of racing skates manufactured.

Let y = the number of figure skates manufactured.

The total profit is: $P = 10x + 12y$. Profit is to be maximized; thus, this is the objective function.

The constraints are:

$x \geq 0, \quad y \geq 0$ A positive number of skates must be manufactured.

$6x + 4y \leq 120$ Only 120 hours are available for fabrication.

$x + 2y \leq 40$ Only 40 hours are available for finishing.

Graph the constraints.

To find the intersection of $6x + 4y = 120$ and $x + 2y = 40$, solve the system:

$$\begin{cases} 6x + 4y = 120 \\ x + 2y = \ 40 \quad \rightarrow \quad x = 40 - 2y \end{cases}$$

Substitute and solve:

$$6(40 - 2y) + 4y = 120 \rightarrow 240 - 12y + 4y = 120$$

$$-8y = -120 \rightarrow y = 15 \rightarrow x = 40 - 2(15) = 10$$

The point of intersection is (10, 15).

The corner points are (0, 0), (0, 20), (20, 0), (10, 15).

Evaluate the objective function:

Vertex	Value of $P = 10x + 12y$
(0, 0)	P = 10(0) + 12(0) = 0
(0, 20)	P = 10(0) + 12(20) = 240
(20, 0)	P = 10(20) + 12(0) = 200
(10, 15)	P = 10(10) + 12(15) = 280

The maximum profit is $280, when 10 racing skates and 15 figure skates are produced.

27. Let x = the number of metal fasteners.
Let y = the number of plastic fasteners.
The total cost is: $C = 9x + 4y$. Cost is to be minimized; thus, this is the objective function.
The constraints are:

   $x \geq 2, \ y \geq 2$ At least 2 of each fastener must be made.
   $x + y \geq 6$ At least 6 fasteners are needed.
   $4x + 2y \leq 24$ Only 24 hours are available.

Graph the constraints.

The corner points are (2, 4), (2, 8), (4, 2), (5, 2).

Evaluate the objective function:

Vertex	Value of $C = 9x + 4y$
(2, 4)	C = 9(2) + 4(4) = 34
(2, 8)	C = 9(2) + 4(8) = 50
(4, 2)	C = 9(4) + 4(2) = 44
(5, 2)	C = 9(5) + 4(2) = 53

The minimum cost is $34, when 2 metal fasteners and 4 plastic fasteners are ordered.

29. Let x = the number of first-class seats.
Let y = the number of coach seats.
The constraints are:

   $8 \leq x \leq 16$ Restriction on first-class seats.
   $80 \leq y \leq 120$ Restriction on coach seats.
(a) $\dfrac{x}{y} \leq \dfrac{1}{12}$ Ratio of seats.

   If $y = 120$, then $\dfrac{x}{120} \leq \dfrac{1}{12} \rightarrow 12x \leq 120 \rightarrow x \leq 10$

The maximum revenue will be obtained with 120 coach seats and 10 first-class seats.
(Note that the first-class seats meet their constraint.)

(b) $\dfrac{x}{y} \le \dfrac{1}{8}$ Ratio of seats.

If $y = 120$, then $\dfrac{x}{120} \le \dfrac{1}{8} \to 8x \le 120 \to x \le 15$

The maximum revenue will be obtained with 120 coach seats and 15 first-class seats. (Note that the first-class seats meet their constraint.)

Systems of Equations and Inequalities

10.R Chapter Review

1. Solve the first equation for y, substitute into the second equation and solve:
$$\begin{cases} 2x - y = 5 & \rightarrow \quad y = 2x - 5 \\ 5x + 2y = 8 \end{cases}$$
$$5x + 2(2x - 5) = 8 \rightarrow 5x + 4x - 10 = 8$$
$$9x = 18 \rightarrow x = 2 \rightarrow y = 2(2) - 5 = 4 - 5 = -1$$
The solution is $x = 2, \ y = -1$.

3. Solve the second equation for x, substitute into the first equation and solve:
$$\begin{cases} 3x - 4y = 4 \\ x - 3y = \dfrac{1}{2} & \rightarrow \quad x = 3y + \dfrac{1}{2} \end{cases}$$
$$3\left(3y + \frac{1}{2}\right) - 4y = 4 \rightarrow 9y + \frac{3}{2} - 4y = 4 \rightarrow 5y = \frac{5}{2} \rightarrow y = \frac{1}{2} \rightarrow x = 3\left(\frac{1}{2}\right) + \frac{1}{2} = 2$$
The solution is $x = 2, \ y = \dfrac{1}{2}$.

5. Solve the first equation for x, substitute into the second equation and solve:
$$\begin{cases} x - 2y - 4 = 0 & \rightarrow \quad x = 2y + 4 \\ 3x + 2y - 4 = 0 \end{cases}$$
$$3(2y + 4) + 2y - 4 = 0 \rightarrow 6y + 12 + 2y - 4 = 0$$
$$8y = -8 \rightarrow y = -1 \rightarrow x = 2(-1) + 4 = 2$$
The solution is $x = 2, \ y = -1$.

7. Substitute the first equation into the second equation and solve:
$$\begin{cases} y = 2x - 5 \\ x = 3y + 4 \end{cases}$$
$$x = 3(2x - 5) + 4 \rightarrow x = 6x - 15 + 4$$
$$-5x = -11 \rightarrow x = \frac{11}{5} \rightarrow y = 2\left(\frac{11}{5}\right) - 5 = -\frac{3}{5}$$
The solution is $x = \dfrac{11}{5}, \ y = -\dfrac{3}{5}$.

9. Multiply each side of the first equation by 5 and each side of the second equation by 30 and add to eliminate y:

$$\begin{cases} x - y + 4 = 0 \xrightarrow{5} 5x - 5y + 20 = 0 \\ \dfrac{1}{2}x + \dfrac{1}{6}y + \dfrac{2}{5} = 0 \xrightarrow{30} \dfrac{15x + 5y + 12 = 0}{20x + 32 = 0} \end{cases}$$

$20x = -32 \rightarrow x = -\dfrac{8}{5}$ Substitute and solve for y: $-\dfrac{8}{5} - y + 4 = 0 \rightarrow y = \dfrac{12}{5}$

The solution of the system is $x = -\dfrac{8}{5}, \; y = \dfrac{12}{5}$.

11. Rewrite each equation and add to eliminate y:

$$\begin{cases} x - 2y - 8 = 0 \longrightarrow x - 2y = 8 \\ 2x + 2y - 10 = 0 \longrightarrow \dfrac{2x + 2y = 10}{3x = 18} \\ x = 6 \end{cases}$$

Substitute and solve for y:

$$6 - 2y = 8 \rightarrow -2y = 2 \rightarrow y = -1$$

The solution of the system is $x = 6, \; y = -1$.

13. Solve the first equation for y, substitute into the second equation and solve:

$$\begin{cases} y - 2x = 11 \;\; \rightarrow \;\; y = 2x + 11 \\ 2y - 3x = 18 \end{cases}$$

$$2(2x + 11) - 3x = 18$$

$$4x + 22 - 3x = 18 \rightarrow x = -4 \rightarrow y = 2(-4) + 11 = 3$$

The solution is $x = -4, \; y = 3$.

15. Multiply each side of the first equation by 2 and each side of the second equation by 3 and add to eliminate y:

$$\begin{cases} 2x + 3y - 13 = 0 \xrightarrow{2} 4x + 6y - 26 = 0 \\ 3x - 2y = 0 \xrightarrow{3} \dfrac{9x - 6y = 0}{13x - 26 = 0} \\ \phantom{3x - 2y = 0 \xrightarrow{3} 9x - 6y} 13x = 26 \\ \phantom{3x - 2y = 0 \xrightarrow{3} 9x - 6y = 0} x = 2 \end{cases}$$

Substitute and solve for y:

$$3(2) - 2y = 0 \rightarrow -2y = -6 \rightarrow y = 3$$

The solution of the system is $x = 2, \; y = 3$.

17. Multiply each side of the second equation by –3 and add to eliminate x:

$$\begin{cases} 3x - 2y = 8 \longrightarrow 3x - 2y = 8 \\ x - \tfrac{2}{3}y = 12 \xrightarrow{-3} \dfrac{-3x + 2y = -36}{0 = -28} \end{cases}$$

The system has no solution, so the system is inconsistent.

19. Multiply each side of the first equation by –2 and add to the second equation to eliminate x; and multiply each side of the first equation by –3 and add to the third equation to eliminate x:

$$\begin{cases} x + 2y - z = 6 \\ 2x - y + 3z = -13 \\ 3x - 2y + 3z = -16 \end{cases}$$

$$\xrightarrow{-2} \quad -2x - 4y + 2z = -12$$
$$\xrightarrow{} \quad \underline{2x - y + 3z = -13}$$
$$\phantom{\xrightarrow{-2}} \quad -5y + 5z = -25 \xrightarrow{-1/5} y - z = 5$$

$$\xrightarrow{-3} \quad -3x - 6y + 3z = -18$$
$$\xrightarrow{} \quad \underline{3x - 2y + 3z = -16}$$
$$\phantom{\xrightarrow{-3}} \quad -8y + 6z = -34$$

Multiply each side of the first result by 8 and add to the second result to eliminate y:

$$y - z = 5 \xrightarrow{8} 8y - 8z = 40$$
$$-8y + 6z = -34 \xrightarrow{} \underline{-8y + 6z = -34}$$
$$-2z = 6$$
$$z = -3$$

Substituting and solving for the other variables:

$$\begin{array}{ll} y - (-3) = 5 & x + 2(2) - (-3) = 6 \\ y = 2 & x + 4 + 3 = 6 \\ & x = -1 \end{array}$$

The solution is $x = -1$, $y = 2$, $z = -3$.

21. $A + C = \begin{bmatrix} 1 & 0 \\ 2 & 4 \\ -1 & 2 \end{bmatrix} + \begin{bmatrix} 3 & -4 \\ 1 & 5 \\ 5 & -2 \end{bmatrix} = \begin{bmatrix} 4 & -4 \\ 3 & 9 \\ 4 & 0 \end{bmatrix}$

23. $6A = 6 \cdot \begin{bmatrix} 1 & 0 \\ 2 & 4 \\ -1 & 2 \end{bmatrix} = \begin{bmatrix} 6 & 0 \\ 12 & 24 \\ -6 & 12 \end{bmatrix}$

25. $AB = \begin{bmatrix} 1 & 0 \\ 2 & 4 \\ -1 & 2 \end{bmatrix} \cdot \begin{bmatrix} 4 & -3 & 0 \\ 1 & 1 & -2 \end{bmatrix} = \begin{bmatrix} 4 & -3 & 0 \\ 12 & -2 & -8 \\ -2 & 5 & -4 \end{bmatrix}$

27. $CB = \begin{bmatrix} 3 & -4 \\ 1 & 5 \\ 5 & -2 \end{bmatrix} \cdot \begin{bmatrix} 4 & -3 & 0 \\ 1 & 1 & -2 \end{bmatrix} = \begin{bmatrix} 8 & -13 & 8 \\ 9 & 2 & -10 \\ 18 & -17 & 4 \end{bmatrix}$

29. Augment the matrix with the identity and use row operations to find the inverse:

$$A = \begin{bmatrix} 4 & 6 \\ 1 & 3 \end{bmatrix} \rightarrow \begin{bmatrix} 4 & 6 & | & 1 & 0 \\ 1 & 3 & | & 0 & 1 \end{bmatrix}$$

$$\rightarrow \begin{bmatrix} 1 & 3 & | & 0 & 1 \\ 4 & 6 & | & 1 & 0 \end{bmatrix} \rightarrow \begin{bmatrix} 1 & 3 & | & 0 & 1 \\ 0 & -6 & | & 1 & -4 \end{bmatrix} \rightarrow \begin{bmatrix} 1 & 3 & | & 0 & 1 \\ 0 & 1 & | & -\frac{1}{6} & \frac{2}{3} \end{bmatrix} \rightarrow \begin{bmatrix} 1 & 0 & | & \frac{1}{2} & -1 \\ 0 & 1 & | & -\frac{1}{6} & \frac{2}{3} \end{bmatrix}$$

Interchange $\quad R_2 = -4r_1 + r_2 \quad R_2 = -\frac{1}{6}r_2 \quad R_1 = -3r_2 + r_1$
r_1 and r_2

$$A^{-1} = \begin{bmatrix} \frac{1}{2} & -1 \\ -\frac{1}{6} & \frac{2}{3} \end{bmatrix}$$

31. Augment the matrix with the identity and use row operations to find the inverse:

$$A = \begin{bmatrix} 1 & 3 & 3 \\ 1 & 2 & 1 \\ 1 & -1 & 2 \end{bmatrix} \rightarrow \begin{bmatrix} 1 & 3 & 3 & | & 1 & 0 & 0 \\ 1 & 2 & 1 & | & 0 & 1 & 0 \\ 1 & -1 & 2 & | & 0 & 0 & 1 \end{bmatrix}$$

$$\rightarrow \begin{bmatrix} 1 & 3 & 3 & | & 1 & 0 & 0 \\ 0 & -1 & -2 & | & -1 & 1 & 0 \\ 0 & -4 & -1 & | & -1 & 0 & 1 \end{bmatrix} \rightarrow \begin{bmatrix} 1 & 3 & 3 & | & 1 & 0 & 0 \\ 0 & 1 & 2 & | & 1 & -1 & 0 \\ 0 & -4 & -1 & | & -1 & 0 & 1 \end{bmatrix} \rightarrow \begin{bmatrix} 1 & 0 & -3 & | & -2 & 3 & 0 \\ 0 & 1 & 2 & | & 1 & -1 & 0 \\ 0 & 0 & 7 & | & 3 & -4 & 1 \end{bmatrix}$$

$R_2 = -r_1 + r_2$ $R_2 = -r_2$ $R_1 = -3r_2 + r_1$

$R_3 = -r_1 + r_3$ $R_3 = 4r_2 + r_3$

$$\rightarrow \begin{bmatrix} 1 & 0 & -3 & | & -2 & 3 & 0 \\ 0 & 1 & 2 & | & 1 & -1 & 0 \\ 0 & 0 & 1 & | & \frac{3}{7} & -\frac{4}{7} & \frac{1}{7} \end{bmatrix} \rightarrow \begin{bmatrix} 1 & 0 & 0 & | & -\frac{5}{7} & \frac{9}{7} & \frac{3}{7} \\ 0 & 1 & 0 & | & \frac{1}{7} & \frac{1}{7} & -\frac{2}{7} \\ 0 & 0 & 1 & | & \frac{3}{7} & -\frac{4}{7} & \frac{1}{7} \end{bmatrix} \longrightarrow A^{-1} = \begin{bmatrix} -\frac{5}{7} & \frac{9}{7} & \frac{3}{7} \\ \frac{1}{7} & \frac{1}{7} & -\frac{2}{7} \\ \frac{3}{7} & -\frac{4}{7} & \frac{1}{7} \end{bmatrix}$$

$R_3 = \frac{1}{7} r_3$ $R_1 = 3r_3 + r_1$

                      $R_2 = -2r_3 + r_2$

33. Augment the matrix with the identity and use row operations to find the inverse:

$$A = \begin{bmatrix} 4 & -8 \\ -1 & 2 \end{bmatrix} \rightarrow \begin{bmatrix} 4 & -8 & | & 1 & 0 \\ -1 & 2 & | & 0 & 1 \end{bmatrix}$$

$$\rightarrow \begin{bmatrix} -1 & 2 & | & 0 & 1 \\ 4 & -8 & | & 1 & 0 \end{bmatrix} \rightarrow \begin{bmatrix} -1 & 2 & | & 0 & 1 \\ 0 & 0 & | & 1 & 4 \end{bmatrix} \rightarrow \begin{bmatrix} 1 & -2 & | & 0 & -1 \\ 0 & 0 & | & 1 & 4 \end{bmatrix}$$

  Interchange $R_2 = 4r_1 + r_2$ $R_1 = -r_1$

     r_1 and r_2

There is no inverse because there is no way to obtain the identity on the left side. The matrix is singular.

35. $\begin{cases} 3x - 2y = 1 \\ 10x + 10y = 5 \end{cases}$ can be written as: $\begin{bmatrix} 3 & -2 & | & 1 \\ 10 & 10 & | & 5 \end{bmatrix}$

$$\rightarrow \begin{bmatrix} 3 & -2 & | & 1 \\ 1 & 16 & | & 2 \end{bmatrix} \rightarrow \begin{bmatrix} 1 & 16 & | & 2 \\ 3 & -2 & | & 1 \end{bmatrix} \rightarrow \begin{bmatrix} 1 & 16 & | & 2 \\ 0 & -50 & | & -5 \end{bmatrix} \rightarrow \begin{bmatrix} 1 & 16 & | & 2 \\ 0 & 1 & | & \frac{1}{10} \end{bmatrix} \rightarrow \begin{bmatrix} 1 & 0 & | & \frac{2}{5} \\ 0 & 1 & | & \frac{1}{10} \end{bmatrix}$$

$R_2 = -3r_1 + r_2$ Interchange $R_2 = -3r_1 + r_2$ $R_2 = -\frac{1}{50} r_2$ $R_1 = -16r_2 + r_1$

                   r_1 and r_2

The solution is $x = \dfrac{2}{5}, y = \dfrac{1}{10}$.

37. $\begin{cases} 5x + 6y - 3z = 6 \\ 4x - 7y - 2z = -3 \\ 3x + y - 7z = 1 \end{cases}$ can be written as $\begin{bmatrix} 5 & 6 & -3 & | & 6 \\ 4 & -7 & -2 & | & -3 \\ 3 & 1 & -7 & | & 1 \end{bmatrix}$

$$\rightarrow \begin{bmatrix} 1 & 13 & -1 & | & 9 \\ 4 & -7 & -2 & | & -3 \\ 3 & 1 & -7 & | & 1 \end{bmatrix} \rightarrow \begin{bmatrix} 1 & 13 & -1 & | & 9 \\ 0 & -59 & 2 & | & -39 \\ 0 & -38 & -4 & | & -26 \end{bmatrix} \rightarrow \begin{bmatrix} 1 & 13 & -1 & | & 9 \\ 0 & 1 & -\frac{2}{59} & | & \frac{39}{59} \\ 0 & -38 & -4 & | & -26 \end{bmatrix}$$

$R_1 = -r_2 + r_1 \qquad R_2 = -4r_1 + r_2 \qquad R_2 = -\frac{1}{59}r_2$
$R_3 = -3r_1 + r_3$

$$\rightarrow \begin{bmatrix} 1 & 0 & -\frac{33}{59} & | & \frac{24}{59} \\ 0 & 1 & -\frac{2}{59} & | & \frac{39}{59} \\ 0 & 0 & -\frac{312}{59} & | & -\frac{52}{59} \end{bmatrix} \rightarrow \begin{bmatrix} 1 & 0 & -\frac{33}{59} & | & \frac{24}{59} \\ 0 & 1 & -\frac{2}{59} & | & \frac{39}{59} \\ 0 & 0 & 1 & | & \frac{1}{6} \end{bmatrix} \rightarrow \begin{bmatrix} 1 & 0 & 0 & | & \frac{1}{2} \\ 0 & 1 & 0 & | & \frac{2}{3} \\ 0 & 0 & 1 & | & \frac{1}{6} \end{bmatrix}$$

$R_1 = -13r_2 + r_1 \qquad R_3 = -\frac{59}{312}r_3 \qquad R_1 = \frac{33}{59}r_3 + r_1$
$R_3 = 38r_2 + r_3 \qquad\qquad\qquad R_2 = \frac{2}{59}r_3 + r_2$

The solution is $x = \frac{1}{2}, y = \frac{2}{3}, z = \frac{1}{6}.$

39. $\begin{cases} x & -2z = 1 \\ 2x + 3y & = -3 \\ 4x - 3y - 4z = 3 \end{cases}$ can be written as $\begin{bmatrix} 1 & 0 & -2 & | & 1 \\ 2 & 3 & 0 & | & -3 \\ 4 & -3 & -4 & | & 3 \end{bmatrix}$

$$\rightarrow \begin{bmatrix} 1 & 0 & -2 & | & 1 \\ 0 & 3 & 4 & | & -5 \\ 0 & -3 & 4 & | & -1 \end{bmatrix} \rightarrow \begin{bmatrix} 1 & 0 & -2 & | & 1 \\ 0 & 1 & \frac{4}{3} & | & -\frac{5}{3} \\ 0 & -3 & 4 & | & -1 \end{bmatrix} \rightarrow \begin{bmatrix} 1 & 0 & -2 & | & 1 \\ 0 & 1 & \frac{4}{3} & | & -\frac{5}{3} \\ 0 & 0 & 8 & | & -6 \end{bmatrix}$$

$R_2 = -2r_1 + r_2 \qquad R_2 = \frac{1}{3}r_2 \qquad\qquad R_3 = 3r_2 + r_3$
$R_3 = -4r_1 + r_3$

$$\rightarrow \begin{bmatrix} 1 & 0 & -2 & | & 1 \\ 0 & 1 & \frac{4}{3} & | & -\frac{5}{3} \\ 0 & 0 & 1 & | & -\frac{3}{4} \end{bmatrix} \rightarrow \begin{bmatrix} 1 & 0 & 0 & | & -\frac{1}{2} \\ 0 & 1 & 0 & | & -\frac{2}{3} \\ 0 & 0 & 1 & | & -\frac{3}{4} \end{bmatrix}$$

$R_3 = \frac{1}{8}r_3 \qquad\qquad R_1 = 2r_3 + r_1$
$R_2 = -\frac{4}{3}r_3 + r_2$

The solution is $x = -\frac{1}{2}, y = -\frac{2}{3}, z = -\frac{3}{4}.$

41. $\begin{cases} x - y + z = 0 \\ x - y - 5z = 6 \\ 2x - 2y + z = 1 \end{cases}$ can be written as: $\begin{bmatrix} 1 & -1 & 1 & | & 0 \\ 1 & -1 & -5 & | & 6 \\ 2 & -2 & 1 & | & 1 \end{bmatrix}$

$$\rightarrow \begin{bmatrix} 1 & -1 & 1 & | & 0 \\ 0 & 0 & -6 & | & 6 \\ 0 & 0 & -1 & | & 1 \end{bmatrix} \rightarrow \begin{bmatrix} 1 & -1 & 1 & | & 0 \\ 0 & 0 & 1 & | & -1 \\ 0 & 0 & -1 & | & 1 \end{bmatrix} \rightarrow \begin{bmatrix} 1 & -1 & 0 & | & 1 \\ 0 & 0 & 1 & | & -1 \\ 0 & 0 & 0 & | & 0 \end{bmatrix} \rightarrow \begin{cases} x = y + 1 \\ z = -1 \end{cases}$$

$R_2 = -r_1 + r_2 \qquad R_2 = -\frac{1}{6}r_2 \qquad R_1 = -r_2 + r_1$
$R_3 = -2r_1 + r_3 \qquad\qquad\qquad R_3 = r_2 + r_3$

The solution is $x = y + 1, z = -1, y$ is any real number..

43. $\begin{cases} x - y - z - t = 1 \\ 2x + y + z + 2t = 3 \\ x - 2y - 2z - 3t = 0 \\ 3x - 4y + z + 5t = -3 \end{cases}$ can be written as $\begin{bmatrix} 1 & -1 & -1 & -1 & | & 1 \\ 2 & 1 & 1 & 2 & | & 3 \\ 1 & -2 & -2 & -3 & | & 0 \\ 3 & -4 & 1 & 5 & | & -3 \end{bmatrix}$

$$\rightarrow \begin{bmatrix} 1 & -1 & -1 & -1 & | & 1 \\ 0 & 3 & 3 & 4 & | & 1 \\ 0 & -1 & -1 & -2 & | & -1 \\ 0 & -1 & 4 & 8 & | & -6 \end{bmatrix} \rightarrow \begin{bmatrix} 1 & -1 & -1 & -1 & | & 1 \\ 0 & -1 & -1 & -2 & | & -1 \\ 0 & 3 & 3 & 4 & | & 1 \\ 0 & -1 & 4 & 8 & | & -6 \end{bmatrix} \rightarrow \begin{bmatrix} 1 & -1 & -1 & -1 & | & 1 \\ 0 & 1 & 1 & 2 & | & 1 \\ 0 & 3 & 3 & 4 & | & 1 \\ 0 & -1 & 4 & 8 & | & -6 \end{bmatrix}$$

$R_2 = -2r_1 + r_2$ Interchange r_2 and r_3 $R_2 = -r_2$
$R_3 = -r_1 + r_3$
$R_4 = -3r_1 + r_4$

$$\rightarrow \begin{bmatrix} 1 & 0 & 0 & 1 & | & 2 \\ 0 & 1 & 1 & 2 & | & 1 \\ 0 & 0 & 0 & -2 & | & -2 \\ 0 & 0 & 5 & 10 & | & -5 \end{bmatrix} \rightarrow \begin{bmatrix} 1 & 0 & 0 & 1 & | & 2 \\ 0 & 1 & 1 & 2 & | & 1 \\ 0 & 0 & 0 & 1 & | & 1 \\ 0 & 0 & 1 & 2 & | & -1 \end{bmatrix} \rightarrow \begin{bmatrix} 1 & 0 & 0 & 1 & | & 2 \\ 0 & 1 & 1 & 2 & | & 1 \\ 0 & 0 & 1 & 2 & | & -1 \\ 0 & 0 & 0 & 1 & | & 1 \end{bmatrix}$$

$R_1 = r_2 + r_1$ $R_3 = -\frac{1}{2}r_3$ Interchange r_3 and r_4
$R_3 = -3r_2 + r_3$ $R_4 = \frac{1}{5}r_4$
$R_4 = r_2 + r_4$

$$\rightarrow \begin{bmatrix} 1 & 0 & 0 & 1 & | & 2 \\ 0 & 1 & 0 & 0 & | & 2 \\ 0 & 0 & 1 & 2 & | & -1 \\ 0 & 0 & 0 & 1 & | & 1 \end{bmatrix} \rightarrow \begin{bmatrix} 1 & 0 & 0 & 0 & | & 1 \\ 0 & 1 & 0 & 0 & | & 2 \\ 0 & 0 & 1 & 0 & | & -3 \\ 0 & 0 & 0 & 1 & | & 1 \end{bmatrix}$$

$R_2 = -r_3 + r_2$ $R_1 = -r_4 + r_1$
$R_3 = -2r_4 + r_3$

The solution is $x = 1, y = 2, z = -3, t = 1$.

45. Evaluating the determinant:

$$\begin{vmatrix} 3 & 4 \\ 1 & 3 \end{vmatrix} = 3(3) - 4(1) = 9 - 4 = 5$$

47. Evaluating the determinant:

$$\begin{vmatrix} 1 & 4 & 0 \\ -1 & 2 & 6 \\ 4 & 1 & 3 \end{vmatrix} = 1 \begin{vmatrix} 2 & 6 \\ 1 & 3 \end{vmatrix} - 4 \begin{vmatrix} -1 & 6 \\ 4 & 3 \end{vmatrix} + 0 \begin{vmatrix} -1 & 2 \\ 4 & 1 \end{vmatrix}$$

$$= 1[2(3) - 6(1)] - 4[-1(3) - 6(4)] + 0[-1(1) - 2(4)]$$
$$= 1(6 - 6) - 4(-3 - 24) + 0(-1 - 8)$$
$$= 1(0) - 4(-27) + 0(-9) = 0 + 108 + 0 = 108$$

49. Evaluating the determinant:

$$\begin{vmatrix} 2 & 1 & -3 \\ 5 & 0 & 1 \\ 2 & 6 & 0 \end{vmatrix} = 2 \begin{vmatrix} 0 & 1 \\ 6 & 0 \end{vmatrix} - 1 \begin{vmatrix} 5 & 1 \\ 2 & 0 \end{vmatrix} + (-3) \begin{vmatrix} 5 & 0 \\ 2 & 6 \end{vmatrix}$$

$$= 2(0 - 6) - 1(0 - 2) + (-3)(30 - 0) = -12 + 2 - 90 = -100$$

51. Set up and evaluate the determinants to use Cramer's Rule:

$$\begin{cases} x - 2y = 4 \\ 3x + 2y = 4 \end{cases}$$

$$D = \begin{vmatrix} 1 & -2 \\ 3 & 2 \end{vmatrix} = 1(2) - 3(-2) = 2 + 6 = 8$$

$$D_x = \begin{vmatrix} 4 & -2 \\ 4 & 2 \end{vmatrix} = 4(2) - 4(-2) = 8 + 8 = 16$$

$$D_y = \begin{vmatrix} 1 & 4 \\ 3 & 4 \end{vmatrix} = 1(4) - 4(3) = 4 - 12 = -8$$

Find the solutions by Cramer's Rule: $x = \dfrac{D_x}{D} = \dfrac{16}{8} = 2$ $y = \dfrac{D_y}{D} = \dfrac{-8}{8} = -1$

53. Set up and evaluate the determinants to use Cramer's Rule:
$$\begin{cases} 2x + 3y = 13 \\ 3x - 2y = 0 \end{cases}$$

$$D = \begin{vmatrix} 2 & 3 \\ 3 & -2 \end{vmatrix} = -4 - 9 = -13$$

$$D_x = \begin{vmatrix} 13 & 3 \\ 0 & -2 \end{vmatrix} = -26 - 0 = -26$$

$$D_y = \begin{vmatrix} 2 & 13 \\ 3 & 0 \end{vmatrix} = 0 - 39 = -39$$

Find the solutions by Cramer's Rule: $x = \dfrac{D_x}{D} = \dfrac{-26}{-13} = 2$ $y = \dfrac{D_y}{D} = \dfrac{-39}{-13} = 3$

55. Set up and evaluate the determinants to use Cramer's Rule:
$$\begin{cases} x + 2y - z = 6 \\ 2x - y + 3z = -13 \\ 3x - 2y + 3z = -16 \end{cases}$$

$$D = \begin{vmatrix} 1 & 2 & -1 \\ 2 & -1 & 3 \\ 3 & -2 & 3 \end{vmatrix} = 1\begin{vmatrix} -1 & 3 \\ -2 & 3 \end{vmatrix} - 2\begin{vmatrix} 2 & 3 \\ 3 & 3 \end{vmatrix} + (-1)\begin{vmatrix} 2 & -1 \\ 3 & -2 \end{vmatrix}$$

$$= 1(-3 + 6) - 2(6 - 9) - 1(-4 + 3) = 3 + 6 + 1 = 10$$

$$D_x = \begin{vmatrix} 6 & 2 & -1 \\ -13 & -1 & 3 \\ -16 & -2 & 3 \end{vmatrix} = 6\begin{vmatrix} -1 & 3 \\ -2 & 3 \end{vmatrix} - 2\begin{vmatrix} -13 & 3 \\ -16 & 3 \end{vmatrix} + (-1)\begin{vmatrix} -13 & -1 \\ -16 & -2 \end{vmatrix}$$

$$= 6(-3 + 6) - 2(-39 + 48) - 1(26 - 16) = 18 - 18 - 10 = -10$$

$$D_y = \begin{vmatrix} 1 & 6 & -1 \\ 2 & -13 & 3 \\ 3 & -16 & 3 \end{vmatrix} = 1\begin{vmatrix} -13 & 3 \\ -16 & 3 \end{vmatrix} - 6\begin{vmatrix} 2 & 3 \\ 3 & 3 \end{vmatrix} + (-1)\begin{vmatrix} 2 & -13 \\ 3 & -16 \end{vmatrix}$$

$$= 1(-39 + 48) - 6(6 - 9) - 1(-32 + 39) = 9 + 18 - 7 = 20$$

$$D_z = \begin{vmatrix} 1 & 2 & 6 \\ 2 & -1 & -13 \\ 3 & -2 & -16 \end{vmatrix} = 1\begin{vmatrix} -1 & -13 \\ -2 & -16 \end{vmatrix} - 2\begin{vmatrix} 2 & -13 \\ 3 & -16 \end{vmatrix} + 6\begin{vmatrix} 2 & -1 \\ 3 & -2 \end{vmatrix}$$

$$= 1(16 - 26) - 2(-32 + 39) + 6(-4 + 3) = -10 - 14 - 6 = -30$$

Find the solutions by Cramer's Rule:

$$x = \dfrac{D_x}{D} = \dfrac{-10}{10} = -1 \qquad y = \dfrac{D_y}{D} = \dfrac{20}{10} = 2 \qquad z = \dfrac{D_z}{D} = \dfrac{-30}{10} = -3$$

57. Find the partial fraction decomposition:

$$\frac{6}{x(x-4)} = \frac{A}{x} + \frac{B}{x-4} \quad \text{(Multiply both sides by } x(x-4)\text{.)}$$

$$6 = A(x-4) + Bx$$

Let $x = 4$: then $6 = A(4-4) + B(4) \;\rightarrow\; 4B = 6 \;\rightarrow\; B = \dfrac{3}{2}$

Let $x = 0$: then $6 = A(0-4) + B(0) \;\rightarrow\; -4A = 6 \;\rightarrow\; A = -\dfrac{3}{2}$

$$\frac{6}{x(x-4)} = \frac{\left(-\dfrac{3}{2}\right)}{x} + \frac{\left(\dfrac{3}{2}\right)}{x-4}$$

59. Find the partial fraction decomposition:

$$\frac{x-4}{x^2(x-1)} = \frac{A}{x} + \frac{B}{x^2} + \frac{C}{x-1} \quad \text{(Multiply both sides by } x^2(x-1)\text{.)}$$

$$x - 4 = Ax(x-1) + B(x-1) + Cx^2$$

Let $x = 1$: then $1 - 4 = A(1)(1-1) + B(1-1) + C(1)^2 \rightarrow -3 = C \rightarrow C = -3$

Let $x = 0$: then $0 - 4 = A(0)(0-1) + B(0-1) + C(0)^2 \rightarrow -4 = -B \rightarrow B = 4$

Let $x = 2$: then $2 - 4 = A(2)(2-1) + B(2-1) + C(2)^2 \rightarrow -2 = 2A + B + 4C$

$$\rightarrow 2A = -2 - 4 - 4(-3) \;\rightarrow\; 2A = 6 \;\rightarrow\; A = 3$$

$$\frac{x-4}{x^2(x-1)} = \frac{3}{x} + \frac{4}{x^2} + \frac{-3}{x-1}$$

61. Find the partial fraction decomposition:

$$\frac{x}{(x^2+9)(x+1)} = \frac{A}{x+1} + \frac{Bx+C}{x^2+9} \quad \text{(Multiply both sides by } (x+1)(x^2+9)\text{.)}$$

$$x = A(x^2+9) + (Bx+C)(x+1)$$

Let $x = -1$: then $-1 = A((-1)^2 + 9) + (B(-1)+C)(-1+1)$

$$\rightarrow -1 = A(10) + (-B+C)(0) \rightarrow -1 = 10A \rightarrow A = -\frac{1}{10}$$

Let $x = 1$:   then $1 = A(1^2+9) + (B(1)+C)(1+1) \;\rightarrow\; 1 = 10A + 2B + 2C$

$$\rightarrow 1 = 10\left(-\frac{1}{10}\right) + 2B + 2C \;\rightarrow\; 2 = 2B + 2C \;\rightarrow\; B + C = 1$$

Let $x = 0$: then $0 = A(0^2+9) + (B(0)+C)(0+1) \;\rightarrow\; 0 = 9A + C$

$$\rightarrow 0 = 9\left(-\frac{1}{10}\right) + C \;\rightarrow\; C = \frac{9}{10} \quad B = 1 - C \;\rightarrow\; B = 1 - \frac{9}{10} \;\rightarrow\; B = \frac{1}{10}$$

$$\frac{x}{(x^2+9)(x+1)} = \frac{-\left(\dfrac{1}{10}\right)}{x+1} + \frac{\left(\dfrac{1}{10}x + \dfrac{9}{10}\right)}{x^2+9}$$

63. Find the partial fraction decomposition:

$$\frac{x^3}{(x^2+4)^2} = \frac{Ax+B}{x^2+4} + \frac{Cx+D}{(x^2+4)^2} \text{ (Multiply both sides by } (x^2+4)^2.)$$

$$x^3 = (Ax+B)(x^2+4) + Cx+D$$

$$x^3 = Ax^3 + Bx^2 + 4Ax + 4B + Cx + D$$

$$x^3 = Ax^3 + Bx^2 + (4A+C)x + 4B + D$$

$$A = 1$$

$$B = 0$$

$$4A+C = 0 \;\rightarrow\; 4(1)+C = 0 \;\rightarrow\; C = -4$$

$$4B+D = 0 \;\rightarrow\; 4(0)+D = 0 \;\rightarrow\; D = 0$$

$$\frac{x^3}{(x^2+4)^2} = \frac{x}{x^2+4} + \frac{-4x}{(x^2+4)^2}$$

65. Find the partial fraction decomposition:

$$\frac{x^2}{(x^2+1)(x^2-1)} = \frac{x^2}{(x^2+1)(x-1)(x+1)} = \frac{A}{x-1} + \frac{B}{x+1} + \frac{Cx+D}{x^2+1}$$

$$\text{(Multiply both sides by } (x-1)(x+1)(x^2+1).)$$

$$x^2 = A(x+1)(x^2+1) + B(x-1)(x^2+1) + (Cx+D)(x-1)(x+1)$$

Let $x = 1$: then $1^2 = A(1+1)(1^2+1) + B(1-1)(1^2+1) + (C(1)+D)(1-1)(1+1)$

$$\rightarrow \; 1 = 4A \;\rightarrow\; A = \frac{1}{4}$$

Let $x = -1$: then

$$(-1)^2 = A(-1+1)((-1)^2+1) + B(-1-1)((-1)^2+1) + (C(-1)$$
$$+ D)(-1-1)(-1+1)$$

$$\rightarrow \; 1 = -4B \;\rightarrow\; B = -\frac{1}{4}$$

Let $x = 0$: then

$$0^2 = A(0+1)(0^2+1) + B(0-1)(0^2+1) + (C(0)+D)(0-1)(0+1)$$

$$\rightarrow \; 0 = A-B-D \;\rightarrow\; 0 = \frac{1}{4}-\left(-\frac{1}{4}\right)-D \;\rightarrow\; D = \frac{1}{2}$$

Let $x = 2$: then

$$2^2 = A(2+1)(2^2+1) + B(2-1)(2^2+1) + (C(2)+D)(2-1)(2+1)$$

$$\rightarrow \; 4 = 15A+5B+6C+3D \;\rightarrow\; 4 = 15\left(\frac{1}{4}\right)+5\left(-\frac{1}{4}\right)+6C+3\left(\frac{1}{2}\right)$$

$$\rightarrow \; 6C = 4 - \frac{15}{4}+\frac{5}{4}-\frac{3}{2} \;\rightarrow\; 6C = 0 \;\rightarrow\; C = 0$$

$$\frac{x^2}{(x^2+1)(x^2-1)} = \frac{x^2}{(x^2+1)(x-1)(x+1)} = \frac{\left(\frac{1}{4}\right)}{x-1} + \frac{-\left(\frac{1}{4}\right)}{x+1} + \frac{\left(\frac{1}{2}\right)}{x^2+1}$$

67. Solve the first equation for y, substitute into the second equation and solve:

$$\begin{cases} 2x + y + 3 = 0 \quad \rightarrow \quad y = -2x - 3 \\ \quad x^2 + y^2 = 5 \end{cases}$$

$$x^2 + (-2x - 3)^2 = 5 \rightarrow x^2 + 4x^2 + 12x + 9 = 5$$

$$5x^2 + 12x + 4 = 0 \rightarrow (5x + 2)(x + 2) = 0$$

$$x = -\frac{2}{5} \quad \text{or} \quad x = -2$$

$$y = -\frac{11}{5} \qquad y = 1$$

Solutions: $\left(-\frac{2}{5}, -\frac{11}{5}\right), (-2, 1)$.

69. Multiply each side of the second equation by 2 and add the equations to eliminate xy:

$$\begin{cases} 2xy + y^2 = 10 \quad \longrightarrow \quad 2xy + y^2 = 10 \\ -xy + 3y^2 = 2 \quad \xrightarrow{\ 2\ } \quad -2xy + 6y^2 = 4 \end{cases}$$

$$7y^2 = 14 \rightarrow y^2 = 2 \rightarrow y = \pm\sqrt{2}$$

If $y = \sqrt{2}$: $\quad 2x(\sqrt{2}) + (\sqrt{2})^2 = 10$

$$2\sqrt{2}x = 8 \rightarrow x = \frac{8}{2\sqrt{2}} = 2\sqrt{2}$$

If $y = -\sqrt{2}$: $\quad 2x(-\sqrt{2}) + (-\sqrt{2})^2 = 10$

$$-2\sqrt{2}x = 8 \rightarrow x = \frac{8}{-2\sqrt{2}} = -2\sqrt{2}$$

Solutions: $\left(2\sqrt{2}, \sqrt{2}\right), \left(-2\sqrt{2}, -\sqrt{2}\right)$

71. Substitute into the second equation into the first equation and solve:

$$\begin{cases} x^2 + y^2 = 6y \\ \quad x^2 = 3y \end{cases}$$

$$3y + y^2 = 6y$$

$$y^2 - 3y = 0$$

$$y(y - 3) = 0$$

$$y = 0 \quad \text{or} \quad y = 3$$

If $y = 0$: $\quad x^2 = 3(0) \rightarrow x^2 = 0 \rightarrow x = 0$

If $y = 3$: $\quad x^2 = 3(3) \rightarrow x^2 = 9 \rightarrow x = \pm 3$

Solutions: $(0, 0), (-3, 3), (3, 3)$

73. Factor the second equation, solve for x, substitute into the first equation and solve:

$$\begin{cases} 3x^2 + 4xy + 5y^2 = 8 \\ \quad x^2 + 3xy + 2y^2 = 0 \quad \rightarrow \quad (x + 2y)(x + y) = 0 \quad \rightarrow \quad x = -2y \text{ or } x = -y \end{cases}$$

Substitute $x = -2y$ and solve:

$$3x^2 + 4xy + 5y^2 = 8$$
$$3(-2y)^2 + 4(-2y)y + 5y^2 = 8$$
$$12y^2 - 8y^2 + 5y^2 = 8$$
$$9y^2 = 8$$
$$y^2 = \frac{8}{9}$$
$$y = \pm\frac{2\sqrt{2}}{3}$$

If $y = \frac{2\sqrt{2}}{3}$: $x = -2\left(\frac{2\sqrt{2}}{3}\right) = \frac{-4\sqrt{2}}{3}$

If $y = \frac{-2\sqrt{2}}{3}$: $x = -2\left(\frac{-2\sqrt{2}}{3}\right) = \frac{4\sqrt{2}}{3}$

Substitute $x = -y$ and solve:

$$3x^2 + 4xy + 5y^2 = 8$$
$$3(-y)^2 + 4(-y)y + 5y^2 = 8$$
$$3y^2 - 4y^2 + 5y^2 = 8$$
$$4y^2 = 8$$
$$y^2 = 2$$
$$y = \pm\sqrt{2}$$

If $y = \sqrt{2}$: $x = -\sqrt{2}$
If $y = -\sqrt{2}$: $x = \sqrt{2}$

Solutions: $\left(\frac{-4\sqrt{2}}{3}, \frac{2\sqrt{2}}{3}\right), \left(\frac{4\sqrt{2}}{3}, \frac{-2\sqrt{2}}{3}\right), \left(-\sqrt{2}, \sqrt{2}\right), \left(\sqrt{2}, -\sqrt{2}\right)$

75.   Multiply each side of the second equation by $-y$ and add the equations to eliminate y:

$$\begin{cases} x^2 - 3x + y^2 + y = -2 \longrightarrow \quad x^2 - 3x + y^2 + y = -2 \\ \dfrac{x^2-x}{y} + y + 1 = 0 \xrightarrow{-y} \quad -x^2 + x - y^2 - y = 0 \end{cases}$$

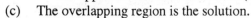

$$-2x \qquad\qquad = -2$$
$$x = 1$$

If $x = 1$ $1^2 - 3(1) + y^2 + y = -2 \rightarrow y^2 + y = 0 \rightarrow y(y+1) = 0$
$$\rightarrow y = 0 \text{ or } y = -1$$

Note that $y \ne 0$ because that would cause division by zero in the original equation.
Solution: $(1, -1)$

77.   Graph the system of linear inequalities:
$$\begin{cases} -2x + y \le 2 \\ x + y \ge 2 \end{cases}$$

(a)   Graph the line $-2x + y = 2$. Use a solid
       line since the inequality uses $\le$.
       Choose a test point not on the line, such as
       $(0, 0)$. Since $-2(0) + 0 \le 2$ is true,
       shade the side of the line containing
       $(0, 0)$.
(b)   Graph the line $x + y = 2$. Use a solid line
       since the inequality uses $\ge$.
       Choose a test point not on the line, such as
       $(0, 0)$. Since $0 + 0 \ge 2$ is false, shade
       the opposite side of the line from $(0, 0)$.
(c)   The overlapping region is the solution.
(d)   The graph is unbounded.

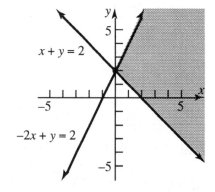

(e)   Find the vertices:
   To find the intersection of $x + y = 2$ and $-2x + y = 2$, solve the system:
   $$\begin{cases} x + y = 2 \quad \rightarrow \quad x = 2 - y \\ -2x + y = 2 \end{cases}$$
   Substitute and solve:
   $$-2(2 - y) + y = 2 \rightarrow -4 + 2y + y = 2 \rightarrow 3y = 6 \rightarrow y = 2$$
   $$x = 2 - 2 = 0$$
   The point of intersection is (0, 2).
   The corner point is (0, 2).

79.   Graph the system of linear inequalities:
   $$\begin{cases} x \geq 0 \\ y \geq 0 \\ x + y \leq 4 \\ 2x + 3y \leq 6 \end{cases}$$

(a)   Graph $x \geq 0$; $y \geq 0$. Shaded region is the first quadrant.

(b)   Graph the line $x + y = 4$. Use a solid line since the inequality uses $\leq$.
   Choose a test point not on the line, such as (0, 0). Since $0 + 0 \leq 4$ is true, shade the side of the line containing (0, 0).

(c)   Graph the line $2x + 3y = 6$. Use a solid line since the inequality uses $\leq$.
   Choose a test point not on the line, such as (0, 0). Since $2(0) + 3(0) \leq 6$ is true, shade the side of the line containing (0, 0).

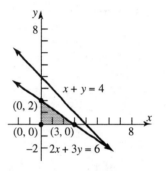

(d)   The overlapping region is the solution.

(e)   The graph is bounded.

(f)   Find the vertices:
   The x-axis and y-axis intersect at (0, 0).
   The intersection of $2x + 3y = 6$ and the y-axis is (0, 2).
   The intersection of $2x + 3y = 6$ and the x-axis is (3, 0).
   The three corner points are (0, 0), (0, 2), and (3, 0).

81.   Graph the system of linear inequalities:
   $$\begin{cases} x \geq 0 \\ y \geq 0 \\ 2x + y \leq 8 \\ x + 2y \geq 2 \end{cases}$$

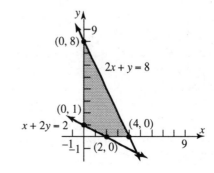

(a)   Graph $x \geq 0$; $y \geq 0$. Shaded region is
the first quadrant.

(b)   Graph the line $2x + y = 8$. Use a solid
line since the inequality uses $\leq$.
Choose a test point not on the line, such
as $(0, 0)$. Since $2(0) + 0 \leq 8$ is true,
shade the side of the line containing $(0, 0)$.

(c)   Graph the line $x + 2y = 2$. Use a solid line since the inequality uses $\geq$.
Choose a test point not on the line, such as $(0, 0)$. Since $0 + 2(0) \geq 2$ is false, shade
the opposite side of the line from $(0, 0)$.

(d)   The overlapping region is the solution.

(e)   The graph is bounded.

(f)   Find the vertices:
The intersection of $x + 2y = 2$ and the y-axis is $(0, 1)$.
The intersection of $x + 2y = 2$ and the x-axis is $(2, 0)$.
The intersection of $2x + y = 8$ and the y-axis is $(0, 8)$.
The intersection of $2x + y = 8$ and the x-axis is $(4, 0)$.
The four corner points are $(0, 1)$, $(0, 8)$, $(2, 0)$, and $(4, 0)$.

83.   Graph the system of inequalities:
$$\begin{cases} x^2 + y^2 \leq 16 \\ x + y \geq 2 \end{cases}$$

(a)   Graph the circle $x^2 + y^2 = 16$. Use a solid line since the
inequality uses $\leq$. Choose a test point not on the circle, such
as $(0, 0)$. Since $0^2 + 0^2 \leq 16$ is true, shade the side of the
circle containing $(0, 0)$.

(b)   Graph the line $x + y = 2$. Use a solid line since the
inequality uses $\geq$. Choose a test point not on the line, such
as $(0, 0)$. Since $0 + 0 \geq 2$ is false, shade the opposite side of
the line from $(0, 0)$.

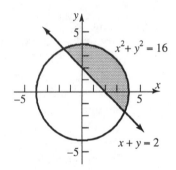

(c)   The overlapping region is the solution.

85.   Graph the system of inequalities:
$$\begin{cases} y \leq x^2 \\ xy \leq 4 \end{cases}$$

(a)   Graph the parabola $y = x^2$. Use a solid line since the
inequality uses $\leq$. Choose a test point not on the parabola,
such as $(1, 2)$. Since $2 \leq 1^2$ is false, shade the opposite side
of the parabola from $(1, 2)$.

(b)   Graph the hyperbola $xy = 4$. Use a solid line since the
inequality uses $\leq$. Choose a test point not on the hyperbola,
such as $(1, 2)$. Since $1 \cdot 2 \leq 4$ is true, shade the same side
of the hyperbola as $(1, 2)$.

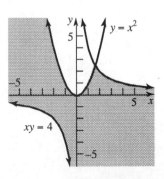

(c)   The overlapping region is the solution.

87.   Maximize $z = 3x + 4y$ Subject to $x \geq 0,\ y \geq 0,\ 3x + 2y \geq 6,\ x + y \leq 8$
Graph the constraints.

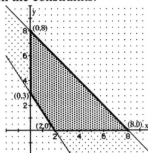

The corner points are $(0, 3), (2, 0), (0, 8), (8, 0)$.

Evaluate the objective function:

Vertex	Value of $z = 3x + 4y$
$(0, 3)$	$z = 3(0) + 4(3) = 12$
$(0, 8)$	$z = 3(0) + 4(8) = 32$
$(2, 0)$	$z = 3(2) + 4(0) = 6$
$(8, 0)$	$z = 3(8) + 4(0) = 24$

The maximum value is 32 at $(0, 8)$.

89.   Minimize $z = 3x + 5y$
Subject to $x \geq 0,\ y \geq 0,\ x + y \geq 1,\ 3x + 2y \leq 12,\ x + 3y \leq 12$
Graph the constraints.

To find the intersection of $3x + 2y = 12$ and $x + 3y = 12$, solve the system:
$$\begin{cases} 3x + 2y = 12 \\ x + 3y = 12 \end{cases} \rightarrow \quad x = 12 - 3y$$

Substitute and solve:

$$3(12 - 3y) + 2y = 12 \rightarrow 36 - 9y + 2y = 12 \rightarrow -7y = -24 \rightarrow y = \frac{24}{7}$$

$$x = 12 - 3\left(\frac{24}{7}\right) = 12 - \frac{72}{7} = \frac{12}{7}$$

The point of intersection is $\left(\frac{12}{7}, \frac{24}{7}\right)$.

The corner points are $(0, 1), (1, 0), (0, 4), (4, 0), \left(\frac{12}{7}, \frac{24}{7}\right)$

Evaluate the objective function:

Vertex	Value of $z = 3x + 5y$
$(0, 1)$	$z = 3(0) + 5(1) = 5$
$(0, 4)$	$z = 3(0) + 5(4) = 20$
$(1, 0)$	$z = 3(1) + 5(0) = 3$
$(4, 0)$	$z = 3(4) + 5(0) = 12$
$\left(\dfrac{12}{7}, \dfrac{24}{7}\right)$	$z = 3\left(\dfrac{12}{7}\right) + 5\left(\dfrac{24}{7}\right) = \dfrac{36}{7} + \dfrac{120}{7} = \dfrac{156}{7} \approx 22.3$

The minimum value is 3 at $(1, 0)$.

91. Maximize $z = 5x + 4y$ Subject to $x \geq 0, \ y \geq 0, \ x + 2y \geq 2, \ 3x + 4y \leq 12, \ y \geq x$
     Graph the constraints.

To find the intersection of $x + 2y = 2$ and $y = x$, substitute and solve:
$$x + 2x = 2 \rightarrow 3x = 2 \rightarrow x = \frac{2}{3} \rightarrow y = \frac{2}{3}$$
The point of intersection is $\left(\dfrac{2}{3}, \dfrac{2}{3}\right)$.

To find the intersection of $y = x$ and $3x + 4y = 12$, substitute and solve:
$$3x + 4x = 12 \rightarrow 7x = 12 \rightarrow x = \frac{12}{7} \rightarrow y = \frac{12}{7}$$
The point of intersection is $\left(\dfrac{12}{7}, \dfrac{12}{7}\right)$.

The corner points are $(0, 1)$, $(0, 3)$, $\left(\dfrac{2}{3}, \dfrac{2}{3}\right)$, $\left(\dfrac{12}{7}, \dfrac{12}{7}\right)$.

Evaluate the objective function:

Vertex	Value of $z = 5x + 4y$
$(0, 1)$	$z = 5(0) + 4(1) = 4$
$(0, 3)$	$z = 5(0) + 4(3) = 12$
$\left(\dfrac{2}{3}, \dfrac{2}{3}\right)$	$z = 5\left(\dfrac{2}{3}\right) + 4\left(\dfrac{2}{3}\right) = \dfrac{18}{3} = 6$
$\left(\dfrac{12}{7}, \dfrac{12}{7}\right)$	$z = 5\left(\dfrac{12}{7}\right) + 4\left(\dfrac{12}{7}\right) = \dfrac{108}{7} \approx 15.43$

The maximum value is $\dfrac{108}{7}$ at $\left(\dfrac{12}{7}, \dfrac{12}{7}\right)$.

93. Multiply each side of the first equation by –2 and eliminate x:

$$\begin{cases} 2x + 5y = 5 \\ 4x + 10y = A \end{cases} \xrightarrow{\;-2\;} \begin{array}{r} -4x - 10y = -10 \\ 4x + 10y = \quad A \\ \hline 0 = A - 10 \end{array}$$

If there are to be infinitely many solutions, the sum in elimination should be $0 = 0$.
Therefore, $A - 10 = 0$ or $A = 10$.

95. when $x = 1,\ y = (1)^2 + b(1) + c = 2 \rightarrow 1 + b + c = 2 \rightarrow b + c = 1$

when $x = -1,\ y = (-1)^2 + b(-1) + c = 3 \rightarrow 1 - b + c = 3 \rightarrow -b + c = 2$

so we have the system $\begin{cases} b + c = 1 \\ -b + c = 2 \end{cases}$

subtracting the first equation from the second equation yields

$$\begin{array}{r} b + c = 1 \\ -(-b + c = 2) \\ \hline 2b = -1 \end{array} \quad \therefore b = -0.5$$

Back-substituting we get: $-0.5 + c = 1 \rightarrow c = 1.5$
Therefore, $y = x^2 - 0.5x + 1.5$, which satisfies the given conditions.

97. $y = ax^2 + bx + c$

At $(0, 1)$ the equation becomes:

$1 = a(0)^2 + b(0) + c$

$c = 1$

At $(1, 0)$ the equation becomes:

$0 = a(1)^2 + b(1) + c$

$0 = a + b + c$

$a + b + c = 0$

At $(-2, 1)$ the equation becomes:

$1 = a(-2)^2 + b(-2) + c = 4a - 2b + c \rightarrow 4a - 2b + c = 1$

The system of equations is:

$$\begin{cases} a + b + c = 0 \\ 4a - 2b + c = 1 \\ \qquad c = 1 \end{cases}$$

Substitute $c = 1$ into the first and second equations and simplify:

$$\begin{cases} a + b + 1 = 0 & \rightarrow & a + b = -1 & \rightarrow & a = -b - 1 \\ 4a - 2b + 1 = 1 & \rightarrow & 4a - 2b = 0 \end{cases}$$

Solve the first equation for a, substitute into the second equation and solve:

$4(-b - 1) - 2b = 0$

$-4b - 4 - 2b = 0$

$$-6b = 4 \rightarrow b = -\frac{2}{3} \rightarrow a = \frac{2}{3} - 1 = -\frac{1}{3}$$

The quadratic function is $y = -\frac{1}{3}x^2 - \frac{2}{3}x + 1$.

99. Let x = the number of pounds of coffee that costs $3.00 per pound.
Let y = the number of pounds of coffee that costs $6.00 per pound.
Then $x + y = 100$ represents the total amount of coffee in the blend.
The value of the blend will be represented by the equation: $3x + 6y = 3.90(100)$.
Solve the system of equations:
$$\begin{cases} x + y = 100 \quad \rightarrow \quad y = 100 - x \\ 3x + 6y = 390 \end{cases}$$
Solve by substitution:
$$3x + 6(100 - x) = 390 \rightarrow 3x + 600 - 6x = 390$$
$$-3x = -210 \rightarrow x = 70$$
$$y = 100 - 70 = 30$$
The blend is made up of 70 pounds of the $3 per pound coffee and 30 pounds of the $6 per pound coffee.

101. Let x = the number of small boxes.
Let y = the number of medium boxes.
Let z = the number of large boxes.
Oatmeal raisin equation: $x + 2y + 2z = 15$
Chocolate chip equation: $x + y + 2z = 10$
Shortbread equation: $y + 3z = 11$
Multiply each side of the second equation by -1 and add to the first equation to eliminate x:
$$\begin{cases} x + 2y + 2z = 15 \quad \longrightarrow \quad x + 2y + 2z = 15 \\ x + y + 2z = 10 \quad \xrightarrow{-1} \quad -x - y - 2z = -10 \\ \quad\quad y + 3z = 11 \quad\quad\quad\quad\quad\quad y \quad\quad = 5 \end{cases}$$
Substituting and solving for the other variables:
$$\begin{array}{ll} 5 + 3z = 11 & x + 5 + 2(2) = 10 \\ 3z = 6 & x + 9 = 10 \\ z = 2 & x = 1 \end{array}$$
1 small box, 5 medium boxes, and 2 large boxes of cookies should be purchased.

103. Let x = the length of the lot.
Let y = the width of the lot.
Perimeter equation: $2x + 2y = 68$
Diagonal equation: $x^2 + y^2 = 26^2$
Solve the system of equations:
$$\begin{cases} 2x + 2y = 68 \quad \rightarrow \quad y = 34 - x \\ x^2 + y^2 = 676 \end{cases}$$
Solve by substitution:
$$x^2 + (34 - x)^2 = 676 \rightarrow x^2 + 1156 - 68x + x^2 = 676$$
$$2x^2 - 68x + 480 = 0 \rightarrow x^2 - 34x + 240 = 0$$
$$(x - 24)(x - 10) = 0 \rightarrow x = 24 \text{ or } x = 10$$
$$y = 10 \text{ or } y = 24$$
The dimensions of the lot are 24 feet by 10 feet.

105. Let x = the length of one leg.
Let y = the length of the other leg.
Perimeter equation: $x + y + 6 = 14$
Pythagorean equation: $x^2 + y^2 = 6^2$
Solve the system of equations:
$$\begin{cases} x + y = 8 & \rightarrow \quad y = 8 - x \\ x^2 + y^2 = 36 \end{cases}$$
Solve by substitution:
$$x^2 + (8-x)^2 = 36 \rightarrow x^2 + 64 - 16x + x^2 = 36$$
$$2x^2 - 16x + 28 = 0 \rightarrow x^2 - 8x + 14 = 0$$
$$x = \frac{8 \pm \sqrt{64-56}}{2} = \frac{8 \pm \sqrt{8}}{2} = \frac{8 \pm 2\sqrt{2}}{2} = 4 \pm \sqrt{2}$$
If $x = 4 - \sqrt{2}$, then $y = 8 - 4 + \sqrt{2} = 4 + \sqrt{2}$
If $x = 4 + \sqrt{2}$, then $y = 8 - 4 - \sqrt{2} = 4 - \sqrt{2}$
The legs are $4 + \sqrt{2}$ and $4 - \sqrt{2}$.

107. Let x = the length of the side of the smaller square.
Then $2x$ = the length of the side of the larger square.
The needed fencing is $4x + 8x = 12x$.
Solve the area equation:
$$x^2 + (2x)^2 = 5000 \rightarrow 5x^2 = 5000 \rightarrow x^2 = 1000 \rightarrow x = 10\sqrt{10}$$
$12x = 120\sqrt{10} \approx 379.5$ feet of fence are needed.

109. Let x = the speed of the boat in still water.
Let y = the speed of the river current.
Let d = the distance from Chiritza to the Flotel Orellana (100 kilometers)

	Rate	Time	Distance
trip downstream	$x + y$	$\dfrac{5}{2}$	100
trip downstream	$x - y$	3	100

The system of equations is:
$$\begin{cases} (x+y)\left(\dfrac{5}{2}\right) = 100 & \rightarrow \quad 5x + 5y = 200 \\ (x-y)(3) = d & \rightarrow \quad 3x - 3y = 100 \end{cases}$$

$$5x + 5y = 200 \xrightarrow{\ 3\ } 15x + 15y = 600$$
$$3x - 3y = 100 \xrightarrow{\ 5\ } \underline{+\ 15x - 15y = 500}$$
$$30x = 1100$$
$$\therefore x = \frac{1100}{30} = \frac{110}{3}$$

$$\rightarrow 3\left(\frac{110}{3}\right) - 3y = 100 \rightarrow 110 - 3y = 100 \rightarrow 10 = 3y \rightarrow y = \frac{10}{3}$$

The speed of the boat $= \dfrac{110}{3} \approx 36.67$ km/hr; the speed of the current $= \dfrac{10}{3} \approx 3.33$ km/hr.

111. Let x = the number of hours for Bruce to do the job alone.
Let y = the number of hours for Bryce to do the job alone.
Let z = the number of hours for Marty to do the job alone.

Then $\dfrac{1}{x}$ represents the fraction of the job that Bruce does in one hour.

$\dfrac{1}{y}$ represents the fraction of the job that Bryce does in one hour.

$\dfrac{1}{z}$ represents the fraction of the job that Marty does in one hour.

The equation representing Bruce and Bryce working together is:

$$\frac{1}{x}+\frac{1}{y}=\frac{1}{\left(\frac{4}{3}\right)}=\frac{3}{4}=0.75$$

The equation representing Bryce and Marty working together is:

$$\frac{1}{y}+\frac{1}{z}=\frac{1}{\left(\frac{8}{5}\right)}=\frac{5}{8}=0.675$$

The equation representing Bruce and Marty working together is:

$$\frac{1}{x}+\frac{1}{z}=\frac{1}{\left(\frac{8}{3}\right)}=0.375$$

Solve the system of equations:
$$\begin{cases} x^{-1}+y^{-1}=0.75 \\ y^{-1}+z^{-1}=0.675 \\ x^{-1}+z^{-1}=0.375 \end{cases}$$

Let $u=x^{-1},\ v=y^{-1},\ w=z^{-1}$
$$\begin{cases} u+v=0.75 & \to & u=0.75-v \\ v+w=0.675 & \to & w=0.675-v \\ u+w=0.375 \end{cases}$$

Substitute into the third equation and solve:
$$0.75-v+0.675-v=0.375 \to -2v=-1 \to v=0.5$$
$$u=0.75-0.5=0.25$$
$$w=0.675-0.5=0.125$$
Solve for x, y, and z: $x=4,\ y=2,\ z=8$ (reciprocals)
Bruce can do the job in 4 hours, Bryce in 2 hours, and Marty in 8 hours.

113. Let x = the number of gasoline engines produced each week.
Let y = the number of diesel engines produced each week.
The total cost is: $C=450x+550y$. Cost is to be minimized; thus, this is the objective function.
The constraints are:
$20 \le x \le 60$ number of gasoline engines needed and capacity each week.
$15 \le y \le 40$ number of diesel engines needed and capacity each week.
$x+y \ge 50$ number of engines produced to prevent layoffs.

Graph the constraints.

The corner points are (20, 30), (20, 40), (35, 15), (60, 15), (60, 40).

Evaluate the objective function:

Vertex	Value of $C = 450x + 550y$
(20, 30)	$C = 450(20) + 550(30) = 25,500$
(20, 40)	$C = 450(20) + 550(40) = 31,000$
(35, 15)	$C = 450(35) + 550(15) = 24,000$
(60, 15)	$C = 450(60) + 550(15) = 35,250$
(60, 40)	$C = 450(60) + 550(40) = 49,000$

The minimum cost is $24,000, when 35 gasoline engines and 15 diesel engines are produced.

The excess capacity is 15 gasoline engines, since only 20 gasoline engines had to be delivered.

Chapter 11

Sequences; Induction; The Binomial Theorem

11.1 Sequences

1. $a_1 = 1,\ a_2 = 2,\ a_3 = 3,\ a_4 = 4,\ a_5 = 5$

3. $a_1 = \dfrac{1}{1+2} = \dfrac{1}{3},\ a_2 = \dfrac{2}{2+2} = \dfrac{2}{4} = \dfrac{1}{2},\ a_3 = \dfrac{3}{3+2} = \dfrac{3}{5},\ a_4 = \dfrac{4}{4+2} = \dfrac{4}{6} = \dfrac{2}{3},$
$a_5 = \dfrac{5}{5+2} = \dfrac{5}{7}$

5. $a_1 = (-1)^{1+1}(1^2) = 1,\ a_2 = (-1)^{2+1}(2^2) = -4,\ a_3 = (-1)^{3+1}(3^2) = 9,$
$a_4 = (-1)^{4+1}(4^2) = -16,\ a_5 = (-1)^{5+1}(5^2) = 25$

7. $a_1 = \dfrac{2^1}{3^1+1} = \dfrac{2}{4} = \dfrac{1}{2},\ a_2 = \dfrac{2^2}{3^2+1} = \dfrac{4}{10} = \dfrac{2}{5},\ a_3 = \dfrac{2^3}{3^3+1} = \dfrac{8}{28} = \dfrac{2}{7},$
$a_4 = \dfrac{2^4}{3^4+1} = \dfrac{16}{82} = \dfrac{8}{41},\ a_5 = \dfrac{2^5}{3^5+1} = \dfrac{32}{244} = \dfrac{8}{61}$

9. $a_1 = \dfrac{(-1)^1}{(1+1)(1+2)} = \dfrac{-1}{2\cdot 3} = \dfrac{-1}{6},\ a_2 = \dfrac{(-1)^2}{(2+1)(2+2)} = \dfrac{1}{3\cdot 4} = \dfrac{1}{12},$
$a_3 = \dfrac{(-1)^3}{(3+1)(3+2)} = \dfrac{-1}{4\cdot 5} = \dfrac{-1}{20},\ a_4 = \dfrac{(-1)^4}{(4+1)(4+2)} = \dfrac{1}{5\cdot 6} = \dfrac{1}{30},$
$a_5 = \dfrac{(-1)^5}{(5+1)(5+2)} = \dfrac{-1}{6\cdot 7} = -\dfrac{1}{42}$

11. $a_1 = \dfrac{1}{e^1} = \dfrac{1}{e},\ a_2 = \dfrac{2}{e^2},\ a_3 = \dfrac{3}{e^3},\ a_4 = \dfrac{4}{e^4},\ a_5 = \dfrac{5}{e^5}$

13. $\dfrac{n}{n+1}$ 15. $\dfrac{1}{2^{n-1}}$ 17. $(-1)^{n+1}$ 19. $(-1)^{n+1} n$

21. $a_1 = 2,\ a_2 = 3+2 = 5,\ a_3 = 3+5 = 8,\ a_4 = 3+8 = 11,\ a_5 = 3+11 = 14$

23. $a_1 = -2,\ a_2 = 2+(-2) = 0,\ a_3 = 3+0 = 3,\ a_4 = 4+3 = 7,\ a_5 = 5+7 = 12$

25. $a_1 = 5,\ a_2 = 2\cdot 5 = 10,\ a_3 = 2\cdot 10 = 20,\ a_4 = 2\cdot 20 = 40,\ a_5 = 2\cdot 40 = 80$

27. $a_1 = 3$, $a_2 = \dfrac{3}{2}$, $a_3 = \dfrac{\left(\dfrac{3}{2}\right)}{3} = \dfrac{1}{2}$, $a_4 = \dfrac{\left(\dfrac{1}{2}\right)}{4} = \dfrac{1}{8}$, $a_5 = \dfrac{\left(\dfrac{1}{8}\right)}{5} = \dfrac{1}{40}$

29. $a_1 = 1$, $a_2 = 2$, $a_3 = 2 \cdot 1 = 2$, $a_4 = 2 \cdot 2 = 4$, $a_5 = 4 \cdot 2 = 8$

31. $a_1 = A$, $a_2 = A + d$, $a_3 = (A + d) + d = A + 2d$, $a_4 = (A + 2d) + d = A + 3d$,
    $a_5 = (A + 3d) + d = A + 4d$

33. $a_1 = \sqrt{2}$, $a_2 = \sqrt{2 + \sqrt{2}}$, $a_3 = \sqrt{2 + \sqrt{2 + \sqrt{2}}}$, $a_4 = \sqrt{2 + \sqrt{2 + \sqrt{2 + \sqrt{2}}}}$,
    $a_5 = \sqrt{2 + \sqrt{2 + \sqrt{2 + \sqrt{2 + \sqrt{2}}}}}$

35. $\displaystyle\sum_{k=1}^{10} 5 = \underbrace{5 + 5 + 5 + \ldots + 5}_{10 \text{ times}} = 50$

37. $\displaystyle\sum_{k=1}^{6} k = 1 + 2 + 3 + 4 + 5 + 6 = 21$

39. $\displaystyle\sum_{k=1}^{5} (5k + 3) = 8 + 13 + 18 + 23 + 28 = 90$

41. $\displaystyle\sum_{k=1}^{3} (k^2 + 4) = 5 + 8 + 13 = 26$

43. $\displaystyle\sum_{k=1}^{6} (-1)^k 2^k = (-1)^1 \cdot 2^1 + (-1)^2 \cdot 2^2 + (-1)^3 \cdot 2^3 + (-1)^4 \cdot 2^4 + (-1)^5 \cdot 2^5 + (-1)^6 \cdot 2^6$
    $= -2 + 4 - 8 + 16 - 32 + 64 = 42$

45. $\displaystyle\sum_{k=1}^{4} (k^3 - 1) = 0 + 7 + 26 + 63 = 96$

47. $\displaystyle\sum_{k=1}^{n} (k + 2) = 3 + 4 + 5 + 6 + \cdots + (n + 2)$

49. $\displaystyle\sum_{k=1}^{n} \dfrac{k^2}{2} = \dfrac{1}{2} + 2 + \dfrac{9}{2} + 8 + \dfrac{25}{2} + \cdots + \dfrac{n^2}{2}$

51. $\displaystyle\sum_{k=0}^{n} \dfrac{1}{3^k} = 1 + \dfrac{1}{3} + \dfrac{1}{9} + \dfrac{1}{27} + \cdots + \dfrac{1}{3^n}$

53. $\displaystyle\sum_{k=0}^{n-1} \dfrac{1}{3^{k+1}} = \dfrac{1}{3} + \dfrac{1}{9} + \dfrac{1}{27} + \cdots + \dfrac{1}{3^n}$

55. $\displaystyle\sum_{k=2}^{n} (-1)^k \ln k = \ln 2 - \ln 3 + \ln 4 - \ln 5 + \cdots + (-1)^n \ln n$

57. $1 + 2 + 3 + \cdots + 20 = \displaystyle\sum_{k=1}^{20} k$

59. $\dfrac{1}{2} + \dfrac{2}{3} + \dfrac{3}{4} + \cdots + \dfrac{13}{13 + 1} = \displaystyle\sum_{k=1}^{13} \dfrac{k}{k + 1}$

61. $1 - \dfrac{1}{3} + \dfrac{1}{9} - \dfrac{1}{27} + \cdots + (-1)^6 \left(\dfrac{1}{3^6}\right) = \displaystyle\sum_{k=0}^{6} (-1)^k \left(\dfrac{1}{3^k}\right)$

63. $\quad 3 + \dfrac{3^2}{2} + \dfrac{3^3}{3} + \cdots + \dfrac{3^n}{n} = \displaystyle\sum_{k=1}^{n} \dfrac{3^k}{k}$

65. $\quad a + (a+d) + (a+2d) + \cdots + (a+nd) = \displaystyle\sum_{k=0}^{n} (a+kd)$

67. $\quad B_1 = 1.01(3000) - 100 = \$2930$

69. $\quad p_1 = 1.03(2000) + 20 = 2080; \qquad p_2 = 1.03(2080) + 20 = 2162.4$

71. $\quad a_1 = 1, \ a_2 = 1, \ a_3 = 2, \ a_4 = 3, \ a_5 = 5, \ a_6 = 8, \ a_7 = 13, \ a_8 = 21, \ a_n = a_{n-1} + a_{n-2}$
$\quad a_8 = a_7 + a_6 = 13 + 8 = 21$
After 7 months there are 21 mature pairs of rabbits.

73. $\quad$ 1, 1, 2, 3, 5, 8, 13 $\qquad$ This is the Fibonacci sequence.

75. To show that $\quad 1 + 2 + 3 + ... + (n-1) + n = \dfrac{n(n+1)}{2}$

$\quad$ Let

$\quad\quad S = 1 + 2 + 3 + \text{............} + (n-1) + n,$ we can reverse the order to get

$\quad\quad \underline{+S = n + (n-1) + (n-2) + ... + 2 + 1,}$ now add these two lines to get

$\quad\quad 2S = [1+n] + [2+(n-1)] + [3+(n-2)] + ... + [(n-1)+2] + [n+1]$

$\quad\quad\quad\quad\quad\quad\quad\quad\quad\quad\quad\quad\quad$ n terms

$\quad$ So we have

$\quad\quad 2S = [1+n] + [1+n] + [1+n] + ... + [n+1] + [n+1] = n \cdot [n+1]$

$\quad\quad\quad\quad\quad\quad\quad\quad\quad\quad\quad$ n terms

$\quad\quad \therefore \ S = \dfrac{n \cdot (n+1)}{2}$

Chapter 11

Sequences; Induction; The Binomial Theorem

11.2 Arithmetic Sequences

1. $d = a_{n+1} - a_n = (n+1+4) - (n+4) = n+5-n-4 = 1$
   $a_1 = 1+4 = 5, \ a_2 = 2+4 = 6, \ a_3 = 3+4 = 7, \ a_4 = 4+4 = 8$

3. $d = a_{n+1} - a_n = (2(n+1) - 5) - (2n-5) = 2n+2-5-2n+5 = 2$
   $a_1 = 2 \cdot 1 - 5 = -3, \ a_2 = 2 \cdot 2 - 5 = -1, \ a_3 = 2 \cdot 3 - 5 = 1, \ a_4 = 2 \cdot 4 - 5 = 3$

5. $d = a_{n+1} - a_n = (6 - 2(n+1)) - (6 - 2n) = 6 - 2n - 2 - 6 + 2n = -2$
   $a_1 = 6 - 2 \cdot 1 = 4, \ a_2 = 6 - 2 \cdot 2 = 2, \ a_3 = 6 - 2 \cdot 3 = 0, \ a_4 = 6 - 2 \cdot 4 = -2$

7. $d = a_{n+1} - a_n = \left(\dfrac{1}{2} - \dfrac{1}{3}(n+1) \right) - \left(\dfrac{1}{2} - \dfrac{1}{3}n \right) = \dfrac{1}{2} - \dfrac{1}{3}n - \dfrac{1}{3} - \dfrac{1}{2} + \dfrac{1}{3}n = -\dfrac{1}{3}$
   $a_1 = \dfrac{1}{2} - \dfrac{1}{3} \cdot 1 = \dfrac{1}{6}, \ a_2 = \dfrac{1}{2} - \dfrac{1}{3} \cdot 2 = -\dfrac{1}{6}, \ a_3 = \dfrac{1}{2} - \dfrac{1}{3} \cdot 3 = -\dfrac{1}{2}, \ a_4 = \dfrac{1}{2} - \dfrac{1}{3} \cdot 4 = -\dfrac{5}{6}$

9. $d = a_{n+1} - a_n = \ln 3^{n+1} - \ln 3^n = (n+1)\ln 3 - n\ln 3 = \ln 3(n+1-n) = \ln 3$
   $a_1 = \ln 3^1 = \ln 3, \ a_2 = \ln 3^2 = 2\ln 3, \ a_3 = \ln 3^3 = 3\ln 3, \ a_4 = \ln 3^4 = 4\ln 3$

11. $a_n = a + (n-1)d = 2 + (n-1)3 = 2 + 3n - 3 = 3n - 1$
    $a_5 = 3 \cdot 5 - 1 = 14$

13. $a_n = a + (n-1)d = 5 + (n-1)(-3) = 5 - 3n + 3 = 8 - 3n$
    $a_5 = 8 - 3 \cdot 5 = -7$

15. $a_n = a + (n-1)d = 0 + (n-1)\dfrac{1}{2} = \dfrac{1}{2}n - \dfrac{1}{2}$
    $a_5 = \dfrac{1}{2} \cdot 5 - \dfrac{1}{2} = 2$

17. $a_n = a + (n-1)d = \sqrt{2} + (n-1)\sqrt{2} = \sqrt{2} + \sqrt{2}n - \sqrt{2} = \sqrt{2}n$
    $a_5 = 5\sqrt{2}$

19. $a_1 = 2, \ d = 2, \ a_n = a + (n-1)d$
    $a_{12} = 2 + (12-1)2 = 2 + 11(2) = 2 + 22 = 24$

21. $a_1 = 1, \ d = -2 - 1 = -3, \ a_n = a + (n-1)d$
$a_{10} = 1 + (10 - 1)(-3) = 1 + 9(-3) = 1 - 27 = -26$

23. $a_1 = a, \ d = (a+b) - a = b, \ a_n = a + (n-1)d$
$a_8 = a + (8 - 1)b = a + 7b$

25. $a_8 = a + 7d = 8 \quad a_{20} = a + 19d = 44$
Solve the system of equations:
$$8 - 7d + 19d = 44$$
$$12d = 36 \rightarrow d = 3 \rightarrow a = 8 - 7(3) = 8 - 21 = -13$$
Recursive formula: $a_1 = -13 \quad a_n = a_{n-1} + 3$

27. $a_9 = a + 8d = -5 \quad a_{15} = a + 14d = 31$
Solve the system of equations:
$$-5 - 8d + 14d = 31 \rightarrow 6d = 36 \rightarrow d = 6$$
$$a = -5 - 8(6) = -5 - 48 = -53$$
Recursive formula: $a_1 = -53 \quad a_n = a_{n-1} + 6$

29. $a_{15} = a + 14d = 0 \quad a_{40} = a + 39d = -50$
Solve the system of equations:
$$-14d + 39d = -50 \rightarrow 25d = -50 \rightarrow d = -2$$
$$a = -14(-2) = 28$$
Recursive formula: $a_1 = 28 \quad a_n = a_{n-1} - 2$

31. $a_{14} = a + 13d = -1 \quad a_{18} = a + 17d = -9$
Solve the system of equations:
$$-1 - 13d + 17d = -9$$
$$4d = -8 \rightarrow d = -2 \rightarrow a = -1 - 13(-2) = -1 + 26 = 25$$
Recursive formula: $a_1 = 25 \quad a_n = a_{n-1} - 2$

33. $S_n = \dfrac{n}{2}(a + a_n) = \dfrac{n}{2}(1 + (2n - 1)) = \dfrac{n}{2}(2n) = n^2$

35. $S_n = \dfrac{n}{2}(a + a_n) = \dfrac{n}{2}(7 + (2 + 5n)) = \dfrac{n}{2}(9 + 5n) = \dfrac{9}{2}n + \dfrac{5}{2}n^2$

37. $a_1 = 2, \ d = 4 - 2 = 2, \ a_n = a + (n-1)d$

$$70 = 2 + (n - 1)2 \rightarrow 70 = 2 + 2n - 2 \rightarrow 70 = 2n \rightarrow n = 35$$
$$S_n = \dfrac{n}{2}(a + a_n) = \dfrac{35}{2}(2 + 70) = \dfrac{35}{2}(72) = 35(36) = 1260$$

39. $a_1 = 5, \ d = 9 - 5 = 4, \ a_n = a + (n-1)d$
$$49 = 5 + (n - 1)4$$
$$49 = 5 + 4n - 4 \qquad\qquad S_n = \dfrac{n}{2}(a + a_n) = \dfrac{12}{2}(5 + 49) = 6(54) = 324$$
$$48 = 4n \rightarrow n = 12$$

41. Using the sum of the sequence feature:

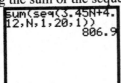

43. $d = 5.2 - 2.8 = 2.4$
     $a = 2.8$
     $36.4 = 2.8 + (n - 1)2.4$
     $36.4 = 2.8 + 2.4n - 2.4$
     $36 = 2.4n$
     $n = 15$
     $a_n = 2.8 + (n - 1)2.4 = 2.8 + 2.4n - 2.4 = 2.4n + 0.4$

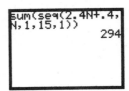

45. $d = 7.48 - 4.9 = 2.58$
     $a = 4.9$
     $66.82 = 4.9 + (n - 1)2.58$
     $66.82 = 4.9 + 2.58n - 2.58$
     $64.5 = 2.58n$
     $n = 25$
     $a_n = 4.9 + (n - 1)2.58 = 4.9 + 2.58n - 2.58$
     $a_n = 2.58n + 2.32$

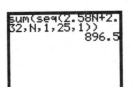

47. Find the common difference of the terms and solve the system of equations:
     $$(2x + 1) - (x + 3) = d \rightarrow x - 2 = d$$
     $$(5x + 2) - (2x + 1) = d \rightarrow 3x + 1 = d$$
     $$3x + 1 = x - 2 \rightarrow 2x = -3 \rightarrow x = -1.5$$

49. The total number of seats is: $S = 25 + 26 + 27 + \cdots$
     This is the sum of an arithmetic sequence with $d = 1$, $a = 25$, and $n = 30$.
     Find the sum of the sequence:
     $$S_{30} = \frac{30}{2}[2(25) + (30 - 1)(1)] = 15(50 + 29) = 15(79) = 1185$$
     There are 1185 seats in the theater.

51. The lighter colored tiles have 20 tiles in the bottom row and 1 tile in the top row. The
     number decreases by 1 as we move up the triangle. This is an arithmetic sequence with
     $a_1 = 20$, $d = -1$, and $n = 20$. Find the sum:
     $$S = \frac{20}{2}[2(20) + (20 - 1)(-1)] = 10(40 - 19) = 10(21) = 210 \text{ lighter tiles.}$$
     The darker colored tiles have 19 tiles in the bottom row and 1 tile in the top row. The
     number decreases by 1 as we move up the triangle. This is an arithmetic sequence with
     $a_1 = 19$, $d = -1$, and $n = 19$.
        Find the sum:
     $$S = \frac{19}{2}[2(19) + (19 - 1)(-1)] = \frac{19}{2}(38 - 18) = \frac{19}{2}(20) = 190 \text{ darker tiles.}$$

53.   Find n in an arithmetic sequence with $a_1 = 10$, $d = 4$, $s_n = 2040$.

$$s_n = \frac{n}{2}\left[2a_1 + (n-1)d\right] \rightarrow 2040 = \frac{n}{2}\left[2(10) + (n-1)4\right]$$

$$4080 = n\left[20 + 4n - 4\right] \rightarrow 4080 = n(4n + 16)$$

$$4080 = 4n^2 + 16n \rightarrow 1020 = n^2 + 4n$$

$$n^2 + 4n - 1020 = 0 \rightarrow (n + 34)(n - 30) = 0 \rightarrow n = -34 \text{ or } n = 30$$

There are 30 rows in the corner section of the stadium.

55.   Answers will vary.

Sequences; Induction; The Binomial Theorem

11.3 Geometric Sequences; Geometric Series

1. $r = \dfrac{3^{n+1}}{3^n} = 3^{n+1-n} = 3$; $\;a_1 = 3^1 = 3,\; a_2 = 3^2 = 9,\; a_3 = 3^3 = 27,\; a_4 = 3^4 = 81$

3. $r = \dfrac{-3\left(\dfrac{1}{2}\right)^{n+1}}{-3\left(\dfrac{1}{2}\right)^{n}} = \left(\dfrac{1}{2}\right)^{n+1-n} = \dfrac{1}{2}$

$a_1 = -3\left(\dfrac{1}{2}\right)^1 = -\dfrac{3}{2},\; a_2 = -3\left(\dfrac{1}{2}\right)^2 = -\dfrac{3}{4},\; a_3 = -3\left(\dfrac{1}{2}\right)^3 = -\dfrac{3}{8},\; a_4 = -3\left(\dfrac{1}{2}\right)^4 = -\dfrac{3}{16}$

5. $r = \dfrac{\left(\dfrac{2^{n+1-1}}{4}\right)}{\left(\dfrac{2^{n-1}}{4}\right)} = \dfrac{2^n}{2^{n-1}} = 2^{n-(n-1)} = 2$

$a_1 = \dfrac{2^{1-1}}{4} = \dfrac{2^0}{2^2} = 2^{-2} = \dfrac{1}{4},\; a_2 = \dfrac{2^{2-1}}{4} = \dfrac{2^1}{2^2} = 2^{-1} = \dfrac{1}{2},\; a_3 = \dfrac{2^{3-1}}{4} = \dfrac{2^2}{2^2} = 1,$

$a_4 = \dfrac{2^{4-1}}{4} = \dfrac{2^3}{2^2} = 2$

7. $r = \dfrac{2^{\left(\frac{n+1}{3}\right)}}{2^{\left(\frac{n}{3}\right)}} = 2^{\left(\frac{n+1}{3} - \frac{n}{3}\right)} = 2^{1/3}$; $\;a_1 = 2^{1/3},\; a_2 = 2^{2/3},\; a_3 = 2^{3/3} = 2,\; a_4 = 2^{4/3}$

9. $r = \dfrac{\left(\dfrac{3^{n+1-1}}{2^{n+1}}\right)}{\left(\dfrac{3^{n-1}}{2^n}\right)} = \dfrac{3^n}{3^{n-1}} \cdot \dfrac{2^n}{2^{n+1}} = 3^{n-(n-1)} \cdot 2^{n-(n+1)} = 3 \cdot 2^{-1} = \dfrac{3}{2}$

$a_1 = \dfrac{3^{1-1}}{2^1} = \dfrac{3^0}{2} = \dfrac{1}{2},\; a_2 = \dfrac{3^{2-1}}{2^2} = \dfrac{3^1}{2^2} = \dfrac{3}{4},\; a_3 = \dfrac{3^{3-1}}{2^3} = \dfrac{3^2}{2^3} = \dfrac{9}{8},$

$a_4 = \dfrac{3^{4-1}}{2^4} = \dfrac{3^3}{2^4} = \dfrac{27}{16}$

11. $\{n+2\}$ Arithmetic

$d = (n+1+2) - (n+2) = n+3-n-2 = 1$

13. $\{4n^2\}$ Examine the terms of the sequence: 4, 16, 36, 64, 100, ...

There is no common difference; there is no common ratio; neither.

15. $\left\{3 - \frac{2}{3}n\right\}$ Arithmetic

$d = \left(3 - \frac{2}{3}(n+1)\right) - \left(3 - \frac{2}{3}n\right) = 3 - \frac{2}{3}n - \frac{2}{3} - 3 + \frac{2}{3}n = -\frac{2}{3}$

17. 1, 3, 6, 10, ...   Neither

There is no common difference or common ratio.

19. $\left\{\left(\frac{2}{3}\right)^n\right\}$ Geometric

$r = \dfrac{\left(\frac{2}{3}\right)^{n+1}}{\left(\frac{2}{3}\right)^n} = \left(\frac{2}{3}\right)^{n+1-n} = \frac{2}{3}$

21. $-1, -2, -4, -8, ...$ Geometric $r = \dfrac{-2}{-1} = \dfrac{-4}{-2} = \dfrac{-8}{-4} = 2$

23. $\{3^{n/2}\}$ Geometric

$r = \dfrac{3^{\left(\frac{n+1}{2}\right)}}{3^{\left(\frac{n}{2}\right)}} = 3^{\left(\frac{n+1}{2} - \frac{n}{2}\right)} = 3^{1/2}$

25. $a_5 = 2 \cdot 3^{5-1} = 2 \cdot 3^4 = 2 \cdot 81 = 162$ $a_n = 2 \cdot 3^{n-1}$

27. $a_5 = 5(-1)^{5-1} = 5(-1)^4 = 5 \cdot 1 = 5$ $a_n = 5 \cdot (-1)^{n-1}$

29. $a_5 = 0 \cdot \left(\frac{1}{2}\right)^{5-1} = 0 \cdot \left(\frac{1}{2}\right)^4 = 0$ $a_n = 0 \cdot \left(\frac{1}{2}\right)^{n-1} = 0$

31. $a_5 = \sqrt{2} \cdot \left(\sqrt{2}\right)^{5-1} = \sqrt{2} \cdot \left(\sqrt{2}\right)^4 = \sqrt{2} \cdot 4 = 4\sqrt{2}$ $a_n = \sqrt{2} \cdot \left(\sqrt{2}\right)^{n-1} = \left(\sqrt{2}\right)^n$

33. $a = 1, \ r = \frac{1}{2}, \ n = 7$ $a_7 = 1 \cdot \left(\frac{1}{2}\right)^{7-1} = \left(\frac{1}{2}\right)^6 = \frac{1}{64}$

35. $a = 1, \ r = -1, \ n = 9$ $a_7 = 1 \cdot (-1)^{9-1} = (-1)^8 = 1$

37. $a = 0.4, \ r = 0.1, \ n = 8$ $a_7 = 0.4 \cdot (0.1)^{8-1} = 0.4(0.1)^7 = 0.00000004$

39. $a = \dfrac{1}{4},\ r = 2;\quad S_n = a\left(\dfrac{1-r^n}{1-r}\right) = \dfrac{1}{4}\left(\dfrac{1-2^n}{1-2}\right) = -\dfrac{1}{4}\left(1-2^n\right)$

41. $a = \dfrac{2}{3},\ r = \dfrac{2}{3};\quad S_n = a\left(\dfrac{1-r^n}{1-r}\right) = \dfrac{2}{3}\left(\dfrac{1-\left(\dfrac{2}{3}\right)^n}{1-\dfrac{2}{3}}\right) = \dfrac{2}{3}\left(\dfrac{1-\left(\dfrac{2}{3}\right)^n}{\dfrac{1}{3}}\right) = 2\left(1-\left(\dfrac{2}{3}\right)^n\right)$

43. $a = -1,\ r = 2;\quad S_n = a\left(\dfrac{1-r^n}{1-r}\right) = -1\left(\dfrac{1-2^n}{1-2}\right) = 1-2^n$

45. Using the sum of the sequence feature:

```
sum(seq(2^N/4,N,
0,14,1))
          8191.75
```

47. Using the sum of the sequence feature:

```
sum(seq((2/3)^N,
N,1,15,1))
        1.995432683
```

49. Using the sum of the sequence feature:

```
sum(seq(-1*2^N,N
,0,14,1))
           -32767
```

51. $a = 1,\ r = \dfrac{1}{3}\quad$ Since $|r| < 1,\ S_n = \dfrac{a}{1-r} = \dfrac{1}{\left(1-\dfrac{1}{3}\right)} = \dfrac{1}{\left(\dfrac{2}{3}\right)} = \dfrac{3}{2}$

53. $a = 8,\ r = \dfrac{1}{2}\quad$ Since $|r| < 1,\ S_n = \dfrac{a}{1-r} = \dfrac{8}{\left(1-\dfrac{1}{2}\right)} = \dfrac{8}{\left(\dfrac{1}{2}\right)} = 16$

55. $a = 2,\ r = -\dfrac{1}{4}\quad$ Since $|r| < 1,\ S_n = \dfrac{a}{1-r} = \dfrac{2}{\left(1-\left(-\dfrac{1}{4}\right)\right)} = \dfrac{2}{\left(\dfrac{5}{4}\right)} = \dfrac{8}{5}$

57. $a = 5$, $r = \dfrac{1}{4}$ Since $|r| < 1$, $S_n = \dfrac{a}{1-r} = \dfrac{5}{\left(1-\dfrac{1}{4}\right)} = \dfrac{5}{\left(\dfrac{3}{4}\right)} = \dfrac{20}{3}$

59. $a = 6$, $r = -\dfrac{2}{3}$ Since $|r| < 1$, $S_n = \dfrac{a}{1-r} = \dfrac{6}{\left(1-\left(-\dfrac{2}{3}\right)\right)} = \dfrac{6}{\left(\dfrac{5}{3}\right)} = \dfrac{18}{5}$

61. Find the common ratio of the terms and solve the system of equations:

$$\dfrac{x+2}{x} = r$$

$$\dfrac{x+3}{x+2} = r$$

$$\dfrac{x+2}{x} = \dfrac{x+3}{x+2} \rightarrow x^2 + 4x + 4 = x^2 + 3x \rightarrow x = -4$$

63. This is a geometric series with $a = \$18,000$, $r = 1.05$, $n = 5$. Find the 5th term:

$$a_5 = 18000(1.05)^{5-1} = 18000(1.05)^4 = \$21,879.11$$

65. (a) Find the 10th term of the geometric sequence:

$a = 2$, $r = 0.9$, $n = 10$ $a_{10} = 2(0.9)^{10-1} = 2(0.9)^9 = 0.775$ feet

(b) Find n when $a_n < 1$:

$$2(0.9)^{n-1} < 1 \rightarrow (0.9)^{n-1} < 0.5$$

$$(n-1)\log 0.9 < \log 0.5 \rightarrow n-1 > \dfrac{\log 0.5}{\log 0.9} \rightarrow n > \dfrac{\log 0.5}{\log 0.9} + 1 = 7.58$$

On the 8th swing the arc is less than 1 foot.

(c) Find the sum of the first 15 swings:

$$S_{15} = 2\left(\dfrac{1-(0.9)^{15}}{1-0.9}\right) = 2\left(\dfrac{1-(0.9)^{15}}{0.1}\right) = 20\left(1-(0.9)^{15}\right) = 15.88 \text{ feet}$$

(d) Find the infinite sum of the geometric series:

$$S = \dfrac{2}{1-0.9} = \dfrac{2}{0.1} = 20 \text{ feet}$$

67. This is an ordinary annuity with $P = \$100$ and $n = (12)(30) = 360$ payment periods.

The interest rate per period is $\dfrac{.12}{12} = .01$. Thus,

$$A = 100\left[\dfrac{\left[1+\dfrac{.12}{12}\right]^{360}-1}{\left(\dfrac{.12}{12}\right)}\right] \approx \$349496.41$$

69. This is an ordinary annuity with $P = \$500$ and $n = (4)(20) = 80$ payment periods. The interest rate per period is $\dfrac{.08}{4} = .02$. Thus,

$$A = 500\left[\dfrac{\left[1 + \dfrac{.08}{4}\right]^{80} - 1}{\left(\dfrac{.08}{4}\right)}\right] \approx \$96885.98$$

71. This is an ordinary annuity with $A = \$50000$ and $n = (12)(10) = 120$ payment periods. The interest rate per period is $\dfrac{.06}{12} = .005$. Thus,

$$50000 = P\left[\dfrac{\left[1 + \dfrac{.06}{12}\right]^{120} - 1}{\left(\dfrac{.06}{12}\right)}\right] \longrightarrow P = 50000\left[\dfrac{\left(\dfrac{.06}{12}\right)}{\left[1 + \dfrac{.06}{12}\right]^{120} - 1}\right] \approx \$305.10$$

73. Both options are geometric sequences:

Option A: $a = \$20,000;\quad r = 1.06;\quad n = 5$
$$a_5 = 20,000(1.06)^{5-1} = 20,000(1.06)^4 \approx \$25,250$$
$$S_5 = 20000\left(\dfrac{1 - (1.06)^5}{1 - 1.06}\right) \approx \$112,742$$

Option B: $a = \$22,000;\quad r = 1.03;\quad n = 5$
$$a_5 = 22,000(1.03)^{5-1} = 22,000(1.03)^4 \approx \$24,761$$
$$S_5 = 22000\left(\dfrac{1 - (1.03)^5}{1 - 1.03}\right) \approx \$116,801$$

Option A provides more money in the 5th year, while Option B provides the greatest total amount of money over the 5 year period.

75. Option 1: Total Salary $= \$2,000,000(7) + \$100,000(7) = \$14,700,000$
    Option 2: Geometric series with: $a = \$2,000,000,\quad r = 1.045,\quad n = 7$
    Find the sum of the geometric series:
$$S = 2,000,000\left(\dfrac{1 - (1.045)^7}{1 - 1.045}\right) \approx \$16,038,304$$
    Option 3: Arithmetic series with: $a = \$2,000,000,\quad d = \$95,000,\quad n = 7$
    Find the sum of the arithmetic series:
$$S_7 = \dfrac{7}{2}(2(2,000,000) + (7 - 1)(95,000)) = \$15,995,000$$

Option 2 provides the most money;　Option 1 provides the least money.

77. This is a geometric sequence with $a = 1$, $r = 2$, $n = 64$.
     Find the sum of the geometric series:

$$S_{64} = 1\left(\frac{1 - 2^{64}}{1 - 2}\right) = \frac{1 - 2^{64}}{-1} = 2^{64} - 1 = 1.845 \times 10^{19} \text{ grains}$$

79. The common ratio, $r = 0.90 < 1$. The sum is: $S = \dfrac{1}{1 - 0.9} = \dfrac{1}{0.10} = 10$.
     The multiplier is 10.

81. This is an infinite geometric series with $a = 4$, and $r = \dfrac{1.03}{1.09}$.

     Find the sum: Price $= \dfrac{4}{\left(1 - \dfrac{1.03}{1.09}\right)} \approx \$72.67$.

83. – 85. Answers will vary.

Chapter 11

Sequences; Induction; The Binomial Theorem

11.4 Mathematical Induction

1. I: $n = 1$: $2 \cdot 1 = 2$ and $1(1+1) = 2$
    II: If $2 + 4 + 6 + \cdots + 2k = k(k+1)$
    then $2 + 4 + 6 + \cdots + 2k + 2(k+1)$
    $$= \left[2 + 4 + 6 + \cdots + 2k\right] + 2(k+1) = k(k+1) + 2(k+1)$$
    $$= (k+1)(k+2)$$
    Conditions I and II are satisfied; the statement is true.

3. I: $n = 1$: $1 + 2 = 3$ and $\dfrac{1}{2} \cdot 1(1+5) = 3$
    II: If $3 + 4 + 5 + \cdots + (k+2) = \dfrac{1}{2} \cdot k(k+5)$
    then $3 + 4 + 5 + \cdots + (k+2) + [(k+1)+2]$
    $$= \left[3 + 4 + 5 + \cdots + (k+2)\right] + (k+3) = \dfrac{1}{2} \cdot k(k+5) + (k+3)$$
    $$= \dfrac{1}{2}k^2 + \dfrac{5}{2}k + k + 3 = \dfrac{1}{2}k^2 + \dfrac{7}{2}k + 3 = \dfrac{1}{2} \cdot \left(k^2 + 7k + 6\right)$$
    $$= \dfrac{1}{2} \cdot (k+1)(k+6)$$
    Conditions I and II are satisfied; the statement is true.

5. I: $n = 1$: $3 \cdot 1 - 1 = 2$ and $\dfrac{1}{2} \cdot 1(3 \cdot 1 + 1) = 2$
    II: If $2 + 5 + 8 + \cdots + (3k - 1) = \dfrac{1}{2} \cdot k(3k+1)$
    then $2 + 5 + 8 + \cdots + (3k - 1) + [3(k+1) - 1]$
    $$= \left[2 + 5 + 8 + \cdots + (3k - 1)\right] + (3k+2) = \dfrac{1}{2} \cdot k(3k+1) + (3k+2)$$
    $$= \dfrac{3}{2}k^2 + \dfrac{1}{2}k + 3k + 2 = \dfrac{3}{2}k^2 + \dfrac{7}{2}k + 2 = \dfrac{1}{2} \cdot \left(3k^2 + 7k + 4\right)$$
    $$= \dfrac{1}{2} \cdot (k+1)(3k+4)$$
    Conditions I and II are satisfied; the statement is true.

7. I: $n = 1$: $2^{1-1} = 1$ and $2^1 - 1 = 1$

II: If $1 + 2 + 2^2 + \cdots + 2^{k-1} = 2^k - 1$

then $1 + 2 + 2^2 + \cdots + 2^{k-1} + 2^{k+1-1}$

$$= \left[1 + 2 + 2^2 + \cdots + 2^{k-1}\right] + 2^k = 2^k - 1 + 2^k$$

$$= 2 \cdot 2^k - 1 = 2^{k+1} - 1$$

Conditions I and II are satisfied; the statement is true.

9. I: $n = 1$: $4^{1-1} = 1$ and $\dfrac{1}{3} \cdot \left(4^1 - 1\right) = 1$

II: If $1 + 4 + 4^2 + \cdots + 4^{k-1} = \dfrac{1}{3} \cdot \left(4^k - 1\right)$

then $1 + 4 + 4^2 + \cdots + 4^{k-1} + 4^{k+1-1}$

$$= \left[1 + 4 + 4^2 + \cdots + 4^{k-1}\right] + 4^k = \dfrac{1}{3} \cdot \left(4^k - 1\right) + 4^k$$

$$= \dfrac{1}{3} \cdot 4^k - \dfrac{1}{3} + 4^k = \dfrac{4}{3} \cdot 4^k - \dfrac{1}{3} = \dfrac{1}{3}\left(4 \cdot 4^k - 1\right) = \dfrac{1}{3} \cdot \left(4^{k+1} - 1\right)$$

Conditions I and II are satisfied; the statement is true.

11. I: $n = 1$: $\dfrac{1}{1(1+1)} = \dfrac{1}{2}$ and $\dfrac{1}{1+1} = \dfrac{1}{2}$

II: If $\dfrac{1}{1\cdot2} + \dfrac{1}{2\cdot3} + \dfrac{1}{3\cdot4} + \cdots + \dfrac{1}{k(k+1)} = \dfrac{k}{k+1}$

then $\dfrac{1}{1\cdot2} + \dfrac{1}{2\cdot3} + \dfrac{1}{3\cdot4} + \cdots + \dfrac{1}{k(k+1)} + \dfrac{1}{(k+1)(k+1+1)}$

$$= \left[\dfrac{1}{1\cdot2} + \dfrac{1}{2\cdot3} + \dfrac{1}{3\cdot4} + \cdots + \dfrac{1}{k(k+1)}\right] + \dfrac{1}{(k+1)(k+2)}$$

$$= \dfrac{k}{k+1} + \dfrac{1}{(k+1)(k+2)} = \dfrac{k}{k+1} \cdot \dfrac{k+2}{k+2} + \dfrac{1}{(k+1)(k+2)}$$

$$= \dfrac{k^2 + 2k + 1}{(k+1)(k+2)} = \dfrac{(k+1)(k+1)}{(k+1)(k+2)} = \dfrac{k+1}{k+2}$$

Conditions I and II are satisfied; the statement is true.

13. I: $n = 1$: $1^2 = 1$ and $\dfrac{1}{6} \cdot 1(1+1)(2\cdot1+1) = 1$

II: If $1^2 + 2^2 + 3^2 + \cdots + k^2 = \dfrac{1}{6} \cdot k(k+1)(2k+1)$

then $1^2 + 2^2 + 3^2 + \cdots + k^2 + (k+1)^2$

$$= \left[1^2 + 2^2 + 3^2 + \cdots + k^2\right] + (k+1)^2 = \dfrac{1}{6}k(k+1)(2k+1) + (k+1)^2$$

$$= (k+1)\left[\dfrac{1}{6}k(2k+1) + k + 1\right] = (k+1)\left[\dfrac{1}{3}k^2 + \dfrac{1}{6}k + k + 1\right]$$

$$= (k+1)\left[\frac{1}{3}k^2 + \frac{7}{6}k + 1\right] = \frac{1}{6}(k+1)\left[2k^2 + 7k + 6\right]$$

$$= \frac{1}{6} \cdot (k+1)(k+2)(2k+3)$$

Conditions I and II are satisfied; the statement is true.

15. I: $n = 1$: $5 - 1 = 4$ and $\frac{1}{2} \cdot 1(9 - 1) = 4$

II: If $4 + 3 + 2 + \cdots + (5 - k) = \frac{1}{2} \cdot k(9 - k)$

then $4 + 3 + 2 + \cdots + (5 - k) + (5 - (k + 1))$

$$= [4 + 3 + 2 + \cdots + (5 - k)] + (4 - k) = \frac{1}{2}k(9 - k) + (4 - k)$$

$$= \frac{9}{2}k - \frac{1}{2}k^2 + 4 - k = -\frac{1}{2}k^2 + \frac{7}{2}k + 4 = -\frac{1}{2} \cdot [k^2 - 7k - 8]$$

$$= -\frac{1}{2} \cdot (k+1)(k - 8) = \frac{1}{2} \cdot (k+1)(8 - k) = \frac{1}{2} \cdot (k+1)[9 - (k+1)]$$

Conditions I and II are satisfied; the statement is true.

17. I: $n = 1$: $1(1 + 1) = 2$ and $\frac{1}{3} \cdot 1(1 + 1)(1 + 2) = 2$

II: If $1 \cdot 2 + 2 \cdot 3 + 3 \cdot 4 + \cdots + k(k + 1) = \frac{1}{3} \cdot k(k + 1)(k + 2)$

then $1 \cdot 2 + 2 \cdot 3 + 3 \cdot 4 + \cdots + k(k + 1) + (k + 1)(k + 1 + 1)$

$$= [1 \cdot 2 + 2 \cdot 3 + 3 \cdot 4 + \cdots + k(k + 1)] + (k + 1)(k + 2)$$

$$= \frac{1}{3} \cdot k(k + 1)(k + 2) + (k + 1)(k + 2) = (k + 1)(k + 2)\left[\frac{1}{3}k + 1\right]$$

$$= \frac{1}{3} \cdot (k + 1)(k + 2)(k + 3)$$

Conditions I and II are satisfied; the statement is true.

19. I: $n = 1$: $1^2 + 1 = 2$ is divisible by 2

II: If $k^2 + k$ is divisible by 2

then $(k + 1)^2 + (k + 1) = k^2 + 2k + 1 + k + 1 = (k^2 + k) + (2k + 2)$

Since $k^2 + k$ is divisible by 2 and $2k + 2$ is divisible by 2, then $(k + 1)^2 + (k + 1)$ is divisible by 2.

Conditions I and II are satisfied; the statement is true.

21. I: $n = 1$: $1^2 - 1 + 2 = 2$ is divisible by 2

II: If $k^2 - k + 2$ is divisible by 2

then $(k + 1)^2 - (k + 1) + 2 = k^2 + 2k + 1 - k - 1 + 2 = (k^2 - k + 2) + (2k)$

Since $k^2 - k + 2$ is divisible by 2 and $2k$ is divisible by 2, then

$(k + 1)^2 - (k + 1) + 2$ is divisible by 2.

Conditions I and II are satisfied; the statement is true.

23. I: $n = 1$: If $x > 1$ then $x^1 = x > 1$.

II: Assume, for any natural number k, that if $x > 1$, then $x^k > 1$.
Show that if $x^k > 1$, then $x^{k+1} > 1$:
$$x^{k+1} = x^k \cdot x > 1 \cdot x = x > 1$$
$$\uparrow$$
$$(x^k > 1)$$

Conditions I and II are satisfied; the statement is true.

25. I: $n = 1$: $a - b$ is a factor of $a^1 - b^1 = a - b$.

II: If $a - b$ is a factor of $a^k - b^k$
Show that $a - b$ is a factor of $a^{k+1} - b^{k+1} = a \cdot a^k - b \cdot b^k$
$$= a \cdot a^k - a \cdot b^k + a \cdot b^k - b \cdot b^k = a(a^k - b^k) + b^k(a - b)$$

Since $a - b$ is a factor of $a^k - b^k$ and $a - b$ is a factor of $a - b$, then
$a - b$ is a factor of $a^{k+1} - b^{k+1}$.

Conditions I and II are satisfied; the statement is true.

27. $n = 1$: $1^2 - 1 + 41 = 41$ is a prime number.
$n = 41$: $41^2 - 41 + 41 = 41^2$ is not a prime number.

29. I: $n = 1$: $ar^{1-1} = a$ and $a\left(\dfrac{1 - r^1}{1 - r}\right) = a$

II: If $a + ar + ar^2 + \cdots + ar^{k-1} = a\left(\dfrac{1 - r^k}{1 - r}\right)$

then $a + ar + ar^2 + \cdots + ar^{k-1} + ar^{k+1-1}$

$$= \left[a + ar + ar^2 + \cdots + ar^{k-1}\right] + ar^k = a\left(\dfrac{1 - r^k}{1 - r}\right) + ar^k$$

$$= \frac{a(1 - r^k) + ar^k(1 - r)}{1 - r} = \frac{a - ar^k + ar^k - ar^{k+1}}{1 - r} = a\left(\frac{1 - r^{k+1}}{1 - r}\right)$$

Conditions I and II are satisfied; the statement is true.

31. I: $n = 4$: The number of diagonals of a quadrilateral is $\dfrac{1}{2} \cdot 4(4 - 3) = 2$

II: Assume that for any integer k the number of diagonals of a convex polygon
with k sides (k vertices) is $\dfrac{1}{2} \cdot k(k - 3)$. A convex polygon with $k + 1$ sides

($k + 1$ vertices) consists of a convex polygon with k sides (k vertices) plus
a triangle for a total of $k + 1$ vertices. The number of diagonals of this
convex polygon consists of the original ones plus $k - 1$ additional ones,

namely, $\dfrac{1}{2} \cdot k(k - 3) + (k - 1) = \dfrac{1}{2}k^2 - \dfrac{3}{2}k + k - 1 = \dfrac{1}{2}k^2 - \dfrac{1}{2}k - 1$

$$= \frac{1}{2} \cdot \left(k^2 - k - 2\right) = \frac{1}{2} \cdot (k+1)(k-2)$$

Conditions I and II are satisfied; the statement is true.

33. Answers will vary.

Chapter 11

Sequences; Induction; The Binomial Theorem

11.5 The Binomial Theorem

1. $\dbinom{5}{3} = \dfrac{5!}{3!\,2!} = \dfrac{5 \cdot 4 \cdot 3 \cdot 2 \cdot 1}{3 \cdot 2 \cdot 1 \cdot 2 \cdot 1} = \dfrac{5 \cdot 4}{2 \cdot 1} = 10$

3. $\dbinom{7}{5} = \dfrac{7!}{5!\,2!} = \dfrac{7 \cdot 6 \cdot 5 \cdot 4 \cdot 3 \cdot 2 \cdot 1}{5 \cdot 4 \cdot 3 \cdot 2 \cdot 1 \cdot 2 \cdot 1} = \dfrac{7 \cdot 6}{2 \cdot 1} = 21$

5. $\dbinom{50}{49} = \dfrac{50!}{49!\,1!} = \dfrac{50 \cdot 49!}{49! \cdot 1} = \dfrac{50}{1} = 50$

7. $\dbinom{1000}{1000} = \dfrac{1000!}{1000!\,0!} = \dfrac{1}{1} = 1$

9. $\dbinom{55}{23} = \dfrac{55!}{23!\,32!} = 1.866442159 \times 10^{15}$

11. $\dbinom{47}{25} = \dfrac{47!}{25!\,22!} = 1.483389769 \times 10^{13}$

13. $(x+1)^5 = \dbinom{5}{0}x^5 + \dbinom{5}{1}x^4 + \dbinom{5}{2}x^3 + \dbinom{5}{3}x^2 + \dbinom{5}{4}x^1 + \dbinom{5}{5}x^0$
$= x^5 + 5x^4 + 10x^3 + 10x^2 + 5x + 1$

15. $(x-2)^6 = \dbinom{6}{0}x^6 + \dbinom{6}{1}x^5(-2) + \dbinom{6}{2}x^4(-2)^2 + \dbinom{6}{3}x^3(-2)^3 + \dbinom{6}{4}x^2(-2)^4$
$+ \dbinom{6}{5}x(-2)^5 + \dbinom{6}{6}x^0(-2)^6$
$= x^6 + 6x^5(-2) + 15x^4 \cdot 4 + 20x^3(-8) + 15x^2 \cdot 16 + 6x \cdot (-32) + 64$
$= x^6 - 12x^5 + 60x^4 - 160x^3 + 240x - 192x + 64$

17. $(3x+1)^4 = \dbinom{4}{0}(3x)^4 + \dbinom{4}{1}(3x)^3 + \dbinom{4}{2}(3x)^2 + \dbinom{4}{3}(3x) + \dbinom{4}{4}$
$= 81x^4 + 4 \cdot 27x^3 + 6 \cdot 9x^2 + 4 \cdot 3x + 1 = 81x^4 + 108x^3 + 54x^2 + 12x + 1$

19. $(x^2 + y^2)^5 = \binom{5}{0}(x^2)^5(y^2)^0 + \binom{5}{1}(x^2)^4(y^2) + \binom{5}{2}(x^2)^3(y^2)^2 + \binom{5}{3}(x^2)^2(y^2)^3$
$$+ \binom{5}{4}x^2(y^2)^4 + \binom{5}{5}(y^2)^5$$
$$= x^{10} + 5x^8y^2 + 10x^6y^4 + 10x^4y^6 + 5x^2y^8 + y^{10}$$

21. $(\sqrt{x} + \sqrt{2})^6 = \binom{6}{0}(\sqrt{x})^6(\sqrt{2})^0 + \binom{6}{1}(\sqrt{x})^5(\sqrt{2})^1 + \binom{6}{2}(\sqrt{x})^4(\sqrt{2})^2 + \binom{6}{3}(\sqrt{x})^3(\sqrt{2})^3$
$$\binom{6}{4}(\sqrt{x})^2(\sqrt{2})^4 + \binom{6}{5}(\sqrt{x})(\sqrt{2})^5 + \binom{6}{6}(\sqrt{x})^0(\sqrt{2})^6$$
$$= x^3 + 6\sqrt{2}x^{5/2} + 15 \cdot 2x^2 + 20 \cdot 2\sqrt{2}x^{3/2} + 15 \cdot 4x + 6 \cdot 4\sqrt{2}x^{1/2} + 8$$
$$= x^3 + 6\sqrt{2}x^{5/2} + 30x^2 + 40\sqrt{2}x^{3/2} + 60x + 24\sqrt{2}x^{1/2} + 8$$

23. $(ax + by)^5 = \binom{5}{0}(ax)^5 + \binom{5}{1}(ax)^4 \cdot by + \binom{5}{2}(ax)^3(by)^2 + \binom{5}{3}(ax)^2(by)^3$
$$+ \binom{5}{4}ax(by)^4 + \binom{5}{5}(by)^5$$
$$= a^5x^5 + 5a^4x^4by + 10a^3x^3b^2y^2 + 10a^2x^2b^3y^3 + 5axb^4y^4 + b^5y^5$$

25. $n = 10, \ j = 4, \ x = x, \ a = 3$
$$\binom{10}{4}x^6 \cdot 3^4 = \frac{10!}{4!6!} \cdot 81x^6 = \frac{10 \cdot 9 \cdot 8 \cdot 7}{4 \cdot 3 \cdot 2 \cdot 1} \cdot 81x^6 = 17{,}010x^6$$
The coefficient of x^6 is 17,010.

27. $n = 12, \ j = 5, \ x = 2x, \ a = -1$
$$\binom{12}{5}(2x)^7 \cdot (-1)^5 = \frac{12!}{5!7!} \cdot 128x^7(-1) = \frac{12 \cdot 11 \cdot 10 \cdot 9 \cdot 8}{5 \cdot 4 \cdot 3 \cdot 2 \cdot 1} \cdot (-128)x^7 = -101{,}376x^7$$
The coefficient of x^7 is $-101{,}376$.

29. $n = 9, \ j = 2, \ x = 2x, \ a = 3$
$$\binom{9}{2}(2x)^7 \cdot 3^2 = \frac{9!}{2!7!} \cdot 128x^7(9) = \frac{9 \cdot 8}{2 \cdot 1} \cdot 128x^7 \cdot 9 = 41{,}472x^7$$
The coefficient of x^7 is $41{,}472$.

31. $n = 7, \ j = 4, \ x = x, \ a = 3$
$$\binom{7}{4}x^3 \cdot 3^4 = \frac{7!}{4!3!} \cdot 81x^3 = \frac{7 \cdot 6 \cdot 5}{3 \cdot 2 \cdot 1} \cdot 81x^3 = 2835x^3$$

33. $n = 9, \ j = 2, \ x = 3x, \ a = -2$
$$\binom{9}{2}(3x)^7 \cdot (-2)^2 = \frac{9!}{2!7!} \cdot 2187x^7 \cdot 4 = \frac{9 \cdot 8}{2 \cdot 1} \cdot 8748x^7 = 314{,}928x^7$$

35. The constant term in $\begin{pmatrix} 12 \\ j \end{pmatrix}(x^2)^{12-j}\left(\dfrac{1}{x}\right)$ occurs when:

$$2(12-j)=j \;\rightarrow\; 24-2j=j \;\rightarrow\; 3j=24 \;\rightarrow\; j=8.$$

Evaluate the 9th term:

$$\begin{pmatrix} 12 \\ 8 \end{pmatrix}(x^2)^4 \cdot \left(\dfrac{1}{x}\right)^8 = \dfrac{12!}{8!\,4!}x^8 \cdot \dfrac{1}{x^8} = \dfrac{12\cdot 11\cdot 10\cdot 9}{4\cdot 3\cdot 2\cdot 1}x^0 = 495$$

37. The x^4 term in $\begin{pmatrix} 10 \\ j \end{pmatrix}(x)^{10-j}\left(\dfrac{-2}{\sqrt{x}}\right)^{j}$ occurs when:

$$10-j-\dfrac{1}{2}j=4 \;\rightarrow\; -\dfrac{3}{2}j=-6 \;\rightarrow\; j=4.$$

Evaluate the 5th term:

$$\begin{pmatrix} 10 \\ 4 \end{pmatrix}(x)^6 \cdot \left(\dfrac{-2}{\sqrt{x}}\right)^4 = \dfrac{10!}{6!\,4!}x^6 \cdot \dfrac{16}{x^2} = \dfrac{10\cdot 9\cdot 8\cdot 7}{4\cdot 3\cdot 2\cdot 1}\cdot 16x^4 = 3360x^4$$

The coefficient is 3360.

39. $(1.001)^5 = \left(1+10^{-3}\right)^5 = \begin{pmatrix} 5 \\ 0 \end{pmatrix}\cdot 1^5 + \begin{pmatrix} 5 \\ 1 \end{pmatrix}\cdot 1^4\cdot 10^{-3} + \begin{pmatrix} 5 \\ 2 \end{pmatrix}\cdot 1^3\cdot \left(10^{-3}\right)^2 + \begin{pmatrix} 5 \\ 3 \end{pmatrix}\cdot 1^2\cdot \left(10^{-3}\right)^3 + \ldots$

$\qquad = 1 + 5(0.001) + 10(0.000001) + 10(0.000000001) + \ldots$

$\qquad = 1 + 0.005 + 0.000010 + 0.000000010 + \ldots$

$\qquad = 1.00501$ (correct to 5 decimal places)

41. $\begin{pmatrix} n \\ n-1 \end{pmatrix} = \dfrac{n!}{(n-1)!(n-(n-1))!} = \dfrac{n!}{(n-1)!(1)!} = n$

$\begin{pmatrix} n \\ n \end{pmatrix} = \dfrac{n!}{n!(n-n)!} = \dfrac{n!}{n!\,0!} = \dfrac{n!}{n!\cdot 1} = \dfrac{n!}{n!} = 1$

43. Show that $\begin{pmatrix} n \\ 0 \end{pmatrix} + \begin{pmatrix} n \\ 1 \end{pmatrix} + \ldots + \begin{pmatrix} n \\ n \end{pmatrix} = 2^n$

$$2^n = (1+1)^n$$

$$= \begin{pmatrix} n \\ 0 \end{pmatrix}\cdot 1^n + \begin{pmatrix} n \\ 1 \end{pmatrix}\cdot 1^{n-1}\cdot 1 + \begin{pmatrix} n \\ 2 \end{pmatrix}\cdot 1^{n-2}\cdot 1^2 + \ldots + \begin{pmatrix} n \\ n \end{pmatrix}\cdot 1^{n-n}\cdot 1^n$$

$$= \begin{pmatrix} n \\ 0 \end{pmatrix} + \begin{pmatrix} n \\ 1 \end{pmatrix} + \ldots + \begin{pmatrix} n \\ n \end{pmatrix}$$

45. $\begin{pmatrix} 5 \\ 0 \end{pmatrix}\left(\dfrac{1}{4}\right)^5 + \begin{pmatrix} 5 \\ 1 \end{pmatrix}\left(\dfrac{1}{4}\right)^4\left(\dfrac{3}{4}\right) + \begin{pmatrix} 5 \\ 2 \end{pmatrix}\left(\dfrac{1}{4}\right)^3\left(\dfrac{3}{4}\right)^2 + \begin{pmatrix} 5 \\ 3 \end{pmatrix}\left(\dfrac{1}{4}\right)^2\left(\dfrac{3}{4}\right)^3$

$\qquad + \begin{pmatrix} 5 \\ 4 \end{pmatrix}\left(\dfrac{1}{4}\right)\left(\dfrac{3}{4}\right)^4 + \begin{pmatrix} 5 \\ 5 \end{pmatrix}\left(\dfrac{3}{4}\right)^5 = \left(\dfrac{1}{4}+\dfrac{3}{4}\right)^5 = (1)^5 = 1$

Sequences; Induction; The Binomial Theorem

11.R Chapter Review

1. $a_1 = (-1)^1 \dfrac{1+3}{1+2} = -\dfrac{4}{3}, \quad a_2 = (-1)^2 \dfrac{2+3}{2+2} = \dfrac{5}{4}, \quad a_3 = (-1)^3 \dfrac{3+3}{3+2} = -\dfrac{6}{5},$

        $a_4 = (-1)^4 \dfrac{4+3}{4+2} = \dfrac{7}{6}, \quad a_5 = (-1)^5 \dfrac{5+3}{5+2} = -\dfrac{8}{7}$

3. $a_1 = \dfrac{2^1}{1^2} = \dfrac{2}{1} = 2, \quad a_2 = \dfrac{2^2}{2^2} = \dfrac{4}{4} = 1, \quad a_3 = \dfrac{2^3}{3^2} = \dfrac{8}{9}, \quad a_4 = \dfrac{2^4}{4^2} = \dfrac{16}{16} = 1, \quad a_5 = \dfrac{2^5}{5^2} = \dfrac{32}{25}$

5. $a_1 = 3, \quad a_2 = \dfrac{2}{3} \cdot 3 = 2, \quad a_3 = \dfrac{2}{3} \cdot 2 = \dfrac{4}{3}, \quad a_4 = \dfrac{2}{3} \cdot \dfrac{4}{3} = \dfrac{8}{9}, \quad a_5 = \dfrac{2}{3} \cdot \dfrac{8}{9} = \dfrac{16}{27}$

7. $a_1 = 2, \quad a_2 = 2 - 2 = 0, \quad a_3 = 2 - 0 = 2, \quad a_4 = 2 - 2 = 0, \quad a_5 = 2 - 0 = 2$

9. $\{n+5\}$ Arithmetic
    $d = (n+1+5) - (n+5) = n+6-n-5 = 1$

    $S_n = \dfrac{n}{2}[6+n+5] = \dfrac{n}{2}(n+11)$

11. $\{2n^3\}$ Examine the terms of the sequence: $2, 16, 54, 128, 250, \ldots$
    There is no common difference; there is no common ratio; neither.

13. $\{2^{3n}\}$ Geometric $r = \dfrac{2^{3(n+1)}}{2^{3n}} = \dfrac{2^{3n+3}}{2^{3n}} = 2^{3n+3-3n} = 2^3 = 8$

    $S_n = 8\left(\dfrac{1-8^n}{1-8}\right) = 8\left(\dfrac{1-8^n}{-7}\right) = \dfrac{8}{7}(8^n - 1)$

15. $0, 4, 8, 12, \ldots$ Arithmetic $d = 4 - 0 = 4$

    $S_n = \dfrac{n}{2}(2(0) + (n-1)4) = \dfrac{n}{2}(4(n-1)) = 2n(n-1)$

17. $3, \dfrac{3}{2}, \dfrac{3}{4}, \dfrac{3}{5}, \dfrac{3}{16}, \ldots$ Geometric

    $r = \dfrac{\left(\dfrac{3}{2}\right)}{3} = \dfrac{3}{2} \cdot \dfrac{1}{3} = \dfrac{1}{2}$

    $S_n = 3\left(\dfrac{1-\left(\dfrac{1}{2}\right)^n}{1-\dfrac{1}{2}}\right) = 3\left(\dfrac{1-\left(\dfrac{1}{2}\right)^n}{\left(\dfrac{1}{2}\right)}\right) = 6\left(1-\left(\dfrac{1}{2}\right)^n\right)$

19. Neither.　　There is no common difference or common ratio.

21. $\displaystyle\sum_{k=1}^{5}(k^2+12)=13+16+21+28+37=115$

23. $\displaystyle\sum_{k=1}^{10}(3k-9)=\sum_{k=1}^{10}3k-\sum_{k=1}^{10}9=3\sum_{k=1}^{10}k-\sum_{k=1}^{10}9=3\left(\frac{10(10+1)}{2}\right)-10(9)=165-90=75$

25. $\displaystyle\sum_{k=1}^{7}\left(\tfrac{1}{3}\right)^k=\frac{1}{3}\left(\frac{1-\left(\frac{1}{3}\right)^7}{1-\frac{1}{3}}\right)=\frac{1}{3}\left(\frac{1-\left(\frac{1}{3}\right)^7}{\frac{2}{3}}\right)=\frac{1}{2}\left(1-\frac{1}{2187}\right)=\frac{1}{2}\cdot\frac{2186}{2187}=\frac{1093}{2187}$

27. Arithmetic;　$a_1=3,\ d=4,\ a_n=a+(n-1)d$
$a_9=3+(9-1)4=3+8(4)=3+32=35$

29. Geometric;　$a=1,\ r=\dfrac{1}{10},\ n=11$　　$a_{11}=1\cdot\left(\dfrac{1}{10}\right)^{11-1}=\left(\dfrac{1}{10}\right)^{10}=\dfrac{1}{10,000,000,000}$

31. Arithmetic;　$a_1=\sqrt{2},\ d=\sqrt{2},\ n=9,\ a_n=a+(n-1)d$
$a_9=\sqrt{2}+(9-1)\sqrt{2}=\sqrt{2}+8\sqrt{2}=9\sqrt{2}$

33. $a_7=a+6d=31\qquad a_{20}=a+19d=96$
Solve the system of equations:
$31-6d+19d=96$
$13d=65$
$d=5$　　　　　　General formula:　$\{5n-4\}$
$a=31-6(5)=31-30=1$

35. $a_{10}=a+9d=0\qquad a_{18}=a+17d=8$
Solve the system of equations:
$-9d+17d=8$
$8d=8$
$d=1$　　　　　　General formula:　$\{n-10\}$
$a=-9(1)=-9$

37.
$$a=3,\ r=\frac{1}{3}\quad \text{Since } |r|<1,\ S_n=\frac{a}{1-r}=\frac{3}{\left(1-\frac{1}{3}\right)}=\frac{3}{\left(\frac{2}{3}\right)}=\frac{9}{2}$$

39.
$$a=2,\ r=-\frac{1}{2}\quad \text{Since } |r|<1,\ S_n=\frac{a}{1-r}=\frac{2}{\left(1-\left(-\frac{1}{2}\right)\right)}=\frac{2}{\left(\frac{3}{2}\right)}=\frac{4}{3}$$

41.

$$a = 4, \ r = \frac{1}{2} \quad \text{Since} \ |r| < 1, \ S_n = \frac{a}{1-r} = \frac{4}{\left(1 - \frac{1}{2}\right)} = \frac{4}{\left(\frac{1}{2}\right)} = 8$$

43. I: $\quad n = 1$: $\ 3 \cdot 1 = 3$ and $\dfrac{3 \cdot 1}{2}(1+1) = 3$

II: $\quad$ If $\quad 3 + 6 + 9 + \cdots + 3k = \dfrac{3k}{2}(k+1)$

$\quad\quad$ then $\quad 3 + 6 + 9 + \cdots + 3k + 3(k+1)$

$$= \left[3 + 6 + 9 + \cdots + 3k\right] + 3(k+1) = \frac{3k}{2}(k+1) + 3(k+1)$$

$$= (k+1)\left(\frac{3k}{2} + 3\right) = \frac{3}{2}(k+1)(k+2)$$

Conditions I and II are satisfied; the statement is true.

45. I: $\quad n = 1$: $\ 2 \cdot 3^{1-1} = 2$ and $3^1 - 1 = 2$

II: $\quad$ If $\quad 2 + 6 + 18 + \cdots + 2 \cdot 3^{k-1} = 3^k - 1$

$\quad\quad$ then $\quad 2 + 6 + 18 + \cdots + 2 \cdot 3^{k-1} + 2 \cdot 3^{k+1-1}$

$$= \left[2 + 6 + 18 + \cdots + 2 \cdot 3^{k-1}\right] + 2 \cdot 3^k = 3^k - 1 + 2 \cdot 3^k$$

$$= 3 \cdot 3^k - 1 = 3^{k+1} - 1$$

Conditions I and II are satisfied; the statement is true.

47. I: $\quad n = 1$: $\ (3 \cdot 1 - 2)^2 = 1$ and $\dfrac{1}{2} \cdot 1(6 \cdot 1^2 - 3 \cdot 1 - 1) = 1$

II: $\quad$ If $\quad 1^2 + 4^2 + 7^2 + \cdots + (3k-2)^2 = \dfrac{1}{2} \cdot k\left(6k^2 - 3k - 1\right)$

$\quad\quad$ then $\quad 1^2 + 4^2 + 7^2 + \cdots + (3k-2)^2 + \left(3(k+1)-2\right)^2$

$$= \left[1^2 + 4^2 + 7^2 + \cdots + (3k-2)^2\right] + (3k+1)^2$$

$$= \frac{1}{2} \cdot k\left(6k^2 - 3k - 1\right) + (3k+1)^2$$

$$= \frac{1}{2} \cdot \left[6k^3 - 3k^2 - k + 18k^2 + 12k + 2\right] = \frac{1}{2} \cdot \left[6k^3 + 15k^2 + 11k + 2\right]$$

$$= \frac{1}{2} \cdot (k+1)\left[6k^2 + 9k + 2\right] = \frac{1}{2} \cdot (k+1)\left[6k^2 + 12k + 6 - 3k - 3 - 1\right]$$

$$= \frac{1}{2} \cdot (k+1)\left[6(k^2 + 2k + 1) - 3(k+1) - 1\right]$$

$$= \frac{1}{2} \cdot (k+1)\left[6(k+1)^2 - 3(k+1) - 1\right]$$

Conditions I and II are satisfied; the statement is true.

49. $(x+2)^5 = \binom{5}{0}x^5 + \binom{5}{1}x^4 \cdot 2 + \binom{5}{2}x^3 \cdot 2^2 + \binom{5}{3}x^2 \cdot 2^3 + \binom{5}{4}x^1 \cdot 2^4 + \binom{5}{5} \cdot 2^5$

$= x^5 + 5 \cdot 2x^4 + 10 \cdot 4x^3 + 10 \cdot 8x^2 + 5 \cdot 16x + 1 \cdot 32$

$= x^5 + 10x^4 + 40x^3 + 80x^2 + 80x + 32$

51. $(2x+3)^5 = \binom{5}{0}(2x)^5 + \binom{5}{1}(2x)^4 \cdot 3 + \binom{5}{2}(2x)^3 \cdot 3^2 + \binom{5}{3}(2x)^2 \cdot 3^3$

$+ \binom{5}{4}(2x)^1 \cdot 3^4 + \binom{5}{5} \cdot 3^5$

$= 32x^5 + 5 \cdot 16x^4 \cdot 3 + 10 \cdot 8x^3 \cdot 9 + 10 \cdot 4x^2 \cdot 27 + 5 \cdot 2x \cdot 81 + 1 \cdot 243$

$= 32x^5 + 240x^4 + 720x^3 + 1080x^2 + 810x + 243$

53. $n = 9, \; j = 2, \; x = x, \; a = 2$

$\binom{9}{2}x^7 \cdot 2^2 = \dfrac{9!}{2! \, 7!} \cdot 4x^7 = \dfrac{9 \cdot 8}{2 \cdot 1} \cdot 4x^7 = 144x^7$

The coefficient of x^7 is 144.

55. $n = 7, \; j = 5, \; x = 2x, \; a = 1$

$\binom{7}{5}(2x)^2 \cdot 1^5 = \dfrac{7!}{5! \, 2!} \cdot 4x^2(1) = \dfrac{7 \cdot 6}{2 \cdot 1} \cdot 4x^2 = 84x^2$

The coefficient of x^2 is 84.

57. This is an arithmetic sequence with $a = 80, \; d = -3, \; n = 25$

(a) $a_{25} = 80 + (25 - 1)(-3) = 80 - 72 = 8$ bricks

(b) $S_{25} = \dfrac{25}{2}(80 + 8) = 25(44) = 1100$ bricks

1100 bricks are needed to build the steps.

59. This is an ordinary annuity with $P = \$200$ and $n = (12)(20) = 240$ payment periods.

The interest rate per period is $\dfrac{.10}{12} = .008\overline{3}$. Thus,

$$A = 200 \left[\frac{\left[1 + \dfrac{.10}{12}\right]^{240} - 1}{\left(\dfrac{.10}{12}\right)} \right] = \$151873.77$$

61.   This is a geometric sequence with $a = 20, \; r = \dfrac{3}{4}$.

(a) After striking the ground the third time, the height is $20\left(\dfrac{3}{4}\right)^3 = \dfrac{135}{16} \approx 8.44$ feet.

(b) After striking the ground the nth time, the height is $20\left(\dfrac{3}{4}\right)^n$ feet.

(c)   If the height is less than 6 inches (i.e. 0.5 feet), then:
$$0.5 = 20\left(\frac{3}{4}\right)^n \rightarrow 0.025 = \left(\frac{3}{4}\right)^n \rightarrow \log 0.025 = n \log\left(\frac{3}{4}\right) \rightarrow n = \frac{\log 0.025}{\log\left(\frac{3}{4}\right)} \approx 12.82$$

The height is less than 6 inches after the 13th strike.

(d)   Since this is a geometric sequence with $|r| < 1$, the distance is the sum of the two infinite geometric series - the distances going down plus the distances going up.

Distance going down:
$$S_{down} = \frac{20}{\left(1-\frac{3}{4}\right)} = \frac{20}{\left(\frac{1}{4}\right)} = 80 \quad \text{feet.}$$

Distance going up:
$$S_{up} = \frac{15}{\left(1-\frac{3}{4}\right)} = \frac{15}{\left(\frac{1}{4}\right)} = 60 \quad \text{feet.}$$

The total distance traveled is 140 feet.

Counting and Probability

12.1 Sets and Counting

1. $A \cup B = \{1, 3, 5, 7, 9\} \cup \{1, 5, 6, 7\} = \{1, 3, 5, 6, 7, 9\}$

3. $A \cap B = \{1, 3, 5, 7, 9\} \cap \{1, 5, 6, 7\} = \{1, 5, 7\}$

5. $(A \cup B) \cap C = (\{1, 3, 5, 7, 9\} \cup \{1, 5, 6, 7\}) \cap \{1, 2, 4, 6, 8, 9\}$
$$= \{1, 3, 5, 6, 7, 9\} \cap \{1, 2, 4, 6, 8, 9\}$$
$$= \{1, 6, 9\}$$

7. $(A \cap B) \cup C = (\{1, 3, 5, 7, 9\} \cap \{1, 5, 6, 7\}) \cup \{1, 2, 4, 6, 8, 9\}$
$$= \{1, 5, 7\} \cup \{1, 2, 4, 6, 8, 9\}$$
$$= \{1, 2, 4, 5, 6, 7, 8, 9\}$$

9. $(A \cup C) \cap (B \cup C)$
$$= (\{1, 3, 5, 7, 9\} \cup \{1, 2, 4, 6, 8, 9\}) \cap (\{1, 5, 6, 7\} \cup \{1, 2, 4, 6, 8, 9\})$$
$$= \{1, 2, 3, 4, 5, 6, 7, 8, 9\} \cap \{1, 2, 4, 5, 6, 7, 8, 9\}$$
$$= \{1, 2, 4, 5, 6, 7, 8, 9\}$$

11. $\overline{A} = \{0, 2, 6, 7, 8\}$

13. $\overline{A \cap B} = \overline{\{1, 3, 4, 5, 9\} \cap \{2, 4, 6, 7, 8\}} = \overline{\{4\}} = \{0, 1, 2, 3, 5, 6, 7, 8, 9\}$

15. $\overline{A} \cup \overline{B} = \{0, 2, 6, 7, 8\} \cup \{0, 1, 3, 5, 9\} = \{0, 1, 2, 3, 5, 6, 7, 8, 9\}$

17. $\overline{A \cap \overline{C}} = \overline{\{1, 3, 4, 5, 9\} \cap \{0, 2, 5, 7, 8, 9\}} = \overline{\{5, 9\}} = \{0, 1, 2, 3, 4, 6, 7, 8\}$

19. $\overline{A \cup B \cup C} = \overline{\{1, 3, 4, 5, 9\} \cup \{2, 4, 6, 7, 8\} \cup \{1, 3, 4, 6\}}$
$$= \overline{\{1, 2, 3, 4, 5, 6, 7, 8, 9\}} = \{0\}$$

21. $\{a\}, \{b\}, \{c\}, \{d\}, \{a, b\}, \{a, c\}, \{a, d\}, \{b, c\}, \{b, d\}, \{c, d\}, \{a, b, c\}, \{a, b, d\},$
$\{a, c, d\}, \{b, c, d\}, \{a, b, c, d\}, \varnothing$

23. $n(A) = 15, n(B) = 20, n(A \cap B) = 10$
$$n(A \cup B) = n(A) + n(B) - n(A \cap B) = 15 + 20 - 10 = 25$$

25. $n(A \cup B) = 50, n(A \cap B) = 10, n(B) = 20$
$$n(A \cup B) = n(A) + n(B) - n(A \cap B)$$
$$50 = n(A) + 20 - 10$$
$$40 = n(A)$$

27. From the figure:
$$n(A) = 15 + 3 + 5 + 2 = 25$$

29. From the figure:
$$n(A \text{ or } B) = n(A \cup B) = n(A) + n(B) - n(A \cap B) = 25 + 20 - 8 = 37$$

31. From the figure:
$$n(A \text{ but not } C) = n(A) - n(A \cap C) = 25 - 7 = 18$$

33. From the figure:
$$n(A \text{ and } B \text{ and } C) = n(A \cap B \cap C) = 5$$

35. Let $A = \{$those who will purchase a major appliance$\}$
$B = \{$those who will buy a car$\}$
$n(U) = 500, \ n(A) = 200, \ n(B) = 150, \ n(A \cap B) = 25$
$n(A \cup B) = n(A) + n(B) - n(A \cap B) = 200 + 150 - 25 = 325$
$n(\text{purchase neither}) = 500 - 325 = 175$
$n(\text{purchase only a car}) = 150 - 25 = 125$

37. Construct a Venn diagram:

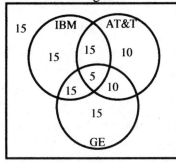

  (a)   15
  (b)   15
  (c)   15
  (d)   25
  (e)   40

39. (a) $n(\text{married}) = n(\text{married, spouse present}) + n(\text{married, spouse absent})$
$$= 54,654 + 3,232 = 57,886 \text{ thousand}$$
  (b) $n(\text{widowed or divorced}) = n(\text{widowed}) + n(\text{divorced})$
$$= 2,686 + 8,208 = 10,894 \text{ thousand}$$
  (c) $n(\text{married, spouse absent or widowed or divorced})$
$$= n(\text{married, spouse absent}) + n(\text{widowed}) + n(\text{divorced})$$
$$= 3,232 + 2,686 + 8,208 = 14,126 \text{ thousand}$$

41.   Answers will vary.

Chapter 12

Counting and Probability

12.2 Permutations and Combinations

1. $P(6, 2) = \dfrac{6!}{(6-2)!} = \dfrac{6!}{4!} = \dfrac{6 \cdot 5 \cdot 4!}{4!} = 30$

3. $P(4, 4) = \dfrac{4!}{(4-4)!} = \dfrac{4!}{0!} = \dfrac{4 \cdot 3 \cdot 2 \cdot 1}{1} = 24$

5. $P(7, 0) = \dfrac{7!}{(7-0)!} = \dfrac{7!}{7!} = 1$

7. $P(8, 4) = \dfrac{8!}{(8-4)!} = \dfrac{8!}{4!} = \dfrac{8 \cdot 7 \cdot 6 \cdot 5 \cdot 4!}{4!} = 1680$

9. $C(8, 2) = \dfrac{8!}{(8-2)!\,2!} = \dfrac{8!}{6!\,2!} = \dfrac{8 \cdot 7 \cdot 6!}{6! \cdot 2 \cdot 1} = 28$

11. $C(7, 4) = \dfrac{7!}{(7-4)!\,4!} = \dfrac{7!}{3!\,4!} = \dfrac{7 \cdot 6 \cdot 5 \cdot 4!}{4! \cdot 3 \cdot 2 \cdot 1} = 35$

13. $C(15, 15) = \dfrac{15!}{(15-15)!\,15!} = \dfrac{15!}{0!\,15!} = \dfrac{15!}{15! \cdot 1} = 1$

15. $C(26, 13) = \dfrac{26!}{(26-13)!\,13!} = \dfrac{26!}{13!\,13!} = 10{,}400{,}600$

17. {*abc, abd, abe, acb, acd, ace, adb, adc, ade, aeb, aec, aed, bac, bad, bae, bca, bcd, bce, bda, bdc, bde, bea, bec, bed, cab, cad, cae, cba, cbd, cbe, cda, cdb, cde, cea, ceb, ced, dab, dac, dae, dba, dbc, dbe, dca, dcb, dce, dea, deb, dec, eab, eac, ead, eba, ebc, ebd, eca, ecb, ecd, eda, edb, edc*}
    $$P(5,3) = \dfrac{5!}{(5-3)!} = \dfrac{5!}{2!} = \dfrac{5 \cdot 4 \cdot 3 \cdot 2!}{2!} = 60$$

19. {123, 124, 132, 134, 142, 143, 213, 214, 231, 234, 241, 243, 312, 314, 321, 324, 341, 342, 412, 413, 421, 423, 431, 432}
    $$P(4,3) = \dfrac{4!}{(4-3)!} = \dfrac{4!}{1!} = \dfrac{4 \cdot 3 \cdot 2 \cdot 1}{1} = 24$$

21. {*abc, abd, abe, acd, ace, ade, bcd, bce, bde, cde*}
    $$C(5,3) = \dfrac{5!}{(5-3)!\,3!} = \dfrac{5 \cdot 4 \cdot 3!}{2 \cdot 1 \cdot 3!} = 10$$

23. $\{123, 124, 134, 234\}$ $C(4,3) = \dfrac{4!}{(4-3)!\,3!} = \dfrac{4 \cdot 3!}{1!\,3!} = 4$

25.   There are 5 choices of shirts and 3 choices of ties; there are $(5)(3) = 15$ combinations.

27.   There are 4 choices for the first letter in the code and 4 choices for the second letter in the code; there are $(4)(4) = 16$ possible two-letter codes.

29.   There are two choices for each of three positions; there are $(2)(2)(2) = 8$ possible three-digit numbers.

31.   To line up the four people, there are 4 choices for the first position, 3 choices for the second position, 2 choices for the third position, and 1 choice for the fourth position. Thus there are $(4)(3)(2)(1) = 24$ possible ways four people can be lined up.

33.   Since no letter can be repeated, there are 5 choices for the first letter, 4 choices for the second letter, and 3 choices for the third letter. Thus, there are $(5)(4)(3) = 60$ possible three-letter codes.

35.   There are 26 possible one-letter names. There are $(26)(26) = 676$ possible two-letter names. There are $(26)(26)(26) = 17576$ possible three-letter names. Thus, there are $26 + 676 + 17576 = 18,278$ possible companies that can be listed on the New York Stock Exchange.

37.   A committee of 4 from a total of 7 students is given by:
$$C(7,4) = \dfrac{7!}{(7-4)!\,4!} = \dfrac{7!}{3!\,4!} = \dfrac{7 \cdot 6 \cdot 5 \cdot 4!}{3 \cdot 2 \cdot 1 \cdot 4!} = 35$$
35 committees are possible.

39.   There are 2 possible answers for each question. Therefore, there are $2^{10} = 1024$ different possible arrangements of the answers.

41.   There are 9 choices for the first digit, and 10 choices for each of the other three digits. Thus, there are $(9)(10)(10)(10) = 9000$ possible four-digit numbers.

43.   There are 5 choices for the first position, 4 choices for the second position, 3 choices for the third position, 2 choices for the fourth position, and 1 choice for the fifth position. Thus, there are $(5)(4)(3)(2)(1) = 120$ possible arrangements of the books.

45.   There are 8 choices for the DOW stocks, 15 choices for the NASDAQ stocks, and 4 choices for the global stocks. Thus, there are $(8)(15)(4) = 480$ different portfolios.

47.   The first person can have any of 365 days, the second person can have any of the remaining 364 days. Thus, there are $(365)(364) = 132,860$ possible ways two people can have different birthdays.

49. Choosing 2 boys from the 4 boys can be done C(4,2) ways, and choosing 3 girls from the 8 girls can be done in C(8,3) ways. Thus, there are a total of:

$$C(4,2) \cdot C(8,3) = \frac{4!}{(4-2)!\,2!} \cdot \frac{8!}{(8-3)!\,3!} = \frac{4!}{2!\,2!} \cdot \frac{8!}{5!\,3!}$$

$$= \frac{4 \cdot 3!}{2 \cdot 1 \cdot 2 \cdot 1} \cdot \frac{8 \cdot 7 \cdot 6 \cdot 5!}{5!\,3!} = 336$$

51. This is a permutation with repetition. There are $\frac{9!}{2!\,2!} = 90,720$ different words.

53. (a) $C(7,2) \cdot C(3,1) = 21 \cdot 3 = 63$

     (b) $C(7,3) = 35$

     (c) $C(3,3) = 1$

Chapter 12

Counting and Probability

12.3 Probability

1. Probabilities must be between 0 and 1, inclusive. Thus, 0, 0.01, 0.35, and 1 are probabilities.

3. All the probabilities are between 0 and 1.
The sum of the probabilities is $0.2 + 0.3 + 0.1 + 0.4 = 1$.
This is a probability model.

5. All the probabilities are between 0 and 1.
The sum of the probabilities is $0.3 + 0.2 + 0.1 + 0.3 = 0.9$.
This is not a probability model.

7. The sample space is: $S = \{HH, HT, TH, TT\}$.

Each outcome is equally likely to occur; so $P(E) = \dfrac{n(E)}{n(S)}$.

The probabilities are: $P(HH) = \dfrac{1}{4}$, $P(HT) = \dfrac{1}{4}$, $P(TH) = \dfrac{1}{4}$, $P(TT) = \dfrac{1}{4}$.

9. The sample space of tossing two fair coins and a fair die is:
$S = \{HH1, HH2, HH3, HH4, HH5, HH6, HT1, HT2, HT3, HT4, HT5,$
$HT6, TH1, TH2, TH3, TH4, TH5, TH6, TT1, TT2, TT3, TT4, TT5, TT6\}$

There are 24 equally likely outcomes and the probability of each is $\dfrac{1}{24}$.

11. The sample space for tossing three fair coins is:
$S = \{HHH, HHT, HTH, THH, HTT, THT, TTH, TTT\}$

There are 8 equally likely outcomes and the probability of each is $\dfrac{1}{8}$.

13. The sample space is:
$S = \{1$ Yellow, 1 Red, 1 Green, 2 Yellow, 2 Red, 2 Green, 3 Yellow, 3 Red,
3 Green, 4 Yellow, 4 Red, 4 Green$\}$

There are 12 equally likely events and the probability of each is $\dfrac{1}{12}$. The probability of

getting a 2 or 4 followed by a Red is $P(2 \text{ Red}) + P(4 \text{ Red}) = \dfrac{1}{12} + \dfrac{1}{12} = \dfrac{1}{6}$.

15. The sample space is:
S = {1 Yellow Forward, 1 Yellow Backward, 1 Red Forward, 1 Red Backward, 1 Green Forward, 1 Green Backward, 2 Yellow Forward, 2 Yellow Backward, 2 Red Forward, 2 Red Backward, 2 Green Forward, 2 Green Backward, 3 Yellow Forward, 3 Yellow Backward, 3 Red Forward, 3 Red Backward, 3 Green Forward, 3 Green Backward, 4 Yellow Forward, 4 Yellow Backward, 4 Red Forward, 4 Red Backward, 4 Green Forward, 4 Green Backward}

There are 24 equally likely events and the probability of each is $\frac{1}{24}$. The probability of getting a 1, followed by a Red or Green, followed by a Backward is

$P(1 \text{ Red Backward}) + P(1 \text{ Green Backward}) = \frac{1}{24} + \frac{1}{24} = \frac{1}{12}$.

17. The sample space is:
S = {1 1 Yellow, 1 1 Red, 1 1 Green, 1 2 Yellow, 1 2 Red, 1 2 Green, 1 3 Yellow, 1 3 Red, 1 3 Green, 1 4 Yellow, 1 4 Red, 1 4 Green, 2 1 Yellow, 2 1 Red, 2 1 Green, 2 2 Yellow, 2 2 Red, 2 2 Green, 2 3 Yellow, 2 3 Red, 2 3 Green, 2 4 Yellow, 2 4 Red, 2 4 Green, 3 1 Yellow, 3 1 Red, 3 1 Green, 3 2 Yellow, 3 2 Red, 3 2 Green, 3 3 Yellow, 3 3 Red, 3 3 Green, 3 4 Yellow, 3 4 Red, 3 4 Green, 4 1 Yellow, 4 1 Red, 4 1 Green, 4 2 Yellow, 4 2 Red, 4 2 Green, 4 3 Yellow, 4 3 Red, 4 3 Green, 4 4 Yellow, 4 4 Red, 4 4 Green}

There are 48 equally likely events and the probability of each is $\frac{1}{48}$. The probability of getting a 2, followed by a 2 or 4, followed by a Red or Green is

$P(2\ 2 \text{ Red}) + P(2\ 4 \text{ Red}) + P(2\ 2 \text{ Green}) + P(2\ 4 \text{ Green}) = \frac{1}{48} + \frac{1}{48} + \frac{1}{48} + \frac{1}{48} = \frac{1}{12}$

19. A, B, C, F   21.   B

23. Let $P(\text{tails}) = x$, then $P(\text{heads}) = 4x$
$$x + 4x = 1 \to 5x = 1 \to x = \frac{1}{5} \quad P(\text{tails}) = \frac{1}{5}, \quad P(\text{heads}) = \frac{4}{5}$$

25. $P(2) = P(4) = P(6) = x \qquad P(1) = P(3) = P(5) = 2x$
$P(1) + P(2) + P(3) + P(4) + P(5) + P(6) = 1$
$$2x + x + 2x + x + 2x + x = 1 \to 9x = 1 \to x = \frac{1}{9}$$
$P(2) = P(4) = P(6) = \frac{1}{9} \qquad P(1) = P(3) = P(5) = \frac{2}{9}$

27. $P(E) = \frac{n(E)}{n(S)} = \frac{n\{1,2,3\}}{10} = \frac{3}{10}$ 29. $P(E) = \frac{n(E)}{n(S)} = \frac{n\{2,4,6,8,10\}}{10} = \frac{5}{10} = \frac{1}{2}$

31. $P(\text{white}) = \frac{n(\text{white})}{n(S)} = \frac{5}{5+10+8+7} = \frac{5}{30} = \frac{1}{6}$

33. The sample space is: S = {BBB, BBG, BGB, GBB, BGG, GBG, GGB, GGG}
$P(3 \text{ boys}) = \frac{n(3 \text{ boys})}{n(S)} = \frac{1}{8}$

35. The sample space is:
S = {BBBB, BBBG, BBGB, BGBB, GBBB, BBGG, BGBG, GBBG, BGGB, GBGB, GGBB, BGGG, GBGG, GGBG, GGGB, GGGG}

$$P(1 \text{ girl}, \ 3 \text{ boys}) = \frac{n(1 \text{ girl}, \ 3 \text{ boys})}{n(S)} = \frac{4}{16} = \frac{1}{4}$$

37. $$P(\text{sum of two die is } 7) = \frac{n(\text{sum of two die is } 7)}{n(S)}$$
$$= \frac{n\{1,6 \text{ or } 2,5 \text{ or } 3,4 \text{ or } 4,3 \text{ or } 5,2 \text{ or } 6,1\}}{n(S)} = \frac{6}{36} = \frac{1}{6}$$

39. $$P(\text{sum of two die is } 3) = \frac{n(\text{sum of two die is } 3)}{n(S)} = \frac{n\{1,2 \text{ or } 2,1\}}{n(S)} = \frac{2}{36} = \frac{1}{18}$$

41. $$P(A \cup B) = P(A) + P(B) - P(A \cap B) = 0.25 + 0.45 - 0.15 = 0.55$$

43. $$P(A \cup B) = P(A) + P(B) = 0.25 + 0.45 = 0.70$$

45. $$P(A \cup B) = P(A) + P(B) - P(A \cap B)$$
$$0.85 = 0.60 + P(B) - 0.05$$
$$P(B) = 0.85 - 0.60 + 0.05 = 0.30$$

47. $$P(\text{not victim}) = 1 - P(\text{victim}) = 1 - 0.253 = 0.747$$

49. $$P(\text{not in } 70's) = 1 - P(\text{in } 70's) = 1 - 0.3 = 0.7$$

51. $$P(\text{white or green}) = P(\text{white}) + P(\text{green}) = \frac{n(\text{white}) + n(\text{green})}{n(S)} = \frac{9+8}{9+8+3} = \frac{17}{20}$$

53. $$P(\text{not white}) = 1 - P(\text{white}) = 1 - \frac{n(\text{white})}{n(S)} = 1 - \frac{9}{20} = \frac{11}{20}$$

55. $$P(\text{strike or one}) = P(\text{strike}) + P(\text{one}) = \frac{n(\text{strike}) + n(\text{one})}{n(S)} = \frac{3+1}{8} = \frac{4}{8} = \frac{1}{2}$$

57. There are 30 households out of 100 with an income of $30,000 or more.
$$P(E) = \frac{n(E)}{n(S)} = \frac{n(30,000 \text{ or more})}{n(\text{total households})} = \frac{30}{100} = \frac{3}{10}$$

59. There are 40 households out of 100 with an income of less than $20,000.
$$P(E) = \frac{n(E)}{n(S)} = \frac{n(\text{less than } \$20,000)}{n(\text{total households})} = \frac{40}{100} = \frac{2}{5}$$

61. (a) $P(1 \text{ or } 2) = P(1) + P(2) = 0.24 + 0.33 = 0.57$
    (b) $P(1 \text{ or more}) = P(1) + P(2) + P(3) + P(4 \text{ or more})$
        $$= 0.24 + 0.33 + 0.21 + 0.17 = 0.95$$
    (c) $P(3 \text{ or fewer}) = P(0) + P(1) + P(2) + P(3) = 0.05 + 0.24 + 0.33 + 0.21 = 0.83$
    (d) $P(3 \text{ or more}) = P(3) + P(4 \text{ or more}) = 0.21 + 0.17 = 0.38$

(e) $P(\text{less than } 2) = P(0) + P(1) = 0.05 + 0.24 = 0.29$

(f) $P(\text{less than } 1) = P(0) = 0.05$

(g) $P(1, 2, \text{ or } 3) = P(1) + P(2) + P(3) = 0.24 + 0.33 + 0.21 = 0.78$

(h) $P(2 \text{ or more}) = P(2) + P(3) + P(4 \text{ or more}) = 0.33 + 0.21 + 0.17 = 0.71$

63.   (a) $P(\text{freshman or female}) = P(\text{freshman}) + P(\text{female}) - P(\text{freshman and female})$

$$= \frac{n(\text{freshman}) + n(\text{female}) - n(\text{freshman and female})}{n(S)} = \frac{18 + 15 - 8}{33} = \frac{25}{33}$$

(b) $P(\text{sophomore or male}) = P(\text{sophomore}) + P(\text{male}) - P(\text{sophomore and male})$

$$= \frac{n(\text{sophomore}) + n(\text{male}) - n(\text{sophomore and male})}{n(S)}$$

$$= \frac{15 + 18 - 8}{33} = \frac{25}{33}$$

65. $P(\text{at least 2 with same birthday}) = 1 - P(\text{none with same birthday})$

$$= 1 - \frac{n(\text{different birthdays})}{n(S)} = 1 - \frac{365 \cdot 364 \cdot 363 \cdot 362 \cdot 361 \cdot 360 \cdot \ldots \cdot 354}{365^{12}} \approx 1 - 0.833 = 0.167$$

67.   The sample space for picking 5 out of 10 numbers in a particular order contains

$$P(10,5) = \frac{10!}{(10-5)!} = \frac{10!}{5!} = 30{,}240 \quad \text{possible outcomes.}$$

One of these is the desired outcome. Thus, the probability of winning is:

$$P(E) = \frac{n(E)}{n(S)} = \frac{n(\text{winning})}{n(\text{total possible outcomes})} = \frac{1}{30240}$$

69.   (a) $P(3 \text{ heads}) = \dfrac{C(5,3)}{2^5} = \dfrac{10}{32} = \dfrac{5}{16}$ (b) $P(0 \text{ heads}) = \dfrac{C(5,0)}{2^5} = \dfrac{1}{32}$

71.   (a) $P(\text{sum } = 7 \text{ three times}) = P(\text{sum } = 7) \cdot P(\text{sum } = 7) \cdot P(\text{sum } = 7)$

$$= \frac{1}{6} \cdot \frac{1}{6} \cdot \frac{1}{6} = \frac{1}{216}$$

(b) $P(\text{sum } = 7 \text{ or } 11 \text{ at least twice})$

$$= P(\text{sum } = 7 \text{ or } 11) \cdot P(\text{sum } = 7 \text{ or } 11) \cdot P(\text{sum } \neq 7 \text{ or } 11) +$$

$$P(\text{sum } = 7 \text{ or } 11) \cdot P(\text{sum } = 7 \text{ or } 11) \cdot P(\text{sum } = 7 \text{ or } 11)$$

$$= \frac{8}{36} \cdot \frac{8}{36} \cdot \frac{28}{36} + \frac{8}{36} \cdot \frac{8}{36} \cdot \frac{8}{36} \approx 0.049$$

73. $P(\text{all 5 defective}) = \dfrac{n(5 \text{ defective})}{n(S)} = \dfrac{1}{C(30,5)} = 7.02 \times 10^{-6}$

$P(\text{at least 2 defective}) = 1 - \left(P(\text{none defective}) + P(\text{one defective}) \right)$

$$= 1 - \left(\frac{C(5,0) \cdot C(25,5)}{C(30,5)} + \frac{C(5,1) \cdot C(25,4)}{C(30,5)} \right) = 1 - 0.817 = 0.183$$

75. $P(\text{one of 5 coins is valued at more than \$10,000}) = \dfrac{C(49,4) \cdot C(1,1)}{C(50,5)} = 0.1$

Chapter 12

Counting and Probability

12.R Chapter Review

1. $A \cup B = \{1, 3, 5, 7\} \cup \{3, 5, 6, 7, 8\} = \{1, 3, 5, 6, 7, 8\}$

3. $A \cap C = \{1, 3, 5, 7\} \cap \{2, 3, 7, 8, 9\} = \{3, 7\}$

5. $\overline{A} \cup \overline{B} = \overline{\{1, 3, 5, 7\}} \cup \overline{\{3, 5, 6, 7, 8\}} = \{2, 4, 6, 8, 9\} \cup \{1, 2, 4, 9\} = \{1, 2, 4, 6, 8, 9\}$

7. $\overline{B \cap C} = \overline{\{3, 5, 6, 7, 8\} \cap \{2, 3, 7, 8, 9\}} = \overline{\{3, 7, 8\}} = \{1, 2, 4, 5, 6, 9\}$

9. $n(A) = 8,\ n(B) = 12,\ n(A \cap B) = 3$
$n(A \cup B) = n(A) + n(B) - n(A \cap B) = 8 + 12 - 3 = 17$

11. From the figure:
$n(A) = 20 + 2 + 6 + 1 = 29$

13. From the figure:
$n(A \text{ and } C) = n(A \cap C) = 1 + 6 = 7$

15. From the figure:
$n(\text{neither in } A \text{ nor in } C) = n(\overline{A \cup C}) = 20 + 5 = 25$

17. $5! = 5 \cdot 4 \cdot 3 \cdot 2 \cdot 1 = 120$

19. $P(8,3) = \dfrac{8!}{(8-3)!} = \dfrac{8!}{5!} = \dfrac{8 \cdot 7 \cdot 6 \cdot 5!}{5!} = 336$

21. $C(8,3) = \dfrac{8!}{(8-3)!\, 3!} = \dfrac{8!}{5!\, 3!} = \dfrac{8 \cdot 7 \cdot 6 \cdot 5!}{5! \cdot 3 \cdot 2 \cdot 1} = 56$

23. There are 2 choices of material, 3 choices of color, and 10 choices of size. The complete assortment would have: $2 \cdot 3 \cdot 10 = 60$ suits.

25. There are two possible outcomes for each game or
$2 \cdot 2 \cdot 2 \cdot 2 \cdot 2 \cdot 2 \cdot 2 = 2^7 = 128$ outcomes for 7 games.

27. Since order is significant, this is a permutation.
$P(9,4) = \dfrac{9!}{(9-4)!} = \dfrac{9!}{5!} = \dfrac{9 \cdot 8 \cdot 7 \cdot 6 \cdot 5!}{5!} = 3024$ ways to seat 4 people in 9 seats

Chapter 12 Counting and Probability

29. Choose 4 runners - order is not significant:
$$C(8,4) = \frac{8!}{(8-4)!\,4!} = \frac{8!}{4!\,4!} = \frac{8\cdot7\cdot6\cdot5\cdot4!}{4\cdot3\cdot2\cdot1\cdot4!} = 70 \text{ ways a squad can be chosen.}$$

31. Choose 14 teams 2 at a time:
$$C(14,2) = \frac{14!}{(14-2)!\,2!} = \frac{14!}{12!\,2!} = \frac{14\cdot13\cdot12!}{12!\cdot2\cdot1} = 91 \text{ ways to pair 14 teams.}$$

33. There are $8\cdot10\cdot10\cdot10\cdot10\cdot10\cdot2 = 1,600,000$ possible phone numbers.

35. There are $24\cdot9\cdot10\cdot10\cdot10 = 216,000$ possible license plates.

37. Since there are repeated letters:
$$\frac{7!}{2!\cdot2!} = \frac{7\cdot6\cdot5\cdot4\cdot3\cdot2\cdot1}{2\cdot1\cdot2\cdot1} = 1260 \text{ different words can be formed.}$$

39. (a) $C(9,4)\cdot C(9,3)\cdot C(9,2) = 126\cdot84\cdot36 = 381,024$ committees can be formed.
 (b) $C(9,4)\cdot C(5,3)\cdot C(2,2) = 126\cdot10\cdot1 = 1260$ committees can be formed.

41. (a) $365\cdot364\cdot363\cdot362\cdot\ldots\cdot348 = 8.634628387 \times 10^{45}$
 (b) $P(\text{no one has same birthday}) = \dfrac{365\cdot364\cdot363\cdot362\cdot\ldots\cdot348}{365^{18}} = 0.6531 = 65.31\%$
 (c) $P(\text{at least 2 have same birthday}) = 1 - P(\text{no one has same birthday})$
$$= 1 - 0.6531 = 0.3469 = 34.69\%$$

43. (a) $P(\text{unemployed}) = 0.054 = 5.4\%$
 (b) $P(\text{not unemployed}) = 1 - P(\text{unemployed}) = 1 - 0.054 = 0.946 = 94.6\%$

45. $P(\$1 \text{ bill}) = \dfrac{n(\$1 \text{ bill})}{n(S)} = \dfrac{4}{9}$

47. Let S be all possible selections, let D be a card that is divisible by 5, and let PN be a 1 or a prime number.
$$n(S) = 100$$
$$n(D) = 20 \quad \text{(There are 20 numbers divisible by 5 between 1 and 100.)}$$
$$n(PN) = 26 \quad \text{(There are 25 prime numbers less than or equal to 100.)}$$
$$P(D) = \frac{n(D)}{n(S)} = \frac{20}{100} = \frac{1}{5} = 0.2$$
$$P(PN) = \frac{n(PN)}{n(S)} = \frac{26}{100} = \frac{13}{50} = 0.26$$

49. (a) $P(5 \text{ heads}) = \dfrac{n(5 \text{ heads})}{n(S)} = \dfrac{C(10,5)}{2^{10}} = \dfrac{\left(\dfrac{10!}{5!\,5!}\right)}{1024} = \dfrac{252}{1024} \approx 0.2461$
 (b) $P(\text{all heads}) = \dfrac{n(\text{all heads})}{n(S)} = \dfrac{1}{2^{10}} = \dfrac{1}{1024} = 0.00098$

A Preview of Calculus: The Limit, Derivative and Integral of a Function

13.1 Finding Limits Using Tables and Graphs

1. $\lim\limits_{x \to 2}\left(4x^3\right)$

X	Y1
1.99	31.522
1.999	31.952
1.9999	31.995
2.0001	32.005
2.001	32.048
2.01	32.482

Y1🔲4X^3

$\lim\limits_{x \to 2}\left(4x^3\right) = 32$

3. $\lim\limits_{x \to 0}\left(\dfrac{x+1}{x^2+1}\right)$

X	Y1
-.01	.9899
-.001	.999
-1E-4	.9999
1E-4	1.0001
.001	1.001
.01	1.0099

Y1🔲(X+1)/(X²+1)

$\lim\limits_{x \to 0}\left(\dfrac{x+1}{x^2+1}\right) = 1$

5. $\lim\limits_{x \to 4}\left(\dfrac{x^2-4x}{x-4}\right)$

X	Y1
3.99	3.99
3.999	3.999
3.9999	3.9999
4.0001	4.0001
4.001	4.001
4.01	4.01

Y1🔲(X²-4X)/(X-4)

$\lim\limits_{x \to 4}\left(\dfrac{x^2-4x}{x-4}\right) = 4$

7. $\lim\limits_{x \to 0}\left(e^x+1\right)$

X	Y1
-.01	1.99
-.001	1.999
-1E-4	1.9999
1E-4	2.0001
.001	2.001
.01	2.0101

Y1🔲e^(X)+1

$\lim\limits_{x \to 0}\left(e^x+1\right) = 2$

9. $\lim\limits_{x \to 0}\left(\dfrac{\cos x-1}{x}\right)$

X	Y1
-.01	.005
-.001	5E-4
-1E-4	5E-5
1E-4	-5E-5
.001	-5E-4
.01	-.005

Y1🔲(cos(X)-1)/X

$\lim\limits_{x \to 0}\left(\dfrac{\cos x-1}{x}\right) = 0$

11. $\lim\limits_{x \to 2} f(x) = 3$

13. $\lim\limits_{x \to 2} f(x) = 4$

15. $\lim\limits_{x \to 3} f(x)$ does not exist because as x gets closer to 3, but is less than 3, $f(x)$ gets closer to 3. However, as x gets closer to 3, but is greater than 3, $f(x)$ gets closer to 6.

17. $f(x) = 3x + 1$

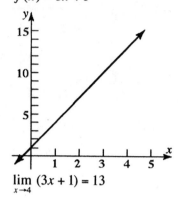

$$\lim_{x \to 4} (3x + 1) = 13$$

19. $f(x) = 1 - x^2$

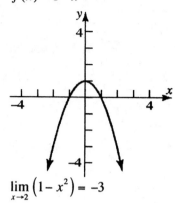

$$\lim_{x \to 2} \left(1 - x^2\right) = -3$$

21. $f(x) = |2x|$

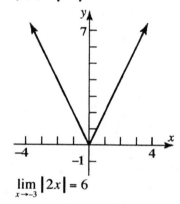

$$\lim_{x \to -3} |2x| = 6$$

23. $f(x) = \sin x$

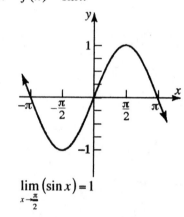

$$\lim_{x \to \frac{\pi}{2}} (\sin x) = 1$$

25. $f(x) = e^x$

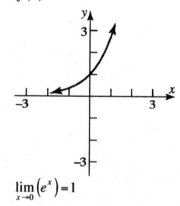

$$\lim_{x \to 0} \left(e^x\right) = 1$$

27. $f(x) = \dfrac{1}{x}$

$$\lim_{x \to -1} \left(\frac{1}{x}\right) = -1$$

646

29. $f(x) = \begin{cases} x^2 & x \geq 0 \\ 2x & x < 0 \end{cases}$

$\lim\limits_{x \to 0} f(x) = 0$

since $\lim\limits_{x \to 0^-} f(x) = \lim\limits_{x \to 0^+} f(x) = 0$

31. $f(x) = \begin{cases} 3x & x \leq 1 \\ x+1 & x > 1 \end{cases}$

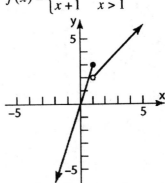

$\lim\limits_{x \to 1} f(x)$ does not exist

since $\lim\limits_{x \to 1^-} f(x) \neq \lim\limits_{x \to 1^+} f(x)$

33. $f(x) = \begin{cases} x & x < 0 \\ 1 & x = 0 \\ 3x & x > 0 \end{cases}$

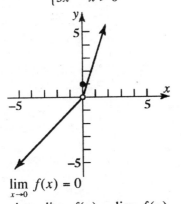

$\lim\limits_{x \to 0} f(x) = 0$

since $\lim\limits_{x \to 0^-} f(x) = \lim\limits_{x \to 0^+} f(x) = 0$

35. $f(x) = \begin{cases} \sin x & x \leq 0 \\ x^2 & x > 0 \end{cases}$

$\lim\limits_{x \to 0} f(x) = 0$

since $\lim\limits_{x \to 0^-} f(x) = \lim\limits_{x \to 0^+} f(x) = 0$

37. $\lim\limits_{x \to 1} \left(\dfrac{x^3 - x^2 + x - 1}{x^4 - x^3 + 2x - 2} \right)$

$\lim\limits_{x \to 1} \left(\dfrac{x^3 - x^2 + x - 1}{x^4 - x^3 + 2x - 2} \right) \approx 0.67$

39. $\lim\limits_{x \to 2} \left(\dfrac{x^3 - 2x^2 + 4x - 8}{x^2 + x - 6} \right)$

X	Y1	
1.99	1.5952	
1.999	1.5995	
1.9999	1.6	
2	ERROR	
2.0001	1.6	
2.001	1.6005	
2.01	1.6048	
Y1=(X^3-2X2+4X-...		

$\lim\limits_{x \to 2} \left(\dfrac{x^3 - 2x^2 + 4x - 8}{x^2 + x - 6} \right) = 1.60$

41. $\displaystyle \lim_{x \to 1}\left(\frac{x^3 + 2x^2 + x}{x^4 + x^3 + 2x + 2}\right)$

$$\lim_{x \to 1}\left(\frac{x^3 + 2x^2 + x}{x^4 + x^3 + 2x + 2}\right) = 0$$

A Preview of Calculus: The Limit, Derivative and Integral of a Function

13.2 Algebra Techniques for Finding Limits

1. $\lim\limits_{x\to 1}(5) = 5$

3. $\lim\limits_{x\to 4}(x) = 4$

5. $\lim\limits_{x\to 2}(3x + 2) = 3(2) + 2 = 8$

7. $\lim\limits_{x\to -1}(3x^2 - 5x) = 3(-1)^2 - 5(-1) = 8$

9. $\lim\limits_{x\to 1}(5x^4 - 3x^2 + 6x - 9) = 5(1)^4 - 3(1)^2 + 6(1) - 9 = 5 - 3 + 6 - 9 = -1$

11. $\lim\limits_{x\to 1}(x^2 + 1)^3 = \left(\lim\limits_{x\to 1}(x^2 + 1)\right)^3 = (1^2 + 1)^3 = 2^3 = 8$

13. $\lim\limits_{x\to 1}\sqrt{5x + 4} = \sqrt{\lim\limits_{x\to 1}(5x + 4)} = \sqrt{5(1) + 4} = \sqrt{9} = 3$

15. $\lim\limits_{x\to 0}\left(\dfrac{x^2 - 4}{x^2 + 4}\right) = \dfrac{\left(\lim\limits_{x\to 0}(x^2 - 4)\right)}{\left(\lim\limits_{x\to 0}(x^2 + 4)\right)} = \dfrac{0^2 - 4}{0^2 + 4} = \dfrac{-4}{4} = -1$

17. $\lim\limits_{x\to 2}\left((3x - 2)^{5/2}\right) = \left(\lim\limits_{x\to 2}(3x - 2)\right)^{5/2} = (3(2) - 2)^{5/2} = 4^{5/2} = 32$

19. $\lim\limits_{x\to 2}\left(\dfrac{x^2 - 4}{x^2 - 2x}\right) = \lim\limits_{x\to 2}\left(\dfrac{(x - 2)(x + 2)}{x(x - 2)}\right) = \lim\limits_{x\to 2}\left(\dfrac{x + 2}{x}\right) = \dfrac{2 + 2}{2} = \dfrac{4}{2} = 2$

21. $\lim\limits_{x\to -3}\left(\dfrac{x^2 - x - 12}{x^2 - 9}\right) = \lim\limits_{x\to -3}\left(\dfrac{(x - 4)(x + 3)}{(x - 3)(x + 3)}\right) = \lim\limits_{x\to -3}\left(\dfrac{x - 4}{x - 3}\right) = \dfrac{-3 - 4}{-3 - 3} = \dfrac{-7}{-6} = \dfrac{7}{6}$

23. $\lim\limits_{x\to 1}\left(\dfrac{x^3 - 1}{x - 1}\right) = \lim\limits_{x\to 1}\left(\dfrac{(x - 1)(x^2 + x + 1)}{x - 1}\right) = \lim\limits_{x\to 1}\left(x^2 + x + 1\right) = 1^2 + 1 + 1 = 3$

25. $\lim\limits_{x\to -1}\left(\dfrac{(x + 1)^2}{x^2 - 1}\right) = \lim\limits_{x\to -1}\left(\dfrac{(x + 1)^2}{(x - 1)(x + 1)}\right) = \lim\limits_{x\to -1}\left(\dfrac{x + 1}{x - 1}\right) = \dfrac{-1 + 1}{-1 - 1} = \dfrac{0}{-2} = 0$

27. $\lim\limits_{x\to1}\left(\dfrac{x^3-x^2+x-1}{x^4-x^3+2x-2}\right)=\lim\limits_{x\to1}\left(\dfrac{x^2(x-1)+1(x-1)}{x^3(x-1)+2(x-1)}\right)=\lim\limits_{x\to1}\left(\dfrac{(x-1)(x^2+1)}{(x-1)(x^3+2)}\right)=\lim\limits_{x\to1}\left(\dfrac{x^2+1}{x^3+2}\right)$

$=\dfrac{1^2+1}{1^3+2}=\dfrac{2}{3}$

29. $\lim\limits_{x\to2}\left(\dfrac{x^3-2x^2+4x-8}{x^2+x-6}\right)=\lim\limits_{x\to2}\left(\dfrac{x^2(x-2)+4(x-2)}{(x+3)(x-2)}\right)=\lim\limits_{x\to2}\left(\dfrac{(x-2)(x^2+4)}{(x+3)(x-2)}\right)=\lim\limits_{x\to2}\left(\dfrac{x^2+4}{x+3}\right)$

$=\dfrac{2^2+4}{2+3}=\dfrac{8}{5}$

31. $\lim\limits_{x\to-1}\left(\dfrac{x^3+2x^2+x}{x^4+x^3+2x+2}\right)=\lim\limits_{x\to-1}\left(\dfrac{x(x^2+2x+1)}{x^3(x+1)+2(x+1)}\right)=\lim\limits_{x\to-1}\left(\dfrac{x(x+1)^2}{(x+1)(x^3+2)}\right)$

$=\lim\limits_{x\to-1}\left(\dfrac{x(x+1)}{x^3+2}\right)=\dfrac{-1(-1+1)}{(-1)^3+2}=\dfrac{-1(0)}{-1+2}=\dfrac{0}{1}=0$

33. $\lim\limits_{x\to2}\left(\dfrac{f(x)-f(2)}{x-2}\right)=\lim\limits_{x\to2}\left(\dfrac{(5x-3)-7}{x-2}\right)=\lim\limits_{x\to2}\left(\dfrac{5x-10}{x-2}\right)=\lim\limits_{x\to2}\left(\dfrac{5(x-2)}{x-2}\right)=\lim\limits_{x\to2}(5)=5$

35. $\lim\limits_{x\to3}\left(\dfrac{f(x)-f(3)}{x-3}\right)=\lim\limits_{x\to3}\left(\dfrac{x^2-9}{x-3}\right)=\lim\limits_{x\to3}\left(\dfrac{(x-3)(x+3)}{x-3}\right)=\lim\limits_{x\to3}(x+3)=3+3=6$

37. $\lim\limits_{x\to-1}\left(\dfrac{f(x)-f(-1)}{x-(-1)}\right)=\lim\limits_{x\to-1}\left(\dfrac{x^2+2x-(-1)}{x+1}\right)=\lim\limits_{x\to-1}\left(\dfrac{x^2+2x+1}{x+1}\right)=\lim\limits_{x\to-1}\left(\dfrac{(x+1)^2}{x+1}\right)$

$=\lim\limits_{x\to-1}x+1=-1+1=0$

39. $\lim\limits_{x\to0}\left(\dfrac{f(x)-f(0)}{x-0}\right)=\lim\limits_{x\to0}\left(\dfrac{3x^3-2x^2+4-4}{x}\right)=\lim\limits_{x\to0}\left(\dfrac{3x^3-2x^2}{x}\right)=\lim\limits_{x\to0}(3x^2-2x)=0$

41. $\lim\limits_{x\to1}\left(\dfrac{f(x)-f(1)}{x-1}\right)=\lim\limits_{x\to1}\left(\dfrac{\left(\frac{1}{x}-1\right)}{(x-1)}\right)=\lim\limits_{x\to1}\left(\dfrac{\left(\frac{1-x}{x}\right)}{(x-1)}\right)=\lim\limits_{x\to1}\left(\dfrac{-1(x-1)}{x(x-1)}\right)=\lim\limits_{x\to1}\left(\dfrac{-1}{x}\right)=\dfrac{-1}{1}=-1$

43. $\lim\limits_{x\to0}\left(\dfrac{\tan x}{x}\right)=\lim\limits_{x\to0}\left(\dfrac{\left(\frac{\sin x}{\cos x}\right)}{(x)}\right)=\lim\limits_{x\to0}\left(\dfrac{\sin x}{x}\cdot\dfrac{1}{\cos x}\right)=\left(\lim\limits_{x\to0}\left(\dfrac{\sin x}{x}\right)\right)\cdot\left(\lim\limits_{x\to0}\left(\dfrac{1}{\cos x}\right)\right)$

$=1\cdot\left(\dfrac{\lim\limits_{x\to0}1}{\lim\limits_{x\to0}(\cos x)}\right)=1\cdot\dfrac{1}{1}=1$

45. $\displaystyle \lim_{x \to 0} \left(\frac{3\sin x + \cos x - 1}{4x} \right) = \lim_{x \to 0} \left(\frac{3\sin x}{4x} + \frac{\cos x - 1}{4x} \right) = \lim_{x \to 0} \left(\frac{3\sin x}{4x} \right) + \lim_{x \to 0} \left(\frac{\cos x - 1}{4x} \right)$

$\displaystyle = \frac{3}{4} \lim_{x \to 0} \left(\frac{\sin x}{x} \right) + \frac{1}{4} \lim_{x \to 0} \left(\frac{\cos x - 1}{x} \right) = \frac{3}{4} \cdot 1 + \frac{1}{4} \cdot 0 = \frac{3}{4}$

Chapter 13

A Preview of Calculus:
The Limit, Derivative and Integral of a Function

13.3 One-Sided Limits; Continuous Functions

1. The domain of $f(x) = 2x + 3$ is all real numbers. Therefore, $f(x)$ is continuous everywhere.

3. The domain of $f(x) = 3x^2 + x$ is all real numbers. Therefore, $f(x)$ is continuous everywhere.

5. The domain of $f(x) = 4 \sin x$ is all real numbers. Therefore, $f(x)$ is continuous everywhere.

7. The domain of $f(x) = 2 \tan x$ is all real numbers except odd integer multiples of $\frac{\pi}{2}$.

   Therefore, $f(x)$ is continuous everywhere except where $x = \frac{k\pi}{2}$ where k is an odd integer.

   $f(x)$ is discontinuous at $x = \frac{k\pi}{2}$ where k is an odd integer.

9. $f(x) = \dfrac{2x+5}{x^2-4} = \dfrac{2x+5}{(x-2)(x+2)}$. The domain of $f(x)$ is all real numbers except
   $x = 2$ and $x = -2$. Therefore, $f(x)$ is continuous everywhere except at $x = 2$ and $x = -2$.
   $f(x)$ is discontinuous at $x = 2$ and $x = -2$.

11. $f(x) = \dfrac{x-3}{\ln x}$. The domain of $f(x)$ is $(0, 1)$ or $(1, \infty)$. Thus, $f(x)$ is continuous on the
    interval $(0, \infty)$ except at $x = 1$. $f(x)$ is discontinuous at $x = 1$.

13. $\lim\limits_{x \to 1^+} (2x + 3) = 2(1) + 3 = 5$

15. $\lim\limits_{x \to 1^-} \left(2x^3 + 5x\right) = 2(1)^3 + 5(1) = 2 + 5 = 7$

17. $\lim\limits_{x \to \frac{\pi}{2}^+} (\sin x) = \sin\left(\dfrac{\pi}{2}\right) = 1$

19. $\lim\limits_{x\to 2^+}\left(\dfrac{x^2-4}{x-2}\right)=\lim\limits_{x\to 2^+}\left(\dfrac{(x+2)(x-2)}{x-2}\right)=\lim\limits_{x\to 2^+}(x+2)=2+2=4$

21. $\lim\limits_{x\to -1^-}\left(\dfrac{x^2-1}{x^3+1}\right)=\lim\limits_{x\to -1^-}\left(\dfrac{(x+1)(x-1)}{(x+1)(x^2-x+1)}\right)=\lim\limits_{x\to -1^-}\left(\dfrac{x-1}{x^2-x+1}\right)=\dfrac{-1-1}{(-1)^2-(-1)+1}=\dfrac{-2}{3}$

23. $\lim\limits_{x\to -2^+}\left(\dfrac{x^2+x-2}{x^2+2x}\right)=\lim\limits_{x\to -2^+}\left(\dfrac{(x+2)(x-1)}{x(x+2)}\right)=\lim\limits_{x\to -2^+}\left(\dfrac{x-1}{x}\right)=\dfrac{-2-1}{-2}=\dfrac{-3}{-2}=\dfrac{3}{2}$

25. $f(x)=x^3-3x^2+2x-6;\ \ c=2$

    1. $f(2)=2^3-3\cdot 2^2+2\cdot 2-6=-6$

    2. $\lim\limits_{x\to 2^-}f(x)=2^3-3\cdot 2^2+2\cdot 2-6=-6$

    3. $\lim\limits_{x\to 2^+}f(x)=2^3-3\cdot 2^2+2\cdot 2-6=-6$

Thus, $f(x)$ is continuous at $c=2$.

27. $f(x)=\dfrac{x^2+5}{x-6};\ \ c=3$

    1. $f(3)=\dfrac{3^2+5}{3-6}=\dfrac{14}{-3}=-\dfrac{14}{3}$

    2. $\lim\limits_{x\to 3^-}f(x)=\dfrac{3^2+5}{3-6}=\dfrac{14}{-3}=-\dfrac{14}{3}$

    3. $\lim\limits_{x\to 3^+}f(x)=\dfrac{3^2+5}{3-6}=\dfrac{14}{-3}=-\dfrac{14}{3}$

Thus, $f(x)$ is continuous at $c=3$.

29. $f(x)=\dfrac{x+3}{x-3};\ \ c=3$

Since $f(x)$ is not defined at $c=3$, the function is not continuous at $c=3$.

31. $f(x)=\dfrac{x^3+3x}{x^2-3x};\ \ c=0$

Since $f(x)$ is not defined at $c=0$, the function is not continuous at $c=0$.

33. $f(x)=\begin{cases}\dfrac{x^3+3x}{x^2-3x}&\text{if }x\neq 0\\[2mm]1&\text{if }x=0\end{cases};\ \ c=0$

    1. $f(0)=1$

    2. $\lim\limits_{x\to 0^-}f(x)=\lim\limits_{x\to 0^-}\left(\dfrac{x^3+3x}{x^2-3x}\right)=\lim\limits_{x\to 0^-}\left(\dfrac{x(x^2+3)}{x(x-3)}\right)=\lim\limits_{x\to 0^-}\left(\dfrac{x^2+3}{x-3}\right)=\dfrac{3}{-3}=-1$

Since $\lim\limits_{x\to 0^-}f(x)\neq f(c)$, the function is not continuous at $c=0$.

Chapter 13   A Preview of Calculus:
The Limit, Derivative and Integral of a Function

35. $f(x) = \begin{cases} \dfrac{x^3 + 3x}{x^2 - 3x} & \text{if } x \neq 0 \\ -1 & \text{if } x = 0 \end{cases}$; $c = 0$

  1. $f(0) = -1$

  2. $\lim\limits_{x \to 0^-} f(x) = \lim\limits_{x \to 0^-}\left(\dfrac{x^3 + 3x}{x^2 - 3x}\right) = \lim\limits_{x \to 0^-}\left(\dfrac{x(x^2 + 3)}{x(x - 3)}\right) = \lim\limits_{x \to 0^-}\left(\dfrac{x^2 + 3}{x - 3}\right) = \dfrac{3}{-3} = -1$

  3. $\lim\limits_{x \to 0^+} f(x) = \lim\limits_{x \to 0^+}\left(\dfrac{x^3 + 3x}{x^2 - 3x}\right) = \lim\limits_{x \to 0^+}\left(\dfrac{x(x^2 + 3)}{x(x - 3)}\right) = \lim\limits_{x \to 0^+}\left(\dfrac{x^2 + 3}{x - 3}\right) = \dfrac{3}{-3} = -1$

  The function is continuous at $c = 0$.

37. $f(x) = \begin{cases} \dfrac{x^3 - 1}{x^2 - 1} & \text{if } x < 1 \\ 2 & \text{if } x = 1 \\ \dfrac{3}{x + 1} & \text{if } x > 1 \end{cases}$; $c = 1$

  1. $f(1) = 2$

  2. $\lim\limits_{x \to 1^-} f(x) = \lim\limits_{x \to 1^-}\left(\dfrac{x^3 - 1}{x^2 - 1}\right) = \lim\limits_{x \to 1^-}\left(\dfrac{(x - 1)(x^2 + x + 1)}{(x - 1)(x + 1)}\right) = \lim\limits_{x \to 1^-}\left(\dfrac{x^2 + x + 1}{x + 1}\right) = \dfrac{3}{2}$

  Since $\lim\limits_{x \to 1^-} f(x) \neq f(c)$, the function is not continuous at $c = 1$.

39. $f(x) = \begin{cases} 2e^x & \text{if } x < 0 \\ 2 & \text{if } x = 0 \\ \dfrac{x^3 + 2x^2}{x^2} & \text{if } x > 0 \end{cases}$; $c = 0$

  1. $f(0) = 2$

  2. $\lim\limits_{x \to 0^-} f(x) = \lim\limits_{x \to 0^-}\left(2e^x\right) = 2e^0 = 2 \cdot 1 = 2$

  3. $\lim\limits_{x \to 0^+} f(x) = \lim\limits_{x \to 0^+}\left(\dfrac{x^3 + 2x^2}{x^2}\right) = \lim\limits_{x \to 0^+}\left(\dfrac{x^2(x + 2)}{x^2}\right) = \lim\limits_{x \to 0^+}(x + 2) = 0 + 2 = 2$

  The function is continuous at $c = 0$.

41. Domain: $\{x \mid -8 \le x < -6 \text{ or } -6 < x < 4 \text{ or } 4 < x \le 6\}$

43. x-intercepts: $-8, -5, -3$

45. $f(-8) = 0$; $f(-4) = 2$

47. $\lim\limits_{x \to -6^-} f(x) = +\infty$

49. $\lim\limits_{x \to -4^-} f(x) = 2$

51. $\lim\limits_{x \to 2^-} f(x) = 1$

53. $\lim\limits_{x \to 4} f(x)$ does exist. $\lim\limits_{x \to 4} f(x) = 0$ since $\lim\limits_{x \to 4^-} f(x) = \lim\limits_{x \to 4^+} f(x) = 0$

55. f is not continuous at -6 because $f(-6)$ does not exist.

57. f is continuous at 0 because $f(0) = \lim\limits_{x \to 0^-} f(x) = \lim\limits_{x \to 0^+} f(x) = 3$

59. f is not continuous at 4 because $f(4)$ does not exist.

61. $R(x) = \dfrac{x-1}{x^2-1} = \dfrac{x-1}{(x-1)(x+1)}$. The domain of R is $\left\{ x \,\middle|\, x \neq -1, x \neq 1 \right\}$. Thus R is

discontinuous at both -1 and 1. $\lim\limits_{x \to -1^-} R(x) = \lim\limits_{x \to -1^-} \left(\dfrac{x-1}{(x-1)(x+1)} \right) = \lim\limits_{x \to -1^-} \left(\dfrac{1}{x+1} \right) = -\infty$ since

when $x < -1$, $\dfrac{1}{x+1} < 0$, and as x approaches -1, $\dfrac{1}{x+1}$ becomes unbounded.

$\lim\limits_{x \to -1^+} \left(\dfrac{1}{x+1} \right) = +\infty$ since when $x > -1$, $\dfrac{1}{x+1} > 0$, and as x approaches -1, $\dfrac{1}{x+1}$ becomes

unbounded. The behavior near $c = 1$ is: $\lim\limits_{x \to 1} R(x) = \lim\limits_{x \to 1} \left(\dfrac{1}{x+1} \right) = \dfrac{1}{2}$. Note there is a hole in

the graph at $\left(1, \dfrac{1}{2} \right)$.

63. $R(x) = \dfrac{x^2+x}{x^2-1} = \dfrac{x(x+1)}{(x-1)(x+1)}$. The domain of R is $\left\{ x \,\middle|\, x \neq -1, x \neq 1 \right\}$. Thus R is

discontinuous at both -1 and 1. $\lim\limits_{x \to 1^-} R(x) = \lim\limits_{x \to 1^-} \left(\dfrac{x(x+1)}{(x-1)(x+1)} \right) = \lim\limits_{x \to 1^-} \left(\dfrac{x}{x-1} \right) = -\infty$ since

when $x < 1$, $\dfrac{x}{x-1} < 0$, and as x approaches 1, $\dfrac{x}{x-1}$ becomes unbounded. $\lim\limits_{x \to 1^+} \dfrac{x}{x-1} = +\infty$

since when $x > 1$, $\dfrac{x}{x-1} > 0$, and as x approaches 1, $\dfrac{x}{x-1}$ becomes unbounded. The

behavior near $c = -1$ is: $\lim\limits_{x \to -1} R(x) = \lim\limits_{x \to -1} \left(\dfrac{x}{x-1} \right) = \dfrac{-1}{-2} = \dfrac{1}{2}$. Note there is a hole in the

graph at $\left(-1, \dfrac{1}{2} \right)$.

65. $R(x) = \dfrac{x^3 - x^2 + x - 1}{x^4 - x^3 + 2x - 2} = \dfrac{x^2(x-1) + 1(x-1)}{x^3(x-1) + 2(x-1)} = \dfrac{(x-1)(x^2+1)}{(x-1)(x^3+2)} = \dfrac{x^2+1}{x^3+2}, \; x \neq 1$

There is a vertical asymptote where $x^3 + 2 = 0$. $x = -\sqrt[3]{2}$ is a vertical asymptote. There is a hole in the graph at $x = 1$.

67. $R(x) = \dfrac{x^3 - 2x^2 + 4x - 8}{x^2 + x - 6} = \dfrac{x^2(x-2) + 4(x-2)}{(x+3)(x-2)} = \dfrac{(x-2)(x^2+4)}{(x+3)(x-2)} = \dfrac{x^2+4}{x+3}, \; x \neq 2$

There is a vertical asymptote where $x + 3 = 0$. $x = -3$ is a vertical asymptote. There is a hole in the graph at $x = 2$.

69. $R(x) = \dfrac{x^3 + 2x^2 + x}{x^4 + x^3 + 2x + 2} = \dfrac{x(x^2 + 2x + 1)}{x^3(x+1) + 2(x+1)} = \dfrac{x(x+1)^2}{(x+1)(x^3+2)} = \dfrac{x(x+1)}{x^3+2}, \; x \neq -1$

There is a vertical asymptote where $x^3 + 2 = 0$. $x = -\sqrt[3]{2}$ is a vertical asymptote. There is a hole in the graph at $x = -1$.

A Preview of Calculus: The Limit, Derivative and Integral of a Function

13.4 The Tangent Problem; The Derivative

1. $f(x) = 3x + 5$ at $(1, 8)$

$$m_{\tan} = \lim_{x \to 1} \left(\frac{f(x) - f(1)}{x - 1} \right) = \lim_{x \to 1} \left(\frac{3x + 5 - 8}{x - 1} \right)$$

$$= \lim_{x \to 1} \left(\frac{3x - 3}{x - 1} \right) = \lim_{x \to 1} \left(\frac{3(x - 1)}{x - 1} \right) = \lim_{x \to 1} (3) = 3$$

Tangent Line: $y - 8 = 3(x - 1)$
$$y - 8 = 3x - 3$$
$$y = 3x + 5$$

3. $f(x) = x^2 + 2$ at $(-1, 3)$

$$m_{\tan} = \lim_{x \to -1} \left(\frac{f(x) - f(-1)}{x + 1} \right) = \lim_{x \to -1} \left(\frac{x^2 + 2 - 3}{x + 1} \right)$$

$$= \lim_{x \to -1} \left(\frac{x^2 - 1}{x + 1} \right) = \lim_{x \to -1} \left(\frac{(x + 1)(x - 1)}{x + 1} \right)$$

$$= \lim_{x \to -1} (x - 1) = -1 - 1 = -2$$

Tangent Line: $y - 3 = -2(x - (-1))$
$$y - 3 = -2x - 2$$
$$y = -2x + 1$$

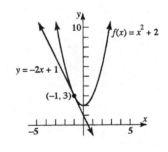

5. $f(x) = 3x^2$ at $(2, 12)$

$$m_{\tan} = \lim_{x \to 2} \left(\frac{f(x) - f(2)}{x - 2} \right) = \lim_{x \to 2} \left(\frac{3x^2 - 12}{x - 2} \right)$$

$$= \lim_{x \to 2} \frac{3((x^2) - 4)}{x - 2} = \lim_{x \to 2} \left(\frac{3(x + 2)(x - 2)}{x - 2} \right)$$

$$= \lim_{x \to 2} (3(x + 2)) = 3(2 + 2) = 12$$

Tangent Line: $y - 12 = 12(x - 2)$
$$y - 12 = 12x - 24$$
$$y = 12x - 12$$

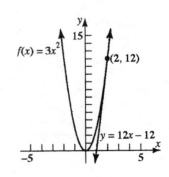

7. $f(x) = 2x^2 + x$ at $(1, 3)$

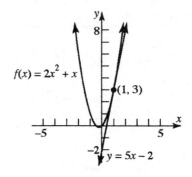

$m_{\tan} = \lim_{x \to 1} \left(\dfrac{f(x) - f(1)}{x - 1} \right) = \lim_{x \to 1} \left(\dfrac{2x^2 + x - 3}{x - 1} \right)$

$= \lim_{x \to 1} \left(\dfrac{(2x + 3)(x - 1)}{x - 1} \right) = \lim_{x \to 1} (2x + 3)$

$= 2(1) + 3 = 5$

Tangent Line: $y - 3 = 5(x - 1)$

$y - 3 = 5x - 5$

$y = 5x - 2$

9. $f(x) = x^2 - 2x + 3$ at $(-1, 6)$

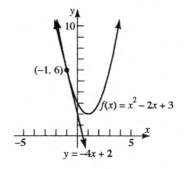

$m_{\tan} = \lim_{x \to -1} \dfrac{f(x) - f(-1)}{x + 1} = \lim_{x \to -1} \dfrac{x^2 - 2x + 3 - 6}{x + 1}$

$= \lim_{x \to -1} \dfrac{x^2 - 2x - 3}{x + 1} = \lim_{x \to -1} \dfrac{(x + 1)(x - 3)}{x + 1}$

$= \lim_{x \to -1} (x - 3) = -1 - 3 = -4$

Tangent Line: $y - 6 = -4(x - (-1))$

$y - 6 = -4x - 4$

$y = -4x + 2$

11. $f(x) = x^3 + x$ at $(2, 10)$

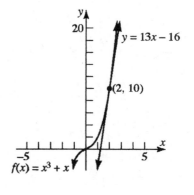

$m_{\tan} = \lim_{x \to 2} \left(\dfrac{f(x) - f(2)}{x - 2} \right) = \lim_{x \to 2} \left(\dfrac{x^3 + x - 10}{x - 2} \right)$

$= \lim_{x \to 2} \left(\dfrac{x^3 - 8 + x - 2}{x - 2} \right)$

$= \lim_{x \to 2} \left(\dfrac{(x - 2)(x^2 + 2x + 4) + x - 2}{x - 2} \right)$

$= \lim_{x \to 2} \left(\dfrac{(x - 2)(x^2 + 2x + 4 + 1)}{x - 2} \right)$

$= \lim_{x \to 2} (x^2 + 2x + 5) = 4 + 4 + 5 = 13$

Tangent Line: $y - 10 = 13(x - 2)$

$y - 10 = 13x - 26$

$y = 13x - 16$

13. $f(x) = -4x + 5$ at 3

$f'(3) = \lim_{x \to 3} \left(\dfrac{f(x) - f(3)}{x - 3} \right) = \lim_{x \to 3} \left(\dfrac{-4x + 5 - (-7)}{x - 3} \right) = \lim_{x \to 3} \left(\dfrac{-4x + 12}{x - 3} \right) = \lim_{x \to 3} \left(\dfrac{-4(x - 3)}{x - 3} \right)$

$= \lim_{x \to 3} (-4) = -4$

15. $f(x) = x^2 - 3$ at 0

$$f'(0) = \lim_{x \to 0}\left(\frac{f(x) - f(0)}{x - 0}\right) = \lim_{x \to 0}\left(\frac{x^2 - 3 - (-3)}{x}\right) = \lim_{x \to 0}\left(\frac{x^2}{x}\right) = \lim_{x \to 0}(x) = 0$$

17. $f(x) = 2x^2 + 3x$ at 1

$$f'(1) = \lim_{x \to 1}\left(\frac{f(x) - f(1)}{x - 1}\right) = \lim_{x \to 1}\left(\frac{2x^2 + 3x - 5}{x - 1}\right) = \lim_{x \to 1}\left(\frac{(2x + 5)(x - 1)}{x - 1}\right) = \lim_{x \to 1}(2x + 5) = 7$$

19. $f(x) = x^3 + 4x$ at -1

$$f'(-1) = \lim_{x \to -1}\left(\frac{f(x) - f(-1)}{x - (-1)}\right) = \lim_{x \to -1}\left(\frac{x^3 + 4x - (-5)}{x + 1}\right) = \lim_{x \to -1}\left(\frac{x^3 + 1 + 4x + 4}{x + 1}\right)$$

$$= \lim_{x \to -1}\left(\frac{(x + 1)(x^2 - x + 1) + 4(x + 1)}{x + 1}\right) = \lim_{x \to -1}\left(\frac{(x + 1)(x^2 - x + 1 + 4)}{x + 1}\right)$$

$$= \lim_{x \to -1}(x^2 - x + 5) = (-1)^2 - (-1) + 5 = 7$$

21. $f(x) = x^3 + x^2 - 2x$ at 1

$$f'(1) = \lim_{x \to 1}\left(\frac{f(x) - f(1)}{x - 1}\right) = \lim_{x \to 1}\left(\frac{x^3 + x^2 - 2x - 0}{x - 1}\right) = \lim_{x \to 1}\left(\frac{x(x^2 + x - 2)}{x - 1}\right)$$

$$= \lim_{x \to 1}\left(\frac{x(x + 2)(x - 1)}{x - 1}\right) = \lim_{x \to 1}(x(x + 2)) = 1(1 + 2) = 3$$

23. $f(x) = \sin x$ at 0

$$f'(0) = \lim_{x \to 0}\left(\frac{f(x) - f(0)}{x - 0}\right) = \lim_{x \to 0}\left(\frac{\sin x - 0}{x - 0}\right) = \lim_{x \to 0}\left(\frac{\sin x}{x}\right) = 1$$

25. Use NDeriv:

27. Use NDeriv:

29. Use NDeriv:

31. Use NDeriv:

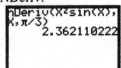

33. Use NDeriv:

```
nDeriv(e^(X)sin(
X),X,2)
      3.643914112
```

35. $V(r) = 3\pi r^2$ at $r = 3$

$$V'(3) = \lim_{r \to 3}\left(\frac{V(r)-V(3)}{r-3}\right) = \lim_{r \to 3}\left(\frac{3\pi r^2 - 27\pi}{r-3}\right) = \lim_{r \to 3}\left(\frac{3\pi(r^2-9)}{r-3}\right)$$

$$= \lim_{r \to 3}\left(\frac{3\pi(r-3)(r+3)}{r-3}\right) = \lim_{r \to 3}\left(3\pi(r+3)\right) = 3\pi(3+3) = 18\pi$$

At the instant $r = 3$ feet, the volume of the cone is increasing at a rate of 18π cubic feet.

37. $V(r) = \frac{4}{3}\pi r^3$ at $r = 2$

$$V'(2) = \lim_{r \to 2}\left(\frac{V(r)-V(2)}{r-2}\right) = \lim_{r \to 2}\left(\frac{\left(\frac{4}{3}\pi r^3 - \frac{32}{3}\pi\right)}{r-2}\right) = \lim_{r \to 2}\left(\frac{\left(\frac{4}{3}\pi(r^3-8)\right)}{r-2}\right)$$

$$= \lim_{r \to 2}\left(\frac{\left(\frac{4}{3}\pi(r-2)(r^2+2r+4)\right)}{r-2}\right) = \lim_{r \to 2}\left(\frac{4}{3}\pi(r^2+2r+4)\right) = \frac{4}{3}\pi(4+4+4) = 16\pi$$

At the instant $r = 2$ feet, the volume of the sphere is increasing at a rate of 16π cubic feet.

39. (a) $-16t^2 + 96t = 0$
    $-16t(t-6) = 0$
    $t = 0$ or $t = 6$
    The ball strikes the ground after 6 seconds.

    (b) $\dfrac{\Delta s}{\Delta t} = \dfrac{s(2)-s(0)}{2-0} = \dfrac{-16(2)^2 + 96(2) - 0}{2} = \dfrac{128}{2} = 64$ feet / sec

    (c) $s'(t) = \lim_{t \to t_0}\left(\dfrac{s(t)-s(t_0)}{t-t_0}\right) = \lim_{t \to t_0}\left(\dfrac{-16t^2 + 96t - \left(-16t_0^2 + 96t_0\right)}{t-t_0}\right)$

    $$= \lim_{t \to t_0}\left(\frac{-16t^2 + 16t_0^2 + 96t - 96t_0}{t-t_0}\right) = \lim_{t \to t_0}\left(\frac{-16\left(t^2-t_0^2\right)+96(t-t_0)}{t-t_0}\right)$$

    $$= \lim_{t \to t_0}\left(\frac{-16(t-t_0)(t+t_0)+96(t-t_0)}{t-t_0}\right) = \lim_{t \to t_0}\left(\frac{(t-t_0)(-16(t+t_0)+96)}{t-t_0}\right)$$

    $$= \lim_{t \to t_0}\left(-16(t+t_0)+96\right) = \left(-16(t_0+t_0)+96\right) = -32t_0 + 96$$

    (d) $s'(2) = -32(2) + 96 = -64 + 96 = 32$ feet / sec

(e) $s'(t) = 0$

$-32t + 96 = 0$

$-32t = -96$

$t = 3$ seconds

(f) $s(3) = -16(3)^2 + 96(3) = -144 + 288 = 144$ feet

(g) $s'(6) = -32(6) + 96 = -192 + 96 = -96$ feet / sec

41.   (a) $\dfrac{\Delta s}{\Delta t} = \dfrac{s(4) - s(1)}{4 - 1} = \dfrac{917 - 987}{3} = \dfrac{-70}{3} = -23\dfrac{1}{3}$ feet / sec

(b) $\dfrac{\Delta s}{\Delta t} = \dfrac{s(3) - s(1)}{3 - 1} = \dfrac{945 - 987}{2} = \dfrac{-42}{2} = -21$ feet / sec

(c) $\dfrac{\Delta s}{\Delta t} = \dfrac{s(2) - s(1)}{2 - 1} = \dfrac{969 - 987}{1} = \dfrac{-18}{1} = -18$ feet / sec

(d) $s(t) = -2.631t^2 - 10.269t + 999.933$

(e) $s'(1) = \lim\limits_{t \to 1} \left(\dfrac{s(t) - s(1)}{t - 1} \right) = \lim\limits_{t \to 1} \left(\dfrac{-2.631t^2 - 10.269t + 999.933 - 987.033}{t - 1} \right)$

$= \lim\limits_{t \to 1} \left(\dfrac{-2.631t^2 - 10.269t + 12.9}{t - 1} \right)$

$= \lim\limits_{t \to 1} \left(\dfrac{-2.631t^2 + 2.631t - 12.9t + 12.9}{t - 1} \right)$

$= \lim\limits_{t \to 1} \left(\dfrac{-2.631t(t - 1) - 12.9(t - 1)}{t - 1} \right) = \lim\limits_{t \to 1} \left(\dfrac{(-2.631t - 12.9)(t - 1)}{t - 1} \right)$

$= \lim\limits_{t \to 1} (-2.631t - 12.9) = -2.631(1) - 12.9 = -15.531$ feet/sec

The instant $t = 1$, the instantaneous speed of the ball is -15.531 feet / sec.

Chapter 13

A Preview of Calculus:
The Limit, Derivative and Integral of a Function

13.5 The Area Problem; The Integral

1. $A \approx f(1)\cdot 1 + f(2)\cdot 1 = 1\cdot 1 + 2\cdot 1 = 1 + 2 = 3$

3. $A \approx f(0)\cdot 2 + f(2)\cdot 2 + f(4)\cdot 2 + f(6)\cdot 2 = 10\cdot 2 + 6\cdot 2 + 7\cdot 2 + 5\cdot 2$
    $= 20 + 12 + 14 + 10 = 56$

5. (a) Graph $f(x) = 3x$:

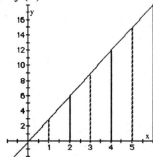

   (b) $A \approx f(0)(2) + f(2)(2) + f(4)(2) = 0(2) + 6(2) + 12(2) = 0 + 12 + 24 = 36$
   (c) $A \approx f(2)(2) + f(4)(2) + f(6)(2) = 6(2) + 12(2) + 18(2) = 12 + 24 + 36 = 72$
   (d) $A \approx f(0)(1) + f(1)(1) + f(2)(1) + f(3)(1) + f(4)(1) + f(5)(1)$
        $= 0(1) + 3(1) + 6(1) + 9(1) + 12(1) + 15(1) = 0 + 3 + 6 + 9 + 12 + 15 = 45$
   (e) $A \approx f(1)(1) + f(2)(1) + f(3)(1) + f(4)(1) + f(5)(1) + f(6)(1)$
        $= 3(1) + 6(1) + 9(1) + 12(1) + 15(1) + 18(1) = 3 + 6 + 9 + 12 + 15 + 18 = 63$
   (f) The actual area is the area of a triangle: $A = \frac{1}{2}(6)(18) = 54$

7.   (a)   Graph $f(x) = -3x + 9$:

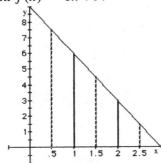

(b) $A \approx f(0)(1) + f(1)(1) + f(2)(1) = 9(1) + 6(1) + 3(1) = 9 + 6 + 3 = 18$

(c) $A \approx f(1)(1) + f(2)(1) + f(3)(1) = 6(1) + 3(1) + 0(1) = 6 + 3 + 0 = 9$

(d) $A \approx f(0)\left(\frac{1}{2}\right) + f\left(\frac{1}{2}\right)\left(\frac{1}{2}\right) + f(1)\left(\frac{1}{2}\right) + f\left(\frac{3}{2}\right)\left(\frac{1}{2}\right) + f(2)\left(\frac{1}{2}\right) + f\left(\frac{5}{2}\right)\left(\frac{1}{2}\right)$

$= 9\left(\frac{1}{2}\right) + \frac{15}{2}\left(\frac{1}{2}\right) + 6\left(\frac{1}{2}\right) + \frac{9}{2}\left(\frac{1}{2}\right) + 3\left(\frac{1}{2}\right) + \frac{3}{2}\left(\frac{1}{2}\right)$

$= \frac{9}{2} + \frac{15}{4} + 3 + \frac{9}{4} + \frac{3}{2} + \frac{3}{4} = \frac{63}{4}$

(e) $A \approx f\left(\frac{1}{2}\right)\left(\frac{1}{2}\right) + f(1)\left(\frac{1}{2}\right) + f\left(\frac{3}{2}\right)\left(\frac{1}{2}\right) + f(2)\left(\frac{1}{2}\right) + f\left(\frac{5}{2}\right)\left(\frac{1}{2}\right) + f(3)\left(\frac{1}{2}\right)$

$= \frac{15}{2}\left(\frac{1}{2}\right) + 6\left(\frac{1}{2}\right) + \frac{9}{2}\left(\frac{1}{2}\right) + 3\left(\frac{1}{2}\right) + \frac{3}{2}\left(\frac{1}{2}\right) + 0\left(\frac{1}{2}\right)$

$= \frac{15}{4} + 3 + \frac{9}{4} + \frac{3}{2} + \frac{3}{4} + 0 = \frac{45}{4}$

(f)   The actual area is the area of a triangle: $A = \frac{1}{2}(3)(9) = \frac{27}{2} = 13.5$

9.   (a)   Graph $f(x) = x^2 + 2$, $[0, 4]$:

(b) $A \approx f(0)(1) + f(1)(1) + f(2)(1) + f(3)(1) = 2(1) + 3(1) + 6(1) + 11(1)$

$= 2 + 3 + 6 + 11 = 22$

(c) $A \approx f(0)\left(\frac{1}{2}\right) + f\left(\frac{1}{2}\right)\left(\frac{1}{2}\right) + f(1)\left(\frac{1}{2}\right) + f\left(\frac{3}{2}\right)\left(\frac{1}{2}\right) + f(2)\left(\frac{1}{2}\right) + f\left(\frac{5}{2}\right)\left(\frac{1}{2}\right)$

$+ f(3)\left(\frac{1}{2}\right) + f\left(\frac{7}{2}\right)\left(\frac{1}{2}\right)$

$= 2\left(\frac{1}{2}\right) + \frac{9}{4}\left(\frac{1}{2}\right) + 3\left(\frac{1}{2}\right) + \frac{17}{4}\left(\frac{1}{2}\right) + 6\left(\frac{1}{2}\right) + \frac{33}{4}\left(\frac{1}{2}\right) + 11\left(\frac{1}{2}\right) + \frac{57}{4}\left(\frac{1}{2}\right)$

$= 1 + \frac{9}{8} + \frac{3}{2} + \frac{17}{8} + 3 + \frac{33}{8} + \frac{11}{2} + \frac{57}{8} = \frac{51}{2}$

(d) $A = \int_0^4 (x^2 + 2)dx$

(e)   Use fnInt function:

```
fnInt(X²+2,X,0,4
)
       29.33333333
```

11.   (a)   Graph $f(x) = x^3$, $[0, 4]$:

(b) $A \approx f(0)(1) + f(1)(1) + f(2)(1) + f(3)(1) = 0(1) + 1(1) + 8(1) + 27(1)$

      $= 0 + 1 + 8 + 27 = 36$

(c) $A \approx f(0)\left(\frac{1}{2}\right) + f\left(\frac{1}{2}\right)\left(\frac{1}{2}\right) + f(1)\left(\frac{1}{2}\right) + f\left(\frac{3}{2}\right)\left(\frac{1}{2}\right) + f(2)\left(\frac{1}{2}\right) + f\left(\frac{5}{2}\right)\left(\frac{1}{2}\right)$

$$+ f(3)\left(\frac{1}{2}\right) + f\left(\frac{7}{2}\right)\left(\frac{1}{2}\right)$$

$$= 0\left(\frac{1}{2}\right) + \frac{1}{8}\left(\frac{1}{2}\right) + 1\left(\frac{1}{2}\right) + \frac{27}{8}\left(\frac{1}{2}\right) + 8\left(\frac{1}{2}\right) + \frac{125}{8}\left(\frac{1}{2}\right) + 27\left(\frac{1}{2}\right) + \frac{343}{8}\left(\frac{1}{2}\right)$$

$$= 0 + \frac{1}{16} + \frac{1}{2} + \frac{27}{16} + 4 + \frac{125}{16} + \frac{27}{2} + \frac{343}{16} = 49$$

(d) $A = \int_0^4 x^3\, dx$

(e)   Use fnInt function:

```
fnInt(X^3,X,0,4)
              64
```

13.   (a)   Graph $f(x) = \frac{1}{x}$, $[1, 5]$:

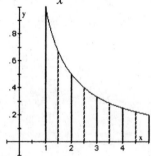

(b) $A \approx f(1)(1) + f(2)(1) + f(3)(1) + f(4)(1) = 1(1) + \frac{1}{2}(1) + \frac{1}{3}(1) + \frac{1}{4}(1)$

$= 1 + \frac{1}{2} + \frac{1}{3} + \frac{1}{4} = \frac{25}{12}$

(c) $A \approx f(1)\left(\frac{1}{2}\right) + f\left(\frac{3}{2}\right)\left(\frac{1}{2}\right) + f(2)\left(\frac{1}{2}\right) + f\left(\frac{5}{2}\right)\left(\frac{1}{2}\right) + f(3)\left(\frac{1}{2}\right) + f\left(\frac{7}{2}\right)\left(\frac{1}{2}\right)$

$+ f(4)\left(\frac{1}{2}\right) + f\left(\frac{9}{2}\right)\left(\frac{1}{2}\right)$

$= 1\left(\frac{1}{2}\right) + \frac{2}{3}\left(\frac{1}{2}\right) + \frac{1}{2}\left(\frac{1}{2}\right) + \frac{2}{5}\left(\frac{1}{2}\right) + \frac{1}{3}\left(\frac{1}{2}\right) + \frac{2}{7}\left(\frac{1}{2}\right) + \frac{1}{4}\left(\frac{1}{2}\right) + \frac{2}{9}\left(\frac{1}{2}\right)$

$= \frac{1}{2} + \frac{1}{3} + \frac{1}{4} + \frac{1}{5} + \frac{1}{6} + \frac{1}{7} + \frac{1}{8} + \frac{1}{9} = \frac{4690}{2520} \approx 1.829$

(d) $A = \int_{1}^{5} \frac{1}{x} dx$ (e)   Use fnInt function:

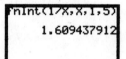

15.   (a)   Graph $f(x) = e^{x}$, $[-1, 3]$:

(b) $A \approx \left(f(-1) + f(0) + f(1) + f(2)\right)(1) = (0.3679 + 1 + 2.7183 + 7.3891)(1)$

$= 11.4753$

(c) $A \approx \left\{f(-1) + f\left(-\frac{1}{2}\right) + f(0) + f\left(\frac{1}{2}\right) + f(1) + f\left(\frac{3}{2}\right) + f(2) + f\left(\frac{5}{2}\right)\right\} \cdot \left(\frac{1}{2}\right)$

$= (0.3679 + 0.6065 + 1 + 1.6487 + 2.7183 + 4.4817 + 7.3891 + 12.1825)(0.5)$

$= 30.3947(0.5) = 15.1974$

(d) $A = \int_{-1}^{3} e^{x} dx$ (e)   Use fnInt function:

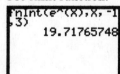

17. (a) Graph $f(x) = \sin x$, $[0, \pi]$:

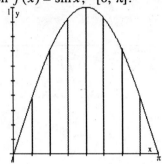

(b) $A \approx \left(f(0) + f\left(\dfrac{\pi}{4}\right) + f\left(\dfrac{\pi}{2}\right) + f\left(\dfrac{3\pi}{4}\right) \right)\left(\dfrac{\pi}{4}\right) = \left(0 + \dfrac{\sqrt{2}}{2} + 1 + \dfrac{\sqrt{2}}{2} \right)\left(\dfrac{\pi}{4}\right)$

$\left(1 + \sqrt{2} \right)\left(\dfrac{\pi}{4}\right) \approx 1.8961$

(c) $A \approx \left(f(0) + f\left(\dfrac{\pi}{8}\right) + f\left(\dfrac{\pi}{4}\right) + f\left(\dfrac{3\pi}{8}\right) + f\left(\dfrac{\pi}{2}\right) + f\left(\dfrac{5\pi}{8}\right) + f\left(\dfrac{3\pi}{4}\right) + f\left(\dfrac{7\pi}{8}\right) \right)\left(\dfrac{\pi}{8}\right)$

$= (0 + 0.3827 + 0.7071 + 0.9239 + 1 + 0.9239 + 0.7071 + 0.3827)(0.3927)$

$= 5.0274(0.3927) = 1.9743$

(d) $A = \int_0^\pi \sin x \ dx$ (e) Use fnInt function:

19. (a) The integral represents the area under the graph of $f(x) = 3x + 1$ from 0 to 4.

      (b)   (c)

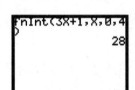

21. (a) The integral represents the area under the graph of $f(x) = x^2 - 1$ from 2 to 5.
     (b)   (c)

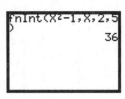

23. (a) The integral represents the area under the graph of $f(x) = \sin x$ from 0 to $\frac{\pi}{2}$.
     (b)   (c)

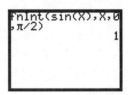

25. (a) The integral represents the area under the graph of $f(x) = e^x$ from 0 to 2.
     (b)   (c)

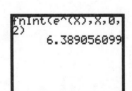

27. Using left endpoints:

$n = 2:$ $0 + 0.5 = 0.5$

$n = 4:$ $0 + 0.125 + 0.25 + 0.375 = 0.75$

$n = 10:$ $0 + 0.02 + 0.04 + 0.06 + \cdots + 0.018 = \dfrac{10}{2}(0 + 0.018) = 0.9$

$n = 100:$ $0 + 0.0002 + 0.0004 + 0.0006 + \cdots + 0.0198 = \dfrac{100}{2}(0 + 0.0198) = 0.99$

Using right endpoints:

$n = 2:$ $0.5 + 1 = 1.5$

$n = 4:$ $0.125 + 0.25 + 0.375 + 0.5 = 1.25$

$n = 10:$ $0.02 + 0.04 + 0.06 + \cdots + 0.20 = \dfrac{10}{2}(0.02 + 0.20) = 1.1$

$n = 100:$ $0.0002 + 0.0004 + 0.0006 + \cdots + 0.02 = \dfrac{100}{2}(0.0002 + 0.02) = 1.01$

A Preview of Calculus:
The Limit, Derivative and Integral of a
Function

13.R Chapter Review

1. $\lim\limits_{x \to 2} \left(3x^2 - 2x + 1\right) = 3(2)^2 - 2(2) + 1 = 12 - 4 + 1 = 9$

3. $\lim\limits_{x \to -2} \left(x^2 + 1\right)^2 = \left(\lim\limits_{x \to -2} \left(x^2 + 1\right)\right)^2 = \left((-2)^2 + 1\right)^2 = 5^2 = 25$

5. $\lim\limits_{x \to 3} \sqrt{x^2 + 7} = \sqrt{\lim\limits_{x \to 3} (x^2 + 7)} = \sqrt{3^2 + 7} = \sqrt{16} = 4$

7. $\lim\limits_{x \to 1^-} \sqrt{1 - x^2} = \sqrt{\lim\limits_{x \to 1^-} (1 - x^2)} = \sqrt{1 - 1^2} = \sqrt{0} = 0$

9. $\lim\limits_{x \to 2} (5x + 6)^{3/2} = \left(\lim\limits_{x \to 2} (5x + 6)\right)^{3/2} = (5(2) + 6)^{3/2} = 16^{3/2} = 64$

11. $\lim\limits_{x \to -1} \left(\dfrac{x^2 + x + 2}{x^2 - 9}\right) = \dfrac{\left(\lim\limits_{x \to -1}(x^2 + x + 2)\right)}{\left(\lim\limits_{x \to -1}(x^2 - 9)\right)} = \dfrac{(-1)^2 + (-1) + 2}{(-1)^2 - 9} = \dfrac{2}{-8} = -\dfrac{1}{4}$

13. $\lim\limits_{x \to 1} \left(\dfrac{x - 1}{x^3 - 1}\right) = \lim\limits_{x \to 1} \left(\dfrac{x - 1}{(x - 1)(x^2 + x + 1)}\right) = \lim\limits_{x \to 1} \left(\dfrac{1}{x^2 + x + 1}\right) = \dfrac{1}{1^2 + 1 + 1} = \dfrac{1}{3}$

15. $\lim\limits_{x \to -3} \left(\dfrac{x^2 - 9}{x^2 - x - 12}\right) = \lim\limits_{x \to -3} \left(\dfrac{(x - 3)(x + 3)}{(x - 4)(x + 3)}\right) = \lim\limits_{x \to -3} \left(\dfrac{x - 3}{x - 4}\right) = \dfrac{-3 - 3}{-3 - 4} = \dfrac{-6}{-7} = \dfrac{6}{7}$

17. $\lim\limits_{x \to -1^-} \left(\dfrac{x^2 - 1}{x^3 - 1}\right) = \lim\limits_{x \to -1^-} \left(\dfrac{(x + 1)(x - 1)}{(x - 1)(x^2 + x + 1)}\right) = \lim\limits_{x \to -1^-} \left(\dfrac{x + 1}{x^2 + x + 1}\right) = \dfrac{-1 + 1}{(-1)^2 + (-1) + 1} = \dfrac{0}{1} = 0$

19. $\lim\limits_{x \to 2} \left(\dfrac{x^3 - 8}{x^3 - 2x^2 + 4x - 8}\right) = \lim\limits_{x \to 2} \left(\dfrac{(x - 2)(x^2 + 2x + 4)}{x^2(x - 2) + 4(x - 2)}\right) = \lim\limits_{x \to 2} \left(\dfrac{(x - 2)(x^2 + 2x + 4)}{(x - 2)(x^2 + 4)}\right)$

$= \lim\limits_{x \to 2} \left(\dfrac{x^2 + 2x + 4}{x^2 + 4}\right) = \left(\dfrac{2^2 + 2(2) + 4}{2^2 + 4}\right) = \dfrac{12}{8} = \dfrac{3}{2}$

21. $$\lim_{x\to3}\left(\frac{x^4-3x^3+x-3}{x^3-3x^2+2x-6}\right)=\lim_{x\to3}\left(\frac{x^3(x-3)+1(x-3)}{x^2(x-3)+2(x-3)}\right)=\lim_{x\to3}\left(\frac{(x-3)(x^3+1)}{(x-3)(x^2+2)}\right)=\lim_{x\to3}\left(\frac{x^3+1}{x^2+2}\right)$$
$$=\frac{3^3+1}{3^2+2}=\frac{28}{11}$$

23. $f(x)=3x^4-x^2+2;\ c=5$
 1. $f(5)=3(5)^4-5^2+2=1852$
 2. $\lim_{x\to5^-}f(x)=3(5)^4-5^2+2=1852$
 3. $\lim_{x\to5^+}f(x)=3(5)^4-5^2+2=1852$

Thus, $f(x)$ is continuous at $c=5$.

25. $f(x)=\dfrac{x^2-4}{x+2};\ c=-2$
Since $f(x)$ is not defined at $c=-2$, the function is not continuous at $c=-2$.

27. $f(x)=\begin{cases}\dfrac{x^2-4}{x+2}&\text{if }x\ne-2\\4&\text{if }x=-2\end{cases};\ c=-2$
 1. $f(-2)=4$
 2. $\lim_{x\to-2^-}f(x)=\lim_{x\to-2^-}\left(\frac{x^2-4}{x+2}\right)=\lim_{x\to-2^-}\left(\frac{(x-2)(x+2)}{x+2}\right)=\lim_{x\to-2^-}(x-2)=-4$

Since $\lim_{x\to-2^-}f(x)\ne f(c)$, the function is not continuous at $c=-2$.

29. $f(x)=\begin{cases}\dfrac{x^2-4}{x+2}&\text{if }x\ne-2\\-4&\text{if }x=-2\end{cases};\ c=-2$
 1. $f(-2)=-4$
 2. $\lim_{x\to-2^-}f(x)=\lim_{x\to-2^-}\left(\frac{x^2-4}{x+2}\right)=\lim_{x\to-2^-}\left(\frac{(x-2)(x+2)}{x+2}\right)=\lim_{x\to-2^-}(x-2)=-4$
 3. $\lim_{x\to-2^+}f(x)=\lim_{x\to-2^+}\left(\frac{x^2-4}{x+2}\right)=\lim_{x\to-2^+}\left(\frac{(x-2)(x+2)}{x+2}\right)=\lim_{x\to-2^+}(x-2)=-4$

The function is continuous at $c=-2$.

31. Domain: $\{x|-6\le x<2\text{ or }2<x<5\text{ or }5<x\le6\}$

35. $f(-6)=2;\ f(-4)=1$ 37. $\lim_{x\to-4^-}f(x)=4$ 39. $\lim_{x\to-2^-}f(x)=-2$

41. $\lim_{x\to2^-}f(x)=-\infty$ 43. $\lim_{x\to0}f(x)$ does not exist. $\lim_{x\to0^-}f(x)=4\ne\lim_{x\to0^+}f(x)=1$

45. f is not continuous at -2 because $\lim\limits_{x \to -2^-} f(x) \neq \lim\limits_{x \to -2^+} f(x)$

47. f is not continuous at 0 because $\lim\limits_{x \to 0^-} f(x) \neq \lim\limits_{x \to 0^+} f(x)$

49. f is continuous at 4 because $f(4) = \lim\limits_{x \to 4^-} f(x) = \lim\limits_{x \to 4^+} f(x)$

51. $R(x) = \dfrac{x+4}{x^2 - 16} = \dfrac{x+4}{(x-4)(x+4)}$. The domain of R is $\left\{ x \mid x \neq -4, x \neq 4 \right\}$. Thus R is

discontinuous at both -4 and 4. $\lim\limits_{x \to 4^-} R(x) = \lim\limits_{x \to 4^-} \left(\dfrac{x+4}{(x-4)(x+4)} \right) = \lim\limits_{x \to 4^-} \left(\dfrac{1}{x-4} \right) = -\infty$ since

when $x < 4$, $\dfrac{1}{x-4} < 0$, and as x approaches 4, $\dfrac{1}{x-4}$ becomes unbounded.

$\lim\limits_{x \to 4^+} \left(\dfrac{1}{x-4} \right) = +\infty$ since when $x > 4$, $\dfrac{1}{x-4} > 0$, and as x approaches 4, $\dfrac{1}{x-4}$ becomes

unbounded. The behavior near $c = -4$ is: $\lim\limits_{x \to -4} R(x) = \lim\limits_{x \to -4} \left(\dfrac{1}{x-4} \right) = \dfrac{1}{-8}$. Note there is a

hole in the graph at $\left(-4, -\dfrac{1}{8} \right)$.

53. $R(x) = \dfrac{x^3 - 2x^2 + 4x - 8}{x^2 - 11x + 18} = \dfrac{x^2(x-2) + 4(x-2)}{(x-9)(x-2)} = \dfrac{(x-2)(x^2+4)}{(x-9)(x-2)} = \dfrac{x^2+4}{x-9}, x \neq 2$

There is a vertical asymptote where $x - 9 = 0$. $x = 9$ is a vertical asymptote. There is a
hole in the graph at $x = 2$.

55. $f(x) = 2x^2 + 8x$ at $(1, 10)$

$$m_{\tan} = \lim_{x \to 1} \left(\frac{f(x) - f(1)}{x - 1} \right) = \lim_{x \to 1} \left(\frac{2x^2 + 8x - 10}{x - 1} \right)$$

$$= \lim_{x \to 1} \left(\frac{2(x + 5)(x - 1)}{x - 1} \right) = \lim_{x \to 1} (2(x + 5))$$

$$= 2(1 + 5) = 12$$

Tangent Line: $y - 10 = 12(x - 1)$
$$y - 10 = 12x - 12$$
$$y = 12x - 2$$

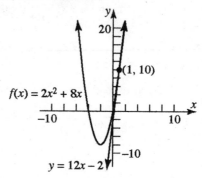

57. $f(x) = x^2 + 2x - 3$ at $(-1, -4)$

$$m_{\tan} = \lim_{x \to -1} \left(\frac{f(x) - f(-1)}{x + 1} \right) = \lim_{x \to -1} \left(\frac{x^2 + 2x - 3 - (-4)}{x + 1} \right)$$

$$= \lim_{x \to -1} \left(\frac{x^2 + 2x + 1}{x + 1} \right) = \lim_{x \to -1} \left(\frac{(x + 1)^2}{x + 1} \right) = \lim_{x \to -1} (x + 1)$$

$$= -1 + 1 = 0$$

Tangent Line: $y - (-4) = 0(x - (-1))$
$$y + 4 = 0$$
$$y = -4$$

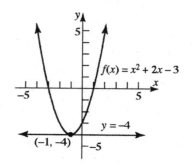

59. $f(x) = x^3 + x^2$ at $(2, 12)$

$$m_{\tan} = \lim_{x \to 2} \left(\frac{f(x) - f(2)}{x - 2} \right) = \lim_{x \to 2} \left(\frac{x^3 + x^2 - 12}{x - 2} \right)$$

$$= \lim_{x \to 2} \left(\frac{x^3 - 2x^2 + 3x^2 - 12}{x - 2} \right)$$

$$= \lim_{x \to 2} \left(\frac{x^2(x - 2) + 3(x - 2)(x + 2)}{x - 2} \right)$$

$$= \lim_{x \to 2} \left(\frac{(x - 2)(x^2 + 3x + 6)}{x - 2} \right)$$

$$= \lim_{x \to 2} (x^2 + 3x + 6) = 4 + 6 + 6 = 16$$

Tangent Line: $y - 12 = 16(x - 2)$
$$y - 12 = 16x - 32$$
$$y = 16x - 20$$

61. $f(x) = -4x^2 + 5$ at 3

$$f'(3) = \lim_{x \to 3} \left(\frac{f(x) - f(3)}{x - 3} \right) = \lim_{x \to 3} \left(\frac{-4x^2 + 5 - (-31)}{x - 3} \right) = \lim_{x \to 3} \left(\frac{-4x^2 + 36}{x - 3} \right)$$

$$= \lim_{x \to 3} \left(\frac{-4(x^2 - 9)}{x - 3} \right) = \lim_{x \to 3} \left(\frac{-4(x - 3)(x + 3)}{x - 3} \right) = \lim_{x \to 3} \left((-4)(x + 3) \right) = -4(6) = -24$$

63. $f(x) = x^2 - 3x$ at 0

$$f'(0) = \lim_{x \to 0} \left(\frac{f(x) - f(0)}{x - 0} \right) = \lim_{x \to 0} \left(\frac{x^2 - 3x - 0}{x} \right) = \lim_{x \to 0} \left(\frac{x(x - 3)}{x} \right) = \lim_{x \to 0} (x - 3) = -3$$

65. $f(x) = 2x^2 + 3x + 2$ at 1

$$f'(1) = \lim_{x \to 1} \left(\frac{f(x) - f(1)}{x - 1} \right) = \lim_{x \to 1} \left(\frac{2x^2 + 3x + 2 - 7}{x - 1} \right) = \lim_{x \to 1} \left(\frac{2x^2 + 3x - 5}{x - 1} \right)$$

$$= \lim_{x \to 1} \left(\frac{(2x + 5)(x - 1)}{x - 1} \right) = \lim_{x \to 1} (2x + 5) = 7$$

67. Use NDeriv:

69. Use NDeriv:

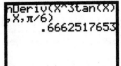

71. (a) $-16t^2 + 96t + 112 = 0$
    $-16(t^2 - 6t - 7) = 0$
    $-16(t + 1)(t - 7) = 0$
    $$t = -1 \text{ or } t = 7$$
    The ball strikes the ground after 7 seconds.

(b) $-16t^2 + 96t + 112 = 112$
    $-16t^2 + 96t = 0$
    $-16t(t - 6) = 0$
    $$t = 0 \text{ or } t = 6$$
    The ball passes the rooftop after 6 seconds.

(c) $\dfrac{\Delta s}{\Delta t} = \dfrac{s(2) - s(0)}{2 - 0} = \dfrac{-16(2)^2 + 96(2) + 112 - 112}{2} = \dfrac{128}{2} = 64 \text{ feet / sec}$

(d) $s'(t) = \lim_{t \to t_0} \left(\dfrac{s(t) - s(t_0)}{t - t_0} \right) = \lim_{t \to t_0} \left(\dfrac{-16t^2 + 96t + 112 - \left(-16t_0^2 + 96t_0 + 112 \right)}{t - t_0} \right)$

$$= \lim_{t \to t_0} \left(\frac{-16t^2 + 16t_0^2 + 96t - 96t_0}{t - t_0} \right) = \lim_{t \to t_0} \left(\frac{-16\left(t^2 - t_0^2 \right) + 96\left(t - t_0 \right)}{t - t_0} \right)$$

$$= \lim_{t \to t_0} \left(\frac{-16(t - t_0)(t + t_0) + 96(t - t_0)}{t - t_0} \right) = \lim_{t \to t_0} \left(\frac{(t - t_0)(-16(t + t_0) + 96)}{t - t_0} \right)$$

$$= \lim_{t \to t_0} \left(-16(t + t_0) + 96 \right) = \left(-16(t_0 + t_0) + 96 \right) = -32t_0 + 96$$

(e) $s'(2) = -32(2) + 96 = -64 + 96 = 32 \text{ feet / sec}$

(f) $s'(t) = 0$

$$-32t + 96 = 0$$

$$-32t = -96$$

$$t = 3 \text{ seconds}$$

(g) $s'(6) = -32(6) + 96 = -192 + 96 = -96 \text{ feet / sec}$

(h) $s'(7) = -32(7) + 96 = -224 + 96 = -128 \text{ feet / sec}$

73. (a) $\dfrac{\Delta R}{\Delta x} = \dfrac{8775 - 2340}{130 - 25} = \dfrac{6435}{105} = \$61.29 \text{ / watch}$

     (b) $\dfrac{\Delta R}{\Delta x} = \dfrac{6975 - 2340}{90 - 25} = \dfrac{4635}{65} = \$71.31 \text{/ watch}$

     (c) $\dfrac{\Delta R}{\Delta x} = \dfrac{4375 - 2340}{50 - 25} = \dfrac{2035}{25} = \$81.40 \text{ / watch}$

     (d) $R(x) = -0.25x^2 + 100.01x - 1.24$

     (e) $R'(25) = \lim\limits_{x \to 25} \left(\dfrac{R(x) - R(25)}{x - 25} \right) = \lim\limits_{x \to 25} \left(\dfrac{-0.25x^2 + 100.01x - 1.24 - 2342.76}{x - 25} \right)$

$$= \lim\limits_{x \to 25} \left(\dfrac{-0.25x^2 + 100.01x - 2344}{x - 25} \right) \quad \text{(Divide)}$$

$$= \lim\limits_{x \to 25} (-0.25x + 93.76) = -0.25(25) + 93.76 = \$87.51\text{/watch}$$

75. (a)   Graph $f(x) = 2x + 3$:

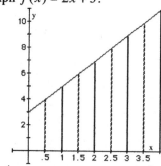

     (b) $A \approx (f(0) + f(1) + f(2) + f(3))(1) = (3 + 5 + 7 + 9)(1) = 24(1) = 24$

     (c) $A \approx (f(1) + f(2) + f(3) + f(4))(1) = (5 + 7 + 9 + 11)(1) = 32(1) = 32$

     (d) $A \approx \left(f(0) + f\left(\dfrac{1}{2}\right) + f(1) + f\left(\dfrac{3}{2}\right) + f(2) + f\left(\dfrac{5}{2}\right) + f(3) + f\left(\dfrac{7}{2}\right) \right) \cdot \left(\dfrac{1}{2}\right)$

$$= (3 + 4 + 5 + 6 + 7 + 8 + 9 + 10)\left(\dfrac{1}{2}\right) = 52\left(\dfrac{1}{2}\right) = 26$$

     (e) $A \approx \left(f\left(\dfrac{1}{2}\right) + f(1) + f\left(\dfrac{3}{2}\right) + f(2) + f\left(\dfrac{5}{2}\right) + f(3) + f\left(\dfrac{7}{2}\right) + f(4) \right) \cdot \left(\dfrac{1}{2}\right)$

$$= (4 + 5 + 6 + 7 + 8 + 9 + 10 + 11)\left(\dfrac{1}{2}\right) = 60\left(\dfrac{1}{2}\right) = 30$$

     (f)   The actual area is the area of a trapezoid:

$$A = \dfrac{1}{2}(3 + 11)(4) = \dfrac{56}{2} = 28$$

77. (a) Graph $f(x) = 4 - x^2$, $[-1, 2]$:

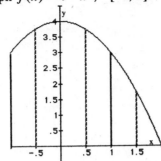

(b) $A \approx \left(f(-1) + f(0) + f(1)\right)(1) = (3 + 4 + 3)(1) = 10(1) = 10$

(c) $A \approx \left(f(-1) + f\left(-\dfrac{1}{2}\right) + f(0) + f\left(\dfrac{1}{2}\right) + f(1) + f\left(\dfrac{3}{2}\right)\right) \cdot \left(\dfrac{1}{2}\right)$

$= \left(3 + \dfrac{15}{4} + 4 + \dfrac{15}{4} + 3 + \dfrac{7}{4}\right)\left(\dfrac{1}{2}\right) = \dfrac{61}{4}\left(\dfrac{1}{2}\right) = \dfrac{61}{8} = 7.625$

(d) $A = \int_{-1}^{2}\left(4 - x^2\right) dx$

(e) Use fnInt function:

79. (a) Graph $f(x) = \dfrac{1}{x^2}$, $[1, 4]$:

(b) $A \approx \left(f(1) + f(2) + f(3)\right)(1) = \left(1 + \dfrac{1}{4} + \dfrac{1}{9}\right)(1) = \dfrac{49}{36}(1) = \dfrac{49}{36} = 1.361$

(c) $A \approx \left(f(1) + f\left(\dfrac{3}{2}\right) + f(2) + f\left(\dfrac{5}{2}\right) + f(3) + f\left(\dfrac{7}{2}\right)\right) \cdot \left(\dfrac{1}{2}\right)$

$= \left(1 + \dfrac{4}{9} + \dfrac{1}{4} + \dfrac{4}{25} + \dfrac{1}{9} + \dfrac{4}{49}\right)\left(\dfrac{1}{2}\right) = 1.024$

(d) $A = \int_{1}^{4}\left(\dfrac{1}{x^2}\right) dx$

(e) Use fnInt function:

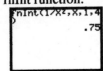

81. (a)   The integral represents the area under the graph of $f(x) = 9 - x^2$ from -1 to 3.

    (b)   (c)

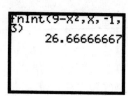

83. (a)   The integral represents the area under the graph of $f(x) = e^x$ from -1 to 1.

    (b)   (c)

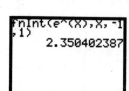

Review

A.1 Algebra Review

1. $x + 2y = -2 + 2 \cdot 3 = -2 + 6 = 4$

3. $5xy + 2 = 5(-2)(3) + 2 = -30 + 2 = -28$

5. $\dfrac{2x}{x-y} = \dfrac{2(-2)}{-2-3} = \dfrac{-4}{-5} = \dfrac{4}{5}$

7. $\dfrac{3x+2y}{2+y} = \dfrac{3(-2)+2(3)}{2+3} = \dfrac{-6+6}{5} = \dfrac{0}{5} = 0$

9. $\dfrac{x^2-1}{x}$ Part (c) must be excluded.
    The value $x = 0$ must be excluded from the domain because it causes division by 0.

11. $\dfrac{x}{x^2-9} = \dfrac{x}{(x-3)(x+3)}$ Part (a) must be excluded.
    The values $x = -3$ and $x = 3$ must be excluded from the domain because they cause division by 0.

13. $\dfrac{x^2}{x^2+1}$ None of the given values are excluded.
    The domain is all real numbers.

15. $\dfrac{x^2+5x-10}{x^3-x} = \dfrac{x^2+5x-10}{x(x-1)(x+1)}$ Parts (b), (c), and (d) must be excluded.
    The values $x = 0$, $x = 1$, and $x = -1$ must be excluded from the domain because they cause division by 0.

17. Since $x = 5$ will cause the denominator of $\dfrac{4}{x-5}$ to equal zero, the domain is the set of all real numbers except 5.

19. Since $x = -4$ will cause the denominator of $\dfrac{x}{x+4}$ to equal zero, the domain is the set of all real numbers except -4.

21.

23. $\dfrac{1}{2} > 0$

25. $-1 > -2$

27. $\pi > 3.14$

29. $\dfrac{1}{2} = 0.5$

31. $\dfrac{2}{3} < 0.67$

33. $x > 0$

35. $x < 2$

37. $x \le 1$

39. Graph on the number line: $x \ge -2$

-2

41. Graph on the number line: $x > -1$

-1

43. $|x + y| = |3 + (-2)| = |1| = 1$

45. $|x| + |y| = |3| + |-2| = 3 + 2 = 5$

47. $\dfrac{|x|}{x} = \dfrac{|3|}{3} = \dfrac{3}{3} = 1$

49. $|4x - 5y| = |4(3) - 5(-2)| = |12 + 10| = |22| = 22$

51. $\big||4x| - |5y|\big| = \big||4(3)| - |5(-2)|\big| = \big||12| - |-10|\big| = |12 - 10| = |2| = 2$

53. $d(C,D) = d(0,1) = |1 - 0| = |1| = 1$

55. $d(D,E) = d(1,3) = |3 - 1| = |2| = 2$

57. $d(A,E) = d(-3,3) = |3 - (-3)| = |6| = 6$

59. $(-4)^2 = (-4)(-4) = 16$

61. $4^{-2} = \dfrac{1}{4^2} = \dfrac{1}{16}$

63. $3^{-6} \cdot 3^4 = 3^{-6+4} = 3^{-2} = \dfrac{1}{3^2} = \dfrac{1}{9}$

65. $\left(3^{-2}\right)^{-1} = 3^{(-2)(-1)} = 3^2 = 9$

67. $\sqrt{25} = \sqrt{5^2} = |5| = 5$

69. $\sqrt{(-4)^2} = |-4| = 4$

71. $\left(8x^3\right)^{-2} = \dfrac{1}{\left(8x^3\right)^2} = \dfrac{1}{8^2 \cdot x^6} = \dfrac{1}{64x^6}$

73. $\left(x^2 y^{-1}\right)^2 = \left(\dfrac{x^2}{y}\right)^2 = \dfrac{x^{2\cdot2}}{y^{1\cdot2}} = \dfrac{x^4}{y^2}$

75. $\dfrac{x^{-2}y^3}{x\,y^4} = \dfrac{x^{-2}}{x}\cdot\dfrac{y^3}{y^4} = x^{-2-1}y^{3-4} = x^{-3}y^{-1} = \dfrac{1}{x^3}\cdot\dfrac{1}{y} = \dfrac{1}{x^3 y}$

77. $\dfrac{(-2)^3 x^4 (yz)^2}{3^2 x\, y^3 z^4} = \dfrac{-8x^4 y^2 z^2}{9x\, y^3 z^4} = \dfrac{-8}{9}x^{4-1}y^{2-3}z^{2-4} = \dfrac{-8}{9}x^3 y^{-1}z^{-2} = \dfrac{-8}{9}x^3\cdot\dfrac{1}{y}\cdot\dfrac{1}{z^2} = -\dfrac{8x^3}{9y\,z^2}$

79. $\left(\dfrac{3x^{-1}}{4y^{-1}}\right)^{-2} = \left(\dfrac{3y^1}{4x^1}\right)^{-2} = \dfrac{1}{\left(\dfrac{3y^1}{4x^1}\right)^2} = \dfrac{1}{\left(\dfrac{3^2 y^{1\cdot2}}{4^2 x^{1\cdot2}}\right)} = \dfrac{1}{\left(\dfrac{9y^2}{16x^2}\right)} = \dfrac{16x^2}{9y^2}$

81.

$A = l \cdot w$

83.

$C = \pi \cdot d$

85.

$A = \dfrac{\sqrt{3}}{4}\cdot x^2$

87.

$V = \dfrac{4}{3}\cdot \pi \cdot r^3$

89.

$V = x^3$

91. $|x - 115| \le 5$

(a) $|x - 115| = |113 - 115| = |-2| = 2 \le 5$ 113 volts is acceptable.

(b) $|x - 115| = |109 - 115| = |-6| = 6 \nleq 5$ 109 volts is not acceptable.

93. $|x - 3| \le 0.01$

(a) $|x - 3| = |2.999 - 3| = |-0.001| = 0.001 \le 0.01$ A radius of 2.999 centimeters is acceptable.

(b) $|x - 3| = |2.89 - 3| = |-0.11| = 0.11 \nleq 0.01$ A radius of 2.89 centimeters is not acceptable.

95. $\dfrac{1}{3} = 0.333333 \ldots > 0.333$, so $\dfrac{1}{3}$ is larger by approximately $0.0003333 \ldots$

97. No.

99. Answers will vary.

Appendix A

Review

A.2 Geometry Review

1. $a = 5,\ b = 12,\quad c^2 = a^2 + b^2 = 5^2 + 12^2 = 25 + 144 = 169 \ \rightarrow\ c = 13$

3. $a = 10,\ b = 24,\quad c^2 = a^2 + b^2 = 10^2 + 24^2 = 100 + 576 = 676 \ \rightarrow\ c = 26$

5. $a = 7,\ b = 24,\quad c^2 = a^2 + b^2 = 7^2 + 24^2 = 49 + 576 = 625 \ \rightarrow\ c = 25$

7. $5^2 = 3^2 + 4^2 \ \rightarrow\ 25 = 9 + 16 \ \rightarrow\ 25 = 25$
   The given triangle is a right triangle. The hypotenuse is 5.

9. $6^2 = 4^2 + 5^2 \ \rightarrow\ 36 = 16 + 25 \ \rightarrow\ 36 \ne 41$
   The given triangle is not a right triangle.

11. $25^2 = 7^2 + 24^2 \ \rightarrow\ 625 = 49 + 576 \ \rightarrow\ 625 = 625$
    The given triangle is a right triangle. The hypotenuse is 25.

13. $6^2 = 3^2 + 4^2 \ \rightarrow\ 36 = 9 + 16 \ \rightarrow\ 36 \ne 25$
    The given triangle is not a right triangle.

15. $A = l \cdot w = 4 \cdot 2 = 8 \text{ in}^2$

17. $A = \dfrac{1}{2} b \cdot h = \dfrac{1}{2}(2)(4) = 4 \text{ in}^2$

19. $A = \pi r^2 = \pi(5)^2 = 25\pi \text{ m}^2 \qquad C = 2\pi r = 2\pi(5) = 10\pi \text{ m}$

21. $V = lwh = 8 \cdot 4 \cdot 7 = 224 \text{ ft}^2$

23. $V = \dfrac{4}{3}\pi r^3 = \dfrac{4}{3}\pi \cdot 4^3 = \dfrac{256}{3}\pi \text{ cm}^3 \qquad S = 4\pi r^2 = 4\pi \cdot 4^2 = 64\pi \text{ cm}^2$

25. $V = \pi r^2 h = \pi(9)^2(8) = 648\pi \text{ in}^3$

27. The diameter of the circle is 2, so its radius is 1. $A = \pi r^2 = \pi(1)^2 = \pi$ square units

29. The diameter of the circle is the length of the diagonal of the square.
    $$d^2 = 2^2 + 2^2 = 4 + 4 = 8 \ \rightarrow\ d = \sqrt{8} = 2\sqrt{2} \qquad r = \sqrt{2}$$
    The area of the circle is: $A = \pi r^2 = \pi\left(\sqrt{2}\right)^2 = 2\pi$ square units

31. The total distance traveled is 4 times the circumference of the wheel.
Total distance $= 4C = 4(\pi d) = 4\pi \cdot 16 = 64\pi = 201.1$ inches

33. Area of the border = area of EFGH − area of ABCD $= 10^2 - 6^2 = 100 - 36 = 64$ ft^2

35. Area of the window = area of the rectangle + area of the semicircle.
$$A = (6)(4) + \frac{1}{2} \cdot \pi \cdot 2^2 = 24 + 2\pi \approx 30.28 \text{ ft}^2$$
Perimeter of the window = 2 heights + width + one-half the circumference.
$$P = 2(6) + 4 + \frac{1}{2} \cdot \pi(4) = 12 + 4 + 2\pi = 16 + 2\pi \approx 22.28 \text{ feet}$$

37. Convert 20 feet to miles, and solve the Pythagorean theorem to find the distance:

$$20 \text{ feet} = 20 \text{ feet} \cdot \frac{1 \text{ mile}}{5280 \text{ feet}} \approx 0.003788 \text{ miles}$$
$$d^2 = (3960 + 0.003788)^2 - 3960^2 = 30$$
$$d \approx 5.477 \text{ miles}$$

39. Convert 100 feet to miles, and solve the Pythagorean theorem to find the distance:
$$100 \text{ feet} = 100 \text{ feet} \cdot \frac{1 \text{ mile}}{5280 \text{ feet}} \approx 0.018939 \text{ miles}$$
$$d^2 = (3960 + 0.018939)^2 - 3960^2 = 150$$
$$d \approx 12.247 \text{ miles}$$
Convert 150 feet to miles, and solve the Pythagorean theorem to find the distance:
$$150 \text{ feet} = 150 \text{ feet} \cdot \frac{1 \text{ mile}}{5280 \text{ feet}} \approx 0.028409 \text{ miles}$$
$$d^2 = (3960 + 0.028409)^2 - 3960^2 = 225$$
$$d \approx 15 \text{ miles}$$

41. Given a rectangle with perimeter = 1000 feet, the largest area will be enclosed by a square with dimensions 250 by 250 feet.   That is, the area $= 250^2 = 62500$ square feet.

A circular pool with circumference = 1000 feet yields the equation : $2\pi r = 1000 \rightarrow r = \dfrac{500}{\pi}$

The area enclosed by the circular pool is: $A = \pi r^2 = \pi \left(\dfrac{500}{\pi}\right)^2 = \dfrac{500^2}{\pi} \approx 79577.47$ square feet

Therefore, a circular pool will enclose the most area.

Review

A.3 Polynomials and Rational Expressions

1. $\left(10x^5 - 8x^2\right) + \left(3x^3 - 2x^2 + 6\right) = 10x^5 + 3x^3 - 10x^2 + 6$

3. $(x+a)^2 - x^2 = x^2 + xa + ax + a^2 - x^2 = 2ax + a^2$

5. $(x+8)(2x+1) = 2x^2 + x + 16x + 8 = 2x^2 + 17x + 8$

7. $\left(x^2 + x - 1\right)\left(x^2 - x + 1\right) = x^4 - x^3 + x^2 + x^3 - x^2 + x - x^2 + x - 1 = x^4 - x^2 + 2x - 1$

9. $(x+1)^3 - (x-1)^3$ can be factored as a difference of cubes.
$$(x+1)^3 - (x-1)^3 = \left[(x+1) - (x-1)\right]\left[(x+1)^2 + (x+1)(x-1) + (x-1)^2\right]$$
$$= \left[x+1-x+1\right]\left[x^2 + 2x + 1 + x^2 - 1 + x^2 - 2x + 1\right]$$
$$= \left[2\right]\left[3x^2 + 1\right] = 6x^2 + 2$$

11. $x^2 - 36 = (x-6)(x+6)$

13. $1 - 4x^2 = (1-2x)(1+2x)$

15. $x^2 + 7x + 10 = (x+2)(x+5)$

17. $x^2 - 2x + 8 = (x+2)(x-4)$

19. $x^2 + 4x + 16$ is prime.

21. $15 + 2x - x^2 = (3+x)(5-x)$

23. $3x^2 - 12x - 36 = 3(x^2 - 4x - 12) = 3(x-6)(x+2)$

25. $y^4 + 11y^3 + 30y^2 = y^2\left(y^2 + 11y + 30\right) = y^2(y+6)(y+5)$

27. $4x^2 + 12x + 9 = (2x+3)(2x+3)$

29. $3x^2 + 4x + 1 = (3x+1)(x+1)$

31. $x^4 - 81 = \left(x^2 + 9\right)\left(x^2 - 9\right) = \left(x^2 + 9\right)(x-3)(x+3)$

33. $x^6 - 2x^3 + 1 = \left(x^3 - 1\right)\left(x^3 - 1\right) = \left(x^3 - 1\right)^2 = \left[(x-1)\left(x^2 + x + 1\right)\right]^2 = (x-1)^2\left(x^2 + x + 1\right)^2$

35. $x^7 - x^5 = x^5\left(x^2 - 1\right) = x^5(x-1)(x+1)$

37. $5 + 16x - 16x^2 = (5 - 4x)(1 + 4x)$

39. $4y^2 - 16y + 15 = (2y - 3)(2y - 5)$

41. $1 - 8x^2 - 9x^4 = -(9x^4 + 8x^2 - 1) = -(9x^2 - 1)(x^2 + 1) = -(3x - 1)(3x + 1)(x^2 + 1)$

43. $x(x + 3) - 6(x + 3) = (x + 3)(x - 6)$

45. $(x + 2)^2 - 5(x + 2) = (x + 2)[(x + 2) - 5] = (x + 2)(x - 3)$

47. $6x(2 - x)^4 - 9x^2(2 - x)^3 = 3x(2 - x)^3[2(2 - x) - 3x] = 3x(2 - x)^3[4 - 2x - 3x]$
$$= 3x(2 - x)^3[4 - 5x]$$

49. $x^3 + 2x^2 - x - 2 = x^2(x + 2) - (x + 2) = (x + 2)(x^2 - 1) = (x + 2)(x - 1)(x + 1)$

51. $x^4 - x^3 + x - 1 = x^3(x - 1) + (x - 1) = (x - 1)(x^3 + 1) = (x - 1)(x + 1)(x^2 - x + 1)$

53. $\dfrac{3x - 6}{5x} \cdot \dfrac{x^2 - x - 6}{x^2 - 4} = \dfrac{3(x - 2)}{5x} \cdot \dfrac{(x - 3)(x + 2)}{(x - 2)(x + 2)} = \dfrac{3(x - 3)}{5x}$

55. $\dfrac{4x^2 - 1}{x^2 - 16} \cdot \dfrac{x^2 - 4x}{2x + 1} = \dfrac{(2x - 1)(2x + 1)}{(x - 4)(x + 4)} \cdot \dfrac{x(x - 4)}{2x + 1} = \dfrac{x(2x - 1)}{x + 4}$

57. $\dfrac{x}{x^2 - 7x + 6} - \dfrac{x}{x^2 - 2x - 24} = \dfrac{x}{(x - 6)(x - 1)} - \dfrac{x}{(x - 6)(x + 4)}$
$$= \dfrac{x(x + 4)}{(x - 6)(x - 1)(x + 4)} - \dfrac{x(x - 1)}{(x - 6)(x + 4)(x - 1)} = \dfrac{x^2 + 4x - x^2 + x}{(x - 6)(x + 4)(x - 1)} = \dfrac{5x}{(x - 6)(x + 4)(x - 1)}$$

59. $\dfrac{4x}{x^2 - 4} - \dfrac{2}{x^2 + x - 6} = \dfrac{4x}{(x - 2)(x + 2)} - \dfrac{2}{(x + 3)(x - 2)}$
$$= \dfrac{4x(x + 3)}{(x - 2)(x + 2)(x + 3)} - \dfrac{2(x + 2)}{(x + 3)(x - 2)(x + 2)}$$
$$= \dfrac{4x^2 + 12x - 2x - 4}{(x - 2)(x + 2)(x + 3)} = \dfrac{4x^2 + 10x - 4}{(x - 2)(x + 2)(x + 3)}$$
$$= \dfrac{2(2x^2 + 5x - 2)}{(x - 2)(x + 2)(x + 3)}$$

Review

A.4 Polynomial Division; Synthetic Division

1. Divide:

$$
\begin{array}{r}
4x^2 - 11x + 23 \\
x+2\overline{\smash{\big)}\,4x^3 - 3x^2 + x + 1} \\
\underline{4x^3 + 8x^2} \\
-11x^2 + x \\
\underline{-11x^2 - 22x} \\
23x + 1 \\
\underline{23x + 46} \\
-45
\end{array}
$$

Check:

$$(x+2)(4x^2 - 11x + 23) + (-45)$$
$$= 4x^3 - 11x^2 + 23x + 8x^2 - 22x + 46 - 45$$
$$= 4x^3 - 3x^2 + x + 1$$

The quotient is $4x^2 - 11x + 23$; the remainder is –45.

3. Divide:

$$
\begin{array}{r}
4x^2 + 13x + 53 \\
x-4\overline{\smash{\big)}\,4x^3 - 3x^2 + x + 1} \\
\underline{4x^3 - 16x^2} \\
13x^2 + x \\
\underline{13x^2 - 52x} \\
53x + 1 \\
\underline{53x - 212} \\
213
\end{array}
$$

Check:

$$(x-4)(4x^2 + 13x + 53) + 213$$
$$= 4x^3 + 13x^2 + 53x - 16x^2 - 52x - 212 + 213$$
$$= 4x^3 - 3x^2 + x + 1$$

The quotient is $4x^2 + 13x + 53$; the remainder is 213.

5. Divide:

$$
\begin{array}{r}
4x - 3 \\
x^2+2\overline{\smash{\big)}\,4x^3 - 3x^2 + x + 1} \\
\underline{4x^3 + 8x} \\
-3x^2 - 7x \\
\underline{-3x^2 - 6} \\
-7x + 7
\end{array}
$$

Check:

$$(x^2 + 2)(4x - 3) + (-7x + 7)$$
$$= 4x^3 - 3x^2 + 8x - 6 - 7x + 7$$
$$= 4x^3 - 3x^2 + x + 1$$

The quotient is $4x - 3$; the remainder is $-7x + 7$.

7.   Divide:

$$2x^3 - 1 \overline{)\begin{array}{l} 2 \\ 4x^3 - 3x^2 + x + 1 \end{array}}$$

$$\begin{array}{r} 4x^3 \phantom{{}- 3x^2 + x} -2 \\ \hline -3x^2 + x + 3 \end{array}$$

The quotient is 2; the remainder is $-3x^2 + x + 3$.

Check:

$(2x^3 - 1)(2) + (-3x^2 + x + 3)$

$= 4x^3 - 2 - 3x^2 + x + 3 = 4x^3 - 3x^2 + x + 1$

9.   Divide:

$$2x^2 + x + 1 \overline{)\begin{array}{l} 2x - \dfrac{5}{2} \\ 4x^3 - 3x^2 + x + 1 \end{array}}$$

$$\begin{array}{r} 4x^3 + 2x^2 + 2x \\ \hline -5x^2 - x \\ -5x^2 - \dfrac{5}{2}x - \dfrac{5}{2} \\ \hline \dfrac{3}{2}x + \dfrac{7}{2} \end{array}$$

The quotient is $2x - \dfrac{5}{2}$; the remainder is $\dfrac{3}{2}x + \dfrac{7}{2}$.

Check:

$(2x^2 + x + 1)(2x - \dfrac{5}{2}) + \left(\dfrac{3}{2}x + \dfrac{7}{2}\right)$

$= 4x^3 - 5x^2 + 2x^2 - \dfrac{5}{2}x + 2x - \dfrac{5}{2} + \dfrac{3}{2}x + \dfrac{7}{2}$

$= 4x^3 - 3x^2 + x + 1$

11.   Divide:

$$4x^2 + 1 \overline{)\begin{array}{l} x - \dfrac{3}{4} \\ 4x^3 - 3x^2 + x + 1 \end{array}}$$

$$\begin{array}{r} 4x^3 \phantom{{}- 3x^2} + x \\ \hline -3x^2 \phantom{{}+ x} + 1 \\ -3x^2 \phantom{{}+ x} - \dfrac{3}{4} \\ \hline \dfrac{7}{4} \end{array}$$

The quotient is $x - \dfrac{3}{4}$; the remainder is $\dfrac{7}{4}$.

Check:

$(4x^2 + 1)\left(x - \dfrac{3}{4}\right) + \dfrac{7}{4}$

$= 4x^3 - 3x^2 + x - \dfrac{3}{4} + \dfrac{7}{4}$

$= 4x^3 - 3x^2 + x + 1$

13. Divide:

$$x-1\overline{\smash{\big)}\,x^4+0x^3+0x^2+0x-1}$$

with quotient x^3+x^2+x+1

$$
\begin{array}{r}
x^3+x^2+x+1 \\
x-1\overline{\smash{\big)}\,x^4+0x^3+0x^2+0x-1} \\
\underline{x^4-\ x^3} \\
x^3 \\
\underline{x^3-\ x^2} \\
x^2 \\
\underline{x^2-\ x} \\
x-1 \\
\underline{x-1} \\
0
\end{array}
$$

Check:

$(x-1)(x^3+x^2+x+1)+0$

$= x^4+x^3+x^2+x-x^3-x^2-x-1$

$= x^4-1$

The quotient is x^3+x^2+x+1; the remainder is 0.

15. Divide:

$$
\begin{array}{r}
x^2+1 \\
x^2-1\overline{\smash{\big)}\,x^4+0x^3+0x^2+0x-1} \\
\underline{x^4-\ x^2} \\
x^2 \\
\underline{x^2-\ 1} \\
0
\end{array}
$$

Check:

$(x^2-1)(x^2+1)+0$

$= x^4+x^2-x^2-1$

$= x^4-1$

The quotient is x^2+1; the remainder is 0.

17. Divide:

$$
\begin{array}{r}
-4x^2-3x-3 \\
x-1\overline{\smash{\big)}\,-4x^3+\ x^2+0x-4} \\
\underline{-4x^3+4x^2} \\
-3x^2 \\
\underline{-3x^2+3x} \\
-3x-4 \\
\underline{-3x+3} \\
-7
\end{array}
$$

Check:

$(x-1)(-4x^2-3x-3)+(-7)$

$= -4x^3-3x^2-3x+4x^2+3x+3-7$

$= -4x^3+x^2-4$

The quotient is $-4x^2-3x-3$; the remainder is -7.

19. Divide:

$$
\begin{array}{r}
x^2-x-1 \\
x^2+x+1\overline{\smash{\big)}\,x^4+0x^3-\ x^2+0x+1} \\
\underline{x^4+\ x^3+\ x^2} \\
-x^3-2x^2 \\
\underline{-x^3-\ x^2-x} \\
-x^2+x+1 \\
\underline{-x^2-x-1} \\
2x+2
\end{array}
$$

Check:

$(x^2+x+1)(x^2-x-1)+2x+2$

$= x^4+x^3+x^2-x^3-x^2-x-x^2-x-1+2x+2$

$= x^4-x^2+1$

The quotient is $x^2 - x - 1$; the remainder is $2x + 2$.

21. Divide:

$$-x^2 + 1)\overline{\smash{\big)}\,x^4 + 0x^3 - x^2 + 0x + 1}$$
$$\underline{x^4 \qquad\;\; - x^2}$$
$$1$$

Check:
$$-x^2(-x^2 + 1) + 1$$
$$= x^4 - x^2 + 1$$

The quotient is $-x^2$; the remainder is 1.

23. Divide:

$$x - a)\overline{\smash{\big)}\,x^3 + 0x^2 + \;\; 0x - a^3}$$
quotient $x^2 + ax + a^2$
$$\underline{x^3 - ax^2}$$
$$ax^2$$
$$\underline{ax^2 - a^2x}$$
$$a^2x - a^3$$
$$\underline{a^2x - a^3}$$
$$0$$

Check:
$$(x - a)(x^2 + ax + a^2) + 0$$
$$= x^3 + ax^2 + a^2x - ax^2 - a^2x - a^3$$
$$= x^3 - a^3$$

The quotient is $x^2 + ax + a^2$; the remainder is 0.

25. Divide:

$$x - a)\overline{\smash{\big)}\,x^4 + 0x^3 + \;\; 0x^2 + \;\; 0x - a^4}$$
quotient $x^3 + ax^2 + a^2x + a^3$
$$\underline{x^4 - ax^3}$$
$$ax^3$$
$$\underline{ax^3 - a^2x^2}$$
$$a^2x^2$$
$$\underline{a^2x^2 - a^3x}$$
$$a^3x - a^4$$
$$\underline{a^3x - a^4}$$
$$0$$

Check:
$$(x - a)(x^3 + ax^2 + a^2x + a^3) + 0$$
$$= x^4 + ax^3 + a^2x^2 + a^3x - ax^3 - a^2x^2 - a^3x - a^4$$
$$= x^4 - a^4$$

The quotient is $x^3 + ax^2 + a^2x + a^3$; the remainder is 0.

27. Use synthetic division:

$$2)\overline{\begin{array}{rrrr} 1 & -1 & 2 & 4 \\ & 2 & 2 & 8 \end{array}}$$
$$\begin{array}{rrrr} 1 & 1 & 4 & 12 \end{array}$$

Quotient: $x^2 + x + 4$ Remainder: 12

29. Use synthetic division:

$$3)\overline{\begin{array}{rrrr} 3 & 2 & -1 & 3 \\ & 9 & 33 & 96 \end{array}}$$
$$\begin{array}{rrrr} 3 & 11 & 32 & 99 \end{array}$$

Quotient: $3x^2 + 11x + 32$ Remainder: 99

31. Use synthetic division:

```
-3)1   0   -4    0    1      0
      -3   9  -15   45  -138
    ─────────────────────────
    1  -3   5  -15   46  -138
```
Quotient: $x^4 - 3x^3 + 5x^2 - 15x + 46$ Remainder: -138

33. Use synthetic division:

```
1)4   0   -3    0    1    0    5
      4   4    1    1    2    2
    ───────────────────────────
    4   4    1    1    2    2    7
```
Quotient: $4x^5 + 4x^4 + x^3 + x^2 + 2x + 2$ Remainder: 7

35. Use synthetic division:

```
-1.1)0.1      0    0.2        0
        -0.11   0.121   -0.3531
     ────────────────────────────
     0.1  -0.11   0.321   -0.3531
```
Quotient: $0.1x^2 - 0.11x + 0.321$ Remainder: -0.3531

37. Use synthetic division:

```
1)1   0   0   0   0  -1
     1   1   1   1   1
   ──────────────────────
   1   1   1   1   1   0
```
Quotient: $x^4 + x^3 + x^2 + x + 1$ Remainder: 0

39. Use synthetic division:

```
2)4  -3  -8   4
     8  10   4
   ──────────────
   4   5   2   8
```
Remainder $= 8 \neq 0$; therefore $x - 2$ is not a factor of $f(x)$.

41. Use synthetic division:

```
2)3  -6   0  -5   10
     6   0   0  -10
   ───────────────────
   3   0   0  -5    0
```
Remainder $= 0$; therefore $x - 2$ is a factor of $f(x)$.

43. Use synthetic division:

```
-3)3   0   0   82   0   0   27
      -9  27  -81  -3   9  -27
    ──────────────────────────────
    3  -9  27    1  -3   9    0
```
Remainder $= 0$; therefore $x + 3$ is a factor of $f(x)$.

45. Use synthetic division:

$$
\begin{array}{r|rrrrrrr}
-4 & 4 & 0 & -64 & 0 & 1 & 0 & -15 \\
   &   & -16 & 64 & 0 & 0 & -4 & 16 \\
\hline
   & 4 & -16 & 0 & 0 & 1 & -4 & 1
\end{array}
$$

Remainder $= 1 \neq 0$; therefore $x + 3$ is not a factor of $f(x)$.

47. Use synthetic division:

$$
\begin{array}{r|rrrrr}
\frac{1}{2} & 2 & -1 & 0 & 2 & -1 \\
   &   & 1 & 0 & 0 & 1 \\
\hline
   & 2 & 0 & 0 & 2 & 0
\end{array}
$$

Remainder $= 0$; therefore $x - \dfrac{1}{2}$ is a factor of $f(x)$.

49. Answers will vary.

Appendix A

Review

A.5 Solving Equations

1.
$$3x = 21$$
$$\frac{3x}{3} = \frac{21}{3}$$
$$x = 7$$

3.
$$5x + 15 = 0$$
$$5x + 15 - 15 = 0 - 15$$
$$5x = -15$$
$$\frac{5x}{5} = \frac{-15}{5} \rightarrow x = -3$$

5.
$$2x - 3 = 5$$
$$2x - 3 + 3 = 5 + 3$$
$$2x = 8$$
$$\frac{2x}{2} = \frac{8}{2} \rightarrow x = 4$$

7.
$$\frac{1}{3}x = \frac{5}{12}$$
$$(3)\left(\frac{1}{3}x\right) = \left(\frac{5}{12}\right)(3)$$
$$x = \frac{5}{4}$$

9.
$$6 - x = 2x + 9$$
$$6 - x - 6 = 2x + 9 - 6$$
$$-x = 2x + 3$$
$$-x - 2x = 2x + 3 - 2x$$
$$-3x = 3$$
$$\frac{-3x}{-3} = \frac{3}{-3} \rightarrow x = -1$$

11.
$$2(3 + 2x) = 3(x - 4)$$
$$6 + 4x = 3x - 12$$
$$6 + 4x - 6 = 3x - 12 - 6$$
$$4x = 3x - 18$$
$$4x - 3x = 3x - 18 - 3x$$
$$x = -18$$

13.
$$8x - (3x + 2) = 3x - 10$$
$$8x - 3x - 2 = 3x - 10$$
$$5x - 2 = 3x - 10$$
$$5x - 2 + 2 = 3x - 10 + 2$$
$$5x = 3x - 8$$
$$5x - 3x = 3x - 8 - 3x$$
$$2x = -8$$
$$\frac{2x}{2} = \frac{-8}{2} \rightarrow x = -4$$

15.
$$\frac{1}{2}x - 4 = \frac{3}{4}x$$
$$\frac{1}{2}x - 4 + 4 = \frac{3}{4}x + 4$$
$$\frac{1}{2}x = \frac{3}{4}x + 4$$
$$\frac{1}{2}x - \frac{3}{4}x = \frac{3}{4}x + 4 - \frac{3}{4}x$$
$$\frac{2}{4}x - \frac{3}{4}x = 4 \rightarrow -\frac{1}{4}x = 4$$
$$(-4)\left(-\frac{1}{4}x\right) = (4)(-4) \rightarrow x = -16$$

17.
$$0.9t = 0.4 + 0.1t$$
$$0.9t - 0.1t = 0.4 + 0.1t - 0.1t$$
$$0.8t = 0.4$$
$$\frac{0.8t}{0.8} = \frac{0.4}{0.8} \to t = 0.5$$

19.
$$\frac{2}{y} + \frac{4}{y} = 3$$
$$y\left(\frac{2}{y} + \frac{4}{y}\right) = y(3)$$
$$2 + 4 = 3y$$
$$6 = 3y$$
$$\frac{6}{3} = \frac{3y}{3}$$
$$y = 2$$

and since y = 2 does not cause a denominator to equal zero, the solution set is {2}.

21.
$$(x+7)(x-1) = (x+1)^2$$
$$x^2 + 6x - 7 = x^2 + 2x + 1$$
$$x^2 + 6x - 7 - x^2 = x^2 + 2x + 1 - x^2$$
$$6x - 7 = 2x + 1$$
$$6x - 7 + 7 = 2x + 1 + 7$$
$$6x = 2x + 8$$
$$6x - 2x = 2x + 8 - 2x$$
$$4x = 8$$
$$\frac{4x}{4} = \frac{8}{4}$$
$$x = 2$$

23.
$$z(z^2 + 1) = 3 + z^3$$
$$z^3 + z = 3 + z^3$$
$$z^3 + z - z^3 = 3 + z^3 - z^3$$
$$z = 3$$

25.
$$x^2 = 9x$$
$$x^2 - 9x = 0$$
$$x(x-9) = 0$$
$$x = 0 \text{ or } x = 9$$
The solution set is {0, 9}.

27.
$$t^3 - 9t^2 = 0$$
$$t^2(t-9) = 0$$
$$t^2 = 0 \to t = 0$$
$$\text{or } t - 9 = 0 \to t = 9$$
The solution set is $\{0, 9\}$

29.
$$\frac{3}{2x-3} = \frac{2}{x+5}$$
$$(2x+3)(x+5)\left(\frac{3}{2x-3}\right) = \left(\frac{2}{x+5}\right)(2x+3)(x+5)$$
$$(x+5)(3) = (2)(2x+3) \to 3x + 15 = 4x + 12$$
$$3x + 15 - 4x = 4x + 12 - 4x \to 15 - 4x = 12$$
$$15 - 4x - 15 = 12 - 15 \to -4x = -3$$
$$\frac{-4x}{-4} = \frac{-3}{-4} \to x = \frac{3}{4}$$

Since $x = \frac{3}{4}$ does not cause any denominator to equal zero, the solution set is $\left\{\frac{3}{4}\right\}$.

31. $(x+2)(3x)=(x+2)(6)$

$3x^2 + 6x = 6x + 12$

$3x^2 + 6x - 6x = 6x + 12 - 6x$

$3x^2 = 12$

$3x^2 - 12 = 12 - 12$

$3x^2 - 12 = 0 \rightarrow 3(x^2 - 4) = 0$

$3(x-2)(x+2) \rightarrow x - 2 = 0 \rightarrow x = 2$

or $x + 2 = 0 \rightarrow x = -2$

Solution set: $\{-2, 2\}$

33.

$$\frac{2}{x-2} + \frac{3}{x+5} = \frac{10}{(x+5)(x-2)}$$

$$(x-2)(x+5)\left(\frac{2}{x-2} + \frac{3}{x+5}\right) = \left(\frac{10}{(x+5)(x-2)}\right)(x-2)(x+5)$$

$$(x-2)(x+5)\left(\frac{2}{x-2}\right) + (x-2)(x+5)\left(\frac{3}{x+5}\right) = 10$$

$$(x+5)(2) + (x-2)(3) = 10$$

$$2x + 10 + 3x - 6 = 10 \rightarrow 5x + 4 = 10$$

$$5x + 4 - 4 = 10 - 4$$

$$5x = 6 \rightarrow \frac{5x}{5} = \frac{6}{5} \rightarrow x = \frac{6}{5}$$

Since $x = \frac{6}{5}$ does not cause any denominator to equal zero, the solution set is $\left\{\frac{6}{5}\right\}$.

35. $|2x| = 6$

$2x = 6$ or $2x = -6$

$x = 3$ or $x = -3$

The solution set is $\{-3, 3\}$.

37. $|2x + 3| = 5$

$2x + 3 = 5$ or $2x + 3 = -5$

$2x = 2$ or $2x = -8$

$x = 1$ or $x = -4$

The solution set is $\{-4, 1\}$.

39. $|1 - 4t| = 5$

$1 - 4t = 5$ or $1 - 4t = -5$

$-4t = 4$ or $-4t = -6$

$t = -1$ or $t = \frac{3}{2}$

The solution set is $\left\{-1, \frac{3}{2}\right\}$.

41. $|-2x| = 8$

$-2x = 8$ or $-2x = -8$

$x = -4$ or $x = 4$

The solution set is $\{-4, 4\}$.

43. $|-2|x = 4$

$2x = 4 \rightarrow x = 2$

The solution set is $\{2\}$.

45. $|x - 2| = -\dfrac{1}{2}$

impossible, since absolute value always yields a non-negative number.

47. $|x^2 - 4| = 0$

$x^2 - 4 = 0$

$x^2 = 4$

$x = \pm 2$

The solution set is $\{-2, 2\}$.

49. $|x^2 - 2x| = 3$

$x^2 - 2x = 3 \rightarrow x^2 - 2x - 3 = 0 \rightarrow (x - 3)(x + 1) = 0 \rightarrow x = 3, x = -1$

or $x^2 - 2x = -3 \rightarrow x^2 - 2x + 3 = 0 \rightarrow x = \dfrac{2 \pm \sqrt{4 - 12}}{2} = \dfrac{2 \pm \sqrt{-8}}{2}$, not real solutions

The solution set is $\{-1, 3\}$.

51. $|x^2 + x - 1| = 1$

$x^2 + x - 1 = 1 \rightarrow x^2 + x - 2 = 0 \rightarrow (x - 1)(x + 2) = 0 \rightarrow x = 1, x = -2$

or $x^2 + x - 1 = -1 \rightarrow x^2 + x = 0 \rightarrow x(x + 1) = 0 \rightarrow x = 0, x = -1$

The solution set is $\{-2, -1, 0, 1\}$.

53.
$x^2 = 4x$

$x^2 - 4x = 0$

$x(x - 4) = 0$

$x = 0$

or $x - 4 = 0 \rightarrow x = 4$

Solution set: $\{0, 4\}$

55.
$z^2 + 4z - 12 = 0$

$(z + 6)(z - 2) = 0$

$z + 6 = 0 \rightarrow z = -6$

or $z - 2 = 0 \rightarrow z = 2$

Solution set: $\{-6, 2\}$

57.
$2x^2 - 5x - 3 = 0$

$(2x + 1)(x - 3) = 0$

$2x + 1 = 0 \rightarrow 2x = -1 \rightarrow x = -\dfrac{1}{2}$

or $x - 3 = 0 \rightarrow x = 3$

Solution set: $\left\{-\dfrac{1}{2}, 3\right\}$

59. $x(x - 7) + 12 = 0$

$x^2 - 7x + 12 = 0$

$(x - 4)(x - 3) = 0$

$x - 4 = 0 \rightarrow x = 4$

or $x - 3 = 0 \rightarrow x = 3$

Solution set: $\{3, 4\}$

61.
$4x^2 + 9 = 12x$

$4x^2 - 12x + 9 = 0$

$(2x - 3)^2 = 0 \rightarrow x = \dfrac{3}{2}$

Solution set: $\left\{\dfrac{3}{2}\right\}$

63.
$6x - 5 = \dfrac{6}{x}$

$6x^2 - 5x = 6 \rightarrow 6x^2 - 5x - 6 = 0$

$(3x + 2)(2x - 3) = 0 \rightarrow x = -\dfrac{2}{3}$ or $x = \dfrac{3}{2}$

Since neither of these values causes a denominator to equal zero, the solution set is $\left\{-\dfrac{2}{3}, \dfrac{3}{2}\right\}$.

65.

$$\frac{4(x-2)}{x-3}+\frac{3}{x}=\frac{-3}{x(x-3)}$$

$$x(x-3)\left(\frac{4(x-2)}{x-3}+\frac{3}{x}\right)=\left(\frac{-3}{x(x-3)}\right)x(x-3)$$

$$x(x-3)\left(\frac{4(x-2)}{x-3}\right)+x(x-3)\left(\frac{3}{x}\right)=-3$$

$$x(4(x-2))+(x-3)(3)=-3$$

$$4x^2-8x+3x-9=-3$$

$$4x^2-5x-6=0$$

$$(4x+3)(x-2)=0$$

$$x=-\frac{3}{4}\ \text{ or }\ x=2$$

Since neither of these values causes a denominator to equal zero, the solution set is $\left\{-\frac{3}{4},2\right\}$.

73. $\left(\frac{8}{2}\right)^2=4^2=16$

77. $\left(\frac{\left(-\frac{2}{3}\right)}{2}\right)^2=\left(-\frac{1}{3}\right)^2=\frac{1}{9}$

67.
$$x^2=25\rightarrow x=\pm\sqrt{25}\rightarrow x=\pm5$$
The solution set is $\{-5,5\}$.

69. $(x-1)^2=4$

$$x-1=\pm\sqrt{4}$$

$$x-1=\pm2$$

$$x-1=2\ \text{ or }\ x-1=-2$$

$$\rightarrow x=3\ \text{ or }\ x=-1$$

The solution set is $\{-1,3\}$.

71. $(2x+3)^2=9$

$$2x+3=\pm\sqrt{9}$$

$$2x+3=\pm3$$

$$2x+3=3\ \text{ or }\ 2x+3=-3$$

$$\rightarrow x=0\ \text{ or }\ x=-3$$

The solution set is $\{-3,0\}$.

75. $\left(\frac{\left(\frac{1}{2}\right)}{2}\right)^2=\left(\frac{1}{4}\right)^2=\frac{1}{16}$

79.
$$x^2+4x=21$$

$$x^2+4x+4=21+4$$

$$(x+2)^2=25$$

$$x+2=\pm\sqrt{25}\rightarrow x+2=\pm5$$

$$x=-2\pm5\rightarrow x=3\ \text{ or }\ x=-7$$

The solution set is $\{-7,3\}$.

81. $x^2 - \dfrac{1}{2}x - \dfrac{3}{16} = 0$

$x^2 - \dfrac{1}{2}x = \dfrac{3}{16}$

$x^2 - \dfrac{1}{2}x + \dfrac{1}{16} = \dfrac{3}{16} + \dfrac{1}{16}$

$\left(x - \dfrac{1}{4}\right)^2 = \dfrac{1}{4}$

$x - \dfrac{1}{4} = \pm\sqrt{\dfrac{1}{4}}$

$x - \dfrac{1}{4} = \pm\dfrac{1}{2}$

$x = \dfrac{1}{4} \pm \dfrac{1}{2} \rightarrow x = \dfrac{3}{4}$

or $x = -\dfrac{1}{4}$

The solution set is $\left\{-\dfrac{1}{4}, \dfrac{3}{4}\right\}$.

83. $3x^2 + x - \dfrac{1}{2} = 0$

$x^2 + \dfrac{1}{3}x - \dfrac{1}{6} = 0$

$x^2 + \dfrac{1}{3}x = \dfrac{1}{6}$

$x^2 + \dfrac{1}{3}x + \dfrac{1}{36} = \dfrac{1}{6} + \dfrac{1}{36}$

$\left(x + \dfrac{1}{6}\right)^2 = \dfrac{7}{36}$

$x + \dfrac{1}{6} = \pm\sqrt{\dfrac{7}{36}}$

$x + \dfrac{1}{6} = \pm\dfrac{\sqrt{7}}{6}$

$x = -\dfrac{1}{6} \pm \dfrac{\sqrt{7}}{6}$

The solution set is $\left\{-\dfrac{1}{6} + \dfrac{\sqrt{7}}{6}, -\dfrac{1}{6} - \dfrac{\sqrt{7}}{6}\right\}$.

85. $x^2 - 4x + 2 = 0$
$a = 1, \quad b = -4, \quad c = 2$

$x = \dfrac{-(-4) \pm \sqrt{(-4)^2 - 4(1)(2)}}{2(1)}$

$= \dfrac{4 \pm \sqrt{16 - 8}}{2} = \dfrac{4 \pm \sqrt{8}}{2}$

$= \dfrac{4 \pm 2\sqrt{2}}{2} = 2 \pm \sqrt{2}$

$\left\{2 - \sqrt{2}, 2 + \sqrt{2}\right\}$

87. $x^2 - 5x - 1 = 0$
$a = 1, \quad b = -5, \quad c = -1$

$x = \dfrac{-(-5) \pm \sqrt{(-5)^2 - 4(1)(-1)}}{2(1)}$

$= \dfrac{5 \pm \sqrt{25 + 4}}{2} = \dfrac{5 \pm \sqrt{29}}{2}$

$\left\{\dfrac{5 + \sqrt{29}}{2}, \dfrac{5 - \sqrt{29}}{2}\right\}$

89. $2x^2 - 5x + 3 = 0$
$a = 2, \quad b = -5, \quad c = 3$

$x = \dfrac{-(-5) \pm \sqrt{(-5)^2 - 4(2)(3)}}{2(2)}$

$= \dfrac{5 \pm \sqrt{25 - 24}}{4} = \dfrac{5 \pm 1}{4}$

$\left\{1, \dfrac{3}{2}\right\}$

91. $4y^2 - y + 2 = 0$
$a = 4, \quad b = -1, \quad c = 2$

$y = \dfrac{-(-1) \pm \sqrt{(-1)^2 - 4(4)(2)}}{2(4)}$

$= \dfrac{1 \pm \sqrt{1 - 32}}{8} = \dfrac{1 \pm \sqrt{-31}}{8}$

No real solution.

93. $4x^2 = 1 - 2x$

$4x^2 + 2x - 1 = 0$

$a = 4, \quad b = 2, \quad c = -1$

$x = \dfrac{-2 \pm \sqrt{2^2 - 4(4)(-1)}}{2(4)}$

$= \dfrac{-2 \pm \sqrt{4 + 16}}{8} = \dfrac{-2 \pm \sqrt{20}}{8}$

$= \dfrac{-2 \pm 2\sqrt{5}}{8} = \dfrac{-1 \pm \sqrt{5}}{4}$

$\left\{ \dfrac{-1 - \sqrt{5}}{4}, \dfrac{-1 + \sqrt{5}}{4} \right\}$

95. $x^2 + \sqrt{3}x - 3 = 0$

$a = 1, \quad b = \sqrt{3}, \quad c = -3$

$x = \dfrac{-(\sqrt{3}) \pm \sqrt{(\sqrt{3})^2 - 4(1)(-3)}}{2(1)}$

$= \dfrac{-\sqrt{3} \pm \sqrt{3 + 12}}{2} = \dfrac{-\sqrt{3} \pm \sqrt{15}}{2}$

$x \approx 1.07, x \approx -2.80$

$\{1.07, -2.80\}$

97. $x^2 - 5x + 7 = 0$

$a = 1, \quad b = -5, \quad c = 7$

$b^2 - 4ac = (-5)^2 - 4(1)(7)$

$= 25 - 28 = -3$

Since the discriminant < 0, we have no real solutions

99. $9x^2 - 30x + 25 = 0$

$a = 9, \quad b = -30, \quad c = 25$

$b^2 - 4ac = (-30)^2 - 4(9)(25)$

$= 900 - 900 = 0$

since the discriminant $= 0$, we have one repeated real solution.

101. $3x^2 + 5x - 8 = 0$

$a = 3, \quad b = 5, \quad c = -8$

$b^2 - 4ac = (5)^2 - 4(3)(-8)$

$= 25 + 96 = 121$

since the discriminant > 0, we have two unequal real solutions

103. $ax - b = c, \quad a \neq 0$

$ax - b + b = c + b$

$ax = c + b$

$\dfrac{ax}{a} = \dfrac{c + b}{a}$

$x = \dfrac{c + b}{a}$

105. $\dfrac{x}{a} + \dfrac{x}{b} = c, \quad a \neq 0, b \neq 0, a \neq -b$

$ab\left(\dfrac{x}{a} + \dfrac{x}{b}\right) = ab \cdot c$

$bx + ax = abc$

$x(a + b) = abc$

$\dfrac{x(a + b)}{a + b} = \dfrac{abc}{a + b}$

$x = \dfrac{abc}{a + b}$

107.
$$\frac{1}{x-a}+\frac{1}{x+a}=\frac{2}{x-1}$$

$$(x-a)(x+a)(x-1)\left(\frac{1}{x-a}+\frac{1}{x+a}\right)=\left(\frac{2}{x-1}\right)(x-a)(x+a)(x-1)$$

$$(x+a)(x-1)(1)+(x-a)(x-1)(1)=(2)(x-a)(x+a)$$

$$x^2-x+ax-a+x^2-x-ax+a=2x^2-2a^2$$

$$2x^2-2x=2x^2-2a^2$$

$$-2x=-2a^2 \to \frac{-2x}{-2}=\frac{-2a^2}{-2} \to x=a^2$$

such that $x \neq \pm a, x \neq 1$

109. Solving for R:

$$\frac{1}{R}=\frac{1}{R_1}+\frac{1}{R_2}$$

$$RR_1R_2\left(\frac{1}{R}\right)=RR_1R_2\left(\frac{1}{R_1}+\frac{1}{R_2}\right)$$

$$R_1R_2=RR_2+RR_1$$

$$R_1R_2=R(R_2+R_1)$$

$$\frac{R_1R_2}{R_2+R_1}=\frac{R(R_2+R_1)}{R_2+R_1}$$

$$\frac{R_1R_2}{R_2+R_1}=R$$

111. Solving for R:

$$F=\frac{mv^2}{R}$$

$$RF=R\left(\frac{mv^2}{R}\right) \to RF=mv^2$$

$$\frac{RF}{F}=\frac{mv^2}{F} \to R=\frac{mv^2}{F}$$

113. Solving for r:

$$S=\frac{a}{1-r}$$

$$S(1-r)=\left(\frac{a}{1-r}\right)(1-r)$$

$$S-Sr=a \to S-Sr-S=a-S$$

$$-Sr=a-S$$

$$\frac{-Sr}{-S}=\frac{a-S}{-S} \to r=\frac{S-a}{S}$$

115. Given a quadratic equation $ax^2+bx+c=0$

the solutions are given by $x_1=\frac{-b+\sqrt{b^2-4ac}}{2a}$ and $x_2=\frac{-b-\sqrt{b^2-4ac}}{2a}$

adding these two values we get

$$\frac{-b+\sqrt{b^2-4ac}}{2a}+\frac{-b-\sqrt{b^2-4ac}}{2a}$$

$$=\frac{-b+\sqrt{b^2-4ac}-b-\sqrt{b^2-4ac}}{2a}=\frac{-2b}{2a}=-\frac{b}{a}$$

117.   The quadratic equation $kx^2 + x + k = 0$ will have a repeated real solution provided the discriminant $= 0$ that is, we need $b^2 - 4ac = 0$ in the given quadratic equation
$$b^2 - 4ac = (1)^2 - 4(k)(k) = 1 - 4k^2$$
so we solve
$$1 - 4k^2 = 0 \rightarrow 1 = 4k^2 \rightarrow \frac{1}{4} = k^2 \rightarrow \pm\sqrt{\frac{1}{4}} = k \rightarrow \pm\frac{1}{2} = k$$

119.   The quadratic equation $ax^2 + bx + c = 0$ has solutions given by
$$x_1 = \frac{-b + \sqrt{b^2 - 4ac}}{2a} \text{ and } x_2 = \frac{-b - \sqrt{b^2 - 4ac}}{2a}$$
The quadratic equation $ax^2 - bx + c = 0$ has solutions given by
$$x_3 = \frac{-(-b) + \sqrt{(-b)^2 - 4ac}}{2a} = \frac{b + \sqrt{b^2 - 4ac}}{2a} = -x_2$$
and
$$x_4 = \frac{-(-b) - \sqrt{(-b)^2 - 4ac}}{2a} = \frac{b - \sqrt{b^2 - 4ac}}{2a} = -x_1$$
so we have the negatives of the first pair of solutions.

121. $x^2 = 9$ is not equivalent to $x = 3$ since $x^2 = 9$ also has $x = -3$ as a solution.
    $x = \sqrt{9}$ is equivalent to $x = 3$ since the equations have equivalent solutions
    $(x-1)(x-2) = (x-1)^2$ is not equivalent to $x - 2 = x - 1$ since the first equation has solution set {1}, but the second equation has no solution.

123 - 127.   Answers will vary.

Appendix A

Review

A.6 Complex Numbers; Quadratic Equations with a Negative Discriminant

1. $(2 - 3i) + (6 + 8i) = (2 + 6) + (-3 + 8)i = 8 + 5i$

3. $(-3 + 2i) - (4 - 4i) = (-3 - 4) + (2 - (-4))i = -7 + 6i$

5. $(2 - 5i) - (8 + 6i) = (2 - 8) + (-5 - 6)i = -6 - 11i$

7. $3(2 - 6i) = 6 - 18i$

9. $2i(2 - 3i) = 4i - 6i^2 = 4i - 6(-1) = 6 + 4i$

11. $(3 - 4i)(2 + i) = 6 + 3i - 8i - 4i^2 = 6 - 5i - 4(-1) = 10 - 5i$

13. $(-6 + i)(-6 - i) = 36 + 6i - 6i - i^2 = 36 - (-1) = 37$

15. $\dfrac{10}{3 - 4i} = \dfrac{10}{3 - 4i} \cdot \dfrac{3 + 4i}{3 + 4i} = \dfrac{30 + 40i}{9 + 12i - 12i - 16i^2} = \dfrac{30 + 40i}{9 - 16(-1)} = \dfrac{30 + 40i}{25}$

$\qquad = \dfrac{30}{25} + \dfrac{40}{25}i = \dfrac{6}{5} + \dfrac{8}{5}i$

17. $\dfrac{2 + i}{i} = \dfrac{2 + i}{i} \cdot \dfrac{-i}{-i} = \dfrac{-2i - i^2}{-i^2} = \dfrac{-2i - (-1)}{-(-1)} = \dfrac{1 - 2i}{1} = 1 - 2i$

19. $\dfrac{6 - i}{1 + i} = \dfrac{6 - i}{1 + i} \cdot \dfrac{1 - i}{1 - i} = \dfrac{6 - 6i - i + i^2}{1 - i + i - i^2} = \dfrac{6 - 7i + (-1)}{1 - (-1)} = \dfrac{5 - 7i}{2} = \dfrac{5}{2} - \dfrac{7}{2}i$

21. $\left(\dfrac{1}{2} + \dfrac{\sqrt{3}}{2}i\right)^2 = \dfrac{1}{4} + 2\left(\dfrac{1}{2}\right)\left(\dfrac{\sqrt{3}}{2}i\right) + \dfrac{3}{4}i^2 = \dfrac{1}{4} + \dfrac{\sqrt{3}}{2}i + \dfrac{3}{4}(-1) = -\dfrac{1}{2} + \dfrac{\sqrt{3}}{2}i$

23. $(1 + i)^2 = 1 + 2i + i^2 = 1 + 2i + (-1) = 2i$

25. $i^{23} = i^{22+1} = i^{22} \cdot i = \left(i^2\right)^{11} \cdot i = (-1)^{11}i = -i$

27. $i^{-15} = \dfrac{1}{i^{15}} = \dfrac{1}{i^{14+1}} = \dfrac{1}{i^{14} \cdot i} = \dfrac{1}{(i^2)^7 \cdot i} = \dfrac{1}{(-1)^7 i} = \dfrac{1}{-i} = \dfrac{1}{-i} \cdot \dfrac{i}{i} = \dfrac{i}{-i^2} = \dfrac{i}{-(-1)} = i$

29. $i^6 - 5 = \left(i^2\right)^3 - 5 = (-1)^3 - 5 = -1 - 5 = -6$

31. $6i^3 - 4i^5 = i^3(6 - 4i^2) = i^2 \cdot i(6 - 4(-1)) = -1 \cdot i(10) = -10i$

33. $(1 + i)^3 = (1 + i)(1 + i)(1 + i) = (1 + 2i + i^2)(1 + i) = (1 + 2i - 1)(1 + i) = 2i(1 + i)$
$= 2i + 2i^2 = 2i + 2(-1) = -2 + 2i$

35. $i^7(1 + i^2) = i^7(1 + (-1)) = i^7(0) = 0$

37. $i^6 + i^4 + i^2 + 1 = \left(i^2\right)^3 + \left(i^2\right)^2 + i^2 + 1 = (-1)^3 + (-1)^2 + (-1) + 1 = -1 + 1 - 1 + 1 = 0$

39. $\sqrt{-4} = 2i$ 41. $\sqrt{-25} = 5i$

43. $\sqrt{(3 + 4i)(4i - 3)} = \sqrt{12i - 9 + 16i^2 - 12i} = \sqrt{-9 + 16(-1)} = \sqrt{-25} = 5i$

45. $x^2 + 4 = 0$
$a = 1, b = 0, c = 4, \quad b^2 - 4ac = 0^2 - 4(1)(4) = -16$

$x = \dfrac{-0 \pm \sqrt{-16}}{2(1)} = \dfrac{\pm 4i}{2} = \pm 2i$ The solution set is $\{-2i, 2i\}$.

47. $x^2 - 16 = 0$
$a = 1, b = 0, c = -16, \quad b^2 - 4ac = 0^2 - 4(1)(-16) = 64$

$x = \dfrac{-0 \pm \sqrt{64}}{2(1)} = \dfrac{\pm 8}{2} = \pm 4$ The solution set is $\{-4, 4\}$.

49. $x^2 - 6x + 13 = 0$
$a = 1, b = -6, c = 13, \quad b^2 - 4ac = (-6)^2 - 4(1)(13) = 36 - 52 = -16$

$x = \dfrac{-(-6) \pm \sqrt{-16}}{2(1)} = \dfrac{6 \pm 4i}{2} = 3 \pm 2i$ The solution set is $\{3 - 2i, 3 + 2i\}$.

51. $x^2 - 6x + 10 = 0$
$a = 1, b = -6, c = 10, \quad b^2 - 4ac = (-6)^2 - 4(1)(10) = 36 - 40 = -4$

$x = \dfrac{-(-6) \pm \sqrt{-4}}{2(1)} = \dfrac{6 \pm 2i}{2} = 3 \pm i$ The solution set is $\{3 - i, 3 + i\}$.

53. $8x^2 - 4x + 1 = 0$
$a = 8, b = -4, c = 1, \quad b^2 - 4ac = (-4)^2 - 4(8)(1) = 16 - 32 = -16$

$x = \dfrac{-(-4) \pm \sqrt{-16}}{2(8)} = \dfrac{4 \pm 4i}{16} = \dfrac{1}{4} \pm \dfrac{1}{4}i$ The solution set is $\left\{\dfrac{1}{4} - \dfrac{1}{4}i, \dfrac{1}{4} + \dfrac{1}{4}i\right\}$.

55. $5x^2 + 2x + 1 = 0$

$a = 5, b = 2, c = 1, \quad b^2 - 4ac = 2^2 - 4(5)(1) = 4 - 20 = -16$

$x = \dfrac{-2 \pm \sqrt{-16}}{2(5)} = \dfrac{-2 \pm 4i}{10} = -\dfrac{1}{5} \pm \dfrac{2}{5}i$ The solution set is $\left\{ -\dfrac{1}{5} - \dfrac{2}{5}i, \ -\dfrac{1}{5} + \dfrac{2}{5}i \right\}$.

57. $x^2 + x + 1 = 0$

$a = 1, b = 1, c = 1, \quad b^2 - 4ac = 1^2 - 4(1)(1) = 1 - 4 = -3$

$x = \dfrac{-1 \pm \sqrt{-3}}{2(1)} = \dfrac{-1 \pm \sqrt{3}\,i}{2} = -\dfrac{1}{2} \pm \dfrac{\sqrt{3}}{2}i$ The solution set is $\left\{ -\dfrac{1}{2} - \dfrac{\sqrt{3}}{2}i, \ -\dfrac{1}{2} + \dfrac{\sqrt{3}}{2}i \right\}$.

59. $x^3 - 8 = 0$

$(x - 2)\left(x^2 + 2x + 4\right) = 0$

$$x - 2 = 0 \rightarrow x = 2$$

or $x^2 + 2x + 4 = 0$

$a = 1, b = 2, c = 4, \quad b^2 - 4ac = 2^2 - 4(1)(4) = 4 - 16 = -12$

$x = \dfrac{-2 \pm \sqrt{-12}}{2(1)} = \dfrac{-2 \pm 2\sqrt{3}\,i}{2} = -1 \pm \sqrt{3}i$ The solution set is $\left\{ -1 - \sqrt{3}i, \ -1 + \sqrt{3}i \right\}$.

61. $x^4 - 16 = 0$

$$\left(x^2 - 4\right)\left(x^2 + 4\right) = 0$$

$(x - 2)(x + 2)\left(x^2 + 4\right) = 0$

$$x - 2 = 0 \rightarrow x = 2$$

or $x + 2 = 0 \rightarrow x = -2$

or $x^2 + 4 = 0$

$a = 1, b = 0, c = 4, \quad b^2 - 4ac = 0^2 - 4(1)(4) = 0 - 16 = -16$

$x = \dfrac{-0 \pm \sqrt{-16}}{2(1)} = \dfrac{\pm 4i}{2} = \pm 2i$ The solution set is $\{-2, \ 2, \ -2i, \ 2i\}$.

63. $x^4 + 13x^2 + 36 = 0$

$\left(x^2 + 9\right)\left(x^2 + 4\right) = 0$

$x^2 + 9 = 0 \rightarrow a = 1, b = 0, c = 9, \quad b^2 - 4ac = 0^2 - 4(1)(9) = 0 - 36 = -36$

$$x = \dfrac{-0 \pm \sqrt{-36}}{2(1)} = \dfrac{\pm 6i}{2} = \pm 3i$$

or $x^2 + 4 = 0 \rightarrow a = 1, b = 0, c = 4, \quad b^2 - 4ac = 0^2 - 4(1)(4) = 0 - 16 = -16$

$$x = \dfrac{-0 \pm \sqrt{-16}}{2(1)} = \dfrac{\pm 4i}{2} = \pm 2i$$

The solution set is $\{-3i, \ 3i, \ -2i, \ 2i\}$.

65. $3x^2 - 3x + 4 = 0$
$a = 3, b = -3, c = 4, \quad b^2 - 4ac = (-3)^2 - 4(3)(4) = 9 - 48 = -39$
The equation has two complex conjugate solutions.

67. $2x^2 + 3x - 4 = 0$
$a = 2, b = 3, c = -4, \quad b^2 - 4ac = 3^2 - 4(2)(-4) = 9 + 32 = 41$
The equation has two unequal real solutions.

69. $9x^2 - 12x + 4 = 0$
$a = 9, b = -12, c = 4, \quad b^2 - 4ac = (-12)^2 - 4(9)(4) = 144 - 144 = 0$
The equation has a repeated real solution.

71.   The other solution is the conjugate of $2 + 3i$, namely, $2 - 3i$.

73. $z + \overline{z} = 3 - 4i + \overline{3 - 4i} = 3 - 4i + 3 + 4i = 6$

75. $z \cdot \overline{z} = (3 - 4i)(\overline{3 - 4i}) = (3 - 4i)(3 + 4i) = 9 + 12i - 12i - 16i^2 = 9 - 16(-1) = 25$

77. $z + \overline{z} = a + bi + \overline{a + bi} = a + bi + a - bi = 2a$
$z - \overline{z} = a + bi - (\overline{a + bi}) = a + bi - (a - bi) = a + bi - a + bi = 2bi$

79. $\overline{z + w} = \overline{(a + bi) + (c + di)} = \overline{(a + c) + (b + d)i} = (a + c) - (b + d)i$
$\qquad\qquad = (a - bi) + (c - di) = \overline{a + bi} + \overline{c + di} = \overline{z} + \overline{w}$

81.   Answers will vary.

Appendix A

Review

A.7 Setting Up Equations; Applications

1. Let A represent the area of the circle and r the radius.
The area of a circle is the product of π times the square of the radius. $\quad A = \pi r^2$

3. Let A represent the area of the square and s the length of a side.
The area of the square is the square of the length of a side. $\quad A = s^2$

5. Let F represent the force, m the mass, and a the acceleration.
Force equals the product of the mass times the acceleration. $\quad F = ma$

7. Let W represent the work, F the force, and d the distance.
Work equals force times distance. $\quad W = Fd$

9. C = total variable cost, x = number of dishwashers manufactured.
$C = 150x$

11.

Amount in Bonds	Amount in CD's	Total
x	$x - 3000$	20,000

$x + x - 3000 = 20000$

$2x - 3000 = 20000$

$2x = 23000 \rightarrow x = 11500$

$11,500 will be invested in bonds. $8,500 will be invested in CD's.

13. Let x represent the score on the final exam and construct the table

	Test1	Test2	Test3	Test4	Test5	Final Exam	Final Exam
score	80	83	71	61	95	x	x
weight	1/7	1/7	1/7	1/7	1/7	1/7	1/7

Compute the final average and set equal to 80.

$$\left(\frac{1}{7}\right)(80 + 83 + 71 + 61 + 95 + x + x) = 80$$

Now solve for x:

$$\left(\frac{1}{7}\right)(390 + 2x) = 80$$

$390 + 2x = 560 \rightarrow 2x = 170 \rightarrow x = 85$

Brooke needs to score an 85 on the final exam to get an average of 80 in the course.

15. l = length, w = width

   $2l + 2w = 60$ Perimeter $= 2l + 2w$

   $l = w + 8$ The length is 8 more than the width.

   $2(w + 8) + 2w = 60$

   $2w + 16 + 2w = 60$

   $4w + 16 = 60 \rightarrow 4w = 44 \rightarrow w = 11$ feet, $l = 19$ feet

17. Let x represent the amount of money invested in bonds.

   Then $50,000 - x$ represents the amount of money invested in CD's.

	Principle	Rate	Time (yrs)	Interest
Bonds	x	0.15	1	$0.15x$
CD's	$50,000 - x$	0.07	1	$0.07(50,000 - x)$

   Since the total interest is to be $6,000, we have:

   $0.15x + 0.07(50,000 - x) = 6,000$

   $(100)(0.15x + 0.07(50,000 - x)) = (6,000)(100)$

   $15x + 7(50,000 - x) = 600,000$

   $15x + 350,000 - 7x = 600,000$

   $8x + 350,000 = 600,000 \rightarrow 8x = 250,000 \rightarrow x = 31,250$

   $31,250 should be invested in bonds at 15% and $18,750 should be invested in CD's at 7%.

19. Let x represent the amount of money loaned at 8%.

   Then $12,000 - x$ represents the amount of money loaned at 18%.

	Principle	Rate	Time (yrs)	Interest
Loan at 8%	x	0.08	1	$0.08x$
Loan at 18%	$12,000 - x$	0.18	1	$0.18(12,000 - x)$

   Since the total interest is to be $1,000, we have:

   $0.08x + 0.18(12,000 - x) = 1,000$

   $(100)(0.08x + 0.18(12,000 - x)) = (1,000)(100)$

   $8x + 18(12,000 - x) = 100,000$

   $8x + 216,000 - 18x = 100,000$

   $-10x + 216,000 = 100,000 \rightarrow -10x = -116,000 \rightarrow x = 11,600$

   $11,600 is loaned at 8% and $400 is loaned at 18%.

21. Let x represent the number of pounds of Earl Gray tea.

   Then $100 - x$ represents the number of pounds of Orange Pekoe tea.

	No. of pounds	Price per pound	Total Value
Earl Gray	x	$5.00	$5x$
Orange Pekoe	$100 - x$	$3.00	$3(100 - x)$
Blend	100	$4.50	$4.50(100)$

   $5x + 3(100 - x) = 4.50(100)$

   $5x + 300 - 3x = 450$

   $2x + 300 = 450 \rightarrow 2x = 150 \rightarrow x = 75$

   75 pounds of Earl Gray tea must be blended with 25 pounds of Orange Pekoe.

23.   Let x represent the number of pounds of cashews.
Then $x + 60$ represents the number of pounds in the mixture.

	No. of pounds	Price per pound	Total Value
cashews	x	$4.00	$4x$
peanuts	60	$1.50	1.50(60)
mixture	$x + 60$	$2.50	$2.50(x + 60)$

$$4x + 1.50(60) = 2.50(x + 60)$$

$$4x + 90 = 2.50x + 150 \rightarrow 1.5x = 60 \rightarrow x = 40$$

40 pounds of cashews must be added to the 60 pounds of peanuts.

25.   Let r represent the speed of the current.

	Rate	Time	Distance
Upstream	$16 - r$	$\dfrac{20}{60} = \dfrac{1}{3}$	$\dfrac{16 - r}{3}$
Downstream	$16 + r$	$\dfrac{15}{60} = \dfrac{1}{4}$	$\dfrac{16 + r}{4}$

Since the distance is the same in each direction:

$$\frac{16 - r}{3} = \frac{16 + r}{4}$$

$$4(16 - r) = 3(16 + r)$$

$$64 - 4r = 48 + 3r \rightarrow 16 = 7r \rightarrow r = \frac{16}{7} \approx 2.286$$

The speed of the current is approximately 2.286 miles per hour.

27.   Let r represent the rate of the Metra commuter train.
Then $r + 50$ represents the rate of the Amtrak train.

	Rate	Time	Distance
Metra train	r	3	$3r$
Amtrak train	$r + 50$	1	$r + 50$

Amtrak distance = Metra distance $- 10$

$$r + 50 = 3r - 10 \rightarrow 60 = 2r \rightarrow r = 30$$

The Metra commuter train travels at a rate of 30 miles per hour.
The Amtrak train travels at a rate of 80 miles per hour.

29.   Let t represent the time it takes to do the job together.

	Minutes to do job	Part of job done in one minute
Trent	30	$\dfrac{1}{30}$
Lois	20	$\dfrac{1}{20}$
Together	t	$\dfrac{1}{t}$

$$\frac{1}{30} + \frac{1}{20} = \frac{1}{t} \rightarrow 2t + 3t = 60 \rightarrow 5t = 60 \rightarrow t = 12$$

Working together, the job can be done in 12 minutes.

31. Let x represent the length of the side of the sheet metal.

$$(x-2)(x-2)(1) = 4$$
$$x^2 - 4x + 4 = 4$$
$$x^2 - 4x = 0$$
$$x(x-4) = 0 \rightarrow x = 0 \text{ or } x = 4$$

Since the side cannot be 0 feet long, the length of a side of the sheet metal is 4 feet. Hence, the dimensions of the sheet metal before the cut are 4 feet by 4 feet.

33. Let r represent the speed of the current.

	Rate	Time	Distance
Upstream	$15-r$	$\dfrac{10}{15-r}$	10
Downstream	$15+r$	$\dfrac{10}{15+r}$	10

Since the total time is 1.5 hours, we have:

$$\frac{10}{15-r} + \frac{10}{15+r} = 1.5$$
$$10(15+r) + 10(15-r) = 1.5(15-r)(15+r)$$
$$150 + 10r + 150 - 10r = 1.5(225 - r^2)$$
$$300 = 1.5(225 - r^2)$$
$$200 = 225 - r^2 \rightarrow r^2 - 25 = 0 \rightarrow (r-5)(r+5) = 0 \rightarrow r = 5 \text{ or } r = -5$$

The speed of the current is 5 miles per hour.

35. Let w represent the width of window.
Then $l = w + 2$ represents the length of the window.
Since the area is 143 square feet, we have: $w(w+2) = 143$

$$w^2 + 2w - 143 = 0 \rightarrow (w+13)(w-11) = 0 \rightarrow w = -13 \text{ which is not practical}$$
$$\text{or } w = 11$$

The width of the rectangular window is 11 feet and the length is 13 feet.

37. Let l represent the length of the rectangle.
Let w represent the width of the rectangle.
The perimeter is 26 meters and the area is 40 square meters.

$$2l + 2w = 26 \quad \rightarrow \quad l + w = 13 \quad \rightarrow \quad w = 13 - l$$
$$lw = 40$$
$$l(13 - l) = 40$$
$$13l - l^2 = 40$$
$$l^2 - 13l + 40 = 0 \rightarrow (l-8)(l-5) = 0$$
$$l = 8 \text{ or } l = 5$$
$$w = 5 \quad w = 8$$

The dimensions are 5 meters by 8 meters.

Appendix A   Review

39.   (a) $s = 96 + 80t - 16t^2$. The ball strikes the ground when the height $= 0$.
      So we solve $0 = 96 + 80t - 16t^2$.
$$0 = -16t^2 + 80t + 96 \rightarrow \frac{0}{-16} = \frac{-16t^2 + 80t + 96}{-16}$$
$$0 = t^2 - 5t - 6 \rightarrow 0 = (t-6)(t+1) \rightarrow t = 6 \text{ or } t = -1, \text{ we discard the negative value}$$
since t represents elapsed time. Therefore, the ball hits the ground after 6 seconds.

      (b) $s = 96 + 80t - 16t^2$. The ball passes the top of the building when the
                                        height $= 96$.
      So we solve $96 = 96 + 80t - 16t^2$.
$$96 = 96 + 80t - 16t^2 \rightarrow 0 = 80t - 16t^2 \rightarrow 0 = 16t(5 - t) \rightarrow t = 0 \text{ or } t = 5$$
      We know that the ball starts ($t = 0$) at a height of 96 feet. Therefore, the ball hits the
      ground after 6 seconds.

41. $l =$ length of the garden
      $w =$ width of the garden
      (a)   The length of the garden is to be twice its width. Thus, $l = 2w$.
            The dimensions of the fence are $l + 4$ and $w + 4$.
            The perimeter is 46 feet, so:
$$2(l+4) + 2(w+4) = 46$$
$$2(2w+4) + 2(w+4) = 46$$
$$4w + 8 + 2w + 8 = 46$$
$$6w + 16 = 46 \rightarrow 6w = 30 \rightarrow w = 5$$
            The dimensions of the garden are 5 feet by 10 feet.
      (b)   Area $= l \cdot w = 5 \cdot 10 = 50$ square feet.
      (c)   If the dimensions of the garden are the same, then the length and width of the fence
            are also the same $(l + 4)$. The perimeter is 46 feet, so:
$$2(l+4) + 2(l+4) = 46$$
$$2l + 8 + 2l + 8 = 46$$
$$4l + 16 = 46 \rightarrow 4l = 30 \rightarrow l = 7.5$$
            The dimensions of the garden are 7.5 feet by 7.5 feet.
      (d)   Area $= l \cdot w = 7.5(7.5) = 56.25$ square feet.

43.   Let x represent the width of the border measured in feet.
      The radius of the pool is 5 feet.
      Then $x + 5$ represents the radius of the circle, including both the pool and the border.
      The total area of the pool and border is $A_T = \pi(x + 5)^2$.
      The area of the pool is $A_P = \pi(5)^2 = 25\pi$.
      The area of the border is $A_B = A_T - A_P = \pi(x + 5)^2 - 25\pi$.
      Since the concrete is 3 inches or 0.25 feet thick, the volume of the concrete in the border is
$$0.25A_B = 0.25\left(\pi(x+5)^2 - 25\pi\right)$$
      Solving the volume equation:
$$0.25\left(\pi(x+5)^2 - 25\pi\right) = 27 \rightarrow \pi\left(x^2 + 10x + 25 - 25\right) = 108 \rightarrow \pi x^2 + 10\pi x - 108 = 0$$

$$x = \frac{-10\pi \pm \sqrt{(10\pi)^2 - 4(\pi)(-108)}}{2(\pi)}$$

$$= \frac{-31.42 \pm \sqrt{2344.1285}}{6.28} = \frac{-31.42 \pm 48.42}{6.28} = 2.71 \text{ or } -12.71$$

The width of the border is approximately 2.71 feet.

45. Let x represent the width of the border measured in feet.

The total area is $A_T = (6 + 2x)(10 + 2x)$.

The area of the garden is $A_G = 6 \cdot 10 = 60$.

The area of the border is $A_B = A_T - A_G = (6 + 2x)(10 + 2x) - 60$.

Since the concrete is 3 inches or 0.25 feet thick, the volume of the concrete in the border is
$$0.25 A_B = 0.25\big((6 + 2x)(10 + 2x) - 60\big)$$

Solving the volume equation:

$$0.25\big((6 + 2x)(10 + 2x) - 60\big) = 27$$

$$60 + 32x + 4x^2 - 60 = 108$$

$$4x^2 + 32x - 108 = 0 \rightarrow x^2 + 8x - 27 = 0$$

$$x = \frac{-8 \pm \sqrt{8^2 - 4(1)(-27)}}{2(1)} = \frac{-8 \pm \sqrt{172}}{2}$$

$$= \frac{-8 \pm 13.11}{2} = 2.56 \text{ or } -10.56 \text{ which is not practical.}$$

The width of the border is approximately 2.56 feet.

47. Let x represent the number of ounces of pure water.

Then $x + 1$ represents the number of gallons in the 60% solution.

	No. of gallons	Conc. of Antifreeze	Pure Antifreeze
water	x	0	0
100% antifreeze	1	1.00	1(1)
60% antifreeze	$x + 1$	0.60	$0.60(x+1)$

$$0 + 1(1) = 0.60(x + 1)$$

$$1 = 0.6x + 0.6 \rightarrow 0.4 = 0.6x \rightarrow x = \frac{4}{6} = \frac{2}{3}$$

$\frac{2}{3}$ gallon of pure water should be added.

49. Let x represent the number of ounces of water to be evaporated.

	No. of ounces	Conc. of Salt	Pure Salt
Water	x	0.00	0
4% Salt	32	0.04	0.04(32)
6% Salt	$32 - x$	0.06	$0.06(32 - x)$

$$0 + 0.04(32) = 0.06(32 - x)$$

$$1.28 = 1.92 - 0.06x \rightarrow 0.06x = 0.64 \rightarrow x = \frac{0.64}{0.06} = \frac{32}{3}$$

32/3 ounces of water need to be evaporated.

51. Let x represent the number of grams of pure gold.
     Then $60 - x$ represents the number of grams of 12 karat gold to be used.

	No. of grams	Conc. of gold	Pure gold
Pure gold	x	1.00	x
12 karat gold	$60 - x$	$\dfrac{1}{2}$	$\dfrac{1}{2}(60-x)$
16 karat gold	60	$\dfrac{2}{3}$	$\dfrac{2}{3}(60)$

$$x + \frac{1}{2}(60 - x) = \frac{2}{3}(60)$$

$$x + 30 - 0.5x = 40 \rightarrow 0.5x = 10 \rightarrow x = 20$$

20 grams of pure gold should be mixed with 40 grams of 12 karat gold.

53. Let t represent the time it takes for Mike to catch up with Dan.

	Time to run mile	Time	Part of mile run in one minute	Distance
Mike	6	t	$\dfrac{1}{6}$	$\dfrac{1}{6}t$
Dan	9	$t + 1$	$\dfrac{1}{9}$	$\dfrac{1}{9}(t + 1)$

Since the distances are the same, we have:

$$\frac{1}{6}t = \frac{1}{9}(t + 1) \rightarrow 3t = 2t + 2 \rightarrow t = 2$$

Mike will pass Dan after 2 minutes, which is a distance of $\frac{1}{3}$ mile.

55. Let t represent the time the auxiliary pump needs to run.

	Time to do job alone	Part of job done in one hour	Time on Job	Part of total job done by each pump
Main Pump	4	$\dfrac{1}{4}$	3	$\dfrac{3}{4}$
Auxiliary Pump	9	$\dfrac{1}{9}$	t	$\dfrac{1}{9}t$

Since the two pumps are emptying one tanker, we have:

$$\frac{3}{4} + \frac{1}{9}t = 1$$

$$27 + 4t = 36$$

$$4t = 9$$

$$t = \frac{9}{4} = 2.25$$

The auxiliary pump must run for 2.25 hours. It must be started at 9:45 a.m.

57. Let t represent the time for the tub to fill with the faucets on and the stopper removed.

	Time to do job alone	Part of job done in one minute	Time on Job	Part of total job done by each
Faucets open	15	$\dfrac{1}{15}$	t	$\dfrac{t}{15}$
Stopper removed	20	$-\dfrac{1}{20}$	t	$-\dfrac{t}{20}$

Since one tub is being filled, we have:

$$\frac{t}{15} + \left(-\frac{t}{20}\right) = 1 \rightarrow 4t - 3t = 60 \rightarrow t = 60 \qquad \therefore \ 60 \text{ minutes is required to fill the tub.}$$

59. Burke's rate is $\dfrac{100}{12}$ meters/sec.

In 9.99 seconds, Burke will run $\dfrac{100}{12}(9.99) = 83.25$ meters.

Lewis would win by 16.75 meters.

61. Let x represent the number of highway miles.

Then $30,000 - x$ represents the number of city miles.

	No. of miles	Gallons per mile	Gallons used
highway	x	1/40	$\dfrac{1}{40}x$
city	$30,000 - x$	1/25	$(30,000 - x)\left(\dfrac{1}{25}\right)$

$$\frac{1}{40}x + (30,000 - x)\left(\frac{1}{25}\right) = 900$$

$$\frac{1}{40}x + 1200 - \frac{1}{25}x = 900$$

$$\frac{1}{40}x - \frac{1}{25}x = -300$$

$$\frac{25 - 40}{(40)(25)}x = -300 \rightarrow \frac{-15}{1000}x = -300 \rightarrow x = -300\left(\frac{1000}{-15}\right) = 20000$$

Therese can allow for 20,000 miles as a business expense.

63. (a) we need to solve the equation $\dfrac{1}{2}n(n - 3) = 65$

$$2\left(\frac{1}{2}\right)n(n - 3) = (65)(2)$$

$$n(n - 3) = 130 \rightarrow n^2 - 3n = 130 \rightarrow n^2 - 3n - 130 = 0 \rightarrow (n + 10)(n - 13) = 0$$

$$n = -10 \ \text{ or } \ n = 13$$

since n must be a positive integer (it represents the number of sides of the polygon), we discard the negative value. Therefore, we conclude that the polygon has 13 sides and 65 diagonals.

(b) we need to solve the equation $\frac{1}{2}n(n-3) = 80$

$$2\left(\frac{1}{2}\right)n(n-3) = (80)(2) \rightarrow n(n-3) = 160 \rightarrow n^2 - 3n = 160 \rightarrow n^2 - 3n - 160 = 0$$

the solutions to this equation are $n = \dfrac{3 \pm \sqrt{9+640}}{2} = \dfrac{3 \pm \sqrt{649}}{2}$,but neither of these

answers is a whole number. Therefore, there cannot exist a polygon having 80 diagonals.

65. The time traveled with the tail wind was:

$$919 = 550t$$

$$t = \frac{919}{550} \approx 1.671 \text{ hours}$$

Since they were 20 minutes early, the time in still air would have been:

$$1.671 \text{ hours} + 20 \text{ minutes} = 1.671 + 0.333 \approx 2 \text{ hours}$$

Thus the rate in still air is:

$$919 = r(2)$$

$$r \approx 460 \text{ nautical miles / hour}$$

The tail wind was $550 - 460 = 90$ nautical miles per hour.

67. Answers will vary.

Review

A.8   Inequalities

1. $[0, 2]$ $0 \le x \le 2$

3. $(-1,2)$ $-1 < x < 2$

5. $[0,3)$ $0 \le x < 3$

7.   (a) $6 < 8$
     (b) $-2 < 0$
     (c) $9 < 15$
     (d) $-6 > -10$

9.   (a) $7 > 0$
     (b) $-1 > -8$
     (c) $12 > -9$
     (d) $-8 < 6$

11. (a) $2x + 4 < 5$
     (b) $2x - 4 < -3$
     (c) $6x + 3 < 6$
     (d) $-4x - 2 > -4$

13. $[0, 4]$

15. $[4, 6)$

17. $[4, \infty)$

19. $(-\infty, -4)$

21. $2 \le x \le 5$

23. $-3 < x < -2$

25. $x \ge 4$

27. $x < -3$

29. If $x < 5$, then $x - 5 < 0$.

31. If $x > -4$, then $x + 4 > 0$.

33. If $x \ge -4$, then $3x \ge -12$.

35. If $x > 6$, then $-2x < -12$.

37. If $x \geq 5$, then $-4x \leq -20$.

39. If $2x > 6$, then $x > 3$.

41. If $-\dfrac{1}{2}x \leq 3$, then $x \geq -6$.

43. $x + 1 < 5$

$x + 1 - 1 < 5 - 1 \rightarrow x < 4$

$\{x \mid x < 4\}$ or $(-\infty, 4)$

4

45. $1 - 2x \leq 3$

$-2x \leq 2 \rightarrow x \geq -1$

$\{x \mid x \geq -1\}$ or $[-1, +\infty)$

-1

47. $3x - 7 > 2$

$3x > 9 \rightarrow x > 3$

$\{x \mid x > 3\}$ or $(3, +\infty)$

3

49. $3x - 1 \geq 3 + x$

$2x \geq 4 \rightarrow x \geq 2$

$\{x \mid x \geq 2\}$ or $[2, +\infty)$

2

51. $-2(x + 3) < 8$

$-2x - 6 < 8$

$-2x < 14$

$x > -7$

$\{x \mid x > -7\}$ or $(-7, +\infty)$

-7

53. $4 - 3(1 - x) \leq 3$

$4 - 3 + 3x \leq 3$

$3x + 1 \leq 3$

$3x \leq 2$

$x \leq \dfrac{2}{3}$

$\left\{x \mid x \leq \dfrac{2}{3}\right\}$ or $\left(-\infty, \dfrac{2}{3}\right]$

$\dfrac{2}{3}$

55. $\frac{1}{2}(x-4) > x + 8$

$\frac{1}{2}x - 2 > x + 8$

$-\frac{1}{2}x > 10$

$x < -20$

$\{x \mid x < -20\}$ or $(-\infty, -20)$

-20

57. $\frac{x}{2} \geq 1 - \frac{x}{4}$

$2x \geq 4 - x$

$3x \geq 4$

$x \geq \frac{4}{3}$

$\left\{x \mid x \geq \frac{4}{3}\right\}$ or $\left[\frac{4}{3}, +\infty\right)$

$\frac{4}{3}$

59. $0 \leq 2x - 6 \leq 4$

$6 \leq 2x \leq 10$

$3 \leq x \leq 5$

$\{x \mid 3 \leq x \leq 5\}$ or $[3, 5]$

$3 \qquad 5$

61. $-5 \leq 4 - 3x \leq 2$

$-9 \leq -3x \leq -2$

$3 \geq x \geq \frac{2}{3}$

$\left\{x \mid \frac{2}{3} \leq x \leq 3\right\}$ or $\left[\frac{2}{3}, 3\right]$

$\frac{2}{3} \qquad 3$

63. $-3 < \frac{2x-1}{4} < 0$

$-12 < 2x - 1 < 0$

$-11 < 2x < 1$

$\frac{-11}{2} < x < \frac{1}{2}$

$\left\{x \mid -\frac{11}{2} < x < \frac{1}{2}\right\}$ or $\left(-\frac{11}{2}, \frac{1}{2}\right)$

$-\frac{11}{2} \qquad \frac{1}{2}$

65. $1 < 1 - \frac{1}{2}x < 4$

$0 < -\frac{1}{2}x < 3$

$0 > x > -6$

$\{x \mid -6 < x < 0\}$ or $(-6, 0)$

$-6 \qquad 0$

67. $(4x+2)^{-1} < 0$

$$\frac{1}{4x+2} < 0 \rightarrow 4x+2 < 0$$

$$x < -\frac{1}{2}$$

$$\left\{ x \,\middle|\, x < -\frac{1}{2} \right\} \text{ or } \left(-\infty, -\frac{1}{2}\right)$$

$$-\frac{1}{2}$$

69. $0 < \dfrac{2}{x} < \dfrac{3}{5}$

$$0 < \frac{2}{x} \quad \text{and} \quad \frac{2}{x} < \frac{3}{5}$$

$$0 < \frac{2}{x} \rightarrow x > 0$$

therefore $\dfrac{2}{x} < \dfrac{3}{5} \rightarrow 10 < 3x \rightarrow \dfrac{10}{3} < x$

$$\left\{ x \,\middle|\, \frac{10}{3} < x \right\} \text{ or } \left(\frac{10}{3}, +\infty\right)$$

$$\frac{10}{3}$$

71. $0 < (2x-4)^{-1} < \dfrac{1}{2}$

$$0 < \frac{1}{2x-4} \quad \text{and} \quad \frac{1}{2x-4} < \frac{1}{2}$$

$$0 < \frac{1}{2x-4} \rightarrow 2x-4 > 0$$

therefore $\dfrac{1}{2x-4} < \dfrac{1}{2}$

$$2 < 2x-4 \rightarrow 3 < x$$

$$\left\{ x \,\middle|\, 3 < x \right\} \text{ or } (3, +\infty)$$

$$3$$

73. $|2x| < 8$

$$-8 < 2x < 8$$
$$-4 < x < 4$$
$$\left\{ x \,\middle|\, -4 < x < 4 \right\} \text{ or } (-4, 4)$$

$$-4 \qquad\qquad 4$$

75. $|3x| > 12$

$$3x < -12 \quad \text{or} \quad 3x > 12$$
$$x < -4 \quad \text{or} \quad x > 4$$
$$\left\{ x \,\middle|\, x < -4 \text{ or } x > 4 \right\} \text{ or}$$
$$(-\infty, -4) \cup (4, +\infty)$$

$$-4 \qquad\qquad 4$$

77. $|2x-1| \le 1$

$$-1 \le 2x-1 \le 1$$
$$0 \le \;\; 2x \;\; \le 2$$
$$0 \le \;\;\; x \;\;\; \le 1$$
$$\left\{ x \,\middle|\, 0 \le x \le 1 \right\} \text{ or } [0,1]$$

$$0 \qquad\qquad 1$$

79. $|1 - 2x| > 3$
$1 - 2x < -3$ or $1 - 2x > 3$
$-2x < -4$ or $-2x > 2$
$x > 2$ or $x < -1$
$\{x \mid x < -1 \text{ or } x > 2\}$ or $(-\infty, -1) \cup (2, \infty)$

-1 2

81. $|-4x| + |-5| \le 9$
$|4x| + 5 \le 9$
$|4x| \le 4$
$-4 \le 4x \le 4$
$-1 \le x \le 1$
$\{x \mid -1 \le x \le 1\}$ or $[-1, 1]$

-1 1

83. $|-2x| \ge |-4|$
$|2x| \ge 4$
$2x \le -4$ or $2x \ge 4$
$x \ge -2$ or $x \ge 2$
$\{x \mid x \le -2 \text{ or } x \ge 2\}$
or $(-\infty, -2] \cup [2, +\infty)$

-2 2

85. $|x - 2| < 0.5$
$-0.5 < x - 2 < 0.5$
$-0.5 + 2 < x < 0.5 + 2$
$1.5 < x < 2.5$
    Solution set: $\{x \mid 1.5 < x < 2.5\}$

87. $|x - (-3)| > 2$
$x - (-3) < -2$ or $x - (-3) > 2$
$x + 3 < -2$ or $x + 3 > 2$
$x < -5$ or $x > -1$
Solution set: $\{x \mid x < -5 \text{ or } x > -1\}$

89. $21 < \text{young adult's age} < 30$

91.   A temperature x that differs from $98.6°$ F by at least $1.5°$
$$|x - 98.6°| \ge 1.5°$$
$$x - 98.6° \le -1.5° \text{ or } x - 98.6° \ge 1.5°$$
$$x \le 97.1° \text{ or } \qquad x \ge 100.1°$$
The temperatures that are considered unhealthy are those that are less than $97.1°$F or greater than $100.1°$F, inclusive.

93.   (a)   An average 25-year-old male can expect to live at least 48.4 more years.
$25 + 48.4 = 73.4$. Therefore, the average age of a 25-year-old male will be ≥ 73.4.
   (b)   An average 25-year-old female can expect to live at least 54.7 more years.
$25 + 54.7 = 79.7$. Therefore, the average age of a 25-year-old female will be ≥ 79.7.
   (c)   By the given information, a female can expect to live 6.3 years longer.

95.   Let P represent the selling price and C represent the commission.
Calculating the commission:
$C = 45,000 + 0.25(P - 900,000) = 45,000 + 0.25P - 225,000 = 0.25P - 180,000$

Calculate the commission range, given the price range:

$$900,000 \le \qquad P \qquad \le 1,100,000$$
$$0.25(900,000) \le \qquad 0.25P \qquad \le 0.25(1,100,000)$$
$$225,000 \le \qquad 0.25P \qquad \le 275,000$$
$$225,000 - 180,000 \le 0.25P - 180,000 \le 275,000 - 180,000$$
$$45,000 \le \qquad C \qquad \le 95,000$$

The agent's commission ranges from \$45,000 to \$95,000, inclusive.

$$\frac{45,000}{900,000} = 0.05 = 5\% \text{ to } \frac{95,000}{1,100,000} = 0.086 = 8.6\%, \text{ inclusive.}$$

As a percent of selling price, the commission ranges from 5% to 8.6%.

97. Let W represent the weekly wage and T represent the withholding tax.
Calculating the tax:

$$T = 69.90 + 0.28(W - 517) = 69.90 + 0.28W - 144.76 = 0.28W - 74.86$$

Calculating the withholding tax range, given the range of weekly wages:

$$525 \le \qquad W \qquad \le 600$$
$$0.28(525) \le \qquad 0.28W \qquad \le 0.28(600)$$
$$147 \le \qquad 0.28W \qquad \le 168$$
$$147 - 74.86 \le 0.28W - 74.86 \le 168 - 74.86$$
$$72.14 \le \qquad T \qquad \le 93.14$$

The amount of withholding tax ranges from \$72.14 to \$93.14, inclusive.

99. Let K represent the monthly usage in kilowatt-hours.
Let C represent the monthly customer bill.
Calculating the bill:

$$C = 0.10494K + 9.36$$

Calculating the range of kilowatt-hours, given the range of bills:

$$80.24 \le \qquad C \qquad \le 271.80$$
$$80.24 \le 0.10494K + 9.36 \le 271.80$$
$$70.88 \le \qquad 0.10494K \qquad \le 262.44$$
$$675.43 \le \qquad K \qquad \le 2500.86$$

The range of usage in kilowatt-hours varied from 675.43 to 2500.86.

101. Let C represent the dealer's cost and M represent the markup over dealer's cost.
If the price is \$8800, then $8800 = C + MC = C(1 + M)$

Solving for C: $C = \dfrac{8800}{1 + M}$

Calculating the range of dealer costs, given the range of markups:

$$0.12 \le \quad M \quad \le 0.18$$
$$1.12 \le 1 + M \le 1.18$$
$$\frac{1}{1.12} \ge \frac{1}{1 + M} \ge \frac{1}{1.18}$$
$$\frac{8800}{1.12} \ge \frac{8800}{1 + M} \ge \frac{8800}{1.18}$$
$$7857.14 \ge \quad C \quad \ge 7457.63$$

The dealer's cost ranged from \$7457.63 to \$7857.14, inclusive.

103. Let T represent the score on the last test and G represent the course grade.
Calculating the course grade and solving for the last test:

$$G = \frac{68 + 82 + 87 + 89 + T}{5} = \frac{326 + T}{5} \rightarrow T = 5G - 326$$

Calculating the range of scores on the last test, given the grade range:

$$80 \leq \quad G \quad < 90$$
$$400 \leq \quad 5G \quad < 450$$
$$74 \leq 5G - 326 < 124$$
$$74 \leq \quad T \quad < 124$$

The fifth test must be greater than or equal to 74.

105.   Since $a < b$

$$\frac{a}{2} < \frac{b}{2} \qquad\qquad \frac{a}{2} < \frac{b}{2}$$

$$\frac{a}{2} + \frac{a}{2} < \frac{a}{2} + \frac{b}{2} \qquad \frac{a}{2} + \frac{b}{2} < \frac{b}{2} + \frac{b}{2} \qquad \text{Thus, } a < \frac{a+b}{2} < b$$

$$a < \frac{a+b}{2} \qquad\qquad \frac{a+b}{2} < b$$

107.   If $0 < a < b$, then $0 < a^2 < ab$ and $0 < ab < b^2$

$$ab - a^2 > 0 \qquad\qquad b^2 - ab > 0$$
$$ab > a^2 > 0 \qquad\qquad b^2 > ab > 0$$
$$\left(\sqrt{ab}\right)^2 > a^2 \qquad\qquad b^2 > \left(\sqrt{ab}\right)^2$$
$$\sqrt{ab} > a \qquad\qquad b > \sqrt{ab}$$

Thus, $a < \sqrt{ab} < b$

109. For $0 < a < b$, $\dfrac{1}{h} = \dfrac{1}{2}\left(\dfrac{1}{a} + \dfrac{1}{b}\right)$

$$h \cdot \frac{1}{h} = \frac{1}{2}\left(\frac{b+a}{ab}\right) \cdot h \quad \rightarrow \quad 1 = \frac{1}{2}\left(\frac{b+a}{ab}\right) \cdot h \quad \rightarrow \quad \frac{2ab}{a+b} = h$$

$$h - a = \frac{2ab}{a+b} - a \qquad\qquad\qquad b - h = b - \frac{2ab}{a+b}$$

$$= \frac{2ab - a(a+b)}{a+b} \qquad\qquad\qquad = \frac{b(a+b) - 2ab}{a+b}$$

$$= \frac{2ab - a^2 - ab}{a+b} \qquad\qquad\qquad = \frac{ab + b^2 - 2ab}{a+b}$$

$$= \frac{ab - a^2}{a+b} \qquad\qquad\qquad\qquad = \frac{b^2 - ab}{a+b}$$

$$= \frac{a(b-a)}{a+b} > 0 \qquad\qquad\qquad = \frac{b(b-a)}{a+b} > 0$$

Therefore, $h > a$. \qquad Therefore, $h < b$. \qquad Thus, $a < h < b$.

Review

A.9 *n*th Roots; Rational Exponents

1. $\sqrt[3]{27} = 3$

3. $\sqrt[3]{-8} = -2$

5. $\sqrt{8} = \sqrt{4 \cdot 2} = 2\sqrt{2}$

7. $\sqrt[3]{-8x^4} = \sqrt[3]{-8 \cdot x^3 \cdot x} = -2x\sqrt[3]{x}$

9. $\sqrt[4]{x^{12}y^8} = \sqrt[4]{(x^3)^4(y^2)^4} = x^3y^2$

11. $\sqrt[4]{\dfrac{x^9y^7}{xy^3}} = \sqrt[4]{x^8y^4} = x^2y$

13. $\sqrt{36x} = 6\sqrt{x}$

15. $\sqrt{3x^2}\sqrt{12x} = \sqrt{36x^2 \cdot x} = 6x\sqrt{x}$

17. $\left(\sqrt{5}\ \sqrt[3]{9}\right)^2 = 5\left(\sqrt[3]{81}\right) = 5\left(\sqrt[3]{27 \cdot 3}\right) = 5 \cdot 3\left(\sqrt[3]{3}\right) = 15\sqrt[3]{3}$

19. $\left(3\sqrt{6}\right)\left(2\sqrt{2}\right) = 6\sqrt{12} = 6\sqrt{4 \cdot 3} = 12\sqrt{3}$

21. $\left(\sqrt{3} + 3\right)\left(\sqrt{3} - 1\right) = \left(\sqrt{3}\right)^2 - \sqrt{3} + 3\sqrt{3} - 3 = 3 + 2\sqrt{3} - 3 = 2\sqrt{3}$

23. $\left(\sqrt{x} - 1\right)^2 = \left(\sqrt{x}\right)^2 - 2\sqrt{x} + 1 = x - 2\sqrt{x} + 1$

25. $3\sqrt{2} - 4\sqrt{8} = 3\sqrt{2} - 4\sqrt{4 \cdot 2} = 3\sqrt{2} - 8\sqrt{2} = -5\sqrt{2}$

27. $\sqrt[3]{16x^4} - \sqrt[3]{2x} = \sqrt[3]{8 \cdot 2 \cdot x^3 \cdot x} - \sqrt[3]{2x} = 2x\sqrt[3]{2x} - \sqrt[3]{2x} = (2x - 1)\sqrt[3]{2x}$

29. $\dfrac{1}{\sqrt{2}} \cdot \dfrac{\sqrt{2}}{\sqrt{2}} = \dfrac{\sqrt{2}}{2}$

31. $\dfrac{-\sqrt{3}}{\sqrt{5}} \cdot \dfrac{\sqrt{5}}{\sqrt{5}} = \dfrac{-\sqrt{15}}{5} = -\dfrac{\sqrt{15}}{5}$

33. $\dfrac{\sqrt{3}}{5 - \sqrt{2}} \cdot \dfrac{5 + \sqrt{2}}{5 + \sqrt{2}} = \dfrac{\sqrt{3}\left(5 + \sqrt{2}\right)}{25 - 2} = \dfrac{\sqrt{3}\left(5 + \sqrt{2}\right)}{23}$

35. $\dfrac{2 - \sqrt{5}}{2 + 3\sqrt{5}} \cdot \dfrac{2 - 3\sqrt{5}}{2 - 3\sqrt{5}} = \dfrac{4 - 6\sqrt{5} - 2\sqrt{5} + 3\left(\sqrt{5}\right)^2}{4 - 9\left(\sqrt{5}\right)^2} = \dfrac{4 - 8\sqrt{5} + 3 \cdot 5}{4 - 9 \cdot 5}$

$$= \dfrac{4 - 8\sqrt{5} + 15}{4 - 45} = \dfrac{19 - 8\sqrt{5}}{-41} = \dfrac{-19 + 8\sqrt{5}}{41}$$

37. $\dfrac{5}{\sqrt[3]{2}} \cdot \dfrac{\left(\sqrt[3]{2}\right)^2}{\left(\sqrt[3]{2}\right)^2} = \dfrac{5\left(\sqrt[3]{2}\right)^2}{\left(\sqrt[3]{2}\right)^3} = \dfrac{4\left(\sqrt[3]{2^2}\right)}{2} = \dfrac{5\sqrt[3]{4}}{2}$

39. $\dfrac{\sqrt{x+h}-\sqrt{x}}{\sqrt{x+h}+\sqrt{x}} \cdot \dfrac{\sqrt{x+h}-\sqrt{x}}{\sqrt{x+h}-\sqrt{x}} = \dfrac{\left(\sqrt{x+h}\right)^2 - 2\sqrt{x}\cdot\sqrt{x+h}+\left(\sqrt{x}\right)^2}{\left(\sqrt{x+h}\right)^2 - \left(\sqrt{x}\right)^2}$

$$= \dfrac{x+h-2\sqrt{x(x+h)}+x}{x+h-x} = \dfrac{2x+h-2\sqrt{x(x+h)}}{h}$$

41. $\sqrt[3]{2t-1} = 2$
$\left(\sqrt[3]{2t-1}\right)^3 = 2^3$

$2t-1 = 8 \rightarrow 2t = 9 \rightarrow t = \dfrac{9}{2}$

Check $t = \dfrac{9}{2}$: $\sqrt[3]{2\left(\dfrac{9}{2}\right)-1} = \sqrt[3]{9-1} = \sqrt[3]{8} = 2$

The solution is $t = \dfrac{9}{2}$.

43. $\sqrt{15-2x} = x$
$\left(\sqrt{15-2x}\right)^2 = x^2$

$15 - 2x = x^2 \rightarrow x^2 + 2x - 15 = 0$
$(x+5)(x-3) = 0 \rightarrow x = -5 \text{ or } x = 3$

Check $x = -5$: $\sqrt{15-2(-5)} = \sqrt{25}$
$= 5 \neq -5$

Check $x = 3$: $\sqrt{15-2(3)} = \sqrt{9} = 3 = 3$

The solution is $x = 3$.

45. $8^{2/3} = \left(2^3\right)^{2/3} = 2^2 = 4$

47. $(-27)^{1/3} = \left((-3)^3\right)^{1/3} = -3$

49. $16^{3/2} = \left(2^4\right)^{3/2} = 2^6 = 64$

51. $9^{-3/2} = \left(3^2\right)^{-3/2} = 3^{-3} = \dfrac{1}{3^3} = \dfrac{1}{27}$

53. $\left(\dfrac{9}{8}\right)^{3/2} = \left(\dfrac{3^2}{2^3}\right)^{3/2} = \dfrac{3^{6/2}}{2^{9/2}} = \dfrac{3^3}{2^{8/2}\cdot 2^{1/2}} = \dfrac{27}{2^4\cdot 2^{1/2}} = \dfrac{27}{16\cdot\sqrt{2}}\cdot\dfrac{\sqrt{2}}{\sqrt{2}} = \dfrac{27\sqrt{2}}{16\cdot 2} = \dfrac{27\sqrt{2}}{32}$

55. $\left(\dfrac{8}{9}\right)^{-3/2} = \left(\dfrac{2^3}{3^2}\right)^{-3/2} = \dfrac{2^{-9/2}}{3^{-6/2}} = \dfrac{2^{-8/2}\cdot 2^{-1/2}}{3^{-3}} = \dfrac{2^{-4}\cdot 2^{-1/2}}{3^{-3}} = \dfrac{27}{16\cdot\sqrt{2}}\cdot\dfrac{\sqrt{2}}{\sqrt{2}} = \dfrac{27\sqrt{2}}{16\cdot 2} = \dfrac{27\sqrt{2}}{32}$

57. $x^{3/4}\cdot x^{1/3}\cdot x^{-1/2} = x^{3/4+1/3-1/2} = x^{(9+4-6)/12} = x^{7/12}$

59. $\left(x^3 y^6\right)^{1/3} = \left(x^3\right)^{1/3}\left(y^6\right)^{1/3} = xy^2$

61. $\left(x^2 y\right)^{1/3}\left(x y^2\right)^{2/3} = x^{2/3}y^{1/3}x^{2/3}y^{4/3} = x^{4/3}y^{5/3}$

63. $\left(16x^2 y^{-1/3}\right)^{3/4} = \left(2^4 x^2 y^{-1/3}\right)^{3/4} = 2^3 x^{3/2} y^{-1/4} = \dfrac{8x^{3/2}}{y^{1/4}}$

65. $\dfrac{x}{(1+x)^{1/2}}+2(1+x)^{1/2}=\dfrac{x+2(1+x)^{1/2}(1+x)^{1/2}}{(1+x)^{1/2}}=\dfrac{x+2(1+x)}{(1+x)^{1/2}}=\dfrac{x+2+2x}{(1+x)^{1/2}}=\dfrac{3x+2}{(1+x)^{1/2}}$

67. $\dfrac{\left(\sqrt{1+x}-x\cdot\dfrac{1}{2\sqrt{1+x}}\right)}{1+x}=\dfrac{\left(\sqrt{1+x}-\dfrac{x}{2\sqrt{1+x}}\right)}{1+x}=\dfrac{\left(\dfrac{2\sqrt{1+x}\sqrt{1+x}-x}{2\sqrt{1+x}}\right)}{1+x}$

$=\dfrac{2(1+x)-x}{2(1+x)^{1/2}}\cdot\dfrac{1}{1+x}=\dfrac{2+x}{2(1+x)^{3/2}}$

69. $\dfrac{(x+4)^{1/2}-2x(x+4)^{-1/2}}{x+4}=\dfrac{\left((x+4)^{1/2}-\dfrac{2x}{(x+4)^{1/2}}\right)}{x+4}=\dfrac{\left((x+4)^{1/2}\cdot\dfrac{(x+4)^{1/2}}{(x+4)^{1/2}}-\dfrac{2x}{(x+4)^{1/2}}\right)}{x+4}$

$=\dfrac{\left(\dfrac{x+4-2x}{(x+4)^{1/2}}\right)}{x+4}=\dfrac{-x+4}{(x+4)^{1/2}}\cdot\dfrac{1}{x+4}=\dfrac{-x+4}{(x+4)^{3/2}}$

71. $(x+1)^{3/2}+x\cdot\dfrac{3}{2}(x+1)^{1/2}=(x+1)^{1/2}\left(x+1+\dfrac{3}{2}x\right)=(x+1)^{1/2}\left(\dfrac{5}{2}x+1\right)=\dfrac{1}{2}(x+1)^{1/2}(5x+2)$

73. $6x^{1/2}(x^2+x)-8x^{3/2}-8x^{1/2}=2x^{1/2}(3(x^2+x)-4x-4)=2x^{1/2}(3x^2-x-4)$
$=2x^{1/2}(3x-4)(x+1)$

75. $x\left(\dfrac{1}{2}\right)(8-x^2)^{-1/2}(-2x)+(8-x^2)^{1/2}=-x^2(8-x^2)^{-1/2}+(8-x^2)^{1/2}=(8-x^2)^{-1/2}\left[-x^2+(8-x^2)\right]$

$=(8-x^2)^{-1/2}\left[-x^2+8-x^2\right]=(8-x^2)^{-1/2}\left[8-2x^2\right]$

$=\dfrac{8-2x^2}{(8-x^2)^{1/2}}=\dfrac{2(4-x^2)}{(8-x^2)^{1/2}}=\dfrac{2(2-x)(2+x)}{(8-x^2)^{1/2}}$

Graphing Utilities

B.1 The Viewing Rectangle

1. (- 1, 4)

3. (3, 1)

5.
$X \min = -6$
$X \max = 6$
$X \operatorname{scl} = 2$
$Y \min = -4$
$Y \max = 4$
$Y \operatorname{scl} = 2$

7.
$X \min = -6$
$X \max = 6$
$X \operatorname{scl} = 2$
$Y \min = -1$
$Y \max = 3$
$Y \operatorname{scl} = 1$

9.
$X \min = 3$
$X \max = 9$
$X \operatorname{scl} = 1$
$Y \min = 2$
$Y \max = 10$
$Y \operatorname{scl} = 2$

11.
$X \min = -11$
$X \max = 5$
$X \operatorname{scl} = 1$
$Y \min = -3$
$Y \max = 6$
$Y \operatorname{scl} = 1$

13.
$X \min = -30$
$X \max = 50$
$X \operatorname{scl} = 10$
$Y \min = -90$
$Y \max = 50$
$Y \operatorname{scl} = 10$

15.
$X \min = -10$
$X \max = 110$
$X \operatorname{scl} = 10$
$Y \min = -10$
$Y \max = 160$
$Y \operatorname{scl} = 10$

Appendix B　　Graphing Utilities

17. $P_1 = (1,3); P_2 = (5,15)$

$$d(P_1, P_2) = \sqrt{(5-1)^2 + (15-3)^2}$$
$$= \sqrt{(4)^2 + (12)^2}$$
$$= \sqrt{16 + 144}$$
$$= \sqrt{160} = 2\sqrt{10}$$

19. $P_1 = (-4,6); P_2 = (4,-8)$

$$d(P_1, P_2) = \sqrt{(4-(-4))^2 + (-8-6)^2}$$
$$= \sqrt{(8)^2 + (-14)^2}$$
$$= \sqrt{64 + 196}$$
$$= \sqrt{260} = 2\sqrt{65}$$

Appendix B

Graphing Utilities

B.2   Using a Graphing Utility to Graph Equations

1. (a) $y = x + 2$
    $X \min = -5$
    $X \max = 5$
    $X \text{ scl} = 1$
    $Y \min = -4$
    $Y \max = 4$
    $Y \text{ scl} = 1$

    (b)
    $X \min = -10$
    $X \max = 10$
    $X \text{ scl} = 1$
    $Y \min = -8$
    $Y \max = 8$
    $Y \text{ scl} = 1$

    (c)
    $X \min = -10$
    $X \max = 10$
    $X \text{ scl} = 2$
    $Y \min = -8$
    $Y \max = 8$
    $Y \text{ scl} = 2$

    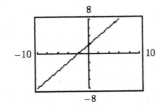

    (d)
    $X \min = -5$
    $X \max = 5$
    $X \text{ scl} = 1$
    $Y \min = -20$
    $Y \max = 20$
    $Y \text{ scl} = 5$

3. (a) $y = -x + 2$

$X \min = -5$

$X \max = 5$

$X \, \text{scl} = 1$

$Y \min = -4$

$Y \max = 4$

$Y \, \text{scl} = 1$

(b)

$X \min = -10$

$X \max = 10$

$X \, \text{scl} = 1$

$Y \min = -8$

$Y \max = 8$

$Y \, \text{scl} = 1$

(c)

$X \min = -10$

$X \max = 10$

$X \, \text{scl} = 2$

$Y \min = -8$

$Y \max = 8$

$Y \, \text{scl} = 2$

(d)

$X \min = -5$

$X \max = 5$

$X \, \text{scl} = 1$

$Y \min = -20$

$Y \max = 20$

$Y \, \text{scl} = 5$

5. (a) $y = 2x + 2$

$X \min = -5$

$X \max = 5$

$X \, \text{scl} = 1$

$Y \min = -4$

$Y \max = 4$

$Y \, \text{scl} = 1$

(b)

$X \min = -10$

$X \max = 10$

$X \, \text{scl} = 1$

$Y \min = -8$

$Y \max = 8$

$Y \, \text{scl} = 1$

(c)

X min $= -10$
X max $= 10$
X scl $= 2$
Y min $= -8$
Y max $= 8$
Y scl $= 2$

(d)

X min $= -5$
X max $= 5$
X scl $= 1$
Y min $= -20$
Y max $= 20$
Y scl $= 5$

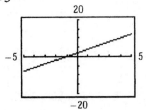

7. (a) $y = -2x + 2$
X min $= -5$
X max $= 5$
X scl $= 1$
Y min $= -4$
Y max $= 4$
Y scl $= 1$

(b)

X min $= -10$
X max $= 10$
X scl $= 1$
Y min $= -8$
Y max $= 8$
Y scl $= 1$

(c)

X min $= -10$
X max $= 10$
X scl $= 2$
Y min $= -8$
Y max $= 8$
Y scl $= 2$

(d)

X min $= -5$
X max $= 5$
X scl $= 1$
Y min $= -20$
Y max $= 20$
Y scl $= 5$

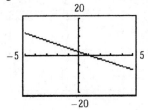

9. (a) $y = x^2 + 2$
   $X\min = -5$
   $X\max = 5$
   $X\operatorname{scl} = 1$
   $Y\min = -4$
   $Y\max = 4$
   $Y\operatorname{scl} = 1$

(b)
   $X\min = -10$
   $X\max = 10$
   $X\operatorname{scl} = 1$
   $Y\min = -8$
   $Y\max = 8$
   $Y\operatorname{scl} = 1$

(c)
   $X\min = -10$
   $X\max = 10$
   $X\operatorname{scl} = 2$
   $Y\min = -8$
   $Y\max = 8$
   $Y\operatorname{scl} = 2$

(d)
   $X\min = -5$
   $X\max = 5$
   $X\operatorname{scl} = 1$
   $Y\min = -20$
   $Y\max = 20$
   $Y\operatorname{scl} = 5$

11. (a) $y = -x^2 + 2$
   $X\min = -5$
   $X\max = 5$
   $X\operatorname{scl} = 1$
   $Y\min = -4$
   $Y\max = 4$
   $Y\operatorname{scl} = 1$

(b)
   $X\min = -10$
   $X\max = 10$
   $X\operatorname{scl} = 1$
   $Y\min = -8$
   $Y\max = 8$
   $Y\operatorname{scl} = 1$

728

(c)

X min $= -10$
X max $= 10$
X scl $= 2$
Y min $= -8$
Y max $= 8$
Y scl $= 2$

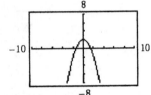

(d)

X min $= -5$
X max $= 5$
X scl $= 1$
Y min $= -20$
Y max $= 20$
Y scl $= 5$

13. (a) $3x + 2y = 6$
X min $= -5$
X max $= 5$
X scl $= 1$
Y min $= -4$
Y max $= 4$
Y scl $= 1$

(b)

X min $= -10$
X max $= 10$
X scl $= 1$
Y min $= -8$
Y max $= 8$
Y scl $= 1$

(c)

X min $= -10$
X max $= 10$
X scl $= 2$
Y min $= -8$
Y max $= 8$
Y scl $= 2$

(d)

X min $= -5$
X max $= 5$
X scl $= 1$
Y min $= -20$
Y max $= 20$
Y scl $= 5$

15. (a) $-3x + 2y = 6$
    $X\min = -5$
    $X\max = 5$
    $X\text{scl} = 1$
    $Y\min = -4$
    $Y\max = 4$
    $Y\text{scl} = 1$

(b)
    $X\min = -10$
    $X\max = 10$
    $X\text{scl} = 1$
    $Y\min = -8$
    $Y\max = 8$
    $Y\text{scl} = 1$

(c)
    $X\min = -10$
    $X\max = 10$
    $X\text{scl} = 2$
    $Y\min = -8$
    $Y\max = 8$
    $Y\text{scl} = 2$

(d)
    $X\min = -5$
    $X\max = 5$
    $X\text{scl} = 1$
    $Y\min = -20$
    $Y\max = 20$
    $Y\text{scl} = 5$

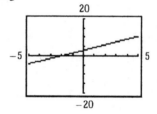

17. $y = x + 2;$ $-3 \le x \le 3$

18.

X	Y₁	
-3	-1	
-2	0	
-1	1	
0	2	
1	3	
2	4	
3	5	

X=3

19. $y = -x + 2;$ $-3 \le x \le 3$

X	Y₁	
-3	5	
-2	4	
-1	3	
0	2	
1	1	
2	0	
3	-1	

X=3

21. $y = 2x + 2;$ $-3 \le x \le 3$

X	Y1	
-3	-4	
-2	-2	
-1	0	
0	2	
1	4	
2	6	
3	8	

X=3

23. $y = -2x + 2;$ $-3 \le x \le 3$

X	Y1	
-3	8	
-2	6	
-1	4	
0	2	
1	0	
2	-2	
3	-4	

X=3

25. $y = x^2 + 2;$ $-3 \le x \le 3$

X	Y1	
-3	11	
-2	6	
-1	3	
0	2	
1	3	
2	6	
3	11	

X=3

27. $y = -x^2 + 2;$ $-3 \le x \le 3$

X	Y1	
-3	-7	
-2	-2	
-1	1	
0	2	
1	1	
2	-2	
3	-7	

X=3

29. $3x + 2y = 6;$ $-3 \le x \le 3$

X	Y1	
-3	7.5	
-2	6	
-1	4.5	
0	3	
1	1.5	
2	0	
3	-1.5	

X=3

31. $-3x + 2y = 6;$ $-3 \le x \le 3$

X	Y1	
-3	-1.5	
-2	0	
-1	1.5	
0	3	
1	4.5	
2	6	
3	7.5	

X=3

Appendix B

Graphing Utilities

B.3   Using a Graphing Utility to Locate Intercepts and Check for Symmetry

1. $y = x^2 + 4x + 2$

The smaller x-intercept is $x \approx -3.41$.

3. $y = 2x^2 + 4x + 1$

The smaller x-intercept is $x \approx -1.71$.

5. $y = 2x^2 - 3x - 1$

The smaller x-intercept is $x \approx -0.28$.

7. $y = x^3 + 3.2x^2 - 16.83x - 5.31$

The positive x-intercept is $x = 3$.

9. $y = x^4 - 1.4x^3 - 33.71x^2 + 23.94x + 292.41$ We zoom in on the positive x-intercept:

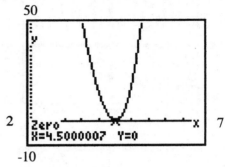

The positive x-intercept is $x \approx 4.50$.

11. $y = \pi x^3 - (8.88\pi + 1)x^2 - (42.066\pi - 8.88)x + 42.066$

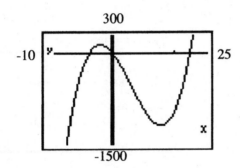

We zoom in on the positive x-intercepts:

The smallest positive x-intercept is $x \approx 0.32$.

We zoom in on the positive x-intercepts:

The largest positive x-intercept is $x \approx 12.3$.

13. $y = x^3 + 19.5x^2 - 1021x + 1000.5$

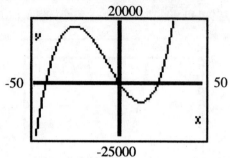

We zoom in on the positive x-intercepts:

The largest positive x-intercept is $x = 23$.

We zoom in on the positive x-intercepts:

The smallest positive x-intercept is $x = 1$.

15. x-intercepts: $(-1.5, 0)$; $(1.5, 0)$

y-intercept: $(0, -2)$

y-axis symmetry

17. x-intercepts: none

y-intercept: none

origin symmetry

Graphing Utilities

B.5   Square Screens

1. yes

    X min = −3

    X max = 3

    X scl = 2

    Y min = −2

    Y max = 2

    Y scl = 2

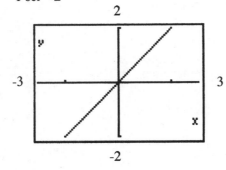

3. yes

    X min = 0

    X max = 9

    X scl = 3

    Y min = −2

    Y max = 4

    Y scl = 2

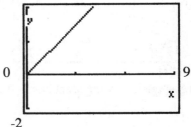

5. no

    X min = −6

    X max = 6

    X scl = 1

    Y min = −2

    Y max = 2

    Y scl = 0.5

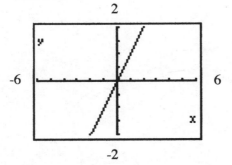

7. yes

    X min = 0

    X max = 9

    X scl = 1

    Y min = −2

    Y max = 4

    Y scl = 1

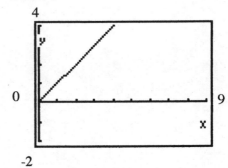

9. One possible answer:

$X \min = -4$

$X \max = 8$

$X \text{scl} = 1$

$Y \min = 4$

$Y \max = 12$

$Y \text{scl} = 1$